Organic Polymers in Energy-Environmental Applications

Organic Polymers in Energy-Environmental Applications

Edited by Ramesh Oraon, Pardeep Singh, Sanchayita Rajkhowa, Sangita Agarwal, and Ravindra Pratap Singh

WILEY-VCH

Editors

Dr. Ramesh Oraon
Central University of Jharkhand
Ratu-Lohardaga Road
Ranchi
Jharkhand 835205
India

Dr. Pardeep Singh
University of Delhi
PGDAV College
New Delhi
India

Dr. Sanchayita Rajkhowa
Haflong Government College
Haflong
Dima Hasao
Assam 788819
India

Dr. Sangita Agarwal
RCC Institute of Information Technology
Canal South Road
Kolkata
West Bengal 700015
India

Dr. Ravindra Pratap Singh
Government of India
New Delhi
India

Library of Congress Card No.: applied for

British Library Cataloguing-in-Publication Data
A catalogue record for this book is available from the British Library.

Bibliographic information published by the Deutsche Nationalbibliothek The Deutsche Nationalbibliothek lists this publication in the Deutsche Nationalbibliografie; detailed bibliographic data are available on the Internet at <http://dnb.d-nb.de>.

Print ISBN: 978-3-527-35237-1
ePDF ISBN: 978-3-527-84279-7
ePub ISBN: 978-3-527-84280-3
oBook ISBN: 978-3-527-84281-0

Typesetting Straive, Chennai, India

Contents

Preface

Polymers, natural or synthetic, have changed the lifestyle of each individual through their immense application in almost every field. The first polymer discovered (early nineteenth century) was completely natural and derived from "Rubber" plants. It was known to have been used by the indigenous people in South America for centuries. Then, in 1839, Charles Goodyear successfully vulcanized the rubber by cross-linking the polymeric chains that impart strength to it. The year 1907 marked a special achievement in producing the first synthetic polymer that was resistant to heat and electric current and used in various materials. It was named after the creator Leo Baekeland as "Bakelite." Since then, there have been a series of remarkable discoveries in this budding field: polyethylene (1930s), nylon (1930s), polyester and PET (1940s), polypropylene (1950s), high-density polyethylene (HDPE) (1950s), and polycarbonate (1950s), to name a few. Since the mid-twentieth century, a significant advancement in polymer science and engineering has occurred, leading to novel and improved polymeric materials and applications. Although these polymers are of great use to humankind, they also have an adverse impact on the environment as they are not biodegradable and have been persistent for centuries. In the late twentieth century, the discovery of biodegradable polymers revolutionizes polymeric research and science. In recent years, research has focused on creating advanced polymers with unique properties, such as conducting polymers for electronics, smart polymers, biodegradable polymers, biopolymers advanced polymer blends and composites, 3D printing, polymeric nanoparticles, and polymers for energy storage that respond to environmental changes. The deployment of organic polymers has surpassed the traditional applications and can be seen in advances in imaging technology such as bioimaging, oil absorption, tissue engineering, and self-healing nanometer coatings in automobile and domestic housing. The recent advancements in organic polymers are expanding the range of applications and addressing important challenges, including environmental sustainability and improved performance across various industries.

The inception of this book was based on numerous applications and a plethora of research advancements in this field. The add-on factor of biodegradability to polymers has motivated us to dive into the depth of the knowledge. With this view, we attempted to comprehend the subject matter for our audience, including students, researchers, academicians, and scientists. The editors Dr. Ramesh Oraon (Central University of Jharkhand), Dr. Pardeep Singh (Delhi University), Dr. Sanchayita Rajkhowa (Haflong Govt. College, Assam, India), Dr. Sangita Agarwal (RCC Institute of Information Technology, Kolkata), and Dr. R.P. Singh (Central Public Works Dept., New Delhi) were keen to publish a handy material on this burning topic with elaborate details starting from its history, development, applications, and future aspects.

Chapter 1 starts with a basic introduction and history of organic polymers. It also offers a comprehensive summary of the basis of polymerization methodologies for synthesizing organic polymers, with recent developments describing the various applications of such materials. Functionalization and characterization of organic polymers demonstrate that the generation of hydrogels and their rheological properties were significantly influenced by the alteration of carboxylates in high molecular mass heparin (HMWH) with various maleimide groups and with thiol-derivatization of PEG crosslinker. In Chapter 2, the progress of porous organic polymers (POPs) as a potential catalyst in various applications like water splitting, CO_2 capture, and degradation of organic pollutants, among others, along with the various synthetic processes of POP catalysts as well as their properties and potential applications, are broadly discussed. The subsequent chapters 3, 4, 10, 14 and 18 comprehensively understand organic polymers in photodetectors, energy storage and solar cells, agriculture, pharmaceuticals, drug delivery systems, etc. There is a great emphasis on green synthesis of organic polymers and their application in environmental remediation (Chapter 11). Another chapter (Chapter 22) provides information about the history, the current research scenario, and the future scope of organic polymer-based adhesives. Owing to their diverse functions and utility, there is a blossoming resurgence of modified polymers in science and technology. In this regard, the methods of polymeric surface modification and their applications, along with the prospects in the future, are also discussed (Chapter 5). A few chapters (Chapters 9, 13, 20, 21 and 24) discuss organic polymers' current and developing trends in various fields, while another includes the future roadmap of POPs (Chapters 19 and 23). Organic polymers are a fundamental part of modern society, and their diverse properties and applications make them a crucial area of scientific and industrial research. Advances in polymer science continue to drive innovation in various industries, from materials engineering to medicine.

Through this journey, we hope to provide our readers with a deeper understanding of organic polymers, their applications, and recent trends that navigate to the future. We aim to make the content accessible, engaging, and relevant.

We extend our heartfelt thanks to the authors, researchers, educators, and all the contributors who have contributed their expertise to this endeavor.

So, dear readers, as you turn the pages of this book, I invite you to embark on this adventure with an open heart and a curious mind. Let us delve into the past, uncover its aspects and applications, and emerge with a deeper appreciation for the incredible journey of the future.

Thank you for joining us on this exploration.

Dr. Ramesh Oraon
Dr. Pardeep Singh
Dr. Sanchayita Rajkhowa
Dr. Sangita Agarwal
Dr. Ravindra Pratap Singh

Acknowledgments

We sincerely thank and appreciate everyone who helped make this book possible. We recognize the arduous labor, devotion, and support of the authors who helped make this project a reality with great pleasure and humility. The authors' passion, creativity, and unwavering commitment to work have made this project an absolute joy to edit. Every page exhibited your talent as a writer, and we feel privileged to have contributed to realizing one's dream.

We express our appreciation to Wiley and the publishing staff for their significant help and direction during the editorial process. This book has really benefited from your experience, professionalism, and unceasing effort.

The quality of a book depends to a large extent on proper reviewing. We gratefully acknowledge the time and expertise devoted to reviewing the manuscripts by a large group of reviewers.

We thank our colleagues and contributors for their invaluable collaboration, insights, and constructive feedback. Your input has been indispensable in refining the manuscripts and ensuring their quality. We also want to acknowledge the contributions of the proofreaders, copyeditors, and designers who worked diligently to make this book visually appealing and error-free. Your attention to detail and dedication to perfection have been truly exceptional.

A special thanks goes to the friends and family who provided emotional support during the long hours and deadlines. Your encouragement and understanding were crucial in helping me see this project through to completion.

Last but not least, we express our appreciation to the readers who will soon embark on this literary journey. Your interest in our work motivates us to continue creating, and we hope you find this book as rewarding to read as it was to edit.

Without the combined efforts of every member of the team, this book would not have been possible. We are incredibly appreciative of the chance to work upon this project and we are looking forward to engaging in many more imaginative partnerships in the futures.

Dr. Ramesh Oraon
Dr. Pardeep Singh
Dr. Sanchayita Rajkhowa
Dr. Sangita Agarwal
Dr. Ravindra Pratap Singh

1

Organic Polymers: Past and the Present

Jyotirmoy Sarma[1], *Subhasish Roy*[1], *Bhaskar Sharma*[1], *Fredy A. Madukkakuzhy*[1], *Monjumoni Das*[2], *and Pallabi Borah*[1]

[1]*Assam Don Bosco University, Department of Chemistry, Tapesia Gardens, Sonapur, Assam 782402, India*
[2]*Sibsagar College, Department of Chemistry, Joysagar, Assam 785665, India*

1.1 Introduction and History of Polymers

The word polymer is derived from the Greek word "polumeros" where "Polus" means "many" and "meros" means "units." Henceforth polymers can be defined as the complex and giant molecules or "**macromolecules**" which are supposed to form by the combination of many small repeating molecules called monomers. Examples of some commercially important polymers and their practical applications have been highlighted in Table 1.1. The most practical distinguishing feature of polymer from its monomer is its huge difference in physical, chemical, and mechanical properties after the polymerization process occurs (Dorel 2008). For example, ethene is a gas but when they combine with each other via the polymerization process, a new class of compound, i.e., polyethene, is formed which differs from its monomer in terms of many physicochemical properties. Monomers being smaller have low molecular weight, while polymers being much larger have very high molecular weight. Compared to simple organic molecules, polymers aren't composed of identical molecules; hence, a polymer sample generally comprises chains of different lengths, which is why their molecular weight is always expressed as an average molecular weight. For instance, the HDPE (high-density polyethylene) molecules are all long-chain carbon chains, but the lengths generally vary by thousands of monomer units. Depending on the type of monomeric units, polymers may be of different types such as homopolymers where all the repeating units (RUs) are same and co-polymers which can be made up of two or more monomer species. For example, in case of homopolymers such as polythene the monomer unit is ethylene, in polyvinylchloride (PVC) the monomer unit is vinyl chloride. Important examples of co-polymers include polyethylene-vinyl acetate (PEVA), nitrile rubber, and acrylonitrile butadiene styrene (ABS) which are formed by the combination of more than one monomer.

Based on the type of backbone chain and composition, polymeric materials are classified into two types, viz. organic polymers and inorganic polymers

Organic Polymers in Energy-Environmental Applications, First Edition.
Edited by Ramesh Oraon, Pardeep Singh, Sanchayita Rajkhowa, Sangita Agarwal, and Ravindra Pratap Singh.

Table 1.1 Some commercially important polymers and their uses.

Name of polymer and structure	Monomer	Practical applications
Polythene 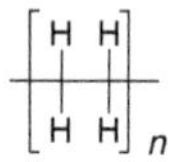	Ethene 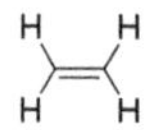	Electrical insulators, packing of materials
Polystyrene 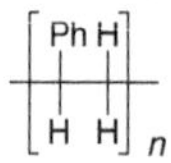	Styrene 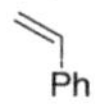	As insulator, wrapping material, for construction of toys
Polyvinyl chloride 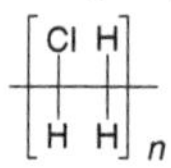	Vinyl chloride 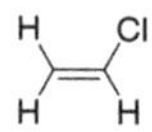	In the manufacture of raincoats, handbags
Polytetrafluoroethylene (Teflon) 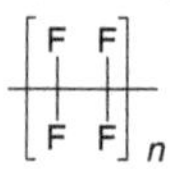	Tetrafluoroethene 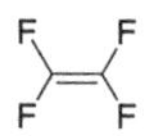	As lubricant insulator in the manufacture of semiconductors, non-stick coating in kitchen cookware and medical devices
Polyacrylonitrile (PAN) 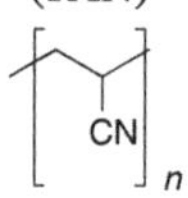	Acrylonitrile 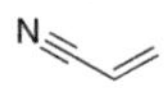	In construction of synthetic fibers and wools
Styrene butadiene rubber (SBS) or Buna-S rubber 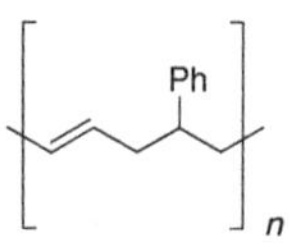	1,3-Butadiene and styrene 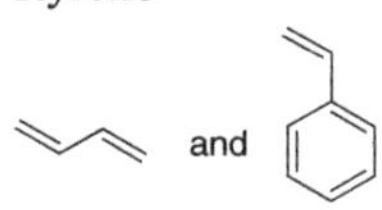	For making of automobile tires and footwear, etc.
Terylene (Dacron), polyester 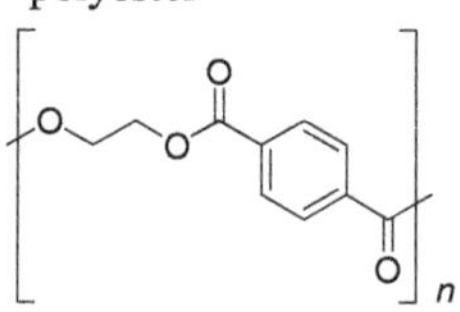	Ethylene glycol and terephthalic acid 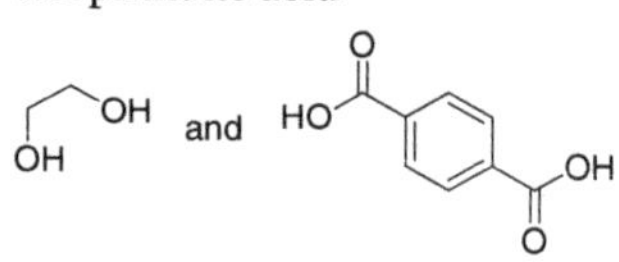	For making fibers, safety belts, plastic bottles, hard wear clothes like dresses

Table 1.1 (Continued)

Name of polymer and structure	Monomer	Practical applications
Nylon-6,6 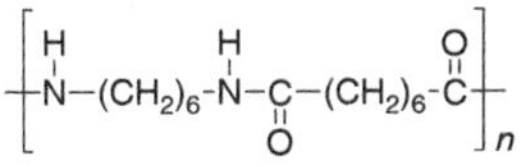	Hexamethylenediamine and adipic acid 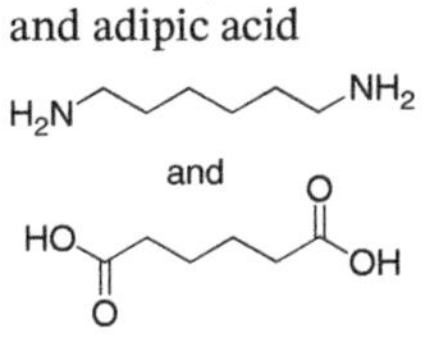	In making brushes, synthetic fibers, water-resistant machine parts
Nylon-6 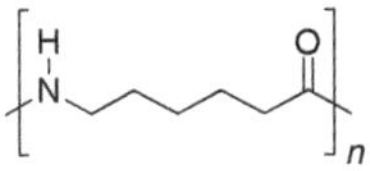	Caprolactum 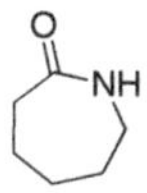	For manufacture of carpets, tire cords, seat belts, parachutes, ropes, etc.

(Currell and Frazer 1969; Gowarikar et al. 2022). Basically, organic polymers are made of carbon-carbon backbone skeleton (Peng et al. 2017), while inorganic polymers do not have carbon-carbon skeleton, rather they have a skeleton like Si–Si for polysilanes, Si–O for polysiloxanes, Si–N for polysilazanes, S–S for polysulfides, B–N for polyborazylenes, and S–N for polythiazyls (Seth 2020; Indra and Shrray 2015).

Cellulose is one of the most abundant organic polymers on Earth, and it is a linear polymer of as many as 10,000 dextro-glucose units joined with each other. Starch, belonging to carbohydrates, can be found in grains and potatoes. Starch is a polymer, also known as a polysaccharide because it is made from glucose as monosaccharide. Starch molecules include two types of glucose polymers, amylose and amylopectin. Amylopectin, being a major starch component, is found in many plants. Amylose belongs to a linear-chain polymer having around two hundred glucose molecules per molecule.

Based on the existence and method of formation, organic polymers may be further categorized as natural or synthetic, and interestingly both of them find equal attention in our day-to-day life. Natural organic polymers can be found in nature or living system and important examples of natural organic polymers include proteins or polypeptides, polynucleotides like DNA and RNA (DNA is a double-stranded polynucleotide chain, while RNA is a single-chain structure of polynucleotides), silk, wool, cellulose, natural rubber. Synthetic polymers are man-made polymers which are being synthesized in the industry or laboratory. Important examples of synthetic polymers include polyethylene (both low-density polyethylene – LDPE and high-density polyethylene – HDPE), polypropene (PP), polyacrylonitrile (PAN), polyaniline (PANI), polystyrene (PS), polyvinylchloride (PVC), tetrafluoroethene (Teflon), polyacetylene, nylon, thermoplastic polyurethane (TPU), and Bakelite. Most of the synthetic polymers possess enhanced lifetime and improved mechanical properties. However due to the absence of a functional group in most of them, they do not have some important physical and chemical properties which limit their synthetic utility in many practical applications. To address this problem post-synthetic functionalization and further modification of polymers have been done in recent times to achieve multinational properties.

During the fifteenth century, Christopher Columbus was involved in the discovery of rubber by isolating it from trees, and later Joseph Priestly observed that the material is helpful for erasing pencil marks on paper. This observation launched the rubber industry. Combining the latex of rubber tree with the morning glory plant juice in various proportions helped to achieve rubber's distinct properties for designing selective products like bouncing balls, various kinds of rubber bands, etc. To gain the advantages and properties of both natural and synthetic polymers, researchers across the globe were always in search for the development of improved semi-synthetic organic polymers. For example, vulcanization of rubber was introduced for enhancing the quality of natural rubber where a small amount of sulfur is added as a cross-linking agent which can enhance the quality and stability of rubber. Vulcanized rubber is comparatively stronger, elastic, more resistant to abrasion and temperature change, and most importantly inert with respect to chemicals and electric current as compared to untreated natural rubber (Brown and Poon 2005).

Likewise, natural resources polymeric materials like cellulose and proteins have been extensively used for making improved polymers via copolymerization techniques.

Henri Braconnot's Braconnot, Christian Schönbein, and coworkers in 1830s first developed the derivatives of cellulose for constructing novel semi-synthetic materials, known as celluloid and cellulose acetate. Later the term "polymer" was introduced in 1833 by Jöns Jakob Berzelius, even though Berzelius could not provide significant contribution for the development of modern polymer science.

In 1909, Leo Baekeland developed Bakelite from cheap and readily available chemicals such as phenol and formaldehyde which opened the door of emerging technology for designing many innovative polymeric products (Baekeland 1909; Wallace 1945).

In spite of the noteworthy developments in polymer synthesis, the molecular nature of polymers was not clear until the concept was introduced by Hermann Staudinger in the year 1922. Earlier formation of polymers was explained via aggregation theory proposed by Thomas Graham in 1861. According to Graham, cellulose and other polymers were supposed to have colloidal nature where aggregation of small molecules having smaller molecular masses was joined by some unknown intermolecular force. Later, Hermann Staudinger first anticipated that polymers consist of long chains of atoms held together by some covalent bonds. Staudinger, for his immense contribution in providing concepts to understand about polymeric structure, was awarded the Nobel Prize in Chemistry in the year 1953.

During the period of World War II there had been an increased demand of natural polymers such as silk and rubber. But due to their limited supply, alternative synthetic polymers such as nylon were introduced to meet the essential requirements.

Then, the invention of advanced polymeric materials like Kevlar and Teflon became eye-catching to launch the robust polymer industry.

The most important properties of a reactant to be a monomer or to take part in polymerization reaction are (i) presence of reactive functional group and (ii) the minimum requirement of bifunctionality. Depending upon the nature of functionality, for example, bi- or tri-functionality, the polymers may be designed as chain polymers, blocked or cross-linked polymers, respectively. For the synthesis of polymers, various methods such as addition polymerization, condensation polymerization, and co-ordination polymerization have been employed. In most of the synthetic procedures use of catalyst, initiator, application of high pressure and temperatures are the important requirements to assist rapid polymerization process.

Due to the great demand of polymeric materials in day-to-day life, many researchers across the globe are working in this field, and hence Nobel Prize has been awarded to many scientists for the development of this particular field. Karl Ziegler (1898–1973) and Giulio Natta (1903–1979) independently developed catalysts which smoothly assist polymerization under ambient condition with high stereospecificity to furnish isotactic polypropylene (i-PP) and HDPE. In the year 1963, Giulio Natta and Karl Ziegler were jointly awarded Nobel Prize in Chemistry for their immense contributions in the area of polymer science. Next, Flory was awarded Nobel Prize in Chemistry in 1974 for his important contribution in both theoretical and experimental development of polymer science. In the year 2000, Alan G. MacDiarmid, Alan J. Heeger, and Hideki Shirakawa were jointly awarded Nobel Prize in Chemistry for their work on conductive polymers which provide significant contributions in the advancement of molecular electronics. John Bennett Fenn, Koichi Tanaka, and Kurt Wüthrich were awarded Nobel Prize in Chemistry in the year 2002 for their innovative methods in identification and structural analysis of biological macromolecules. Robert Grubbs, Richard Schrock, and Yves Chauvin were awarded Nobel Prize in Chemistry in the year 2005 for the development of olefin metathesis also known as ring-opening metathesis polymerization reaction (ROMP). The use of Ru-based Grub's catalyst for polymerization gained significant popularity in recent past. Some of the major breakthroughs that evolved during the development of polymer science have been depicted in Table 1.2.

Even though polymeric materials have important applications, their non-biodegradability always poses a threat to the surroundings. For example, polyethene compounds cannot be broken down by the bacteria present in the soil, and due to constant accumulation of these materials soil and water pollution arises. To solve this particular problem, extensive study is going on around the world for the development of biodegradable polymers which are also considered as green polymers. One important example of biodegradable polymers is poly β-hydroxybutyrate-*co*-β-hydroxy valerate (PHBV) which can be prepared by the combination of 3-hydroxy butanoic acid and 3-hydroxy pentanoic acid.

Recently polymer matrix composites (PMCs) have been introduced which possess a wide range of potential applications (Tewatia et al. 2017; Ramakrishna et al. 2001). PMCs are polymers reinforced with additives such as glass fibers, carbon fibers, carbon nanofibers, graphene, and carbon nanotubes. They have been

Table 1.2 Major breakthroughs in the history and discovery of polymer science.

Year	Name of the scientists and contribution
1833	Berzelius for the introduction of the polymer terminology
1839	Edward Simon for the synthesis of polystyrene
1859	A.V. Lourenço and Charles Adolphe Wurtz for the production of polyethylene glycol
1900	Hermann Staudinger (Father of Modern Polymer Science) for providing the idea about the structure of polymers (long-chain molecules), awarded Nobel Prize award in 1953
1938	W.H. Carothers for the synthesis of nylon which laid the foundation of the synthetic fiber industry
1963	Giulio Natta and Karl Ziegler awarded Nobel Prize in Chemistry for the metathesis ring-opening polymerization reaction
1974	Paul Flory awarded Nobel Prize in Chemistry for providing significant contributions in the area of experimental and theoretical polymer research
1984	Bruce Merrifield for the development of solid-phase protein synthesis awarded Nobel Prize in Chemistry
2000	Alan J. Heeger, Alan G. MacDiarmid, and Hideki Shirakawa awarded Nobel Prize in Chemistry for the discovery and development of electrically conductive polymers
2002	John Bennett Fenn, Koichi Tanaka, and Kurt Wüthrich for identification and structure analysis of biological macromolecules, awarded Nobel Prize in Chemistry
2005	Grubbs and coworkers awarded Nobel Prize in Chemistry for the introduction of metathesis tactics which provided significant advancement in synthetic chemistry

widely employed because of the magical enhancements of mechanical, thermal, and electrical properties associated with this special type of material. Presently, researchers are looking for intelligent polymers which are specially designed materials that can extend the scope of polymer science.

1.2 Classification of Organic Polymers

Polymers are divided primarily into organic polymers and inorganic polymers based on the nature of backbone chain. Primarily, organic polymers are made up of carbon backbone structures, while inorganic polymers do not have carbon atoms in their backbone. Organic polymers are suitable for several applications because of their simple structures and biodegradable nature, which make them environmentally friendly. Nevertheless hybrid polymers like "silicones," which are organosilicon polymers or polysiloxanes, are also well known because of their potential industrial applicability. Silicones found wide applications because of the presence of strong carbon-silicon (C—Si) bonds. The formation takes place in a few steps, starting from the alkylation/arylation of a silicon tetrachloride and finally resulting in silicones, $R_2(SiO)_n$. Here "R" represents an alkyl/aryl group. Silicones found commercial applications mainly under three category namely fluids, elastomers, and resins (Zielecka et al. 2002).

It is well known that most of the polymers are built up via two organic-chemistry-based mechanisms. One is addition/chain and the another is condensation/step-growth polymerization mechanism. Both of these mechanisms involve a combination of small monomers or repeating units. The mechanism of polymer formation results in either addition or chain polymers. Apart from this, polymers are broadly classified based on their source, structure, molecular forces operating, and thermal behavior (Kariduraganavar et al. 2014).

(a) Originally polymers were classified as either "natural or synthetic polymers." Natural polymers occur naturally, whereas synthetic polymers can be synthesized artificially. Natural polymers are obtained from animals and plants. A few examples include

  (i) *Proteins*: Wool and natural silk are protein-based products. Proteins are the building blocks of animal cells. The proteins are formed by the repetition of α-amino acids (Figure 1.1).

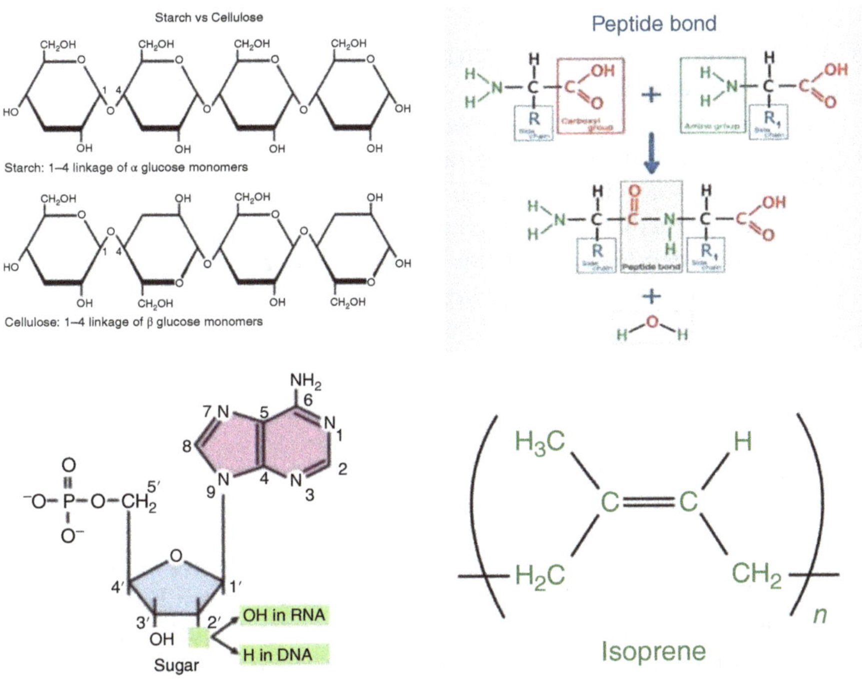

Figure 1.1 Building blocks of a few natural polymers. Source: Adapted from https://d1hj4to4g9ba46.cloudfront.net/questions/1536297_1144437_ans_9129a4f3e7f84e03932353ae379c21a0.png.

(ii) *Carbohydrate polysaccharides*: Repetition of smaller monosaccharides like "glucose" leads to the formation of macromolecules like cellulose and starch, which constructs the plant cell. These biological molecules also supply energy to animal cells (Figure 1.1).

(iii) *Nucleic acids*: "RNA" and "DNA" are the bio-polymers formed by the repetition of nucleotides. These are the primary constituents of all living cells (Figure 1.1).

(iv) *Natural rubber*: This is a well-known plant-extracted component. It is present in the latex of rubber plant. It is also known as *cis*-polyisoprene, which is a polymer of isoprene. However the trans-isomer is known as "gutta-percha" (Figure 1.1).

On the other hand, most of the daily-use polymers are synthetic ones. This covers a series of polyvinyl-based chain polymers like "polyvinylchloride" and functional-group-based condensed polymers like "polyethyleneterephthalate." A few examples are PVC, polyethylene, polystyrene, Teflon, synthetic rubber, nylon-6, and nylon-66. Most of these are organic polymers. Moreover "semi-synthetic polymers" also do exist. These are obtained by chemical modification of natural polymers. For example, one of the important materials, cellulose diacetate, is synthesized from cellulose by reacting with acetic anhydride in the presence of sulfuric acid (Kamide 2005). It is formed by the acetylation reaction. Some of the hydrogen atoms in cellulose are substituted by acetyl ($-COCH_3$) groups.

(b) Polymers are divided into three structural categories, namely, linear, branched, and cross-linked polymers.

(i) Linear polymers are long straight-chain molecules with highly ordered structures. Importantly, the randomness is comparatively less as the monomers are linked together in linear fashion. Because of the linear arrangement their structure is densely packed. Some of the characteristics, like high densities, melting temperatures, and mechanical strengths, are the result of this. Polyester and polyethylene belong to this category.

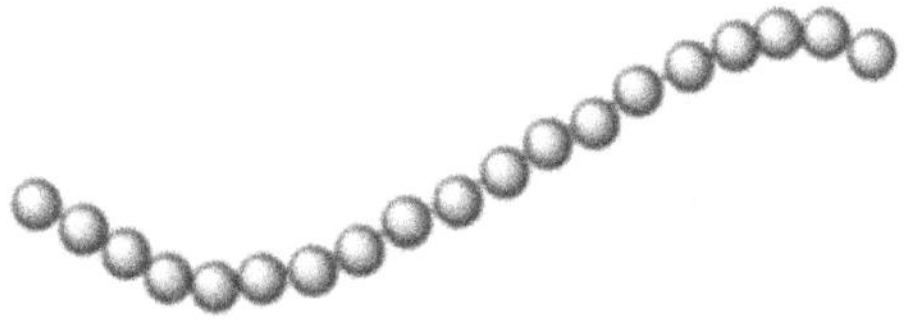

(ii) Branched one possesses secondary polymer chains connected to a primary backbone, resulting in a branching-tree-like structure. The chains are less well packed due to branching, which results in much lower melting temperatures, densities, and mechanical strengths (Abuzreda 2023). Branch-chain polymers can be created via condensation or addition polymerization methods. Examples include amylopectin, glycogen, LDPE, and any rubbers that have been vulcanized.

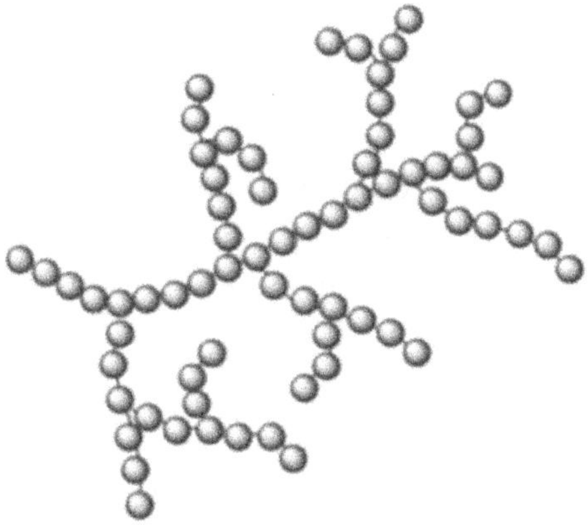

(iii) Network or more specifically the cross-linked polymers are formed by connecting two or more chains by cross-linking agents. The initially formed linear polymeric chains are joined together to form a three-dimensional network structure. The chain linkages give these polymers their hard, stiff, and brittle characteristics. Condensation polymers are invariably cross-linked polymers. A group of cross-linked polymers is known as resins.

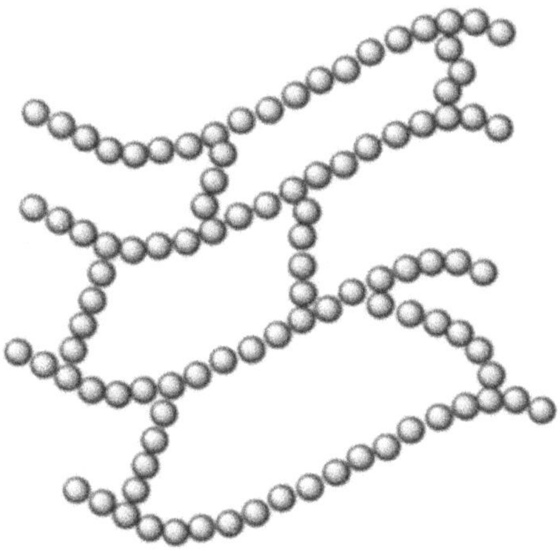

(c) Depending upon the mechanism of formation, polymers are classified into addition and condensation polymers. Addition or chain reaction polymers are formed by successive addition of monomers (small molecules). The linked monomers are called "repeating units" (RUs). The RUs are joined predominantly by covalent bonds. Their mechanism of formation is primarily dependent upon the three steps: initiation, propagation, and termination. The propagation step is responsible for getting high-molecular-weight polymers. The molecular weight is an important property in polymer characterization. "Polyethylene" is a chain polymer formed by the repetition of ethylene monomers.

$$n(CH_2{=}CH_2) \rightarrow (CH_2 - CH_2)_n$$

Alternatively, condensation reaction between monomers bearing typical functional groups like hydroxyl (–OH), carboxyl (–COOH), ester (–COOR), and amide (–$CONH_2$) results in the formation of condensed polymers. During condensation reaction some small molecules like water or mineral acids are released from the monomers by condensation. Polymerization proceeds by successive formation of dimers, trimers, and tetramers. However the extent of polymerization between them is important for having a high-molecular-weight condensation polymer (Kariduraganavar et al. 2014). Nylon-6,6 is a common example of condensation polymer. It is formed by the condensation of hexamethylene diamine, NH_2–$(CH_2)_6$–NH_2, and adipic acid, COOH–$(CH_2)_4$–COOH. The reaction takes place between a diamine and a dibasic acid. The two monomers are linked by a peptide bond as shown in Figure 1.1. One of the advantages here is that no initiator is required for the reaction.

(d) Elastomers and fibers are two subcategories of polymers that are determined by the molecular forces between individual chains.
An elastomer is plastic and its polymeric chains are bounded by very weak van der Waal forces. Because of such weak operating molecular forces, these polymers are stretchable as well as compressible. They can be reverted back to their earlier form. It is a randomly oriented amorphous polymer. Vulcanized rubber is an elastomer that exhibits cross-linking by sulfur atoms. Such cross-linking prevents the chains from slipping over one another.
However, due to the highly organized geometry and potent intermolecular forces of attraction, fibers possess the highest tensile strength and negligible elasticity. Fibers have a high melting point and low solubility, and are crystalline in nature. Fibers include polymers like cellulose, nylon, wool, and silk (Abuzreda 2023).

(e) Polymers are divided into thermoplastics and thermosetting plastics based on heat treatment. Interestingly most of organic macromolecules belong to these categories.
Thermoplastics are a type of polymer that softens when heated and may take on any shape when cooled. Thermoplastic polymers can undergo transition from hardness to softness and vice versa. At normal temperature conditions, they are hard. With increase in temperature, hardness decreases and they become soft and viscous. The polymeric chains can slip over one another during heating. When cooled, this material stiffens up after being soft and viscous. Hence depending on temperature, thermoplastic polymers can be molded into various shapes and products. Without affecting the chemical identity and mechanical merits of the plastic, the heating, softening, and cooling procedures can be performed as frequently as needed (Carothers 1929). They have highly ordered crystalline as well as less ordered amorphous regions. Polyethylene, polycarbonate, polypropylene, Teflon, and PVC are a few typical examples.

When thermosetting polymers are heated, they undergo chemical transformation and cross-linking takes place, and when they are cooled, they become permanently hard and infusible. Repeating heat treatment doesn't cause them to soften; instead, they degrade. These polymers undergo setting to only one form during heat treatment. It cannot be converted back to the original form. They become hard on heating and turn into a solid that cannot be re-melted by changing the temperature (Carothers 1929). These polymers set into a solid structure because of the existence of cross-linking. With rise in the degree of cross-linking the rigidity also increases. Upon heating the hardness increases due to extensive cross-linking. Moreover, cross-linking reduces the mobility of the polymer chains, causing them to be relatively brittle. Common examples of thermosetting polymers are phenol-formaldehyde resin and urea-formaldehyde resin (Figure 1.2).

(f) In addition to these, organic polymers also categorized as conducting polymers as they found promising applications in organic light-emitting diode (OLED), semiconductors, and superconductors. Interestingly, the first conducting polymer is an organic polymer, polypyrrole, as reported by Donald Weiss and coworkers in 1960. Polypyrrole is a heterocyclic polymer containing carbon and nitrogen atoms in its rings. Because of the conjugation of electrons in their rings, they can act as conducting polymers. Conducive polymers' most important benefit is their processability, which is mostly attained through dispersion (Saldívar-Guerra and Vivaldo-Lima 2013).

Conducting polymers can be further classified into intrinsically conducting polymers, extrinsically conducting polymers or doped polymers.

Intrinsically conducting polymers: This class of polymers conducts electricity due to the presence of delocalized electron pairs or conjugated π-bonds in

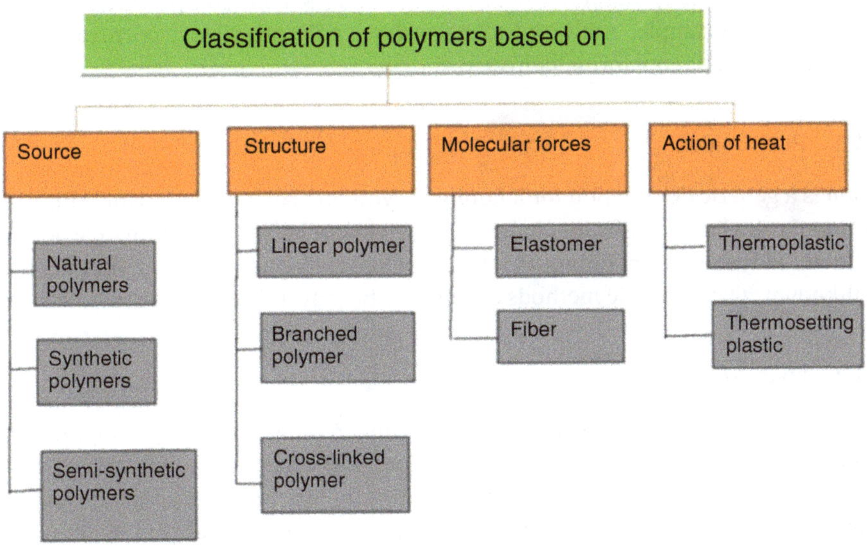

Figure 1.2 General classification of polymers (self-made).

the structural frame. In a polymeric material with an electric field, the delocalized electrons can move across it. Coupled bands, valence bands, and conduction bands are produced as a result of overlapping orbitals.

Semiconducting or impure polymers: Doping is a process by which conductivity of materials can be enhanced by adding or doping some impurity. Impurity means certain elements in their chemical form. Doping creates charge carriers in the polymer chain by either oxidation or reduction processes. The common doping processes are p-doping and n-doping. Both involve creation of positive holes and negative excess electrons by the redox processes.

1.3 Synthesis and Properties of Polymers

Polymer can be synthesized by two ways: (i) polymerization of the monomeric units and (ii) chemical modification of monomeric side chain of the polymer. Through the first method molecular weight can be increased, whereas the second method will change the structure of the molecule without molecular weight variation.

The chemical process which converts molecules into polymer macromolecules is known as polymerization. This process is the combination of different reactions which determine the features of the product. The properties of the obtained macromolecules are also dependent on the chemical composition of the starting materials. This chapter will mainly cover the synthesis of polymers by using different polymerization processes. Basically there are two types of polymerization processes: viz. **step growth** or **condensation polymerization** and **addition polymerization**.

Step Growth Polymerization is one type of polymerization process in which bi-functional or multifunctional monomers react to form first dimers, then trimers, longer oligomers, and eventually long-chain polymers. Through this step growth polymerization, many naturally occurring as well as synthetic polymers can be obtained. For example, polyamides, polyesters, polycarbonates, polyurethanes, epoxy resin, and phenol formaldehyde resin.

1.3.1 Polyamides

Nylon is a generic designation for a family of synthetic polyamides. Nylon, silk-like thermoplastics have significant applications in fabric, fibers, shapes, films, etc. Many synthetic polyamides, including nylon-6,6, nylon-6, nylon-6,10, and nylon-11, are well-known. The synthetic methods of some of the polyamides are discussed below.

1.3.2 Nylon-6,6

Hexamethylene diamine and adipic acid (Hexan-1,6-dioic acid) are co-polymers in it. Benzene, a byproduct of petroleum refining and cracking, is the primary raw material used in the manufacture of nylon-6,6. Adipic acid serves as the precursor once more in the creation of hexamethylene diamine. The synthesis of nylon-6,6 is shown in Scheme 1.1 (Gowarikar et al. 2022).

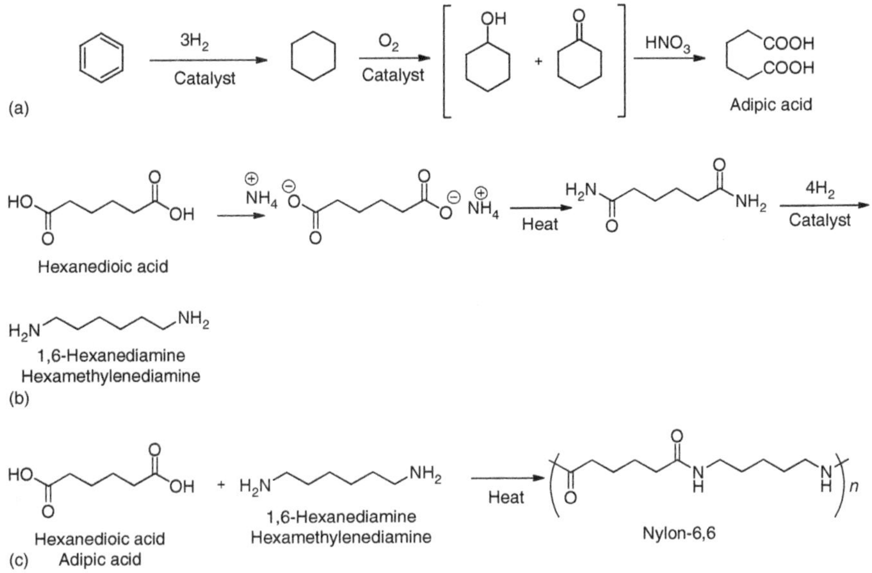

Scheme 1.1 (a) Synthesis of adipic acid from benzene, (b) synthesis of hexamethylenediamine from adipic acid, and (c) synthesis of nylon-6,6.

During fabrication, nylon fibers are cold-drawn to about four times their original length, which increases crystallinity, tensile strength, and stiffness.

1.3.3 Nylon-6

Nylon-6 or polycaprolactam, a semicrystalline polyamide, is synthesized from caprolactam. Fibers, brush bristles, high-impact moldings, and tire cords are all made from nylon-6. The synthesis of nylon-6 is depicted in the Scheme 1.2.

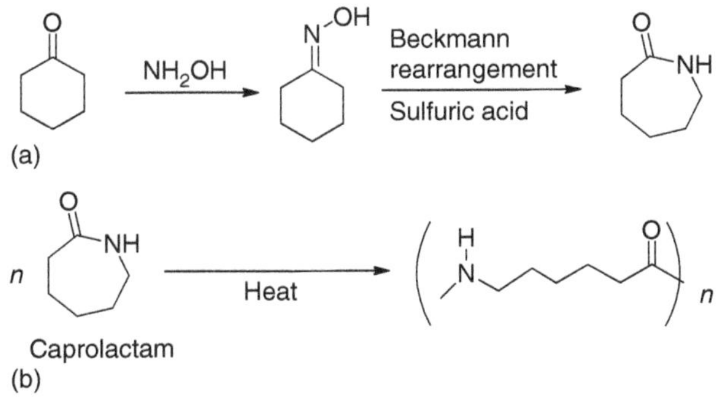

Scheme 1.2 (a) Synthesis of caprolactam and (b) synthesis of nylon 6.

Properties of nylon:

- ➢ With their parallel arrangement of molecules and hydrogen bonds holding them together, nylon fibers have a linear structure. Strong intermolecular forces give these fibers a more crystalline structure and provide them great strength, flexibility, and a high melting point.

- Nylons are chemically stable and resistant to abrasion.
- The yarn is smooth, long lasting, and can be spun into fabric.
- The fabric can withstand heat and steam and is strong, glossy, moisture-resistant, simple to dye, and retains color.

1.3.4 Polyesters

Polyester is a class of polymer in which an ester functional group is present in every repeating unit of the polymer chain. Some polyesters are naturally occurring compounds found in plants and insects, some are synthetic compounds like polybutyrate. But it most frequently refers to a type of substance known as polyethylene terephthalate as a specific material (PET). Biodegradable polyesters include some synthetic as well as some natural varieties; however, the majority of synthetic polyesters are not biodegradable. Synthetic polyester is extensively used in cloth industry.

1.3.5 Polyethylene Terephthalate (PET)

The most popular thermoplastic polymer of the polyester family is PET. This is used in clothing fibers, food and liquid storage containers, thermoforming for manufacturing, and in combination with glass fiber for engineering resins (Cann and Connelly 2000). There are two strategies for the synthesis of PET. The first method is the direct reaction between terephthalic acid and ethylene glycol which is mentioned below in Scheme 1.3a. PET is commercially known as Terylene.

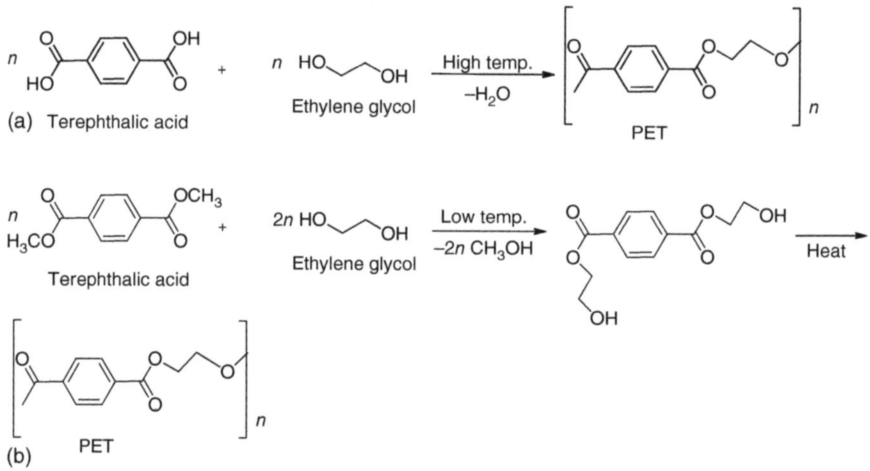

Scheme 1.3 (a) Synthesis of PET by direct method and (b) synthesis of PET by trans esterification method.

This method is a typical Fisher-type esterification reaction where an acid reacts with an alcohol by following the usual mechanism. The second method is the ester exchange of a diester and a diol. This is a transesterification reaction in which one ester is transformed into another.

By this method, PET is synthesized by the reaction between dimethylterephthalate and ethylene glycol as shown in Scheme 1.3b (Gowarikar et al. 2022).

Terylene's properties:

- ➢ It offers excellent mechanical strength and stability in its dimensions.
- ➢ It has a very high strength and will only slightly lose strength when wet.
- ➢ It is elastic by nature, stable in the temperature range of −40 to 100 °C, and has good insulating qualities as well as creep and abrasion resistance.
- ➢ It exhibits little water absorption and is resistant to water, diluted acids, salts, aliphatic and aromatic hydrocarbons, and alcohols at room temperature.

1.3.6 Polycarbonates

Polycarbonates are a type of thermoplastic polymers featuring a carbonate group in their structure. As this type of polymer can be worked, molded, and thermoformed easily, so it has a variety of applications. The very important polycarbonate Lexan is produced by the reaction of bisphenol A (BPA) and phosgene $COCl_2$ in dichloromethane by using phase transfer catalyst as shown in Scheme 1.4. High impact and tensile strength, tough translucent polymer Lexan retains its shape over a wide temperature range. It is utilized in athletic goods like catcher's masks for baseball and hockey as well as bicycle, football, and snowmobile helmets. Moreover, it is employed in the manufacture of safety and unbreakable windows.

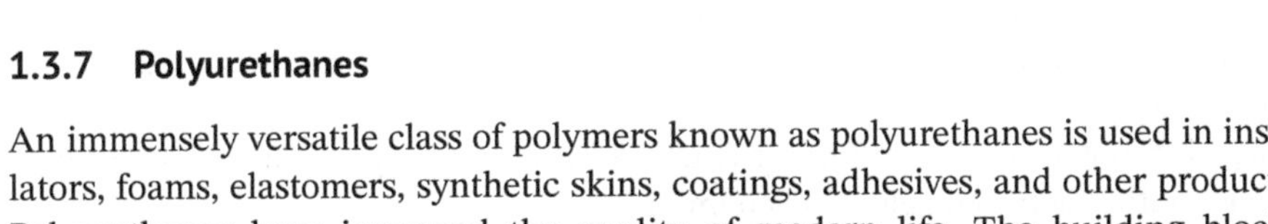

Scheme 1.4 Synthesis of Lexan (Ghosh 2011).

1.3.7 Polyurethanes

An immensely versatile class of polymers known as polyurethanes is used in insulators, foams, elastomers, synthetic skins, coatings, adhesives, and other products. Polyurethanes have improved the quality of modern life. The building blocks of polyurethanes alternate between hard urethane units and flexible polyester or polyether units. The more flexible urethane blocks are made from low-MW polyesters or polyethers that include –OH groups at the ends of each polymer chain, whereas the rigid urethane blocks are made from a diisocyanate. Dr. Otto Bayer and his coworkers first synthesized polyurethane by simple diisocyanate polyaddition procedures (Janik et al. 2014). It achieved industrial-scale synthesis in 1937, and by the 1950s, it had become a market staple. A urethane linkage distinguishes polyurethanes, and they are easy to synthesize via an addition reaction between alcohol and an isocyanate. The synthesis of polyurethanes is shown in Scheme 1.5.

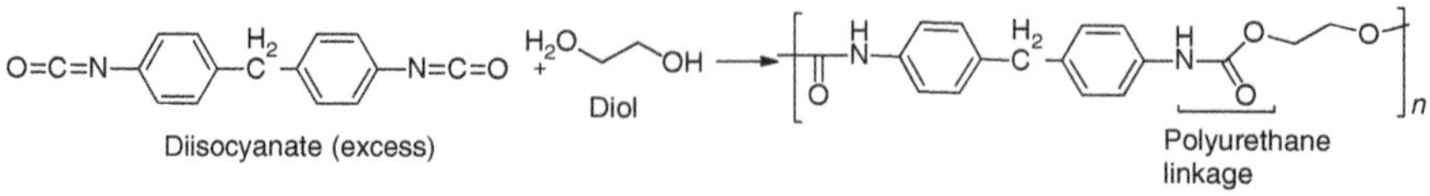

Scheme 1.5 Synthesis of a polyurethane.

1.3.7.1 Properties of Polyurethanes

Molecular weight between cross-links and the number of segments in a stiff sequence are the key factors affecting polyurethanes' thermal characteristics. Long alkyl chains, aromatic groups, branching and cross-linking, as well as the strength of the secondary bonding forces, all have a significant impact on mechanical qualities like elongation break, rip strength, and tensile strength. The types of polyols and isocyanates that are used to create polyurethane have a significant impact on its chemical characteristics.

Some key properties of polyurethanes are mentioned below:

- ➢ Good abrasion resistance
- ➢ Good impact resistance
- ➢ Good toughness
- ➢ Low viscosity
- ➢ High elongation
- ➢ Good flexibility
- ➢ Good tear strength
- ➢ Low shrinkage
- ➢ Hydrolytically stable

1.3.8 Epoxy Resin

Epoxy resins, a very significant group of thermosetting polymers, are extensively used in aircraft structures because of their low shrinkage and low release of volatiles during curing, high strength, and good durability in hot and moist environments (Januszewski et al. 2021). This type of resins can be produced in different forms ranging from low-viscosity liquids to high-melting solids. The most widely used epoxide monomer, diepoxide, is prepared by the treatment of 1 mol of bisphenol A with 2 mol of epichlorohydrin. This on further reaction with diamine gives the resin. The synthetic route is depicted below in Scheme 1.6.

1.3.8.1 Properties of Epoxy Resin

Epoxies are compatible with the majority of substrates. They are particularly well suited for composite applications since they have a tendency to moisten surfaces easily. Epoxy resin can also be used to modify some polymers like polyurethane or unsaturated polyesters in order to improve their physical and chemical properties. Some other properties are listed below:

- ➢ High strength
- ➢ Low shrinkage

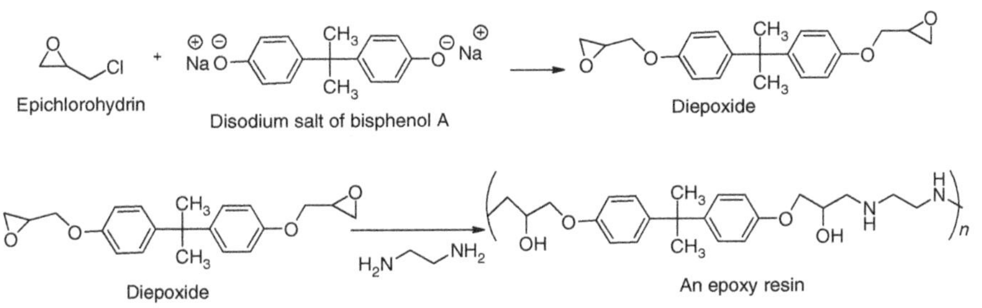

Scheme 1.6 Synthesis of an epoxy resin.

- ➢ Effective electric insulation
- ➢ Its glass transition temperature ranges from 150 to 220 °C
- ➢ Excellent adhesion to various substrates
- ➢ Chemical and solvent resistance

1.3.9 Phenol Formaldehyde Resin

The condensation polymer phenol-formaldehyde resins can be synthesized by the polycondensation reaction between phenol and formaldehyde in the presence of acidic or basic catalyst. These are thermosetting polymers. They were the first commercial synthetic resins prepared by Baekeland, an American chemist who gave the name Bakelite. Electrophilic aromatic substitution reaction between phenol and formaldehyde in the presence of a mineral acid like HCl produces *o*-hydroxymethyl phenol and *p*-hydroxymethylphenol, which are the actual monomers. This is the initial stage of the mechanism of polymerization. A linear (straight-chain) polymer known as Novolak is produced when only one type of monomers, such as *o*-hydroxymethylphenol, react together in a head-to-tail fashion. However, if cross-polymerization occurs with a certain amount of polysubstituted phenol, Bakelite – a strongly cross-linked polymer – is produced. The reaction for preparing Bakelite is depicted in Scheme 1.7

1.3.9.1 Properties of Phenol Formaldehyde Resin

Phenol formaldehyde is a synthetic resin made from phenol and formaldehyde. It is thermosetting in nature. It is used to make products such as adhesives, coatings, and molded products. Some of the key properties of phenol-formaldehyde resin are high strength, resistance to chemicals and solvents, low flammability, and good thermal stability. Novolak is a solid and stable resin. It has good solvent and chemical resistance. It is electrically insulated with good dielectric resistance.

Chain growth or addition polymerization is a type of polymerization process which involves adding monomers sequentially that are either unsaturated or have additional reactive functional groups. Carbocations, carbanions, radicals, and organometallic complexes are just a few examples of the reaction intermediates that are produced during this kind of polymerization. This chain growth polymerization proceeds through chain initiation, propagation, and termination steps. The synthesis of some of the important addition polymers is mentioned below.

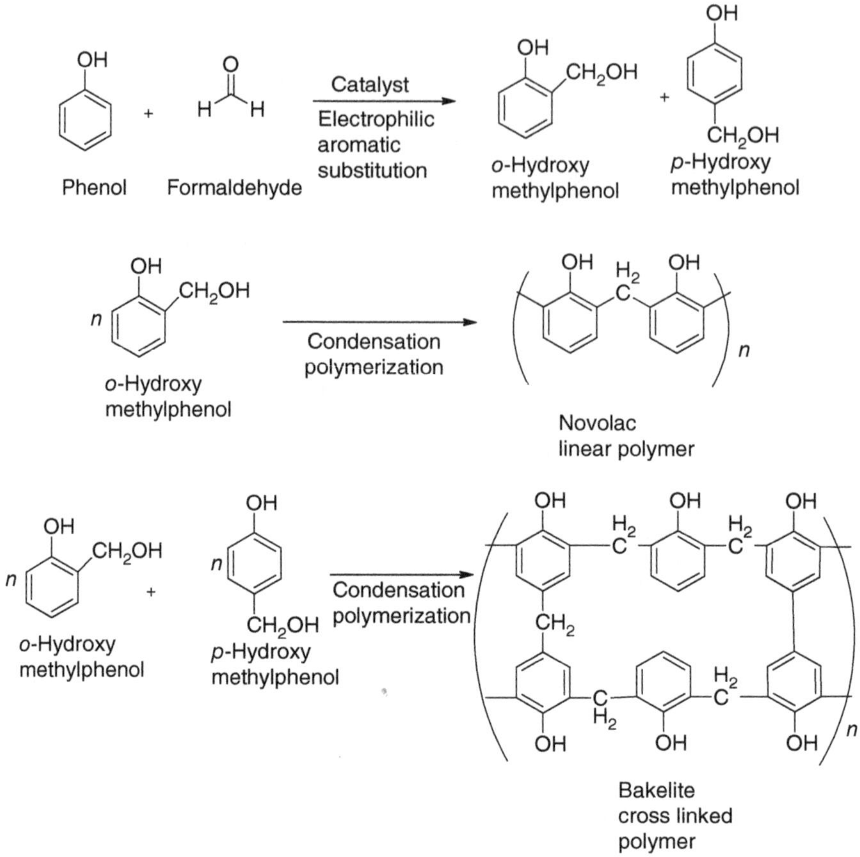

Scheme 1.7 Synthesis of phenol formaldehyde resin.

1.3.10 Polyethene

Polyethene, a simple and extensively used homopolymer, is synthesized by the addition polymerization of single monomer ethylene. The reaction for preparation of polyethene is shown in Scheme 1.8. Polyethene can also be synthesized by coordination polymerization by using Ziegler–Natta catalyst.

$$n\ H_2C{=}CH_2 \xrightarrow{\text{Addition polymerization}} (H_2C{-}CH_2)_n$$

Scheme 1.8 Synthesis of polyethene.

1.3.11 Coordination Polymerization

Ziegler–Natta catalysts are used in coordination polymerization. Combination of triethylaluminum and titanium or vanadium tetrachloride is the most used Ziegler–Natta catalyst. This catalyst is highly stereoselective. Polyethene can be used to explain the mechanism. The ethene monomer produces a π-complex by first attacking the catalyst surface (Ti). It then produces a transition state by partially using its π-electrons and interacting with the alkyl group. In the end, the Ti-alkyl link breaks, leading to the formation of bonds between the monomer and the alkyl group and between the monomer and Ti. Schemes 1.9 and 1.10 describe the process and mechanism of synthesizing polyethylene utilizing Ziegler–Natta catalysts.

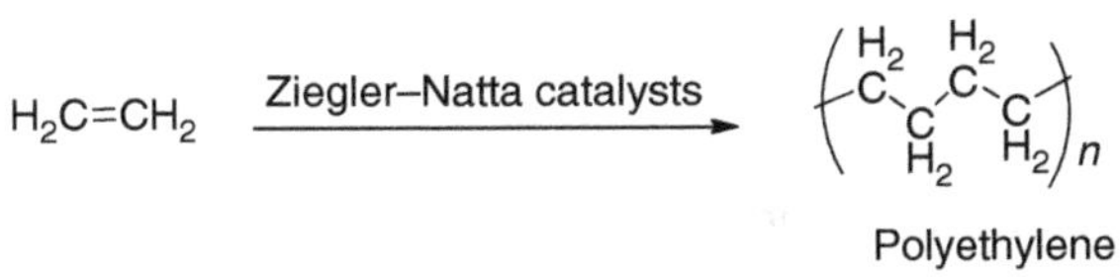

Scheme 1.9 Synthesis of polyethylene by Ziegler–Natta catalysis.

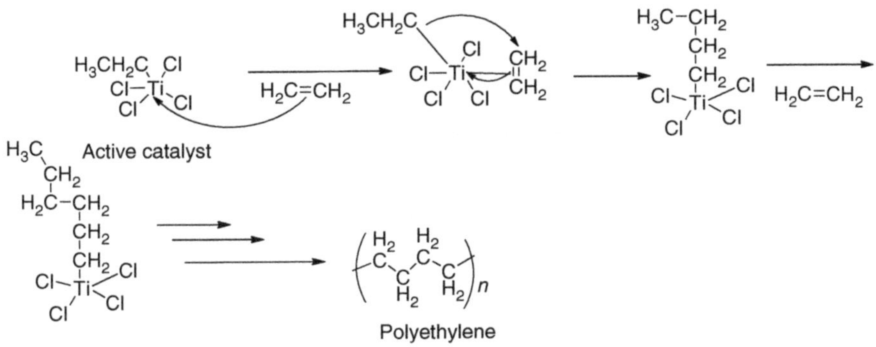

Scheme 1.10 Mechanism of synthesis of polyethylene by coordination polymerization.

Polyethenes can be broadly categorized into two types

(i) LDPE
(ii) HDPE

Very-LDPE is produced using a free radical process at 80–350 °C, high pressures of 1000–3000 atm, and the presence of oxygen or peroxide.

Properties:

- It is a transparent waxy substance with superior electrical insulating qualities, a low brittleness temperature, film transparency, and strong impact strength.
- It is chemically inert and has good resistance to acids and alkalis. However, it swells and is permeable to oils.
- It melts in temperature range of 107–120 °C. Its glass transition temperature (*Tg*) value is −120 °C.
- It is flexible over a wide temperature range.

HDPE can be prepared by the coordination polymerization using Ziegler–Natta catalyst, but it can also be prepared by the ionic mechanism at low temperatures of 60–70 °C and very low pressures of 6–7. HDPE is a linear polymer with little to no branching.

Properties:

- ➢ It is opaque and has less impact strength, but better barrier properties than LDPE.
- ➢ It has better chemical resistance than LDPE.
- ➢ It melts in temperature range of 130–138 °C. Its glass transition temperature is −20 °C.
- ➢ It exhibits better stiffness, toughness, tensile strength, and heat resistance.

1.3.12 Polyvinylchloride

PVC, a thermoplastic homo-polymer, can be prepared by the addition polymerization of single monomer vinyl chloride. The reaction for preparation of PVC is shown in Scheme 1.11.

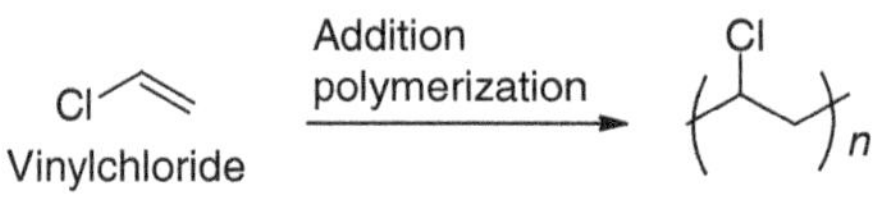

Scheme 1.11 Synthesis of polyvinyl chloride.

PVC can be classified into two classes:

(a) Unplasticized PVC
(b) Plasticized PVC

PVC-R or PVC-U, often known as rigid or unplasticized PVC, is a strong, rigid material that is also chemically resistant and naturally flame retardant. To make plasticized polyvinyl chloride (PVC-P) more flexible, several chemical substances known as plasticizers have been added to the polymer.

Properties:

- ➢ It is a substance that is chemically inert, colorless, odorless, and very resistant to acids and alkalis. Plasticized PVC is less chemically resistant than unplasticized PVC, nevertheless.
- ➢ PVC is now replacing conventional building materials like wood, metal, and concrete because of its versatile properties, including its lightweight nature, durability, and ease of processing. It is also resistant to atmospheric oxygen or oxidation. It also has greater stiffness and rigidity than polyethene.

1.3.13 Polytetrafluoroethylene (PTFE)

Polytetrafluoroethylene (PTFE), also known as Teflon, is a homopolymer and its monomer unit is tetrafluoroethylene. It is also a thermoplastic polymer. This is typically synthesized through the use of peroxide or ammonium persulfate as initiators in the emulsion polymerization of tetrafluoroethylene. The procedure for preparing Teflon is depicted in Scheme 1.12.

Properties:

- ➢ It is a polymer with a high degree of crystallinity (93–98%).
- ➢ It is extremely resistant to chemicals.
- ➢ Almost none of the solvents can dissolve it, and neither water nor oil can moisten it.
- ➢ It has a low coefficient of friction and maintains its slickness over a broad temperature range (−40 to 300 °C).
- ➢ It has great temperature stability and is temperature resistant.
- ➢ It possesses superb electrical insulating qualities.

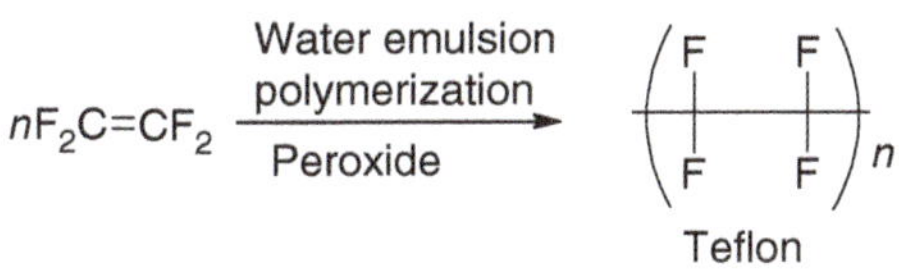

Scheme 1.12 Synthesis of Teflon.

General applications of organic polymers are discussed in the following section.

1.3.14 Application of Natural Polymers

Organic polymers, either synthetic or natural, are crucial for everyday life. The natural polymers primarily belong to proteins and polysaccharides. Both categories are either plant based or animal based. The existing polysaccharides however are also bacteria based (e.g., Dextran), fungi based (e.g., pullulan), and algae based (e.g., alginate) (McChesney et al. 2007) (Figure 1.3).

Natural polymers are the building blocks of most of the living things providing basic structural materials and participating in vital life processes. For example, the polymers cellulose and starch are the primary construction units of a plant. Natural fibers such as cotton are made up of cellulose. Cellulose is a linear polymer of glucose units, an abundant organic polymer on Earth. Cellulose has gained immense technological importance in recent days. Cellulose as such being water insoluble finds some technological barrier in applications (Guo et al. 1998). So numerous chemical modifications have been carried out upon cellulose, resulting in cellulose derivatives.

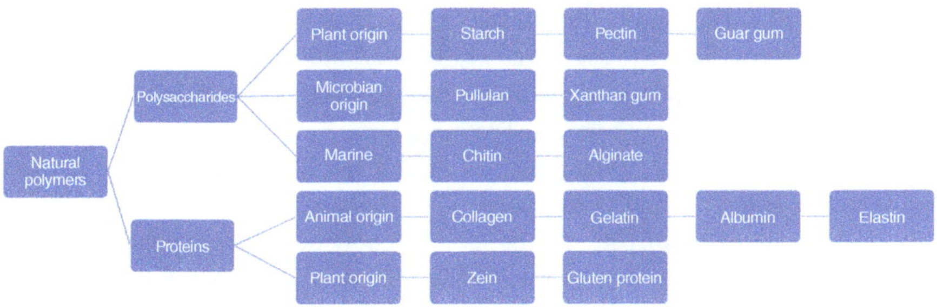

Figure 1.3 A flowchart of natural polymers (self-made).

These derivatives found good industrial importance because of their water-soluble properties. As an example, etherification results in cellulose ethers, which are water soluble in nature. Cellulose ethers are important materials in pharmaceutical industry because of their hydrophilicity and good swelling properties. Apart from this, cellulose and its derivatives are versatile components in many industries such as veterinary foods, wood and paper, fibers and clothes, body care, cosmetics, paint, and coatings (Dumitriu 1996). Cellulose ethers and cellulose esters, which are derivatives of cellulose, exhibit different physicochemical and mechanical properties. Information about a few important semi-synthetic derivatives like methyl cellulose, cellulose nitrates, hydroxypropyl methylcellulose, and ethyl (hydroxyethyl) cellulose is noteworthy to mention here because of their tremendous applications.

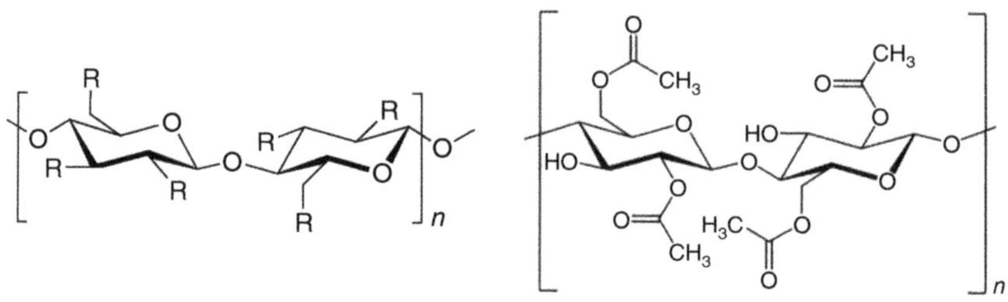

Cellulose ether   Cellulose ester

The carbohydrate, starch, is a natural polymer formed by the repetition of glucose units. The breakdown of starch in our body releases energy. At the same time, starch is a raw material for bioplastics in industry. Processing of starch with plasticizers can increase its mechanical strength ((Gajre)Kulkarni et al. 2012). In general, starch is an important gradient of food/bakery industry, paper, and textile mills. Polysaccharides are important analogues of cosmetic industries. Starch nanoparticles possess anti-inflammatory properties, which are essential in drug designing.

Natural-polymer-based hydrogels seek research attention for their applications in food industry and agriculture. Hydrogels that can absorb large amount of water can exist either naturally or synthesized. Natural-polymer-loaded drugs find effective treatment of various kinds of diseases. As an example, chitosan-loaded ascorbic acid was found effective against cervical cancer. Polymeric nanoparticles especially derived from natural polymers are mostly used in controlled drug-delivery systems because of their certain pharmaceutically acceptable properties like biocompatibility, bioadhesion capacity, low cost, availability, and lack of toxicity (Pandey and Khuller 2004). The polymeric nanoparticles can act in controlled drug-release systems in addition to drug carriers. Proteins, for example, collagen and gelatin are suitable candidates for development of drug-delivery systems because of their stability under different conditions of pH and ionic strength.

Our food primarily consists of the natural polymers, polysaccharides, and proteins. A small amount of polynucleotides (DNA and RNA) is also involved in the cellular part. Proteins are the source of amino acids in our body and have high water-holding capacity. The creation and stabilization of food structures is dependent upon the polymers used. Moreover the physiological and biological function of food also relies upon the nature of polymer used. Most of the emulsifiers and stabilizers used in food industry are natural polymers (Adelmann et al. 2012). During food formulation, they stabilize the emulsion droplets and foams. Polymers are also applied as thickeners in food processing to increase the viscosity of the food. A few of the natural polymers can act as gelling agents or can form gels. A gel is a 3D network formed by cross-linking of polymer chains, trapping the water molecules and making them flow resistant. The textural properties of food (such as chewy or creamy) depend on the type and concentration of gel-forming polymers used during food formulation. Interestingly, natural polymers can also develop the bioavailability of vitamins (Guo et al. 1998). Essential nutrients undergo degradation by the action of heat, light, and oxygen. Polymers in this context can protect those by enhancing stability and strength. The antioxidative and antimicrobial functions of certain natural polymers place them in food formulations.

One of the important natural polymers is “natural rubber” that finds enormous application in our daily life. The polymerization of isoprene results in natural rubber. Isoprene is chemically 2-methyl-1,3-butadiene. Its polymerization results in two geometrical isomers. One is “cis” and the other is “trans.” These are chemistry terms. The “cis” isomer is known as “natural rubber” or *cis*-1,4-polyisoprene (Flory 1953). Interestingly one another isomer “neoprene” is a synthetic rubber.

Commonly natural rubber is also known as latex, which is cultivated across the producers. The modern application of natural rubber in refrigeration seals, freon/air conditioning, motor mounts, engine coolants, automotive gaskets, and seals originates from the elastic nature of their molecular chains (Abuzreda 2023).

1.3.15 Application of Synthetic Polymers

Synthetic organic polymers are synthesized in laboratories by various chemical modifications. The modification results in different compositions of the polymeric chains. This in turn helps in improvement of the polymeric properties. Properties like tensile strength, gelation, and chain stiffness are important for possible industrial application of synthesized polymers (Flory 1953). They are commonly found in a variety of consumer products. They find technological use as adhesives, fibers, paints, plastics, and rubbers. We have mentioned about cellulose as natural polymer; at the same time cellulose derivatives like hydroxyl propyl methyl cellulose (HPMC), sodium carboxymethyl cellulose fall under synthetic category (Dumitriu 1996). So derivatives of natural polymers belong to synthetic category. Generally organic polymers are flexible, have a low density, and are stable over a wide pH range.

Most of the synthetic polymers are classified based on their mechanism of formation. Some of them are chain polymers, and others are condensation polymers. One of the simplest but highly utilized synthetic chain polymers is polyethylene (PE). This is formed by the repetition of ethylene units. However, two important varieties of PE are LDPE and HDPE which found very good industrial applications. HDPE is more rigid than LDPE due to higher crystallinity (Felgel-Farnholz et al. 2023). A linear polyethylene is a HDPE, whereas a branched polyethylene is a LDPE.

A series of other polymers like PVC, polypropylene (PP), polystyrene (PS), PTFE or Teflon, polyvinyl alcohol (PVA), polyvinyl acetate (PVAc), polymethyl acrylate (PMA), polymethyl methacrylate (PMMA), and PAN can be also formed in a similar fashion. A vinyl group $CH_2{=}CH–$ undergoes substitution with several atoms/groups like –chloro, phenyl, cyano, and alcohol to yield these polymers (Flory 1953). For example, the polymer PVC was formed by the repetition of vinyl chloride monomers. "PVC" is a strong plastic that can be made softer and more flexible by adding plasticizers.

$$n(CH_2{=}CH - Cl)\text{ Vinyl chloride} \rightarrow \ (-CH_2 - CHCl) -_n \text{ PVC}$$

Nevertheless, the polymer Teflon is widely used because of its non-reactive and hydrophobic nature. It does not interact with water as well as water-containing materials, so it is used mostly in cookware like non-stick pans. Non-reactivity arises from the strength of the carbon–fluorine bond (Hanford and Joyce 1946).

On the other hand, a range of synthetic polymers fall under the condensation polymer category. For example, polyester or PET, a synthetic fiber, is a condensation polymer of ethylene glycol (alcohol) and terephthalic acid. These polymers found application based upon their broad classification as thermosets, thermoplastics, elastomers, and synthetic fibers. Thermosets are mostly used in adhesives as they don't undergo structural change after getting set. An elastomer being a flexible polymer found several suitable applications. Apart from polyester, a few other important synthetic condensation polymers include nylon, Bakelite, melamine, polyurethanes, polycarbonates, and epoxies (Carothers 1929).

Most of the synthesized organic polymers have carbon backbone in their structure. The insertion of the heterochain (e.g., oxygen, sulfur) in the polymeric structure can change the polymeric properties for achieving required applications. Some examples of heterochain polymers are polyamides, polyesters, polyurethanes, polysulfides, and polycarbonates that have other elements in their chain (Flory 1953).

A few common applications of some synthetic polymers are presented in the following table.

	Polymers	**Applications**
1.	Polyethylene (PE)	Used in plastic bags, plastic films, containers, bottles, food packaging, flexible pipes, toys, housewares, etc.
2.	LDPE	Used in manufacturing various containers, dispensing bottles, wash bottles, tubing, plastic parts of computer components, molded laboratory equipment. Also used in packaging like foils, trays, plastic bags, etc.
3.	HDPE	Used in manufacturing toys, chemical containers, shampoo bottles, pipe systems, milk jugs, recycling bins, grocery bags, cereal box liners, flower pots, etc.

(Continued)

	Polymers	Applications
4.	Polypropylene (PP)	Used in appliances such as refrigerators, radios, and TVs. Also used in package films, pipes, storage tanks, set covers, dishwasher-safe plates, trays, cups, and many toys.
5.	Polyvinyl chloride (PVC)	Manufacturing of adhesives, coatings, and fibers. Raw material in automotive under body coating.
6.	Polystyrene (PS)	Used in protective packaging, containers, lids, bottles, trays, tumblers, disposable cutlery, making of models, toys, etc.
7.	Polytetrafluoroethylene (PTFE) or Teflon	Used in coating frying pans, baking sheets, other cooking utensils, automotive parts like gears, slide plates, seal gaskets, plain bearings, medical devices, and insulation.
8.	Polyvinyl alcohol (PVA)	Used in paper making, textile warp sizing, as a thickener, as emulsion stabilizer, and in biomedical devices such as in the manufacturing of contact lenses, artificial heart surgery, drug-delivery systems, wound dressings, etc.
9.	Polyacrylonitrile (PAN)	Used for making textile fibers, military aircraft, tennis rackets, tents, high-tech bicycle, pressure vessels, fishing rods, carpets, etc. It is also used as electrolyte separators in batteries.
10.	Polyvinyl acetate (PVA)	Emulsified polyvinyl acetate used as water-based adhesives including pastes and glues, bookbinding, lamination of metal foils. Non-emulsified polyvinyl acetate is used as a thermosetting adhesive. Used as a raw material for the manufacture of other polymers.
11.	Polymethyl acrylate (PMA)	Used in acrylics and in sheets for laminated safety glass as a spinning solvent, designing of LED bulbs.
12.	Polymethyl methacrylate (PMMA)	Used in making basins, bathroom fixtures, sink. In medicine, used in contact lenses, bone cements, and membrane dialysis. PMMA is also used to make attractive signboards and durable lenses for automobile lighting in buildings for decorative purposes.
13.	Nylon	Used for making textiles like sarees, shirts, neckties, socks, curtains, sleeping bags, rain coats, swim wear, cycle wear, etc. It is also used in making ropes, car seat belts, fishing nets, tire cords, strings for sports rackets and musical instruments, paint brushes, etc.
14.	Polyester	Used in textiles such as clothing, upholstery, curtains, blended with other fibers such as cotton and wool. Also used in automotive components like bumpers, door panels, dashboards, as substituent for metal. Used as insulating material in buildings, vehicles, electrical equipment.
15.	Polyurethanes	Used as coatings, films, foams, adhesives, sealants, elastomers, foundation garments, and swim suits. They are also used as a leather substitute such as cushion materials, carpet underlays, furniture, and packaging. Protector for laptops, outer cases of mobile electronic devices, sporting goods, medical devices, etc.

(Continued)

Polymers	Applications
16. Polycarbonates	Used to make a variety of lenses, such as those for automobile headlights, safety glass lenses, SCUBA masks, and sunglasses or eyeglasses. Windscreens of small motorized vehicles such as motor cycles, helicopters, small airplanes are made up of polycarbonates. The light-weight polycarbonates used as displays include some LCD screens, mobile screens, plasma screens by replacing glass. Aliphatic polycarbonates are used to create nanomedicine materials with increased biocompatibility and degradability.
17. Polyamides	Synthetic polyamides are commonly used in textiles, automotive industry, carpets, kitchen utensils, sportswear, etc. In industry, it is used to synthesize different kinds of substances such as tents, fishing nets, ropes, and tennis rackets. For example, nylon-6,6 is used for making swimwear, hoses, machine parts due to its waterproof properties and nylon-6 is widely used in gears, fitting, and bearings in automotive industry. Kevler, another synthetic polyamide, is widely used in the manufacturing of motorcycle helmets, brakes, clutch lining, etc.
18. Polyimides	Used in aerospace, flexible printed circuits, automotive, heaters, bar code labels, pressure-sensitive tape, electrical insulation, safety, loudspeaker components, high-density electronic interconnects, high-temperature mechanical seals, photovoltaic modules, etc. They are used as surface coating in supersonic aircraft.
19. Bakelite	Bakelite is used in industrial electrical applications such as electronics, power generation, and aerospace industries as well as for wire insulation and other automobile components like brake pads. Bakelite also finds its use as non-conducting components of phones, radios, and other electrical equipment, including insulators, bases, and sockets for light bulbs and electron tubes.
20. Epoxy resins	Epoxy resins are extensively used because of their excellent adhesion, chemical and heat resistance, good-to-excellent mechanical properties, and good electrical insulating properties. Epoxy adhesives are used in the construction of aircraft, automobiles, bicycles, boats, golf clubs, snowboards, etc. It is used in the electronic industry for printed wire boards, insulators, switchgear, motors, generators, and transformers. Nevertheless, for embedding electron microscopes, water-soluble epoxides, named "durcupan," are frequently utilized.
21. Silicones	Used in surface coatings, manufacture of laminates, as foaming agents in airplanes and missiles, as elastomers, and as a filler to reinforce silicone rubbers. Used in many products such as electrical (e.g., insulation), electronics (e.g., coatings), household (e.g., sealants and cooking utensils), textiles and papers (e.g., coating), automobile (e.g., gaskets), and office machines (e.g., keyboard pads).

Polyurethane, which is a thermoplastic, is also considered as one of the most applicable polymers due to some of its inherent properties like elasticity, transparency, and resistance to oil, grease, and abrasion. It is a hydrophilic polymer that can react with water. This thermoplastic elastomer consists of urethane groups (–NH–CO–O–) in the main polymer chain (Akindoyo et al. 2016). However incorporation of some other groups like ester, ether, amide, and urea produces some copolymers with suitable industrial applications.

Due in large part to their excellent tensile strength, synthetic co-polymers have found a wide range of uses. Because two or more distinct monomers are combined, co-polymers are able to withstand pressure. The atoms of different monomers can form strong bond between them because of electronegativity differences. They are generally categorized into linear, alternate, block, and graft co-polymers depending upon the arrangement of chains (Saldívar-Guerra and Vivaldo-Lima 2013). Several types of acrylic, vinyl acetate, and styrene-based co-polymers have found extensive applications in industries.

1.3.16 Recent Advances in Organic Polymers

In the modern era, with the advancement of technology, polymeric materials have potential countless applications. These include applications in electronics (Shaw and Seidler 2001), water treatment (Donga et al. 2019), hydrogen storage (Kato and Nishide 2018), drug-delivery systems (Liechty et al. 2010), solar cells (Brabec et al. 2003), sequestration of carbon dioxide (Wang et al. 2017), and energy storage capacitors (Feng et al. 2022). Due to their diverse applications, many efforts have been made recently in the design and study of novel polymeric materials. The classification of polymers falls under various categories because of their structures, broad applications, and varying properties. Thus instead of classifying them under one category they are classified as synthetic, semi-synthetic, or natural; organic and inorganic; linear, branched, and cross-linked polymers; homomer and co-polymers; addition and condensation polymers; and elastomers, fibers, thermosetting, and thermoplastics. This section gives a brief account of the recent developments in the field of polymer science particularly organic polymers.

Clean energy has become a necessity of modern society, and organic polymers may play a pivotal role in the design of novel materials to harvest clean energy. The major challenge in the sustainable development of society is the increasing population which invariably puts pressure on natural resources. The ever-increasing population results in an increase in the consumption of fossil fuels and hence increases air pollution and global warming. There has been an exhaustive effort to pave the way toward clean energy either by using solar cells or designing novel photocatalysts for water splitting (Takanabe 2017).

Although photolysis of water is considered to be the holy grail approach for hydrogen production toward sustainable energy, there was very limited success in the field until the employment of graphitic carbon nitride polymers (g-C_3N_4) (Wang et al. 2021). Conjugated organic polymers due to their tunable properties have attracted much attention in the pursuit of sustainable hydrogen production from water as compared to their inorganic counterparts. One such organic polymer is graphitic carbon nitride, a layered conjugated organic polymer, which has become the center of attraction since the last decade due to its various unique properties such as ease of synthesis, being metal-free, and having an environmentally benign nature, and a band gap of 2.7 eV. These properties make the polymer the most suitable candidate for hydrogen production. Despite the abovementioned properties, the pristine g-C_3N_4 has some bottlenecks such as limited active sites, high charge carrier recombination, and low specific surface areas which have to be overcome before it is used as catalyst in full scale for water splitting. Engineering the g-C_3N_4 is one of the many possibilities to make the material more efficient for water splitting. Elemental doping is one of the engineering techniques where a heteroatom is incorporated in g-C_3N_4. In one such study, the bulk g-C_3N_4 was doped with elemental cobalt and it was observed that the incorporation of Co not only results in the narrowing of the band gap but also increases the specific surface areas, thereby making the separation of photoinduced carriers more efficient (Chen et al. 2017). In another attempt g-C_3N_4 was doped with elemental iron and it was observed that the Fe-doped g-C_3N_4 nanosheets have a broader absorption spectrum than the bulk g-C_3N_4. Moreover, the prepared material is more efficient than the bulk g-C_3N_4 in case of hydrogen-evolving reaction (Gao et al. 2017). Non-metal-doped g-C_3N_4 has also been observed to be more efficient than the pristine g-C_3N_4. In one such study, Jiang et al. (2019) observed that the photocatalytic performance was significantly increased after it was doped with oxygen. Apart from g-C_3N_4 there are porous organic polymers that can be a platform to carry out photocatalytic water splitting. Porous organic polymers are generally defined as a group of covalently connected polymer materials with high porosity and specific surface area. They are made of different light elements (carbon, boron, hydrogen, oxygen, and nitrogen) and strong covalent bonds. Porous organic polymers gained popularity due to their versatile properties such as being tunable, i.e., their geometry and dimensions can be controlled during the synthesis process. Since porous organic polymers are connected through strong covalent bonds they have high chemical and thermal stability. There have been continuous efforts to prepare porous organic polymers to carry out the hydrogen and oxygen evolution reactions (Xu et al. 2018). Porous organic polymers can be grouped into three classes: conjugated microporous polymers (Jiang et al. 2008), covalent triazine frameworks (Jiang et al. 2018) and covalent organic frameworks (Cote 2005). The application of porous organic polymers for water splitting is still in the infancy period, and the material has to compete with the state-of-art materials such as metal oxides and graphitic carbon nitrides. There is no doubt that the addition of porous organic polymers into the family of photocatalysts will give us more options to explore the myriad landscape of photocatalysts and help the scientific community to design environmentally benign novel photocatalysts.

Hydrogen storage is another area where the organic polymer has tremendous potential. Hydrogen (H_2) due to its properties such as energy density and clean combustion, an abundance in the form of water is going to become the major energy source and will replace the non-renewable petroleum energy resources.

The major bottlenecks in using H_2 as energy source are its storage and production (Cousins and Zhang 2019). Hydrogen can be stored using highly porous organic polymers through adsorption. These approaches have various advantages such as low cost, easy to scale up, metal-free, and environmentally friendly. In this case four groups of highly porous polymers have been extensively used for hydrogen storage: (i) hyper cross-linked polymers (HCPs) (Wood et al. 2007), (ii) Polymers of Intrinsic Microporosity (PIMs), (iii) Conjugated Microporous Polymers (CMPs), and (iv) Porous Aromatic Frameworks (PAFs) (Germain et al. 2009).

Ease of functionalizing pore environments, large accessible surface areas, and high porosities make solid-state porous materials promising materials for hydrogen (Chen et al. 2022). Similarly amorphous POPs, because of their ease of processibility and good mechanical stability, make them viable candidates for hydrogen storage. The processing cost is relatively low and free of heavy metal making the materials and processes friendly to the environment.

Conducting polymers have been extensively investigated as a potential hydrogen storage due to the low cost, ease of synthesis, and processability to achieve the desired morphological and microstructural architecture, ease of doping and composite formation, chemical stability, and functional properties.

Organic polymers are being extensively used for targeted drug-delivery systems and controlled drug-delivery systems (Mahato et al. 2020). Drug delivery refers to a combination of methods, technologies, formulations, or processes used to deliver the therapeutic agent into the body to achieve the desired therapeutic effect (Aimetti et al. 2009). When a drug is administered to the body because of various factors only a small amount reaches the sites of action. The various factors include metabolism of the therapeutic agent or the drug while en route to the site, reaching wrong sites (tissues or organs), or quick removal of the drug from the site. This affects the efficiency of any drug greatly. The above shortcomings can be overcome if targeted drug-delivery approaches or smart drug-delivery approaches or controlled drug-delivery approaches are applied. In this method, to minimize the side effects and to attain the maximum efficacy of the drugs, the drugs are administered at an appropriate time, dosage, and location (Zishan et al. 2017). Recently, in medicine organic polymers have become an inseparable and indispensable part as drug-delivery systems. The objective of a sustainable drug-delivery system is to control drug release, reduce drug frequency, thereby increasing the efficiency and decreasing the side effects of the drug (Allen and Cullis 2004). Considering the efficiency of the targeted drug-delivery system over the conventional systems many studies were done on carriers of the drug. The vesicular carrier systems, microparticulate systems, monoclonal antibodies and fragments, and polymeric carriers are the most commonly used carriers for targeted drug delivery. The advancement in polymeric science helped scientists and pharmacologists to design new drug carriers, thereby making polymers an inseparable part of drug-delivery systems. The development of novel drug-delivery systems or controlled drug-delivery systems depends on the carriers that are used to deliver the drugs to the specific site. It is observed that polymer-based drug nanocarriers can increase the solubility of hydrophobic drugs, decrease their cytotoxicity toward normal tissue, prolong drug availability, favorably alter biodistribution, improve utilization efficiency (Mahajan et al. 2007) and thereby may show promising pharmacokinetics both at the cellular and whole body levels, e.g., polymeric nanocarrier poly(D,L-lactide-*co*-glycolide). Various organic polymers utilized in drug delivery are poly(D,L-lactide-*co*-glycolide), polyglycolic acid, poly-L-glutamic acid, polylactic acid, poly(*N*-isopropylacrylamide), dextran, dendrimers, chitosan, and so on (Mathiowitz et al. 1997). Amphiphilic polymers such as micelles, dendrimers, and nanomicelles, containing both hydrophilic and hydrophobic parts, have been extensively studied as they are thought to be good candidates as polymeric carriers in drug delivery (Srivastava et al. 2016). Smart polymers are another class that changes their properties according to the environment such as pH. In summary, it can be concluded that the future of drug delivery is very bright, and the inclusion of technologies in treating diseases will become more and more efficient.

In general, when the term good conductors comes to our mind, we generally picture metals, salt solutions, and fused salts and usually treat organic solids and liquids as good insulators. However, this notion is changing since the discovery of conjugated organic polymers capable of conducting electricity. In 2000 Heeger, MacDiarmid, and Shirakawa were awarded the Nobel Prize in Chemistry for their work on the discovery and development of conductive polymers (Yang and Pierstorff 2012). The two branches of materials science which have witnessed exponential growth in the twentieth century and almost touched all aspects of modern life are semiconductors and polymers. The history of conjugated polymers dates back to polypyrrole and PANI. Synthetic metals are defined as organic polymers which possess both the properties of metals (such as electronic, electric, magnetic, and optical) and those of polymers (such as mechanical properties and processibility). Moreover, with the discovery that the conductivity of conjugated organic polymers can be easily tuned or controlled, there has been exponential growth in the application of conjugated organic polymers in organic photovoltaic devices, light-emitting diodes, sensors, and field effect transistors (MacDiarmid 2001). The most commonly studied conjugated organic polymers are polyacetylene, polythiophene, polyphenylene-vinylene, polypyrrole, polyfuran, polyphenylene, and PANI. It is observed that synthetic metals exhibit high electrical conductivity in their oxidized and reduced states. Porous organic polymers have a very large surface area and as a result, make them an ideal candidate for gas trapping and hence can be used as chemical sensors for heavy metals, explosives, biological molecules, sensors for volatile organic compounds and solvents, and sensors for anions (Bao and Lovinger 1999). Engineering polymers represent another class of organic polymers that can withstand high temperatures and have high physical and mechanical properties which make them good candidates for their use in aerospace electronics and robotic systems (Mittal 2011). Stretchable electronics are another area where organic polymers have potential applications. Stretchable polymers, because of their unique physicochemical properties, are envisioned as the future of electronic devices. Stretchable electronics consist of electronic materials integrated with an elastomer or stretchable material. Stretchable polymers have the potential to be applied in biomedical engineering, healthcare, and human–machine interface (Lim et al. 2015).

Thus it is seen that in the near future polymers will become the epicenter of research and development. Moreover, it is believed that organic polymers will play an important part in deciphering the solutions to the problems that arise after the indiscriminate use of resources.

1.4 Conclusion and Future Scope

It is well known that nature from the beginning selected biopolymers for survival and growth. Likewise, humans also started their journey from Stone Age and passed through several stages to reach the present age where polymer material is always one of the best choices of interest. Polymer technology is a branch of science where extensive research is going on for the introduction of novel material which can meet our daily requirements in a sustainable fashion. Presently, researchers are looking for intelligent polymers which are specially designed materials having multidimensional applications that can extend the scope of polymer science. In this chapter, a thorough development in the field of polymer science has been nicely described in a concise manner. Starting from the origin of polymers, their history and discovery, types and mode of classification, the recent advancements in polymer sciences, and their important practical applications have been properly incorporated with suitable examples. This book also explains about the scope and limitations of various techniques used in the synthesis and study of properties of various types of organic polymers. In addition, important applications of polymers in material science such as in OLEDs, semiconductors, superconductors, and PMCs have been presented. The modification of polymers and the research involved for the same will definitely help mankind in future.

References

Abuzreda, A. (2023). Nanopolymers types, classification, properties, and uses. *Advance in Environmental Waste Management & Recycling* 6: 402–408.

Adelmann, H., Binks, B.P., and Mezzenga, R. (2012). Oil powders and gels from particle-stabilized emulsions. *Langmuir* 28 (3): 1694–1697.

Aimetti, A.A., Machen, A.J., and Anseth, K.S. (2009). Poly(ethylene glycol) hydrogels formed by thiol-ene photopolymerization for enzyme-responsive protein delivery. *Biomaterials* 30: 6048–6054.

Akindoyo, J.O., Beg, M.D.H., Ghazali, S. et al. (2016). Polyurethane types, synthesis and applications – a review. *RSC Advances* 6: 114453–114482.

Allen, T. and Cullis, P. (2004). Drug delivery systems: entering the mainstream. *Science* 303: 1818–1822.

Baekeland, L.H. (1909). The synthesis, constitution, and uses of bakelite. *Industrial and Engineering Chemistry* 1: 149–161.

Bao, Z.N. and Lovinger, A. (1999). Soluble regioregular polythiophene derivatives as semiconducting materials for field-effect transistors. *Chemistry of Materials* 11 (9): 2607–2612.

Brabec, C., Dyakonov, V., Parisi, J., and Serdar, N. (2003). *Organic Photovoltaics*. Berlin: Springer Verlag.

Brown, W.H. and Poon, T. (2005). *Introduction to Organic Chemistry*, 3e. Wiley.

Cann, M.C. and Connelly, M.E. (2000). Dupont Petretec polyester regeneration technology. In: *Real-World Cases in Green Chemistry*, 25–31. American Chemical Society.

Carothers, W.H. (1929). Studies on polymerization and ring formation. I. An introduction to the general theory of condensation polymers. *Journal of the American Chemical Society* 51 (8): 2548–2559.

Chen, P.-W., Li, K., Yu, Y.-X., and Zhang, W.-D. (2017). Cobalt-doped graphitic carbon nitride photocatalysts with high activity for hydrogen evolution. *Applied Surface Science* 392: 608–615.

Chen, Z., Kirlikovali, K.O., Idrees, K.B. et al. (2022). Porous materials for hydrogen storage. *Chem* 8 (3): 693–716. https://doi.org/10.1016/j.chempr.2022.01.012.

Cote, A.P. (2005). Porous, crystalline, covalent organic frameworks. *Science* 310: 1166–1170. https://doi.org/10.1126/science.1120411.

Cousins, K. and Zhang, R. (2019). Highly porous organic polymers for hydrogen fuel storage. *Polymers* 11 (4): 690. https://doi.org/10.3390/polym11040690.

Currell, B.R. and Frazer M, J. (1969). Inorganic polymers. *Royal Institute of Chemistry Reviews* 2: 13–40.

Donga, C., Mabape, K.I.S., Mishra, S.B., and Mishra, A.K. (2019). Chapter 8: Polymer-based engineering materials for removal of nanowastes from water. In: *Emerging and Nanomaterial Contaminants in Wastewater* (ed. A.K. Mishra, H.M. Anawar, and N. Drouiche), 217–243. Elsevier.

Dorel, F. (2008). Polymer history. *Designed Monomers and Polymers* 11 (1): 1–15.

Dumitriu, S. (1996). Cellulose and its derivatives: structures, reactions and medical uses. In: *Polysaccharides in Medicinal Applications* (ed. D.N.-S. Hon), 87–106. New York, NY: Marcel Dekker, Inc.

Felgel-Farnholz, A., Schweighuber, A., Klampfl, C.W., and Fischer, J. (2023). Comparative study on the degradation of HDPE, LLDPE and LDPE during multiple extrusions. *Polymer Degradation and Stability* 216: 110486.

Feng, Q.-K., Zhong, S.-L., Pei, J.-Y. et al. (2022). Recent progress and future prospects on all-organic polymer dielectrics for energy storage capacitors. *Chemical Reviews* 122 (3): 3820–3878.

Flory, P.J. (1953). *Principles of Polymer Chemistry*. Ithaca, NY: Cornell University Press.

(Gajre)Kulkarni, V., Butte, K., and Rathod, S. (2012). Natural polymers – a comprehensive review. *International Journal of Research in Pharmaceutical and Biomedical Sciences* 3: 1597–1613.

Gao, J., Wang, Y., Zhou, S. et al. (2017). A facile one-step synthesis of Fe-doped g-C_3N_4 nanosheets and their improved visible light photocatalytic performance. *ChemCatChem* 9 (9): 1708–1715.

Ghosh, P. (2011). *Polymer Science and Technology: Plastics, Rubbers, Blends and Composites. 3rd ed.* New York: McGraw-Hill Education. https://www.accessengineeringlibrary.com/content/book/9780070707047.

Germain, J., Frechet, J.M.J., and Svec, F. (2009). Nanoporous polymers for hydrogen storage. *Small* 5 (10): 1098–1111. https://doi.org/10.1002/smll.200801762.

Gowarikar, V.R., Vishwanathan, N.V., and Jayadev, S. (2022). *Textbook of Polymer Chemistry*, 4e. Wiley.

Guo, J., Skinner, G.W., Harcum, W.W., and Barnum, P.E. (1998). Pharmaceutical applications of naturally occurring water-soluble polymers. *Pharmaceutical Science & Technology Today* 1: 254–261.

Hanford, W.E. and Joyce, R.M. (1946). Polytetrafluoroethylene. *Journal of the American Chemical Society* 68: 2082–2085.

Indra, V.I.G. and Shrray, S. (2015). Recent developments in Inorganic polymers: a review with focus on Si-Al based inorganic polymers. *Oriental Journal of Chemistry* 31 (4): 2475–2480.

Janik, H., Sienkiewicz, M., and Kucinska-Lipka, J. (2014). Chapter 9: Polyurethanes. In: *Handbook of Thermoset Plastics*, 3e (ed. H. Dodiuk), 253–295. Elsevier.

Januszewski, R., Dutkiewicz, M., Nowicki, M. et al. (2021). Synthesis and properties of epoxy resin modified with novel reactive liquid rubber-based systems. *Industrial and Engineering Chemistry Research* 60 (5): 2178–2186.

Jiang, J.-X., Su, F., Trewin, A. et al. (2008). Synthetic control of the pore dimension and surface area in conjugated microporous polymer and copolymer networks synthetic control of the pore dimension and surface area in conjugated microporous polymer and copolymer networks. *Journal of the American Chemical Society* 130 (24): 7710–7720.

Jiang, Q., Sun, L., Bi, J. et al. (2018). MoS_2 quantum dots-modified covalent triazine-based frameworks for enhanced photocatalytic hydrogen evolution. *ChemSusChem* 11: 1108–1113.

Jiang, Y., Sun, Z., Tang, C. et al. (2019). Enhancement of photocatalytic hydrogen evolution activity of porous oxygen doped g-C_3N_4 with nitrogen defects induced by changing electron transition. *Applied Catalysis, B: Environmental* 240: 30–38.

Kamide, K. (2005). *Cellulose and Cellulose Derivatives: Molecular Characterization and its Applications*. Amsterdam, The Netherlands: Elsevier.

Kariduraganavar, M.Y., Kittur, A.A., and Kamble, R.R. (2014). Chapter 1: Polymer synthesis and processing. In: *Natural and Synthetic Biomedical Polymers* (ed. S.G. Kumbar, C.T. Laurencin, and M. Deng), 1–31. Elsevier. ISBN 9780123969835.

Kato, R. and Nishide, H. (2018). Polymers for carrying and storing hydrogen. *Polymer Journal* 50: 77–82.

Liechty, W.B., Kryscio, D.R., Slaughter, B.V., and Peppas, N.A. (2010). Polymers for drug delivery systems. *Annual Review of Chemical and Biomolecular Engineering* 1: 149–173.

Lim, S., Son, D., Kim, J. et al. (2015). Transparent and stretchable interactive human machine interface based on patterned graphene heterostructures. *Advanced Functional Materials* 25 (3): 375–383.

MacDiarmid, A.G. (2001). Synthetic metals: a novel role for organic polymers. *Angewandte Chemie International Edition* 40: 2581–2590.

Mahajan, H.S., Patil, S.B., Gattani, S.G., and Kuchekar, B.S. (2007). Targeted drug delivery systems. *Pharmacy Times* 39 (2): 19–21.

Mahato, N., Jang, H., Dhyani, A., and Cho, S. (2020). Recent progress in conducting polymers for hydrogen storage and fuel cell applications. *Polymers* 12 (11): 2480. https://doi.org/10.3390/polym12112480.

Mathiowitz, E., Jacob, J.S., Jong, Y.S. et al. (1997). Biologically erodable microsphere as potential oral drug delivery system. *Nature* 386 (6623): 410–414.

McChesney, J.D., Venkataraman, S.K., and Henri, J.T. (2007). Plant natural products: back to the future or into extinction? *Phytochemistry* 68: 2015–2022.

Mittal, V. (2011). *High Performance Polymers and Engineering Plastics*. Wiley.

Pandey, R. and Khuller, G.K. (2004). Polymer based drug delivery systems for mycobacterial infections. *Current Drug Delivery* 1: 195–201.

Peng, S., Wang, D., Lu, J. et al. (2017). A review on organic polymer-based thermoelectric materials. *Journal of Polymers and the Environment* 25: 1208–1218.

Ramakrishna, S., Mayer, J., Wintermantel, E., and Leong, K.W. (2001). Biomedical applications of polymer-composite materials: a review. *Composites Science and Technology* 61 (9): 1189–1224.

Saldívar-Guerra, E. and Vivaldo-Lima, E. (2013). Chapter 1: Introduction to polymers and polymer types. In: *Handbook of Polymer Synthesis, Characterization, and Processing* (ed. E. Saldívar-Guerra and E. Vivaldo-Lima), 1–14. Wiley.

Seth, C.R. (2020). Conjugated and conducting organic polymers: the first 150 years. *ChemPlusChem* 85 (7): 1412–1429.

Shaw, J.M. and Seidler, P.F. (2001). Organic electronics: introduction. *IBM Journal of Research and Development* 45 (1): 3–9.

Srivastava, A., Yadav, T., Sharma, S. et al. (2016). Polymers in drug delivery. *Journal of Biosciences and Medicines* 4 (1): 69–84. https://doi.org/10.4236/jbm.2016.41009.

Takanabe, K. (2017). Photocatalytic water splitting: quantitative approaches toward photocatalyst by design. *ACS Catalysis* 7 (11): 8006–8022.

Tewatia, A., Hendrix, J., Nosker, T., and Lynch-Branzoi, J. (2017). Multi-scale carbon (micro/nano) fiber reinforcement of polyetheretherketone using high shear melt-processing. *Fibers* 5: 3.

Wallace, P.C. (1945). Leo Hendrik Baekeland, 1863–1944: an appreciation. *Chemical & Engineering News* 23: 228–229.

Wang, W., Zhouab, M., and Yuan, D. (2017). Carbon dioxide capture in amorphous porous organic polymers. *Journal of Materials Chemistry A* 5: 1334–1347.

Wang, S., Zhang, J., Li, B. et al. (2021). Engineered graphitic carbon nitride-based photocatalysts for visible-light-driven water splitting: a review. *Energy & Fuels* 35: 6504–6526.

Wood, C.D., Tan, B., Trewin, A. et al. (2007). Hydrogen storage in microporous hypercrosslinked organic polymer networks. *Chemistry of Materials* 19 (8): 2034–2048. https://doi.org/10.1021/cm070356a.

Xu, C., Zhang, W., Tang, J. et al. (2018). Porous organic polymers: an emerged platform for photocatalytic water splitting. *Frontiers in Chemistry* 6: 592. https://doi.org/10.3389/fchem.2018.00592.

Yang, W.W. and Pierstorff, E. (2012). Reservoir based polymer drug delivery systems. *Journal of Laboratory Automation* 17: 50–58. https://doi.org/10.1177/2211068211428189.

Zielecka, M., Kozakiewicz, J., and Przybylski, J. (2002). Studies on the surface properties of silicone-containing hybrid systems. *Surface Coatings International Part B: Coatings Transactions* 85: 215–220.

Zishan, M., Zeeshan, A., Faisal, S. et al. (2017). Vesicular drug delivery system used for liver diseases. *World Journal of Pharmaceutical Sciences* 5 (4): 28–35.

2

Basics of Polymerizations and Application Toward Organic Materials

Abhijit Mahanta[1], Arpita Joarder[2], Neha Khetawat[2], Dipjyoti Dutta[3], Panchanan Puzari[2], and Rekha R. Dutta[4]

[1] *Digboi College, Department of Chemistry, Tinisukia, NH 38, 786171, India*
[2] *The Assam Kaziranga University, Department of Chemistry, Koraikhowa, Jorhat, NH 37, Assam, 785006, India*
[3] *CSIR-North East Institute of Science and Technology (NEIST), Material Sciences and Technology Division (MSTD), Jorhat, Assam, 785006, India*
[4] *The Assam Kaziranga University, Department of Chemistry, Koraikhowa, Jorhat, NH 37, Assam 785006, India*

2.1 Introduction

Since the earth's origin, elements like carbon, oxygen, and hydrogen have been uniting to form composite polymer molecules. A polymer served as the material basis for the origin of life, which is the most exciting and intriguing process that could have ever been brought about by such a combination. In nature, this polymer known as protein was created from straightforward chemical constituents like CO_2, CH_4, and NH_3 (Errahali et al. 2014). However, despite being a fundamental component of the natural world and having appeared in everyday items like starch, wood, cotton, cellulose, and natural rubber, humanity did not fully understand polymers until almost the end of World War II. In nineteenth century, Leo Baekeland (1935) developed a resin from two very common chemicals phenol and formaldehyde, which was later named after him as Bakelite. Bakelite was a forerunner of many synthetic polymers. For instance, Jacques Brandenburger developed a famous transparent material called cellophane (Hisano 2017). Within a decade, several new forms of polymers with better properties had emerged from laboratories in different parts of the world.

In recent times, the applications of polymers have gained much attention in diverse areas of chemistry and industry. Contemporary applications of polymer material extend from electronic devices, adhesives, foams, coatings, and packaging materials to industrial fibers, silicone heart valves, textile, Teflon-coated cookware, and composites (Namazi 2017). They have also massive application in optical devices and biomedical devices, and they are considered as precursors for many high-tech ceramics. The list is practically never-ending. These assorted practices of polymers have enormously aided to advance our living standards. Polymers are attained via chemical reaction of monomers. The polymer chain is formed by the

Organic Polymers in Energy-Environmental Applications, First Edition.
Edited by Ramesh Oraon, Pardeep Singh, Sanchayita Rajkhowa, Sangita Agarwal, and Ravindra Pratap Singh.

reaction of monomer molecule with alternative molecule from a similar type or another kind in the appropriate condition (Namazi 2017). This practice in nature has led to the development of natural polymers, whereas the synthetic polymers are man-made. Natural and synthetic polymers comprise inorganic and organic polymers; elastomer, blends, fibers, ceramics, coatings, plastics, and composites (Patil et al. 2017). Herein, in this chapter we will emphasize only about the basis and application of organic polymer.

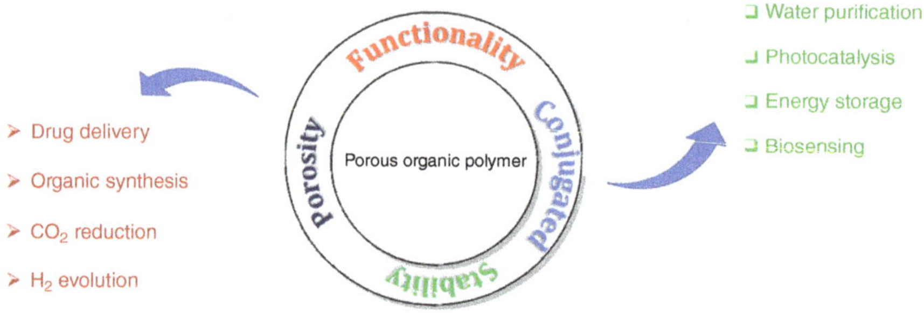

Organic polymers are made up of carbon backbone structure. Some common examples of organic polymers are polyamides, polycarbonates, polyesters, and polyurethanes (Patil et al. 2017). Organic polymers have found their immense applications in some of the exciting areas like water purification, molecular recognition, holography, biomedical applications, drug delivery system, biosensor devices, tissue engineering, cosmetics aircraft, aerospace, and sports equipment (Patil et al. 2017). Bio-based polymers are also included in this category, e.g. polylactic acid, polyhydroxy alkanolate, and polypropylene which are mainly made up from renewable resources (Babu et al. 2013). Bio-based polymers have the benefit of reducing carbon footprint, since the former are degraded into their component monomers with the aid of fungi and bacteria (microorganisms) and produce CO_2, CH_4, and water without creating any toxic material (Laurichesse and Avérous 2014). On the other hand, plastics/materials which are petroleum based like polyethylene, PVC, polystyrene, and poly-butyrene have very destructive effects on the environment as they are nonbiodegradable (Avella et al. 2009; Nakajima et al. 2017). To solve this matter, bio-based polymers have arose as an auxiliary to traditional (petroleum) plastics. In current years the chemistry of porous organic polymer (POP) is considered as a fascinating area of research. POPs are a type of organic macromolecules which are constructed by strong covalent linkages of different organic moieties (Zhang and Riduan 2012). On the ground of their structural uniqueness and degree of long-range order, POPs can be classified into different categories, e.g., polymer of intrinsic microporosity (PIMs), conjugated microporous polymers (CMPs), hyper cross-linked polymers (HCPs), covalent triazine frameworks (CTFs), covalent organic frameworks (COFs), and porous aromatic frameworks (PAFs). All the mentioned POPs are amorphous materials, with the exception of CTFs and COFs, which are crystalline having ordered structures and synthesized under thermodynamic control (Zhang et al. 2017a,b). POPs feature high surface area expectedly between 100 and 1000 m^2 g^{-1} which are greater than those of porous carbons,

MOFs. A 3D-COF-103 with a high Brunauer–Emmett–Teller (BET) surface area of 4210 m^2 g^{-1} was reported by Yaghi and coworkers in 2007 (El-Kaderi et al. 2007; Ahn et al. 2006). Ben and coworkers developed a PAF (PAF-1) with a high specific area (4210 m^2 g^{-1}) using the Yamamoto reaction. POPs also display greater chemical and thermal stabilities, extended π-conjugations, light weight and high contents of C, H, N. The said excellence of POPs contributes to it becoming an emerging material (Li and Yang 2017). Additionally, POPs-based catalysts also exhibit high degree of polymerization, poor solubility in organic solvent, and rigid chemical bonds, making them easier to separate from reaction medium and recycle (Wong et al. 2016). Their surface areas pore size and porous structure can be designed and easily modified by introducing definite functional components. Owing to the aforementioned characteristics, POPs are used in amazing potential applications in diverse areas like gas storage/separation, photocatalytic reactions, energy storage, chemical and bio-sensing, degradation of organic pollutants, and H_2 evolution (Zhang et al. 2017a,b). In addition to these, POPs have a tremendous application in the field of heterogeneous catalysis. Incorporation of active catalytic species into POPs via metal–ligand coordination or covalent bonding and adjusting pore structures of the solids improves the catalytic properties of POPs (El-Kaderi et al. 2007). Hence, this process makes the heterogeneous catalytic system of POPs suitable for diverse applications. On the contrary, COF, which represents crystalline POP, has been receiving widespread application in the field of heterogenous catalysis as it experiences uniform nanopores along with extended and ordered skeleton which enables specific incorporation of functional units (El-Kaderi et al. 2007). Quite a lot of COFs with various topologies have been thoroughly investigated in catalysis, gas storage and separation, and energy storage since the pioneering work reported by O.M. Yaghi and coworkers (El-Kaderi et al. 2007; Fontanals et al. 2015). Moreover, CTFs, a distinct subclass of COFs, have triazine rings that are often synthesized via the trimerization reaction of aromatic nitriles (Zhao et al. 2017; Planells et al. 2013; Ghosh et al. 2015). Due to low crystallinity, greater chemical stability, and high nitrogen content, most reported CTFs are hence well suited for photocatalysis applications. Nevertheless, the harsh reaction conditions for synthesis (such as use of high temperature of over 400 °C) and irreversible production of aromatic 1,3,5-triazine rings restrict the development of CTFs in comparison to COFs containing imine, azine linkage, boroxine, or boronate ester (Kuhn et al. 2008). Recently, polycondensation approach under low temperature was established by B. Tan and coworkers which was a great success (Wang et al. 2017).

In this chapter, we offer an overview of preparation of POPs (including COFs, CMPs, CTFs, and HCPs), their synthesis methodologies, and various applications in the construction of organic materials and future development.

2.2 Preparation of Covalent Organic Framework (COF)

Organic polymers are gaining attention due to their innumerable advantages. Organic polymers are the macromolecules that are composed of carbon backbone structures, with the repetition of monomer units. We can synthesize polymers on the basis of their applications. The various types of organic polymers that can be synthesized are COF, HCPs, POPs, organic polymeric membranes, etc. Polymerization is a process in which three-dimensional networks or polymer chains are formed, by combining reactant monomers together under suitable reaction conditions. Though we can prepare organic polymers by different methods the basics of polymerization are mainly classified into two types: (i) addition polymerization and (ii) condensation polymerization. In addition polymerization, the monomers combine with each other without loss of any atom or molecules whereas in condensation polymerization, condensation polymers are formed by the combination of two different monomers, with the loss of small molecules like water. The reversible condensation of building blocks containing light elements and connected by covalent bonds resulted in the formation of a new type of porous crystalline organic polymers which were called COF (Cote et al. 2005). It resulted in the long-range-order crystalline structures due to their thermodynamically controlled covalent bonding and self-healing ability. Their crystallinity and ordered structures are maintained because of the ability to withstand harsh condition and chemical stability in organic solvents (Diercks and Yaghi 2017; El-Kaderi et al. 2007). The reason for the good strength of the framework is pi stacking interactions and hydrogen bonding which can further protect them from solvation and hydrolysis (Lohse and Bein 2018; Ding and Wang 2013).

The preparation of certain light elements containing COFs is described below.

2.2.1 Hydrazone-Based COF

The co-condensation of hydrazides and aldehydes in the presence of a catalyst, i.e., acetic acid, results in the formation of COF that are hydrazone linked (Figure 2.1a). The ethoxy unit at the ortho position of the building units has been integrated to form the essential hydrazine-linked COF which includes the formation of intramolecular hydrogen bonding. The construction of the 2D conformation using the solvothermal conditions has also been attempted which results in the enhancement of thermal stability and crystallinity (Abuzeid et al. 2021). For example COF_{42} and COF_{43} were synthesized using aromatic aldehydes and 2,5-diethoxyterephthalohydrazide (Wang et al. 2020).

2.2.1.1 Boron-Containing COF

Boronated anhydrides and boronated esters have been used for the formation of a wide range of boron-containing COF. Based on the condensation process two categories of COF are formed (i) through co-condensation of catechol derivatives and boronic acids to yield a five-membered ring boronate esters (Figure 2.1b). For example, solvothermal co-condensation of 4,4-diphenyl butadiynebis(boronic acid) (DPB) and 2,3,6,7,10,11-hexahydroxytriphenylene (HHTP) yielded HHTP-DPB COFP which was synthesized by Feriante et al. (2020), (ii) through self-condensation of a single building block leading to the formation for cyclic six-membered boroxine-linked COFP (Figure 2.1c). For example, 2D COF was synthesized

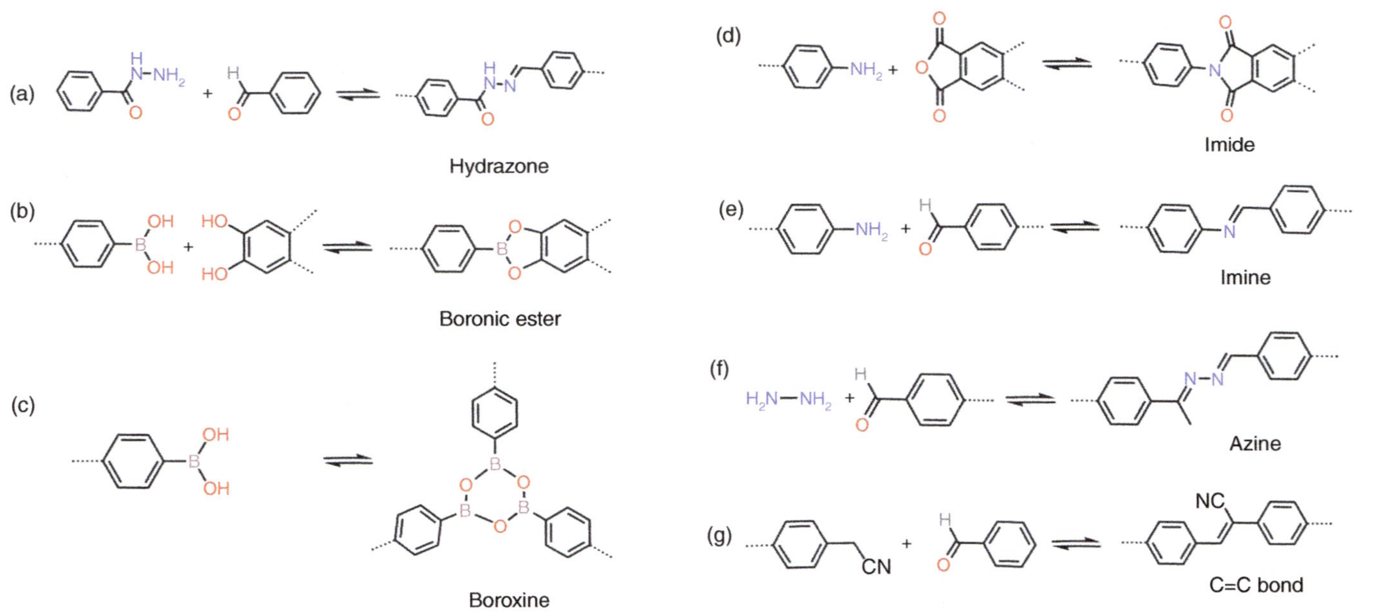

Figure 2.1 Condensation of various linkages for COF formation: (a) hydrazone, (b) boronic ester, (c) boroxine, (d) imide, (e) imine, (f) azine, (g) C=C bond. Source: Adapted from Jin et al. (2018).

through self-condensation of 1,4-benzenediboronic acid, and the self-condensation of TBPS resulted in the synthesis of 3D COF (Spitler et al. 2011).

2.2.1.2 Imide-Based COF

At high temperatures, the condensation of an acetic acid with amine yields imide linkages (Figure 2.1d) Under the solvothermal condition in methylene/*N*-methyl-2-pyrrolidone mixture and isoquinoline as catalyst, the reaction between triamine building units and aromatic anhydrides leads to the formation of 2D polyimide COFs (Fang et al. 2014). The 3D COF were formed at lower temperature for over five days (Fang et al. 2015). Further, synthetic methods such as ionothermal conditions and catalyst and solvent-free methods have been used for the synthesis of such polymer (Wu et al. 2015; Abuzeid et al. 2021).

2.2.1.3 Imine-Based COF

The Schiff base reaction results in the formation of imine-based COFs. This takes place by means of co-condensation of aldehydes and aromatic amines in the presence of a catalyst which can be either an organic acid or a Lewis acid (Figure 2.1e). These imine-based COF are categorized into five classes based on their topology (i) hexagonal, (ii) trigonal, (iii) tetragonal, (iv) rhombic, (v) kagome structures (Abuzeid et al. 2021).

2.2.1.4 Azine-Based COF

A large number of azine (Figure 2.1f)-based COF are synthesized because the two aldehydes are connected by the short hydrazine monomer resulting in the formation of polygon structures. Triphenyl triazine, substituted benzene, and a number of knots have been explored which leads to diverse topologies (Stegbauer et al. 2015; Alahakoon et al. 2017; Li et al. 2019; Lu et al. 2019; Guan et al. 2019; Alahakoon et al. 2016).

2.2.1.5 sp^2-Hybridized Carbon-Conjugated COF

The Knoevenagel condensation of aldehydes and benzyl cyanides in the presence of a base results in the synthesis of fully pie conjugated COF (Figure 2.1g). The condensation of tetrakis(4-formylphenyl)pyrene and 1,4-phenylenediacetonitrile in mesitylene/1,4-dioxirane in the presence of aqueous NaOH as catalyst resulted in the formation of C=C linked COF which was first proposed by Jiang and coworkers (Jin et al. 2017). The condensation 1,4-phenylenediacetonitrile and 1,3,5-tri(4-formylphenyl)benzene with cesium carbonate as the catalyst in the formation of a 2D poly(phenylenevinylene) framework polymer (Feng et al. 2016). A mixture of $CHCl_3$ and 1,4-dioxirane was used to construct a 2D COF which constituted triazine central planar units bridged by sp^2-hybridized carbon linkers (Zhao et al. 2019). The reaction of linkers of various lengths with tetrakis(4-formylphenyl)pyrene resulted in the synthesis of various sp^2-hybridized carbon-conjugated COF (Jin et al. 2018).

2.2.2 Preparation of Covalent Triazine Framework (CTF)

The first examples of CTF-containing triazine molecules and aromatic rings were reported by Thomas and his coworkers (Mohamed et al. 2020). These polymers are amorphous porous partially crystalline containing layered and conjugated structures (Qian et al. 2021). CTF has captivated attention due to its promising implementation in organosynthesis, reduction of carbon dioxide, hydrogen gas, and hydrogen peroxide production, degradation of pollutant, organic semiconductor devices, and splitting of water (Li et al. 2020; Liu et al. 2018,2019; Jiang et al. 2015; Xu et al. 2020; Puthiaraj et al. 2016). CTF has been prepared by a number of methods which include superacid-catalyzed trimerization, catalysis of P_2O_5 at 400 °C, Suzuki cross-coupling, cyclotrimerization of nitrile groups in the presence of zinc chloride, and nickel-catalyzed Yamamoto coupling as described in Figure 2.2 (Qian et al. 2021). For the preparation of highly crystalline structures of CTF, monomers with planar characteristics are used. If nonplanar monomers are used, it leads to the formation of semicrystalline or amorphous polymers (Bojdys et al.

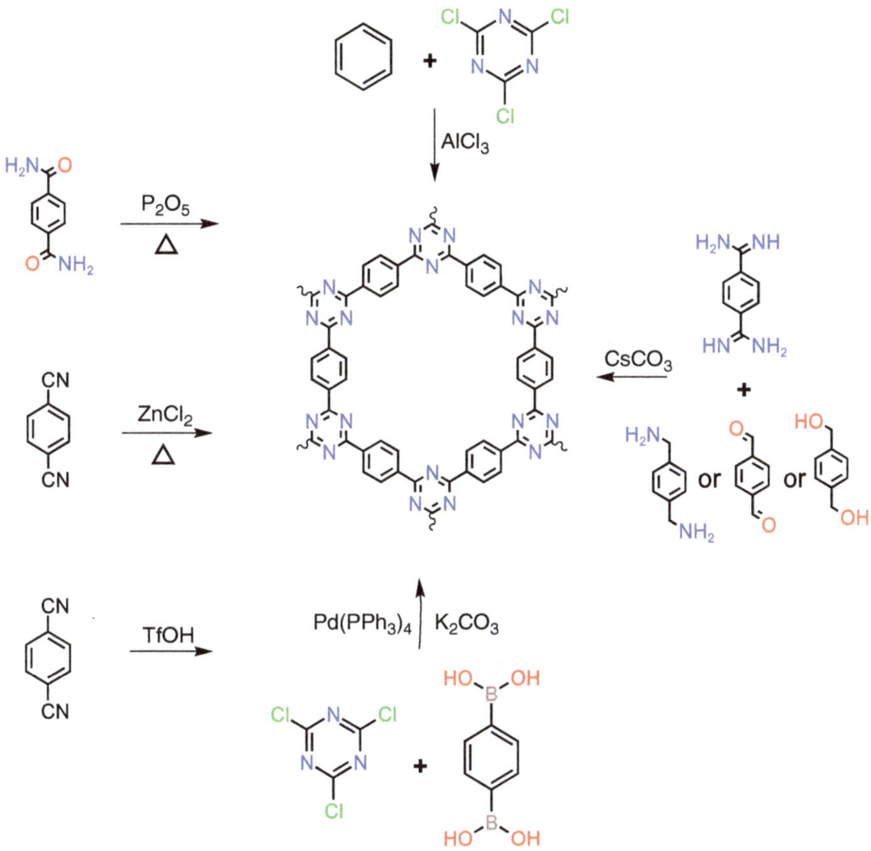

Figure 2.2 Formation of CTF reproduced from Abuzeid et al. (2021).

2010; Katekomol et al. 2013; Zhang et al. 2010; Kuhn et al. 2009). A well-stacked 2D CTF structure was prepared within 10 minutes at 25 °C by Xu. There was a TfOH-catalyzed interfacial reaction with refining techniques applied after the trimerization reaction (Abuzeid et al. 2021). Using this approach AB-staggered crystalline CTF were formed, whereas AA stacking crystalline CTF were formed by the ionothermal method.

2.2.3 Preparation of Hyper Cross-linked Polymers (HCP)

The porous polymers that have fascinated much curiosity are the HCP for properties like high surface area, high hydrothermal, mechanical, and chemical stability, extended conjugation, variation in synthetic methods, and tunable porosities. To establish their configuration, characteristics, structure, and stabilities, a number of methods have been used. The Friedel Crafts reaction containing a reaction solvent, Lewis acid catalyst, electrophile, and nucleophile that is arene based resulted in the construction of HCP (Gu et al. 2021). The reactions were carried out under inert atmosphere to eliminate the formation of by-products and obtain pure polymer (Gu et al. 2021). The HCP-containing metal species can be formulated through post alteration, i.e., either by using a metal complex as a monomer in the knitting process or by adding a metal complex or salt (Gu et al. 2021). For example, the Friedel Crafts alkylation of polystyrene results in the formation of HCP through crosslinking as shown in Figure 2.3a (Davankov and Tsyurupa 1990; Jiang and Cooper 2009). Also the direct polycondensation of small monomers like dichloroxylene (ortho, para, and meta isomer) (DCX), 4,4′-bis(chloromethyl)-1,1′-biphenyl (BCMBP) and

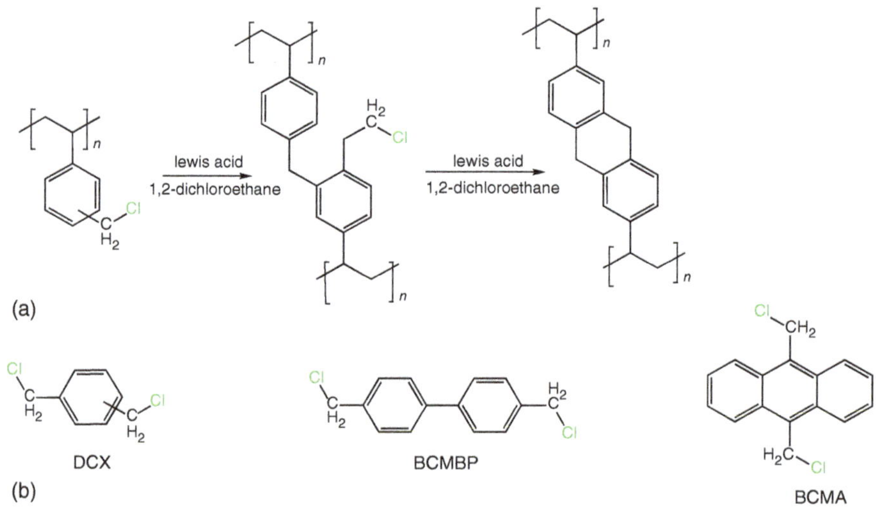

Figure 2.3 (a) Friedel Crafts alkylation reaction for the formation of HCP (54), (b) three bischloromethyl monomers used for the preparation of HCP by direct polycondensation. Source: Wood et al. (2007)/American Chemical Society.

bis(chloromethyl) anthracene (BCMA) yields HCP (Figure 2.3b (Jiang and Cooper 2009; Wood et al. 2007).

2.2.4 Preparation of Nitrogen-Containing Porous Organic Polymers

Synthesis of nitrogen-containing POPs takes place by the following methods (i) the porous carbon polymers are made to react with nitrogen-containing materials, (ii) N-containing materials having nitrogen-free precursors are co-carbonized with POPs containing carbon (Wu et al. 2017).

2.2.5 Preparation of Melamine-Formaldehyde (MF) Spheres

In the synthesis of melamine formaldehyde spheres, melamine was allowed to react with formaldehyde solution followed by addition of lysine (Liu et al. 2017a,b).

2.2.6 Preparation of Nitrogen-Rich Microporous Carbon

Melamine formaldehyde was mixed with water and ethanol. After that, hexamethylenetetramine (HMT), resorcinol, and citric acid were added. The formed nitrogen-rich microporous polymer was washed to remove impurities (Liu et al. 2012, 2017a,b). It can also be synthesized by using resorcinol and HMT in water, followed by adding citric acid (Liu et al. 2017a,b).

2.2.7 Preparation of Conjugated Microporous Polymer (CMP)

Because of the diversity, modularity, large surface area, and outstanding physiochemical stabilities, CMP is making an appearance in this field. These polymers are extensively used for photocatalysis, luminescent materials, absorbents, and light-harvesting materials. The engrossing properties of these polymers are mainly due to two reasons. First, their highly cross-linked polymeric network structures and high surface area; second, their expanded conjugated structures along the chains. The most widely used methods for the preparation of CMP are electropolymerization, Suzuki–Miyaura and Sonogashira–Hagihara cross-coupling, Schiff base formation, phenazine ring fusion, and alkyne metathesis. These approaches allow a range of different substrates, polysynthetic modification, good selectivities, suitable reaction condition, and high yield production (Luo et al. 2021; Sheng et al. 2021; Abuzeid et al. 2021; Chaoui et al. 2017).

2.2.8 Synthesis of Organic Polymeric Membranes

Using trimesoylchloride (TMC) with the addition of bisfunctional or multifunctional phenol monomers in water medium, cross-linked polyarylates are synthesized which is a porous membrane and a good example of polymers of intrinsic microporosity (Jimenez-Solomon et al. 2016; Ulbricht 2020).

2.3 Application Toward Organic Materials

2.3.1 Application of Porous Organic Polymers for Separation of Organic Gas

POPs have been broadly utilized for the storage adsorption of gases like hydrogen, methane as well as carbon dioxide with enhanced capacity and selectivity. Among the three gases, hydrogen is considered as an ideal fresh energy carrier because of its carbon-free nature, emission of only water as byproduct during its application in engine or fuel cell. Besides these, the efficiency of the fuel cell can be enhanced compared to internal combustion engine. Despite so many advantages, the lack of suitable, safe, and low-cost method causes limited application of hydrogen as fuel. In this regard, porous materials with high surface area have been considered as potential applicant in the sorption of H_2. These POPs are considered owing to their suitable composition of light element, available binding sites for hydrogen, and possibility of large-scale production value.

Similarly POPs could be utilized in the adsorption of carbon dioxide gas which is discharged into the atmosphere due to the increasing consumption of fossil fuels. This increasing amount of released carbon dioxide becomes a major issue of concern in prospect of global warming. Despite so many efforts, the practical solution of this problem is yet to be completed. To avoid the energy-exhausted CO_2 capture, the physisorption methods are more reliable as compared to chemical adsorption. But these techniques are suffering from low CO_2 selectivity, interference by tuned functional groups, and missing link in improving postcombustion process. In this aspect, porous materials are considered as dynamically efficient and precise. Till date a number of POPs have been described for the adsorption of CO_2 under suitable pre- and postcombustion capture. Due to the high porosity and pore surface functionality, the CO_2 uptake capacity is increasing making it a potential applicant for this purpose. Among the recently reported literature, the work of Yavuz and Coskun on azobridged nanoporous covalent organic polymers, work of Hani and his group on POPs and Dai group on heaxaaztriphenylene-based CFTs has been considered as significant for CO_2 adsorption.

The primary component of natural gas methane is considered as plentiful energy resource. The application of POPs for CH_4 gas adsorption and storage is quite less as compared to other gases, though organic porous materials are promising candidates for the same. The work of Cooper and coworkers is considered for storage of methane by utilizing series of HCPs. Among the developed series, 4,4′-bis(chloromethyl)-1,10-biphenyl (BCMBP) showed highest methane adsorption capacity.

In recent development, POPs are utilized in the separation of ethylene from ethane, one of the most efficient feedstock molecules used in petrochemical industry. For the large-scale chemical separation, similar volatility and molecular size create major problem. Utilization of cryogenic distillation method under high pressure and low temperature is being developed recently; still it is not efficient owing to the high cost and energy consumption. Organic-solvent-based adsorbent and porous absorbents are considered as potential users in this separation technique owing to their low cost and low-energy consumption value. On the other hand, POPs with strong covalent character can show high chemical and physical as well as moisture stability. Adornment of metal-like copper (catecholate) can improve the ethane-ethylene adsorption selectivity. One recent development of organic complexation with pops can not only provide suitable adsorption site but also enhance the stability and separation (Zhang et al. 2017a,b).

2.3.2 Catalytic Application of POPs

POPs and their functionalized derivatives like metal-free POP, chiral POP, and POP supported nano-particles, metal-free POPs establish as a new dimensional nanoreactor in recent development owing to their unique skeleton and pores. Molecules like amine, catechol, and pyridine groups have high affinity for metal ions. POPs are manufactured with a high density organic content in their framework, thus providing a high potential for metal ion. This property provides a high demand of this POPs functionalized material in the field of heterogeneous catalysis (Zhang et al. 2017a,b) which are summarized in following section.

2.3.2.1 Porphyrin-Based Functionalized POPs for Catalytic Application

Metal-porphyrin and their co-compounds can provide a large number of important biological processes which have a numerous number of catalytic applications. Macrocyclic porphyrin can fit in the organic framework through their interlinks to form POPs. This framework is adequate for metal ions to utilize in heterogeneous metalloporphyrin catalysis. The development of metal–organic CMP with Ir was reported in 2011. The metal–organic CMP was utilized in the catalytic reduction of amine. The catalyst showed high catalytic activity with 90% yield of the product. In 2009, Deng and his coworker developed a class of cobalt-aluminum coordinated CMP which can show excellent CO_2 capture at atmospheric pressure and room temperature. The cobalt-coordinated microporous polymers can also act as heterogenous catalyst for the reaction of CO_2 and propylene oxide at room temperature (Zhang et al. 2017a,b).

2.3.2.2 Application of Chiral Porous Organic Polymer in Asymmetric Catalysis

Only minimum numbers of reports are there to address the application of chiral POP catalysts. There are some criteria for the potential application of POP catalysts. To fulfill the criteria, the synthesis should be cost effective; the catalyst should be reusable with maximum catalytic efficiency. The development of chiral cross-linked polymer network by utilizing 1,10-binaphthyl and titanium ions is one example of this application. The developed network shows fair stability and different enantioselectivity for asymmetric catalysis, synthesis of three chiral porous through cross coupling polycondensation of diene. Herein they have utilized porous polymer as chiral stationary phase (CSP) in the chromatographic separation of race mates (Zhang et al. 2017a,b).

2.3.3 Application of POPs as Adsorbent for Volatile Organic Compounds

Volatile organic compounds, emitted from different kinds of industries, can be harmful to human health and environment. Owing to its low cost and high efficiency, adsorption method can be used for the treatment of VOCs. Activated carbons and mesoporous materials are mostly used to remove VOCs because of their plentiful porosity and surface area. Despite these advantages, the hydrophilic nature and low desorption value can cause limiting commercial application of these materials which should be more advanced. In this regard, =POPs structure possess high specific area, fair stability, and high degree of hydrophobicity and have emerged as excellent adsorbing materials for VOC. In this purpose, different kinds of POPs materials are utilized. HCPs, CMPs, PAFs are commonly used POPs materials for VOC adsorption. The abundant porosity in the surfaces of the POPs plays a crucial rule in adsorption. An example included in the recent development is the synthesis of a new kind of PAF material, by Ullmann's reaction which has excellent adsorption capacity toward methanol, benzene, and toluene at saturated vapor pressure showing the values of 29.16, 11.74, and 11.53 $mmol\,g^{-1}$, respectively. Construction of porphyrin-based porous covalent porphyrin network with large surface area and great hydrophobicity shows excellent adsorption capacity toward saturated hydrocarbons like *n*-pentane, *n*-hexane, *n*-heptane, *n*-octane, cyclopentane, and cyclohexane. In 2015, Wang and his group synthesized hydrophobic HCPs with large surface area through Friedel Crafts alkylation reaction by utilizing benzyl chloride monomer. The prepared polymer was selective for benzene adsorption when water is present. In their work, benzene adsorption capacity was 86% in dry condition and 80% in the presence of humidity (Lu et al. 2021).

The application of composite material of POP and other material can be more advantageous for the adsorption of volatile organic compounds. Recently Liu and his group have prepared grapheme-based hyper cross-linked porous carbon composite which shows outstanding adsorption capacity for 2,4-dichlorophenol. During this application, low yield of production and poor structural density of POPs led to lengthy production leading to limited application (Lu et al. 2021).

2.3.4 Application of POPs for Chemical Sensor

Π-Conjugated POPs with light-emitting properties are utilized as attractive materials in the construction of chemical sensor for the detection of various chemicals. The photo physical and electronic properties of POPs can be tuned by modifying the optical band-gap as they possess large surface area which helps them to provide large interface for analyte interaction. The first example of development of chemosensing CMP was reported by Jiang and his group. The synthesized CMP exhibits blue florescence behavior with large surface area by condensation reaction of carbazole derivative TCB utilized to detect the arene vapors. Compared to linear polymer, the porous TCB-CMP showed not only advanced detection sensitivity toward arene vapors but also provided faster detection capability. Another

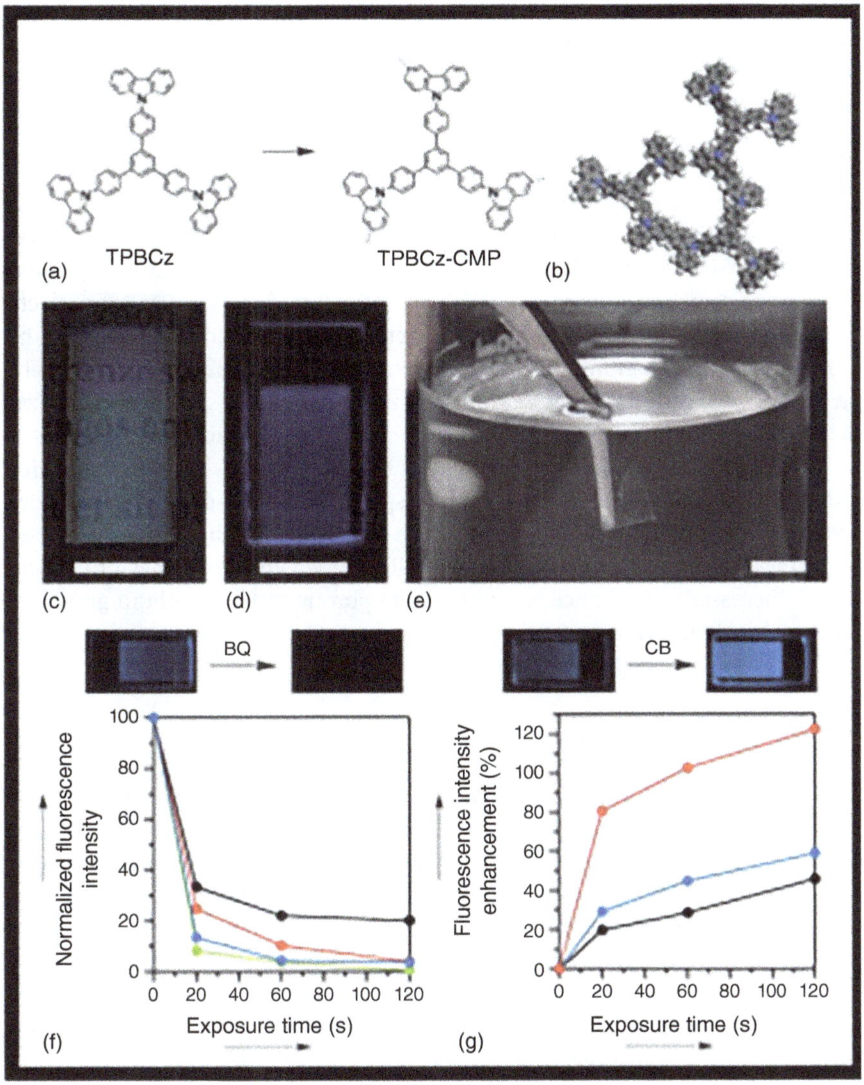

Figure 2.4 Schematic depiction of the synthesis of TPBCz-CMP films by electrochemical method. (b) Elementary pore of the film. (c) A film on an ITO substrate ($0.5 \times 1.5\ cm^2$), (d) film under UV light, and (e) a 100 nm thick freestanding film ($0.5 \times 1\ cm^2$) in ACN solvent. Scale bars: 0.5 cm, (f) Standardized fluorescence intensity of the CMP films upon exposure to vapors of benzonitrile (BN, black line), 1,2-dinitrobenzene (DNB, red line), hexafluorobenzene (HFB, blue line), and 1,4-benzoquinone (BQ, green line) for different periods of time. Photos: fluorescence quenching by BQ vapor, (g) enhancement of fluorescence intensity of the CMP films upon exposure to vapors of toluene (black line), benzene (blue line), and chlorobenzene (CB, red line) for different periods of time. Photos: enhancement of fluorescence intensity upon exposure to CB vapor. Source: Zhang et al. (2017a,b)/Reproduced with permission from Royal Society of Chemistry.

application of COPs as chemical sensor was to detect nitroaromatic explosives which are considered as one of the serious pollutants. Among all the forms, 2,4,6-trinitrophenol (PA) has become popular as it is widely used in many industries like glass, dye, fireworks, and leather industries. In the period of commercialization, PA could be released into the environment, leading to contamination of soil and aquatic bio systems. Considering their major threats toward environment, the development of an efficient and consistent sensor for their detection is very serious issue of concern. In this regard is the work of CaO and his group to synthesize two porous luminescent COPs by copolymerization. The developed sensor showed good detection capability with high sensitivity toward nitroaromatic explosives with a limit of detection 1 ppm. The development of light-emitting CMP/poly(lactic acid) nanofibrous films by electrospinning method was reported by Copper and his group. The fabricated nanofibrous films show high flexibility and high porosity with large surface area. The prepared CMP-based nanofibrous films were then utilized to detect nitroaromatic vapors. The sensor could detect benzoquinone vapors as well. The application of porous CMP-based hybrid materials in super capacitor construction has also been reported. In this aspect porous two-dimensional carbon nanosheets sandwiched with grapheme were developed by Zhang and his group. The synthesis nanomaterial showed outstanding supercapacitor behavior capacity with 48% increased capacitance as compared to porous carbons without grapheme (Zhang et al. 2017a,b) (Figure 2.4).

2.4 Conclusions

This chapter focuses on the various synthetic and wide application routes of POPs in various developments of organic materials. POPs have numerous advantages like amazing tailorability in terms of internal structures, compositions, and properties which make conceivable their rational design for any kind of precise application. The synthesis and application part of POPs are interdisciplinary with the integration of different fields like chemistry, physics, engineering as well as material science, while in the design of products for particular probable application is included. From this chapter the basic knowledge of useful POPs, their synthetic methodologies as well as applications in various fields have been discussed. Future work should be continued to develop methodologies that allow ascendable synthesis of POPs through environmental-friendly and low-cost method since large amount of production is necessary for many emerging applications in future.

References

Abuzeid, H.R., EL-Mahdy, A.F., and Kuo, S.W. (2021). Covalent organic frameworks: design principles, synthetic strategies, and diverse applications. *Giant* 6: 100054.

Ahn, J.H., Jang, J.E., Oh, C.G. et al. (2006). Rapid generation and control of microporosity, bimodal pore size distribution, and surface area in Davankov-type hyper-cross-linked resins. *Macromolecules* 39 (2): 627–632.

Alahakoon, S.B., Thompson, C.M., Nguyen, A.X. et al. (2016). An azine-linked hexaphenylbenzene based covalent organic framework. *Chemical Communications* 52 (13): 2843–2845.

Alahakoon, S.B., McCandless, G.T., Karunathilake, A.A. et al. (2017). Enhanced structural organization in covalent organic frameworks through fluorination. *Chemistry – A European Journal* 23 (18): 4255–4259.

Avella, M., Buzarovska, A., Errico, M.E. et al. (2009). Eco-challenges of bio-based polymer composites. *Materials* 2 (3): 911–925.

Babu, R.P., O'connor, K., and Seeram, R. (2013). Current progress on bio-based polymers and their future trends. *Progress in Biomaterials* 2 (1): 1–16.

Baekeland, L.H. (1935). Impress of chemistry upon industry bakelite, an example. *Industrial and Engineering Chemistry* 27 (5): 538–543.

Bojdys, M.J., Jeromenok, J., Thomas, A., and Antonietti, M. (2010). Rational extension of the family of layered, covalent, triazine-based frameworks with regular porosity. *Advanced Materials* 22 (19): 2202–2205.

Chaoui, N., Trunk, M., Dawson, R. et al. (2017). Trends and challenges for microporous polymers. *Chemical Society Reviews* 46 (11): 3302–3321.

Cote, A.P., Benin, A.I., Ockwig, N.W. et al. (2005). Porous, crystalline, covalent organic frameworks. *Science* 310 (5751): 1166–1170.

Davankov, V.A. and Tsyurupa, M.P. (1990). Structure and properties of hypercrosslinked polystyrene—the first representative of a new class of polymer networks. *Reactive Polymers* 13 (1–2): 27–42.

Diercks, C.S. and Yaghi, O.M. (2017). The atom, the molecule, and the covalent organic framework. *Science* 355 (6328): eaal1585.

Ding, S.Y. and Wang, W. (2013). Covalent organic frameworks (COFs): from design to applications. *Chemical Society Reviews* 42 (2): 548–568.

El-Kaderi, H.M., Hunt, J.R., Mendoza-Cortés, J.L. et al. (2007). Designed synthesis of 3D covalent organic frameworks. *Science* 316 (5822): 268–272.

Errahali, M., Gatti, G., Tei, L. et al. (2014). Microporous hyper-cross-linked aromatic polymers designed for methane and carbon dioxide adsorption. *The Journal of Physical Chemistry C* 118 (49): 28699–28710.

Fang, Q., Zhuang, Z., Gu, S. et al. (2014). Designed synthesis of large-pore crystalline polyimide covalent organic frameworks. *Nature Communications* 5 (1): 1–8.

Fang, Q., Wang, J., Gu, S. et al. (2015). 3D porous crystalline polyimide covalent organic frameworks for drug delivery. *Journal of the American Chemical Society* 137 (26): 8352–8355.

Feng, X., Zhuang, X., Zhao, W. et al. (2016). A two-dimensional conjugated polymer framework with fully sp^2-bonded carbon skeleton. *Polymer Chemistry* 7 (25): 4176–4181.

Feriante, C., Evans, A.M., Jhulki, S., et al. (2020). New mechanistic insights into the formation of imine-linked two-dimensional covalent organic frameworks. *Journal of the American Chemical Society* 142 (43): 18637–18644.

Fontanals, N., Marcé, R.M., Borrull, F., and Cormack, P.A.G. (2015). Hypercrosslinked materials: preparation, characterisation and applications. *Polymer Chemistry* 6 (41): 7231–7244.

Ghosh, S., Kouamé, N.A., Ramos, L. et al. (2015). Conducting polymer nanostructures for photocatalysis under visible light. *Nature Materials* 14 (5): 505–511.

Gu, Y., Son, S.U., Li, T., and Tan, B. (2021). Low-cost hypercrosslinked polymers by direct knitting strategy for catalytic applications. *Advanced Functional Materials* 31 (12): 2008265.

Guan, P., Qiu, J., Zhao, Y. et al. (2019). A novel crystalline azine-linked three-dimensional covalent organic framework for CO_2 capture and conversion. *Chemical Communications* 55 (83): 12459–12462.

Hisano, A. (2017). Selling food in clear packages: the development of cellophane and the expansion of self-service merchandising in the United States, 1920s–1950s. *International Journal of Food Design* 2 (2): 153–166.

Jiang, J.X. and Cooper, A.I. (2009). Microporous organic polymers: design, synthesis, and function. In: *Functional Metal-Organic Frameworks: Gas Storage, Separation and Catalysis*, vol. 128 (1), 1–33.

Jiang, X., Wang, P., and Zhao, J. (2015). 2D covalent triazine framework: a new class of organic photocatalyst for water splitting. *Journal of Materials Chemistry A* 3 (15): 7750–7758.

Jimenez-Solomon, M.F., Song, Q., Jelfs, K.E. et al. (2016). Polymer nanofilms with enhanced microporosity by interfacial polymerization. *Nature Materials* 15 (7): 760–767.

Jin, E., Asada, M., Xu, Q. et al. (2017). Two-dimensional sp^2 carbon–conjugated covalent organic frameworks. *Science* 357 (6352): 673–676.

Jin, E., Li, J., Geng, K. et al. (2018). Designed synthesis of stable light-emitting two-dimensional sp^2 carbon-conjugated covalent organic frameworks. *Nature Communications* 9 (1): 1–10.

Katekomol, P., Roeser, J., Bojdys, M. et al. (2013). Covalent triazine frameworks prepared from 1,3,5-tricyanobenzene. *Chemistry of Materials* 25 (9): 1542–1548.

Kuhn, P., Antonietti, M., and Thomas, A. (2008). Porous, covalent triazine-based frameworks prepared by ionothermal synthesis. *Angewandte Chemie International Edition* 47 (18): 3450–3453.

Kuhn, P., Thomas, A., and Antonietti, M. (2009). Toward tailorable porous organic polymer networks: a high-temperature dynamic polymerization scheme based on aromatic nitriles. *Macromolecules* 42 (1): 319–326.

Laurichesse, S. and Avérous, L. (2014). Chemical modification of lignins: towards biobased polymers. *Progress in Polymer Science* 39 (7): 1266–1290.

Li, Z. and Yang, Y.W. (2017). Creation and bioapplications of porous organic polymer materials. *Journal of Materials Chemistry B* 5 (47): 9278–9290.

Li, X., Qi, Y., Yue, G. et al. (2019). Solvent-and catalyst-free synthesis of an azine-linked covalent organic framework and the induced tautomerization in the adsorption of U (VI) and Hg (II). *Green Chemistry* 21 (3): 649–657.

Li, J., Liu, P., Huang, H. et al. (2020). Metal-free 2D/2D black phosphorus and covalent triazine framework heterostructure for CO_2 photoreduction. *ACS Sustainable Chemistry & Engineering* 8 (13): 5175–5183.

Liu, L., Deng, Q.F., Hou, X.X., and Yuan, Z.Y. (2012). User-friendly synthesis of nitrogen-containing polymer and microporous carbon spheres for efficient CO_2 capture. *Journal of Materials Chemistry* 22 (31): 15540–15548.

Liu, J., Zan, W., Li, K. et al. (2017a). Solution synthesis of semiconducting two-dimensional polymer via trimerization of carbonitrile. *Journal of the American Chemical Society* 139 (34): 11666–11669.

Liu, L., Xie, Z.H., Deng, Q.F. et al. (2017b). One-pot carbonization enrichment of nitrogen in microporous carbon spheres for efficient CO_2 capture. *Journal of Materials Chemistry A* 5 (1): 418–425.

Liu, M., Huang, Q., Wang, S. et al. (2018). Crystalline covalent triazine frameworks by in situ oxidation of alcohols to aldehyde monomers. *Angewandte Chemie International Edition* 57 (37): 11968–11972.

Liu, M., Jiang, K., Ding, X. et al. (2019). Controlling monomer feeding rate to achieve highly crystalline covalent triazine frameworks. *Advanced Materials* 31 (19): 1807865.

Lohse, M.S. and Bein, T. (2018). Covalent organic frameworks: structures, synthesis, and applications. *Advanced Functional Materials* 28 (33): 1705553.

Lu, J., Lin, F., Wen, Q. et al. (2019). Large-scale synthesis of azine-linked covalent organic frameworks in water and promoted by water. *New Journal of Chemistry* 43 (16): 6116–6120.

Lu, S., Liu, Q., Han, R. et al. (2021). Potential applications of porous organic polymers as adsorbent for the adsorption of volatile organic compounds. *Journal of Environmental Sciences* 105: 184–203.

Luo, S., Zeng, Z., Wang, H. et al. (2021). Recent progress in conjugated microporous polymers for clean energy: synthesis, modification, computer simulations, and applications. *Progress in Polymer Science* 115: 101374.

Mohamed, M.G., Ebrahium, S.M., Hammam, A.S. et al. (2020). Enhanced CO_2 capture in nitrogen-enriched microporous carbons derived from polybenzoxazines containing azobenzene and carboxylic acid units. *Journal of Polymer Research* 27 (8): 1–12.

Nakajima, H., Dijkstra, P., and Loos, K. (2017). The recent developments in biobased polymers toward general and engineering applications: polymers that are upgraded from biodegradable polymers, analogous to petroleum-derived polymers, and newly developed. *Polymers* 9 (10): 523.

Namazi, H. (2017). Polymers in our daily life. *BioImpacts: BI* 7 (2): 73.

Patil, A., Patel, A., and Purohit, R. (2017). An overview of polymeric materials for automotive applications. *Materials Today: Proceedings* 4 (2): 3807–3815.

Planells, M., Abate, A., Hollman, D.J. et al. (2013). Diacetylene bridged triphenylamines as hole transport materials for solid state dye sensitized solar cells. *Journal of Materials Chemistry A* 1 (23): 6949–6960.

Puthiaraj, P., Lee, Y.R., Zhang, S., and Ahn, W.S. (2016). Triazine-based covalent organic polymers: design, synthesis and applications in heterogeneous catalysis. *Journal of Materials Chemistry A* 4 (42): 16288–16311.

Qian, Z., Wang, Z.J., and Zhang, K.A. (2021). Covalent triazine frameworks as emerging heterogeneous photocatalysts. *Chemistry of Materials* 33 (6): 1909–1926.

Sheng, X., Shi, H., Yang, L. et al. (2021). Rationally designed conjugated microporous polymers for contaminants adsorption. *Science of the Total Environment* 750: 141683.

Spitler, E.L., Koo, B.T., Novotney, J.L. et al. (2011). A 2D covalent organic framework with 4.7-nm pores and insight into its interlayer stacking. *Journal of the American Chemical Society* 133 (48): 19416–19421.

Stegbauer, L., Hahn, M.W., Jentys, A. et al. (2015). Tunable water and CO_2 sorption properties in isostructural azine-based covalent organic frameworks through polarity engineering. *Chemistry of Materials* 27 (23): 7874–7881.

Ulbricht, M. (2020). Design and synthesis of organic polymers for molecular separation membranes. *Current Opinion in Chemical Engineering* 28: 60–65.

Wang, K., Yang, L.M., Wang, X. et al. (2017). Covalent triazine frameworks via a low-temperature polycondensation approach. *Angewandte Chemie International Edition* 56 (45): 14149–14153.

Wang, Y., Liu, Y., Li, H. et al. (2020). Three-dimensional mesoporous covalent organic frameworks through steric hindrance engineering. *Journal of the American Chemical Society* 142 (8): 3736–3741.

Wong, Y.L., Tobin, J.M., Xu, Z., and Vilela, F. (2016). Conjugated porous polymers for photocatalytic applications. *Journal of Materials Chemistry A* 4 (48): 18677–18686.

Wood, C.D., Tan, B., Trewin, A. et al. (2007). Hydrogen storage in microporous hypercrosslinked organic polymer networks. *Chemistry of Materials* 19 (8): 2034–2048.

Wu, S., Gu, S., Zhang, A. et al. (2015). A rational construction of microporous imide-bridged covalent–organic polytriazines for high-enthalpy small gas absorption. *Journal of Materials Chemistry A* 3 (2): 878–885.

Wu, J.Y., Mohamed, M.G., and Kuo, S.W. (2017). Directly synthesized nitrogen-doped microporous carbons from polybenzoxazine resins for carbon dioxide capture. *Polymer Chemistry* 8 (36): 5481–5489.

Xu, C., Xie, Q., Zhang, W. et al. (2020). A vinylene-bridged conjugated covalent triazine polymer as a visible-light-active photocatalyst for degradation of methylene blue. *Macromolecular Rapid Communications* 41 (7): 2000006.

Zhang, Y. and Riduan, S.N. (2012). Functional porous organic polymers for heterogeneous catalysis. *Chemical Society Reviews* 41 (6): 2083–2094.

Zhang, W., Li, C., Yuan, Y.P. et al. (2010). Highly energy-and time-efficient synthesis of porous triazine-based framework: microwave-enhanced ionothermal polymerization and hydrogen uptake. *Journal of Materials Chemistry* 20 (31): 6413–6415.

Zhang, W., Aguila, B., and Ma, S. (2017a). Retracted Article: Potential applications of functional porous organic polymer materials. *Journal of Materials Chemistry A* 5 (19): 8795–8824.

Zhang, W., Aguila, B., and Ma, S. (2017b). Retraction: Potential applications of functional porous organic polymer materials. *Journal of Materials Chemistry A* 5 (35): 18896–18896.

Zhao, Z., Su, H., Zhang, P. et al. (2017). Polyyne bridged AIE luminogens with red emission: design, synthesis, properties and applications. *Journal of Materials Chemistry B* 5 (8): 1650–1657.

Zhao, Y., Liu, H., Wu, C. et al. (2019). Fully conjugated two-dimensional sp^2-carbon covalent organic frameworks as artificial photosystem I with high efficiency. *Angewandte Chemie* 131 (16): 5430–5435.

3

Organic Polymers Fabrication for Solar Cells

Sudeshna Kalita[1,2], *Hemanta Singhania*[3], *and Anup Singhania*[1,2]

[1] *Council of Scientific and Industrial Research–North East Institute of Science and Technology (CSIR–NEIST), Chemical Sciences and Technology Division, Jorhat, Assam 785006, India*
[2] *Academy of Scientific and Innovative Research (AcSIR), Ghaziabad, Uttar Pradesh 201002, India*
[3] *Numaligarh Refinery Limited, Numaligarh, Assam 78569, India*

3.1 Introduction

Increasing trends in global warming and the energy crisis have demanded alternative sources of energy supply. The limited sources of fossil fuels as well as carbon emissions are other associated issues (Delucchi 2010). Hence, in the aspect of the environment, the paradigm has shifted toward green energy. Continuous investigations are ongoing to find superior solutions or make the existing methods greener (Dincer and Acar 2015). Various alternatives are explored (Li et al. 2018a,b,c; Singhania et al. 2020), out of which solar cells are one of the currently trending research areas (Kannan and Vakeesan 2016). To generate electricity, scientists have put continuous efforts into designing and developing new materials that can be used in large-scale production and effectively convert light energy into electricity or solar fuels at a cheaper cost while maintaining sustainability (Muñoz-García et al. 2021). There is space for significant growth in the use of renewable energy sources because they now make up just 10% of the overall energy supply (29% of the total electricity supply) (Hannah et al. 2022).

The sun is the core energy source for nature, providing the world with 173,000 TW of energy each year (Archer 2011). The research in the energy sector has been moving toward finding sustainable renewable energy sources that could be achieved by developing efficient, low-cost photovoltaic devices. The world's fastest-growing energy technology is photovoltaic electricity generation, which has increased by more than, on average, 34% annually during the past 10 years (Newell and Raimi 2020). But barely 1% of the world's electricity is produced by solar cells. Between 2019 and 2025, renewable electricity output is expected to increase by 50%, according to the International Energy Agency (IEA). With the consumers' ability to manufacture their own energy increasing rapidly, new opportunities are presented.

The solar cell is one of the effective ways to harness the available light energy, and organic polymer as a component of solar cells makes it easier by providing a

Organic Polymers in Energy-Environmental Applications, First Edition.
Edited by Ramesh Oraon, Pardeep Singh, Sanchayita Rajkhowa, Sangita Agarwal, and Ravindra Pratap Singh.

simple method and more affordable by lowering the cost. Organic-polymer-based materials are used practically in every element of modern civilization due to their ability to chemically alter the material properties of polymers and their several quick and inexpensive processing methods (Osaka 2015).

3.1.1 Parameter that Describes the Effectiveness of Solar Cells

We discuss several accounts on the performance advancement of photovoltaics where organic polymers are fabricated and also the latter's role in solar devices. Hence the functionality of solar cells is described with some quantified parameters; thereby, we here define those parameters, mainly with two quantified parameters - the current–voltage (J–V) measurement and power conversion efficiency (PCE) (Furkan and Mehmet Emin 2010). The current–voltage (J–V) measurement is the foremost tool for determining the solar cell's performance as it demonstrates the various fluctuations in power and voltage generated by a photovoltaic device. The second factor that affects how well a solar cell performs is its PCE, defined as the proportion of power generated by the solar cell to the optical power of incident light. The graph of J–V of a solar cell is plotted between the current density J–V curves by changing the external load under sun irradiation. Notably, this parameter includes energy loss owing to the difference in externally applied voltage and that of the emitted photons, where internal voltage drop is also considered. Thus, PCE will always be less than 100%. Another term, power density or flux, tells us about the power generation of each square meter of a cell, whereas PCE describes the solar cell's total power. The output energy of the solar cells is described in terms of three other parameters, namely short-circuit current density (J_{SC}), open-circuit voltage (V_{OC}), and fill factor (FF). The active layer's ability to absorb photons, the efficacy of free charge carrier creation, and the charge collection all affect the J_{SC} or the device photocurrent at zero bias. The difference between the highest occupied molecular orbital (HOMO) of the donor material and the lowest unoccupied molecular orbital (LUMO) of the electron acceptor fullerene derivative, known as the V_{OC} or photovoltage at zero current density, determines the photovoltage (Mi et al. 2014). Moreover, the parameter *FF* determines the maximum power from a solar cell with respect to V_{OC} and J_{SC}, which hence, determines the maximum power from a solar cell (Jain and Kapoor 2004; Green 1981). The *FF* can be described as the ratio of the maximum power from the solar cell to the product of V_{OC} and J_{SC}. The active layer's balanced charge transport and recombination characteristics are necessary for the FF to function. The standard illumination has an intensity of $1000\,W\,m^{-2}$ (AM1.5 G).

3.1.2 Generation of Solar Cells

Solar cell devices are technological marvels that generate free charges when moving through semiconductors by means of sunlight, converting the light energy into electricity (Bagher et al. 2015). In literature, various forms of solar cells are reported depending on the materials used to manufacture them (Soga 2006). First-generation solar cells are silicon-based crystalline devices. The method was first identified in 1839, and later in 1954, Chaplin, Fuller, and Pearson manufactured the first solar cell (Perlin 2004) and introduced it to the world. This generation of solar cells was created using monocrystalline or polycrystalline silicon crystals on silicon wafers. However, high manufacturing and process costs are the main drawback of this type of solar cell type (Razykov et al. 2011). In terms of cost, the polycrystalline form was cheap but less efficient among single-crystal, amorphous, and polycrystalline silicon. These silicon-based photovoltaic devices are long-lasting compared to other non-silicon-based solar devices. As with the rise in temperature, the solar panels lose some of their efficiency, so, as a necessary consequence, these panels are being arranged in such a way that they get better air circulation. Monocrystalline silicon solar cells exhibit maximum efficiency of up to 25.6% (Masuko et al. 2014; Jošt et al. 2022), whereas polycrystalline ones show conversion of up to 19.3%.

In 1970, second-generation solar cells were introduced; these were thin films composed of layers of semiconductor materials having only a few micrometers of thickness, mainly composed of cadmium telluride (CdTe) (Hubbard et al. 2008) and copper indium gallium diselenide (CIGS) (Chirilă et al. 2011) and on coupling with n-type materials they resulted in p-n junctions (Shah et al. 1999). CdTe exhibits a bandgap of 1.5 eV, allowing in a matching solar spectrum absorption. According to a report, the maximum power conversion of CIGS cells was 22.6%, whereas the GaAs solar cells obtained 28.8% (Ren et al. 2017) in all single-junction solar cells containing a bandgap of 1.42 eV (Green et al. 2018, 2021).

Recently, third-generation solar cells have been reported based on new materials besides silicon, such as organic polymers and dyes. Under this generation, organic solar cells (OSCs), dye-sensitized solar cells (DSSCs), quantum dot solar cells (QDSCs), and perovskite solar cells (PSCs) are included (NREL 2019). These photovoltaic devices have been explored extensively owing to their several beneficial properties such as low manufacturing cost, light weight, flexibility, and large area; features that allow them to be applied in wearable devices, calculators, building photovoltaics, and other domains. According to a report, PSCs have obtained a PCE of 24.2% (Jung et al. 2019) and still progressing (25.8%) (George et al. 2022; Kratzenberg et al. 2019). OSCs comprised polymer donor moieties, and non-fullerene acceptors obtained PCE of more than 19%, but their poor absorption below 550 nm limited the functionality (Cui et al. 2021). Herein, we will focus on the OSCs or polymer OSCs, also known as plastic solar cells, due to the plastic nature of polymeric materials.

3.2 Organic Solar Cells

3.2.1 General Considerations

OSCs or organic photovoltaic (OPV) solar cells are the technologies used in third-generation solar cell systems on which extensive research and development are being carried out (Wöhrle and Meissner 1991; Li et al. 2018a,b,c). They are named as such because, unlike traditional crystalline solar cells that use silicon as an absorber of sun rays, OSCs make use of carbon-based organic polymers or small molecules and organic electronics having desirable semiconducting properties. These compounds are printed onto thin plastic sheets (thickness less than 100 nm) by typically dissolving in ink. As a result (Clarke and Durrant 2010), these solar cells are much superior to the former silicon photovoltaics in terms of greater mechanical flexibility, lesser weight, more color variations, compactness, and capability to generate high power even under low light intensities. Various future applications of the OPVs can be envisioned owing to these properties, such as generating energy by incorporating them in sunroofs of hybrid and electric cars, on the windows of our homes and offices, achieving small tasks of lighting and charging cell phones by integrating them in fabrics of tents and bag packs, generating electricity to power up a medication and vaccine refrigerator in remote areas with no access to electricity (Khan et al. 2018; Badea et al. 2018). In contrast to large silicon solar panels occupying the entire length of a terrace or that of a large solar field, the thin and flexible OPVs thus have the potential to generate power in places that is otherwise quite impossible.

Additionally, they have cheap manufacturing costs and can also pay back the energy consumed during their manufacture in much lesser time. The fundamental core of OSCs involves using light-sensitive organic compounds such as conjugated polymers, dyes, or small organic moieties as semiconductors operating through the photovoltaic effect to generate solar power (Lu et al. 2021). Although the power generation ability of the OSCs is not at par with the crystalline silicon solar cells, improvements are being made (generating efficiency of up to 19%) by altering the polymer's geometry, absorption properties, and synthetic and self-assembly protocols (Green et al. 2018, 2021).

Heeger, MacDiarmid, and Shirakawa, in the year 2000, received the Nobel Prize (Rasmussen 2018) in Chemistry for the "discovery and development of conducting polymers," a completely new class of conducting organic material having great potential of being fabricated into organic optoelectrical systems like photovoltaics and organic light-emitting diodes (OLEDs). Semiconductors based on organic compounds (Chen 2018; Sariciftci et al. 1992) have several benefits and attractive properties in modern days because of their numerous appealing qualities, such as their availability, lightweight design, low cost of production, processing at lower temperatures, along with mechanical flexibility. This chapter mainly discusses solar cells designed and manufactured using organic polymers. The several roles played by various organic polymers in photovoltaic performance will be elaborated, as well as their preparation strategies.

What makes these materials applicable in solar cells for power generation is that the conducting polymers comprise the sp^2-hybridized conjugated carbon chain, and the conjugation renders the π-electronic cloud of these materials highly delocalized and polarized. This, in turn, provides the materials with visible light-absorbing capability in addition to transporting electrical current – two fundamental requirements for materials to be fabricated in solar cells (Hoppe and Sariciftci 2007). The alternating electron-donating and -accepting interactions within the molecules decrease their bandgap energy resulting in intense and broad absorption bands with high absorption coefficients. The bands, therefore, cover an array of wavelengths, encompassing those of the visible spectrum. Hence, they can easily harvest visible light even with a thickness of 100–200 nm only (Brédas et al. 2009). Conjugated polymers with dissolvable side chain derivatives are soluble in water or other organic solvents and can be used as ink to deposit over thin, homogeneous, and flexible films (Günes et al. 2007; Hoppe and Sariciftci 2007). Some examples of polymers commonly used in PSCs are diketopyrrolopyrrole (DPP) (Li et al. 2016), B ← N bridged bipyridine systems like P-BNBP-fBT (Long et al. 2016, 2019), indacenodithieno[3,2-*b*]thiophene (IDTT) (Li et al. 2019a,b), and bithiophene derivatives such as 3,3′-difluoro-2,2′-bithiophene (Figure 3.1).

3.2.2 Working Mechanism

In general, the OPVs operate by the photovoltaic effect of transforming the radiant light energy captured from sun rays directly into electric power. First, the absorption of sunlight by the organic absorber takes place. Through light absorption, the OPV knocks electrons loose from the organic semiconductor. Electron acceptors used together with the absorber accept these free electrons and transfer them into the electrode to generate direct current from light radiation. A detailed four-step mechanism of polymer-fabricated solar cells is described below:

(a) *Photogeneration of exciton*: Light energy in packets of energy called photons is absorbed by the OSCs, which comprise an electron-donating and an electron-accepting compound in conjunction with each other. Upon absorbing a photon, an electron belonging to the HOMO of the donor polymer gets excited to its LUMO, leaving behind an electron-hole. Thus, an electrically neutral exciton is created.
(b) *Migration of exciton*: The bound electron–hole pair or exciton moves to the donor–acceptor interfacial region.
(c) *Charge separation*: At the juncture of the donor–acceptor system, the exciton gets separated into the individual charges – free electron and hole. Dissociation of the electron–hole pair occurs only at the heterojunction owing to their contrasting electron-accepting and -losing tendencies. Here, the electron on the donor's LUMO gets transported to that of the acceptor, while the hole stays back in its original site of generation, i.e., the donor's LUMO.
(d) *Collection of charge*: The charge carriers formed now move toward their respective electrodes by the influence of an internal field produced by the variation in the work function of the electrodes and get accumulated.

Hence, in conclusion, the excited hole–electron pairs, or excitons, created in solar cells split apart into free carriers at the electrical junction between the electron-accepting and -donating layers before moving through charge separation, all of which are controlled by an existing electric field (Figure 3.2) (Zhu et al. 2009; Andersson 2020). The free charges then accumulate in the respective anode and cathode, which are finally transferred to an external circuit to generate electricity from solar light.

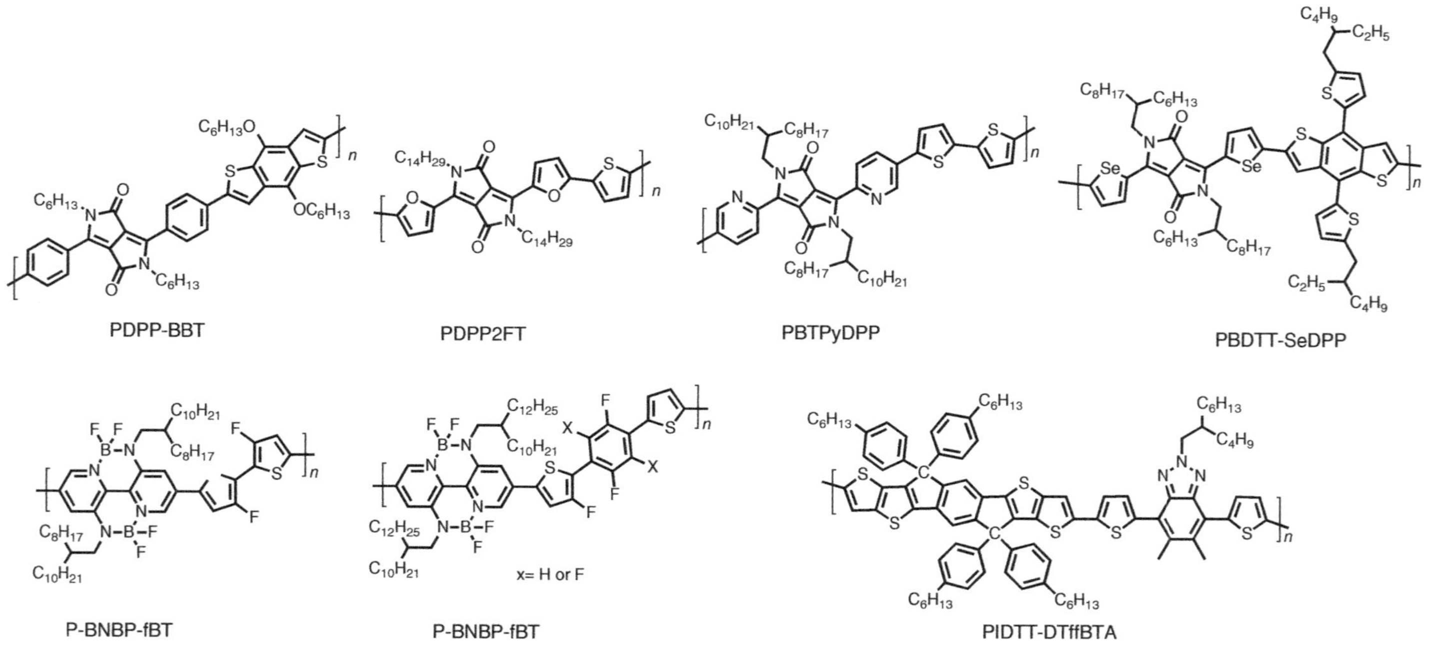

Figure 3.1 Examples of DPP-based organic-polymer-bearing aromatic substituents that comprise B ← N bridged bipyridine and IDTT.

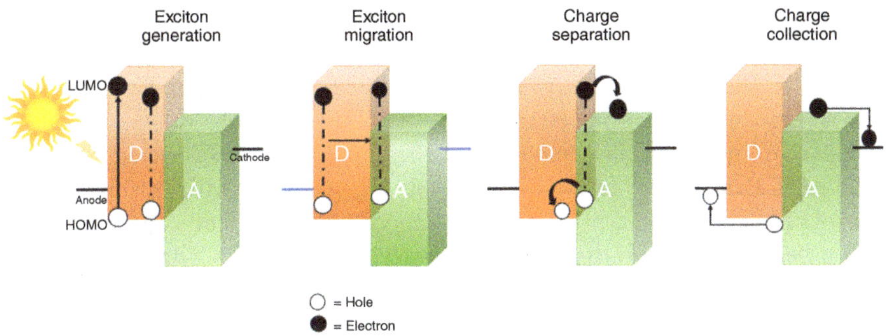

Figure 3.2 A general working mechanism of OSCs comprises the donor (D) and acceptor (A).

3.2.3 Architecture of OSCs

Solar cells that perform well need to be developed, and in this regard, the architectural designs of OSCs are essential to address their issues. The design improvements of solar cells can provide better charge transfer, absorption, stability, and high PCE (Venkataraman et al. 2010). In addition, the module design and processing technologies need to be explored for large-scale application of OSCs (Liao et al. 2010) (Figure 3.3): (a) single or monolayer, (b) bilayer heterojunction, (c) bulk heterojunction (BHJ), and (d) tandem solar cells. In addition, the design of graded heterojunction and continuous heterojunction solar cells have also been reported.

The era of polymer solar cells began with just a single-layer architecture where the organic layer was stuffed in between a layer of Indium tin oxide (ITO), used as the anode, and a layer of Al, Mg, or Ca metal, performing as the cathode. Here, a conductor connects the two electrodes to an external circuit, and an electrical field is created in the layers of the OSCs due to the difference in the electrode's functions. The PCE of single-layer OSCs is limited owing to the lesser diffusion length of the photogenerated excitons, which hence causes greater electron–hole recombination. However, the <1% quantum efficiency as well as <0.1% PCE achieved by these devices paved the way for better devices with bilayer planar heterojunction and BHJ architectures. These second generations of polymeric solar architectures incorporate electron donor (D) and electron acceptor (A) materials in their device structure.

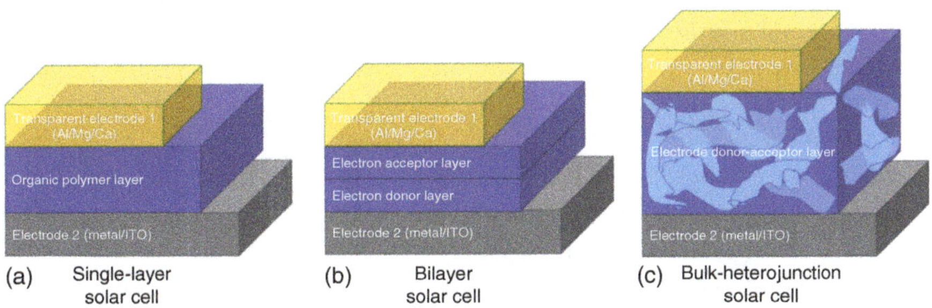

Figure 3.3 Device architecture of the OSCs device.

The electrostatic force generated at the heterojunction between the two materials, as a result of the variation in their ionization potentials, causes the excitons to dissociate much more efficiently than that achieved in the homojunction of a single-layer solar cell (Nelson 2002).

Bilayer heterojunction solar cells were the first photovoltaic systems to have incorporated this concept in 1986 (Tang 1986). A CuPc/PV bilayer heterojunction photovoltaic device was synthesized with copper phthalocyanine (CuPc) as the D and a perylene tetracarboxylic derivative (PV) as the A, sandwiched amidst an anodic ITO electrode and Ag cathode resulting in PCE of approximately 1%. The charge separation induced at the heterojunction provided a distinct advantage of not having to rely on an external electric field for the same (Tang 1986). Many more bilayer devices were synthesized over the years, but they suffered the limitation of exciton diffusion length (Menke and Holmes 2014). This exciton diffusion length is related to the distance between the excitons and the heterojunction. Since the excitons need to travel to the junction within their lifetime to get separated into their charges, only those excitons get successfully dissociated, which are photogenerated near the donor-acceptor interface of the bilayer heterojunction, while the others recombine. As excitons generated in organic polymers live for a very limited amount of time, their maximum diffusion length is confined to only 20 nm. Correspondingly, an extremely small percentage of the excitons generated gets converted into their charge carriers (Li et al. 2018a,b,c). The outcome of this limitation was the BHJ polymeric solar cells in which both A and D are blended such that the junction of exciton creation is not confined to a particular position. Charge separation can thus effectively occur at any position of the photoactive layer, thereby leading to a more electric current generation.

3.2.4 Preparation Techniques

Different strategies are adopted to produce organic polymer solar cells based on thin films. One of the most commonly used tactics is vacuum evaporation which is for low molecular-weight polymers; however, solution processing is used for higher molecular-weight materials (Hoppe and Sariciftci 2004). It can be attributed to the fact that higher molecular weight materials or polymers decompose at high temperatures, resulting in the formation of greater molar masses, which makes evaporation difficult. Hence solution processing is advantageous as it performs under low temperatures. The standard fabrication techniques with examples are discussed below.

3.2.4.1 Spin Coating

This laboratory solution-based technique is a low-cost method that deposits a homogeneous layer on the surface of a solid through the application of centrifugal force while also necessitating a liquid–vapor interface (Mammeri 2019; Yilbas et al. 2019). Generally, a drop of the material is placed at the center of a circular surface, and upon rotation, the drop turns into a uniform thick film of 1–10 μm (Liu and Kelly 2014). The centrifugal force upon rotation induces the liquid to spread into a film, and the solvate rapidly evaporates. Depending on the required film thickness, the rotational speed is regulated at a given value. For instance, on 125-mm silicon wafers with 0.5–1.5 m thickness, spin coating of poly(1*H*,1*H*-perfluorooctyl methacrylate) solutions has been carried out. The measurement of the thickness of the film can be investigated by these factors: solvent evaporation rate, the velocity of the spinning substrate (spin rate), and the viscosity of the coating solution. Accordingly, controlling or monitoring these parameters is essential for better film formation. Burmann et al. (2014) described that solvent evaporation affects the film or membrane casting, and slow evaporation is preferable to defeat free membranes; however, the rotational time has no significant role. The inexpensive spin-coating approach is often used to make BHJ films for large-area organic cells. In comparison to films produced without technology, these function better and are simpler to duplicate. In 2014, Kelly et al. reported hybrid PSCs that achieved more than 10% PCE upon the solution-processed spin-coated method (Liu and Kelly 2014). Zhang and co-workers employed a PTB7 polymer in a solution-processed PSC that obtained a PCE of 13.29% under ambient air compared to its counterparts which needed a more sophisticated environment (Du et al. 2015a,b). The results suggested that the spin-coating technique enhanced absorption capacity and maintained the quality of the film, thus improving its efficiency.

3.2.4.2 Blade Coating

Doctor-blade coating or blade coating is one of the popular fabrication techniques in which a sharp blade is positioned at a set distance from the substrate surface (Sankaran et al. 2016; Hsu et al. 2022). In front of the blade lies the ink composition. After that, a wet film is left on the substrate and is dried as the blade is moved linearly across it. The gap between the substrate and the blade is smaller than the final thickness. The viscoelastic characteristics of the solution and the coating rate impact the wet film's eventual thickness. Blade coating is more resource-efficient than spin coating. It can be used for the scaling up of OSCs layer deposition because the drying conditions are closer to those produced with industrial processing The reproducibility of the wet layer film's thickness is, however, weak. This occurs as a result of the solution's shearing rate affecting the ultimate film thickness. This method works well to produce thicker films from a viscous solution as well as coatings on a big scale. Its restriction is that it cannot achieve uniformity at the nanoscale and in extremely thin layers that spin coating can.

Some reports also suggested that the small-scale laboratory-based device performance could not deliver the same result when implemented on a large scale. Thus, more investigations need to be done to achieve effective results on a large scale by improvement in scale-up techniques. Consequently, fabrication techniques and device stability need to improve, which could cause a substantial increase in efficiency. These issues are explored from time to time by researchers. For example, based on this spin-coating method and upon scaling-up using blade coating, Son et al. described the mechanism of PBDB-T (poly[2,6-(4,8-bis(5-(2-ethylhexyl) thiophen-2-yl)benzo[1,2-*b*:4,5-*b'*]dithiophene)-*co*-(1,3-di(5-thiophene-2-yl)-5,7-bis (2-ethylhexyl)-benzo[1,2-*c*:4,5-*c'*] dithiophene-4,8-dione)]) based BHJ solar cells with fullerene or non-fullerene acceptor using the blade-coating technique (Park et al. 2020). This polymer with PC71BM in another BHJ film-based solar cells obtains similar morphology with an active surface area of 1.35 cm^2 irrespective of coating temperature. However, the processing affected non-fullerene ITIC-based BHJ films, which was evident by the increase in the blade coating's processing temperature, which increased their efficiency by around 4% at 50 °C. Because of the ITIC's extensive aggregation and its resulting slow evaporation, there was a substantial decrease in the power conversion when the coating was performed at room temperature. This outcome which ranks among the best of large-area Perovskite cells offers research requirements for a smooth transfer from small-area coverage spin-coating technology to big-area (here active area 58.5 cm^2) covering roll-to-roll technology.

Spin-coating techniques to integrate OSC are limited in the laboratory and unsuitable for industrial scale (Diao et al. 2015; Wu et al. 2017; Hoang et al. 2016). The doctor blade coating was used for large-scale production that can monitor the morphology of the film. In comparison to spin-coating techniques, blade coating offers a better rate of bleeding, higher crystallization, and greater molecular orientation, resulting in improved photovoltaic performance. Reports on blade-coating methods being adopted in semitransparent OSCs (Zhang et al. 2018a; Gu et al. 2016) can be found. Zhu et al. used two fabrication methods to study semitransparent OSCs using low-bandgap polymer and a non-fullerene moiety (Zhu et al. 2020a). They employed the ternary process by introducing a third component resulting in a PCE of 12.86% via spin coating. The spin-coating method of semitransparent OSCs was 10.46%, while that obtained through blade coating was 9.52% PCE.

The small-molecule-based solar cells offer potential advantages over PSC; however, the hurdle is to regulate the solution viscosity and film thickness. Thus, adding a third component could possibly erase these issues. Based on blade-coating work on ternary BHJ solar cells in 2017, Guo et al. outlined a work where a polymer was presented as the third component to regulate the viscosity, morphology, and function of the small molecule (Wu et al. 2017). The work mainly described the effect of coating speed as well as the role of the third component. The polymer p-DTS$(FBTTh_2)_2$ with PCPDTBT employed as the third component increased the absorption range and enhanced external quantum efficiency. Two regimes are identified that are affected by coating speed. The film thickness increases upon an increase in speed that could be categorized as Landau–Levich regime blade coating. Additionally, the surface morphology improves with increased coating speed as crystal size increases in both binary and ternary blends. This could be attributed to the third component that enables better connectivity of small crystals with better drying. Subsequently, this results in the improvement of exciton dissociation and hole transport. Moreover, the phase segregation of the acceptor was improvised with the coating speed with the inclusion of a third component. Overall, the work depicted the effect that the blade-coating method, with its speed, has on nanoscale morphology, along with the addition of a conjugated polymer that could enhance photovoltaic performance. In another report based on the same strategy, Wei Ma and coworkers created printed OSCs that could achieve a PCE of 12.02% in an ambient environmental setting (Zhang et al. 2018a). This blade-coating method demonstrates excellent compatibility in printing large areas by the roll-to-roll process while assisting in slot-die coating as a prototype tool. When manufactured through this blade-coating strategy, the ternary organic solar devices of PBDB-T:PTB7-Th:FOIC provided one of the highest PCE (12.02%) ideals for printed organic solar panels in the ambient environment. Using blade coating could accelerate the crystallization and charge carrier mobility of the non-fullerene acceptor polymer FOIC and the polymer donors PBDB-T and PTB7-Th; however, their degree of induction varied in the order: FOIC > PBDB-T > PTB7-Th. As such, the FOIC devices manufactured through blade coating yield greater crystalline behavior than that of spin-coated FOIC films, but this crystallinity combined with the electron mobility displays a decreasing behavior on increasing the amount of PTB7-Th in the PBDB-T:FOIC polymer blends, owing to the fact that PTB7-Th is much more miscible with FOIC compared to PBDB-T. The method of adding PTB7-th boosts the photocurrent while still sustaining the crystalline phase very well. The manufacturing of greatly efficient OSCs to cover large areas has a lot of potential, all thanks to the excellent synergism in shape optimization provided by the combined techniques of blade coating with the ternary strategies. The PCE might be improved to the results shown here, and the whole fabrication procedure represents a valuable new strategy for making highly efficient perovskite-based photovoltaic systems.

3.2.4.3 Inkjet Printing

It is ideal to transfer the lab-scale spin-coating method to bulk-scale industrial processes like roll-to-roll coating and printing since polymers and other necessary components for solution processing are readily available (Peng et al. 2017). Among all the different kinds of fabrication processes, inkjet printing has attracted a lot of attention in both academic research and industrial applications and is thought to be one of the most promising ways to achieve these needs (Karunakaran et al. 2019). Inkjet printing is a flexible approach to deposit functional layers of organic semiconductors made of homogeneous molecular or colloidal liquid-phase dyes. The working mechanism relies on the exact deposition and fixation of ink droplets with regulated properties that are ejected from the nozzle onto a target substrate. Inkjet printing uses two most frequently used methods to create ink droplets, i.e., continuous inkjet printing and drop-on-demand inkjet printing (Basaran et al. 2013).

Conjugated organic polymers have a given advantage of getting easily dissolved in water and other solvents when engineered with soluble side chains. This unique property enables them to be put in service as "ink" for depositing in solar films through an inkjet printer (Schlisske et al. 2021). The inkjet printing technique surpasses the limitation of device geometry as it can print the polymer "inks" at every layer of the solar device. The science behind this technique depends on releasing tiny droplets from the printer nozzle by overcoming the intermolecular forces of attraction between the cartridge and ink and also on the drying procedure of the solvent in which the polymer is dissolved. Using this printing technique, ultrathin, lightweight, and flexible OSCs were developed for harvesting light energy.

The preparation techniques control the solar panel's transparency, which can open up greater opportunities for the future. For instance, controlled digital printing of photopolymerizable *N*-vinylcaprolactam monomers at desired areas of a PSC appeared as translucent holes and resulted in a semitransparent PSC that showed a PCE of about 11.2% (Pendyala et al. 2021). Producing solar cells by inkjet printing technique is remarkably cheap as the desired polymer solutions can be filled in refillable printer cartridges; the substrates on which the polymer is to be deposited are fabricated on a compact disc. This disc is then inserted in the printer, and the polymer inks are printed out on the desired substrate using the original printing software. This method thus enables the printing of multiple solar films at a time and allows the reprinting of a single film to result in a multilayer solar structure. Apart from being cheap, inkjet printing boasts several other advantages, including being non-contact, requiring no photomask, employing more material, and being scalable and in accordance with roll-to-roll printing (Karunakaran et al. 2019). The advantage of not using a photomask is reflected in the liberty obtained in the solar cell. For example, different artistic shapes of OSCs were demonstrated on large areas of solar panels through industrial inkjet printing, which confirms that this technique allows product customization by giving desired shapes and sizes (Eggenhuisen et al. 2015). Overall, creating an ink formula with the right viscosity, wettability, and stability is the main problem encountered while using inkjet printing to create solar cells. Due to the variation in solubility, solvent engineering is necessary to filter out a suitable solvent or a combination of solvents for a better ink formula. The requirement of a smaller nozzle for printing in high-resolution becomes another significant barrier to the widespread usage of inkjet printing as it causes nozzle clogging. Hence, the optimization of inkjet printing parameters needs to be done to resolve these issues.

Here, we have discussed only three deposition techniques of organic materials for the solar cell's device. However, several other printing techniques, including screen printing, Gravure printing, slot-die coating, and spray coating, are also available. We refer to some dedicated literature reports (Gutoff and Cohen 2006; Krebs 2009b) for more detailed studies.

3.3 The Role of Organic Polymers in Solar Cells and Their Recent Progress

Various polymeric semiconducting materials are designed and developed, further investigating their conducting properties in the solar energy sector. The polymers are incorporated with different functionalized groups or molecules such as DPP, isoindigo (IID) units, naphthalene diimide (NDI), perylene diimide (PDI), and thieno[2′,3′:5′,6′]pyrido[3,4-g]thieno[3,2-*c*]isoquinoline-5,11(4*H*,10*H*)-dione. Some reported organic polymers that have remarkable solar performance are shown in Figure 3.4 (Zhou et al. 2021; Hou et al. 2019). Based on the properties or design of organic polymers, they are named differently, for instance, conjugated organic polymers, asymmetric organic polymers, and porous organic polymers. Depending

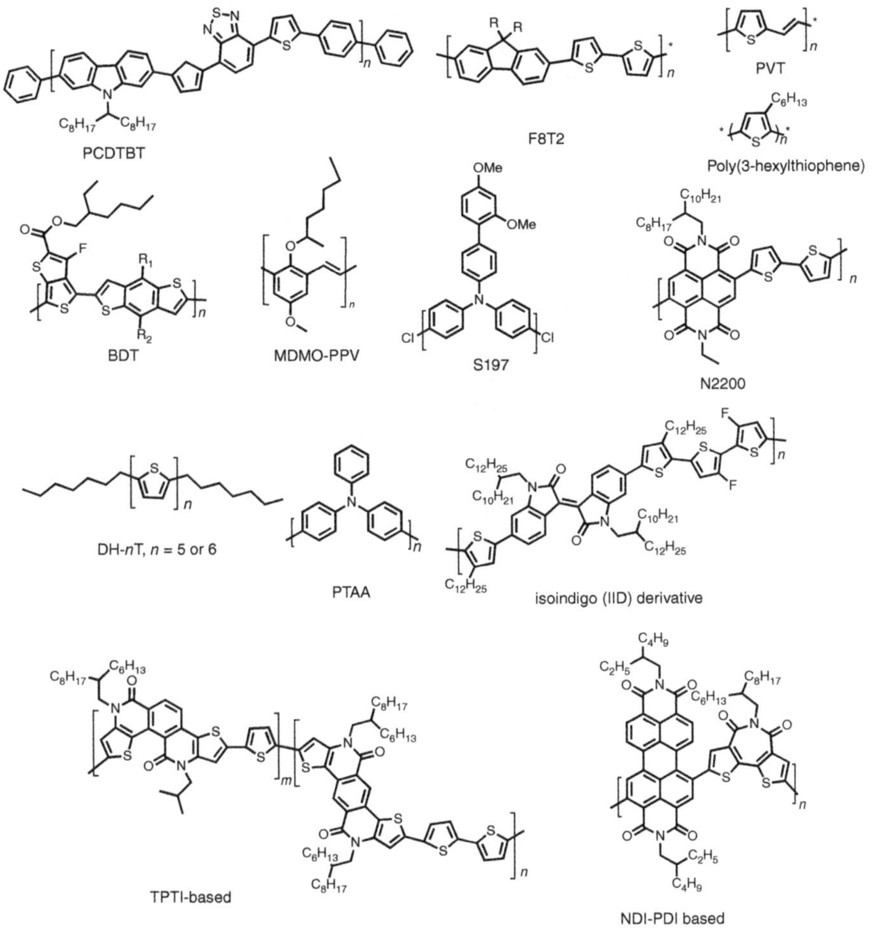

Figure 3.4 Examples of reported polymers used in solar cells.

upon the bandgap, two types of polymers are found that are implemented in solar cells; wide bandgap and low-bandgap organic polymers. The prime goal of these polymers in solar cells is to maximize the solar cells' efficacy or improvise V_{OC}, FF, etc., among other factors.

DSSCs are constructed using a TiO_2 (titanium dioxide) membrane that promotes electron transfer to the conductive layer and onto the electric circuit. The smaller size of these nanoparticles provides more significant benefit, yet these systems are agglomerated or poorly dispersed owing to the van der Waals forces, resulting in poor performance. Thus, tremendous research on developing more efficient TiO_2 nanoparticles with a combination of various organic polymers has been found. According to certain reports, polymers are used as templates or dispersants (Lee et al. 2017). For instance, polyethylene glycol (PEG) is used as a dispersant. In a report, the two different molecular weight-bearing PEG chains were used as a binder with TiO_2 to prepare bi-layered electrodes to improve the efficiency of

DSSCs (Lee et al. 2006). The low-molecular-weight P1 (20,000) and P2 with higher molecular weight (200,000) of the PEG chain is combined with itself or each other to make four systems, among which P1/P2 system shows better performance in ideal condition. The greater surface area, along with extra-large pore size, results in better electron transport, which thereby helps in maximizing energy conversion (up to 9.04%).

Li et al. reported a method for the nanoporous film preparation by using a copolymer consisting of styrene and butyl acrylate. When these films were employed in solar cells (Li et al. 2009), the device performed better by a factor of 2.7 in terms of per weight. In another report, the PEG polymer gel with molecular weights 1000 and 4000 was used as an electrolyte to overcome the disadvantages associated with a liquid electrolyte, such as stability in the long term. In addition to these polymeric gels, poly(vinyl alcohol) (PVA) 60,000 was also used, which showed better performance (0.083%) among all, with a *FF* of 59.84.

Moreover, these polymeric electrolytes enhance the current density-voltage of DSSC compared to other liquid electrolytes (Seni and Puspitasari 2017). Nevertheless, the dispersion issue of TiO_2 can be resolved with PEG polymer, yet the reports suggested that it can also trap the photo-excited electron, which could be another issue. Apart from this, polymers are also used with mesoporous TiO_2 film to prepare an organized film that enhances the light-scattering effect of metal oxide (Ahn et al. 2011; Kim et al. 2013). For instance, polystyrene (PS), PVC-g-POEM (comprising poly(vinyl chloride) backbone and poly(oxyethylene methacrylate)), and PLMA-POEM (comprising poly(lauryl methacrylate) [PLMA] and hydrophilic poly(oxyethylene methacrylate) [POEM]) (Figure 3.5).

Another novel class of solar cells generated from quantum dots, termed quantum-dot-sensitized solar cells (QDSCs), were the utmost regular polymer is polysulfide/sulfide (Sn^{2-}/S^{2-}), was used as an electrolyte (Chakrapani et al. 2011). This electrolyte offers high stability to QDs and increases the high open-circuit voltage, but it requires high redox potential that decreases the overall V_{OC}. Other alternatives were investigated but failed to match the expectations. Afrooz et al.

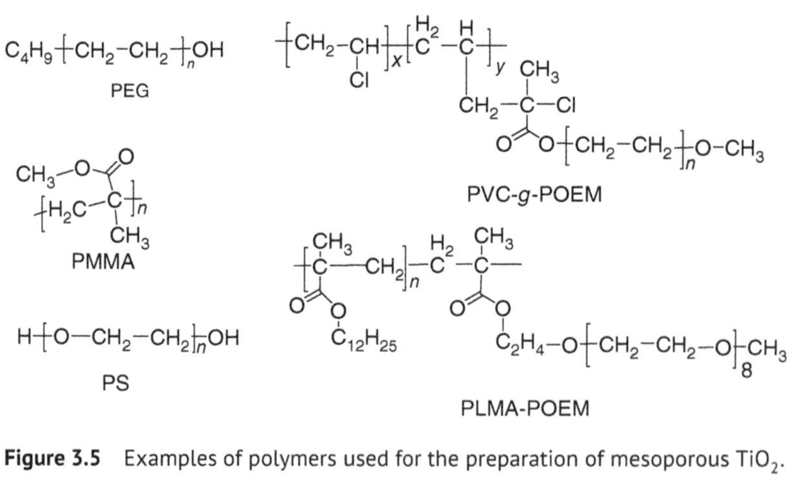

Figure 3.5 Examples of polymers used for the preparation of mesoporous TiO_2.

modified the polysulfide electrolyte by incorporating aromatic or linear amine as additives to enhance the functionality of QDSCs (Beiraghdar et al. 2021). The same group also employed thiourea and urea moiety in the electrolyte. The investigation revealed that among all aromatic amines, 1,8-naphthalenediamine shows better results, and diethylamine and thiourea are other moieties that present better power conversion. Polymers are exploited as additives with polysulfide electrolytes. These polymer additives in polysulfide electrolytes are one of the better methods that control charge recombination occurring at the interface between the photoanode and electrolyte, yielding greater PCEs (Dürr et al. 2006). The additives are adsorbed on TiO_2 nanoparticles and lower the charge recombination by protecting surface traps of TiO_2. Zhong and coworkers utilized a method in which PEG polymer in combination with polysulfide was adopted to suppress the charge recombination in the surface of TiO_2/QDs/electrolyte compared to polysulfide electrolyte (Du et al. 2015a,b). Due to the steric factor, the recombination of charges did not occur, and it subsequently improvised the PCE from 5.80% to 6.74%. This work is not efficient apart from CdSe-sensitized QDSCs. Thus, more exploration will be needed. Later, in another report, the same group used a functional polymer, poly(vinyl pyrrolidone) (PVP), increasing the *FF* and the V_{OC}. Also, this additive was compatible with a wide range of QDSCs. This is because PVP provides a barrier to the photoanode that hinders the surface defect of TiO_2, along with suppressing charge recombination. Overall, the PCE in $CdSe_xTe_{1-x}$ of 9.77% was afforded with PVC additive (Jiang et al. 2016). Xinhua Zhong (Sun et al. 2018) evaluated the effect of different PEG polymers having molecular weights ranging from 300 to 20,000 in CdSe, and Zn–Cu–In–Se QDSCs. The polymers mentioned above have shown affirmative response on V_{OC}; however, the higher generation PEG exhibits an unfavorable impact on *FF* and short-circuit density (J_{SC}).

Electrolytes made from gel polymer work as a replacement for the usual electrolytes in QDSCs owing to their minimal cost, lightness, and flexible nature. Poly(acrylamide-*co*-acrylic acid) (PAAm-PAA) is compatible with the electrolyte; hence, Buraidah and his team employed this polymer in QDSC (Lee et al. 2020). In 2020, when Dassanayakee and his group infused PbS QDs in polymeric gel electrolytes to investigate their role in solar cells (Dissanayake et al. 2020), they found that the polysulfide liquid, when incorporated with CdS QD, increased the solar cell's efficiency from 1.19% to 1.51%, whereas PbS QD-based nanofiber gel exhibited enhancement in PCE from 0.94% to 1.46%. The reason is that the conductivity of sulfide anion increased owing to indirect ionic dissociation facilitated by the quantum dots. Improving hole-extracting capacity by the better redox electrolyte could be another way of boosting the power conversion in QDSCs. In this regard, Maitia et al. demonstrated a negatively charged polyurethane gel electrolyte that can transmit ions, which is explored in 3-sulfanylpropanoic acid functionalized CdS QDSCs (Kumar et al. 2022). Generally, polyurethanes consist of hard and soft segments, and here aliphatic polyurethanes with pedant groups are used. The system's oxidized hole was trapped with the electrolyte's redox active group. The nature of polyurethane chains controlled the electrical conductivity as well as thermal stability, and this conductivity could further increase upon an increase in negative ion concentration. Hence, the photovoltaic conversion was investigated by changing chain composition (by varying soft segment and pedant moiety), and urethane moiety achieved a maximum conversion of about 1.16%. Besides organic polymers, natural biopolymers or modifications of these polymers are also used to modulate polysulfide electrolytes (Rasal et al. 2021).

A new type of organic polymer, porous organic polymers or POPs, has attracted a great deal of attention due to their excellent pore size and surface area (Zhang et al. 2020; Lee et al. 2020) with other excellent properties such as thermal and chemical stability. The electrical tunability and the host–guest system of these polymers could be advantageous for charge transport as well as separation. These porous systems are composed of linkers and nodes to form networks of monomeric building blocks, resulting in various topologies. Also, thanks to their high dimensional geometric structures, they can actively participate in electron/hole transportation irrespective of direction, subsequently finding applications in a broad disciplinary range. One such role is their use in a diversity of solar cells. Different types of POP materials are reported based on their amorphous or crystalline nature. The amorphous POPs include microporous (pore size <2 nm) hyper-crosslinked polymers (HCPs), conjugated microporous polymers (CMPs) where an extended π-conjugation and three-dimensional microporous skeleton are in combination, porous aromatic frameworks or PAFs that are constructed with aromatic building blocks linked by carbon-carbon bonds, and polymers of intrinsic microporosity (PIMs) where microporosity can be seen without any covalently linked network. While on the other hand, covalent organic frameworks or COFs represent the crystalline POPs. The robustness, tunable topology, and pore size are a few of the advantages that make POPs exploited in photovoltaic devices. Better charge transport and ultra-long charge carrier lifetime are used in the advancement of solar cell technology that leads toward green energy. On that account, the photovoltaic applications by solar energy harnessing via POPs need a better understanding of photophysical properties and device design.

PSCs are another group of photovoltaic cells that has sparked a lot of interest as they can provide a maximum of 25.2% conversion efficiency, compared to OSCs, which give 20% (Jeong et al. 2021; Cui et al. 2021). In 2013, Bein and his coworkers reported crystalline COFs that exhibit donor–acceptor frameworks when used with electron-accepting fullerene molecules and light-directed charge transfer exhibited by [6,6]-phenyl-C_{61}-butyricacid methyl ester (PCBM). In 2011, Wang et al. reported an inverted polymer solar cell (IPSC) comprising a buffer layer of transition metal oxide with poly[*N*-9″-hepta-decan yl-2,7-carbazolealt-5,5-(4′,7′-di-thienyl-2′,1′,3′-benzothiadiazole)]:[6,6]-phenyl C_{70}-butyric acid methyl ester (PCDTBT:$PC_{70}BM$) serve like photoactive layer (Liu et al. 2012). The cathode buffer layer is MoO_3-Al, and a dual layer of this serves as an anode. The Al quantity in the composite controlled the photovoltaic performance by adjusting interfacial property at cathode buffer/BHJs resulting in an increase in V_{OC} to 55% Al content. Overall, the device exhibits 0.88 V of $V_{OC,}$ with a 70.7% *FF* and a maximum PCE of 6.77% under light irradiation of 49.8 mW cm^{-2}. Jiang et al. developed a technique for obtaining a π-extended CMP on an electrode interface. Their film's conducting properties are investigated with fullerenes for energy conservation (Gu et al. 2015). This electropolymerization method mainly used thiophene-based monomers which were employed as a photoactive layer with different concentrations of C_{60} molecule. The results were a PCE of 5.02% with $V_{OC} = 0.732$ V and $FF = 0.682$.

Many other groups have designed, synthesized, and explored several polymers in terms of charge-selective separation, wide band gap, etc. (Wang et al. 2022; Cai et al. 2016; Li et al. 2019a,b; Ma et al. 2017; Huo et al. 2017). The wide-bandgap polymers show low PCE owing to a finite absorption window. To tackle this hurdle, Xiaobo Wan and coworkers developed a building block of dibenzonaphthyridinedione (DBND) that could be used to synthesize polymers with efficient PCE. The resultant two polymers were mixed with phenyl-C_{71}-butyric acid methyl ester ($PC_{71}BM$) in OSCs to achieve a PCE of up to 5.75% and 6.32% (Cai et al. 2016). In another report, the same group reported the role of fluorine in a different position (*para and ortho-*) in a conjugated polymer backbone, which overall affects the PCE in solar cells (Cai et al. 2017). In a similar type of report, the group investigated the O- and N-alkylation of the polymer donor tetracyclic lactam in OSCs.

Improvements in the PCE of the OSCs have been achieved by improving the donor polymers in BHJ structure, which is an intermix of two types of system – donor and acceptor. The donor participates in the processes of excitation and dissociation, and hence, by increasing donor substances' functionality, the OSCs' efficiency could be enhanced. One way to achieve high-performance donor/acceptor polymer is by tuning energy levels via engineering electron-rich or -deficient molecules in a polymer chain. Benzodithiophene (BDT) is used as donor moiety in OSCs owing to their excellent π–π stacking capacity and hole mobility. However, the available two types of symmetrical BDTs, one-dimensional (1D) BDT and two-dimensional (2D) BDT, exhibit some disadvantages. Thus, researchers came up with a solution to address these disadvantages by combining these two together, known as the "symmetry-breaking" strategy resulting in asymmetric organic polymers (Liu et al. 2016), introduced by Yang et al., termed asymmetric organic polymers, have garnered attention in the field of OSCs in recent decades (Zhao et al. 2022).

In recent times, different asymmetric backbone-bearing polymers have been reported (Lin et al. 2019; Wang et al. 2016; Qin et al. 2017; Fan et al. 2014; Qiu et al. 2015; He et al. 2016; Yu et al. 2016). The alkyl and alkoxy groups offer different benefits; as for the tuning energy level, alkyl side chains are preferable, but J_{SC} and FF values are prominent in alkoxy side chains. Thus, Hoang et al. (2015) reported and developed asymmetric BDT derivatives bearing polymers for the BHT solar cells that omitted the issue created by the alkyl group. Basically, the BDT moiety was modified with alkyl and alkoxy chains on two sides, resulting in a suitable HOMO energy level and effective J_{SC} and FF. An efficiency of 7.64% was achieved with 4-octyl-8-octyloxy-BDT moiety based on this strategy outperforming counterparts containing symmetric dioctyl-BDT showing PCE of 6.48% (Figure 3.6). Further structural changes in side chains subsequently afford an 8.12% PCE owing to the improvement in hole mobility and balance of hole/electron mobility.

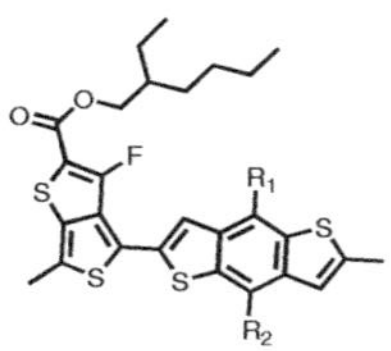

P1: $R_1 = R_2$ = octyloxy, PCE = 7.18%
P2: R_1 = octyl, R_2 = octyloxy, PCE = 7.64%
P3: $R_1 = R_2$ = octyl, PCE = 6.48%
P4: R_1 = octyl, R_2 = butoxydiethoxy, PCE = 8.12%

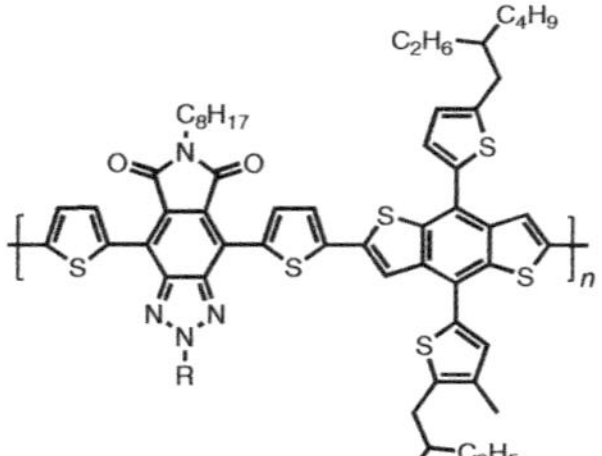

R = 2-ethylhexyl
P5, PCE = 10.2%

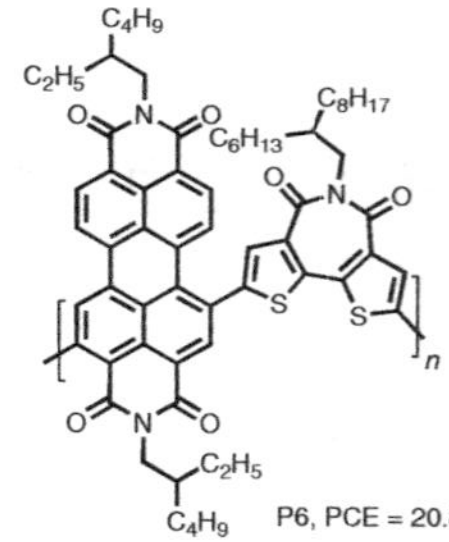

P6, PCE = 20.8%

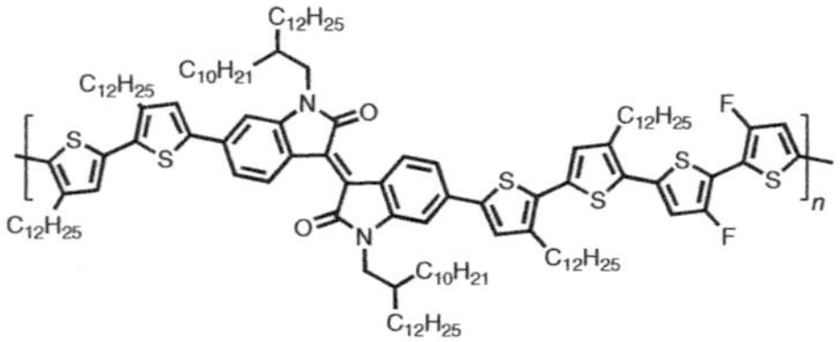

P7, PCE = 10.7%

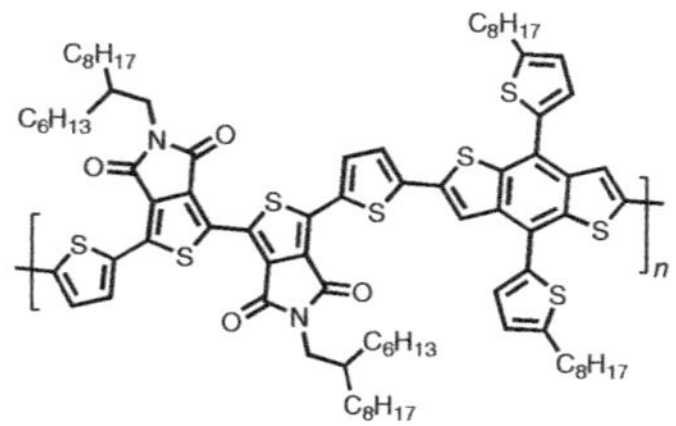

P8, PCE = 14.2%

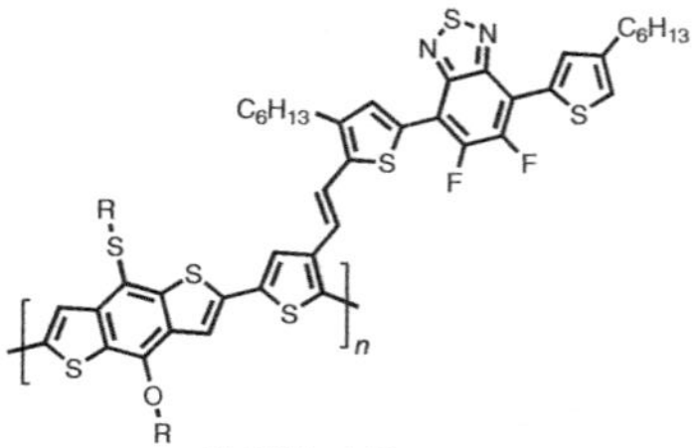

P9, PCE = 4.4%

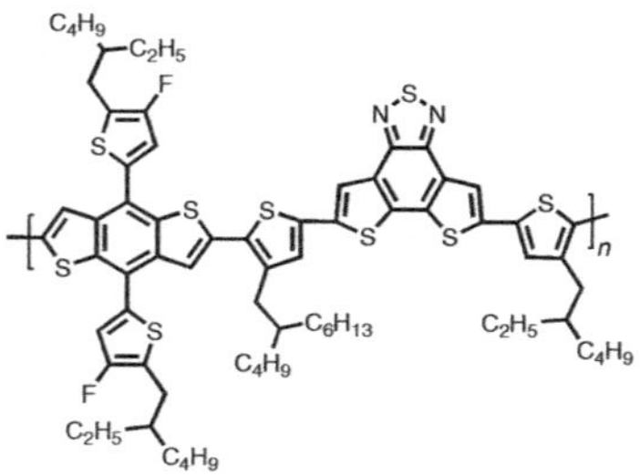

P10, PCE = 18.22%

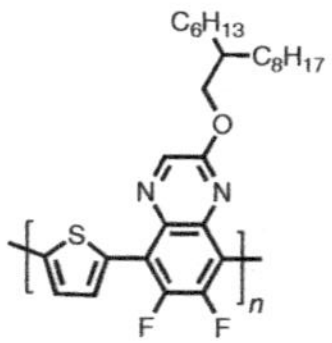

P11, PCE = 17.7%

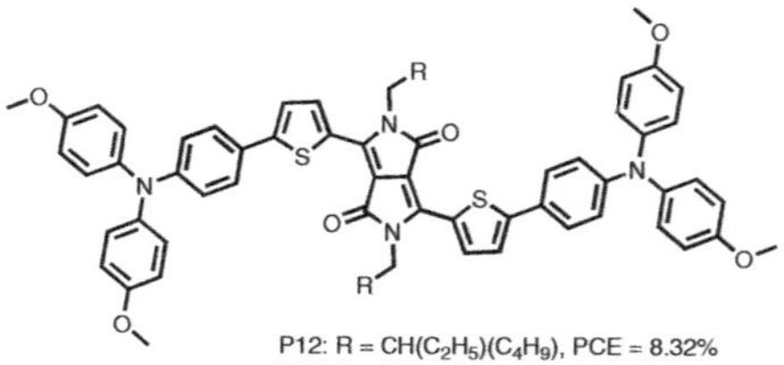

P12: R = $CH(C_2H_5)(C_4H_9)$, PCE = 8.32%
P13: R = C_5H_{11}, PCE = 11.16%
P14: R = $CH_2(OCH_2CH_2)_2OCH_3$, PCE = 12.05%
P15: R = $CONHCH_2CH_2(OCH_2CH_2)_2OCH_3$, PCE = 20.19%

Figure 3.6 Reported various types of polymer with their PCE values.

Cirpan and coworkers have modified conjugated polymers composed of BDT and benzothiadiazole with selenophene and fluorine units for the BHJ solar cells (Aslan et al. 2021). The fluorine substitute polymers showed greater negative homo energy and achieved PCE of 2.63% with J_{SC} of 7.24 mA cm^{-2}, V_{OC} of 0.72 V, and *FF* of 50.6%. In a report, a PCE of 3.41% was achieved under AM1.5 G illumination was achieved with a difluoroquinoxaline and diketopyrrolopyrrole-based conjugated polymers in OSCs. These low-bandgap polymers consist of electron-deficient difluoroquinoxaline and diketopyrrolopyrrole substrates and electron-rich thiophene or bithiophene units that result in a push–pull system with an increase in HOMO energy level. Further, the effective polymer was blended with the synthesized copolymers with the fullerene $PC_{71}BM$ resulting in high-performance PSCs (Keum et al. 2022). In terms of the low band gap or low-energy HOMO of a conjugated polymer, the group of Youngeup Jin introduced a fused phenyl ring in the polymer, with 2-fluorophenazine serving as an acceptor moiety (Lee et al. 2022).

The hydrophobic alkyl side chains are vividly used in active materials for solar cells because of their controlled self-assembly and strong ability to increase solubility. However, recently hydrophilic oligo(ethylene glycol) (OEG) chains have been widely employed owing to their flexibility and less steric factor. The work reported by Wang et al. revealed that the integration of the OEG side chain in the conjugated polymer (Meng et al. 2015), poly[2,7-fluorene-*alt*-5,5-(4,7-di-2-thienyl-2,1,3-benzothiadiazole)] results in lowering the distance of π–π stacking in addition to increasing hole mobility. Additionally, the film showed a bathochromic shift spectrum and low bandgap compared to the side chains of alkyl groups. A PCE of 4.04% was achieved from this PSC device upon performing with a polar nonhalogenated solvent that was ascribed to the polar nature of the OEG chain. However, increasing the OEG chain's length limited the hole mobility; that being the case, a suitable length of the OEG chain with a higher charge carrier, crystallinity, and narrower bandgap are critical features that call for maintenance in high-performance device development. Mainly, the OEG side chain in polymer donor are exploited with fullerene blends; however, limited investigations are found with non-fullerene. You et al. examined the role of the OEG chains with non-fullerene molecular acceptors and compared them to the alkyl chain. The authors used triethylene glycol (TEG) and decyl chain in a conjugated polymer DICTF with a non-fullerene acceptor blended with a polymer donor, resulting in PCE of 3.09% and 6.93%, respectively (Zhang et al. 2018b). The decrease in PCE for the TEG side chain is unexpected, which was accounted for by the difference in the system's morphology. Thus, the effect of the OEG chain on the polymer backbone with a non-fullerene acceptor needs investigation with more suitable engineering.

Incorporating hydrogen-bonding units in the side chain of the polymer for solar cells could effectively improve the orientation and morphology. Effective hole mobility is a way to develop high-performance solar cells. In these regards, Zhang et al. demonstrated that urea containing alkyl chain and branching alkyl chain in diketopyrrolopyrrole-quaterthiophene conjugated polymer offered higher hole mobility than without urea-functionalized polymers (Yao et al. 2016). The three polymers with three different ratios of urea-based alkyl chains vs branching alkyl chains were prepared, which could lead to 13.1 cm^2 (V s)$^{-1}$ and, when intermixed with $PC_{71}BM$, results in a maximum of 6.8% PCE. These outcomes are due to the lamellar packing arrangement of the alkyl chain, which enhances the interlayer packing and thin-film morphology. A report by Gu et al. suggested that careful selection of hydrogen bonding in a conjugated polymer is necessary as it offers a strategy to enhance mechanical performance, whereas it could be unfavorable for device performance (Galuska et al. 2022). Rondeau-Gagné et al. have introduced amine moiety in the alkyl chain capable of showing hydrogen bonding which governs the conjugated polymers' solid-state structure and self-assembly (Ocheje et al. 2018). This diketopyrrolopyrrole-based polymer is incorporated with 5–30 mol% of amides that advance the charge carrier's mobility up to 2.46 cm^2 (V s)$^{-1}$ owing to the formation of lamellar packing at the expense of intramolecular hydrogen bonding without hampering conjugation. The account on the conjugated polymer that consists of carbazole and cyanoacetic acid results in donor–acceptor type system employed in DSSCs and OPV cells are demonstrated. This multifunctional system achieved PCE with a value of 4.11% in DSSCs (Chang et al. 2011).

Some of the lactam, imide-functionalized, and asymmetric polymers are shown in Figure 3.6 that hold remarkable conversion efficiencies (P5: Abdulahi et al. 2019; P6: Shi et al. 2020; P7: Liao et al. 2019; P8: Zhao et al. 2020; P9: Huang et al. 2015; P10: Liu et al. 2020a; P11: Chai et al. 2021). An exceptional photovoltaic work reported in 2010 by the team of Luping Yu described that benzodithiophene polymer reached PCE up to 7.4% upon blending with $PC_{71}BM$ in polymer BHJ solar cells (Liang et al. 2010). These innovative semiconducting polymers exhibit a low bandgap of about 1.6 eV that shows absorption at the red-light region. The good structural design of the polymer offered various advantages, such as decent hole mobility, good solubility, and worthy V_{OC} (Sharma et al. 2021). Lee and coworkers created a material with a hole-transporting ability for use in PSCs. The material was based on a diketopyrrolopyrrole system, which was further engineered with different side chains. As a result, the PCE could be enhanced from 8.32% to 17.64%, and this maximum value could be increased to 20.19% with an *FF* of 73.87% upon adding a dopant (P12–P15, Figure 3.6).

The negative electronic moieties are explored in side chains of the polymer that are found to be synchronized with the aggregation ability and energy level. The literature report revealed that such structural modification on the donor part (for instance, in the BDT unit) improved the functionality of PSC (Hou et al. 2008; Zhang et al. 2015; Huang et al. 2018; Jia et al. 2019; Hoang et al. 2015; Vohra et al. 2015). A table with other PV parameters (Table 3.1) is shown. Recently, in 2020, a polymer fluorinated donor polymer was synthesized and explored in solar cells. The system was designed based on fluorination and sulfurization on BDT-composed phenyl chains resulting in a new polymer PBDB-PS2F that showed a PCE of 13.56%. The role of fluorine was supported by the fact that the non-fluorinated polymer exhibited a PCE of 10.36%. The fluorinated PBDB-PS2F system lowers the HOMO energy level and provides a perfect phase separation film morphology.

The electronic band gap (E_g) of organic semiconductors materials is an essential parameter that could be investigated for advanced device performance (Scharber

Table 3.1 The photovoltaic performance of D–A polymers.

Name of polymer	V_{OC} (V)	J_{SC} (mA cm^{-2})	*FF* (%)	*PCE* (%)	References
P1	0.66	16.29	67	7.18	Hoang et al. (2015)
P2	0.73	15.23	69	7.64	Hoang et al. (2015)
P3	0.78	12.48	66	6.48	Hoang et al. (2015)
P4	0.73	15.78	71	8.12	Hoang et al. (2015)
P5	0.84	17.78	69	10.24	Abdulahi et al. (2019)
P6	1.12	23.30	79.4	20.8	Shi (2020)
P7	0.81	21.97	60.82	10.7	Liao et al. (2019)
P8	0.83	25.60	66.7	14.2	Zhao et al. (2020)
P9	0.86	8.24	62.0	4.4	Huang et al. (2015)
P10	0.85	27.70	76.6	18.22	Liu et al. (2020a)
P11	0.88	25.30	79.3	17.7	Chai, et al. (2021)
P12	0.93	14.37	60.23	8.32	Sharma et al. (2021)
P14	0.92	17.69	63.16	11.16	Sharma et al. (2021)
P15	0.94	19.75	64.03	12.05	Sharma et al. (2021)
P16	1.14	22.94	73.87	20.19	Sharma et al. (2021)

et al. 2021). These band gaps are explained with the help of the HOMO and LUMO. Hence, molecular engineering in semiconducting materials designing is necessary to tune the band gap by keeping other parameters in mind. While tuning the band gap, other factors often become inefficient, affecting the device's performance. A lot of attention and research is growing in organic and polymer electronics due to variables other than electrical and optical characteristics, such as light weight, mechanical flexibility, and low manufacturing costs. Pomerantz et al. categorized that low band gap polymer as having a band gap less than 1.5 eV using polyacetylene as a reference point. However, a band gap with 2 eV can be considered low band gap material in solar cells (references). The D-A-based conjugated polymer with a narrower band gap ($E_g < 1.6$ eV) allows exploring the NIR range wavelength in solar cells. However, these polymers usually exhibit low open-circuit voltage (V_{OC}), which hampers the device's performance. Thus, the challenge is to develop a low band gap conjugated polymer with high V_{OC} that could retain an effective absorption range. The most widely used low-bandgap polymers fall under the NIR regime, having between 1.59 and 0.4 eV. For the structural designs of better conjugated polymer for solar cells, it needs to address a few factors such as donor–acceptor role, alternation of bond length, conjugated structure's planarity, the effect of different substituents, intermolecular interactions, and length of π-conjugation.

The conjugated polymers function as donor moieties in photovoltaic applications and are mixed with acceptor units, resulting in photoinduced electron transfer. This is known as bulk heterojunction (BHJ). Yang et al. developed an efficient method to develop the polymers PTBTz-Cl and PTBTz-Me by thiazole-induced method to afford high-performance solar cells materials (Kang et al. 2019). The chlorine or methyl group-based thiazole unit was attached to the backbone of the quinoid polymer, increasing the π–π stacking along with a decreased HOMO energy gap that could assist in retaining redshift absorption. However, owing to the electron-pulling effect of the Cl unit in the PTBTz-Cl polymer better results in terms of a reduced HOMO, greater crystallinity, and PCE of 10.35% could be achieved. Owing to the noncovalent interaction between –Cl and –S, the planarity of the backbone is enhanced; interestingly, the V_{OC} of 0.94 V was achieved compared to other synthesized polymers that could be beneficial in semitransparent PSCs and tandem solar cells. Many efforts have been made to develop the advancement of various materials with these features. Queisser et al. have calculated and found that a maximum PCE of around 33% could be afforded by using optimum bandgap polymers under standard conditions. Zhu et al. reported another low-bandgap polymer based on cyclopentadithiophene and benzothiadiazole (≈1.5 eV) gave PCE of >5%. Due to enhanced charge-transport properties and energy level, this efficiency was achieved with $PC_{71}BM$ (Mühlbacher et al. 2006). The literature data shows that PCE increased from ≈ 4.8% to ≈17.4% along with short-circuit voltage in the last two decades in BHJ solar cells. Fullerene and non-fullerene-based acceptors are generally used in BHJ solar cells. The tuning of the HOMO of the donor and LUMO energy level of the acceptor is important, as well as morphology in the nanodomain. Various reviews are available in the literature that describes the progress and mechanism of low-bandgap solar devices (Parida et al. 2011; Polman et al. 2016). Organic materials are a promising candidate as PCE values are increased in a short period of time; however, compared to inorganic photovoltaic performance, organic-based solar cells are still lagging behind (Scharber and Sariciftci 2021; Xiao et al. 2017; Yoshikawa et al. 2017). The main reason for this might be the low charge mobility of organic systems that limit sunlight absorption (Würfel et al. 2015). In the year of 2018, Meng et al. developed and reported a remarkable photovoltaic work to achieve a conversion efficiency of 17.3% in tandem solar cells (Meng et al. 2018). Apart from this, various strategies have been developed to increase conversion efficiency (Zhu et al. 2020a,b; Liu et al. 2020a,b). Gang Li and coworkers lowered the electronic energy level of HOMO of the PBDTTT-based polymer by a structural modification that enhanced conversion efficiency. The introduction of ketone instead of an ester unit results in a PCE of 6.58%, V_{OC} of 0.70 V, a J_{SC} of 14.7 mA cm^{-2}, and an *FF* of 0.64 (Hou et al. 2009). Jianhui Hou and coworkers explored the 2D conjugated BDT-based polymer with different functionalization for photovoltaic applications (Huo et al. 2011). The introduction of 2-alkylthienyl groups in the side chain of the aforementioned polymers results in wider absorption bands and higher hole units compared to alkoxyl substituted units, which results in conversion efficiency up to 7.59% and 8.79% with advancement in V_{OC} and J_{SC}. In 2014, the same group introduced three different side units, furan, thiophene, and selenophene, in the polymer composed of BDT and thieno[3,4-*b*]thiophene moieties (Zhang et al. 2014). The absorption band, crystallinities, and aggregation sizes of the polymers were affected by the dihedral angles between the BDT unit and the aforementioned side chain. The thiophene- and selenophone-based polymers exhibit PCEs of 9.0% and 8.78%, respectively. The steric factor of the side chain and that of the polymer participates in tuning the performance of the solar cells.

According to a report published by Jin et al., the synthesis of π-conjugated polymers by introducing 2D conjugation was studied. The conjugation was a consequence of using 4,8-dithienylbenzo[1,2-*b*:4,5-*b*′]dithiophene (DTBDT) as the donor unit (Chakravarthi et al. 2015). This divinylbenzene-directed polymer achieved a maximum PCE of 7.10% by inducing intramolecular charge transfer (ICT) and maintaining HOMO energy levels (−5.21 to −5.28 eV). The ternary techniques are used to obtain high-performance solar devices. Based on this, Uddin et al. reported that the ternary device consists of two conjugated polymer donor systems and non-fullerene acceptor material, IEICO-4F (Duan et al. 2020). A PCE of 12.1% was obtained, which is 16% more than its binary counterpart. Yang et al. described a semi-empirical method with a combination of available materials to achieve a PCE of around 20% (Chen et al. 2020). Conductive organic polymers are exploited in PSCs as a hole transfer material; for example, the polymer PEDOT:PSS are employed for this purpose owing to their better film-forming abilities. Another example is polyaniline (PANI) which requires a relatively lesser price and exhibits multiple oxidation states, which is why they are employed in solid DSSCs (Xiao et al. 2014).

Apart from TiO_2, zinc oxide (ZnO) is another potential alternative that finds its use as an electron transport layer (ETL) in solar cells (Chen et al. 2012; Li et al. 2014), especially in IPSCs. It is environmentally sustainable, possesses good electron mobility, and is greatly transparent in the Vis as well as near-IR regions (E_g = ~3.3 eV). As electron entrapment leads to higher mobility of holes compared to electrons in most organic semiconductors, it leads to poor electron mobility. But the higher VB energy level and lower CB energy level of ZnO overcome this limitation by blocking holes (Ryu et al. 2017). The simple solution-based deposition process and high natural abundance also provide an added advantage to this material. Even so, certain conditions limit the performance of ZnO as an ETL. One such obstacle is the below-par interlayer connectivity between the ZnO film and the solar cell's active layer, which in turn causes resistance in electron mobility. This causes lower J_{SC} and *FF* values. The oxygen surface defects in the ZnO layer further substantially decrease the stability of the device; hence PCE decreases, as it plays the main role in charge recombination. Methods like sol–gel used to prepare these films result in their lower conductivity, further degrading the material's electron-transporting ability.

Addressing these issues is crucial to increase their impact as ETL and consequently increase the solar device's performance for their large-scale applications. In this context, both organic and inorganic dopants have been used. Inorganic Li_2CO_3 doping on ZnO ETL has achieved the highest reported PCE of 10.08% in PTB7:PC_{71}BM-based IPSC (Qiu et al. 2018a,b). The XPS study also concluded that doping alkali metal fluorides mitigate the oxygen defects in the ZnO structures. Comparison between fluorides of sodium (NaF), potassium (KF), and cesium (CsF) showed that NaF is the most suitable dopant for ZnO in accordance with the PCE order of 8.64%, 8.39%, and 8.28% achieved in the ZnO:NaF, ZnO:KF, and

ZnO:CsF systems, respectively. Doping ZnO thin films with organic water-soluble materials has proved beneficial in blocking holes and yielding much better electron transportation. When dye molecules of perylene bisimide functionalized with sulfonic acid (PBI–SO_3H) were incorporated into ZnO thin films, a homogeneous hybrid ZnO: PBI–SO_3H thin film was produced having electrostatic bonds between the Zn atom of inorganic ZnO and the SO_3H groups of the organic dye molecules. These electrostatic forces of attraction between the two facilitate electron conductivity and a higher hole-blocking ability of the hybrid film. On being used as a cathode interlayer in IPSCs, an efficiency of 13.6% could be accomplished with PBDB-T-2Cl:ITIC-4F and 15.4% with PM6:Y6 as the active layers (Luo et al. 2020).

Increasing the interfacial area has also proved to be valuable in increasing the PCE of inverted solar cells with ZnO as the ETL. For instance, the use of nano-ridged ZnO structures, also termed ZnO-R, results in increased efficiency of 8.38% from 7.41% and also causes a reduction in electric power leakage (Beek et al. 2005). Another example is employing nanowire structures of ZnO (ZnO NW), which addressed the photogenerated electron–hole recombination and increased electron transport efficiency due to its highly organized structure and large surface area (Lee et al. 2019). Massive groups of hydroxyl ions on the surface of ZnO-NPs are to be blamed for their insufficient contact with the active layer, decreasing the overall PCE of the IPSCs. This can be combatted by acid treatment of the ZnO thin films. By treating the inverted PM6:Y6 solar cell with 2-phenylethylmercaptan or PET, the efficiency of the solar cell could be increased up to 16.46% (Han et al. 2021). Another way to decrease surface defects in the nanoparticle film is through surface modification of the same by silane compounds such as 3-aminopropyltrimethoxysilane (APTMS) to form ZnO@APTMS nanoparticle wherein the hydroxyl groups of ZnO film are capped by APTMS (Wei et al. 2018). This reduces the oxygen defects to increase device stability and efficiency.

ITO is employed as a transparent conductive electrode (TCE) (Schmidt et al., 2009), yet due to a few limitations (for example, chemically unstable, availability), graphene has become an alternative choice for researchers (Xu and Liu 2016). Graphene is a promising material that serves as TCE in OSCs (Ma and Zhi 2019; Rana et al. 2014). The novel properties of graphene electrode, such as excellent optical, electrical, and mechanical, offers great opportunities to develop flexible optoelectronic devices (such as wearable devices). For this kind of flexible device, an annealing-free fabrication process is explored that inhibits or reduces the chance of damaging the substrate or active component of the device. This carbon material has been widely explored in OSCs, and PCE is more than 8% achieved (Liu et al. 2015; Park et al. 2014). For instance, Park and coworkers described an approach of an inverted OSC comprising a poly[[4,8-bis[(2-ethylhexyl)oxy]benzo[1,2-*b*,4,5-*b*′]-dithiophene-2,6-diyl][3-fluoro-2-[(2-ethylhexyl)carbonyl]-thieno[3,4-*b*]thiophene diyl]] (PTB7) as well as a hybrid material graphene/annealing-free ZnO-NP, where ZnO serve as electron-transporting layers (Jung et al. 2018a,b). Here, n-doped graphene electrodes resulted from integrating the NPs in the hydrophobic surface of graphene. The effective charge transfer by ZnO to graphene electrode results in a PCE of 8.16% on the glass, whereas on a flexible substrate (polyethylene terephthalate) a PCE of 7.41% was achieved. This low-cost and low-temperature processing approach was based on hybrid material beneficial for OSCs. The same group reported another work based on the modification of graphene with a thin layer of polynorepinephrine derivative resulting a PCE of 7.93% in OSC (Jung et al. 2018a).

To obtain high-performance OSCs with graphene, suitable deposition techniques that depend on processing conditions need to be developed (Garg et al. 2017). Among various techniques, chemical vapor deposition is one of the promising techniques. The silver nanowires (AgNWs) electrode exhibits remarkable properties in OSCs; however, some issues, like the rough surface, hamper device performance. These disadvantages are overcome by developing hybrid material with graphene as TCE exhibiting good mechanical flexibility (Lee et al. 2013). Zhang et al. (Zhang et al. 2019) synthesized a hybrid electrode comprising graphene and AgNWs. Graphene: AgNW compos, which was further modified with poly(3,4-ethylenedioxythiophene):poly-(styrenesulfonate) (PH1000). This electrode was employed with non-fullerene flexible OSCs that achieved a PCE of 13.44% with V_{OC} of 0.829 V, J_{SC} of 23.20 mA cm^{-2}, and FF of 69.82%. Notably, this device retains 84.6% of the initial PCE after bending multiple times. Ion migration of PSCs hampers the device performance. In this regard, graphene derivatives have been exploited in PSCs by providing better stability (Su et al. 2020). Wang et al. described various sizes of amino-functionalized graphene quantum dots that serve as cathode interfacial layers in OSCs with fullerene and nonfullerene (Wang et al. 2019). Among all, the medium-sized QDs result in a PCE of 11.87 % and 12.81% with PBDB-T:ITIC and J71:ITIC. Capilla and coworkers reviewed the role of graphene materials in OSCs (Velasco Davoise et al. 2022). In 2020, the group of Park reported a PCE of more than 15% with polyimide-integrated graphene electrodes in OSCs (Koo et al. 2020). Here, polyimide plays the dual role of carrier film and substrate, resulting in a thermally stable electrode exhibiting high optical transmittance of over 92%. The graphene or its derivative played various roles in OSCs, such as an active layer, electrode, and transport layers. However, the PCE of graphene-based OSCs still needs to increase to match conventional OSCs.

3.4 Conclusion

The OSCs, DSSCs, and PSCs are just a few solar technologies that have extensively utilized polymers in their structures. We have discussed the various roles of the aforementioned polymers in solar cells. Polymers are widely applied as active layers in OPVs to affect light harvesting effectiveness and performance characteristics. The polymers are used as dispersants, additives in mesoporous film perpetration, and also in some solar cells exploited as hole transport layers owing to excellent hole mobility. The various designing approaches by introducing various moieties in polymer could be one of the ways to make them better-performing agents in solar cells. In DSSC, polymers can be employed to provide flexibility as pore- and film-forming photoanode materials and film materials. In addition, because of their high catalytic activity, conductive polymers and their derivatives serve in generating platinum-free counter-electrode materials.

The polymer with a long chain and suitable functional groups can solidify a liquid electrolyte resulting in a quasi-solid state of the electrolyte. In perovskite films, the organic semiconducting materials generate nucleation, govern the crystallization, and improvise device stability. However, the low conductivity of quasi-solid-state electrolytes is one issue that needs to be addressed. Symmetry-breaking approaches in designing novel donor polymers that adjust energy levels and improve solar absorption range are some key features that assist in high-performance OSCs. However, the synthetic route might be complex and expensive, which needs to be solved. The advantages of polymers in OSCs are that they have improvised device performance by serving various roles. However, the PCE value needs to be improvised more, keeping in mind the disadvantages associated with the polymer, such as degradation at high temperatures. Moreover, large-scale production needs to be done as laboratory results are not matched upon bulk-scale production.

References

Abdulahi, B.A., Li, X., Mone, M. et al. (2019). Structural engineering of pyrrolo[3,4-*f*]benzotriazole-5,7(2*H*,6*H*)-dione-based polymers for non-fullerene organic solar cells with an efficiency over 12%. *Journal of Materials Chemistry A* 7 (33): 19522–19530.

Ahn, S.H., Jeon, H., Son, K.J. et al. (2011). Efficiency improvement of dye-sensitized solar cells using graft copolymer-templated mesoporous TiO_2 films as an interfacial layer. *Journal of Materials Chemistry* 21 (6): 1772–1779.

Andersson, A. (2020). Electrical performance study of organic photovoltaics for indoor applications: with potential in Internet of Things devices.

Archer, D. (2011). *Global Warming: Understanding the Forecast*. Wiley.

Aslan, S.T., Cevher, D., Bolayır, E. et al. (2021). Synthesis of selenophene substituted benzodithiophene and fluorinated benzothiadiazole based conjugated polymers for organic solar cell applications. *Electrochimica Acta* 398: 139298.

Badea, G., Felseghi, R.A., Varlam, M. et al. (2018). Design and simulation of Romanian solar energy charging station for electric vehicles. *Energies* 12 (1): 74.

Bagher, A.M., Vahid, M.M.A., and Mohsen, M. (2015). Types of solar cells and application. *American Journal of Optics and Photonics* 3 (5): 94–113.

Basaran, O.A., Gao, H., and Bhat, P.P. (2013). Nonstandard inkjets. *Annual Review of Fluid Mechanics* 45: 85–113.

Beek, W.J., Wienk, M.M., Kemerink, M. et al. (2005). Hybrid zinc oxide conjugated polymer bulk heterojunction solar cells. *The Journal of Physical Chemistry B* 109 (19): 9505–9516.

Beiraghdar, N., Dehghani, H., and Afrooz, M. (2021). Modification of polysulfide electrolyte by applying various amines, thiourea and urea as efficient additives to improve photovoltaic performance of quantum dot-sensitized solar cells. *Solar Energy* 220: 384–393.

Brédas, J.L., Norton, J.E., Cornil, J., and Coropceanu, V. (2009). Molecular understanding of organic solar cells: the challenges. *Accounts of Chemical Research* 42 (11): 1691–1699.

Burmann, P., Zornoza, B., Tellez, C., and Coronas, J. (2014). Mixed matrix membranes comprising MOFs and porous silicate fillers prepared via span coating for gas separation. *Chem. Eng. Sci.* 107: 66–75.

Cai, M., Bao, X., Wang, X. et al. (2016). From isoindigo to dibenzonaphthyridinedione: a building block for wide-bandgap conjugated polymers with high power conversion efficiency. *Chemistry of Materials* 28 (17): 6196–6206.

Cai, M., Bao, X., Liu, Y.F. et al. (2017). Unexpected opposite influences of para vs ortho backbone fluorination on the photovoltaic performance of a wide-bandgap conjugated polymer. *Chemistry of Materials* 29 (21): 9162–9170.

Chai, G., Chang, Y., Zhang, J. et al. (2021). Fine-tuning of side-chain orientations on nonfullerene acceptors enables organic solar cells with 17.7% efficiency. *Energy & Environmental Science* 14 (6): 3469–3479.

Chakrapani, V., Baker, D., and Kamat, P.V. (2011). Understanding the role of the sulfide redox couple (S2−/S n 2−) in quantum dot-sensitized solar cells. *Journal of the American Chemical Society* 133 (24): 9607–9615.

Chakravarthi, N., Gunasekar, K., Kim, C.S. et al. (2015). Synthesis, characterization, and photovoltaic properties of 4,8-dithienylbenzo[1,2-*b*:4,5-*b*′]dithiophene-based donor–acceptor polymers with new polymerization and 2D conjugation extension pathways: a potential donor building block for high performance and stable inverted organic solar cells. *Macromolecules* 48 (8): 2454–2465.

Chang, D.W., Ko, S.J., Kim, J.Y. et al. (2011). Multifunctional conjugated polymers with main-chain donors and side-chain acceptors for dye sensitized solar cells (DSSCs) and organic photovoltaic cells (OPVs). *Macromolecular Rapid Communications* 32 (22): 1809–1814.

Chen, F.C. (2018). Organic semiconductors. In: *Encyclopedia of Modern Optics* (ed. B.D. Guenther and D.G. Steel), 220–231. Elsevier.

Chen, S., Small, C.E., Amb, C.M. et al. (2012). Inverted polymer solar cells with reduced interface recombination. *Advanced Energy Materials* 2 (11): 1333–1337.

Chen, X., Jia, Z., Chen, Z. et al. (2020). Efficient and reproducible monolithic perovskite/organic tandem solar cells with low-loss interconnecting layers. *Joule* 4 (7): 1594–1606.

Chirilă, A., Buecheler, S., Pianezzi, F. et al. (2011). Highly efficient Cu(In,Ga)Se_2 solar cells grown on flexible polymer films. *Nature Materials* 10 (11): 857–861.

Clarke, T.M. and Durrant, J.R. (2010). Charge photogeneration in organic solar cells. *Chemical Reviews* 110 (11): 6736–6767.

Cui, Y., Xu, Y., Yao, H. et al. (2021). Single-junction organic photovoltaic cell with 19% efficiency. *Advanced Materials* 33 (41): 2102420.

Delucchi, M.A. (2010). Impacts of biofuels on climate change, water use, and land use. *Annals of the New York Academy of Sciences* 1195 (1): 28–45.

Diao, Y., Zhou, Y., Kurosawa, T. et al. (2015). Flow-enhanced solution printing of all-polymer solar cells. *Nature Communications* 6 (1): 7955.

Dincer, I. and Acar, C. (2015). A review on clean energy solutions for better sustainability. *International Journal of Energy Research* 39 (5): 585–606.

Dissanayake, M.A.K.L., Liyanage, T., Jaseetharan, T. et al. (2020). Effect of PbS quantum dot-doped polysulfide nanofiber gel polymer electrolyte on efficiency enhancement in CdS quantum dot-sensitized TiO_2 solar cells. *Electrochimica Acta* 347: 136311.

Du, J., Meng, X., Zhao, K. et al. (2015a). Performance enhancement of quantum dot sensitized solar cells by adding electrolyte additives. *Journal of Materials Chemistry A* 3 (33): 17091–17097.

Du, Y., Cai, H., Ni, J. et al. (2015b). Air-processed, efficient $CH_3NH_3PbI_{3-x}Cl_x$ perovskite solar cells with organic polymer PTB7 as a hole-transport layer. *RSC Advances* 5 (82): 66981–66987.

Duan, L., Zhang, Y., Deng, R. et al. (2020). Balance between energy transfer and exciton separation in ternary organic solar cells with two conjugated polymer donors. *ACS Applied Energy Materials* 3 (6): 5792–5803.

Dürr, M., Yasuda, A., and Nelles, G. (2006). On the origin of increased open circuit voltage of dye-sensitized solar cells using 4-*tert*-butyl pyridine as additive to the electrolyte. *Applied Physics Letters* 89 (6): 061110.

Eggenhuisen, T.M., Galagan, Y., Biezemans, A.F.K.V. et al. (2015). High efficiency, fully inkjet printed organic solar cells with freedom of design. *Journal of Materials Chemistry A* 3 (14): 7255–7262.

Fan, L., Cui, R., Guo, X. et al. (2014). A new two-dimensional donor/acceptor copolymer based on 4,8-bis(2′-ethylhexylthiophene)thieno[2,3-*f*]benzofuran for high-performance polymer solar cells. *Journal of Materials Chemistry C* 2 (28): 5651–5659.

Furkan, D. and Mehmet Emin, M. (2010). Critical factors that affecting efficiency of solar cells. *Smart Grid and Renewable Energy* 1 (1): 1947.

Galuska, L.A., Ocheje, M.U., Ahmad, Z.C. et al. (2022). Elucidating the role of hydrogen bonds for improved mechanical properties in a high-performance semiconducting polymer. *Chemistry of Materials* 34 (5): 2259–2267.

Garg, R., Elmas, S., Nann, T., and Andersson, M.R. (2017). Deposition methods of graphene as electrode material for organic solar cells. *Advanced Energy Materials* 7 (10): 1601393.

George, J., Joseph, A.P., and Balachandran, M. (2022). Perovskites: emergence of highly efficient third-generation solar cells. *International Journal of Energy Research* 46 (15): 21856–21883.

Green, M.A. (1981). Solar cell fill factors: general graph and empirical expressions. *Solid State Electronics* 24 (8): 788–789.

Green, M.A., Hishikawa, Y., Dunlop, E.D. et al. (2018). Solar cell efficiency tables (version 52). *Progress in Photovoltaics: Research and Applications* 26 (7): 427–436.

Green, M., Dunlop, E.D., Hohl-Ebinger, J. et al. (2021). Solar cell efficiency tables (version 57). *Progress in Photovoltaics: Research and Applications* 29 (1): 3–15.

Gu, C., Huang, N., Chen, Y. et al. (2015). π-conjugated microporous polymer films: designed synthesis, conducting properties, and photoenergy conversions. *Angewandte Chemie* 127 (46): 13798–13802.

Gu, X., Yan, H., Kurosawa, T. et al. (2016). Comparison of the morphology development of polymer–fullerene and polymer–polymer solar cells during solution-shearing blade coating. *Advanced Energy Materials* 6 (22): 1601225.

Günes, S., Neugebauer, H., and Sariciftci, N.S. (2007). Conjugated polymer-based organic solar cells. *Chemical Reviews* 107 (4): 1324–1338.

Gutoff, E.B. and Cohen, E.D. (2006). *Coating and Drying Defects: Troubleshooting Operating Problems*. Wiley.

Han, Y., Dong, H., Pan, W. et al. (2021). An efficiency of 16.46% and a T_{80} lifetime of over 4000 h for the PM6:Y6 inverted organic solar cells enabled by surface acid treatment of the zinc oxide electron transporting layer. *ACS Applied Materials & Interfaces* 13 (15): 17869–17881.

Hannah, R., Max, R. and Pablo, R. (2022). Energy. Published online at Our World In Data.org. https://ourworldindata.org/energy (accessed 10 December 2022).

He, D., Qiu, L., Zhang, Z. et al. (2016). Two new fluorinated copolymers based on thieno[2,3-*f*]benzofuran for efficient polymer solar cells. *RSC Advances* 6 (67): 62923–62933.

Hoang, Q.V., Song, C.E., Moon, S.J. et al. (2015). Asymmetric electron-donating 4-alkyl-8-alkoxybenzo[1,2-*b*:4,5-*b*′]dithiophene unit for use in high-efficiency bulk heterojunction polymer solar cells. *Macromolecules* 48 (12): 3918–3927.

Hoang, Q.V. et al. (2016). Poly(vinyl pyrrolidone): a superior and general additive in polysulfide electrolytes for high efficiency quantum dot sensitized solar cells. *Journal of Materials Chemistry A* 4 (29): 11416–11421.

Hoppe, H. and Sariciftci, N.S. (2004). Organic solar cells: an overview. *Journal of Materials Research* 19 (7): 1924–1945.

Hoppe, H. and Sariciftci, N.S. (2007). Polymer solar cells. *Photoresponsive Polymers* II: 1–86.

Hou, J., Park, M.H., Zhang, S. et al. (2008). Bandgap and molecular energy level control of conjugated polymer photovoltaic materials based on benzo[1,2-*b*:4,5-*b*′]dithiophene. *Macromolecules* 41 (16): 6012–6018.

Hou, J., Chen, H.Y., Zhang, S. et al. (2009). Synthesis of a low band gap polymer and its application in highly efficient polymer solar cells. *Journal of the American Chemical Society* 131 (43): 15586–15587.

Hou, W., Xiao, Y., Han, G., and Lin, J.Y. (2019). The applications of polymers in solar cells: a review. *Polymers* 11 (1): 143.

Hsu, H.C., Wu, S.H., Tung, Y.L., and Shih, C.F. (2022). Long-term stable perovskite solar cells prepared by doctor blade coating technology using bilayer structure and non-toxic solvent. *Organic Electronics* 101: 106400.

Huang, Y., Wu, F., Zhang, M. et al. (2015). Synthesis and photovoltaic properties of conjugated polymers with an asymmetric 4-(2-ethylhexyloxy)-8-(2-ethylhexylthio)benzo[1,2-*b*:4,5-*b*′]dithiophene unit. *Dyes and Pigments* 115: 58–66.

Huang, G., Jiang, H., Zhang, J. et al. (2018). Two-dimensional copolymers based on an alkylthionaphthyl-substituted benzo[1,2-*b*:4,5-*b*′]dithiophene for high-efficiency polymer solar cells. *ACS Applied Energy Materials* 1 (4): 1506–1511.

Hubbard, S.M., Cress, C.D., Bailey, C.G. et al. (2008). Effect of strain compensation on quantum dot enhanced GaAs solar cells. *Applied Physics Letters* 92 (12): 123512.

Huo, L., Zhang, S., Guo, X. et al. (2011). Replacing alkoxy groups with alkylthienyl groups: a feasible approach to improve the properties of photovoltaic polymers. *Angewandte Chemie International Edition* 50 (41): 9697–9702.

Huo, Z., Zhang, P., Li, J. et al. (2017). Wide bandgap conjugated polymers based on bithiophene and benzotriazole for bulk heterojunction solar cells: thiophene versus thieno[3,2-*b*]thiophene as π-conjugated spacers. *Journal of Macromolecular Science, Part A* 54 (9): 565–574.

Jain, A. and Kapoor, A. (2004). Exact analytical solutions of the parameters of real solar cells using Lambert W-function. *Solar Energy Materials and Solar Cells* 81 (2): 269–277.

Jeong, J., Kim, M., Seo, J. et al. (2021). Pseudo-halide anion engineering for α-$FAPbI_3$ perovskite solar cells. *Nature* 592 (7854): 381–385.

Jia, Y., Zhang, Y., Jiang, W. et al. (2019). A fused-ring non-fullerene acceptor based on a benzo[1,2-*b*:4,5-*b′*]dithiophene central core with a thieno[3,2-*b*]thiophene side-chain for highly efficient organic solar cells. *Journal of Materials Chemistry A* 7 (18): 10905–10911.

Jiang, G., Pan, Z., Ren, Z. et al. (2016). Poly (vinyl pyrrolidone): a superior and general additive in polysulfide electrolytes for high efficiency quantum dot sensitized solar cells. *Journal of Materials Chemistry A* 4 (29): 11416–11421.

Jošt, M., Köhnen, E., Al-Ashouri, A. et al. (2022). Perovskite/CIGS tandem solar cells: from certified 24.2% toward 30% and beyond. *ACS Energy Letters* 7 (4): 1298–1307.

Jung, S., Kim, H., Lee, J. et al. (2018a). Bio-inspired catecholamine-derived surface modifier for graphene-based organic solar cells. *ACS Applied Energy Materials* 1 (11): 6463–6468.

Jung, S., Lee, J., Seo, J. et al. (2018b). Development of annealing-free, solution-processable inverted organic solar cells with n-doped graphene electrodes using zinc oxide nanoparticles. *Nano Letters* 18 (2): 1337–1343.

Jung, H.S., Han, G.S., Park, N.G., and Ko, M.J. (2019). Flexible perovskite solar cells. *Joule* 3 (8): 1850–1880.

Kang, X., Zhou, D., Wang, Q. et al. (2019). Rational design of low band gap polymers for efficient solar cells with high open-circuit voltage: the profound effect of Me and Cl substituents with a similar van der Waals radius. *ACS Applied Materials & Interfaces* 11 (51): 48155–48161.

Kannan, N. and Vakeesan, D. (2016). Solar energy for future world: a review. *Renewable and Sustainable Energy Reviews* 62: 1092–1105.

Karunakaran, S.K., Arumugam, G.M., Yang, W. et al. (2019). Recent progress in inkjet-printed solar cells. *Journal of Materials Chemistry A* 7 (23): 13873–13902.

Keum, S., Lee, J., Song, S. et al. (2022). Design and photovoltaic properties of conjugated polymers based on quinoxaline and diketopyrrolopyrrole for OSCs. *Synthetic Metals* 285: 117016.

Khan, S., Ahmad, A., Ahmad, F. et al. (2018). A comprehensive review on solar powered electric vehicle charging system. *Smart Science* 6 (1): 54–79.

Kim, D.J., Kim, S.J., Roh, D.K., and Kim, J.H. (2013). Synthesis of low-cost, rubbery amphiphilic comb-like copolymers and their use in the templated synthesis of mesoporous TiO_2 films for solid-state dye-sensitized solar cells. *Physical Chemistry Chemical Physics* 15 (19): 7345–7353.

Koo, D., Jung, S., Seo, J. et al. (2020). Flexible organic solar cells over 15% efficiency with polyimide-integrated graphene electrodes. *Joule* 4 (5): 1021–1034.

Kratzenberg, M.G., Rüther, R., and Rambo, C.R. (2019). Towards the maximum efficiency design of a perovskite solar cell by material properties tuning: a multidimensional approach. *Solar Energy* 194: 499–509.

Krebs, F.C. (2009b). Fabrication and processing of polymer solar cells: a review of printing and coating techniques. *Solar Energy Materials and Solar Cells* 93 (4): 394–412.

Kumar, S., Prakash, R., and Maiti, P. (2022). Redox mediation through integrating chain extenders in active ionomer polyurethane hard segments in CdS quantum dot sensitized solar cell. *Solar Energy* 231: 985–1001.

Lee, K.M., Suryanarayanan, V., and Ho, K.C. (2006). The influence of surface morphology of TiO_2 coating on the performance of dye-sensitized solar cells. *Solar Energy Materials and Solar Cells* 90 (15): 2398–2404.

Lee, M.S., Lee, K., Kim, S.Y. et al. (2013). High-performance, transparent, and stretchable electrodes using graphene–metal nanowire hybrid structures. *Nano Letters* 13 (6): 2814–2821.

Lee, C.P., Li, C.T., and Ho, K.C. (2017). Use of organic materials in dye-sensitized solar cells. *Materials Today* 20 (5): 267–283.

Lee, Y.H., Ha, M., Song, I. et al. (2019). High-performance hybrid photovoltaics with efficient interfacial contacts between vertically aligned ZnO nanowire arrays and organic semiconductors. *ACS Omega* 4: 9996–10002.

Lee, Y.C., Buraidah, M.H., and Woo, H.J. (2020). Poly(acrylamide-*co*-acrylic acid) gel polymer electrolyte incorporating with water-soluble sodium sulfide salt for quasi-solid-state quantum dot-sensitized solar cell. *High Performance Polymers* 32 (2): 183–191.

Lee, S., Yang, H.S., Song, S. et al. (2022). A low bandgap conjugated polymer bearing a phenazine moiety for application in organic solar cells. *Synthetic Metals* 289: 117114.

Li, J., Wang, L., Kong, X. et al. (2009). A simple and efficient method using polymer dispersion to prepare controllable nanoporous TiO_2 anodes for dye-sensitized solar cells. *Langmuir* 25 (18): 11162–11167.

Li, P., Jiu, T., Tang, G. et al. (2014). Solvents induced ZnO nanoparticles aggregation associated with their interfacial effect on organic solar cells. *ACS Applied Materials & Interfaces* 6 (20): 18172–18179.

Li, W., Hendriks, K.H., Wienk, M.M., and Janssen, R.A. (2016). Diketopyrrolopyrrole polymers for organic solar cells. *Accounts of Chemical Research* 49 (1): 78–85.

Li, G., Budiawan, W., Wang, P.C., and Chu, C.W. (2018a). Conjugated polymer-based solar cells. In: *Encyclopedia of Modern Optics* (ed. B.D. Guenther and D. Steel), 256–269. Elsevier.

Li, M., Lu, J., Chen, Z., and Amine, K. (2018b). 30 Years of lithium-ion batteries. *Advanced Materials* 30 (33): 1800561.

Li, Y., Xu, G., Cui, C., and Li, Y. (2018c). Flexible and semitransparent organic solar cells. *Advanced Energy Materials* 8 (7): 1701791.

Li, C., Cai, M., Bao, X. et al. (2019a). N-Alkylation vs. O-alkylation: influence on the performance of the photovoltaic cells based on a tetracyclic lactam polymer donor. *RSC Advances* 9 (22): 12310–12318.

Li, F., Tang, A., Zhang, B., and Zhou, E. (2019b). Indacenodithieno[3,2-*b*] thiophene-based wide bandgap D-π-A copolymer for nonfullerene organic solar cells. *ACS Macro Letters* 8 (12): 1599–1604.

Liang, Y., Xu, Z., Xia, J. et al. (2010). For the bright future—bulk heterojunction polymer solar cells with power conversion efficiency of 7.4%. *Advanced Materials* 22 (20): E135–E138.

Liao, K.S., Yambem, S.D., Haldar, A. et al. (2010). Designs and architectures for the next generation of organic solar cells. *Energies* 3 (6): 1212–1250.

Liao, S.F., Lu, C.F., Fenta, A.D. et al. (2019). High face-on ratio isoindigo copolymers with extended nano-fibrillar networks in fullerene-based thick (>300 nm) photovoltaics achieving a high efficiency of 10.7%. *Journal of Materials Chemistry A* 7 (37): 21309–21320.

Lin, H., Zhang, S., Zhang, T. et al. (2019). Simultaneous cobalt and phosphorous doping of MoS_2 for improved catalytic performance on polysulfide conversion in lithium–sulfur batteries. *Advanced Energy Materials* 9 (38): 1902096.

Liu, D. and Kelly, T.L. (2014). Perovskite solar cells with a planar heterojunction structure prepared using room-temperature solution processing techniques. *Nature Photonics* 8 (2): 133–138.

Liu, J., Shao, S., Fang, G. et al. (2012). High-efficiency inverted polymer solar cells with transparent and work-function tunable MoO_3-Al composite film as cathode buffer layer. *Advanced Materials* 24 (20): 2774–2779.

Liu, Z., You, P., Liu, S., and Yan, F. (2015). Neutral-color semitransparent organic solar cells with all-graphene electrodes. *ACS Nano* 9 (12): 12026–12034.

Liu, D., Zhu, Q., Gu, C. et al. (2016). High-performance photovoltaic polymers employing symmetry-breaking building blocks. *Advanced Materials* 28 (38): 8490–8498.

Liu, Q., Jiang, Y., Jin, K. et al. (2020a). 18% Efficiency organic solar cells. *Science Bulletin* 65 (4): 272–275.

Liu, S., Yuan, J., Deng, W. et al. (2020b). High-efficiency organic solar cells with low non-radiative recombination loss and low energetic disorder. *Nature Photonics* 14 (5): 300–305.

Long, X., Ding, Z., Dou, C. et al. (2016). Polymer acceptor based on double B←N bridged bipyridine (BNBP) unit for high-efficiency all-polymer solar cells. *Advanced Materials* 28 (30): 6504–6508.

Long, X., Yao, J., Cheng, F. et al. (2019). Double B←N bridged bipyridine-containing polymer acceptors with enhanced electron mobility for all-polymer solar cells. *Materials Chemistry Frontiers* 3 (1): 70–77.

Lu, H., Li, X., and Lei, Q. (2021). Conjugated conductive polymer materials and its applications: a mini-review. *Frontiers in Chemistry* 9: 732132.

Luo, Y., Fang, S., Zheng, N. et al. (2020). Increased electron transport and hole blocking in an aqueous solution processed dye-doped ZnO cathode interlayer for high performance organic solar cells. *ACS Applied Energy Materials* 3 (2): 1694–1701.

Ma, Y. and Zhi, L. (2019). Graphene-based transparent conductive films: material systems, preparation and applications. *Small Methods* 3 (1): 1800199.

Ma, X., Li, J., Lv, J. et al. (2017). Efficiency boost significantly of ternary organic solar cells by doping low bandgap polymer. *Molecular Crystals and Liquid Crystals* 650 (1): 117–124.

Mammeri, F. (2019). Nanostructured flexible PVDF and fluoropolymer-based hybrid films. In: *Frontiers of Nanoscience*, vol. 14 (ed. M. Benelmekki and A. Erbe), 67–101. Elsevier.

Masuko, K., Shigematsu, M., Hashiguchi, T. et al. (2014). Achievement of more than 25% conversion efficiency with crystalline silicon heterojunction solar cell. *IEEE Journal of Photovoltaics* 4 (6): 1433–1435.

Meng, B., Song, H., Chen, X. et al. (2015). Replacing alkyl with oligo(ethylene glycol) as side chains of conjugated polymers for close π–π stacking. *Macromolecules* 48 (13): 4357–4363.

Meng, L., Zhang, Y., Wan, X. et al. (2018). Organic and solution-processed tandem solar cells with 17.3% efficiency. *Science* 361 (6407): 1094–1098.

Menke, S.M. and Holmes, R.J. (2014). Exciton diffusion in organic photovoltaic cells. *Energy and Environmental Science* 7 (2): 499–512.

Mi, D., Kim, J.H., Kim, H.U. et al. (2014). Fullerene derivatives as electron acceptors for organic photovoltaic cells. *Journal of Nanoscience and Nanotechnology* 14 (2): 1064–1084.

Mühlbacher, D., Scharber, M., Morana, M. et al. (2006). High photovoltaic performance of a low-bandgap polymer. *Advanced Materials* 18 (21): 2884–2889.

Muñoz-García, A.B., Benesperi, I., Boschloo, G. et al. (2021). Dye-sensitized solar cells strike back. *Chemical Society Reviews* 50: 12450–12550.

Nelson, J. (2002). Organic photovoltaic films. *Current Opinion in Solid State and Materials Science* 6 (1): 87–95.

Newell, R.G. and Raimi, D. (2020). Global energy outlook comparison methods: 2020 update. *Resources for the Future: Washington, DC, USA. Report 20-06*.

NREL (2019). Best research-cell efficiencies, National Renewable Energy Laboratory – Photovoltaic Research Homepage. https://www.nrel.gov/pv/cell-efficiency.html (accessed 10 January 2023).

Ocheje, M.U., Charron, B.P., Cheng, Y.H. et al. (2018). Amide-containing alkyl chains in conjugated polymers: effect on self-assembly and electronic properties. *Macromolecules* 51 (4): 1336–1344.

Osaka, I. (2015). Semiconducting polymers based on electron-deficient π-building units. *Polymer Journal* 47 (1): 18–25.

Parida, B., Iniyan, S., and Goic, R. (2011). A review of solar photovoltaic technologies. *Renewable and Sustainable Energy Reviews* 15 (3): 1625–1636.

Park, H., Chang, S., Zhou, X. et al. (2014). Flexible graphene electrode-based organic photovoltaics with record-high efficiency. *Nano Letters* 14 (9): 5148–5154.

Park, S.H., Park, S., Lee, S. et al. (2020). Development of highly efficient large area organic photovoltaic module: effects of nonfullerene acceptor. *Nano Energy* 77: 105147.

Pendyala, N.K., Magdassi, S., and Etgar, L. (2021). Fabrication of perovskite solar cells with digital control of transparency by inkjet printing. *ACS Applied Materials & Interfaces* 13 (26): 30524–30532.

Peng, X., Yuan, J., Shen, S. et al. (2017). Perovskite and organic solar cells fabricated by inkjet printing: progress and prospects. *Advanced Functional Materials* 27 (41): 1703704.

Perlin, J. (2004). Silicon solar cell turns 50. *No. NREL/BR-520-33947*. National Renewable Energy Laboratory, Golden, CO, USA.

Polman, A., Knight, M., Garnett, E.C. et al. (2016). Photovoltaic materials: present efficiencies and future challenges. *Science* 352 (6283): aad4424.

Qin, T., Zang, Y., Bai, W.Y. et al. (2017). The influence of oxygen atoms on conformation and π–π stacking of ladder-type donor-based polymers and their photovoltaic properties. *Macromolecular Rapid Communications* 38 (16): 1700156.

Qiu, B., Cui, R., Yuan, J. et al. (2015). Synthesis and photovoltaic properties of two new alkoxylphenyl substituted thieno[2,3-*f*]benzofuran based polymers. *Physical Chemistry Chemical Physics* 17 (27): 17592–17600.

Qiu, J., Guo, B., Zhang, H. et al. (2018a). Insights into working mechanism of alkali metal fluorides as dopants of ZnO films in inverted polymer solar cells. *The Journal of Physical Chemistry C* 122 (43): 24542–24549.

Qiu, X., Li, Y., Qian, Y. et al. (2018b). Long-acting and safe sunscreens with ultrahigh sun protection factor via natural lignin encapsulation and synergy. *ACS Applied Bio Materials* 1 (5): 1276–1285.

Rana, K., Singh, J., and Ahn, J.H. (2014). A graphene-based transparent electrode for use in flexible optoelectronic devices. *Journal of Materials Chemistry C* 2 (15): 2646–2656.

Rasal, A.S., Korupalli, C., Getachew, G. et al. (2021). Towards green, efficient and stable quantum-dot-sensitized solar cells through nature-inspired biopolymer modified electrolyte. *Electrochimica Acta* 391: 138972.

Rasmussen, S.C. (2018). Early history of conductive organic polymers. In: *Conductive Polymers* (ed. Z. Zhang, M. Rouabhia, and S.E. Moulton), 1–22. CRC Press.

Razykov, T.M., Ferekides, C.S., Morel, D. et al. (2011). Solar photovoltaic electricity: current status and future prospects. *Solar Energy* 85 (8): 1580–1608.

Ren, Z., Liu, H., Liu, Z. et al. (2017). The GaAs/GaAs/Si solar cell – towards current matching in an integrated two terminal tandem. *Solar Energy Materials and Solar Cells* 160: 94–100.

Ryu, S.Y., Seo, J.H., Hafeez, H. et al. (2017). Effects of the wrinkle structure and flat structure formed during static low-temperature annealing of ZnO on the performance of inverted polymer solar cells. *The Journal of Physical Chemistry C* 121 (17): 9191–9201.

Sankaran, S., Glaser, K., Gärtner, S. et al. (2016). Fabrication of polymer solar cells from organic nanoparticle dispersions by doctor blading or ink-jet printing. *Organic Electronics* 28: 118–122.

Sariciftci, N.S., Smilowitz, L., Heeger, A.J., and Wudl, F. (1992). Photoinduced electron transfer from a conducting polymer to buckminsterfullerene. *Science* 258 (5087): 1474–1476.

Scharber, M.C. and Sariciftci, N.S. (2021). Low band gap conjugated semiconducting polymers. *Advanced Materials Technologies* 6 (4): 2000857.

Schlisske, S., Rosenauer, C., Rödlmeier, T. et al. (2021). Ink formulation for printed organic electronics: investigating effects of aggregation on structure and rheology of functional inks based on conjugated polymers in mixed solvents. *Advanced Materials Technologies* 6 (2): 2000335.

Schmidt, H., Flügge, H., Winkler, T. et al. (2009). Efficient semitransparent inverted organic solar cells with indium tin oxide top electrode. *Applied Physics Letters* 94 (24).

Seni, R.S. and Puspitasari, N. (2017). Effect of the addition of PEG and PVA polymer for gel electrolytes in dye-sensitized solar cell (DSSC) with chlorophyll as dye sensitizer. *IOP Conference Series: Materials Science and Engineering*, vol. 214(1). IOP Publishing. 012011. 2nd Materials Research Society of Indonesia Meeting, MRS-Id 2016 - Bandung, West Java, Indonesia, 24 October 2016–26 October 2016.

Shah, A., Torres, P., Tscharner, R. et al. (1999). Photovoltaic technology: the case for thin-film solar cells. *Science* 285 (5428): 692–698.

Sharma, A., Singh, R., Kini, G.P. et al. (2021). Side-chain engineering of diketopyrrolopyrrole-based hole-transport materials to realize high-efficiency perovskite solar cells. *ACS Applied Materials & Interfaces* 13 (6): 7405–7415.

Shi, Y., Chen, W., Wu, Z. et al. (2020). Imide-functionalized acceptor–acceptor copolymers as efficient electron transport layers for high-performance perovskite solar cells. *Journal of Materials Chemistry A* 8 (27): 13754–13762.

Singhania, A., Ghosh, I., Sahoo, P. et al. (2020). Radio waveguide–double Ratchet rotors work in unison on a surface to convert heat into power. *Nano Letters* 20 (9): 6891–6898.

Soga, T. (2006). Fundamentals of solar cell. In: *Nanostructured Materials for Solar Energy Conversion* (ed. T. Soga), 3–43. Elsevier.

Su, H., Wu, T., Cui, D. et al. (2020). The application of graphene derivatives in perovskite solar cells. *Small Methods* 4 (10): 2000507.

Sun, Y., Jiang, G., Zhou, M. et al. (2018). Origin of the effects of PEG additives in electrolytes on the performance of quantum dot sensitized solar cells. *RSC Advances* 8 (52): 29958–29966.

Tang, C.W. (1986). Two-layer organic photovoltaic cell. *Applied Physics Letters* 48 (2): 183–185.

Velasco Davoise, L., Díez-Pascual, A.M., and Peña Capilla, R. (2022). Application of graphene-related materials in organic solar cells. *Materials* 15 (3): 1171.

Venkataraman, D., Yurt, S., Venkatraman, B.H., and Gavvalapalli, N. (2010). Role of molecular architecture in organic photovoltaic cells. *The Journal of Physical Chemistry Letters* 1 (6): 947–958.

Vohra, V., Kawashima, K., Kakara, T. et al. (2015). Efficient inverted polymer solar cells employing favourable molecular orientation. *Nature Photonics* 9 (6): 403–408.

Wang, X., Tang, A., Chen, Y. et al. (2016). Effect of fluorination and symmetry on the properties of polymeric photovoltaic materials based on an asymmetric building block. *RSC Advances* 6 (93): 90051–90060.

Wang, S., Li, Z., Xu, X. et al. (2019). Amino-functionalized graphene quantum dots as cathode interlayer for efficient organic solar cells: quantum dot size on interfacial modification ability and photovoltaic performance. *Advanced Materials Interfaces* 6 (3): 1801480.

Wang, X., Yang, J., Shi, X. et al. (2022). Electrosynthesis of ionic covalent organic frameworks for charge-selective separation of molecules. *Small* 18 (15): 2107108.

Wei, J.J., Ji, G., Zhang, C. et al. (2018). Silane-capped ZnO nanoparticles for use as the electron transport layer in inverted organic solar cells. *ACS Nano* 12 (6): 5518–5529.

Wöhrle, D. and Meissner, D. (1991). Organic solar cells. *Advanced Materials* 3 (3): 129–138.

Wu, X., Lan, S., Zhang, G. et al. (2017). Morphology of a ternary blend solar cell based on small molecule: conjugated polymer: fullerene fabricated by blade coating. *Advanced Functional Materials* 27 (40): 1703268.

Würfel, U., Neher, D., Spies, A., and Albrecht, S. (2015). Impact of charge transport on current–voltage characteristics and power-conversion efficiency of organic solar cells. *Nature Communications* 6 (1): 6951.

Xiao, Y., Han, G., Chang, Y. et al. (2014). An all-solid-state perovskite-sensitized solar cell based on the dual function polyaniline as the sensitizer and p-type hole-transporting material. *Journal of Power Sources* 267: 1–8.

Xiao, Z., Jia, X., and Ding, L. (2017). Ternary organic solar cells offer 14% power conversion efficiency. *Scientific Bulletin* 62 (23): 1562–1564.

Xu, Y. and Liu, J. (2016). Graphene as transparent electrodes: fabrication and new emerging applications. *Small* 12 (11): 1400–1419.

Yao, J., Yu, C., Liu, Z. et al. (2016). Significant improvement of semiconducting performance of the diketopyrrolopyrrole–quaterthiophene conjugated polymer through side-chain engineering via hydrogen-bonding. *Journal of the American Chemical Society* 138 (1): 173–185.

Yilbas, B., Al-Sharafi, A., and Ali, H. (2019). *Self-Cleaning of Surfaces and Water Droplet Mobility*. Elsevier.

Yoshikawa, K., Kawasaki, H., Yoshida, W. et al. (2017). Silicon heterojunction solar cell with interdigitated back contacts for a photoconversion efficiency over 26%. *Nature Energy* 2 (5): 1–8.

Yu, Y., Wu, Y., Zhang, A. et al. (2016). Diketopyrrolopyrrole polymers with thienyl and thiazolyl linkers for application in field-effect transistors and polymer solar cells. *ACS Applied Materials & Interfaces* 8 (44): 30328–30335.

Zhang, S., Ye, L., Zhao, W. et al. (2014). Side chain selection for designing highly efficient photovoltaic polymers with 2D-conjugated structure. *Macromolecules* 47 (14): 4653–4659.

Zhang, M., Guo, X., Ma, W. et al. (2015). A large-bandgap conjugated polymer for versatile photovoltaic applications with high performance. *Advanced Materials* 27 (31): 4655–4660.

Zhang, L., Xu, X., Lin, B. et al. (2018a). Achieving balanced crystallinity of donor and acceptor by combining blade-coating and ternary strategies in organic solar cells. *Advanced Materials* 30 (51): 1805041.

Zhang, S., Gao, J., Wang, W. et al. (2018b). Effect of replacing alkyl side chains with triethylene glycols on photovoltaic properties of easily accessible fluorene-based non-fullerene molecular acceptors: improve or deteriorate? *ACS Applied Energy Materials* 1 (3): 1276–1285.

Zhang, W., Song, W., Huang, J. et al. (2019). Graphene: silver nanowire composite transparent electrode based flexible organic solar cells with 13.4% efficiency. *Journal of Materials Chemistry A* 7: 22021–22028.

Zhang, T., Xing, G., Chen, W., and Chen, L. (2020). Porous organic polymers: a promising platform for efficient photocatalysis. *Materials Chemistry Frontiers* 4 (2): 332–353.

Zhao, J., Li, Q., Liu, S. et al. (2020). Bithieno[3,4-*c*]pyrrole-4,6-dione-mediated crystallinity in large-bandgap polymer donors directs charge transportation and recombination in efficient nonfullerene polymer solar cells. *ACS Energy Letters* 5 (2): 367–375.

Zhao, Y., Zhang, S., Bi, F. et al. (2022). Progress review of asymmetric polymers for organic solar cells. *Journal of Materials Chemistry C* 10 (25): 9419–9440.

Zhou, Y., Zhang, W., and Yu, G. (2021). Recent structural evolution of lactam-and imide-functionalized polymers applied in organic field-effect transistors and organic solar cells. *Chemical Science* 12 (20): 6844–6878.

Zhu, X.Y., Yang, Q., and Muntwiler, M. (2009). Charge-transfer excitons at organic semiconductor surfaces and interfaces. *Accounts of Chemical Research* 42 (11): 1779–1787.

Zhu, C., Huang, H., Jia, Z. et al. (2020a). Spin-coated 10.46% and blade-coated 9.52% of ternary semitransparent organic solar cells with 26.56% average visible transmittance. *Solar Energy* 204: 660–666.

Zhu, C., Yuan, J., Cai, F. et al. (2020b). Tuning the electron-deficient core of a non-fullerene acceptor to achieve over 17% efficiency in a single-junction organic solar cell. *Energy & Environmental Science* 13 (8): 2459–2466.

4

Supercapacitor Energy Storage Incorporating Conjugated Microporous Polymer

Benjamin Raj[1,2], *Arun Kumar Padhy*[3], *Atimanyu Dey*[3], *Ramesh Oraon*[3], *and Mamata Mohapatra*[2]

[1] *Radha Govind University, Department of Chemistry, Lalki Ghati, Ramgarh, 829122, India*
[2] *CSIR-Institute of Minerals and Materials Technology, Hydro and Electrometallurgy Department, Bhubaneswar, Orissa, 751013, India*
[3] *Central University of Jharkhand, Department of Chemistry, Cheri-Manatu, Ranchi, 835222, India*

4.1 Introduction

The pressing demand for uninterrupted power supply and to mitigate the use of fossil fuels and minimize the CO_2 emissions ensures considerable research interest for the scientific community to develop renewable sources of an energy storage system (Fan et al. 2020; Chen et al. 2018; Costa et al. 2021; Akcil 2019). The use of petroleum products leads to a negative impact over the environment, causing severe issues of global warming by polluting the air and water (Mohamed and Allam 2020; Zeng et al. 2015). In the year 2020, global CO_2 emission from fossil fuels combustion was around 30.6 Gt CO_2 which is almost 8% lower than 2019 (IEA U 2020). It is expected that the earth will run into petroleum catastrophe in the coming decade due to the excessive depletion of fossils. To mitigate the use of the above-mentioned fossils, lithium-ion batteries (LIBs) play a vital role as an alternative energy source from small portable devices to large-scale electric vehicles (from an electric vehicle to a microchip) (Xiao et al. 2019; Yao et al. 2018; Zhan et al. 2020; Chen et al. 2018). As a result, there is an urgent need for high-energy storage and specific power devices, such as batteries and supercapacitors. Supercapacitors (also known as ultracapacitors or electric double-layer capacitors) are energy storage devices with high efficiency and low internal resistance that can deliver more energy than batteries due to their energy storage mechanism. It entails a straightforward charge separation between the electrode and the electrolyte (Stoller et al. 2008; Raj et al. 2020). Supercapacitors function as a link between batteries and traditional electrolytic capacitors. Supercapacitors are broadly classified into two types based on their charge storage mechanism, namely electrical double-layer capacitors (EDLCs) and pseudocapacitors. Pseudocapacitors store charge at the interfaces using the redox process, whereas EDLCs work on the nonfaradaic surface interface of the electrode and electrolyte (Sarkar et al. 2017; Yu et al. 2011). EDLCs, in particular, operate in a

Organic Polymers in Energy-Environmental Applications, First Edition.
Edited by Ramesh Oraon, Pardeep Singh, Sanchayita Rajkhowa, Sangita Agarwal, and Ravindra Pratap Singh.

short time span, and their power densities are determined by their physical energy storage mechanism and the dynamics of this storage process (Leistenschneider et al. 2020). This can then be tuned by selecting an electrode with versatile electrode properties and adjusting the applied voltage to the device in relation to the chemical stability of the electrolyte. Research has shifted its attention toward the microporous structural moiety because it provides more active sites for electron transfer and other physicochemical reaction. As conjugated microporous polymers (CMPs) have extended polymeric structures, homogeneous micropores, good thermal and chemical stability, the ease of functionalization with various organic groups are more feasible. In lieu of this many researchers and scientific community developed the composite material for supercapacitor application. Jiang and coworkers (Kou et al. 2011) reported first ever utilization of CMP as a promising material for energy storage application, although previously conjugated polymers are reported to be a good cathode material for the LIBs (Han et al. 2007).

Supercapacitors are energy storage devices which possesses high-power density and low-energy density as compared to battery. But if optimization should be done in a systematic way conjugated microporous materials could be the potential candidate to possess high-energy density like supercapacitors and high-power density like battery. Jiang and coworkers (Kou et al. 2011) ionothermally synthesized Aza-CMPs that have suitable pore size, high surface area, and low interfacial resistance which facilitate the charge accumulation between the electrode and the electrolyte which in terms provide outstanding capacitive property, high-energy density, good cycle life, and better stability. The material they prepared turned out to be an encouraging material for the formation of electrostatic charge separation layer. They observed that the fused aza-microporous framework conducts electricity faster, the high surface area and the porosity helped the fast diffusion of ions during the charge–discharge cycles. It delivered the capacity of $947\,F\,g^{-1}$ which seems to be the best capacity for any material.

4.1.1 Challenges

Although the popularity of CMPs has been increasing linearly in the field of energy storage application due to their numerous captivating properties, there are many challenges which restrict their commercial applications. Some of these difficulties are thoroughly discussed below.

- Because of their limited theoretical capacity, CMPs face significant challenges. As a result, the researcher or scientific community faces a significant challenge in designing a porous system with high theoretical capacity.
- The energy density as well as redox potential is low as compared to the inorganic electrode material. The inorganic electrode material possesses potential 4 V vs Li/Li^+, but in case of organic maximum potential confined to 2.5 V vs Li/Li^+. So, it is another important challenge for the researcher to develop new materials/methods to compete with inorganic electrode.
- CMPs are mostly an insulator or have very wide bandgap semiconductor that possesses very little electronic conductivity with poor conductivity.

- Another significant challenge for conjugated porous organic materials is low active material utilization because of low conductivity.
- Another significant challenge is the issue of solubility. The majority of CMPs are not dispersed in NMP or other electrode-processing solvents containing other additives. As a result, achieving high-quality and uniform distribution of active materials is still a challenge for the scientific community.

4.2 Microporous Polymer Material

The solids that contain interconnected pores or void having the molecular dimensions of size less than 2 nm and high surface area are termed as microporous material (Budd et al. 2004). As per International Union of Pure and Applied Chemistry (IUPAC) the porous materials are categorized in terms of the size of pores: microporous materials (pore size: $d < 2\,\text{nm}$), mesoporous materials (pore size: $2 < d < 50\,\text{nm}$), and macroporous materials (pore size: $d > 50\,\text{nm}$). High surface area and microporosity of these materials play a vital role in the science of adsorption as well as catalysis, gas storage, sensing, and energy storage applications. There structures are similar to zeolites, covalent organic framework, and metal organic framework. Microporous polymers are polymers ranging from highly crosslinked networked structures to linear contorted ones (Chaoui et al. 2017). CMPs are a class of porous materials whose structural moiety is similar to that of covalent organic framework and molecular organic framework having extended π-conjugation (Xie et al. 2011). Triphenylamine derivatives are good electrode materials for SC devices as they have high charge mobilities, unique electronic properties, and high redox activity (Samy et al. 2023). Various researchers have tried to develop some of the microporous polymers for the supercapacitor application. In this regard Ahmed Abdel Nazeer et al. (2020) have demonstrated the hydroxyl-functionalized microporous polymer with twisted carbazole units by $FeCl_3$-faciliated chemical oxidation for enhanced CO_2 uptake and efficient super-capacitor energy storage. The synthesized material was further characterized with various physicochemical techniques. They have reported that the material exhibited high surface area of $1015\,\text{m}^2\,\text{g}^{-1}$, rich nitrogen content, presence of polar CO_2-philic hydroxyl groups in the polymeric network, and remarkable porosity of 0.8 nm. The electrochemical properties of the material were analyzed by employing conventional three-electrode system. The graphite sheet (1 cm × 1 cm) coated with the slurry was used as working electrode, Pt and saturated calomel electrode as counter and reference electrode, respectively. To evaluate the electrochemical performances techniques like cyclic voltammetry (CV) and charge discharge were carried out with the potential range of 0–1 V using 1 M H_2SO_4 as electrolytic solution. The material revealed excellent electrochemical properties of specific capacitance $434\,\text{F}\,\text{g}^{-1}$ at scan rate of $2\,\text{mV}\,\text{s}^{-1}$ and $367\,\text{F}\,\text{g}^{-1}$ at current density of $0.5\,\text{A}\,\text{g}^{-1}$, respectively. They have also found 91% retention after the 2000 charge/discharge cycle at a current density of $10\,\text{A}\,\text{g}^{-1}$. Piyali Bhanja et al. (2018) demonstrated the preparation of new porous polymer TPDA-1, by taking the precursor 2,4,6-trihydroxyisophthalaldehyde and

1,3,5-tris(4-aminophenyl)triazine by using solvothermal Schiff base condensation reaction. The successful synthesis of material was characterized by using various physicochemical techniques. The phase analysis and purity of the sample was done by X-ray diffraction (XRD) analysis. The presence of functional group was analyzed by FTIR and was validated by the UV-Visible spectroscopic techniques. The surface morphology by FESEM and surface area of 545 $m^2 g^{-1}$ and pore volume (0.2546 $cm^3 g^{-1}$) was characterized by Brunauer-Emmett-Teller (BET) surface area. Due to well-organized distribution of pores in 2D layer of covalent organic polymer and presence of interstitial space for transportation of electrolytic ions thereby creating more surface-active sites, they suggested EDLC charge storage mechanism for the material TPDA-1. According to the report the porous material possesses excellent electrochemical properties of capacitance of 469.4 $F g^{-1}$ at 2 $mV s^{-1}$. At 5 $A g^{-1}$, 95% retention properties were observed after 1000 cycles. The graphical representation of synthetic route and its electrochemical properties is shown in Figure 4.1.

Muhamad Gamal et al. (2022) demonstrated the synthesis of two pyrene-4,5,9,10-tetraone (PT)-linked conjugated microporous polymers (PT-CMPs) as organic electrode precursors as shown in Figure 4.2 as probable electrode in energy storage applications. The materials were prepared through the Sonogashira polycondensation reaction of ethynyl pyrene (Py-T)/tetraphenylethene (TPE-T) as common units with brominated pyrene tetraene (PT-Br_2) as a redox-active unit. The electrochemical properties of both the synthesized materials were done by using conventional three-electrode system in which glassy carbon was used as working electrode, platinum and Hg/HgO as counter and reference electrode, respectively. It was evaluated by using cyclic voltammetry (CV) and galvanostatic charge–discharge (GCD) profiles with different scan rate ranges from 2 to 200 $mV s^{-1}$ within the potential window −1.0 to 0.0 V using 1 M KOH as electrolytic solution. The electrochemical results high cycling stability of up to 80% and specific capacitance of 440 $F g^{-1}$ due to its interesting suitable porosity properties, presence of PT moiety, and rod-shaped morphologies. They have also fabricated the symmetric coin cell and found excellent electrochemical properties.

Mostafa Ahmed et al. (2022) demonstrated the successful preparation of two carbazole-tethered CMPs, BC-Py-CMP and BC-BF-CMP, through the coupling reaction of the novel BC-4Bpin with Py-4Br and BF-4Br, respectively. The chemical structure of the prepared CMPs was evaluated by FTIR and solid-state NMR spectroscopy which had excellent thermal stabilities up to T_{d10} values of around 694 °C and char yields of 85.50%. The electrochemical performances of these two synthesized materials were evaluated by conventional three-electrode system using 1 M KOH as electrolytic solution. The cyclic voltammetry was measured with different scan rate ranges from 2 to 200 $mV s^{-1}$ within the potential window of −1 to 0 V and charge discharge as well. The material possess optimum specific capacitance of 260 $F g^{-1}$ at 0.5 $A g^{-1}$ with excellent specific retention of about 89.60% after 600 cycles. The electrochemical capacitance of as prepared sample could be due to the high surface area and greater aromaticity of molecules. The schematic diagram and electrochemical performances is shown in Figure 4.3.

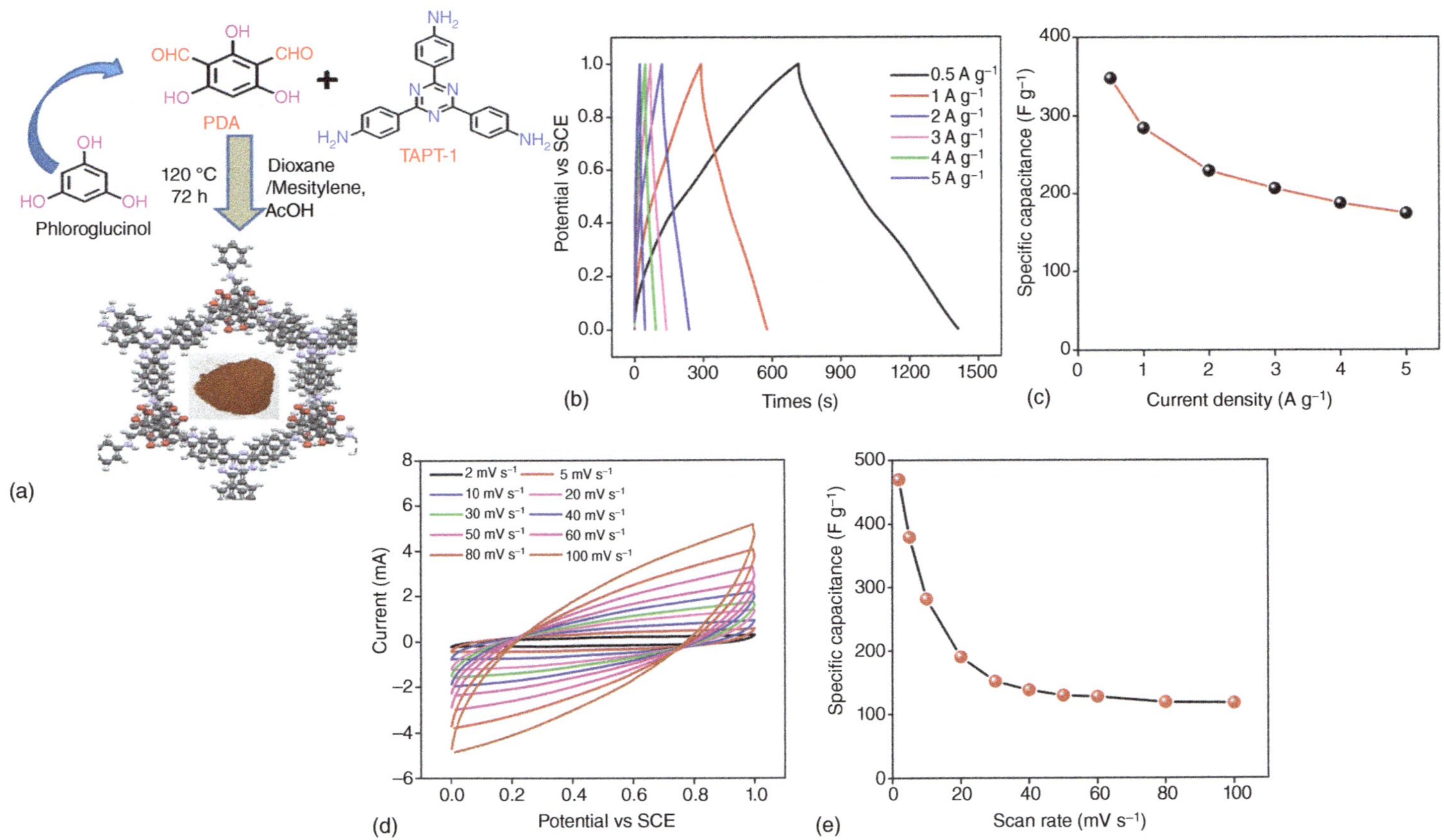

Figure 4.1 (a) Graphical representation of synthetic route, (b) Galvanostatic charge–discharge plots, (c) specific capacitance at different current densities for TPDA-1 electrode measured in 1 M H_2SO_4 electrolyte, (d) cyclic voltammetry in different scan rate, (e) specific capacitance with different scan rate. Source: Bhanja et al. (2018)/American Chemical Society.

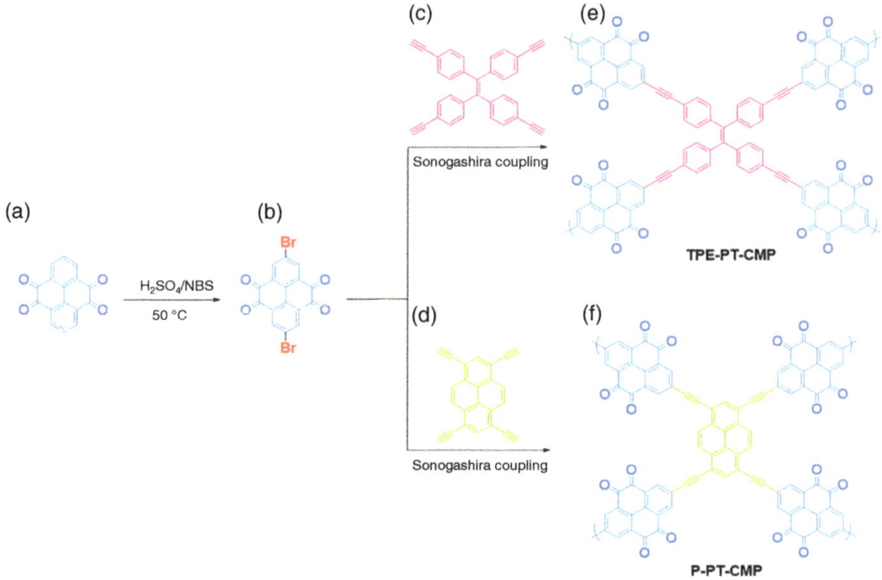

Figure 4.2 Synthesis of PT-Br_2 (b), TPE-PT-CMP (e), and P-PT-CMP (f) from PT (a), TPE-T(c), and Py-T (d). Source: Muhamad Gamal et al. (2022)/American Chemical Society.

Maha Mohamed Samy et al. (2023) prepared two triphenylamine-linked CMPs TPA-TAB and TPA-TBN by single-step Suzuki coupling. The structure of the polymer was characterized by solid-state ^{13}C NMR spectroscopy, and functional group was determined by FT-IR. The thermal stability was measured by thermo-gravimetric analysis. TGA shows exceptional thermal stability of more than 400 °C by applying a heating rate of 20 °C min^{-1} under a nitrogen environment. The high surface area of 490 and 100 m^2 g^{-1} was determined by BET surface area analyzer. The morphology of the synthesized material was characterized by SEM, TEM, and EDS. Electrochemical studies were carried out in the potential range of 0 to −1.0 V. CV of both the materials by sweeping the scan rates which revealed a steady humped rectangular area that shows its EDLC behavior. Galvanometric charge discharge was carried out within the current density of 0–0.5 A g^{-1}. It also indicates pseudo-capacitive and EDLC properties from its bent-triangular shape. The TPA-TAB CMP exhibited high specific capacitance of 684 F g^{-1} at 0.5 A g^{-1} and with the capacitance retention 99.5% after 5000 cycles at 10 A g^{-1}. They also demonstrated the practical application TPA-CMPs cells in the lighting of light-emitting diode. Jie Tong et al. (2023) developed low-cost electrode material for the supercapacitor application. Using benzoquinone and melamine as monomer units, a novel synthetic polymer (SPMA-BQ)-derived material with monomer units combined through C—N single bonds is developed. At the required temperatures (700–900 °C), the SPMA-BQ materials were further carbonized and activated. They evaluated the electrochemical performance using a conventional three-electrode system after successfully forming the required electrode material. They reported that the optimum specific capacitance measured at 800 °C was 414.6 F g^{-1} in a three-electrode system. They have also

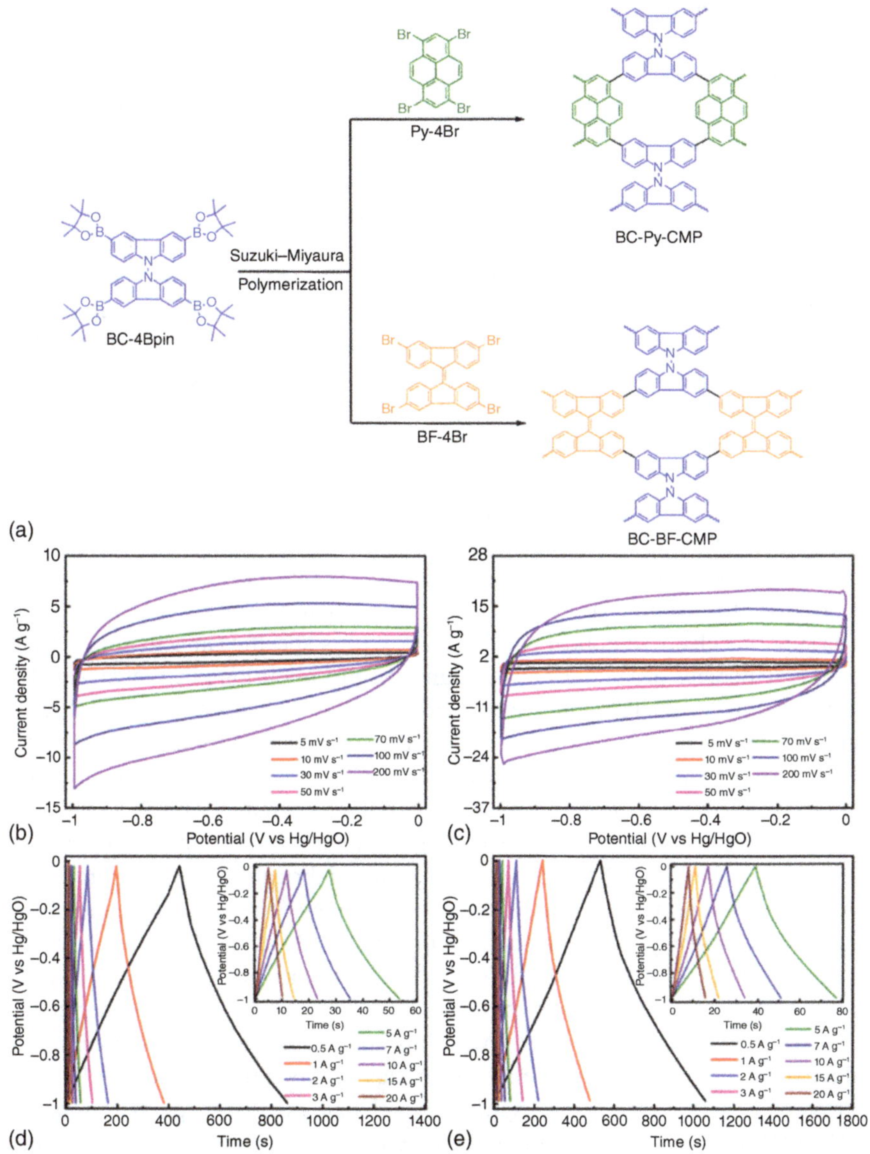

Figure 4.3 Schematic strategies for the synthesis of BC-Py-CMP and BC-BF-CMP (a), Cyclic voltammetry curves of (b) BC-Py-CMP and (c) BC-BF-CMP at various scan rates ($mV\,s^{-1}$). Galvanostatic charge–discharge curves of (d) BC-Py-CMP and (e) BC-BF-CMP at various current densities ($A\,g^{-1}$). Source: Mostafa Ahmed et al. (2022)/with permission of Elsevier.

fabricated solid-state supercapacitor using PVA and KOH and found capacitance of $254\,F\,g^{-1}$ at $0.5\,A\,g^{-1}$. The SSCs also reveal high-energy density, long-term cycling stability, and good rate capability, suggesting their application potential in energy storage fields. Thus, it was proposed that it can be used as a potential electrode material for the supercapacitor in near future. Nan Meng et al. (2022) developed self-templating pyridyl-functionalized conjugated microporous polytriphenylamine for supercapacitor application. Nitrogen-enriched pyridyl CMPs were creatively designed and synthesized as self-templates for the fabrication of nitrogen-doped porous carbon materials (NCs) via a one-step pyrolysis. The synthesized NCs have a high nitrogen content (4.6–11.5 wt%) and a large surface area (400–1059 $m^2\ g^{-1}$). As a result, the NCs have an efficient CO_2 capture capacity of 14.0–17.0 wt% at 273 K and 1 bar, as well as excellent electrochemical energy storage properties with a high specific capacitance (up to 238 $F\,g^{-1}$ at 0.5 $A\,g^{-1}$) and good cyclability (retained almost 100% after 1000 cycles). Furthermore, the NCs were used to build symmetric supercapacitors with a high-energy density (27.7 $Wh\,kg^{-1}$) and power density (7000 $W\,kg^{-1}$). Tzu-Hsin Weng et al. (2022) demonstrated three-dimensional (3D) CMPs (Try-Ph-Th, Try-Ph-Py, and Try-Ph-TPE) containing triptycene (Try) moieties connected to thiophene (Th), pyrene (Py), and tetraphenylethene (TPE) units were prepared using one-pot polycondensation via Suzuki cross-coupling. Thermal stabilities of Try-Ph-Py CMP (T_{d10} = 605 °C; char yield = 80 wt%) and Try-Ph-TPE CMP (T_{d10} = 517 °C; char yield = 71 wt%) were higher than those of Try-Ph-Th CMP (T_{d10} = 471 °C; char yield = 65 wt%) and other CMP materials, according to TGA. Based on N_2 adsorption/desorption analyses, the Try-Ph-TPE CMP sample had a high specific surface area (up to 700 $m^2\ g^{-1}$) and pore volume (0.45 $cm^3\ g^{-1}$), as well as superior electrochemical performance, with a specific capacity of 245 $F\,g^{-1}$ at a current density of 0.5 $A\,g^{-1}$. The high-capacity retentions of the Try-Ph-Th, Try-Ph-Py, and Try-Ph-TPE CMPs were 75.00%, 85.71%, and 92.85%, respectively. When used as real supercapacitors, these CMPs provided high specific capacitances (53, 84.2, and 166 $F\,g^{-1}$, respectively) in addition to their exceptional three-electrode performance. Deok-Ho Roh et al. (2021) demonstrated sono-cavitation and nebulization-based synthesis (SNS) of conjugated microporous polymers for energy storage applications. The SNS allows for the scalable fabrication of microporous and redox-active CMP films via oxidative CC coupling polymerization of the monomer precursor. Surprisingly, the ultrasonic frequency used during SNS has a strong influence on the synthesis of CMP films, with a 30% improvement in reaction yields and 1.3–1.7 times increased surface areas (336–542 $m^2\,g^{-1}$) at 180 kHz compared to 120 kHz. Furthermore, we layer CMP films and CNTs via SNS to create highly conductive, three-dimensional porous electrodes (CMP/carbon nanotube [CNT]). ^{13}C cross-polarization magic angle spinning solid-state nuclear magnetic resonance (CP-MAS NMR) was used to examine the crosslinked structures of the CMP films. Following material confirmation, the electrochemical performances were evaluated using a standard three-electrode system. A working electrode of $1.0 \times 1.0\ cm^2$ CMP film on ITO glass, Pt wire as counter electrode (CE), Ag/AgCl as reference electrode (RE), and degassed 0.1 M Bu_4NPF_6 in acetonitrile were used as electrolyte. The optimal specific capacitance value was observed around 251 $F\,g^{-1}$. Finally, an asymmetric supercapacitor with a CMP/CNT cathode and a carbon anode exhibits a high specific capacitance of 477 $F\,g^{-1}$ at 1 $A\,g^{-1}$, a wide working potential window (0–1.4 V), and robust cycling stability with 94.4% retention after 10,000 cycles. The synthetic route and electrochemical properties of the material is shown in Figures 4.4 and 4.5.

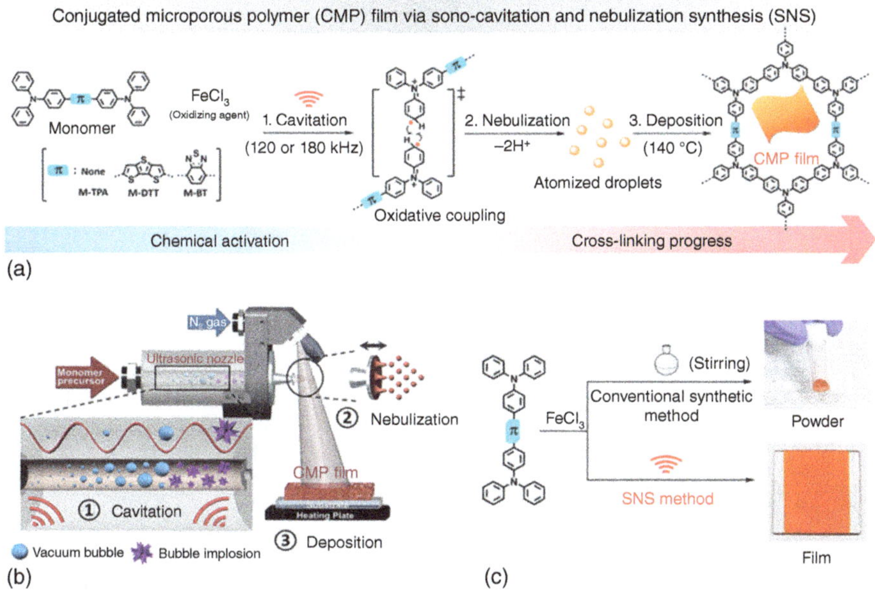

Figure 4.4 SNS reaction process and mechanism. (a) Chemical structures of the monomers and the synthesis of CMP film via the SNS method. (b) Schematic of the three-step SNS process is as follows: (1) Cavitation: the monomer precursor (0.40 mM monomer and 8.0 mM $FeCl_3$ in chloroform/ethanol [v/v = 1 : 1]) is injected into the ultrasonic nozzle using a syringe pump (0.20 ml min^{-1}), where acoustic cavitation occurs in the ultrasonic nozzle (120 or 180 kHz). The high energy of cavitation implosion accelerates the reaction of monomers with oxidizing agents ($FeCl_3$). (2) Nebulization: the activated precursor is nebulized to atomized droplets (8.3–12 μm diameter) at the ultrasonic nozzle tip. (3) Deposition: the atomized droplets are deposited on the target substrate using N_2 carrier gas (3.0 psi), and the deposited CMP films are further polymerized on the substrate by thermal energy (140 °C). (c) Synthetic scheme for CMPs using either the conventional synthetic method of producing CMP powders or the SNS method yielding CMP films.

Similarly Siyu Liu et al. (2023) proposed *p*-phenylenediamine (PPD) as a bridging agent to encapsulate redox *p*-benzoquinone (PBQ) into microporous carbon frameworks. Due to synergistic effects with redox PBQ species, PPD diamine groups can not only covalently link PBQ molecules to the carbon surface, greatly simplifying quinone grafting but also contribute additional pseudocapacitance and accelerate charge transfer. Taking advantage of these appealing properties, the as-synthesized carbon electrodes (i.e., AC-PPD-PBQ) provided an extremely high specific capacitance of 377 F g^{-1} at a current density of 0.5 A g^{-1} and retained 276 F g^{-1} with superior rate performance at 100 A g^{-1}. Furthermore, they demonstrated a maximum energy density of 70.9 Wh kg^{-1} with a wide voltage window of 1.8 V and good cycling stability with 84.4% initial capacitance after 5000 cycles using AHS with NiCoAl-layered double hydroxide/carbon nanotubes (NiCoAl-LDH@CNT) and AC-PPD-PBQ as positive and negative electrodes.

Wei Lyu et al. (2022) reported a unique work of selecting a building block and blending the block with the carbon nanotube to prepare a high-surface area

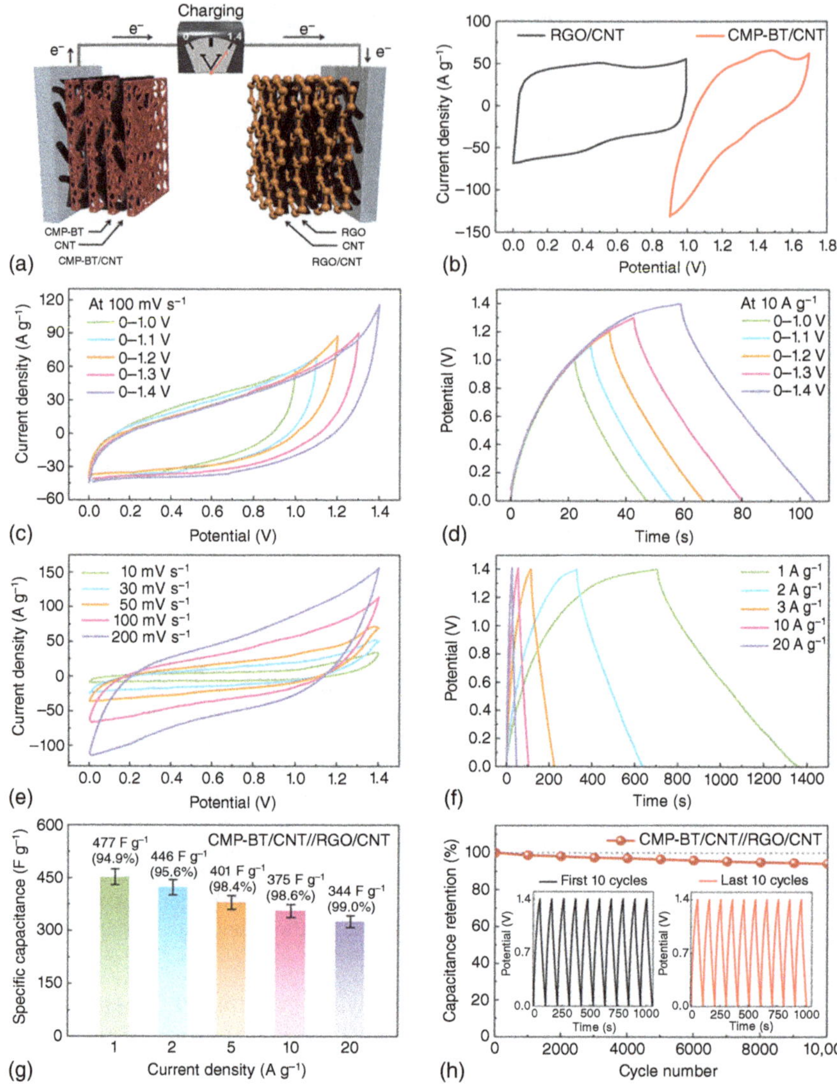

Figure 4.5 (a) Schematic of the ASC, positive (cathode) electrode: CMP-BT/CNT and negative (anode) electrode: RGO/CNT (b) CV curves in two-electrode system, (c) CV with different potential window, (d) charging–discharging with different potential, (e) CV with different scan rate, (f) charging–discharging with different scan rate, (g) average capacitance, and (h) cycling stability.

nitrogen-containing CMP-based composite materials for energy storage application. They put forward a solution to the low electrical conductivity and poor redox activity of CMPs by grafting a nitrogen-containing porous polymer (spirobifluorene bromide core and *p*-phenylenediamine [SACMP]) onto amino-functionalized multiwalled carbon nanotubes (MWNT) by in situ polymerization. The porous polymer MWNT@SACMP was prepared by using Buchwald–Hartwig coupling reaction. The existence of imine functionality was done by FTIR which was further validated by RAMAN spectroscopic techniques. The surface areas and microporosity of prepared material were characterized by N_2 adsorption and desorption isotherm at 77 K. The BET surface area was found to be 660 m^2 g^{-1}. The electrochemical performance was first carried out by using conventional three-electrode system by using 0.5 M H_2SO_4 as electrolytic solution. Techniques like CV, EIS, and GCD show the pseudo-capacitive nature of the material. At 1.0 A g^{-1} the galvanometric charge–discharge curve reveals a maximum specific capacitance of 594 F g^{-1}. Furthermore, they have fabricated a symmetric supercapacitor device with two-electrode configuration to evaluate the practical application of as-synthesized composite. The specific capacitance calculated from GCD was found to be 254 F g^{-1}. The authors concluded that the porous structure of nanotubes provides a diffusion channel for the efficient transportation of electrolyte ions and is the major reason of high capacitance beside the π-stacking interactions between the SACMP and the MWNT, the high redox activity of the porous polymer (SACMP), the high conductivity of the carbon nanotubes, and the surface area of the composite which are the major advantages for the prepared material to be a promising candidate for energy storage devices having high conductivity and stability. Yu-Jen Shao et al. (2023) developed non-conjugated triarylamine-based intrinsic microporous polyamides for electrochromic supercapacitor (ECS). They incorporated the TB scaffolds into the backbone of triarylamine-based polyamides (TPPA-TB and TPPA-Me-TB) to generate microporous structure in the resulting highly transparent polymer thin films cast on ITO-coated glass substrate electrode. The successful formation was confirmed by XRD, BET measurement, and density testing kit. In addition to the conjugated polymers mentioned in the ECS materials, the triarylamine-based polyamides, the non-conjugated polymers, are solution processable and exhibit excellent charge–discharge ability with two-stage color change (green to blue), demonstrating that the intrinsic micropores are caused by carefully incorporating TB scaffolds and the methyl substituents (TPPA-Me-TB). The electrochemical performance of synthesized material was reported by using 0.1 M $TBABF_4$/MeCN as electrolytic solution with a scan rate of 50 mV s^{-1}. The material revealed the excellent charge–discharge ability with a high specific capacitance of 165.3 F g^{-1} at current density of 1.0 A g^{-1}. Based on the above finding, triarylamine-based polyamides are considered as a suitable and potential candidate for ECS materials. Hua Wang et al. (2019) demonstrated the successful fabrication of MWNT@CMPs and carbonized MWNT@CMPs for supercapacitors application as shown in Figure 4.6. For the formation of MWNT@CMPs, conjugated mesoporous/microporous polymers (CMPs) were incorporated onto the surfaces of MWNTs. The morphologies of MWNT@CMP composites changed as the starting monomers for polymer preparation were changed. The surface morphology of as-synthesized material was evaluated by the SEM, TEM spectroscopic techniques. The existence of functional group was confirmed by the FTIR analysis and was further validated with RAMAN spectroscopic techniques. The purity of the sample was analyzed by powder XRD and the confirmation of the oxidation state and relevant binding energy of elements was done by XPS analytical techniques. The material

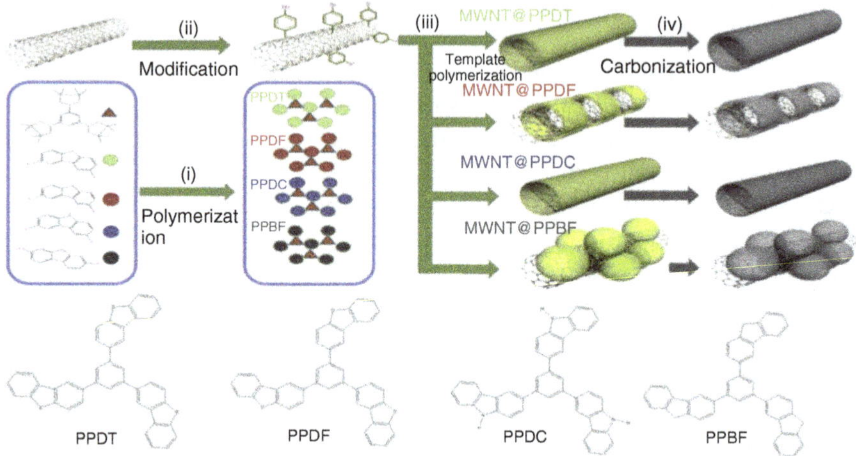

Figure 4.6 Synthesis of CMPs, MWNT@CMPs and the corresponding porous carbon materials. (i) Nitrogen, $Pd(PPh_3)_4$, 100 °C, 24 hours. (ii) Nitrogen, 4-Bromoaniline, 80 °C, 24 hours. (iii) Nitrogen, $Pd(PPh_3)_4$, 100 °C, 24 hours. (iv) Nitrogen, pyrolyzed at 450 °C for 30 minutes then carbonization at 800 °C for 120 minutes. Source: Hua Wang et al. (2019/with permission of Elsevier.

possesses high surface area of 1000 m^2 g^{-1} and pore volume of these core–shell MWNT@CMPs composite materials is large (1.3 cm^3 g^{-1}) which directly influences the electrochemical properties of the material. For electrochemical studies they have analyzed in conventional three-electrode system where glassy carbon coated with material was taken as working electrode, platinum as reference electrode, and Ag/AgCl as reference electrode. The cyclic voltammetry was measured in the potential range from −0.8 to −0.2 V at different scan rates and galvanometric charge–discharge with different current density by using 6 M KOH electrolytic solution, and the highest specific capacitance of these composite materials is 248.4 F g^{-1}. It was 269% greater than the control porous carbon. Carbonization of CMPs moieties on MWNT@CMPs resulted in N- and S-doped MWNT@CMP-C composite materials. The carbonization treatment significantly increased capacitance due to an increase in electrochemical sites during calcination. This study proposes a method for improving the electric performance of CMPs-based materials, as well as a supercapacitor mechanism.

Yuyu Dai et al. (2023) reported the synthesis of three porous imide-based polymeric composites with carbon nanotubes (pNTCDA-TAPB@5–50%CNT, pNTCDATAPM@5–50%CNT, pNTCDA-TAPT@5–50%CNT) by condensation reaction for energy storage application. The porous polymeric materials (pNTCDA-TAPB, pNTCDA-TAPM, pNTCDA-TAPT) have different pore size as well as different central moiety of triphenyl benzene, tetraphenyl methane, and triphenyl triazine. The structure of the composite with corresponding polymer was characterized by FTIR and powder XRD. The result of XRD reveals that the polymer having the core moiety of triphenyl benzene exhibits a better crystallinity property. Further the surface area of these three is determined by nitrogen adsorption–desorption test and the high surface area of 453.46, 9.87, and 4.50 $m^2 g^{-1}$ was calculated, respectively. The morphology of the polymer and the composite was illustrated by SEM and TEM. The HR-TEM reveals no distinct morphology for the three polymers. Tube-type structures with smooth outer surface morphology of composite were confirmed by SEM. The electrochemical performance was firstly studied by conventional three-electrode system by using 1 M Na_2SO_4 solution as the electrolyte, platinum sheet as counter electrode, and Ag/AgCl as reference electrode. The galvanostatic charge–discharge (GCD) technique was applied to evaluate the energy storage capacitance. From GCD the highest specific capacitance of 217.4 $F g^{-1}$ at 0.5 $A g^{-1}$ for pNTCDA-TAPT polymer was calculated among the three polymers. Further to know the mechanism FT-IR analysis was done for pNTCDA-TAPT at varying voltages. From the FT-IR analysis the author concluded that along with imide units of the polymer, the triazine units of the central moiety also participate in the reduction energy storage, electron transport, ion migration and gives a better capacitance than pure polymeric material. The synthetic approach and electrochemical analysis are presented in Figure 4.7a–e.

Ahmed F. Saber et al. (2022) reported the synthesis of three conjugated carbazole-based microporous polymer by making carbazole-based crosslinker (Cz-3BO) react with monomer of 3,6-dibromo-9-(4-bromophenyl)-9*H*-carbazole (Cz-3Br) tris(4-bromophenyl) amine (TPA-3Br) and 5,7-dibromothieno[3,4-*b*] pyrazine (TP-2Br). Polymer of Cz-Cz CMP, Cz-TPA CMP, and Cz-TP CMP was synthesized through Suzuki–Miyaura chemical reaction. The structure of the synthesized polymers was affirmed through FTIR and NMR techniques. The structure concluded the presence of high degree of cross-linking and stiffness which resulted in the reduction of solubility and enhancement of thermal stability. The thermal stability of the as-synthesized polymers is determined by thermogravimetric analysis under the nitrogen atmosphere temperature ranging from ambient temperature to 800 °C. Cz-TP CMP turns out to be the most thermally stable polymer with negligible weight loss of 10% up to 616 °C as a consequence of organic groups decomposition, whereas Cz-Cz CMP and Cz-TPA CMP have a thermal stability of around 560 and 597 °C, respectively. The surface area was determined by BET surface area analyzer. The surface area of developed polymer Cz-TP CMP, CzTPA CMP, and Cz-Cz CMP was found to be 412, 618, and 623 $m^2 g^{-1}$, respectively. The morphology of the synthesized material is studied by SEM and TEM; both the microscopic techniques displayed the microporous structure of the polymer with the internal diameter range of 100–500 nm with high cross-linkage. The result of HR-TEM also suggested both the amorphous and porous properties of the as-synthesized material. Electrochemical properties were studied by the conventional three-electrode system taking 1M H_2SO_4 as an electrolytic solution. The CV was taken in the potential range of 0.1–0.6 V at various scan rates ranging from 5 to 200 $mV s^{-1}$. The rectangular shape of voltagram suggests a typical electric double layer. The specific capacitance was thus calculated for the three materials. The highest capacitance was found to be 271.82 $F g^{-1}$ at 0.5 $A g^{-1}$ with a capacitance retention of 99% for 2000 cycles at the current density of 10 $F g^{-1}$ for Cz-TPA CMP, whereas a low capacitance of 43.70 and 67.38 $F g^{-1}$ was calculated for Cz-Cz and

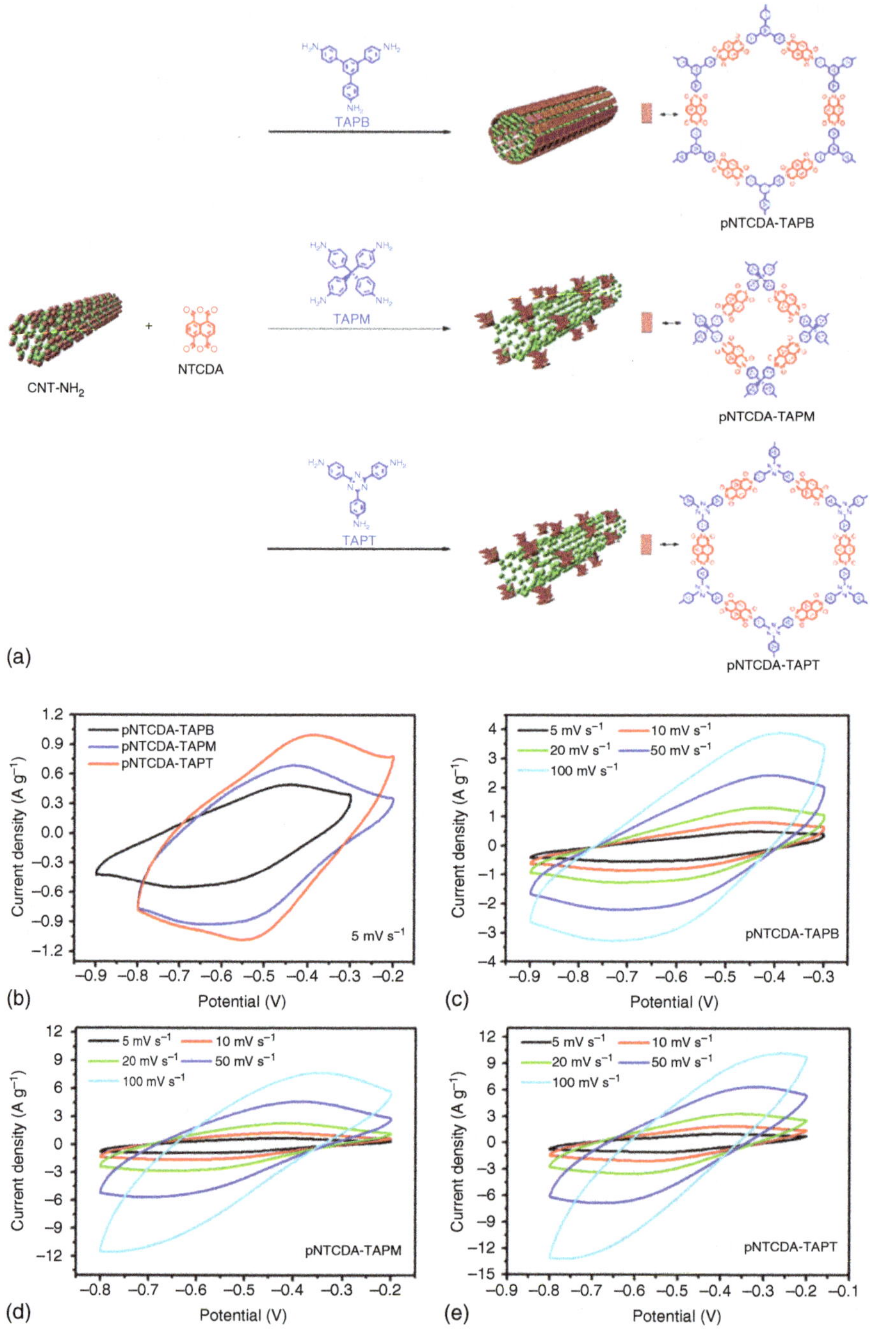

Figure 4.7 (a) The schematic diagram of imide-based CMPs with different core structure and their corresponding composites, (b) CV curves of pNTCDA-TAPB, pNTCDA-TAPM, and pNTCDA-TAPT at 5 mV s^{-1}; CV curves of (c) pNTCDA-TAPB, (d) pNTCDA-TAPM, and (e) pNTCDA-TAPT at different current density. Source: Ahmed F. Saber et al. (2022/with permission of Elsevier.

Cz-TP CMPs. The author concluded that as Cz-Cz CMP has the highest surface area of 623 m^2 g^{-1}, the largest pore size of 1.40 nm than the other, the penetration of ions to the electrode surface facilitates, as a result of which N-rich electroactive CMP material comes out to be the best material for energy storage applications.

4.3 Conclusion

The incorporation of conjugated microporous polymers (CMPs) into supercapacitor energy storage systems presents a promising avenue for advancing energy storage technologies. Through their unique properties such as high surface area, tunable pore size, and excellent electrical conductivity, CMPs offer significant advantages in enhancing the performance and efficiency of supercapacitors. The synthesis and optimization of CMPs for tailored applications in energy storage have demonstrated remarkable progress, showing great potential for addressing challenges associated with traditional energy storage devices. Moving forward, continued research efforts aimed at further understanding the structure–property relationships of CMPs, along with advancements in fabrication techniques and device integration, will be crucial for realizing the full capabilities of supercapacitor energy storage incorporating conjugated microporous polymers, thus paving the way toward sustainable and efficient energy storage solutions for diverse applications.

Conflicts of Interest

The authors do have not any conflict of interest.

References

Ahmed, M., Kotp, M.G., Mansoure, T.H. et al. (2022). Ultrastable carbazole-tethered conjugated microporous polymers for high-performance energy storage. *Microporous and Mesoporous Materials* 333: 111766.

Akcil, A. (2019). *Critical and Rare Earth Elements: Recovery from Secondary Resources*. CRC Press.

Bhanja, P., Das, S.K., Bhunia, K. et al. (2018). A new porous polymer for highly efficient capacitive energy storage. *ACS Sustainable Chemistry & Engineering* 6: 202–209.

Budd, P.M., Makhseed, S.M., Ghanem, B.S. et al. (2004). Microporous polymeric materials. *Materials Today* 7: 40–46.

Chaoui, N., Trunk, M., Dawson, R. et al. (2017). Trends and challenges for microporous polymers. *Chemical Society Reviews* 46: 3302–3321.

Chen, H., Ling, M., Hencz, L. et al. (2018). Exploring chemical, mechanical, and electrical functionalities of binders for advanced energy-storage devices. *Chemical Reviews* 118: 8936–8982.

Chen, Y., Liu, N., Hu, F. et al. (2018). Thermal treatment and ammoniacal leaching for the recovery of valuable metals from spent lithium-ion batteries. *Waste Management* 75: 469–476.

Costa, C.M., Barbosa, J.C., Gonçalves, R. et al. (2021). Recycling and environmental issues of lithium-ion batteries: advances, challenges and opportunities. *Energy Storage Materials* 37: 433–465.

Dai, Y., Wang, Y., Xu, H. et al. (2023). Structure, morphology and energy storage properties of imide conjugated microporous polymers with different cores and the corresponding composites with CNT. *Electrochimica Acta* 441: 141820.

Fan, E., Li, L., Wang, Z. et al. (2020). Sustainable recycling technology for Li-ion batteries and beyond: challenges and future prospects. *Chemical Reviews* 120: 7020–7063.

Han, X., Chang, C., Yuan, L. et al. (2007). Aromatic carbonyl derivative polymers as high-performance Li-ion storage materials. *Advanced Materials* 19: 1616–1621. https://doi.org/10.1002/adma.200602584.

IEA U (2020). Global energy review 2020. Ukr https//www.iea.org/countries/ukraine (accessed 10 September 2020).

Kou, Y., Xu, Y., Guo, Z., and Jiang, D. (2011). Supercapacitive energy storage and electric power supply using an aza-fused π-conjugated microporous framework. *Angewandte Chemie* 123: 8912–8916.

Kou, Y., Xu, Y., Guo, Z., and Jiang, D. (2011). Supercapacitive energy storage and electric power supply using an aza-fused π-conjugated microporous framework. *Angewandte Chemie International Edition* 50: 8753–8757. https://doi.org/10.1002/anie.201103493.

Leistenschneider, D., Heß, L.H., Balducci, A., and Borchardt, L. (2020). Solid-state transformation of aqueous to organic electrolyte – enhancing the operating voltage window of 'in situ electrolyte' supercapacitors. *Sustainable Energy & Fuels* 4: 2438–2447. https://doi.org/10.1039/d0se00180e.

Liu, S., Yang, J., Wang, M. et al. (2023). Encapsulation of redox *p*-benzoquinone into microporous carbon frameworks by a diamine covalent-grafted strategy for aqueous hybrid supercapacitors. *ACS Applied Energy Materials* 6 (5): 2989–2998.

Lyu, W., Yan, C., Chen, Z. et al. (2022). Spirobifluorene-based conjugated microporous polymer-grafted carbon nanotubes for efficient supercapacitive energy storage. *ACS Applied Energy Materials* 5: 3706–3714.

Meng, N., Li, H., Liu, Y., and Liao, Y. (2022). Self-templating synthesis of nitrogen-rich porous carbons using pyridyl functionalized conjugated microporous polytriphenylamine for electrochemical energy storage. *Electrochimica Acta* 402: 139531.

Mohamed, N. and Allam, N.K. (2020). Recent advances in the design of cathode materials for Li-ion batteries. *RSC Advances* 10: 21662–21685.

Mohamed, M.G., Chaganti, S.V., Sharma, S.U. et al. (2022). Constructing conjugated microporous polymers containing the pyrene-4,5,9,10-tetraone unit for energy storage. *ACS Applied Energy Materials* 5: 10130–10140.

Nazeer, A.A., Husain, A.A., Samuel, J. et al. (2020). Hydroxyl-functionalized microporous polymer for enhanced CO_2 uptake and efficient super-capacitor energy storage. *Reactive and Functional Polymers* 154: 104670.

Raj, B., Oraon, R., Mohapatra, M. et al. (2020). Development of SnO_2@rGO hybrid nanocomposites through complexometric approach for multi-dimensional electrochemical application. *Journal of the Electrochemical Society* 167: 167518.

Roh, D.-H., Shin, H., Kim, H.-T., and Kwon, T.-H. (2021). Sono-cavitation and nebulization-based synthesis of conjugated microporous polymers for energy storage applications. *ACS Applied Materials & Interfaces* 13: 61598–61609.

Saber, A.F., Sharma, S.U., Lee, J.-T. et al. (2022). Carbazole-conjugated microporous polymers from Suzuki–Miyaura coupling for supercapacitors. *Polymer (Guildf)* 254: 125070.

Samy, M.M., Mohamed, M.G., Sharma, S.U. et al. (2023). An ultrastable tetrabenzonaphthalene-linked conjugated microporous polymer functioning as a high-performance electrode for supercapacitors. *Journal of the Taiwan Institute of Chemical Engineers* 104750.

Samy, M.M., Mohamed, M.G., Sharma, S.U. et al. (2023). Constructing conjugated microporous polymers containing triphenylamine moieties for high-performance capacitive energy storage. *Polymer (Guildf)* 264: 125541.

Sarkar, S., Howli, P., Das, B. et al. (2017). Novel quaternary chalcogenide/reduced graphene oxide-based asymmetric supercapacitor with high energy density. *ACS Applied Materials & Interfaces* 9: 22652–22664.

Shao, Y.-J., Yen, T.-C., Hu, C.-C., and Liou, G.-S. (2023). Non-conjugated triarylamine-based intrinsic microporous polyamides for electrochromic supercapacitor: diffusion dynamics and charge-discharge studies. *Journal of Materials Chemistry A* 11 (4): 1877–1885.

Stoller, M.D., Park, S., Zhu, Y. et al. (2008). Graphene-based ultracapacitors. *Nano Letters* 8: 3498–3502.

Tong, J., Wang, J., Xu, P., and Zhang, S. (2023). Nitrogen and oxygen codoped porous carbon based on a synthetic polymer for high-performance solid-state supercapacitors. *Journal of Energy Storage* 58: 106349.

Wang, H., Liang, L., Duan, L. et al. (2019). Fabrication of MWNT@ CMPs and carbonized MWNT@ CMPs for supercapacitors. *Materials Chemistry and Physics* 226: 309–317.

Weng, T.-H., Mohamed, M.G., Sharma, S.U. et al. (2022). Ultrastable three-dimensional triptycene-and tetraphenylethene-conjugated microporous polymers for energy storage. *ACS Applied Energy Materials* 5: 14239–14249.

Xiao, J., Li, J., and Xu, Z. (2019). Challenges to future development of spent lithium ion batteries recovery from environmental and technological perspectives. *Environmental Science & Technology* 54: 9–25.

Xie, Z., Wang, C., deKrafft, K.E., and Lin, W. (2011). Highly stable and porous cross-linked polymers for efficient photocatalysis. *Journal of the American Chemical Society* 133: 2056–2059.

Yao, Y., Zhu, M., Zhao, Z. et al. (2018). Hydrometallurgical processes for recycling spent lithium-ion batteries: a critical review. *ACS Sustainable Chemistry & Engineering* 6: 13611–13627.

Yu, G., Hu, L., Vosgueritchian, M. et al. (2011). Solution-processed graphene/MnO_2 nanostructured textiles for high-performance electrochemical capacitors. *Nano Letters* 11 (7): 2905–2911.

Zeng, X., Li, J., and Liu, L. (2015). Solving spent lithium-ion battery problems in China: opportunities and challenges. *Renewable and Sustainable Energy Reviews* 52: 1759–1767.

Zhan, R., Payne, T., Leftwich, T. et al. (2020). De-agglomeration of cathode composites for direct recycling of Li-ion batteries. *Waste Management* 105: 39–48:https://doi.org/10.1016/j.wasman.2020.01.035.

5

Modification of Surface Properties of Polymeric Materials: Methodological Approaches and Applications

Akhi Das, Swrangsi Goyary*, Sukanya Gogoi*, Swaraj Pathak*, and Nilamoni Nath*

Department of Chemistry, Gauhati University, Gopinath Bordoloi Nagar, Guwahati, Assam 781014, India

5.1 Introduction

Polymers are a class of natural or artificial compounds that are made up of massive molecules known as macromolecules, which are multiples of monomers. These monomers are the smaller chemical building blocks. Whether a polymer is natural or man-made depends on the source and the synthesis process. Polymers can be crosslinked networks of monomers in two or three dimensions, or they can be lengthy chains of unbranched or branched monomers. They might have flexible or stiff backbones. Polymers are being frequently utilized and explored in several technical fields. Natural polymers which are highly degradable, for example, collagen and chitosan, also find their use in biomedical applications (Paul et al. 2016) and in engineering. The role of polymers in the evolution of technology, applied sciences, and engineering is quite noteworthy. They have replaced their metallic equivalents in electrical, medicinal, building, and energy applications. This is because they are inexpensive to produce and to buy. They have a broad range of strength, conductivity, degradability, and elasticity and are adaptable to changing production methods, and it is very simple to modify their properties (Govindarajan and Shandas 2014). Although polymer has a vast range of uses in both the chemical and everyday worlds, its future development has been constrained by their low surface hydrophilicity and wear resistance. The polymer's surface has to be modified in order to enhance its surface qualities. Surface modification in polymers is the process of altering a polymer's surface at the nanoscale level while leaving the material's core qualities unaffected and introduce properties which are new and beneficial on its surface. Some newly introduced properties are hydrophilicity and scratch resistance which are of prime importance on the surface of the polymers.

The main purpose behind the surface treatment of polymers is modification at the outermost layer of the polymer to improve its wettability, sealability, printability, barrier properties, dye uptake, adhesion to other materials, and resistance to

*Equal contribution.

Organic Polymers in Energy-Environmental Applications, First Edition.
Edited by Ramesh Oraon, Pardeep Singh, Sanchayita Rajkhowa, Sangita Agarwal, and Ravindra Pratap Singh.

glazing or interaction with the biological environment, while preserving the desired important properties of the polymer. This is accomplished by adding certain functional groups to the surface of the polymer.

There are many techniques for altering the surface of polymers. Some of such important methods, namely, physical, chemical, mechanical, thermal, and photolytic processes are extensively studied to generate various products with particular morphologies and structural compositions. The physicochemical properties are also of importance in pharmaceuticals, electronics, microsensing, and bioengineering industries. There are also some other common surface modification techniques which include solution processing, plasma treatment, irradiation treatment, surface grafting. Here, a detailed introduction of the particular polymer material, modification method, and modification mechanism is provided. Due to their increased ability to be altered for specific needs, which frequently involves modification of their surface properties, polymers have successfully entered advanced application fields over the past 15 years. As a result, surface modification techniques have drawn growing interest and now make up more than half of the worldwide scientific output relevant to polymers and polymer modifications, as demonstrated by an examination of the scientific output from 2000 to 2016. The stable bulk characteristics of polymers, including resistance to heat degradation, solvent treatments, and unaltered mechanical strength after undergoing treatment, make them practical substitutes for traditional metal-based devices. Polymer surfaces, however, prevent their vast variety of planned applications, necessitating modification through a number of means. For instance, the presence of hydroxyl groups causes surfaces to be hydrophilic, which is not always the case with polymeric surfaces. Therefore, it is essential that surfaces that are naturally nonconforming undergo treatments to enable their usage in a variety of industrial applications and to compensate for their inherent limitations.

Surfaces are frequently modified using polymers to provide desirable effects which are beneficial in the long run for their effective functionality. Due to the adjustable bulk material characteristics of polymers, such as elasticity, tensile strength, and density, surface treatments not only increase the efficacy of the present material but also expand the application range of the material. These surface-enhanced polymers are widely used in microfluidics (Sinton 2014), biomedical applications (Menard et al. 2007; Goddard and Hotchkiss 2007; Ikada 1994), textiles (Nikolaev et al. 2014; Hu 2016), electronics (Behl et al. 2002), water-treatment (Saqib and Aljundi 2016), and energy industries (Yang et al. 2013).

When the polymer is subjected to chemical, ionic, or light-based treatment processes that can adjust the surface energies by adding functional groups to the material (Chen et al. 2016), the most popular route is to modulate the surface energy of the material (Ferrero and Periolatto 2015) to alter its adhesive (Sojoudi et al., 2018), absorbing, wetting, or releasing properties for the better. Another route is to mold the surface roughness (Encinas et al. 2010) of the polymer by chemical or mechanical abrasion processes in which the topmost layers are altered by being treated with different types of abrasives. Additionally, the polymer surface is subjected to nano- and micro-scale roughness, which allows for the manipulation of antifouling and anti-wetting capabilities. Surface energy of the polymer and its roughness can be managed by a range of non-invasive techniques such as ultra-violet (UV) radiation, patterning induced by laser, and corona/plasma treatment. In this chapter, we want to highlight the differences between property modification approaches that alter the chemical structure of the polymer without changing its intrinsic properties. Due to the wide variety of chemical and physical properties offered by many families of polymeric materials, choosing and modifying polymers for each specific application is made easier. For instance, certain polymers are used more frequently in cardiovascular stents (Yakacki et al. 2007) because they have a lower density than metals and are more flexible than many other materials. Owing to its cheaper rate, simplicity of manufacture, oxygen permeability, and optical transparency (Zhou et al. 2010; Boxshall et al. 2006), poly(dimethylsiloxane) (PDMS) is frequently utilized in microfluidic devices. The biodegradable aliphatic polyesters are employed in tissue engineering (Morent et al. 2011) owing to their least toxicity, high mechanical characteristics, and low immunogenicity. Due to the many polar hydroxyl groups that are found in cotton fiber's polymers, it is particularly absorbent when used in textiles. By preferentially treating the polymer surface with fluorine-based chemical reagents that may produce functional surfaces because of their low surface energy, the hydrophilicity which is extreme for the cellulose structures can be changed. Classes of silanes have been used to modify cellulose nanofibers so that they function as hydrophobic materials (Baidya et al. 2017).

This chapter explores the benefits and drawbacks, longevity, and durability of induced characteristics for each approach in order to highlight and offer a thorough overview of the various ways for treatment of polymer on their surfaces with an equal emphasis on their field of application.

5.2 Physical Treatment for Polymer Surface Modification

Surface modification using physical techniques involves straightforward, affordable, and scalable pathways. They involve no chemicals and hence they are environment friendly. They are intended to give preferred surface requirements and also, durability of the polymer surface is essential for manufacturing purposes (Li et al. 2007; Yao et al. 2011). When compared to other types of treatments, this method provides polymer surfaces that are stronger and more adverse to abrasion. This physical modification is carried out as a step in the surface manufacturing process, unlike other surface treatment techniques where the variation is applied to a polymer surface that has already been produced. As mentioned before, physical modification is done prior to the manufacture of polymer. It can done using injection molding, (Weng et al. 2018; Zhou et al. 2018; Maghsoudi et al. 2020) (Figure 5.1) extrusion (Okulova et al. 2017; Sollogoub et al. 2011) (Figure 5.3), and lamination template technique (Figure 5.2) (Xu et al. 2011).

Although the physical alteration of surface does not change its in-built chemical composition, it can influence the wettability of the surface by generating roughness onto the polymer on micro- and nanolevels. Scientists have tuned the roughness of the polymer surface for achieving superhydrophobic and omniphobic qualities. The polymer surface can be simply made more hydrophobic by increasing the

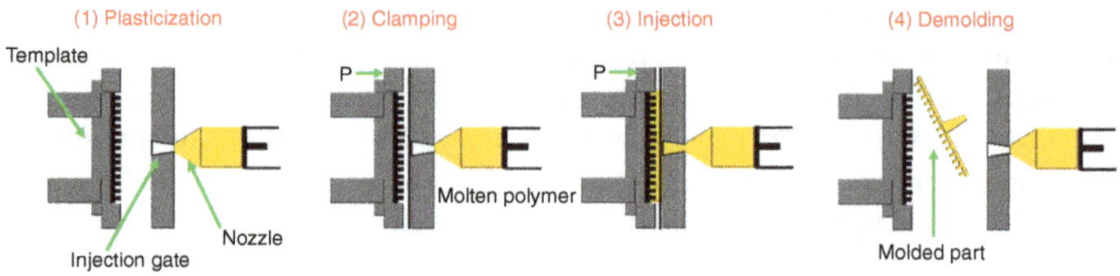

Figure 5.1 Pictorial illustration of injection molding process. Source: Adapted from Maghsoudi et al. (2020).

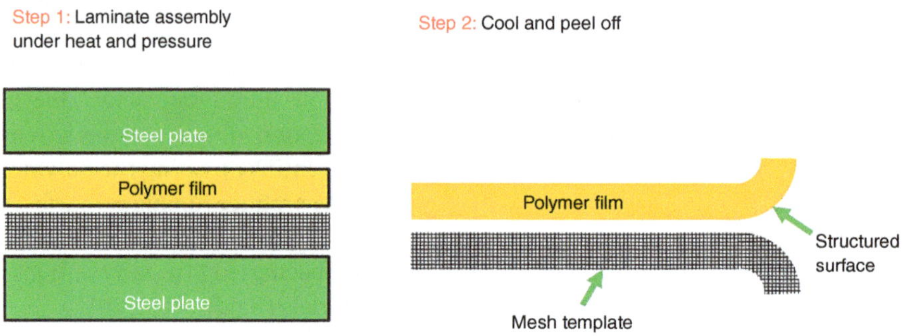

Figure 5.2 Pictorial illustration of lamination template techniques. Source: Adapted from Xu et al. (2011).

surface roughness (Liu and Kim 2014). Using the lamination template approach, Xu et al. (2011) created superhydrophobic polymer surfaces with exceptional abrasion resistance by laminating polyethylene (PE) films beside mesh templates. The mesh was then removed from the PE films, leaving a 3D pattern of orderly microposts on the polymer surface that matched the mesh's pitch and apertures.

Correspondingly, with microinjection compression molding technique, Chen and Huang had given a quick, adaptable approach for replicating polypropylene (PP) surfaces with producing T-shaped micropillars (Chen and Huang 2016). They used the roll-to-roll extrusion coating (R2R-EC) technique by simultaneously examining the replication of the pillars under various conditions (Chen and Huang 2016). The temperature of the cooling zone had a significant impact on the replication process, followed by the extruder output, stripper roll, and pressure.

For commercial purpose, there are several ways to create sustainable, superhydrophobic, antifouling, and ice-phobic surfaces (Xue et al. 2010; Sojoudi et al. 2016). The fundamental benefit of physical methods over other techniques for creating superhydrophobic surfaces, aside from its low cost, is that the use of hazardous fluorine-based polymers is not a requisite. Furthermore, because the created surface roughness occasionally affects the polymer surface chemistry, polymers reformed through physical means remain durable and do not age or degrade.

The physical modification techniques are always preferred, unless otherwise necessary. The method is only applicable to thermoplastic polymers that are able

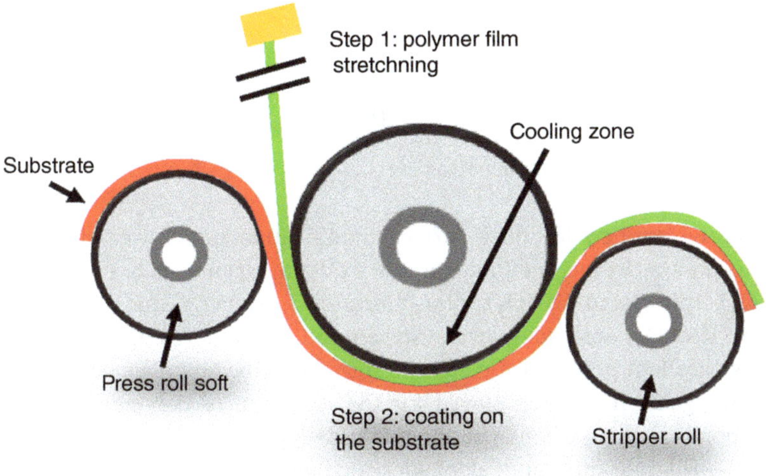

Figure 5.3 Pictorial illustration of extrusion process. Source: Adapted from Sollogoub et al. (2011).

to be treated when solid or molten without losing their inherent qualities like mechanical power and elasticity.

5.3 Chemical Treatment for Polymer Surface Modification

Chemical surface modification of polymeric material is very useful in biomedical applications such as enhancement of the polymer properties without modifying the surface roughness, where standard physical approaches are insufficient. In this wet chemical approach, the polymers are treated with chemicals to improve their surface qualities and to remove microbes and debris to create a sterile environment on the surface, ideal for biomedical applications. Wet chemical treatments comprise several techniques for changing the surface of textile fibers, such as dissolving, swelling, etching, and the creation of adhesion-promoting layers. In polymer fiber material, surface modification through the wet chemical method is mainly subdivided into halogenation, oxidation, hydrolysis, complexation, and the creation of adhesion-promoting layers (Gleissner et al. 2022).

The halogenation method is based on substitution or addition of hydrogen atom by halogen atoms such F, Cl, Br, and I. For the activation of fibers material by this method, chlorination is mostly preferred. Chlorinating agents such as sodium hypochlorite and sodium dichloroisocyanurate solutions are often used for this treatment (Sun and Sun 2004; Barassi and Borrmann 2012). For example, when polyamide (PA) is treated with sodium hypochlorite solution, an O-chlorinated intermediate is formed which further follows rearrangement resulting in a N-chlorinated stable product (Hardy and Robson 1967). The mechanistic pathways involved in this reaction are shown in Figure 5.4.

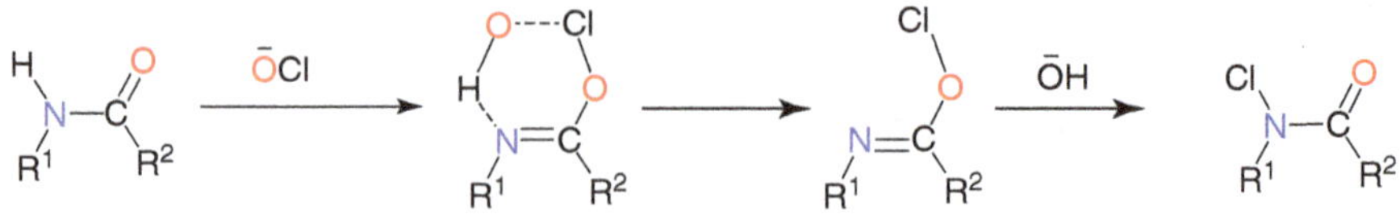

Figure 5.4 Halogenation mechanism in polyamide.

The next crucial method for fiber surface modification under wet chemical treatment is oxidation. In this method, various oxidizing agents such as chromic acid ($H_2Cr_2O_4$), nitric acid (HNO_3), phosphoric acid (H_3PO_4), and potassium permanganate ($KMnO_4$) are widely used (Steven and Burke 2003). Out of these, $KMnO_4$ shows excellent potential for high-performance fibers surface activation. These oxidizing agents mainly employed for the oxidation of polymer material contain weak bonds such as π-bonds in alkynes and alkenes, weak C—C bonds in glycols, weak C—H bonds present in aldehydes, alcohols, and substituted aromatic rings (Montazer and Komeily Nia 2015). This method results in the formation of new functional moieties such as carboxyl, carbonyl, and hydroxyl groups on the polymer surface.

In the hydrolysis method, the polymeric material is treated with alkali to separate individual fiber from the bundles of fibers. This process creates roughness in the fiber material by increasing the aspect ratio of the individual fiber particle, and this technique improves the interfacial bonding of fiber material with the matrix site (Jones et al. 2017). This technique is best suited for materials that contain ester bonds. The breakdown of the polymer is accelerated by harsh circumstances and rising levels of alcohols or certain amines as by products, which weakens the substrate (Tavanai 2009; Musale and Shukla 2017). The hydrolysis of poly(lactic acid) (PLA) by alkali treatment is shown in Figure 5.5 (Abdul Hamid 2014).

Out of all alkaline agents, NaOH is widely used for modifying the surface properties of polymer fiber material, for example, polyethylene terephthalate (PET) (Han et al. 2016; Vecchiato et al. 2017). The mechanistic pathway leads to the formation of an anion as intermediate, which further facilitates the scission of chain length followed by the formation of hydroxyl and carboxylic end group (Zeronian and Collins 1989). The formation of surface carboxylic groups is directly proportional to the NaOH concentration. Another route of hydrolysis is treating polymeric fibers with enzymes. Enzyme proteins are more appropriate method for the hydrolysis since they act as bio-catalyst which may modify and completely degrade polymer material. This is a controlled degradation method for surface modification of fiber material as well as increased roughness and formation of desirable

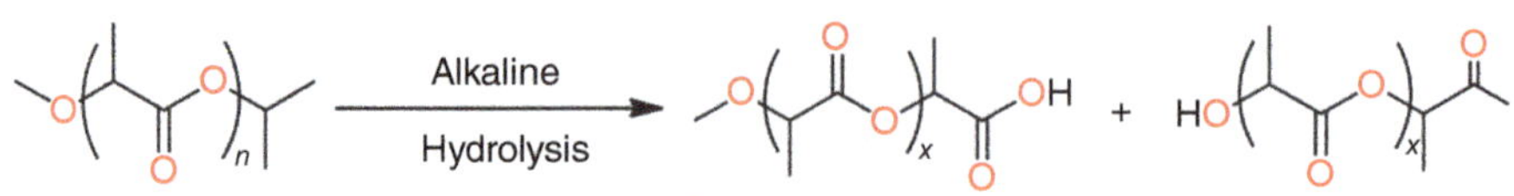

Figure 5.5 Hydrolysis of ester bond of poly(lactic acid) by alkali treatment. Source: Abdul Hamid (2014)/Trans Tech Publications Ltd.

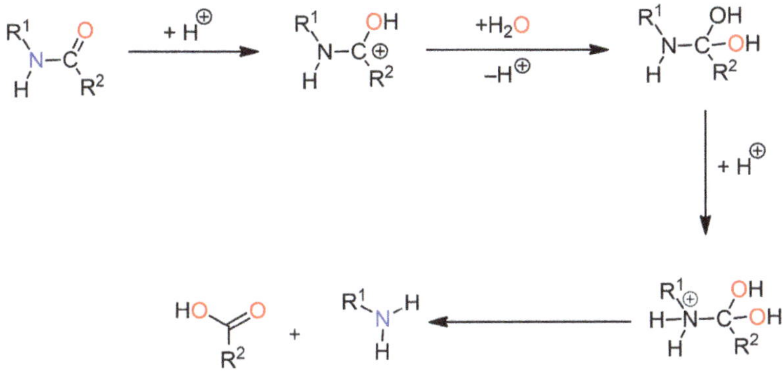

Figure 5.6 Formation of carboxyl and amine groups due to enzymatic treatment on polyamide (Begum et al. 2016; Kanelli et al. 2017; Kim and Seo 2013).

functional groups on the surface. There are various enzymes are being used for the treatment. For example, proteases and peptidases are used for the cleavage of amide bonds in polyamide for the formation of carboxyl and amino groups on the surface. The mechanistic pathway is shown in Figure 5.6 (Begum et al. 2016; Kanelli et al. 2017; Kim and Seo 2013). The activity of enzyme can be optimized by varying different parameters such as pH value, treatment time, concentration, and temperature (Begum et al. 2016; Kanelli et al. 2017; Kim and Seo 2013).

Another crucial method under wet chemical treatment for surface modification of polymer fibers is complexation. In this method the interaction between complexation agents with polymer material is Lewis acid–base interaction, so, the polymer which is appropriate for this method must contain a specific Lewis base site. Generally, polymer chains with non-bonding electron pairs or side groups with nitrogen, phosphorous, oxygen, or sulfur moieties are suitable for this treatment (Jenekhe and Roberts 1993). However, these groups should be involved in the inter- or intramolecular hydrogen bonding or any kind of dispersive forces for the effective surface activation through complexation. Several moderate-to-strong Lewis acids are used for this method. For example, in case of aromatic or aliphatic polyamides $AlCl_3$, $CaCl_3$, and $GaCl_3$ are used as a complexation agent (Vasanthan et al. 2004; Roberts and Jenekhe 1990, 1991).

And the last and most common approach for surface modification of polymer fibers by wet chemical approaches is the layer deposition method. In this method, most commonly, the grafting of siloxane and silanes is used to increase adhesion between aromatic polyamide (PA) fibers and in composites or functional coatings (Jia et al. 2020). There are several reports published to improve the adhesion capacity of aromatic PA fibers. Recently Jia et al. reported a grafting process in aromatic PA initiated by gamma irradiation using 3-aminopropyltriethoxysilane and 1,4-dichlorobutane. They observed the formation of siloxane bonds between 3-aminopropyltriethoxysilane and substrate as a result of the grafting of 1,4-dichlorobutane and its hydroxylation (Jia et al. 2020). Another process suitable for direct silane grafting is the sol–gel method (Landsiedel et al. 2020). For example, the sol–gel method in an aqueous solution using acidic prehydrolysis of alkoxysilane (TESPSA: (3-triethoxysilylpropyl)succinic anhydride) leads to the formation of a layer of porous polysiloxane with evenly dispersed hydroxyl and succinic acid end groups (Landsiedel et al. 2020). The layer deposition is also achieved via a reaction between polymer fibers with different types of amines, for example, chitosan, ethylenediamine, or dopamine (Yu et al. 2015; Wang et al. 2013). Several reports also focus on self-polymerization and deposition of dopamine on the surface of the polymer to improve the adhesion property. For example, Wang et al. reported a successful method for polydopamine deposition on the aromatic PA, which was further verified by comparing the experimental nitrogen-to-carbon (N/C) ratio (0.11) with theoretical N/C ratio (0.125) (Wang et al. 2013).

In the case of plastics, the sol–gel method is mainly utilized for the formation of organic–inorganic hybrid coating and UV-curable coating. This method has many advantages including the requirement of milder reaction conditions and flexibility in terms of formulation. UV-curable coating presents many advantageous features over traditional solvent-based coating systems in terms of eco-friendly and energy-saving curing methods since it is a solvent-free treatment (Schwalm 2006). Various types of coating material with diverse properties such as antibacterial, antistatic, corrosion resistance, and antifouling can be obtained from the sol–gel method (Sangermano et al. 2015). Hybrid organic–inorganic material is an interesting material where an organic phase and an inorganic phase are interconnected. There is a two-step method, which is hydrolysis and condensation reaction, involved in the formation of organic–inorganic hybrid coating, illustrated in Figure 5.7 (Malucelli 2016). This is a conventional bottom-up approach, where a metal alkoxide, $M(OR)_x$ (e.g., tetraethoxysilane), acts as starting reagent. The experimental conditions play an important role in the synthesis of these materials for the synthesis of composite structures with dispersed phase dimensions under 100 nm (nanocomposites). The morphology of these hybrid materials highly depends on various factors such as molecular weight, the number of the reactive functional moieties

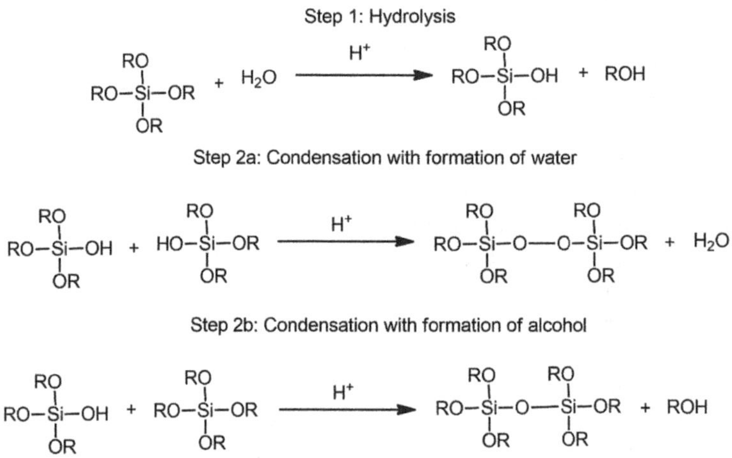

Figure 5.7 Hydrolysis and condensation method involved in sol–gel method. Source: Adapted from Malucelli (2016).

present in the organic polymer/monomer/oligomer, and polymer's solubility in the sol–gel medium.

5.4 Plasma Treatment for Polymer Surface Modification

Plasma treatment has attracted considerable attention due to its straightforward methodology; additionally, this method doesn't require any solvent or hazardous chemicals. An inert gas, such as nitrogen, oxygen, and hydrogen, is used in the surface plasma technique to dissociate and interact with the substrate to modify its characteristics like adhesion, wettability, and printability (Yasuda and Gazicki 1982; Zhou et al. 2010). A plasma is produced when a gaseous mixture comprising particles like neutral particles, UV photons, vacuum ultra-violet photons (VUV), radicals, ions, and free electrons interacts with an electric field or radiation with a low-to-high electric field of energy. Some of these particles got excited due to this collision, and when they move back to their ground energy state by photon emission, it causes plasma luminosity. Based on the gas temperature, plasmas can be divided into two categories, one is thermal and other one is non-thermal or cold plasma (Fridman 2008; Desmet et al. 2009). Since polymers are heat-sensitive materials, thermal plasma cannot be used for this method. So, in the case of polymer surface modification by plasma treatment generally nonthermal plasma is employed (Denes and Manolache 2004). Based on the plasma operation there is a wide range of electron densities and electron temperature (Fridman and Kennedy 2004). There are two types of plasma techniques mainly employed for surface modification of polymers; one is plasma polymerization and the other one is plasma ablation (Luo and Van Ooij 2002). In plasma polymerization polymeric coating is applied on the surface and in plasma ablation, the polymeric layer is removed from the surface. These two methods are described in Figure 5.8.

When the elementary particles present in the plasma (i.e., ions and radicals) interact with the polymer surface, different phenomena may occur. The interacting particles can be modulated by various factors like plasma operating parameters, reactor design, polymer substrate, and gas chemistry (Kulkarni 2019). The polymer surface is continually bombarded by excited plasma species, changing the chemistry and properties of the surface morphology and/or structure (Latag and Vasquez 2018). The interaction between the plasma particles and polymer surface may lead to a variety of surface phenomena, including surface functionalization, roughness/etching, and crosslinking. These surface phenomena are shown in Figure 5.9. All of these processes have an impact on how well the polymer adheres to different substrates, whether they occur in isolation or a synergistic manner (Kulkarni 2019). In this section, various methods associated with surface modification by plasma treatment are discussed (Yoshida et al. 2013).

5.4.1 Addition of Functional Group on Polymer Surfaces

Plasma treatment is a widely used method by polymer scientists to add new functional moieties on the polymer substrate. Typical plasma treatment by

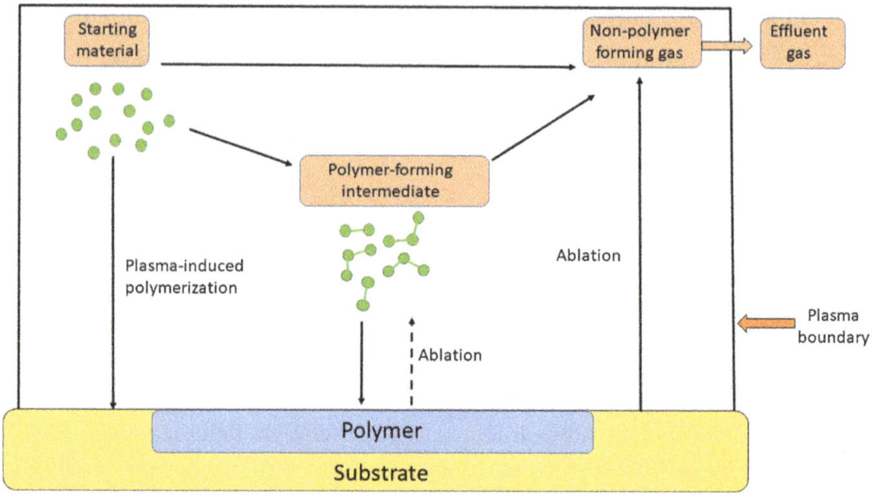

Figure 5.8 A schematic diagram of competitive polymerization–ablation technique. Source: Yasuda (1985)/with permission of Elsevier.

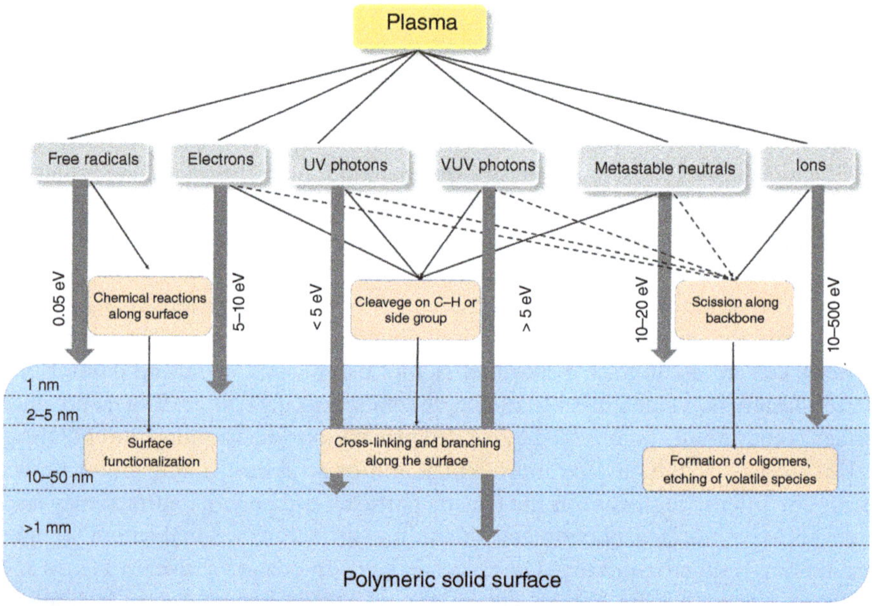

Figure 5.9 Different plasma species interacting with a polymeric substance. Source: Adapted from Taaca et al. (2022).

ammonia, oxygen, or with air can produce amine or carboxyl groups onto the polymer surface (Kull et al. 2005; Sanchis et al. 2006; Lai et al. 2006). So, in order to introduce various functional moieties on the surface of the polymer, it is very important to carefully choose the type of plasma source to change various properties like surface adhesion properties, biocompatibility, and wettability of the polymer material (Grace and Gerenser 2003; Hegemann et al. 2003). Further, this method can be used for the subsequent covalent immobilization for a different type of bio-polymers (Sathish et al. 2019). For example, it was discovered that plasma treatments could immobilize a variety of extracellular matrix protein components including laminin, gelatin, fibronectin, and collagens onto the polymer surface. This significantly increased the proliferation rate and adhesion force of the cells (Nardulli et al. 2010; Huang et al. 2007). Additionally, air or argon-based plasma treatment increases the hydrophilicity of various biopolymers (Luque-Agudo et al. 2021; Mozetič 2020).

5.4.2 Introduction of Roughness on the Polymer Surface

When the polymers are treated with plasma, it is well known that plasma etching and plasma cleaning occur on the surface of the polymer. Plasma etching is mainly employed for the microfabrication of integrated circuits. It is now a most popular and demanding area of plasma-assisted modification technique for scaling up the usage of polymeric material and applications of circuits. Plasma etching has also been explored specifically for improving the adhesion between organic and material surfaces, such as the surface of the cell by altering the biological surface's roughness and wettability, which is in direct contact with the surface of the material (Mirmohammadi et al. 2012). Furthermore, plasma etching would be a good contender for biomedical material surface modification since it allows for efficient surface micro-fabrication. In order to investigate the connection between surface topography and cell adhesion and proliferation, to date several specific surfaces were developed (Zhu et al. 2021).

5.4.3 Formation of Crosslinking

When a polymer material interacts with highly energetic particles like ions or radiations, that is, ultra-violet (UV), VUV, X-ray, and γ-rays, that could produce crosslinking on the polymer surface (Clark and Dilks 1977). When polymers, for example, polypropylene (PP) and polyethylene (PE), interact with activated plasma particles, which results in the formation of radicals on the polymeric surface. This leads to the radicals-induced crosslinking of the polymer chains and eventually the molecular weight of the polymer chain increases (Chapiro 1995). Moreover, crosslinking formation leads to the formation of a 3-D network structure on the polymer surface, making the polymer layer an insoluble substance (Molnar et al. 2020; Kolarova Raskova et al. 2018). Additionally, this molecular network further increases the mechanical strength of the polymer surface (Taaca et al. 2022). The plasma-initiated crosslinking during hydrogel synthesis is illustrated in Figure 5.10.

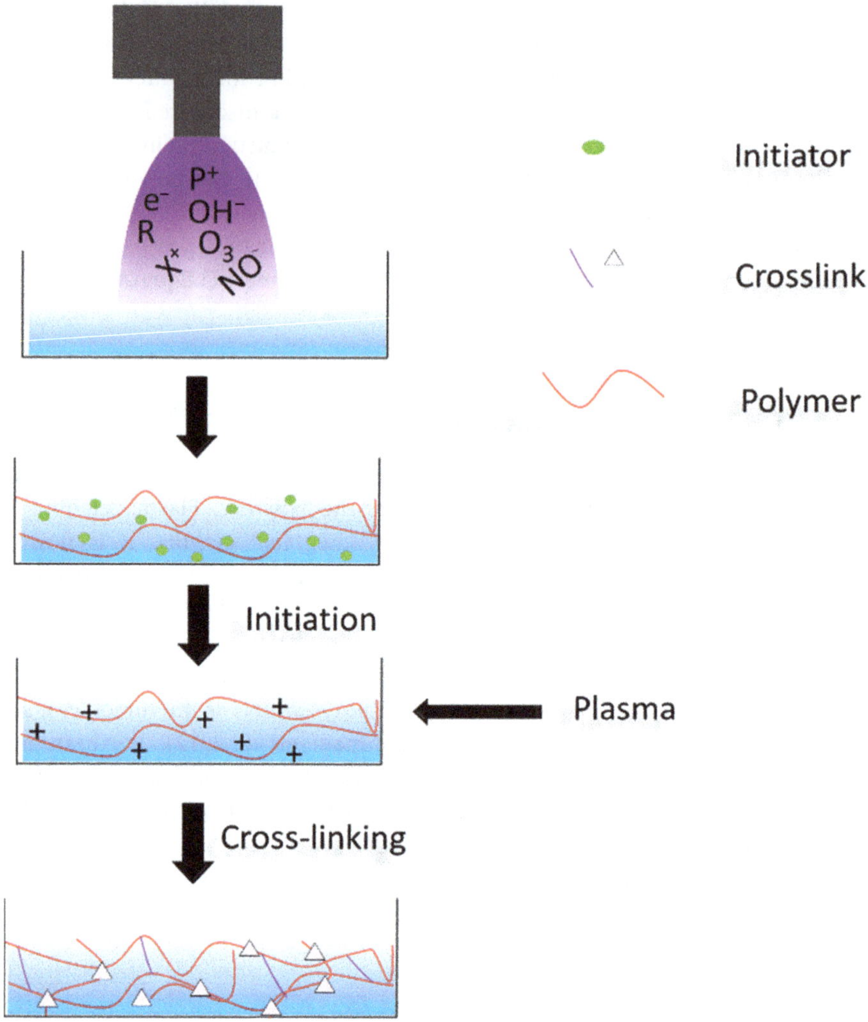

Figure 5.10 Formation of crosslinking due to plasma treatment. Source: Adapted from Taaca et al. (2022).

5.5 Corona Treatment for Polymer Surface Modification

The corona discharge is an immensely utilized technique to modify the top layer of polymeric materials before employing it for adhesive laminating, coating, printing, and metalizing. Surface treatment is used to create a functional group that has a high binding capacity. This allows interaction with various functional groups to form bindings (Gooch 2011b; Tuominen et al. 2010; Zenkiewicz 2008). Corona discharge is a process of non-local thermodynamic equilibrium (LTE) plasma that can be commenced in the atmosphere and air of several active and inert gases where it generates polar groups that are responsible for improving the wettability, adhesive properties,

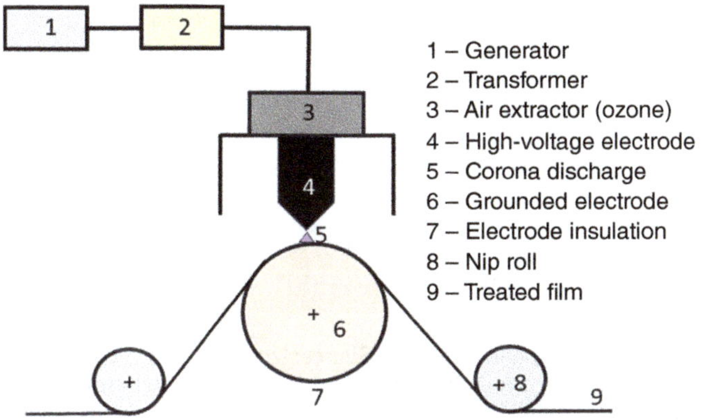

Figure 5.11 Schematic diagram of corona discharge activator. Source: Adapted from Chan (1999), Żenkiewicz (2001), and Izdebska (2016).

and surface energy (Tendero et al. 2006). The surface roughness is affected by the improvement of surface energy unintentionally. The surface energy alters surface morphology by varying the surface roughness (Zenkiewicz 2008; Tuominen et al. 2010; Gooch 2011b). Corona treatment has many applications in various fields of science. Applications include disinfection and sterilization of packaging materials in medical science, resin adhesives in the fabric industry (Brzezinski et al. 2009), and separation of emulsions (Sadeghi et al. 2013; Moghimifar et al. 2014), where ozone is used as a gaseous material. Applications of corona discharge are limited not only to two-dimensional materials but also to three-dimensional objects such as polymer fibers, pipes, and plastic materials (Zenkiewicz 2008). In the 1960s, the first corona discharge was implemented to modify polymer surface. It was employed in the modification of polyethylene (PE) surfaces (Kim 1996; Stepczyńska 2014; Żenkiewicz 2001).

The surface modification of the upper layer of polymers with corona discharge is a physical method mostly used in the industrial sector. The setup of the corona treatment consists of a transformer, a set of electrodes, and a generator, all these devices are known as activators (Figure 5.11). To furnish resistance to heat and ozone obtrusions, the electrode is coated with insulating materials such as ceramics, vulcanized rubbers, polyesters, glass fiber or resin, and epoxy coatings (Chan et al. 1999; Żenkiewicz 2001). The electrode is mainly a grounded electrode and a high-voltage electrode. High-frequency discharge serves as the basis for corona discharge. Corona discharges release oxygen atoms, oxygen free radicals, and ozone (Figure 5.12). These oxidants form oxidizing groups with the free radicals of the surface materials. They create oxidizing groups like ester, carboxyl, carbonyl, and hydroxyl. These electrodes have different potentials and are filled with gas under atmospheric pressures, similar to a partial discharge phenomena. The atmosphere becomes ionized as a result of discharges between the electrodes. As a result of the formation of ionized air (plasma), a blue mist, and light are emitted (Figures 5.11 and 5.12b). The emission of a source of light in the form of a single point and a frustum cone having a tiny cone close to the edge might also be attached to

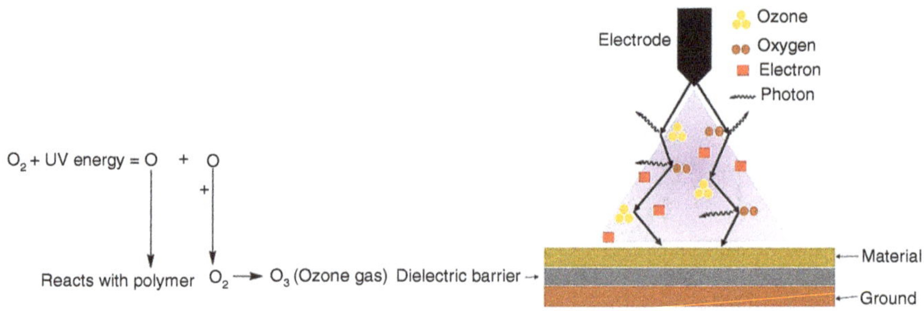

Figure 5.12 Schematic of (a) chemical phase reaction. Source: Izdebska (2016)/with permission of Elsevier and (b) set up of corona discharge in air. Source: Adapted from Goldman et al. (1985) and Izdebska (2016).

the corona radiation in a single point (Koller et al. 2004). High-speed electrons are discharged at the substrate, which is in an area between the grounded and high-voltage electrodes. The surface modification results from the breaking of the chemical bonds of material by hitting the material with the electrons having higher energy than its binding energy. The low-molecular-weight oxidized materials (LMWOM) can appear over the top layers of various polymers. If this is not firmly bonded to the substrate, this can result in issues in blocking and adhesion when the material is printed (Kusano 2014; Chan 1999; Tuominen et al. 2010). The breakage of the polymer chains is caused by the very active electrons (more powerful than ions). This disturbance generates chemical reactions that lead to crosslinking, degradation, and oxidation. It is because of the insertion of the polar groups, producing pits, and formation of electrets while removing infirm border layers. The Fiber and film breakages can be avoided by the little heat capacity (Zenkiewicz 2008; Ding et al. 2014). Studies on the pulsed corona demonstrate that corona discharge strongly relies on the polarity of the pulse and generates various sizes and forms of pennants. In the corona, a sizable portion of pennant fills the gap between the two electrodes. As a result, the streamers' length rises in an increase in the voltage, which might cause a glimmer (Stishkov et al. 2010; Zhang et al. 2008).

Some of the primary objectives of corona discharges are mentioned in the points below (Brzezinski et al. 2009; Ozdemir et al. 1999; Pekárek 2009; Ragoubi et al. 2012; Sparavigna and Wolf 2006; Stepczyńska 2014; Zenkiewicz 2008).

1. Oxidation of the top layer of the polymer surface and the emergence of the polar groups on the surface of the polymer.
2. Polishing the material's surface and enhancing the surface roughness and modification of the structure of the material owing to high surface energy.
3. Production of a suitable surface without affecting the mechanical characteristics of the entire material.
4. Creating a polymer suitable for graft polymerization.
5. In disinfection and sterilization of packaging materials in medical science.

The effectiveness of corona discharges on various polymers has been discussed in recent years. The polymers are mainly PLA, polypropylene (PE), PET, ethylene vinyl acetate (EVA), polystyrene (PS), and polyvinyl chloride (PVC). PE and PP are the two polyolefins most commonly implemented in the commercial sector. Surface energy for PP increases rapidly at $1.2 < E < 4\,\text{kJ m}^{-2}$ and chemical bonds begin to dissolve at $4 < E < 5\,\text{kJ m}^{-2}$. Following this, the structures start to change, causing roughness from ablation and the emergence of LMWOM (Cáceres et al. 2012; Guimond et al. 2002). LMWOM dissolves in inks with polar liquids to enhance the adhesiveness of the polymer films, while it becomes lower with nonpolar solvents (Zenkiewicz 2008; Strobel et al. 2003). The low surface energy of PE needs to be increased energy before printing (Novak and Florián 2001).

5.5.1 Modification in the Top Layer

The polymer's top surfaces are chemically modified because of corona treatment. During this corona treatment chemical and physical engraving occurs. Engraving leads to ablation and cleans the surface to remove the pollutants. Cleaning up enhances the adhesiveness of the material's surface. Moreover, ablation increases surface roughness and eliminates the elements that are weak in nature. (Kusano 2014; Marat-Mendes et al. 1992). After the corona discharge impulse the electrons are responsible for the conversion of particles of oxygen and water into O^- and OH^-. These negative ions interrelate with the positive ions to form a neutral substance (Strobel et al. 2003; Strobel and Lyons 2003).

The relative humidity in the air, the gas amount, and the corona discharge polarity all affect in formation of ozone (Kusano 2014). Ozone affects the surface by creating cuts or fractures, which are released during the discharges (Gooch 2011a). Chemical bonds of the polymer chain (C–H and C–C) break because of electron collision (Figure 5.13) and generate free radicals. These free radicals participate in the crosslinking, unsaturation, recombination, and branching. This crosslinking results in the enchantment of the bond strength of the polymer chain (Kusano 2014).

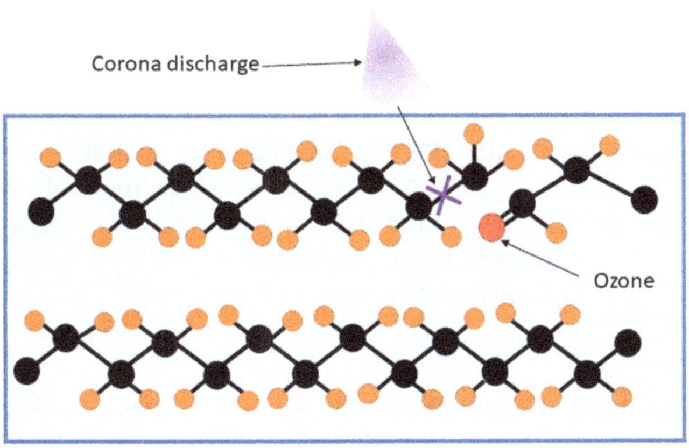

Figure 5.13 Breakage of chemical bonds in the chain of polymer because of corona treatment causes a change in the material's top layer. Source: Adapted from Izdebska (2016) and Kusano (2014).

5.5.2 Factors Affecting the Course and Effectiveness of Corona Treatment

5.5.2.1 Parameters Controlling Optimum Treatment

The quantity of energy carried by ions significantly impacts the corona's effectiveness. These ions are the function of the speed of material shifts, the function of power, and the length of discharge electrodes (Moraczewski et al. 2014). Nevertheless, having fewer free electrons, positive corona discharges are favored over negative discharge for surface treatments because they produce less ozone regarding polarity (van Veldhuizen and Rutgers 2001). The thickness of the polymeric films has a notable effect in the treatment of polymers containing lubricant, thickset materials demanding high energy and endeavor to go through significant surface transformations. Less susceptible to surface treatment are materials with increased surface energy as a result of more stretch. Another essential aspect affecting the efficiency of the treatments is humidity. The polymer's roughness varies even at low applied voltages with increased humidity (80%), highlighting the significance of moistness regulation in the surface corona treatment (Chan 1999).

5.5.2.2 Remedies and Causes of Treatment by Reverse Lateral

The polymers are in porous essence, and the air may become trapped between the corona roller and the polymer films. That subsequently experiences ionization at both high and low levels of power. The corona rollers are glazed in silicone latex, have a small range of elasticity, and are prone to dents and grooves. It is because of the venture effect, which is rapidly operating film carriage air multitudes with a lower and upper side at low power. On account of localized heating brought on by high temperatures, non-oriented polypropylene (PP) is observed to exhibit persistent teardrop fluctuations at higher power. The treatment roller can be cooled by dispersing the power across the sizable surface areas, and the overwarming of the substrates may be constructively controlled.

5.5.2.3 Effects of Additives in Corona Treatment

In corona treatment, various stabilizers are incorporated into polymers for inhibiting the oxidation, preserving the tincture, and maintaining the melting indexes like biphenylene-diphosphonite (P-EPQ), irganox-1010, and tris-nonyl phenyl-phosphite (TNPP). Where irganox-1010 and butylate hydroxy toluene (BHT) are free radical scavengers' primary antioxidants while P-EPQ and TNPP are used to decompose hydrogen peroxide (Sun et al. 1999). Irganox-1010 films have no carbonyl groups, while P-EPQ films have relatively few carboxyl groups.

5.5.2.4 Treatment Time

The production of carbonyls occurs throughout the corona treatment initially in an excessively rapid reaction, and then, in a subsequent long reaction, these carbonyls are converted to the ether groups. The issue of warmth fastening on the material will result from immoderate treatment, which will result in nonpolar ethers with low surface tension. Additionally, it has been observed that excessive treatment degrades polyolefin's wettability and adhesion because of bumps. However, very little treatment is not enough to get suitable surface energy (Kim et al. 1971).

5.5.2.5 Aging

The rate of degradation of the treated film's surface energy with storage relies on the mobility of the polar groups, the ambient temperature, and humidity. This degradation is most likely caused by their reversion and diffusion. In contrast to PP, which aged the fastest, polyimide (PI) and PE were shown to be more resistant to aging, while polystyrene and PET exhibited modest stability (Klemberg-Sapieha et al. 1993).

5.6 UV Treatment for Polymer Surface Modification

Light-sensitive polymers are cured using UV treatment. It is because it enhances the surface adhesiveness (Kraus et al. 2016; Heitz et al. 1996) and functionalization of the surface by photosensitizer/photoinitiator (Shin et al. 2016). A photochemical reaction is instigated by UV light onto the surface of the photo-polymers. This photochemical reaction creates a crosslinked chain of polymers (Figure 5.3) (Carroll et al. 2011). The main objective of UV treatment is to solidify the substance during the course of fabrication. It is not to change its surface properties. The UV treatment on a polymer surface is only useful for the already-hardened surface of the polymer to vary the properties of adhesiveness and wettability.

In UV treatment, polymerization is responsible for curing the surface instead of evaporation since this is a solvent-free, low-temperature, and high-speed process (Figure 5.14). It is a phase in the fabrication process for polymer networks (Decker 2002), imprint lithography (Fader et al. 2012), and polymer composites (Decker 2002). In UV treatment, the photoinitiator plays a crucial role in the rise in concentration. The intensity of irradiation increases the degree of the conversion of

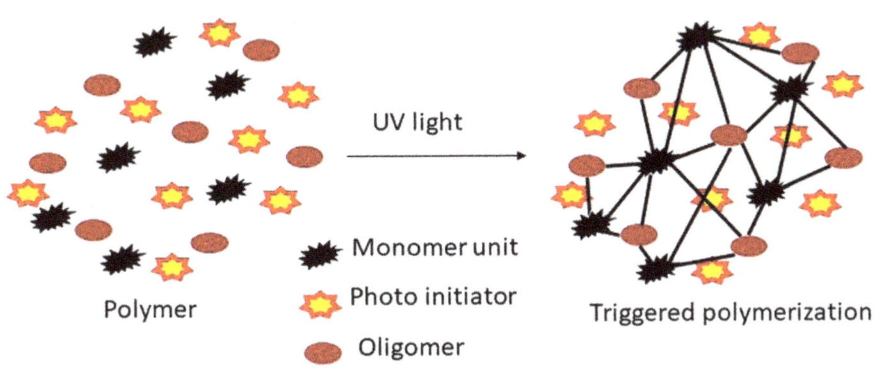

Figure 5.14 UV light treatment for the modification of the surface of the polymer with the photoinitiator. Source: Adapted from Nemani et al. (2018a), Fader et al. (2012), and Decker (2002).

polymerization. It is because of the existence of free radicals on the upper layer of the polymer. The polymer surface having thick layers shows a low polymerization conversion rate because of the lower penetration of the UV light into the layer. The depth of the polymer surface and intensity of UV light have a crucial role in the last luster of UV treatment. The material is cured at high intensities of UV light and leads to a higher value, but this can be lowered by enhancing the thickness of the surface. The conversion rate of polymerization is decreased by increasing the oxygen in the air. This is because the creation of peroxy radicals results from the scavenging of free radicals by the oxygen molecules. These peroxy radicals are not reactive enough to activate the double bonds of acrylate to participate in polymerization (Studer et al. 2003).

Once the sample is unveiled in front of UV light, it releases free radicals by oxidizing the polymer. These interact with the ozone or oxygen in the air resulting in hydrophilic carboxyl and carbonyl species. The degree of modification of the surface may be managed by maintaining the solvent, the photoinitiator, the irradiation period, and the concentration of the monomer (Liu et al. 2016). Polymers' low surface energy leads to poor adhesive quality, which restricts their usage. Enhancing the adhesive property (hydrophilicity) of polymers using a UV-ozone (UVO) or UV oxidation process has been proven to be successful (Liu et al. 2016; Mathieson and Bradley 1996). UV treatments are utilized to promote cell adhesion and proliferation in the biological aspect (Formosa et al. 2008). To increase the adhesiveness between the metal films and the polymer substrates at the nanoscale, Liu et al. (2016) employed the UV-ozone treatments. They were able to create metal–polymer composite films, which could bear prolonged liquid penetration for the extended periods of time (Figure 5.15). Grafting is the technique of covalently bonded monomers on the polymer's surface. Wang (2006) used UV grafting to enhance adherence to PE. Hu et al. (2002) described a single-step process for modification of PDMS surface using UV grafting.

For biomaterial usage, Yang and Rubner (2002) employed thermal/UV curing to stabilize polyacrylamide (PAAm) and poly(acrylic acid) (PAA) multilayers at high pH. At high pH levels to make the PAA/PAAm multilayers more stable, heat energy is employed for creating thermal crosslinking inside them via a

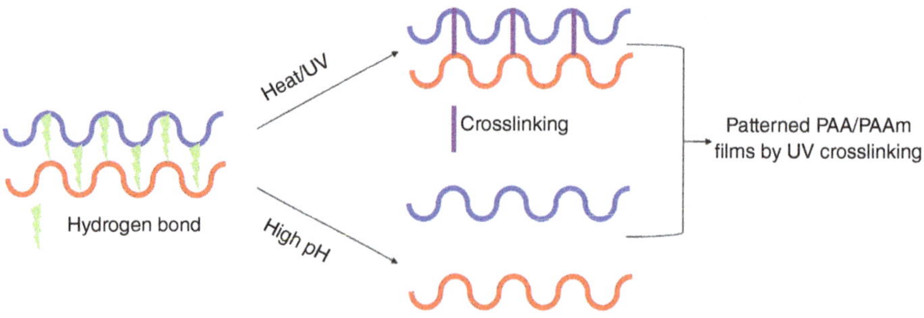

Figure 5.15 Schematic diagram of crosslinking of hydrogen bonded multilayers for the formation of polymeric films by UV treatment. Source: Adapted from Yang and Rubner (2002) and Nemani et al. (2018a).

thermal imidization reaction. UV light is employed for generating photo-crosslinks inside the PAA/PAAm multilayers. It is accomplished by the existence of a photoinitiator-labeled poly(acrylic acid) copolymer (PI-PAA) (Figure 5.15). Its controlled and targeted stability enables water-based subtractive patterning method such as photolithography and inkjet printing for producing multilayer patterned thin films. These films may be utilized for regulating the adherence of cells of mammalian and building light-emitting patterned devices.

5.7 Surface Patterning Treatment for Polymer Surface Modification

Patterned surfaces of the polymers are used widely in the domains of optics, electronics and biomedicine. An organic display was developed that showed luminescent character on applying electric current. They used structured, light-emitting conducting polymers for displaying this vivid full-color visuals (Shimoda et al. 2003). Compared to the original polymer surface, laser-induced periodic surface structures (LIPSS) exhibit better gluing capabilities. Some researchers created inexpensive polymer substrates for optical sensors using the LIPSS approach to lessen the amount of metal required (Rebollar et al. 2012). On the other hand scientists have also demonstrated that stable ink solutions may be used to print higher resolution, structured, and multicolored thin films of luminous colloidal quantum dot (QD) polymer composites, adding to the development of a reliable and effective device architecture (Wood et al. 2009). Their attributes can be adjusted via tuning the thickness of the composite layer of the QD polymer. The polymer surface has showed improvised photoluminescence efficiency, but its electrical capabilities were intact (Aldred et al. 2005).

Block copolymers can be utilized to construct patterned polymer structures as well as magnetic and metallic pillars or arrays for data storage devices (Park et al. 2001). Printed polymeric surfaces with patterns are used in organic light-emitting diodes (OLEDs) and organic thin-film transistors (OTFTs). Likewise, structured nanocrystals are used in memory devices because of their promising durability and longevity features. Patterned polymers have also been used in optics, creating masks, and templates (Julthongpiput et al. 2007; Valkama et al. 2004).

5.7.1 Instability-Induced Polymer Patterning

This patterning technique utilizes the in-built qualities of the polymers that are classified into two kinds: the metastable films and the unstable films. The unstable films have instability mainly because of their inherent chemistry. The metastable films have instability mainly because of external forces like thermal, electrical, and some forms of automated forces (Müller-Buschbaum et al. 2005; Rodríguez-Hernández 2015). Spinodal dewetting is one type of instability that is triggered by electrostatic forces or van der Waals forces. Homogeneous and heterogeneous nucleated dewetting are the two types of the instabilities caused by substrate deformations. When external forces act on thin liquid polymer films, they can cause changes in charge density, surface tension, or buoyancy that lead to instability in the metastable films (Bodenschatz et al. 2000). These cause the surface to get deformed and cause localized liquid flow. The thin film gets ruptured and causes patterning. The different types of molecular interactions or forces along with surface properties rule the morphology and amount of patterning (Mukherjee et al. 2011).

5.7.2 Patterning by Dewetting

The free energy of the system controls the propagation of unstable (or metastable) films to a steady state, which typically results in dewetting which means breaking of the film generating random patterns (Xie et al. 1998; Higgins and Jones 2000; Peter 2003). Whenever this unwanted phenomenon manifests itself on uniform surfaces, bi-continuous structures, droplets, random holes and polygons have been seen, among other things (Reiter 1992). Thin-film polymers subjected to dewetting processes have demonstrated morphological growth, self-organization with subsequent pattern generation (Xing et al. 2007; Bäumchen and Jacobs 2010; Xu et al. 2017) (Figure 5.16). The polymers that have the property of interconverting between liquid and solid states can undergo dewetting process. They interchange between the states by merely adjusting the temperature above or below the glass transition point. This process, which begins with the production of holes, eventually enlarge, merge, and produce isotropic droplets, is controlled by the initial conditions of the film-like film thickness (Hyun et al. 2016).

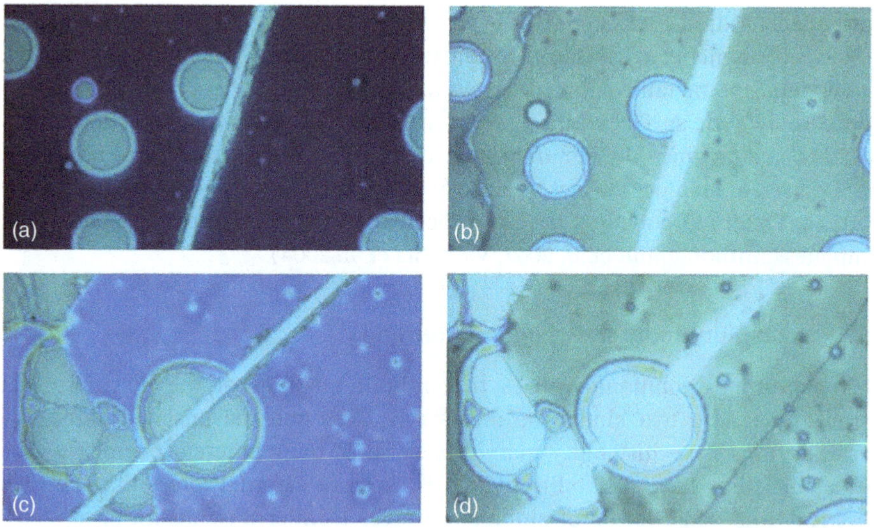

Figure 5.16 Surface morphologies of PMMA films/PS films/Si substrate after dewetting. (a,c) Before etching via acetic acid and (b,d) after etching. Source: Xu et al. (2017)/Royal Society of Chemistry/CC BY 3.0.

The uneven distribution of structures, short-range order, and poor resolution of generated structures are some of the drawbacks of the polymer-patterning approach (Hyun et al. 2016). Another drawback is that the spontaneous dewetting method is plausible in extremely thin, thermodynamically unstable polymer films (Mukherjee et al. 2011). Over these years, patterned structures have found applications in optics, electronic devices and storage devices etc.

5.7.3 Patterning by Evaporation

It is possible to create distinctive patterns by letting non-volatile substances assemble after solvents from thin polymer films have evaporated (Bodenschatz et al. 2000; Khanal and Zubarev 2007). Large-scale evaporation has been used to create hierarchically built amphiphilic diblock copolymer micelles (Li et al. 2014). Like the dewetting process, nonuniform dispersion in the evaporation procedure may result in uneven formations (Watanabe and Yanagisawa 2022) (Figure 5.17). Rigid geometry has been applied to pattern the polymer surface in order to get around this problem (Tao et al. 2008). Biodegradable polymers have been utilized to create honeycomb-like porous polymer films by means of an evaporation-induced self-assembly technique in a humid environment (Kawano et al. 2012). The micro-topography of these honeycomb polymer films for particular diameter encourages adhesion and managed cellular shape.

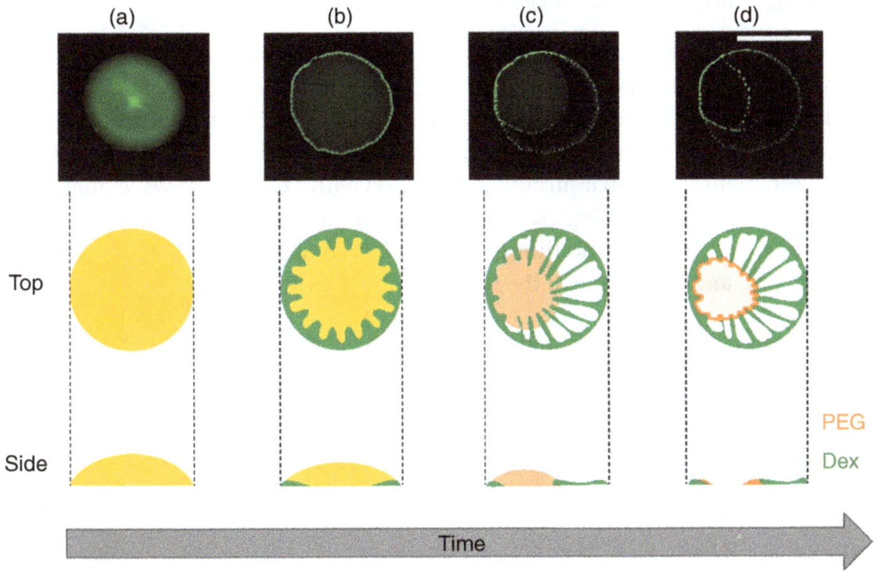

Figure 5.17 Representation of the creation of an evaporation pattern over time. (a) Homogeneous solution immediately following the droplet's deposition, (b) phase separation progressing along the contact line, (c) string production as the PEG (poly(ethylene glycol))-rich phase moves toward the droplet's core, and (d) full drying. Source: Watanabe and Yanagisawa (2022)/MDPI/CC BY 4.0.

5.7.4 Patterning by Electric Field Gradient

Whenever an external electric potential is employed to the polymer film sandwiched amid two electrodes, the film will be disrupted, leading to the creation of patterned structures (Hyun et al. 2016; Cargill et al. 2011) (Figure 5.18). Patterned electrodes with regulated electric fields are used for highly ordered arrays (Deshpande et al. 2004). Further, if higher electric field with high temperature is used across the polymer film, then it can produced micro- or nanosized patterned features (Schaeffer et al. 2000). Also, electrostatic forces acting between polymer and air interfaces can cause capillary and thermal variations. This can create sideways pressure and form protrusions. The ultimate outcome depends on the electric field, film width, air–polymer interface, and the film surface (Mukherjee et al. 2011; Schaeffer et al. 2000). The viscosity has a considerable impact on how long it takes to generate a patterned structure (Wu and Russel 2009). Some research show that the difference in viscosity in polymeric bilayers can cause instability in the electric field which results in uneven patterning (Xu et al. 2016).

This process of patterning technique using electric field is a very popular and is industrial friendly (Liu et al. 2013). Using this method, polymeric organo-rods that resemble gecko nanohairs in terms of hydrophobicity and improved adherence have been produced (Northen and Turner 2005). Due to the electric halt, the pattern resolution produced from variation of electric-field instability is constrained (Liu et al. 2013; Dickey et al. 2008).

5.7.5 Patterning by Thermal Gradient

For electric field patterning, electric filed is used. Similarly, for thermal gradient patterning, heat gradient is used to destabilize the surface of the polymers (Schäffer et al. 2003). In this procedure, polymer film is placed amid the substrate and the top plate having different temperatures with little air spacing dividing them. This gradient between them results in a surface deformation with a distinctive wavelength that gives rise to various morphologies. The exact morphology is determined by the initial film conditions, including the film thickness, the air spacing, the temperatures of the two plates, and the degree of the thermal gradient (Mukherjee et al. 2011).

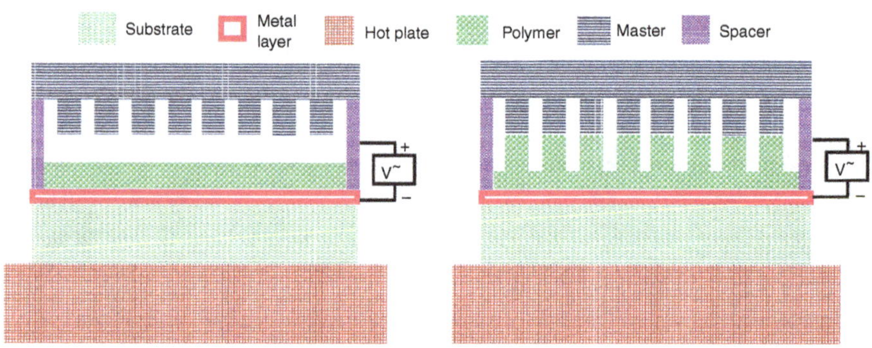

Figure 5.18 Schematic illustration of the patterning technique by electric gradient. Source: Adapted from Cargill et al. (2011).

5.7.6 Photolithography

Photolithography is currently an economically advantageous fabrication technique that can be utilized to generate surface relief features on micro- and nanoscale level in polymer industries without actually losing their inbuilt characteristics. As the name suggests, this technique uses light to transfer geometric pattern on polymer films which has chemical, electrical, and other response to light (Handrea-Dragan and Botiz 2021) (Figure 5.19). This process then dissolves a chosen portion of the film in the right solvent and removes it from the film. The radiation causes photo-crosslinking, functionalization, photopolymerization, and decomposition (Nie and Kumacheva 2008). Nevertheless, large areas with high resolution cannot be patterned by this technique (Acikgoz et al. 2011). For the fabrication of textured surfaces via photolithography, a polymeric film or a template is required, and the setup is costly.

Light-emitting displays (LEDs) (Müller et al. 2003), sensors and actuators (Hoffmann et al. 1999), polymer-dispersed liquid-crystal displays (LCDs) (Penterman et al. 2002; Kato et al. 1994), three-dimensional photonic crystals (Campbell et al. 2000; Moon et al. 2006), and optical components (Wu et al. 2002), and are all made using photolithographically patterned polymers. It is evident that one of

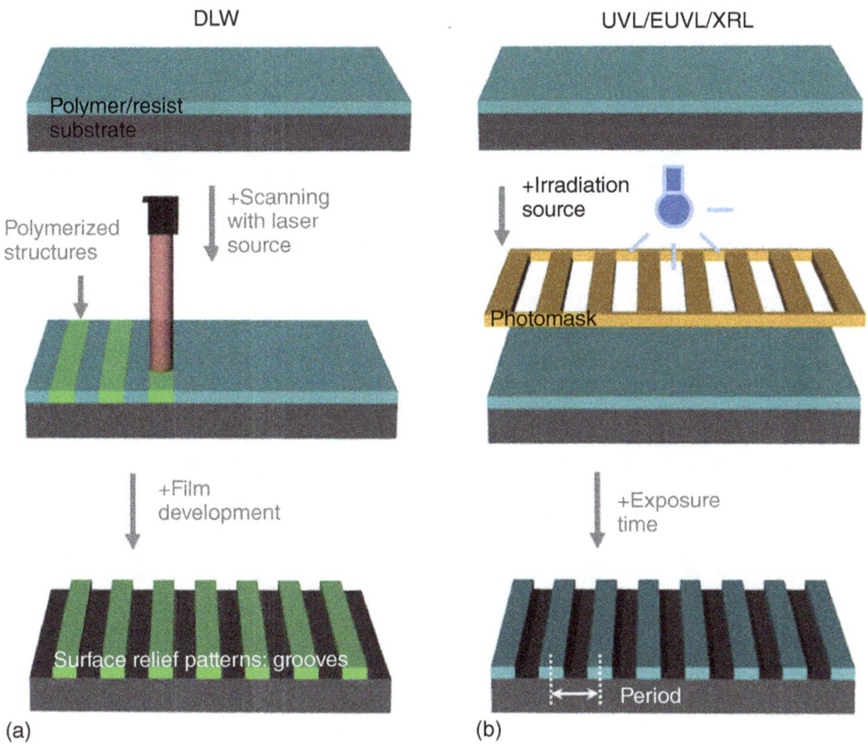

Figure 5.19 Pictorial illustration of different photolithographic systems: (a) direct laser writing and (b) UV, extreme UV, and X-ray lithographies. Source: Handrea-Dragan and Botiz (2021)/MDPI/CC BY 4.0.

the key prerequisites for innovative creation of applications in different areas of nanotechnology is the production of comparatively tiny structured features and customized properties with good picture clarity. Patterned surfaces made using this method find applications in biomedical devices (Tian et al. 2014), cell growth (Castillejo et al. 2012), or organic optoelectronics (Su et al. 2019; Yun et al. 2019) to name a few examples.

5.7.7 Patterning by Block Copolymers

Block copolymers (BCPs) are made through carefully controlling polymerization of one monomer, then extending the strand with a discrete monomer to generate copolymers (Matyjaszewski 2002). Correspondingly, blocks of two or more polymeric chains that remains chemically distinct and covalently bonded to one another are created side by side (Segalman 2005). Instead of being distributed randomly or alternately across the polymer chain, distinct monomers are clustered in different homogeneous regions of the polymer strand (Kim and Hinsberg 2008). The simplest method to direct pattern alignment in a BCP thin film it to equalize domain surface energy which is achieved by spin-coating a random copolymer with identical monomer units between substrate and BCP (Figure 5.20). Phase separation occurs, producing micro- and nanoscale patterned patterns, as a result of the different affinities of these block copolymers toward the interface (Xue et al. 2012). The film thickness and composition, its temperature and level of block incompatibility (such as the Flory–Huggins interaction parameter) all affect the patterned features (Kim and Hinsberg 2008; Nie and Kumacheva 2008). After the patterns are created, mostly annealing is conducted. The size of these copolymers is customizable and can be organized into ordered displays of lines and dots. BCPs are utilized as blueprints for the creation of high-density information storage medium, ceramic nanodots and wires, and nanoscale metal (Lodge 2003). The BCP method's use is constrained by its high cost and low likelihood of producing a sizable region of defect-free patterns (Xue et al. 2012).

5.7.8 Microcontact Printing

It is a very successful method for patterning large polymer surfaces and small pattern range (Qiu et al. 2021) (Figure 5.21). Microcontact printing is a simple

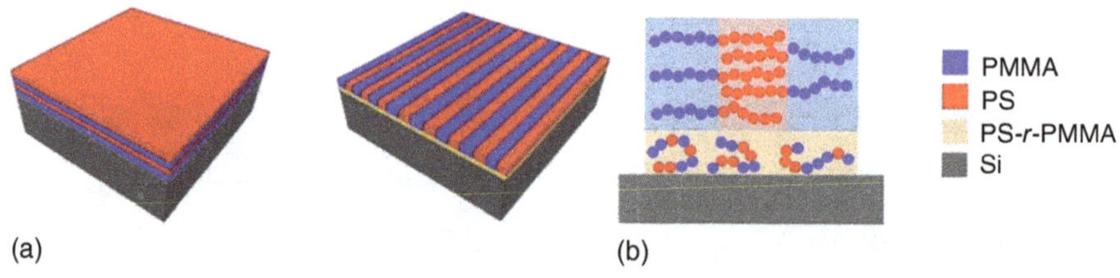

Figure 5.20 llustration of the self-assembly of lamellar PS-*b*-PMMA (polystyrene-block-poly(methyl methacrylate)). (a) Self-assembly of the lamellae and (b) a copolymer brush (PS-*r*-PMMA) forcing the generation of vertical PS and PMMA lamellae.
Source: Pinto-Gómez et al. (2020)/MDPI/CC BY 4.0.

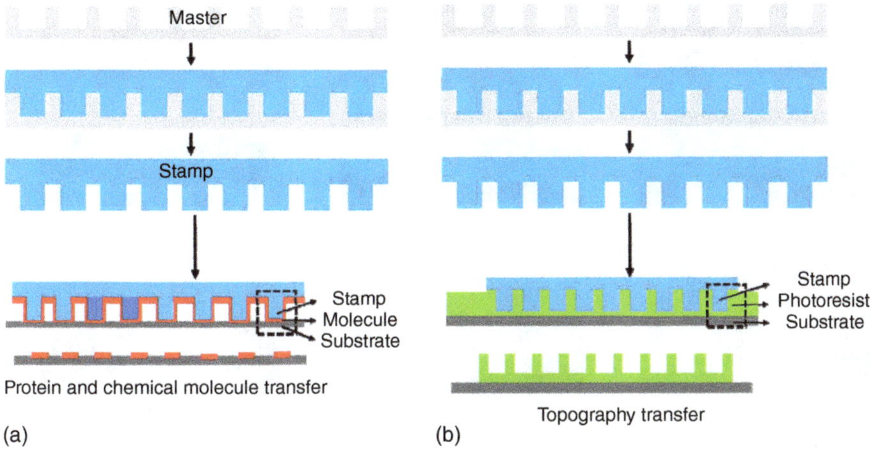

Figure 5.21 Schematic representation of the microcontact printing (μCP) fabrication. (a) PDMS (polydimethylsiloxane) stamp creation and molecule transfer or (b) PDMS stamp creation and topography transfer. Source: Qiu et al. (2021)/with permission of Elsevier.

and economical process where the polymer is transferred to the substrate using an elastic or rigid polymeric stamp with relief features. Due to its transparency, elasticity, and capacity to create conformal contact with difficult surfaces, PDMS is the most used microcontact stamp (Kaufmann and Ravoo 2010). The benefits of this technique are shorter processing periods, minimal ink consumption, and the potential to produce high-resolution patterns on large surface areas. The fabrication of organic electronics, LEDs, circuits, electronic sheets, and computer chips mostly uses the microcontact printing technology (Li and Guo 2006; Menard et al. 2007).

5.7.9 Nanoimprinting Lithography

Nanoimprinting lithography is another patterning technique that creates patterned structures using mechanical deformation of the resist materials. This technique is processed under regulated temperature and pressure. Later, curing method is carried out either by heating thermally or by exposing to UV radiations (Hongbo 2013). One of its major advantage is that it involves a simple equipment with short processing period which makes it a cost-effective approach and provides high-resolution patterns (Chou et al. 1997; Hua et al. 2004). Nevertheless, high viscous polymers pose a challenge for this technique because high temperature can make the polymer shrink during curing process. Recent reports show that this technique was used onto curved lens which made it waterproof and antireflective without altering its inherent chemical composition (Dickson et al. 2017; Handrea-Dragan and Botiz 2021) (Figure 5.22).

5.7.10 Laser Surface Texturing (LST)

A quick and ecologically responsible way to alter the surface of polymers, laser surface texturing (LST) offers good control over the size and shape of the details

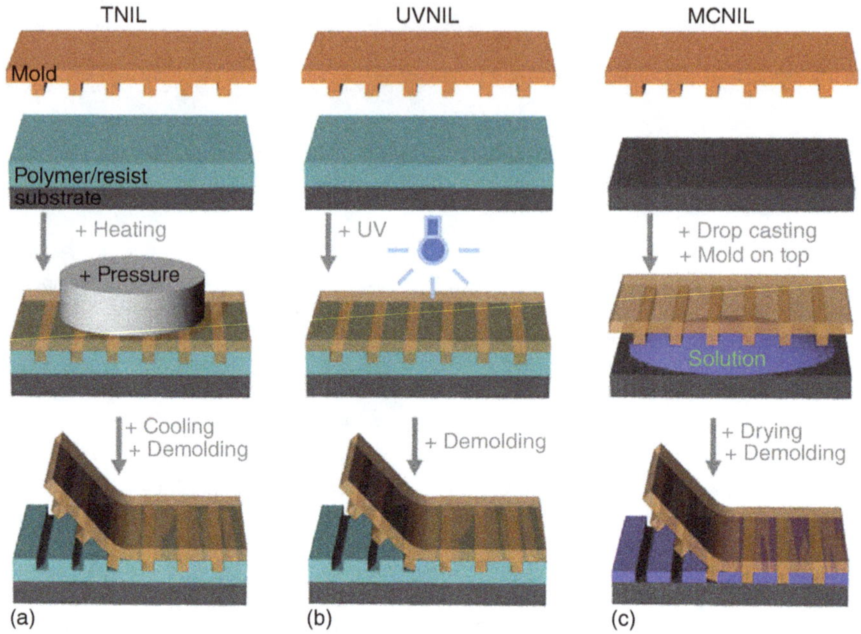

Figure 5.22 Schematic illustration of different nanoimprint lithography (NIL) methods: thermal nanoimprint lithography (TNIL) (a), UV nanoimprint lithography (UVNIL) (b), and molding in capillaries nanoimprint lithography (MCNIL) (c). Source: Handrea-Dragan and Botiz (2021)/MDPI/CC BY 4.0.

which enables the production of ideal patterns (Garrison and Srinivasan 1985; Riveiro et al. 2018) (Figures 5.23 and 5.24). In order to perform LST, a targeted laser beam must be directed at the polymer's surface, where radiation hits the upper layer and produces heat. The surface is altered as an outcome of the heat. Moreover, this heat causes melting and vaporization of the surface. Additionally, if the laser beam is very intense, the chemical bonds on the surface will be disrupted. Even such high-intensity beam also causes modification of the polymer surface. Comparatively, UV radiation is chosen for LST operations over IR radiation. It is because the IR light can cause ionization and breakdown polymers without melting. LST may polarize polymers and change their surface properties, considerably improving the polymer's hydrophilic or hydrophobic character (Garrison and Srinivasan 1985).

One of the best methods for texturing polymers with excellent quality, throughput, and flexibility at the nanoscale scale is direct laser interference patterning (DLIP) (Alamri and Lasagni 2017). As soon as two or more coherent laser beams collide on the polymer surface, DLIP results. This phenomenon can be controlled by adjusting the pitch, sample geometry, and period. The use of DLIP could improve the formation of nanorods, pattern graphene oxide, create organic solar cells, and many other things (Alamri and Lasagni 2017; Müller-Meskamp et al. 2012). For consistent findings when employing the DLIP approach, it is crucial to carry out studies under standardized, controlled settings (Alamri and Lasagni 2017).

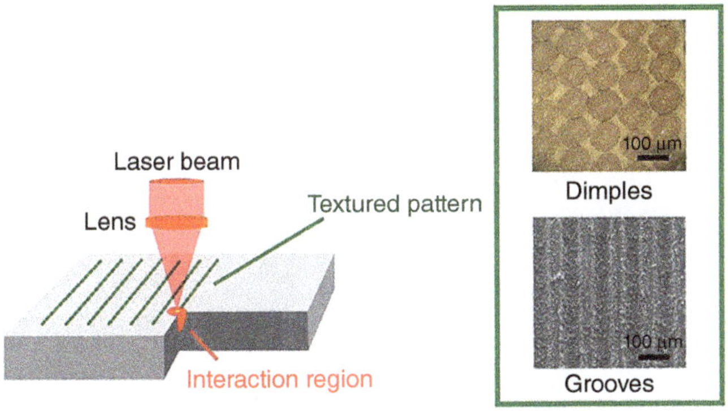

Figure 5.23 Schematic diagram for operating of laser surface texturing (LST). Source: Riveiro et al. (2018)/Frontiers Media S.A./CC BY 4.0.

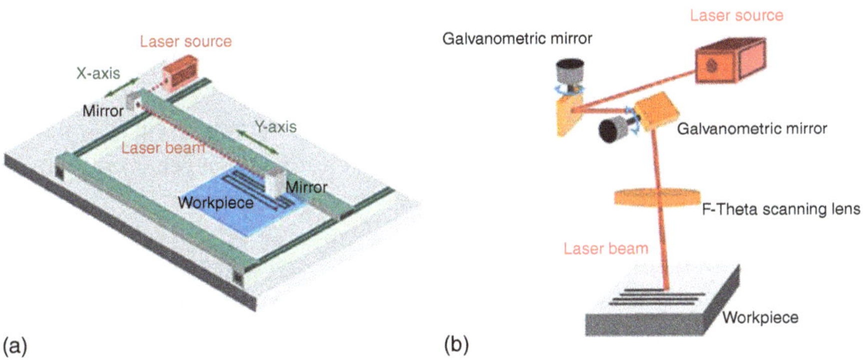

Figure 5.24 Schematic representation of laser-surface-texturing method (a) cartesian, or (b) a scanning system. Source: Riveiro et al. (2018)/Frontiers Media S.A./CC BY 4.0.

Laser surface modification is a low-cost method that can be employed in mass production (Zhai et al. 2011). LST has the ability to alter the qualities of a material in order to advance its adhesion properties (Kromer et al. 2018), increase its load-carrying capacity (Brizmer and Kligerman 2012), modify its friction coefficients (Schille et al. 2015), and lower its wear rate (Jones and Schmid 2016).

5.8 Thermal Annealing Treatment for Polymer Surface Modification

Thermal annealing is a typical procedure for strengthening a solid, such as glass or a metal by increasing, maintaining, and then gradually lowering its temperature. Though it is common, it is an equally important method for improving the surface properties in interwoven polymeric fibers and thin-film polymers. Surface modification through annealing finds its uses in the electrical, clothing, and packaging industries. Additionally, its utility as a post-processing step to stop polymer matrix

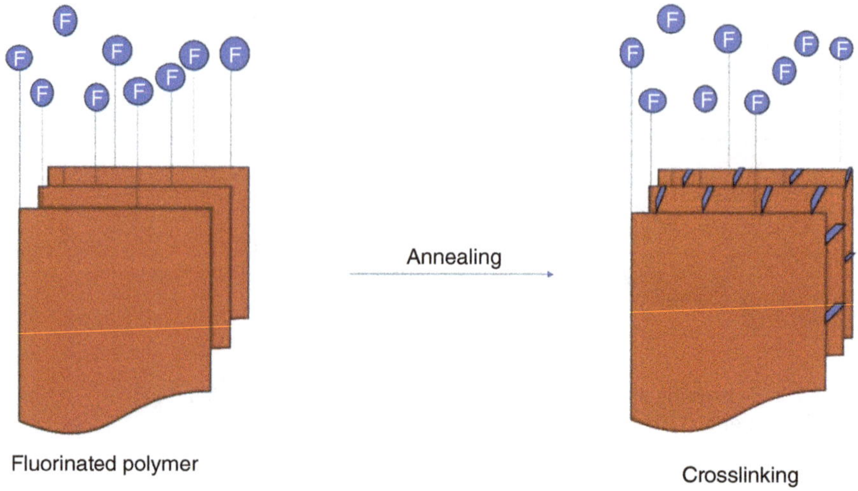

Figure 5.25 Annealing leads to crosslinking in fluorinated polymer. Source: Adapted from Gu et al. (2017).

surface rebuilding and degradation (Yagüe and Gleason 2013) is worth mentioning. The material needs to be subjected to annealing to increase the mechanical strength of polymers. Electrospinning in polymers increases its mechanical strength (Li et al. 2013; Gu et al. 2017) (Figure 5.25).

In annealing, the substrate is put in the oven and undergoes some baking processes where the constant heating causes the polymer fibers to go through morphological changes in their structure. Due to the fusing and overlaying of the delicate fibers, this causes a vital change in the structural qualities, such as the integrity of the membrane (You et al. 2006; Burger et al. 2006; Lee et al. 2008). Additionally, annealing can change the molecular structure and result in high crystallinity (Migliaresi et al. 1991), which raises the elasticity of Young's modulus (Cho et al. 2011; Tan and Lim 2006). Annealing also prevents partial dewetting of the polymer films and induces crosslinking of fluorinated polymer species. In conductive polymers annealing enhances the stability of the host species (Nemani et al. 2018b) by increasing the conductivity during the stabilization of the wettability of chemical vapor deposition (CVD) polymer films (Zhao et al. 2017; Sojoudi et al. 2017). The bulk characteristics of the polymer are significantly altered by annealing, which is a significant negative for applications like bioimplant devices and tailored drug-delivery systems. Thermal annealing has been used to modify conductivity in photonic, optoelectronic, and electronic devices with success, although its usefulness is limited to circumstances in which property changes at the bulk are not as important.

5.9 Conclusion

In the beginning, the primary application of polymer composites was as stable building elements. Modern times, however, are witnessing a rise in interest in sophisticated polymer materials with much improved surface properties. Lately, various surface modification approaches have been coming to light with the advancement in conventional techniques. This clears the way for strong and durable polymer surfaces. UV and laser methods focus on activation of the polymer surface and surface polymerization to attain modified surfaces. However CVD techniques adopt a more complex approach by generating thin coatings of polymer to address surface modification. Primarily, owing to its affordability factor, corona discharge methods remain most popular surface treatment techniques in the packaging sector.

Even though complex surface modification techniques like lithiographic patterning treatments are technically conceivable, their commercial viability is typically determined by economic concerns. The financial and technological should be kept in mind while selecting a suitable surface treatment technique for the desired purpose. The basic research and knowledge regarding a particular method is advantageous, but its economic advantages typically depend on industry.

Herein we have discussed popular surface modification treatments like physical, chemical, and optical methods. We have tried to emphasize on the strategies that modify the surface characteristics and morphologies of polymer surfaces. Nano- and micropatterning has also been studied as a considerable surface modification technique. For every method, a general explanation of the surface techniques, their strategies, and the influence of these processes on polymer surface have been covered. In most cases, we analyze these methods from application standpoint, whereas their mechanisms also require a lot of attention. A comparative study of these methods is also covered in this chapter. Alternative methods can also modify the surface with comparable properties, but the pace with which the modification occurs, varies. Moreover the disadvantages of each method are discussed, and ways to mitigate them are being addressed. This chapter summarizes overall techniques related to surface modification with a spotlight of the main purpose.

Abbreviations

BCPs	block copolymers
BHT	butylate hydroxytoluene
CVD	chemical vapor deposition
DLIP	direct laser interference patterning
LCD	liquid-crystal displays
LDPE	low-density polyethylene
LED	light-emitting displays
LIPSS	laser-induced periodic surface structures
LMWOM	low-molecular-weight oxidized materials
LST	laser surface texturing
LTE	local thermodynamic equilibrium
MCNIL	molding in capillaries nanoimprint lithography
NIL	nanoimprint lithography

OLED	organic light-emitting diodes
OTFTs	organic thin-film transistors
PA	polyamide
PAA	poly(acrylic acid)
PAAm	polyacrylamide
PDMS	poly(dimethylsiloxane)
PE	polyethylene
PEG	poly(ethylene glycol)
P-EPQ	biphenylene-diphosphonite
PET	polyethylene terephthalate
PI	polyimide
PLA	polylactic acid
PMMA	poly(methyl methacrylate)
PP	polypropylene
PS	polystyrene
PS-*b*-PMMA	polystyrene-*block*-poly(methyl methacrylate)
PVC	polyvinyl chloride
QD	quantum dot
R2R-EC	roll-to-roll extrusion coating
TNIL	thermal nanoimprint lithography
TNPP	tris-nonyl phenyl-phosphite
UV	ultra-violet
UVNIL	UV nanoimprint lithography
UVO	UV-ozone
VUV	vacuum ultra-violet
μcp	microcontact printing

References

Abdul Hamid, Z. (2014). Surface modification of poly (lactic acid) (PLA) via alkaline hydrolysis degradation. *Advanced Materials Research* 970: 324–327.

Acikgoz, C., Hempenius, M.A., Huskens, J., and Vancso, G.J. (2011). Polymers in conventional and alternative lithography for the fabrication of nanostructures. *European Polymer Journal* 47: 2033–2052.

Alamri, S. and Lasagni, A.F. (2017). Development of a general model for direct laser interference patterning of polymers. *Optics Express* 25: 9603–9616.

Aldred, M.P., Contoret, A.E.A., Farrar, S.R. et al. (2005). A full-color electroluminescent device and patterned photoalignment using light-emitting liquid crystals. *Advanced Materials* 17: 1368–1372.

Baidya, A., Ganayee, M.A., Jakka Ravindran, S. et al. (2017). Organic solvent-free fabrication of durable and multifunctional superhydrophobic paper from waterborne fluorinated cellulose nanofiber building blocks. *ACS Nano* 11: 11091–11099.

Barassi, G. and Borrmann, T. (2012). N-Chlorination and Orton rearrangement of aromatic polyamides, revisited. *Journal of Membrane Science and Technology* 2: 115.

Bäumchen, O. and Jacobs, K. (2010). Slip effects in polymer thin films. *Journal of Physics: Condensed Matter* 22: 033102.

Begum, S., Wu, J., Takawira, C., and Wang, J. (2016). Surface modification of polyamide 6,6 fabrics with an alkaline protease – subtilisin. *Journal of Engineered Fibers and Fabrics* 11: 64–74.

Behl, M., Seekamp, J., Zankovych, S. et al. (2002). Towards plastic electronics: patterning semiconducting polymers by nanoimprint lithography. *Advanced Materials* 14: 588–591.

Bodenschatz, E., Pesch, W., and Ahlers, G. (2000). Recent developments in Rayleigh-Bénard convection. *Annual Review of Fluid Mechanics* 32: 709–778.

Boxshall, K., Wu, M.H., Cui, Z. et al. (2006). Simple surface treatments to modify protein adsorption and cell attachment properties within a poly(dimethylsiloxane) micro-bioreactor. *Surface and Interface Analysis: An International Journal Devoted to the Development and Application of Techniques for the Analysis of Surfaces, Interfaces and Thin Films* 38 (4): 198–201.

Brizmer, V. and Kligerman, Y. (2012). A laser surface textured journal bearing. *Journal of Tribology* 134: 031702.

Brzezinski, S., Polowinski, S., Kowalczyk, D., and Malinowska, G. (2009). Effect of corona discharge treatment on the surface strength and performance properties of synthetic fibre textiles. *Fibres & Textiles in Eastern Europe* 17: 62–68.

Burger, C., Hsiao, B.S., and Chu, B. (2006). Nanofibrous materials and their applications. *Annual Review of Materials Research* 36: 333–368.

Cáceres, C.A., Mazzola, N., França, M., and Canevarolo, S.V. (2012). Controlling in-line the energy level applied during the corona treatment. *Polymer Testing* 31: 505–511.

Campbell, M., Sharp, D.N., Harrison, M.T. et al. (2000). Fabrication of photonic crystals for the visible spectrum by holographic lithography. *Nature* 404: 53–56.

Cargill, S., Desmulliez, M.P.Y., and Yu, W. (2011). Electro-hydrodynamic instability patterning of polymers. *Journal of Physics: Conference Series* 301: 012032.

Carroll, G.T., Devon, T.L., Moscatelli, A. et al. (2011). Photogeneration of gelatinous networks from pre-existing polymers. *Journal of Applied Polymer Science* 122: 168–174.

Castillejo, M., Rebollar, E., Oujja, M. et al. (2012). Fabrication of porous biopolymer substrates for cell growth by UV laser: the role of pulse duration. *Applied Surface Science* 258: 8919–8927.

Chan, C.-M. (1999). Surface treatment of polypropylene by corona discharge and flame. In: *Polypropylene* (ed. J. Karger-Kocsis), 800–805. Springer.

Chapiro, A. (1995). General consideration of the radiation chemistry of polymers. *Nuclear Instruments and Methods in Physics Research Section B: Beam Interactions with Materials and Atoms* 105: 5–7.

Chen, A.-F. and Huang, H.-X. (2016). Rapid fabrication of t-shaped micropillars on polypropylene surfaces with robust Cassie–Baxter state for quantitative droplet collection. *The Journal of Physical Chemistry C* 120: 1556–1561.

Chen, N., Kim, D.H., Kovacik, P. et al. (2016). Polymer thin films and surface modification by chemical vapor deposition: recent progress. *Annual Review of Chemical and Biomolecular Engineering* 7: 373–393.

Cho, A.-R., Shin, D.M., Jung, H.W. et al. (2011). Effect of annealing on the crystallization and properties of electrospun polylactic acid and nylon 6 fibers. *Journal of Applied Polymer Science* 120: 752–758.

Chou, S.Y., Krauss, P.R., Zhang, W. et al. (1997). Sub-10 nm imprint lithography and applications. *Journal of Vacuum Science & Technology, B: Microelectronics and Nanometer Structures-Processing, Measurement, and Phenomena* 15: 2897–2904.

Clark, D.T. and Dilks, A. (1977). ESCA applied to polymers. XV. RF glow-discharge modification of polymers, studied by means of ESCA in terms of a direct and radiative energy-transfer model. *Journal of Polymer Science, Polymer Chemistry Edition* 15: 2321–2345.

Decker, C. (2002). Light-induced crosslinking polymerization. *Polymer International* 51: 1141–1150.

Denes, F.S. and Manolache, S. (2004). Macromolecular plasma-chemistry: an emerging field of polymer science. *Progress in Polymer Science* 29: 815–885.

Deshpande, P., Pease, L.F. III, Chen, L. et al. (2004). Cylindrically symmetric electrohydrodynamic patterning. *Physical Review E* 70: 041601.

Desmet, T., Morent, R., De Geyter, N. et al. (2009). Nonthermal plasma technology as a versatile strategy for polymeric biomaterials surface modification: a review. *Biomacromolecules* 10: 2351–2378.

Dickey, M.D., Raines, A., Collister, E. et al. (2008). High-aspect ratio polymeric pillar arrays formed via electrohydrodynamic patterning. *Journal of Materials Science* 43: 117–122.

Dickson, M.N., Tsao, J., Liang, E.I. et al. (2017). Conformal reversal imprint lithography for polymer nanostructuring over large curved geometries. *Journal of Vacuum Science & Technology B* 35: 021602.

Ding, L., Wang, L., Shao, L. et al. (2014). The water-dependent decay mechanism of biaxially-oriented corona-treated polyethylene terephthalate films. *RSC Advances* 4: 54805–54809.

Encinas, N., Pantoja, M., Abenojar, J., and Martínez, M.A. (2010). Control of wettability of polymers by surface roughness modification. *Journal of Adhesion Science and Technology* 24: 1869–1883.

Fader, R., Schmitt, H., Rommel, M. et al. (2012). Novel organic polymer for UV-enhanced substrate conformal imprint lithography. *Microelectronic Engineering* 98: 238–241.

Ferrero, F. and Periolatto, M. (2015). Modification of surface energy and wetting of textile fibers. In: *Wetting and Wettability* (ed. M. Aliofkhazraei), 140–168.

Formosa, F., Anfuso, C.D., Satriano, C. et al. (2008). UV-O_3-treated and protein-coated polymer surfaces facilitate endothelial cell adhesion and proliferation mediated by the PKCα/ERK/cPLA2 pathway. *Microvascular Research* 75: 330–342.

Fridman, A. (2008). *Plasma Chemistry*. Cambridge: Cambridge University Press.

Fridman, A. and Kennedy, L.A. (2004). *Plasma Physics and Engineering*. CRC Press.

Garrison, B.J. and Srinivasan, R. (1985). Laser ablation of organic polymers: microscopic models for photochemical and thermal processes. *Journal of Applied Physics* 57: 2909–2914.

Gleissner, C., Landsiedel, J., Bechtold, T., and Pham, T. (2022). Surface activation of high performance polymer fibers: a review. *Polymer Reviews* 62: 757–788.

Goddard, J.M. and Hotchkiss, J.H. (2007). Polymer surface modification for the attachment of bioactive compounds. *Progress in Polymer Science* 32: 698–725.

Goldman, M., Goldman, A., and Sigmond, R.S. (1985). The corona discharge, its properties and specific uses. *Pure and Applied Chemistry* 57: 1353–1362.

Gooch, J. (2011a). *Cauchy's Equation in Encyclopedic Dictionary of Polymers*. New York: Springer.

Gooch, J.W. (2011b). *Encyclopedic Dictionary of Polymers-Hagen-Poiseuille Equation*. New York, NY: Springer.

Govindarajan, T. and Shandas, R. (2014). A survey of surface modification techniques for next-generation shape memory polymer stent devices. *Polymers* 6: 2309–2331.

Grace, J.M. and Gerenser, L.J. (2003). Plasma treatment of polymers. *Journal of Dispersion Science and Technology* 24: 305–341.

Gu, Z.-X., Cheng, J., Zhang, M.-Z. et al. (2017). Effect of sequence structure on wetting behaviors of fluorinated methacrylate polymers based on perfluorohexylethyl methacrylate and stearyl acrylate. *Chinese Journal of Polymer Science* 35: 1061–1072.

Guimond, S., Radu, I., Czeremuszkin, G. et al. (2002). Biaxially oriented polypropylene (BOPP) surface modification by nitrogen atmospheric pressure glow discharge (APGD) and by air corona. *Plasmas and Polymers* 7: 71–88.

Han, M.S., Park, Y., and Park, C.H. (2016). Development of superhydrophobic polyester fabrics using alkaline hydrolysis and coating with fluorinated polymers. *Fibers and Polymers* 17: 241–247.

Handrea-Dragan, M. and Botiz, I. (2021). Multifunctional structured platforms: from patterning of polymer-based films to their subsequent filling with various nanomaterials. *Polymers (Basel)* 13: 445.

Hardy, F.E. and Robson, P. (1967). The formation and hydrolysis of substituted *N*-chloro-*N*-methylbenzamides in aqueous alkali. *Journal of the Chemical Society B: Physical Organic* 1151–1154.

Hegemann, D., Brunner, H., and Oehr, C. (2003). Plasma treatment of polymers for surface and adhesion improvement. *Nuclear Instruments and Methods in Physics Research Section B: Beam Interactions with Materials and Atoms* 208: 281–286.

Heitz, J., Niino, H., and Yabe, A. (1996). Chemical surface modification on polytetrafluoroethylene films by vacuum ultraviolet excimer lamp irradiation in ammonia gas atmosphere. *Applied Physics Letters* 68: 2648–2650.

Higgins, A.M. and Jones, R.A. (2000). Anisotropic spinodal dewetting as a route to self-assembly of patterned surfaces. *Nature* 404: 476–478.

Hoffmann, J., Plötner, M., Kuckling, D., and Fischer, W.-J. (1999). Photopatterning of thermally sensitive hydrogels useful for microactuators. *Sensors and Actuators, A: Physical* 77: 139–144.

Hongbo, L. (2013). Soft UV nanoimprint lithography and its applications. In: *Updates in Advanced Lithography* (ed. H. Sumio), 170–195. Rijeka: IntechOpen.

Hu, J. (2016). *Active Coatings for Smart Textiles*. Woodhead Publishing.

Hu, S., Ren, X., Bachman, M. et al. (2002). Surface modification of poly(dimethyl-siloxane) microfluidic devices by ultraviolet polymer grafting. *Analytical Chemistry* 74: 4117–4123.

Hua, F., Sun, Y., Gaur, A. et al. (2004). Polymer imprint lithography with molecular-scale resolution. *Nano Letters* 4: 2467–2471.

Huang, Y.C., Huang, C.C., Huang, Y.Y., and Chen, K.S. (2007). Surface modification and characterization of chitosan or PLGA membrane with laminin by chemical and oxygen plasma treatment for neural regeneration. *Journal of Biomedical Materials Research. Part A* 82: 842–851.

Hyun, D.C., Park, M., and Jeong, U. (2016). Micropatterning by controlled liquid instabilities and its applications. *Journal of Materials Chemistry C* 4: 10411–10429.

Ikada, Y. (1994). Surface modification of polymers for medical applications. *Biomaterials* 15: 725–736.

Izdebska, J. (2016). Corona treatment. In: *Printing on Polymers: Fundamentals and Applications* (ed. S. Thomas), 123–139.

Jenekhe, S.A. and Roberts, M.F. (1993). Effects of intermolecular forces on the glass transition of polymers. *Macromolecules* 26: 4981–4983.

Jia, C., Zhang, R., Yuan, C. et al. (2020). Surface modification of aramid fibers by amino functionalized silane grafting to improve interfacial property of aramid fibers reinforced composite. *Polymer Composites* 41: 2046–2053.

Jones, K. and Schmid, S.R. (2016). Experimental investigation of laser texturing and its effect on friction and lubrication. *Procedia Manufacturing* 5: 568–577.

Jones, D., Ormondroyd, G.O., Curling, S.F. et al. (2017). Chapter 2: Chemical compositions of natural fibres. In: *Advanced High Strength Natural Fibre Composites in Construction* (ed. M. Fan and F. Fu). Woodhead Publishing.

Julthongpiput, D., Zhang, W., Douglas, J.F. et al. (2007). Pattern-directed to isotropic dewetting transition in polymer films on micropatterned surfaces with differential surface energy contrast. *Soft Matter* 3: 613–618.

Kanelli, M., Vasilakos, S., Ladas, S. et al. (2017). Surface modification of polyamide 6.6 fibers by enzymatic hydrolysis. *Process Biochemistry* 59: 97–103.

Kato, K., Tanaka, K., Shinji Tsuru, S.T., and Shigenobu Sakai, S.S. (1994). Reflective color display using polymer-dispersed cholesteric liquid crystal. *Japanese Journal of Applied Physics* 33: 2635.

Kaufmann, T. and Ravoo, B.J. (2010). Stamps, inks and substrates: polymers in microcontact printing. *Polymer Chemistry* 1: 371–387.

Kawano, T., Iwama, N., Ishihata, H. et al. (2012). Preparation and biomedical application of self-organized honeycomb-patterned polymer films. In: *Interface Oral Health Science 2011* (ed. K. Sasaki, O. Suzuki, and N. Takahashi), 22–26. Springer.

Khanal, B.P. and Zubarev, E.R.J. (2007). Rings of nanorods. *Angewandte Chemie International Edition* 46: 2195–2198.

Kim, S.W. (1996). Surface modification of polypropylene in an impulse corona discharge. *Korean Journal of Chemical Engineering* 13: 97–100.

Kim, H.-C. and Hinsberg, W.D. (2008). Surface patterns from block copolymer self-assembly. *Journal of Vacuum Science & Technology, A: Vacuum, Surfaces, and Films* 26: 1369–1382.

Kim, H.R. and Seo, H.Y. (2013). Enzymatic hydrolysis of polyamide fabric by using acylase. *Textile Research Journal* 83: 1181–1189.

Kim, C.Y., Evans, J., and Goring, D.A.I. (1971). Corona-induced autohesion of polyethylene. *Journal of Applied Polymer Science* 15: 1365–1375.

Klemberg-Sapieha, J.E., Martinu, L., Sapieha, S., and Wertheimer, M.R. (1993). Control and modification of surfaces and interfaces by corona and low pressure plasma. In: *The Interfacial Interactions in Polymeric Composites* (ed. G. Akovali), 201–222. Dordrecht: Springer Netherlands.

Kolarova Raskova, Z., Stahel, P., Sedlarikova, J. et al. (2018). The effect of plasma pretreatment and cross-linking degree on the physical and antimicrobial properties of nisin-coated PVA films. *Materials (Basel)* 11: 1451.

Koller, J., Kříha, V., Píchal, J., and Aubrecht, L. (2004). Corona discharge light emission and electric field intensity comparison. *Czechoslovak Journal of Physics* 54: C810–C815.

Kraus, E., Baudrit, B., Heidemeyer, P. et al. (2016). Surface treatment of polymers by an ultraviolet laser to improve adhesion quality. *Polymer Science, Series D* 9: 5–12.

Kromer, R., Costil, S., Verdy, C. et al. (2018). Laser surface texturing to enhance adhesion bond strength of spray coatings – cold spraying, wire-arc spraying, and atmospheric plasma spraying. *Surface and Coating Technology* 352: 642–653.

Kulkarni, S. (2019). Chapter 3: Plasma assisted polymer synthesis and processing. In: *Non-thermal Plasma Technology for Polymeric Materials* (ed. S. Thomas, M. Mozetič, U. Cvelbar, et al.), 67–93. Elsevier.

Kull, K.R., Steen, M.L., and Fisher, E.R. (2005). Surface modification with nitrogen-containing plasmas to produce hydrophilic, low-fouling membranes. *Journal of Membrane Science* 246: 203–215.

Kusano, Y. (2014). Atmospheric pressure plasma processing for polymer adhesion: a review. *The Journal of Adhesion* 90: 755–777.

Lai, J., Sunderland, B., Xue, J. et al. (2006). Study on hydrophilicity of polymer surfaces improved by plasma treatment. *Applied Surface Science* 252: 3375–3379.

Landsiedel, J., Root, W., Schramm, C. et al. (2020). Tunable colors and conductivity by electroless growth of Cu/Cu_2O particles on sol-gel modified cellulose. *Nano Research* 13: 2658–2664.

Latag, G.V. and Vasquez, M.R. (2018). Effects of RF plasma modification on the thermal and mechanical properties of electrospun chitosan/poly(vinyl alcohol) nanofiber mats. *Journal of Vacuum Science & Technology B* 36: 04I101.

Lee, S.J., Oh, S.H., Liu, J. et al. (2008). The use of thermal treatments to enhance the mechanical properties of electrospun poly(ε-caprolactone) scaffolds. *Biomaterials* 29: 1422–1430.

Li, D. and Guo, L.J. (2006). Micron-scale organic thin film transistors with conducting polymer electrodes patterned by polymer inking and stamping. *Applied Physics Letters* 88: 063513.

Li, X.-M., Reinhoudt, D., and Crego-Calama, M. (2007). What do we need for a superhydrophobic surface? A review on the recent progress in the preparation of superhydrophobic surfaces. *Chemical Society Reviews* 36: 1350–1368.

Li, L., Hashaikeh, R., and Arafat, H.A. (2013). Development of eco-efficient micro-porous membranes via electrospinning and annealing of poly (lactic acid). *Journal of Membrane Science* 436: 57–67.

Li, B., Han, W., Jiang, B., and Lin, Z. (2014). Crafting threads of diblock copolymer micelles via flow-enabled self-assembly. *ACS Nano* 8: 2936–2942.

Liu, T.L. and Kim, C.J. (2014). Repellent surfaces. Turning a surface superrepellent even to completely wetting liquids. *Science* 346: 1096–1100.

Liu, G., Yu, W., Li, H. et al. (2013). Microstructure formation in a thick polymer by electrostatic-induced lithography. *Journal of Micromechanics and Microengineering* 23: 035018.

Liu, J., He, L., Wang, L. et al. (2016). Significant enhancement of the adhesion between metal films and polymer substrates by UV–ozone surface modification in nanoscale. *ACS Applied Materials & Interfaces* 8: 30576–30582.

Lodge, T.P. (2003). Block copolymers: past successes and future challenges. *Macromolecular Chemistry and Physics* 204: 265–273.

Luo, S. and Van Ooij, W.J. (2002). Surface modification of textile fibers for improvement of adhesion to polymeric matrices: a review. *Journal of Adhesion Science and Technology* 16: 1715–1735.

Luque-Agudo, V., Hierro-Oliva, M., Gallardo-Moreno, A.M., and González-Martín, M.L. (2021). Effect of plasma treatment on the surface properties of polylactic acid films. *Polymer Testing* 96: 107097.

Maghsoudi, K., Vazirinasab, E., Momen, G., and Jafari, R. (2020). Advances in the fabrication of superhydrophobic polymeric surfaces by polymer molding processes. *Industrial & Engineering Chemistry Research* 59: 9343–9363.

Malucelli, G. (2016). Hybrid organic/inorganic coatings through dual-cure processes: state of the art and perspectives. *Coatings* 6: 10.

Marat-Mendes, J.N., Raposo, M., and Ribeiro, P.A. (1992). Corona polarization effects on polymer surfaces. In: *Equilibrium Structure and Properties of Surfaces and Interfaces* (ed. A. Gonis and G.M. Stocks), 353–354. Boston, MA: Springer US.

Mathieson, I. and Bradley, R.H. (1996). Improved adhesion to polymers by UV/ozone surface oxidation. *International Journal of Adhesion and Adhesives* 16: 29–31.

Matyjaszewski, K. and Davis, K.A. (2002). *Statistical, Gradient, Block and Graft Copolymers by Controlled/Living Radical Polymerizations* (ed. K. Matyjaszewski and K.A. Davis). Berlin, Heidelberg: Springer Berlin Heidelberg.

Menard, E., Meitl, M.A., Sun, Y. et al. (2007). Micro- and nanopatterning techniques for organic electronic and optoelectronic systems. *Chemical Reviews* 107: 1117–1160.

Migliaresi, C., Cohn, D., De Lollis, A., and Fambri, L. (1991). Dynamic mechanical and calorimetric analysis of compression-molded PLLA of different molecular weights: effect of thermal treatments. *Journal of Applied Polymer Science* 43: 83–95.

Mirmohammadi, S.A., Khorasani, M.T., Mirzadeh, H., and Irani, S. (2012). Investigation of plasma treatment on poly(3-hydroxybutyrate) film surface: characterization and in vitro assay. *Polymer-Plastics Technology and Engineering* 51: 1319–1326.

Moghimifar, V., Raisi, A., and Aroujalian, A. (2014). Surface modification of polyethersulfone ultrafiltration membranes by corona plasma-assisted coating TiO_2 nanoparticles. *Journal of Membrane Science* 461: 69–80.

Molnar, K., Jozsa, B., Barczikai, D. et al. (2020). Plasma treatment as an effective tool for crosslinking of electrospun fibers. *Journal of Molecular Liquids* 303: 112628.

Montazer, M. and Komeily Nia, Z. (2015). Conductive nylon fabric through in situ synthesis of nano-silver: preparation and characterization. *Materials Science and Engineering: C* 56: 341–347.

Moon, J.H., Ford, J., and Yang, S. (2006). Fabricating three-dimensional polymeric photonic structures by multi-beam interference lithography. *Polymers for Advanced Technologies* 17: 83–93.

Moraczewski, K., Stepczynska, M., and Zenkiewicz, M. (2014). Effect of corona discharge on the selected properties of polylactide surface layer intended for autocatalytic metallization/Wplyw wyladowan koronowych na wybrane wlasciwosci warstwy wierzchniej polilaktydu przeznaczonego do autokatalitycznego metalizowania. *Polimery* 59: 482–488.

Morent, R., De Geyter, N., Desmet, T. et al. (2011). Plasma surface modification of biodegradable polymers: a review. *Plasma Processes and Polymers* 8: 171–190.

Mozetič, M. (2020). Plasma-stimulated super-hydrophilic surface finish of polymers. *Polymers (Basel)* 12: 2489.

Mukherjee, R., Sharma, A., and Steiner, U. (2011). Surface instability and pattern formation in thin polymer films. In: *Generating Micro- and Nanopatterns on Polymeric Materials* (ed. A. del Campo and E. Arzt), 217–265. Wiley.

Müller, C.D., Falcou, A., Reckefuss, N. et al. (2003). Multi-colour organic light-emitting displays by solution processing. *Nature* 421: 829–833.

Müller-Buschbaum, P., Bauer, E., Wunnicke, O., and Stamm, M. (2005). The control of thin film morphology by the interplay of dewetting, phase separation and microphase separation. *Journal of Physics: Condensed Matter* 17: S363.

Müller-Meskamp, L., Kim, Y.H., Roch, T. et al. (2012). Efficiency enhancement of organic solar cells by fabricating periodic surface textures using direct laser interference patterning. *Advanced Materials* 24: 906–910.

Musale, R.M. and Shukla, S.R. (2017). Weight reduction of polyester fabric using sodium hydroxide solutions with additives cetyltrimethylammonium bromide and [BMIM]Cl. *The Journal of the Textile Institute* 108: 467–471.

Nardulli, M., Belviso, M., Favia, P. et al. (2010). The study of specific and nonspecific hepatoma cells behavior by means of plasma-treated substrates. *Journal of Biomedical Materials Research. Part B, Applied Biomaterials* 94: 97–107.

Nemani, S.K., Annavarapu, R.K., Mohammadian, B. et al. (2018a). Surface modification of polymers: methods and applications. *Advanced Materials Interfaces* 5: 1801247.

Nemani, S.K., Chen, D., Mohamed, M.H., and Sojoudi, H. (2018b). Stretchable and hydrophobic electrochromic devices using wrinkled graphene and PEDOT: PSS. *Journal of Nanomaterials* 8 (2): 3230293.

Nie, Z. and Kumacheva, E. (2008). Patterning surfaces with functional polymers. *Nature Materials* 7: 277–290.

Nikolaev, A.G., Yushkov, G.Y., Oks, E.M. et al. (2014). Modification of anti-bacterial surface properties of textile polymers by vacuum arc ion source implantation. *Applied Surface Science* 310: 51–55.

Northen, M.T. and Turner, K.L. (2005). A batch fabricated biomimetic dry adhesive. *Nanotechnology* 16: 1159.

Novak, I. and Florián, Š. (2001). Investigation of hydrophilicity of polyethylene modified by electric discharge in the course of aging. *Journal of Materials Science Letters* 20: 1289–1291.

Okulova, N., Johansen, P., Christensen, L., and Taboryski, R. (2017). Replication of micro-sized pillars in polypropylene using the extrusion coating process. *Microelectronic Engineering* 176: 54–57.

Ozdemir, M., Yurteri, C.U., and Sadikoglu, H. (1999). Physical polymer surface modification methods and applications in food packaging polymers. *Critical Reviews in Food Science and Nutrition* 39: 457–477.

Park, M., Chaikin, P.M., Register, R.A., and Adamson, D.H. (2001). Large area dense nanoscale patterning of arbitrary surfaces. *Applied Physics Letters* 79: 257–259.

Paul, U.C., Fragouli, D., Bayer, I.S., and Athanassiou, A. (2016). Functionalized cellulose networks for efficient oil removal from oil–water emulsions. *Polymers* 8: 52.

Pekárek, S. (2009). DC corona discharge ozone production enhanced by magnetic field. *The European Physical Journal D* 56: 91.

Penterman, R., Klink, S.I., De Koning, H. et al. (2002). Single-substrate liquid-crystal displays by photo-enforced stratification. *Nature* 417: 55–58.

Peter, M.-B. (2003). Dewetting and pattern formation in thin polymer films as investigated in real and reciprocal space. *Journal of Physics: Condensed Matter* 15: R1549.

Pinto-Gómez, C., Pérez-Murano, F., Bausells, J. et al. (2020). Directed self-assembly of block copolymers for the fabrication of functional devices. *Polymers* 12: 2432.

Qiu, S., Ji, J., Sun, W. et al. (2021). Recent advances in surface manipulation using micro-contact printing for biomedical applications. *Smart Materials in Medicine* 2: 65–73.

Ragoubi, M., George, B., Molina, S. et al. (2012). Effect of corona discharge treatment on mechanical and thermal properties of composites based on miscanthus fibres and polylactic acid or polypropylene matrix. *Composites Part A: Applied Science and Manufacturing* 43: 675–685.

Rebollar, E., Sanz, M., Pérez, S. et al. (2012). Gold coatings on polymer laser induced periodic surface structures: assessment as substrates for surface-enhanced Raman scattering. *Physical Chemistry Chemical Physics* 14: 15699–15705.

Reiter, G. (1992). Dewetting of thin polymer films. *Physical Review Letters* 68: 75–78.

Riveiro, A., Maçon, A.L.B., Del Val, J. et al. (2018). Laser surface texturing of polymers for biomedical applications. *Frontiers in Physics* 6: 16.

Roberts, M.F. and Jenekhe, S.A. (1990). Lewis acid complexation of polymers: gallium chloride complex of nylon 6. *Chemistry of Materials* 2: 224–226.

Roberts, M.F. and Jenekhe, S.A. (1991). Site-specific reversible scission of hydrogen bonds in polymers: an investigation of polyamides and their Lewis acid-base complexes by infrared spectroscopy. *Macromolecules* 24: 3142–3146.

Rodríguez-Hernández, J. (2015). Wrinkled interfaces: taking advantage of surface instabilities to pattern polymer surfaces. *Progress in Polymer Science* 42: 1–41.

Sadeghi, I., Aroujalian, A., Raisi, A. et al. (2013). Surface modification of polyethersulfone ultrafiltration membranes by corona air plasma for separation of oil/water emulsions. *Journal of Membrane Science* 430: 24–36.

Sanchis, M.R., Blanes, V., Blanes, M. et al. (2006). Surface modification of low density polyethylene (LDPE) film by low pressure O_2 plasma treatment. *European Polymer Journal* 42: 1558–1568.

Sangermano, M., Roppolo, I., and Messori, M. (2015). UV-cured functional coatings. *RSC Smart Materials* 2015: 121–133.

Saqib, J. and Aljundi, I.H. (2016). Membrane fouling and modification using surface treatment and layer-by-layer assembly of polyelectrolytes: state-of-the-art review. *Journal of Water Process Engineering* 11: 68–87.

Sathish, S., Ishizu, N., and Shen, A.Q. (2019). Air plasma-enhanced covalent functionalization of poly(methyl methacrylate): high-throughput protein immobilization for miniaturized bioassays. *ACS Applied Materials & Interfaces* 11: 46350–46360.

Schaeffer, E., Thurn-Albrecht, T., Russell, T.P., and Steiner, U. (2000). Electrically induced structure formation and pattern transfer. *Nature* 403: 874–877.

Schäffer, E., Harkema, S., Roerdink, M. et al. (2003). Morphological instability of a confined polymer film in a thermal gradient. *Macromolecules* 36: 1645–1655.

Schille, J., Ullmann, F., Schneider, L. et al. (2015). Experimental study on laser surface texturing for friction coefficient enhancement. *Journal of Laser Micro/Nano-engineering* 10: 245–253.

Schwalm, R. (2006). UV coatings: basics, recent developments and new applications. *Pigment & Resin Technology* 37.

Segalman, R.A. (2005). Patterning with block copolymer thin films. *Materials Science and Engineering: R: Reports* 48: 191–226.

Shimoda, T., Morii, K., Seki, S., and Kiguchi, H. (2003). Inkjet printing of light-emitting polymer displays. *MRS Bulletin* 28: 821–827.

Shin, J., Liu, X., Chikthimmah, N., and Lee, Y.S. (2016). Polymer surface modification using UV treatment for attachment of natamycin and the potential applications for conventional food cling wrap (LDPE). *Applied Surface Science* 386: 276–284.

Sinton, D. (2014). Energy: the microfluidic frontier. *Lab on a Chip* 14: 3127–3134.

Sojoudi, H., Wang, M., Boscher, N.D. et al. (2016). Durable and scalable icephobic surfaces: similarities and distinctions from superhydrophobic surfaces. *Soft Matter* 12: 1938–1963.

Sojoudi, H., Kim, S., Zhao, H. et al. (2017). Stable wettability control of nanoporous microstructures by iCVD coating of carbon nanotubes. *ACS Applied Materials & Interfaces* 9: 43287–43299.

Sojoudi, H., Arabnejad, H., Raiyan, A. et al. (2018). Scalable and durable polymeric icephobic and hydrate-phobic coatings. *Soft Matter* 14: 3443–3454.

Sollogoub, C., Montmitonnet, P., Demay, Y. et al. (2011). Origin of the bubble defect in the extrusion coating process. *Polymer Engineering & Science* 51: 347–357.

Sparavigna, A. and Wolf, R.A. (2006). Electron and ion densities in corona plasma. *Czechoslovak Journal of Physics* 56: B1062–B1067.

Stepczyńska, M. (2014). Research of biocidal effect of corona discharges on poly(lactic acid) packaging films. *Journal of Food Engineering* 126: 56–61.

Steven, D. and Burke, R.L.D. (2003). *Handbook of Reagents for Organic Synthesis*. Chichester: Wiley. Polymer Reviews 781.

Stishkov, Y.K., Samusenko, A.V., Subbotskii, A.S., and Kovalev, A.N. (2010). Experimental study of pulsed corona discharge in air. *Technical Physics* 55: 1569–1576.

Strobel, M. and Lyons, C.S. (2003). The role of low-molecular-weight oxidized materials in the adhesion properties of corona-treated polypropylene film. *Journal of Adhesion Science and Technology* 17: 15–23.

Strobel, M., Jones, V., Lyons, C.S. et al. (2003). A comparison of corona-treated and flame-treated polypropylene films. *Plasmas and Polymers* 8: 61–95.

Studer, K., Decker, C., Beck, E., and Schwalm, R. (2003). Overcoming oxygen inhibition in UV-curing of acrylate coatings by carbon dioxide inerting, Part I. *Progress in Organic Coatings* 48: 92–100.

Su, Z., Bedolla-Valdez, Z.I., Wang, L. et al. (2019). High-speed photothermal patterning of doped polymer films. *ACS Applied Materials & Interfaces* 11: 41717–41725.

Sun, Y. and Sun, G. (2004). Novel refreshable N-halamine polymeric biocides: N-chlorination of aromatic polyamides. *Industrial & Engineering Chemistry Research* 43: 5015–5020.

Sun, C., Zhang, D., and Wadsworth, L.C. (1999). Corona treatment of polyolefin films – a review. *Advances in Polymer Technology* 18: 171–180.

Taaca, K.L.M., Prieto, E.I., and Vasquez, M.R. (2022). Current trends in biomedical hydrogels: from traditional crosslinking to plasma-assisted synthesis. *Polymers* 14: 2560.

Tan, E.P.S. and Lim, C.T. (2006). Effects of annealing on the structural and mechanical properties of electrospun polymeric nanofibres. *Nanotechnology* 17: 2649.

Tao, A.R., Huang, J., and Yang, P. (2008). Langmuir–Blodgettry of nanocrystals and nanowires. *Accounts of Chemical Research* 41: 1662–1673.

Tavanai, H. (2009). A new look at the modification of polyethylene terephthalate by sodium hydroxide. *The Journal of the Textile Institute* 100: 633–639.

Tendero, C., Tixier, C., Tristant, P. et al. (2006). Atmospheric pressure plasmas: a review. *Spectrochimica Acta Part B: Atomic Spectroscopy* 61: 2–30.

Tian, X., Jin, H., Sainio, J. et al. (2014). Droplet and fluid gating by biomimetic janus membranes. *Advanced Functional Materials* 24: 6023–6028.

Tuominen, M., Lahti, J., Lavonen, J. et al. (2010). The influence of flame, corona and atmospheric plasma treatments on surface properties and digital print quality of extrusion coated paper. *Journal of Adhesion Science and Technology* 24: 471–492.

Valkama, S., Kosonen, H., Ruokolainen, J. et al. (2004). Self-assembled polymeric solid films with temperature-induced large and reversible photonic-bandgap switching. *Nature Materials* 3: 872–876.

Vasanthan, N., Kotek, R., Jung, D.-W. et al. (2004). Lewis acid–base complexation of polyamide 66 to control hydrogen bonding, extensibility and crystallinity. *Polymer* 45: 4077–4085.

Vecchiato, S., Ahrens, J., Pellis, A. et al. (2017). Enzymatic functionalization of HMLS-polyethylene terephthalate fabrics improves the adhesion to rubber. *ACS Sustainable Chemistry & Engineering* 5: 6456–6465.

van Veldhuizen, E.M. and Rutgers, W.R. (2001). Corona discharges: fundamentals and diagnostics. *Invited paper, Proceedings of Frontiers in Low Temperature Plasma,*

Rolduc Conference Centre, The Netherlands, 25 March 2001. Eindhoven University of Technology. pp. 40–49.

Wang, H. (2006). Improving the adhesion of polyethylene by UV grafting. *The Journal of Adhesion* 82: 731–745.

Wang, W., Li, R., Tian, M. et al. (2013). Surface silverized meta-aramid fibers prepared by bio-inspired poly(dopamine) functionalization. *ACS Applied Materials & Interfaces* 5: 2062–2069.

Watanabe, C. and Yanagisawa, M. (2022). Evaporation patterns of dextran–poly(ethylene glycol) droplets with changes in wettability and compatibility. *Life* 12: 373.

Weng, C., Wang, F., Zhou, M. et al. (2018). Fabrication of hierarchical polymer surfaces with superhydrophobicity by injection molding from nature and function-oriented design. *Applied Surface Science* 436: 224–233.

Wood, V., Panzer, M.J., Chen, J. et al. (2009). Inkjet-printed quantum dot–polymer composites for full-color AC-driven displays. *Advanced Materials* 21: 2151–2155.

Wu, N. and Russel, W.B. (2009). Micro- and nano-patterns created via electrohydrodynamic instabilities. *Nano Today* 4: 180–192.

Wu, H., Odom, T.W., and Whitesides, G.M. (2002). Reduction photolithography using microlens arrays: applications in gray scale photolithography. *Analytical Chemistry* 74: 3267–3273.

Xie, R., Karim, A., Douglas, J.F. et al. (1998). Spinodal dewetting of thin polymer films. *Physical Review Letters* 81: 1251–1254.

Xing, R., Luo, C., Wang, Z., and Han, Y. (2007). Dewetting of polymethyl methacrylate on the patterned elastomer substrate by solvent vapor treatment. *Polymer Chemistry* 48: 3574–3583.

Xu, Q.F., Mondal, B., and Lyons, A.M. (2011). Fabricating superhydrophobic polymer surfaces with excellent abrasion resistance by a simple lamination templating method. *ACS Applied Materials & Interfaces* 3: 3508–3514.

Xu, L., Zhang, H., and Shi, T. (2016). Liquid–liquid interfacial slip induced layer instability in a thin polymer bilayer. *Polymer Chemistry* 99: 185–192.

Xu, L., Chen, Z., and Zou, Z. (2017). Dewetting of a pre-patterned thin polymer bilayer: influence of the instability mode. *RSC Advances* 7: 21459–21466.

Xue, C.-H., Jia, S.-T., Zhang, J., and Ma, J.-Z. (2010). Large-area fabrication of superhydrophobic surfaces for practical applications: an overview. *Science and Technology of Advanced Materials* 11: 033002.

Xue, L., Zhang, J., and Han, Y. (2012). Phase separation induced ordered patterns in thin polymer blend films. *Progress in Polymer Science* 37: 564–594.

Yagüe, J.L. and Gleason, K.K. (2013). Enhanced cross-linked density by annealing on fluorinated polymers synthesized via initiated chemical vapor deposition to prevent surface reconstruction. *Macromolecules* 46: 6548–6554.

Yakacki, C.M., Shandas, R., Lanning, C. et al. (2007). Unconstrained recovery characterization of shape-memory polymer networks for cardiovascular applications. *Biomaterials* 28: 2255–2263.

Yang, S.Y. and Rubner, M.F. (2002). Micropatterning of polymer thin films with pH-sensitive and cross-linkable hydrogen-bonded polyelectrolyte multilayers. *Journal of the American Chemical Society* 124: 2100–2101.

Yang, Y., Zhu, G., Zhang, H. et al. (2013). Triboelectric nanogenerator for harvesting wind energy and as self-powered wind vector sensor system. *ACS Nano* 7: 9461–9468.

Yao, X., Song, Y., and Jiang, L. (2011). Applications of bio-inspired special wettable surfaces. *Advanced Materials* 23: 719–734.

Yasuda, H. (1985). *Plasma Polymerization*. Orlando: Academic Press.

Yasuda, H. and Gazicki, M. (1982). Biomedical applications of plasma polymerization and plasma treatment of polymer surfaces. *Biomaterials* 3: 68–77.

Yoshida, S., Hagiwara, K., Hasebe, T., and Hotta, A. (2013). Surface modification of polymers by plasma treatments for the enhancement of biocompatibility and controlled drug release. *Surface and Coatings Technology* 233: 99–107.

You, Y., Won Lee, S., Jin Lee, S., and Park, W.H. (2006). Thermal interfiber bonding of electrospun poly(L-lactic acid) nanofibers. *Materials Letters* 60: 1331–1333.

Yu, D., Mu, S., Liu, L., and Wang, W. (2015). Preparation of electroless silver plating on aramid fiber with good conductivity and adhesion strength. *Colloids and Surfaces A: Physicochemical and Engineering Aspects* 483: 53–59.

Yun, C., Han, J.W., Kang, M.H. et al. (2019). Effect of laser-induced direct micropatterning on polymer optoelectronic devices. *ACS Applied Materials & Interfaces* 11: 47143–47152.

Żenkiewicz, M. (2001). Some effects of corona discharge treatment of biaxially-oriented polypropylene film. *The Journal of Adhesion* 77: 25–41.

Zenkiewicz, M. (2008). Corona discharge in an air as a method of modification of polymeric materials' surface layers. *Polimery* 53: 3–13.

Zeronian, S.H. and Collins, M.J. (1989). Surface modification of polyester by alkaline treatments. *Textile Progress* 20: 1–26.

Zhai, T., Zhang, X., Pang, Z., and Dou, F. (2011). Direct writing of polymer lasers using interference ablation. *Advanced Materials* 23: 1860–1864.

Zhang, Z.L., He, Z.H., Deng, H.M., et al. (2008). Experimental study on DC corona discharge in air-solid two phase mixture. *2008 International Conference on High Voltage Engineering and Application*. Washington (9–12 November 2008), Seattle, Washington, International Information and Engineering Technology Association (IIETA). pp. 420–421.

Zhao, J., Wang, M., and Gleason, K.K. (2017). Stabilizing the wettability of initiated chemical vapor deposited (iCVD) polydivinylbenzene thin films by thermal annealing. *Advanced Materials Interfaces* 4: 1700270.

Zhou, J., Ellis, A.V., and Voelcker, N.H. (2010). Recent developments in PDMS surface modification for microfluidic devices. *Electrophoresis* 31: 2–16.

Zhou, M., Xiong, X., Jiang, B., and Weng, C. (2018). Fabrication of high aspect ratio nanopillars and micro/nano combined structures with hydrophobic surface characteristics by injection molding. *Applied Surface Science* 427: 854–860.

Zhu, G., Wang, G., and Li, J.J. (2021). Advances in implant surface modifications to improve osseointegration. *Materials Advances* 2: 6901–6927.

6

Organic Polymers as Potential Catalysts

Saponjeet Borah[1], Rajkanya Das[2], Bitap R. Thakuria[2], Pulakesh Das[2], and Bedanta Gogoi[2,3]

[1] *Gauhati University, Department of Physics, Gopinath Bordoloi Nagar, Jalukbari, Guwahati, Assam, 781014, India*
[2] *Gauhati University, Department of Chemistry, Gopinath Bordoloi Nagar, Jalukbari, Guwahati, Assam, 781014, India*
[3] *Tezpur University, Department of Chemical Sciences, Napaam, Assam 784028, India*

6.1 Introduction

Numerous chemical processes depend on catalysis, and the choice between homogeneous and heterogeneous catalysts can have a substantial effect on how a reaction turns out. Catalysts are materials that speed up chemical reactions without being consumed. A reaction is said to be homogeneous if both the catalyst and the reactants are in the same phase. On the other hand, the reaction is said to be heterogeneous if the catalyst and the reactants are in two separate phases. The choice of catalyst relies on the particular requirements of the process. Both homogeneous and heterogeneous catalysis have benefits and drawbacks. Selecting the optimum catalytic system for a particular reaction requires consideration of a number of elements, including required reaction rate, product yield, and the viability of catalyst recovery and reuse. In this chapter, we emphasize the use of porous organic polymers (POPs) containing active sites, and how they are applied as catalysts for various transformation reactions such as photocatalysis, coupling reactions, oxidation reactions, and reduction reactions.

6.1.1 Organic Polymer Photocatalysis (OPP)

To address different issues such as environmental pollution, greenhouse effect, energy storage devices, the researchers of the planet have been engaged in making an alternative solution for clean and sustainable energy and technology. Among them, solar energy to chemical energy converted to visible-light-driven photocatalytic technique is a unique one.

Organic Polymers in Energy-Environmental Applications, First Edition.
Edited by Ramesh Oraon, Pardeep Singh, Sanchayita Rajkhowa, Sangita Agarwal, and Ravindra Pratap Singh.

In order to accelerate a chemical reaction, the techniques of photochemistry and catalysis are combined in a process known as photocatalysis (Saravanam et al. 2017). Fujishima and Honda (1972) made the first observation in 1972 about the photocatalytic phenomenon by splitting of water on TiO_2 electrode under ultraviolet (UV) light. The photocatalysis process is subdivided into two types: one is homogeneous and the other is heterogeneous photocatalysis process (Fujishima and Honda 1972; Rajeshwar et al. 2008). Most frequently, metal complexes (transition metals like iron, copper, and chromium) are used as catalysts in homogeneous photocatalytic processes. In this process, under the photon and thermal condition, the higher oxidation state of metal ion complexes is generated hydroxyl radicals. Then, these hydroxyl radicals react with organic matter which leads to destruction of toxic matters (Tahir et al. 2020) In order to carry out a desired reaction, heterogeneous photocatalysis uses a light-activated catalyst at room temperature. Methane, ethane, ethylene, propylene, and carbon monoxide are examples of simple hydrocarbons that can be completely oxidatively destroyed by hydroxyl radicals produced when the n-type semiconductor oxide titanium dioxide (TiO_2) is illuminated in the presence of molecular oxygen (Ollis 1996). Because of its better stability, ease of separation, low cost, and capacity to reuse photocatalysts the heterogeneous photocatalysis typically exceeds than homogeneous photocatalysis (Tahir et al. 2020). The key advantages of photocatalysis are:

(i) Because it can use clean, renewable solar energy, it is a better alternative for current technologies that consume a lot of energy.
(ii) It can be employed in various wastewater management processes to destroy a number of dangerous chemicals.
(iii) A less amount of chemical input is needed, the reaction conditions for photocatalysis are mild, and the reaction period is brief.

6.1.2 The Mechanism of Photocatalysis

As shown in Figure 6.1, an intrinsically semi-conductive POP will be activated to produce photoexcited charge carriers, such as photogenerated electrons in the conduction band (CB) and holes in the valence band, when visible light with energy equal to or greater than the bandgap falls on the POPs (VB). Then a surface redox reaction occurs between POPs and substrates by efficient charge separation/migration and appropriate band potential alignment. The majority of photoinduced electrons and holes may merge, which lower the photocatalysis's effectiveness. A number of techniques have been investigated for POPs to increase the separation of charges and prevent recombination, including the use of noble metals, asymmetrical structures, high crystallinity polymers, high coplanarity polymers, and structural designs with electron donor–acceptor systems (Byun and Zhang 2020). Both excellent photostability and high activity are maintained by metal-free POPs in prolonged reactions. The lack of metal reduces the material's cost and removes the possibility of metal contamination, but the research effort is

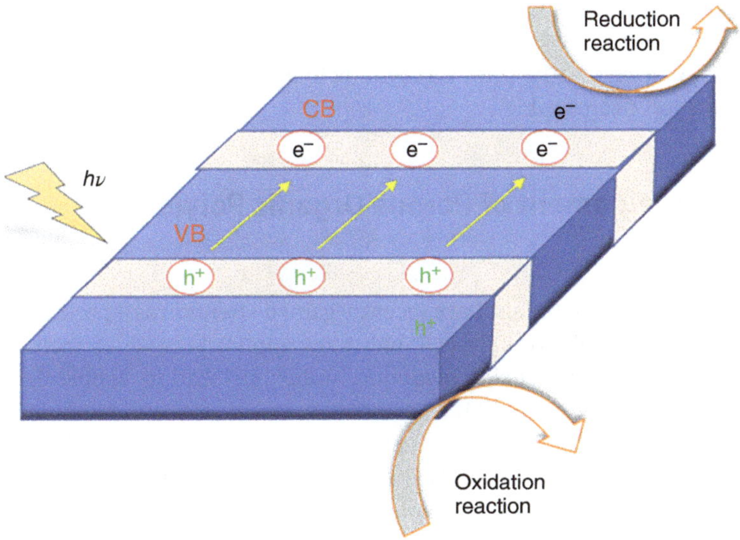

Figure 6.1 Pictorial representation of the photocatalytic reaction.

hampered by the need to design and optimize structures and reaction conditions (Wang et al. 2020).

The holes and electrons that are generated after striking the light on the semiconductive POPs could undergo successive oxidation and reduction reactions with any species, which might be adsorbed on the surface of the semiconductor to give the necessary products. The oxidation and reduction mechanism of photocatalysis is given below:

(i) *Oxidation mechanism:* Water molecules which are present on the photocatalyst's surface are oxidized by positive holes formed in the valence band as a result of the electron's shift to the conduction band, which produces hydroxyl (OH) radicals (agents which have strong oxidative decomposition power). The hydroxyl radicals react with organic matter present in the dyes. If oxygen is present when this process happens, the intermediate radicals in the organic compounds along with the oxygen molecules can experience radical chain reactions and consume oxygen in some cases. In such a case, the organic matter finally decomposes ultimately becoming carbon dioxide and water (Rajeshwar et al. 2008). In these conditions, organic molecules can interact directly with the positive holes leading to oxidative decomposition.
(ii) *Reduction mechanism:* Since oxygen is a material that can be reduced simply, it is reduced instead of being converted into hydrogen. Superoxide anions are produced when the conduction band electrons react with dissolved oxygen species, which causes material breakdown. These superoxide anions bind to the oxidative reactions of intermediate products, producing peroxide or transforming to hydrogen peroxide before becoming water. With organic

stuff as opposed to water, the decrease is likely to happen more quickly. As a result, there are more positive holes when there is a higher concentration of organic stuff. As a result, the photocatalytic activity is increased and carrier recombination gets decreased.

6.2 Recent Development of Porous Organic Polymer (POP)

POPs, a new class of porous materials, have recently gained attention as a promising platform for heterogeneous photocatalysis. It has many applications which include gas and water splitting, carbon dioxide reduction, water purification, visible-light photocatalyst, high porosity H_2, and CO_2 device.

Due to the rapid consumption of fossil fuels in industries, the carbon dioxide concentration is increased from 280 to 415 ppm (1750–2020) in our atmosphere which led to rise in the average temperature of the planet (Shen et al. 2021). It needs to be reduced by using a CO_2 photoreduction reaction (CO_2PRR). The CO_2PRR is an energy-efficient mechanism that converts CO_2 value-added chemicals and fuels using solar energy which is achieved by developing a highly active and selective photocatalysis by using various material catalysts like TiO_2, MoS_2, and MXene (Cheng et al. 2021).

The diverse uses of functional materials with a clearly defined pore structure were further enhanced by the development of metal organic frameworks. The POPs are highly stable both thermally and hydrothermally because they have strong C—C covalent bonds. S. Bandyopadhyay et al. (2016) designed and fabricated a novel series of 4,4-difluoro-4-bora-3*a*,4*a*-diaza-*s*-indacenes (BODIPY)-based POPs exhibiting high carbon dioxide and hydrogen uptake. A 10-fold increase in surface area from 73 $m^2\ g^{-1}$ in BDT1a to 1010 $m^2\ g^{-1}$ in BDT3 was achieved by judicious tuning of reaction conditions and monomer structures. They found the gas adsorption behavior of the polymers revealed a high uptake capacity of CO_2 (up to 16.5 wt%) at 273 K and 1 bar.

Generating POPs for photocatalytic water splitting is a tough but new potential effort in the area of research. When semiconductors are used as photocatalysts, the minimal band gap necessary for thermodynamic water splitting is 1.23 eV, assuming that all of the energy used to split the water came from electrical activity. Splitting of water into hydrogen and oxygen also requires a change in standard Gibb's free energy (ΔG°). The fundamental principle behind photocatalytic water splitting is that the electrons in the highest occupied molecular orbital (HOMO), or valence band (VB) are peacefully moved to the lowest unoccupied molecular orbital (LUMO)when an energetic photon ($h\nu$) stimulates the semiconductor and overcomes the band gap's barrier (CB). During this process, the electron (e^-) would emerge in the LUMO, while the hole (h+) would remain in the HOMO. The majority of the electron–hole pairs merge quickly in the "photo-excited" state. A portion of the electron–hole pairs move to the surface and are absorbed by water, which causes the water splitting effectively (Xu et al. 2018).

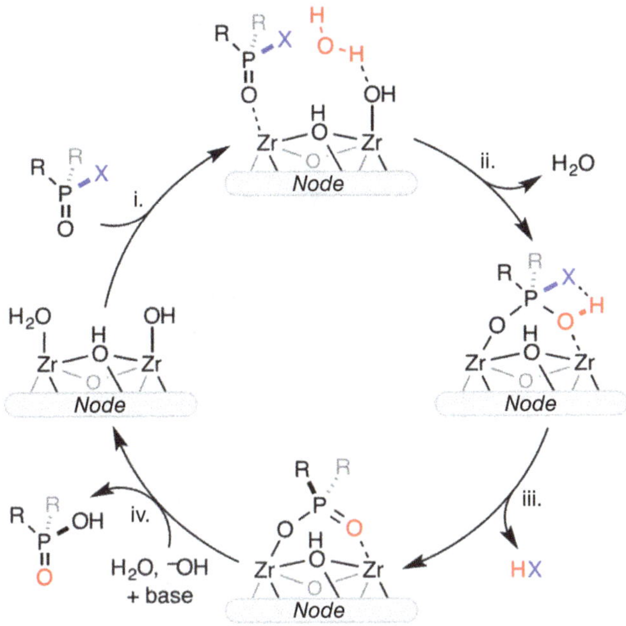

Figure 6.2 Proposed mechanisms for Zr-MOFs-based hydrolysis of organophosphorus substrates (Kirlikovali et al. 2020). Source: Copyright permission is received for reusing the figure.

6.3 Metal–Organic Framework (MOF)-Based Heterogeneous Catalysis

6.3.1 MOFs Chemistry

MOFs are a family of compounds formed by the coordination of metal ions or metal clusters to organic ligands (Figure 6.2). This class of compounds shows promising properties such as adjustable pore structure, high porosity, crystallinity, high surface area, variety of chemical compositions, and versatile functionality (Zhou et al. 2012; Furukawa et al. 2013; Kitagawa et al. 2004). These properties have sparked intense research attention in a number of disciplines, particularly catalysis.

MOFs as catalysts possess unique benefits over conventional porous materials (such as mesoporous silicon, clays, or zeolites), most notably the following (Li et al. 2019a):

(1) Clearly defined structures of MOFs help to understand the chemical reactions at the molecular level.
(2) It is possible to develop MOF-based catalysts by introducing multiple catalytic sites into a single MOF due to the structural diversity and tunability of all MOF components (nodes, linkers, and pores).
(3) The numerous catalytic sites and cooperative microenvironments provide evidence for synergistic catalysis.

(4) The tunable pore environments (hydrophilic and hydrophobic) support the identification and transportation of substrates and products.

According to recent studies, MOFs show extremely useful properties as heterogeneous catalysts for many chemical reactions. The strengths of MOFs as heterogeneous catalyst lie in their homogeneous and customizable pore shape and size, which are very important for size-selective catalysis. Most MOFs are microporous, having pores sizes in the range of 0–9.8 nm, appropriate for the majority of the reaction substrates that are used in industry. Again, because of the large surface area and porosity, substrates can be adsorbed on the surface of MOFs, which improves the interaction between the reactants and the catalytic sites and increases the catalytic efficiency. In order to understand the underlying catalytic mechanism, it is very important to understand the link between the well-defined tunable structure of MOFs and catalytic performance (Li et al. 2019b).

6.3.2 Active Sites of MOFs for Heterogeneous Catalysis

The MOFs containing metal nodes or clusters show excellent catalytic behavior compared to conventional catalysts. MOF-based single-metal sites catalysts have many benefits because of the complete isolation and dispersion of the metal sites. Metal site in different topological MOFs with tunable spatial positions and orientations enables synergistic catalysis and cascade reactions. Organic linkers and other potential terminal ligands like water and solvent molecules are coordinated with the metal ions in the metal nodes of MOFs. These molecules can be eliminated by thermal or vacuum processing without disturbing the host framework structures. As a result, the metal ions in the metal nodes/clusters that are exposed have vacant orbitals that can interact with substrates or other molecules through lone-pair electrons. Thus, these exposed metal sites can function as Lewis acid sites which can absorb and energize the reactant molecules to facilitate charge transfer as they transform into products (Jiao et al. 2018). Férey et al. (2005) reported a MOF named MIL-101(Cr) with chemical formula ($Cr_3X(H_2O)_2O(1,4\text{-bdc})_3$; X = F, OH; bdc = benzene-1,4-dicarboxylate). Due to its Lewis acid activity, it is a suitable MOF for use in heterogeneous catalysis. The water ligand gets easily eliminated under vacuum at a higher temperature. The exposed Cr(III) metal ions are available for potential reactants providing a significantly higher catalytic activity. Over the last 10 years, zirconium-based metal–organic frameworks have become a cutting-edge catalyst for detoxifying nerve agents. Farha, Hupp, and coworkers reported zirconium-based metal–organic framework MOF (NU-1000) for the catalytic hydrolysis of organophosphorus nerve agents and their stimulants (Mondloch et al. 2015). These MOFs have strong Zr(IV)—O bonds, which contribute to their exceptional chemical and thermal stability. The Lewis acidic behavior of Zr(IV) active sites facilitates the faster hydrolysis of organophosphorus substrates. The phosphoryl oxygen of the organophosphorus substrate coordinated to the Zr(IV) active site in the first step of the mechanism, as shown in Figure 6.2 (Kirlikovali et al. 2020).

In the second step, the hydroxyl group attached to the Zr-MOF acts as a nucleophile and attacks the phosphorous atom of the organophosphorus substrate. This nucleophilic attack can also occur from the attack of the water molecule coming from solvents. Step three involves the elimination of a leaving group from the organophosphorus substrate. Then in the final step, the hydrolyzed product gets removed from the catalyst surface by one hydroxo and one aqua ligand, and catalytic cycles get completed. It can be easily understood from the catalytic cycle that the rate of hydrolysis significantly depends on the accessibility of the organophosphorus substrate to the Zr(IV) active (Kirlikovali et al. 2020). The advantages of Lewis acid sites of MOFs are also explored in gas-phase reactions. Xu and coworkers developed a MOF Cu(mipt) (mipt = 5-methylisophthalate), which is microporous with open Cu(II) sites on channel walls. This MOF exhibits excellent and consistent catalysis for the oxidation of CO (Zou et al. 2007).

6.3.3 Active Sites at Organic Linkers

The presence of organic linkers (such as chiral molecules, salens, and metalloporphyrin) in MOFs makes the MOFs much more efficient for various catalysis processes (Cho et al. 2006). The modification of the functional groups present in organic linkers makes it possible to create various active sites within the MOFs. Examples of such types of functional groups are sulfonic acid, amide, sulfoxy, amino, pyridyl, bipyridyl, etc. (Zou et al. 2007). On the other hand, these functional groups can also be incorporated into the organic linkers by functionalized modification approach. The acidic or basic nature of functional groups has influenced the catalytic activity of the MOFs (Zhu et al. 2017; Jiang et al. 2013).

Many research groups have tried to functionalize MOFs with strong Bronsted acids like sulfonic acids (Jiang and Yaghi 2015). For the first time, Zhou et al. (2014) developed ~100% sulfonic acid functionalized MIL-101(Cr) by a hydrothermal process followed by a facile post-synthetic HCL treatment. The resulting MIL-101-SO_3H demonstrated high activity, outstanding selectivity, and recyclability toward the alcoholysis of epoxides under ambient conditions. On the other hand, basic sites can also be carefully added onto organic linkers by simply using N-containing ligands as building blocks (Dong et al. 2013). Hartmann and coworkers demonstrated the activity of NH_2 functionalized MOFs Fe-MIL-101-NH_2 and Al-MIL-101-NH_2, by testing in the Knoevenagel condensation reaction between benzaldehyde and malononitrile. The 90% yield of the expected condensation products showed the high catalytic activity of the NH_2-functionalized MOFs (Hartmann, and Fischer 2012). Other N-containing ligands such as pyridyl and amide are also used to develop MOF basic catalysts (Hasegawa et al. 2007).

Metalloporphyrins and metallosalens are also grafted into MOFs. A porous metal metalloporphyrin structure PCN-222(Fe) was developed by Zhou and his coworkers, and it shows outstanding peroxidase-like catalytic activity (Feng et al. 2012). Later, the same group also reported a number of extremely stable Zr-porphyrinic MOFs, such as PCN-223, PCN-224, and PCN-225, which exhibit promising catalytic potential (Feng et al. 2014, 2013). On the other hand, metallosalens are potential starting materials for the synthesis of porous MOF catalysts, which have been explored for heterogeneous catalytic applications. For example alkene epoxidation, cyclopropanation, and hydrolytic kinetic resolution of epoxides with excellent shape, size, and enantio-selectivities (Falkowski et al. 2011; Zhu et al. 2012).

6.3.4 MOFs-Based Composites for Heterogeneous Catalysis

The catalytic applications of pure MOFs are limited by their poor activity from having fewer active sites and low thermal, mechanical, and chemical stability from the fragility of the metal linker coordinative bond (Yuan et al. 2018; Burtch et al. 2018). Many strategies have been put out to improve the catalytic characteristics of MOFs. One such approach is based on the combination MOFs with a range of functional materials to develop new MOF composites. The MOF composites are composed of two parts, viz. MOF molecule and functional material. The diverse functional materials such as molecular species, metal nanoparticles (NP), polyoxometalates (POMs), and enzymes have all been combined with MOFs to form MOF composites (Zhu and Xu 2014; Chen et al. 2017). Following sections discuss catalytic activities of various reported MOF-composites.

6.3.5 MOF-Molecular Species Composites

Molecular species such as salen complexes, dye molecules, and porphyrin are reportedly enclosed in MOF pores via covalent or non-covalent interactions. The encapsulation through non-covalent interactions preserves the structure and characteristics of molecular species by not disrupting their coordination sphere. In addition, direct encapsulation can keep the mobility of the molecular species to a certain extent, enabling them to incorporate with the MOF backbone (Bogaerts et al. 2013). The effective encapsulation of molecular species in MOFs can lead to better performance in a variety of heterogeneous catalytic reactions (Juan-Alcaniz et al. 2012). Alkordi et al. (2008) reported encapsulation of a porphyrin, 5,10,15,20-tetrakis-(1-methyl-4-pyridinio) porphyrin tetra(*p*-toluenesulfonate) ([H_2TMPyP] [*p*-tosyl]$_4$), in an indium imidazole dicarboxylate-based rho-ZMOF. An enantioselective MOF-composite was reported by Speybroeck and his coworkers, consisting of a chiral Mn(III)salen complex entrapped in the MIL-101 metal–organic framework (Bogaerts et al. 2013).

6.3.6 MOF–Metal Nanoparticle Composites

Metal nanoparticles have received a wide range of interests because of their high chemical activities. The high surface-to-volume ratio with a high surface energy of metal nanoparticles leads to aggregation. The incorporation of metal nanoparticles into the MOFs with large surface area and porosity can restrict the aggregation to a certain extent (Chen and Xu 2019). Metal NPs typically serve as active centers for catalytic transformation when MOF–metal NP composites are used in a catalytic process, while MOFs serve as a stabilizer to protect metal NPs with improved stability (Chen et al. 2014, 2015). Yuan et al. reported a MOF-nanoparticle composite where palladium NPs were supported on MIL-101(Cr). The resulting Pd/MIL-101(Cr) was used as a heterogeneous catalyst with high catalytic activity and negligible metal leaching for water-mediated Suzuki–Miyaura and Ullmann coupling reactions of aryl chlorides (Yuan et al. 2010). Immobilization of ultrafine Pt nanoparticles inside the pores of a MOF MIL-101 was done by Xu and his coworker using a double solvent method. They were able to synthesis Pt@MIL-101 composite without aggregation of Pt nanoparticles on the surface of the MOF. The composite showed higher activity for catalytic reaction in all three phases: solid-phase ammonia borane thermal dehydrogenation, liquid-phase ammonia borane hydrolysis, and gas-phase CO oxidation (Aijaz et al. 2012).

6.3.7 MOF–Polyoxometalate Composites

POMs are a type of distinct, anionic metal–oxygen clusters, have a variety of features that make them interesting candidates for prospective use in the field of catalysis. Properties such as tailorable size and shapes, strong acidities, redox potentials and solubilities offer excellent possibilities for different catalytic transformations, although the small specific surface area and insufficient stability of POMs restricted their applications (Chen and Xu 2019). A useful way of optimizing and stabilizing POMs to improve their catalytic capabilities is immobilizing them into MOFs (Ye and Wu 2016; Du et al. 2014). Maksimchuk et al. reported POM–MOF composites by inserting polyoxotungstates $[PW_4O_{24}]^{3-}$ (PW_4) and $[PW_{12}O_{40}]^{3-}$ (PW_{12}) into nanocages of the metal organic framework MIL-101. The composite materials PW_x/MIL-101 ($x = 4$ or 12) were used as heterogeneous catalysts for the epoxidation of alkenes with aqueous hydrogen peroxide under moderate reaction conditions (Maksimchuk et al. 2010).

POM–MOF composites also find their applications in heterogeneous asymmetric catalysis. Han et al. developed POM–MOF composites by encapsulating $[BW_{12}O_{40}]^{5-}$ and the chiral group, L- or D-pyrrolidin-2-ylimidazole (PYI), into MOFs, results in two enantiomorphs Ni-PYI1 and Ni-PYI2 via self-assembly method, respectively. The resultant composites facilitated the asymmetric dihydroxylation of aryl olefins with high stereoselectivity by acting as amphipathic catalysts (Han et al. 2013).

6.3.8 MOF–Enzyme Composites

Enzymes are a group of extremely effective biological catalysts, capable of catalyzing a wide range of reactions with high specificity and selectivity under very moderate reaction conditions. They are made up of linear amino acid chains that fold into precise and intricate tertiary structures (Benkovic and Hammes-Schiffer 2003; Dongen et al. 2014). These catalysts have undergone thousands of years of evolutionary fine-tuning and exhibit remarkable efficiency, chemo-, enantio-, and regioselectivity, although most of the catalytic applications are severely hindered by their low thermal stability, low tolerance to organic solvents and metal ions, narrow optimum pH ranges. Furthermore, the contamination of enzymes in the products necessitates intensive separation and purification procedures. The immobilization of enzymes on the MOF substrates can protect the enzymes from deactivation

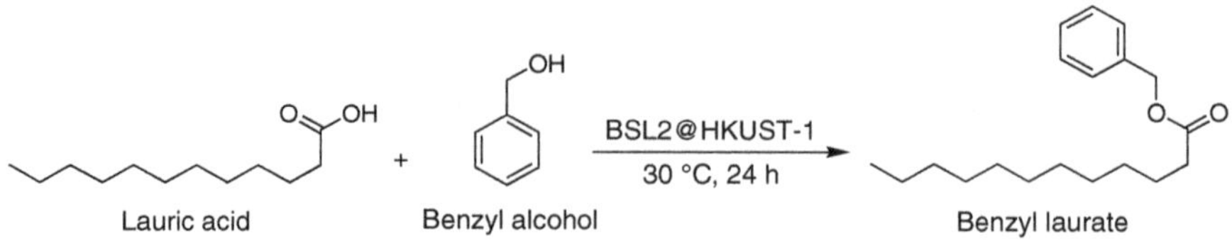

Figure 6.3 Esterification of lauric acid and benzyl alcohol using BSL2@HKUST-1 as a catalyst.

reaction conditions, enhance their recyclability, and reduce their contamination in the desired product (Lykourinou et al. 2011; Drout et al. 2019). The interactions between the MOF's functional linker and inorganic nodes with enzymes can stabilize enzymes from leaching (Lian et al. 2017). Enzyme viz. *Bacillus subtilis lipase* (BSL2) was immobilized into a Cu-BTC-based hierarchically porous metal–organic framework HKUST-1 by Liu and his coworkers. The catalytic performance of the composite catalyst BSL2@HKUST-1 was studied by using the catalyst in the esterification reaction between lauric acid and benzyl alcohol (Figure 6.3). The results showed remarkably enhanced enzymatic activity as compared to the free enzyme in isooctane solvent (Cao et al. 2016).

The effective dispersion of the BSL2 on the mesopores of the MOF HKUST-1 prevents the aggregation of the enzyme and showed a 17 times higher initial rate as compared to free BSL2. Additionally, the BSL2@HKUST-1 showed higher reusability with only an insignificant drop in conversion after 10 cycles (Cao et al. 2016).

An interesting example of MOF-based biohybrid catalyst was reported by Wu and his coworker. By stepwise immobilization of Pd nanoparticles and *Candida antarctica lipase B* (CalB) onto a MOF UiO-66-NH_2 they developed the biohybrid catalyst and utilized it for the synthesis of benzyl hexanoate from benzaldehyde and ethyl hexanoate (Wang et al. 2019). The resulting heterogeneous biohybrid catalyst showed excellent catalytic efficiency for one-pot cascade reactions.

6.3.9 MOF-Based Heterogeneous Polymerization Catalysts

For the manufacture of polymers, using MOFs as heterogeneous polymerization catalysts has been found to be an effective strategy. The ability to make structural modification and highly tailorable pores of MOFs provides enormous possibilities for the polymerization reaction. An in-depth understanding of the polymerization mechanism is also possible because of the clearly defined structures of MOFs (Goetjen et al. 2020). MOFs act as a promising catalyst in reversible deactivation radical polymerization (RDRP) as well as coordination polymerization reactions (Mochizuki et al. 2018).

6.4 Reversible Deactivation Radical Polymerization (RDRP)

MOFs can act as efficient catalysts for RDRP reactions by replacing small metal complexes. Use of Fe- and Cu-based MOF as heterogeneous catalyst for polymerization reactions was found to be an excellent alternative option for metal catalysts (Schmidt 2019).

Lee et al. demonstrated how to activate [Cu_2bdc_2DABCO] thermally using amine compounds, such as added DABCO (bdc, benzene dicarboxylate; DABCO, diaminobicyclooctane). Cu(II) MOF, DABCO, initiator, and monomer were mixed and heated to begin the polymerization. The monomers, such as styrene, benzyl methacrylate (BMA), 4-vinyl pyridine (4VP), or isoprene, were used in the process along with ethyl α-bromoisobutyrate (EBiB) as an initiator. The catalyst showed good catalytic activity for synthesis of polymers such as polystyrene, poly(benzyl methacrylate), poly(4-vinylpyridine), polyisoprene in a controlled way, which was previously difficult to do using conventional methods (Lee et al. 2016).

6.5 Coordination Polymerization

Another polymerization technique that heavily depends on metal complexes, particularly metal–monomer complexes, is coordination polymerization. As a result, MOFs have been used often as coordination polymerization catalysts. The polymerization mechanism and product characteristics, such as molar mass or stereoselectivity, can be greatly influenced, particularly by the defined crystal structure and surroundings of the catalytic center. Ethylene is one of the major monomers that has typically been polymerized using coordination polymerization (Schmidt 2019). Liu et al. reported ethylene polymerization reaction by using Cr(III)-based heterogeneous catalyst (IRMOF-3-SI-Cr). Salicylideneimine was added to the surface of isoreticular metal–organic framework-3 (IRMOF-3, [$Zn_4O(ATA)_3$]; ATA = 2-aminoterephthalate) to instigate a condensation reaction with the ATA and enable it to bind to Cr(III) ions (Liu et al. 2014). Coordination polymerization of butadiene is also studied by many research groups. Dincă and coworkers reported a stereoselective polymerization of butadiene using bis(1*H*-1,2,3-triazolo[4,5-b],[4′,5′-i])dibenzo[1,4]dioxin)-based MOFs, Co(II) ion as metal sites (Dubey et al. 2017). In this approach, poly(1,3-butadiene) was produced with 99% selectivity for 1,4-cis propagation. The catalyst was also reusable, and little leaching was observed.

6.6 Covalent Organic Frameworks (COFs)-Based Heterogeneous Catalysis

6.6.1 COFs Chemistry

COFs are a family of porous crystalline molecular frameworks that allow the covalent incorporation of organic groups into an organized polymer skeleton to construct two (2D)- and three-dimensional (3D) porous structures. The potential of COFs to integrate various structural characteristics, such as stability, designability, porosity, and tenability in a single material, makes them significant (Shinde et al.

2015; Xu et al. 2016). Again, because of their structural orderings, COFs enable the precise insertion of molecular catalysts to produce a clearly defined structure with a unique local interface that includes spatial confinement, electron density, and orientation. COFs possess distinct structural characteristics over metal–organic frameworks (MOFs) in three main ways. They are as follows (Guo and Jiang 2020):

1. The covalent linkages within the framework maintain sufficient stability for keeping crystalline porous structures in a variety of solvents, such as dimethyl sulfoxide, imidazole, coordinative pyridine, strong acids (concentrated HCl), and bases (NaOH or KOH) (Geng et al. 2020).
2. The intralayer π–π interactions between the layered structures of COFs can only be exfoliated with significant external power and produces few-layer nanosheets with retention of the fundamental structural characteristic of COFs.
3. Almost no catalytic sites that are covalently attached to the frameworks degrade.

6.6.2 Structural Origins of COFs for Various Heterogeneous Catalyses

COFs can be constructed as heterogeneous catalysts by utilizing different chemistries because of the diversity of frameworks and pores. Guo and Jiang (2020) examined four potential structural origins for designing COFs for heterogeneous catalysis, i.e., (i) skeleton design, (ii) pore surface engineering, (iii) pore confinement, and (iv) systematic organization.

Usually, catalytic active sites are added to the frameworks of porous materials to design them as catalysts. In fact, the active catalytic centers could also be found in the skeleton of pure COFs. For some specific catalytic reactions, the linkages of COFs can function as acidic or basic active sites (Yusran et al. 2020a). The incorporation of catalytic sites on the pore walls can be done by fixing the side groups of monomers used to develop the COFs. This method allows pore surface engineering of various heterogeneous catalytic systems. Since pore size, pore shape, and pore environment can all be fully designed and controlled, the open pores of COFs allow for the physical loading of a variety of catalysts, such as metal nanoparticles and enzymes. These catalysts create catalytic systems by investigating confinement effects in the clearly defined pores with specific local environments. COFs are also incredibly potent and can be used to create multielectron catalytic systems for many chemical reactions, which enables the effective transport of electrons to the reaction centers (Guo and Jiang 2020).

6.6.3 Classification of COFs-Based Catalyst

COF-based catalysts are classified into four categories based on their structures and chemical compositions: (i) COFs with reactive skeletons, (ii) COFs with reactive pendant groups, (iii) COFs with reactive metals (including metal nanoparticles), and (iv) COFs with molecular catalysts. The skeleton of pure COFs itself can act as an active catalytic center (Yusran et al. 2020b). COFs can also incorporate reactive pendant groups such as thiophene, methoxy, hydroxyl, sulfoacids, and chiral

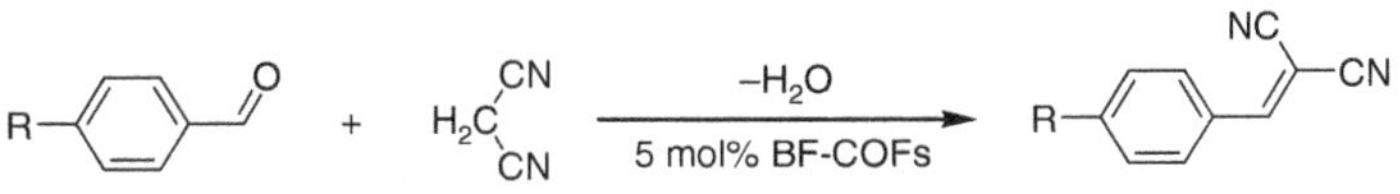

Figure 6.4 Heterogeneous catalysis reaction using BF-CPFs.

pyrrolidine moieties (Peng et al. 2015; Huang et al. 2019) as catalytic sites. Contrary to MOFs, which are made up of metal clusters that operate as reactive catalytic sites, adding reactive metals (Ding et al. 2011; Aiyappa et al. 2016; Yan et al. 2019) such as Mn, Co, Cu, Pd, and reactive metal nanoparticles to COFs involves a particular design of the COF structure as a scaffold. COFs are considered ideal confinement materials because of their excellent porosity and extensive capacity to accommodate incoming groups from small-to-large sizes (Yusran et al. 2020b). This enables excellent opportunities to incorporate various molecular catalysts (such as metal clusters and polyoxometalates) (Kibsgaard et al. 2014) to construct a COF-based porous catalyst with improved catalytic performance.

6.6.4 Applications of COFs as Heterogeneous Catalyst

6.6.4.1 COFs with Reactive Skeletons for Heterogeneous Catalysis

The active catalytic centers in the skeleton or linkages of COFs can catalyze specific chemical reactions. For example, in 2014, Yan and other coworkers developed two 3D base-functionalized microporous COFs, namely BF-COF-1, and BF-COF-2, as heterogeneous base catalysts (Figure 6.4). The imine linkages arose from the 3,5,7-tetraaminoadamantane (TAA) building blocks within the BF-COFs possessed strong basic sites. On the other hand, the microporous nature of COFs can offer a platform for size-selective catalysis. The catalytic performance of BF-COFs was investigated using the conventional base-catalyzed Knoevenagel condensation reactions (Fang et al. 2014).

The COFs exhibited promising catalytic performance with significant conversions. For example, for reactants with relatively small molecular diameters, BF-COF-1 and BF-COF-2 demonstrated conversion rates of up to 96% and 98%, respectively. On the other hand, catalysts showed lower conversion rates with reactants with relatively larger molecular diameters. This work made it clearly evident how highly effective size-selective catalysis could be carried out using BF-COFs with uniform pores (Fang et al. 2014).

Another excellent example of a COF catalyst with a reactive skeleton is Py-An COF, which is a two-dimensional COF with π-frameworks. This COF was prepared by Jiang and his team and utilized it under ambient conditions in catalyzing Diels–Alder reactions (Figure 6.5). The catalyst contains anthracene (An) backbone on the edges and pyrene (Py) backbone at the vertices.

The excellent catalytic activity of Py-An COF promotes the Diels–Alder reaction between 9-hydroxymethylanthracene and N-substituted maleimide derivatives under ambient temperature. This finding shows that COFs with reactive skeletons provide promising chemical catalysts that are simple to prepare without additional post-synthetic metallation steps (Wu et al. 2015).

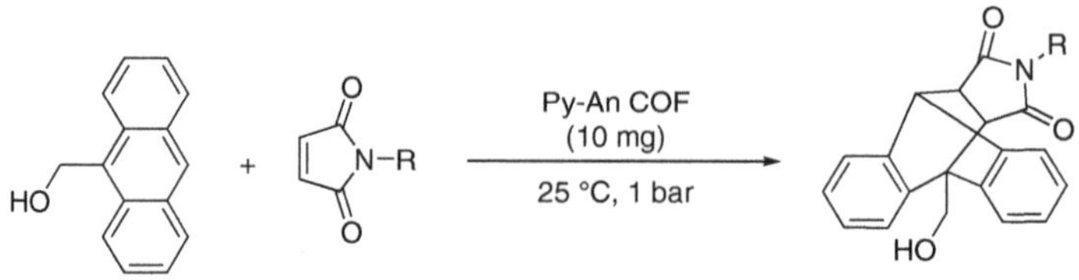

Figure 6.5 Py-An COF-based catalytic Diels–Alder reaction between 9-hydroxy-methylanthracene and N-substituted maleimide under ambient conditions. Source: Wu et al. (2015)/Royal Society of Chemistry.

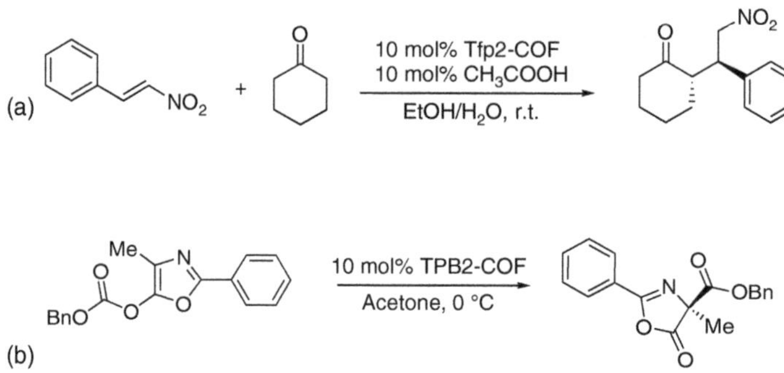

Figure 6.6 (A) Asymmetric Michael addition reactions catalyzed by Tfp2-COF, (B) asymmetric catalysis by COF-based catalyst.

6.6.4.2 COFs with Reactive Pendant Groups for Heterogeneous Catalysis

In addition to reactive skeletons, COFs with reactive pendant groups have also been widely developed for heterogeneous catalysis. The pendant groups with properties like electron-rich and chirality are catalytically reactive to several chemical reactions. For example, Zhang et al. (2019) developed two chiral COF catalysts named Tfp2-COF and TPB2-COF (Figure 6.6).

These catalysts were made of chiral pyrrolidine and DHIP moieties, respectively. The catalyst Tfp2-COF was employed to catalyze the Michael addition reaction, while the TPB2-COF catalyst catalyzed the asymmetric Steglich rearrangement reaction. Both the catalysts outperform their amorphous and homogeneous analogues in terms of stereoselectivity and diastereoselectivity (Zhang et al. 2019). Therefore, these chiral COFs can be considered effective heterogeneous catalysts for asymmetric catalysis.

Electron-rich organic pendant groups may also function in COFs as a reactive catalytic center to catalyze specific chemical reactions. For example, an electron-rich organic pendant group squaramide (SQ) was incorporated into a COF via a bottom-up synthesis method by Li and other coworkers. The synthesized 2D COF (COF-SQ) was then used as a heterogeneous catalyst for hydrogen-bonding organocatalysis reactions (Figure 6.7). High catalytic efficiency and good recyclability were achieved while using COF-SQ to catalyze the Michael-addition reaction. In the catalytic reaction between β-nitrostyrene and 2,4-pentanedione, 10 mol% of

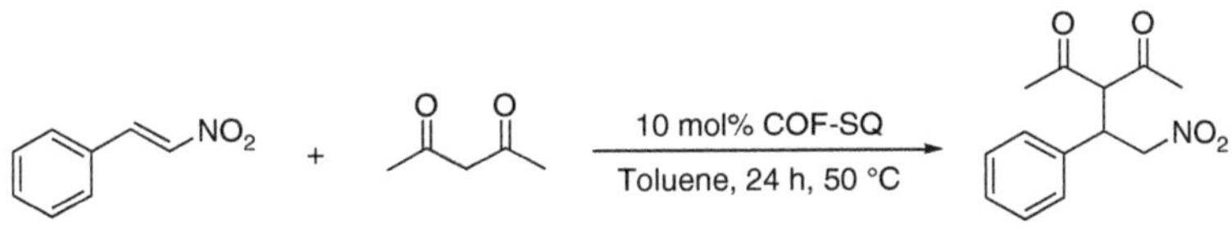

Figure 6.7 The Michael addition reaction between β-nitrostyrene and 2,4-pentanedione using COF-SQ as the heterogeneous catalyst. Source: Li et al. (2019a)/Royal Society of Chemistry.

COF-SQ was able to produce 95% yield using toluene as solvent at 50 °C for 24 hours (Li et al. 2019b).

6.6.4.3 COFs with Reactive Metals for Heterogeneous Catalysis

The earliest report of a COF with a reactive-metal-based catalyst for heterogeneous catalysis was reported by Ding et al. in the year 2011. They intercalated Pd(II) metal into COF-LZU1 by treating COF-LZU1 with $Pd(OAc)_2$ using dichloromethane as solvents at room temperature. The resultant catalyst Pd/COF-LZU1 was then used to catalyze the Suzuki–Miyaura coupling reaction between the bromobenzene substrate and phenylboronic acid (Figure 6.8) (Ding et al. 2011). The product yield of the reaction was 97%, with a very small amount of 0.5 mol% catalyst used. The catalyst surpassed the Pd-containing MOF analogues with their high catalytic performance under identical conditions (Llabrés i Xamena et al. 2007).

Metal nanoparticles can also be intercalated into COFs by reduction processing using strong reducing agents. Nevertheless, intercalation is also possible without the use of severe reducing agents, making them more effective and environmentally friendly. For instance, Mullangi et al. reported a triazine(trzn) COF with the loading of Pd nanoparticles into it. The resultant Pd(0)@trzn-COF catalyst was synthesized in a single-step approach using alcohol as a mild reducing agent instead of a severe reducing agent such as $NaBH_4$, $LiAlH_4$, or hydrazine (Figure 6.9) (Mullangi et al. 2015).

Remarkably, Pd(0)@trzn-COF catalyst can catalyze numerous Heck and C–C coupling reactions with excellent activity. Mullangi et al. reported a Heck coupling reaction between styrene and 1,4-dibromobenzene using Pd(0)@trzn-COF as a

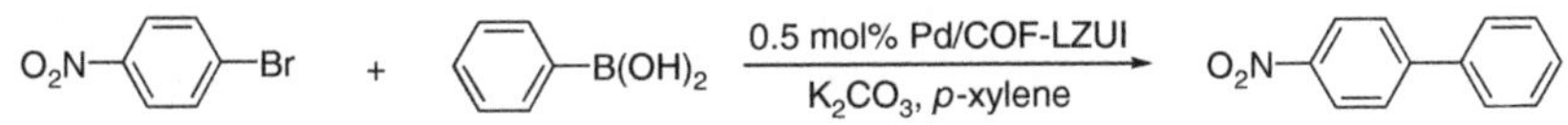

Figure 6.8 Suzuki–Miyaura cross-coupling reaction between *p*-nitrobromobenzene and phenylboronic acid by using 0.5 mol% Pd/COF-LZUI as the catalyst. Source: Ding et al. (2011)/American Chemical Society.

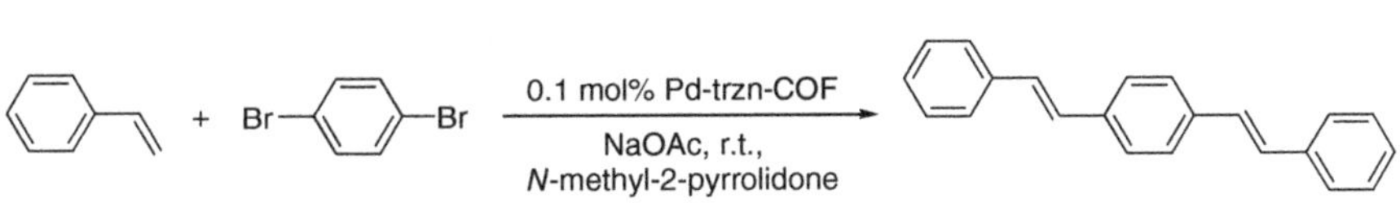

Figure 6.9 Heck coupling reaction between styrene and 1,4-dibromobenzene using Pd(0)@trzn-COF as the catalyst.

catalyst. Even at a very low catalyst dosage, the reaction gave 94% product yield, which shows excellent catalytic activity of the catalyst. Moreover, the catalyst also demonstrated good recyclability with minimal metal leaching (Mullangi et al. 2015).

6.6.4.4 COFs with Molecular Catalysts for Heterogeneous Catalysis

The immobilization of molecular catalysts into COFs has proven to be a promising way of increasing the catalytic efficiency of COF-based catalysts. For instance, Gao et al. synthesized three COFs composites by immobilization of varying amounts of phosphomolybdic acid (PMA) onto COF-300. The resultant three composites were denoted as PMA@COF-300a, PMA@-COF-300b, and PMA@COF-300c. The catalytic characteristics of these composites were examined in the epoxidation of olefins with *t*-BuOOH as the oxidizing agent (Gao et al. 2015).

As mentioned in the above table, PMA@COF-300c, which has higher amounts of PMA, showed higher catalytic efficiency with 65% and 91% epoxidation of *cis*-cyclooctene after three hours at 50 and 60 °C respectively, as compared to the other two composites. However, the homogeneous PMA catalyst showed the highest catalytic efficiency with 99% conversion after three hours at 60 °C. Again, the relatively inert 1-octene was converted to 1,2-epoxyoctane by PMA@COF-300c with 77% conversion after nine hours at 60 °C. Therefore, these findings indicate that all three PMA@COF-300 composites are active and selective catalysts for a variety of olefin epoxidation (Gao et al. 2015).

6.7 Polymer-Based Homogeneous Catalysis

The mainstream homogeneous catalysts have well-defined chemical structures and have the ability of breaking, formation, and rearrangement of chemical bonds (Van Leeuwen 2006). Chemical reactions catalyzed by homogeneous catalysts are selective and specific, producing just the desired product. It is possible to precisely manipulate the reaction conditions. Additionally, such catalysts can be used in solution phase which allows for greater mixing and mass transfer. However, it is reported that homogeneous catalysts are difficult to remove from reaction mixtures that result into poorer catalyst utilization, catalyst poisoning, and low efficiency. Other drawbacks of homogeneous catalysts include their susceptibility to impurities, pH, and other reaction conditions (Ali et al. 2014). There are very limited number of articles demonstrating the development of polymers that can act as homogeneous catalysis. Below are a couple of examples of homogeneous catalysts based on polymers.

Xie et al. have performed asymmetric transfer hydrogenation (ATH) of various ketones by means of a copolymer of dimethyl aminopropyl acrylamide (DMAPM), and *N*-*p*-styrenesulfonyl-1,2-diphenylethylenediamine (V-TsDPEN). The required copolymer was synthesized under solvothermal conditions. The monomer required for the reaction V-TsDPEN was synthesized from 1,2-diphenylethylene-diamine and *p*-styrenesulfonyl chloride followed by the coordination of $\{RuCl_2(p\text{-cymene})\}_2$.

The resulting polymer is a pH-responsive soluble polymer and was employed as an efficient catalyst. The polymer reduced acetophenone using the R-version catalyst under identical reaction conditions, and enantioselectivity was unaffected (Xie et al. 2016).

Using a biodegradable, homogeneous, and reusable polymer-supported catalyst called PEG-SO_3H, Paul et al. created a highly effective and affordable synthetic process for the production of coumarin and uracil-fused pyrrole derivatives. The Michael addition and intramolecular cyclisation was catalyzed by using PEG-SO_3H in methanol. The catalytic reaction results in the removal of the nitro group. Additionally, the catalyst can be used for five cycles (Paul and Das 2012).

6.8 Conclusion

In this chapter, an effort has been made to discuss the recent progress in the development of organic polymers for homogeneous as well as heterogeneous catalysis. From the aforementioned discussion, it can be stated that polymer-based catalysis finds a wide variety of chemical reactions that have industrial as well as small-scale production. One of the advantages of using polymer-based catalyst is the ease of separation of the catalyst after the chemical reaction. In general, isolation of the product after the catalysis does not require sophisticated instrumentation in the case of polymer-based heterogeneous catalysis. In addition, the catalytic properties of the polymers can be tuned by preparing composites with different other materials such as metal oxides and nanoparticles. It is also well known that polymers provide matrix-like structures or porous network that facilitates the catalysis. The solubility of network structured polymers is relatively less and hence can be easily used as heterogeneous catalysts. However, there are some challenges of using polymers as catalyst from the perspective of synthesizing the polymers as well as using them as catalysts. The reported polymer catalysts are non-biodegradable in nature that make a permanent waste. Additionally, the synthesis of polymers generally requires large quantities of monomers, radicals, and solvents. Therefore, the production of polymers is not cost-effective. In literature, it was found that most of the polymer-catalyzed reactions are performed at ambient temperature as the catalyst cannot withstand higher temperatures and leads to different toxic and degradation products. A great deal of scientific research and innovation is required to address these issues in the future on the ground of polymer-based catalysis.

References

Aijaz, A., Karkamkar, A., Choi, Y.J. et al. (2012). Immobilizing highly catalytically active Pt nanoparticles inside the pores of metal–organic framework: a double solvents approach. *Journal of the American Chemical Society* 134: 13926–13929.

Aiyappa, H.B., Thote, J., Shinde, D.B. et al. (2016). Cobalt-modified covalent organic framework as a robust water oxidation electrocatalyst. *Chemistry of Materials* 28: 4375–4379.

Ali, M., Rahman, M., Sarkar, M., and Abd Hamid, S.B. (2014). Heterogeneous metal catalysts for oxidation reactions. *Journal of Nanomaterials* 2014: 1–23.

Alkordi, M.H., Liu, Y., Larsen, R.W. et al. (2008). Zeolite-like metal–organic frameworks as platforms for applications: on metalloporphyrin-based catalysts. *Journal of the American Chemical Society* 130: 12639–12641.

Bandyopadhyay, S., Anil, A.G., James, A., and Patra, A. (2016). Multifunctional porous organic polymers: tuning of porosity, CO_2, and H_2 storage and visible-light-driven photocatalysis. *ACS Applied Materials and Interfaces* 8: 27669–27678.

Benkovic, S.J. and Hammes-Schiffer, S. (2003). A perspective on enzyme catalysis. *Science* 301: 1196–1202.

Bogaerts, T., Van Yperen-De Deyne, A., Liu, Y.Y. et al. (2013). Mnsalen@MIL101(al): a heterogeneous, enantioselective catalyst synthesized using a 'bottle around the ship' approach. *Chemical Communications* 49: 8021–8023.

Burtch, N.C., Heinen, J., Bennett, T.D. et al. (2018). Mechanical properties in metal–organic frameworks: emerging opportunities and challenges for device functionality and technological applications. *Advanced Materials* 30: 1704124.

Byun, J. and Zhang, K.A.I. (2020). Designing conjugated porous polymers for visible light-driven photocatalytic chemical transformations. *Materials Horizons* 7: 15–31.

Cao, Y., Wu, Z., Wang, T. et al. (2016). Immobilization of *Bacillus subtilis* lipase on a Cu-BTC based hierarchically porous metal-organic framework material: a biocatalyst for esterification. *Dalton Transactions* 45: 6998–7003.

Chen, L. and Xu, Q. (2019). Metal-organic framework composites for catalysis. *Matter* 1: 57–89.

Chen, L., Chen, H., Luque, R., and Li, Y. (2014). Metal–organic framework encapsulated Pd nanoparticles: towards advanced heterogeneous catalysts. *Chemical Science* 5: 3708–3714.

Chen, L., Chen, X., Liu, H., and Li, Y. (2015). Encapsulation of mono- or bimetal nanoparticles inside metal–organic frameworks via in situ incorporation of metal precursors. *Small* 11: 2642–2648.

Chen, L., Luque, R., and Li, Y. (2017). Controllable design of tunable nanostructures inside metal–organic frameworks. *Chemical Society Reviews* 46: 4614–4630.

Cheng, Y., Ding, X., and Han, B. (2021). Porous organic polymers for photocatalytic carbon dioxide reduction. *ChemPhotoChem* 5: 406–417.

Cho, S.H., Ma, B., Nguyen, S.T. et al. (2006). A metal–organic framework material that functions as an enantioselective catalyst for olefin epoxidation. *Chemical Communications* 2563–2565.

Ding, S.Y., Gao, J., Wang, Q. et al. (2011). Construction of covalent organic framework for catalysis: Pd/COF-LZU1 in Suzuki–Miyaura coupling reaction. *Journal of the American Chemical Society* 133: 19816–19822.

Dong, X.W., Liu, T., Hu, Y.Z. et al. (2013). Urea postmodified in a metal–organic framework as a catalytically active hydrogen-bond-donating heterogeneous catalyst. *Chemical Communications* 49: 7681–7683.

Dongen, S.F.M.v., Elemans, J.A.A.W., Rowan, A.E., and Nolte, R.J.M. (2014). Processive catalysis. *Angewandte Chemie International Edition* 53: 11420–11428.

Drout, R.J., Robison, L., and Farha, O.K. (2019). Catalytic applications of enzymes encapsulated in metal–organic frameworks. *Coordination Chemistry Reviews* 381: 151–160.

Du, D.Y., Qin, J.S., Li, S.L. et al. (2014). Recent advances in porous polyoxometalate-based metal–organic framework materials. *Chemical Society Reviews* 43: 4615–4632.

Dubey, R.J.C., Comito, R.J., Wu, Z. et al. (2017). Highly stereoselective heterogeneous diene polymerization by Co-MFU-4l: a single-site catalyst prepared by cation exchange. *Journal of the American Chemical Society* 139: 12664–12669.

Falkowski, J.M., Wang, C., Liu, S., and Lin, W. (2011). Actuation of asymmetric cyclopropanation catalysts: reversible single-crystal to single-crystal reduction of metal-organic frameworks. *Angewandte Chemie International Edition* 50: 8674–8678.

Fang, Q., Gu, S., Zheng, J. et al. (2014). 3D Microporous base-functionalized covalent organic frameworks for size-selective catalysis. *Angewandte Chemie International Edition* 53: 2878–2882.

Feng, D., Gu, Z.Y., Li, J.R. et al. (2012). Zirconium-metalloporphyrin PCN-222: mesoporous metal-organic frameworks with ultrahigh stability as biomimetic catalysts. *Angewandte Chemie International Edition* 51: 10307–10310.

Feng, D., Chung, W.C., Wei, Z. et al. (2013). Construction of ultrastable porphyrin Zr metal-organic frameworks through linker elimination. *Journal of the American Chemical Society* 135: 17105–17110.

Feng, D., Gu, Z.Y., Chen, Y.P. et al. (2014). A highly stable porphyrinic zirconium metal–organic framework with shp-a topology. *Journal of the American Chemical Society* 136: 17714–17717.

Férey, G., Mellot Draznieks, C., Serre, C. et al. (2005). A chromium terephthalate-based solid with unusually large pore volumes and surface area. *Science* 309: 2040–2042.

Fujishima, A. and Honda, K. (1972). Electrochemical photolysis of water at a semiconductor electrode. *Nature* 238: 37–38.

Furukawa, H., Cordova, K.E., O'Keeffe, M., and Yaghi, O.M. (2013). The chemistry and applications of metal-organic frameworks. *Science* 341: 1230444.

Gao, W., Sun, X., Niu, H. et al. (2015). Phosphomolybdic acid functionalized covalent organic frameworks: structure characterization and catalytic properties in olefin epoxidation. *Microporous and Mesoporous Materials* 213: 59–67.

Geng, K., He, T., Liu, R. et al. (2020). Covalent organic frameworks: design, synthesis, and functions. *Chemical Reviews* 120: 8814–8933.

Goetjen, T.A., Liu, J., Wu, Y. et al. (2020). Metal–organic framework (MOF) materials as polymerization catalysts: a review and recent advances. *Chemical Communications* 56: 10409–10418.

Guo, J. and Jiang, D. (2020). Covalent organic frameworks for heterogeneous catalysis: principle, current status, and challenges. *ACS Central Science* 6: 869–879.

Han, Q., He, C., Zhao, M. et al. (2013). Engineering chiral polyoxometalate hybrid metal-organic frameworks for asymmetric dihydroxylation of olefins. *Journal of the American Chemical Society* 135: 10186–10189.

Hartmann, M. and Fischer, M. (2012). Amino-functionalized basic catalysts with MIL-101 structure. *Microporous and Mesoporous Materials* 164: 38–43.

Hasegawa, S., Horike, S., Matsuda, R. et al. (2007). Three-dimensional porous coordination polymer functionalized with amide groups based on tridentate ligand: selective sorption and catalysis. *Journal of the American Chemical Society* 129: 2607–2614.

Huang, W., He, Q., Hu, Y., and Li, Y. (2019). Molecular heterostructures of covalent triazine frameworks for highly enhanced photocatalytic hydrogen production. *Angewandte Chemie International Edition* 58: 8676–8680.

Jiang, J. and Yaghi, O.M. (2015). Brønsted acidity in metal-organic frameworks. *Chemical Reviews* 115: 6966–6997.

Jiang, H.L., Feng, D., Wang, K. et al. (2013). An exceptionally stable, porphyrinic Zr metal–organic framework exhibiting pH-dependent fluorescence. *Journal of the American Chemical Society* 135: 13934–13938.

Jiao, L., Wang, Y., Jiang, H.L., and Xu, Q. (2018). Metal-organic frameworks as platforms for catalytic applications. *Advanced Materials* 30: 1703663.

Juan-Alcaniz, J., Gascon, J., and Kapteijn, F. (2012). Metal–organic frameworks as scaffolds for the encapsulation of active species: state of the art and future perspectives. *Journal of Materials Chemistry* 22: 10102–10118.

Kibsgaard, J., Jaramillo, T.F., and Besenbacher, F. (2014). Building an appropriate active-site motif into a hydrogen-evolution catalyst with thiomolybdate $[Mo_3S_{13}]_2^-$ clusters. *Nature Chemistry* 6: 248–253.

Kirlikovali, K.O., Chen, Z., Islamoglu, T. et al. (2020). Zirconium-based metal–organic frameworks for the catalytic hydrolysis of organophosphorus nerve agents. *ACS Applied Materials and Interfaces* 12: 14702–14720.

Kitagawa, S., Kitaura, R., and Noro, S. (2004). Functional porous coordination polymers. *Angewandte Chemie International Edition* 43: 2334–2375.

Lee, H.-C., Antonietti, M., and Schmidt, B.V.K.J. (2016). A Cu(II) metal-organic framework as a recyclable catalyst for ARGET ATRP. *Polymer Chemistry* 7: 7199–7203.

Li, D., Xu, H.Q., Jiao, L., and Jiang, H.L. (2019a). Metal-organic frameworks for catalysis: state-of-the-art, challenges, and opportunities. *EnergyChem* 1: 100005.

Li, X., Wang, Z., Sun, J. et al. (2019b). Squaramide-decorated covalent organic framework as a new platform for biomimetic hydrogen-bonding organocatalysis. *Chemical Communications* 55: 5423–5426.

Lian, X., Fang, Y., Joseph, E. et al. (2017). Enzyme-MOF (metal–organic framework) composites. *Chemical Society Reviews* 46: 3386–3401.

Liu, B., Jie, S., Bu, Z., and Li, B.G. (2014). A MOF-supported chromium catalyst for ethylene polymerization through post-synthetic modification. *Journal of Molecular Catalysis A: Chemical* 387: 63–68.

Llabrés i Xamena, F.X., Abad, A., Corma, A., and Garcia, H. (2007). MOFs as catalysts: activity, reusability, and shape-selectivity of a Pd-containing MOF. *Journal of Catalysis* 250: 294–298.

Lykourinou, V., Chen, Y., Wang, X.S. et al. (2011). Immobilization of MP-11 into a mesoporous metal–organic framework, MP-11@mesoMOF: a new platform for enzymatic catalysis. *Journal of the American Chemical Society* 133: 10382–10385.

Maksimchuk, N.V., Kovalenko, K.A., Arzumanov, S.S. et al. (2010). Hybrid polyoxotungstate/MIL-101 materials: synthesis, characterization, and catalysis of H_2O_2-based alkene epoxidation. *Inorganic Chemistry* 49: 2920–2930.

Mochizuki, S., Kitao, T., and Uemura, T. (2018). Controlled polymerizations using metal-organic frameworks. *Chemical Communications* 54: 11843–11856.

Mondloch, J.E., Katz, M.J., Isley, W.C. III, et al. (2015). Destruction of chemical warfare agents using metal–organic frameworks. *Nature Materials* 14: 512–516.

Mullangi, D., Nandi, S., Shalini, S. et al. (2015). Pd loaded amphiphilic COF as catalyst for multi-fold Heck reactions, C-C couplings, and CO oxidation. *Scientific Reports* 5: 10876.

Ollis, D.F. (1996). Heterogeneous photocatalytic oxidation of atmospheric trace contaminants. NASA. *Technical Reports Server (NTRS) 19970015392.*

Paul, S. and Das, A.R. (2012). A new application of polymer supported, homogeneous and reusable catalyst PEG–SO_3H in the synthesis of coumarin and uracil fused pyrrole derivatives. *Catalysis Science & Technology* 2 (6): 1130–1135.

Peng, Y., Hu, Z., Gao, Y. et al. (2015). Synthesis of a sulfonated two-dimensional covalent organic framework as an efficient solid acid catalyst for biobased chemical conversion. *ChemSusChem* 8: 3208–3212.

Rajeshwar, K., Osugi, M.E., Chanmanee, W. et al. (2008). Heterogeneous photocatalytic treatment of organic dyes in air and aqueous media. *Journal of Photochemistry and Photobiology C* 9: 171–192.

Saravanam, R., Gracia, F., and Stephen, A. (2017). Chapter 2: Basic principles, mechanism, and challenges of photocatalysis. In: *Nanocomposites for Visible Light-Induced Photocatalysis* (ed. M.M. Khan, D. Pradhan, and Y. Sohn), 19–40. Springer International Publishing AG.

Schmidt, B.V.K.J. (2019). Metal-organic frameworks in polymer science: polymerization catalysis, polymerization environment, and hybrid materials. *Macromolar Rapid Communications* 41: 1900333.

Shen, H., Peppel, T., Strunk, J., and Sun, Z. (2021). Photocatalytic reduction of CO_2 by metal-free-based materials: recent advances and future perspective. *Solar RRL* 2020: 1900546.

Shinde, D.B., Kandambeth, S., Pachfule, P. et al. (2015). Bifunctional covalent organic frameworks with two-dimensional organocatalytic micropores. *Chemical Communications* 51: 310–313.

Tahir, B., Iqbal, T., Rafique, M. et al. (2020). Chapter 5: Nanomaterials for photocatalysis. In: *Nanotechnology and Photocatalysis for Environmental Applications: A Volume in Micro and Nano Technologies* (ed. M.B. Tahir, M. Rafique, and M.S. Rafique), 65–76. Elsevier.

Van Leeuwen, P.W. (2006). *Homogeneous Catalysis: Understanding the Art*. Springer Science & Business Media.

Wang, Y., Zhang, N., Zhang, E. et al. (2019). Heterogeneous metal–organic-framework-based biohybrid catalysts for cascade reactions in organic solvent. *Chemistry – A European Journal* 25: 1716–1721.

Wang, T.X., Liang, H.P., Anito, D.A. et al. (2020). Emerging applications of porous organic polymers in visible-light photocatalysis. *Journal of Materials Chemistry A* 8: 7003–7034.

Wu, Y., Xu, H., Chen, X. et al. (2015). A π-electronic covalent organic framework catalyst: π-walls as catalytic beds for Diels–Alder reactions under ambient conditions. *Chemical Communications* 51: 10096–10098.

Xie, Y., Wang, M., Wu, X. et al. (2016). A pH-responsive soluble-polymer-based homogeneous ruthenium catalyst for highly efficient asymmetric transfer hydrogenation (ATH). *ChemPlusChem* 81: 541–549.

Xu, H.S., Ding, S.Y., An, W.K. et al. (2016). Constructing crystalline covalent organic frameworks from chiral building blocks. *Journal of the American Chemical Society* 138: 11489–11492.

Xu, C., Zhang, W., Tang, J. et al. (2018). Porous organic polymers: an emerged platform for photocatalytic water splitting. *Frontiers in Chemistry* 6: 592.

Yan, S., Guan, X., Li, H. et al. (2019). Three-dimensional salphen-based covalent organic frameworks as catalytic antioxidants. *Journal of the American Chemical Society* 141: 2920–2924.

Ye, J.J. and Wu, C.D. (2016). Immobilization of polyoxometalates in crystalline solids for highly efficient heterogeneous catalysis. *Dalton Transactions* 45: 10101–10112.

Yuan, B., Pan, Y., Li, Y. et al. (2010). A highly active heterogeneous palladium catalyst for the Suzuki–Miyaura and Ullmann coupling reactions of aryl chlorides in aqueous media. *Angewandte Chemie International Edition* 49: 4054–4058.

Yuan, S., Feng, L., Wang, K. et al. (2018). Stable metal–organic frameworks: design, synthesis, and applications. *Advanced Materials* 30: 1704303.

Yusran, Y., Li, H., Guan, X. et al. (2020a). Covalent organic frameworks for catalysis. *EnergyChem* 2: 100035.

Yusran, Y., Guan, X., Li, H. et al. (2020b). Postsynthetic functionalization of covalent organic frameworks. *National Science Review* 7: 170–190.

Zhang, J., Han, X., Wu, X. et al. (2019). Chiral DHIP- and pyrrolidine-based covalent organic frameworks for asymmetric catalysis. *ACS Sustainable Chemistry and Engineering* 7: 5065–5071.

Zhou, H.C., Long, J.R., and Yaghi, O.M. (2012). Introduction to metal-organic frameworks. *Chemical Reviews* 112: 673–674.

Zhou, Y.X., Chen, Y.Z., Hu, Y. et al. (2014). MIL-101-SO_3H: a highly efficient Brønsted acid catalyst for heterogeneous alcoholysis of epoxides under ambient conditions. *Chemistry – A European Journal* 20: 14976–14980.

Zhu, Q.L. and Xu, Q. (2014). Metal–organic framework composites. *Chemical Society Reviews* 43: 5468–5512.

Zhu, C., Yuan, G., Chen, X. et al. (2012). Chiral nanoporous metal–metallosalen frameworks for hydrolytic kinetic resolution of epoxides. *Journal of the American Chemical Society* 134: 8058–8061.

Zhu, L., Liu, X.Q., Jiang, H.L., and Sun, L.B. (2017). Metal-organic frameworks for heterogeneous basic catalysis. *Chemical Reviews* 117: 8129–8176.

Zou, R.Q., Sakurai, H., Han, S. et al. (2007). Probing the Lewis acid sites and CO catalytic oxidation activity of the porous metal–organic polymer [Cu(5-methylisophthalate)]. *Journal of the American Chemical Society* 129: 8402–8403.

7

Environmental Fate of Water-Soluble Cellulosic-Polymer-Based Composites

Bugra K. Debbarma, Sanghamitra Deka, Darshan Pegu, Golam Mohiuddin, and Prasanta Baishya

University of Science & Technology Meghalaya, Department of Chemistry, Killing Road, Ri-Bhoi, Meghalaya 793101, India

7.1 Introduction

In current days, the usage of polymers plays a vital role in every aspect of human life. Their applications vary from everyday packaging materials, biomedical applications, adhesives to various large-scale uses such as in construction and manufacturing textiles. The modifiable structure, light weight, weather-resistant properties, low cost are the key advantages of using the polymers (Godavitarne et al. 2017). Polymer history started with the development of celluloid in 1850s–1860s (Freinkel 2011). The term "polymer" was first coined by Swedish chemist J.J. Berzelius (Feldman, 2008). Polyvinyl chloride (PVC) was synthesized in 1920s following the development of Bakelite in 1907 (Geyer 2020). Due to the versatile and durable nature of plastic, it became the key element for the production of warfare materials during World War II. The use of newly invented plastics in the postwar period increased to a greater extent with the global annual production of two million metric tons (Mt) by 1950 (Geyer 2020). Plastic pollution is now emerging as a global crisis which has negative impacts in every step right from the extraction, disposal till its incineration (Ramesh Kumar et al. 2020). The plastic particles being insoluble and toxic are a major marine pollutant and can be ingested by marine organisms, leading to the accumulation of microplastics and causing harm to many oceanic ecosystems (Koelmans et al. 2015). The conventional synthetic polymers usually originate from petroleum and are non-biodegradable (Lu et al. 2009; Hazarika et al. 2015). In the process of polymerization including their synthesis, medium of reaction and chemical feedstock induce toxicity. Petroleum as a resource is limited, and serious environmental issues are arising due to the excessive use of synthetic polymers which have led to the research of alternative materials to mitigate such problems (Lu et al. 2009; Hazarika et al. 2015). To overcome these drawbacks green routes are adapted for more efficient modeling of polymeric substances (Erdmenger et al. 2010). Moreover, governments around the world have decided to make stringent regulations to control environmental issues and raise awareness (Hazarika et al. 2015). Therefore, attempts

Organic Polymers in Energy-Environmental Applications, First Edition.
Edited by Ramesh Oraon, Pardeep Singh, Sanchayita Rajkhowa, Sangita Agarwal, and Ravindra Pratap Singh.

are being made to generate new authentic and high-performance materials within a rational budget in the last few years. Simpler chemical processes and innovative product designs from chemical industries for future generations can assist in creating a lesser environmental impact. Researches have been going on to produce a biocompatible alternative to plastic which causes no more threat to the nature and can be disposed of easily.

The use of biodegradable polymers is a good alternative and more beneficial in terms of different aspects than synthetic polymers. Biodegradation is the key factor for considering the utilization of natural polymers in different applications in place of synthetic-polymeric-based products. Natural polymers are easily available and have a low cost (Jiang et al. 2020; Rahman et al. 2021). Along with their low-cost and eco-friendly nature, they are a desirable choice for the development of materials for replacing synthetic polymers. Naturally available, water-soluble polymers are given more importance in the recent years to form a substitute for the synthetic polymers as well as other natural polymers. Water-soluble cellulosic polymers are easily available, low in cost, easy to synthesize, and more environment friendly as compared to other polymers (George and Sabapathi 2015).

Water-soluble polymers have the property to dissolve, disperse, or swell in water, resulting in a physical modification through gelation, thickening, emulsification, or stabilization (Kadajji and Betageri 2011). The backbone of such polymeric chain contains hydrophilic groups which may be nonionic, cationic, anionic, or amphoteric in nature (Kadajji and Betageri 2011). Water-soluble polymers can be either synthetic or from a natural source (Kadajji and Betageri 2011). Polymers that are water soluble when extracted from natural sources are termed as biopolymers, e.g., polysaccharides, protein, polynucleotides, and polypeptides. These biopolymers impart specific biological properties, have distinct microstructures, and exhibit perfect monodispersity (McCormick et al. 2002). Water-soluble polymers can be either homo- or co-polymer where properties like selectivity, solubility, and sorption capacity of metal ions can be controlled or improved by the selection of comonomers (Rivas et al. 2018).

Solvents play a significant role in the preparation of different polymeric-based substances. Various steps like dissolving substances, maintaining viscosity, heat transportation, purification, and separation processes are involved when a solvent is used in the preparation of a material (Cvjetko Bubalo et al. 2015). According to the green chemistry principle an ideal solvent eases the transfer of mass but does not solvate the reactants (Li and Trost 2008). The use of water as a solvent has multiple advantages like being renewable, eco-friendly, and inexpensive compared to the other solvents. Therefore, it is considered as the most ideal solvent available among all the green solvents (Hazarika et al. 2015). However there are certain issues like water impurity and insolubility of organic compounds for using water as a solvent.

The applications of water-soluble polymers are immense. Water-soluble polymers can be applied to a wide range of industrial applications, where pharmaceuticals, food industries, water treatment, and textiles are few to mention but are not limited to these alone. They have found variety of applications in oilfield treatments, coating agents, cosmetic products, medical products, gel formations, textile industries, etc.

In water treatment, water-soluble polymers are used with ultrafiltration membranes in a technique called LPR (liquid-phase polymer-based retention). This technique is used for the removal of metal or metalloid ions in water (Rivas et al. 2018). However, limitations are faced due to the poor mechanical properties, heat resistivity, and easy moisture absorption properties of natural polymers. Several studies have been done for enhancing the properties of natural polymers.

In this chapter, we have tried to focus on the future aspects of two naturally available water-soluble cellulosic polymer, starch and carboxymethyl cellulose (CMC), which can be a preferable keystone substitute to synthetic polymers.

7.2 Starch: A Widely Known Water-Soluble Polymer

Starch is one of the most abundant plant polysaccharides. It is also a major source of carbohydrates and energy. Cassava (*Manihot esculenta*), potato (*Solanum tuberosum*), corn/maize (*Zea mays*), wheat (*Triticum aestivum*), and sweet potato (*Ipomoea batatas*) are the major sources of starch (Ratnayake and Jackson 2009). It is naturally regenerated in plants through photosynthesis. Starch can be hydrolyzed into glucose by the action of microorganisms and enzymes, through which it is then metabolized into water and carbon dioxide. It is then recycled into starch again in plants (Lu et al. 2009). It is a suitable substance for developing completely biodegradable, renewable, and low-cost sustainable materials. There has been an effort to lower the utilization of plastic and solve the problem caused by non-biodegradable materials. Starch is a good alternate material that can replace the non-biodegradable material.

Starch has numerous applications and can be used in synthesizing biodegradable packaging materials for their odorless, non-toxic, and physiologically inert properties. They are widely used in hybrid materials suitable for pharmaceutical applications such as food grade thickeners, wound dressing, tissue engineering, emulsion stabilizers, moisture binders in food products, and drug delivery (Jiang et al. 2020; Shah et al. 2016; Hemamalini and Dev 2018). Starch films are used for protecting food. Starch can be used as an alternative polymeric film for packaging of food materials which is environment friendly (Molavi et al. 2021). Starch films keep the food fit for consumption for longer periods of time. Starch films are obtained from native starch or from any of its component. With appropriate modifications and improvements, natural polymers become a favorable material for any commercial, industrial, or other required purposes. They can be obtained either by solution casting or by thermoplastic processing, i.e., wet method and dry method, respectively (Jimenez et al. 2012). Pareta and Edirisinghe (2006) have developed a novel method for the preparation of starch films and coating.

7.2.1 Structure of Starch

Starch is mainly stored as carbohydrates in plants in the form of insoluble, semi-crystalline granules (Copeland et al. 2009). It is the second most abundant polysaccharide after cellulose found in plants. It is found deposited in different parts of the plant such as the stem, tuberous tissues, seeds, and even in different algae (Hemamalini and Dev 2018). The molecular structure of starch is mainly composed of amylose and amylopectin which are two different polymers of D-glucopyranose.

Amylose is made up of α (1,4)-linked glucopyranose units with a molecular mass ranging from 105 to 107 g mol^{-1}. While generally amylose is considered as a linear chain molecule, most amylose molecules when prepared contain two to five branches (Whistler and Daniel 2000). Amylopectin contains a backbone of α (1,4)-linked glucopyranose units with α (1,6) branch points at every 20–30 glucose units. Its molecular mass ranges from 107 to 109 g mol^{-1}. The degree of polymerization of amylose and amylopectin is found to be around 6000 and 2,000,000, respectively. Amylose mostly forms the amorphous region, while amylopectin constitutes the crystalline region of the starch granule (Hemamalini and Dev 2018; Thakur et al. 2019). The percentage of amylose and amylopectin content basically depends on the source of starch. However, approximately 10–20% of amylose and 80–90% of amylopectin are generally found to be present in any natural source of starch. Amylose is water soluble and has a helical structure. Amylopectin forms a helical structure and due to its crystalline nature, starch forms discrete granules (Lu et al. 2009).

The presence of numerous hydroxyl groups on the starch chain results in the strong hydrophilic nature of the starch molecule. In the structure of amylose (Figure 7.1), two secondary hydroxyl groups are attached each at the C-2 and C-3 position of the glucose residue, and the C-6 position contains one non-bonded primary hydroxyl group. These free hydroxyl groups can be easily reduced, oxidized, form hydrogen bonds, and also take part in ether and ester formation (Lu et al. 2009). Variation in size, chemical composition, shape, and structure of starch granules can be observed depending on their source of origin (Molavi et al. 2021). The structure of amylose and amylopectin is given in Figures 7.1 and 7.2.

7.2.2 Advantages and Drawbacks of Starch as a Polymer

Starch is an attractive alternative to conventional plastics due to its inherent biodegradability, high availability, easy renewability, low cost, etc. (Jiang et al. 2020). Non-sustainable, disposable plastic products available are problematic for creating large amounts of plastic waste and environmental problems. Therefore, new biodegradable-starch-based polymeric substances are being researched and developed seeking for a better alternative. Starch-based materials demand a very

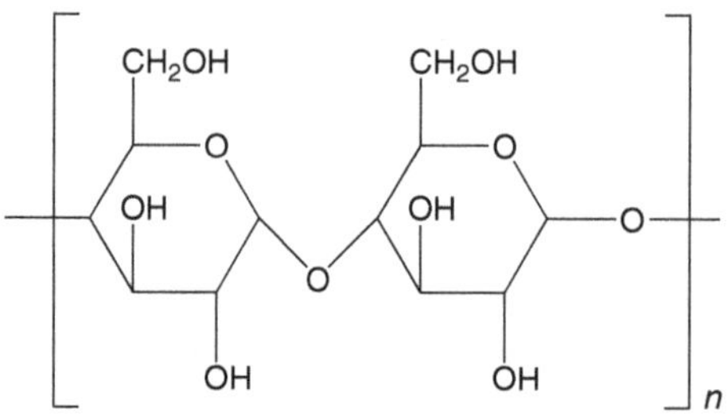

Figure 7.1 Structure of amylose.

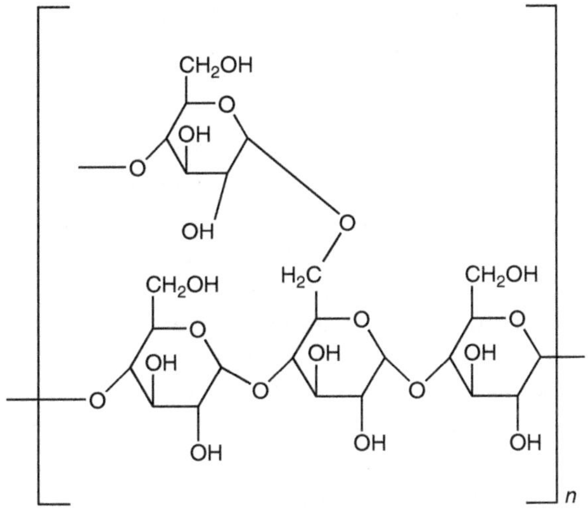

Figure 7.2 Structure of amylopectin.

low material cost and can be processed easily by utilizing green routes giving rise to eco-friendly and biodegradable materials (Jiang et al. 2020).

Various properties and characteristics of starch are utilized for creating films, packaging, and coating materials. Starch is easily available, biodegradable, and biocompatible and gives a high yield on extraction. Safe edible materials have also been made possible due to the functional properties of starch. It is odorless, tasteless, and generally non-toxic with nutritional values (Shah et al. 2016; Abe et al. 2021). Starch can form coatings and films due to the linear nature of amylose. Starch-based films are semi-permeable to gases like oxygen and carbon dioxide, lipids, moisture as well as flavor components (Shah et al. 2016). Depending on the application, color and transparency are other important features of starch-based polymers. During modification of starch, color and transparency should be considered as an important term as it reflects the utilization, consumer acceptance, and overall appearance of the material (Abe et al. 2021).

However the poor physical properties of starch such as high hydrophilic nature, low temperature of decomposition, and elevated recrystallization behavior limit the use of native starch for the preparation of end products (Baishya and Maji 2014). The products prepared from native starch are brittle in character. Native starch shows inadequate mechanical strength, dimensional firmness, and processability (Lu et al. 2009). Thus, modifying starch with proper chemical and physical treatment is necessary for improved applications and to achieve a broader range of functionality. Starch can be modified by chemical or physical treatment.

7.2.3 Modification of Starch

Many of the naturally occurring polymers are not suitable enough to be used for applications directly due to their low thermal stability, low water resistance, poor mechanical properties, and chemical resistance properties of their end products. Thus, it is necessary to modify starch in order to obtain the desired properties for advanced applications (Baishya et al. 2017). The different chemical or physical modifications methods include blending, grafting, and use of crosslinkers to improve the properties (Lu et al. 2009). Polylactic acid (PLA) is a biodegradable polyester that can be used to develop PLA/starch blends to improve their mechanical and thermal properties (Zhang and Sun 2004). PLA, being hydrophobic, interacts poorly with hydrophilic starch granules. To increase the compatibility between hydrophilic starch and hydrophobic PLA different compatibilizers were used (Lu et al. 2009; Zhang and Sun 2004). Introducing plasticizing agents such as glycerol to the PLA/starch blend can convert the blend into thermoplastic starch (TPS) which improves their dispersion and interfacial affinity (Lu et al. 2009; Ribba et al. 2017). TPS is prepared by heating at a temperature between 140 and 160 °C at high pressure and high shear. TPS can be repeatedly softened and molded in the desired shape (Carvalho 2008). The starch obtained from natural source has to undergo gelatinization process in the presence of excess water where it becomes irreversible change. Swelling of granules depending on the quantity of water is known as starch gelatinization (Jimenez et al. 2012). Yam starch films is one of the examples of TPS produced by gelatinization (Mali et al. 2002). In cold water, strong hydrogen bonds (intermolecular and intramolecular) hold the starch chain together causing it to be insoluble. When starch gets converted to TPS the hydrogen bonds of the starch molecule are broken down which results in the degradation of the crystallinity of the starch granule (Lu et al. 2009; Ribba et al. 2017). This allows the reaction of water molecules with the hydroxyl groups of the starch granules making it partially soluble in water (Ribba et al. 2017). Modification of starch chemically can improve the barrier and physiochemical properties, but at the same time it reduces its biodegradation property to a certain level. The key factor for lower biodegradation rate is due to the reduced degree of hydrophilicity of the modified starch. When starch is treated through chemical derivatization, it becomes less soluble in water as the interactions between the starch chain and water molecule reduce. Moisture is important for microorganisms to multiply and break down a starch bioplastic (Abe et al. 2021). One of the best and encouraging products is to manufacture products that are eco-friendly and less dependent on fossil fuels. Few of the ways by which the properties of polymers can be modified are grafting, blending, crosslinking, etc. (Hazarika et al. 2015).

7.2.3.1 Grafting

Grafting is often considered a very important technique to alter the chemical as well as the physical properties of starch. Different functional groups are incorporated in the polymers for improving the properties of the final products. Various techniques are used to graft copolymerize depending on the monomer of the polymeric backbones (Baishya and Maji 2016a). In grafting, the monomers are bonded covalently onto the main long chain of the polymer, and further polymerization takes place on this chain (Haroon et al. 2016). Grafting can be carried out by different techniques such as free radical polymerization, coupling reactions, and controlled radical polymerization (Meimoun et al. 2018; Chen et al. 2017). In order to graft, an active site has to be formed in the mid of the polymeric chain, where the grafting of the polymer takes place when exposed to the monomer that can bind to it rather than its end point. Starch is more reactive as more number of hydroxyl groups are attached to it. Grafting of renewable polymers like starch, cellulose, and polysaccharides with synthetic polymers are able to produce hybrid products with induced control properties. Djordjevic et al. in their paper reported how grafting of starch with acrylic acid in the presence of various initiators like AIBN, KSP, and BPO gives different results (Haroon et al. 2016; Djordjevic et al. 2013).

Naturally occurring polymers are not water repellent and do not have good mechanical strength to make use of them for better applications. Thus it is essential to modify their native structure by grafting with synthetic polymer to achieve molded innovative products with controlled properties and structure. Novel renewable-based products have been designed through ring-opening polymerization, controlled radical polymerization, and click chemistry giving interesting new opportunities (Meimoun et al. 2018; Uliniuc et al. 2013). Radical polymerization is often opted to produce graft copolymers. The chain transfer reactions are initiated by the elimination of hydrogen atoms in most of the cases. Lately, graft copolymers synthesized from natural polymers have been significantly increasing in the industrial applications. Therefore, grafting has become a potential technique in the modification of the properties of natural polymers by using different monomers (Hazarika et al. 2015). Principally, there are three techniques used for grafting to synthesize co-polymers, viz., grafting onto, grafting from, and grafting through. Grafting onto process takes place between two different polymers' functional groups. In grafting from, the grafting takes place within the specific functional group which triggers the polymerization of vinyl monomers. Whereas grafting through takes place in the copolymerization of macromolecules. Among these three types of technique the most preferred one is the grafting from, as it produces high yield. The reaction pathways that grafting follows are free-radical, ionic path, and living polymerization (Haroon et al. 2016).

7.2.3.2 Blending

Blending is another method used for the modification of starch which improves the properties of starch to an utmost level. Blending is a physical method carried out to modify the physical properties of starch. Blending is preferable more pertinently used as it is chemically safer and low in cost. Starch of various concentrations can also be blended with urea formaldehyde (UF) resin by esterification and oxidation. The produced blend of UF-starch shows improved water resistant properties, good insulation, low brittleness, higher temperature resistivity, chemical stability, oil resistance, mildew resistance, slower aging, and also non-toxic to the environment (Gadhave et al. 2017). The wheat gluten protein and modified potato starch (70/30 ratio) were used to prepare the composites using extrusion method, and it exhibited excellent gas barrier properties for packaging applications (Muneer et al. 2015). Starch can be blended with petrochemical-based polymers to improve their biodegradability (Bien et al. 2001). Naturally occurring starch found in potatoes is modified with allylglycidyl ether (AGE), which is used in the crosslinked composites preparation (Baishya et al. 2017). One of the techniques used in blending is heat-moisture treatment (HMT) (da Rosa Zavareze et al. 2012). By this technique, the functional properties of starch improve exceedingly and so also its durability (Goel et al. 2020).

PLA is another eco-friendly polyester which is used for various purposes like packaging of food (Muller et al. 2017), agriculture films, and medical devices. However, the rate of its biodegradability is very low and gets delayed for recycling into the ecosystem. To overcome this challenge, PLA can be blended with starch and wood flour to increase the rate of its biodegradability and reduce its cost. Blending of PLA with starch has shown to improve tremendously in accelerating the rate of its biodegradability without generating any toxic substances that are harmful to the environment (Gadhave et al. 2017). The blending of starch with other polymers with proper ratio in the preparation of composites may provide some new alternatives to the development of bio-based composites with enhanced properties. Starch was physically blended with polyolefin as a filler by Griffin with low concentrations ranging from 6% to 15% to induce the degradation rate of the vinyl polymers through the introduction of starch to the carbon-carbon matrix (Griffin 1977; Bikiaris et al. 1998).

7.2.3.3 Crosslinking Agent

Crosslinking is the chemical modification of starch which is carried out to improve its performance for different applications (Baishya and Maji 2016a; Reddy and Yang 2010). Crosslinking agents are best used for the treatment of two incompatible or semi-compatible materials to improve their properties for the advanced applications. The agents are mostly bifunctional compounds, in which one functional is attached to polymer and the other component is the reinforcing agent. Crosslinking alters the intra- and interhydrogen bond of starch chains with stronger and permanent covalent bonds.

The functional group plays an important role in deciding the properties of modified starches. Various crosslinking agents like dimethyloldihydroxy-ethyleneurea (DMDHEU), epichlorohydrin, *N*-methylol acrylamide (NMA), and 2-hydroxyethyl methacrylate (HEMA) are used to enhance the properties of the starch-based composites (Baishya and Maji 2014; Moad 2011; Hazarika and Maji 2013). DMDHEU is widely used as crosslinking agent in the cotton and wood industries (Emmerich et al. 2019). *N,N*-dimethylformamide also shows high degree of crosslinking (Haroon et al. 2016). Epichlorohydrin (EPI) is the most commonly used crosslinking agent of starch. Jyothi et al. (2006) in their work on cassava starch reported using EPI as crosslinking agent. The modified starch showed better water-resistance capacities than native starch. Starch is also modified by AGE which gives several crosslinked matrixes that reduce the excessive motion of macromolecules and the hydrophilicity of starch (Duanmu et al. 2007a). Crosslinking agents also play an immense role in improving the viscoelasticity and increasing the strength of composites (Zeng et al. 2019). Baishya et al. (2018) have synthesized a nanocomposite blended with methyl methacrylate-grafted starch and wheat gluten using citric acid as the crosslinker and reported the improved properties of the starch-based composites. The mechanism of modification starch with the crosslinker and grafting is explained in Figure 7.3 (Baishya and Maji 2016b).

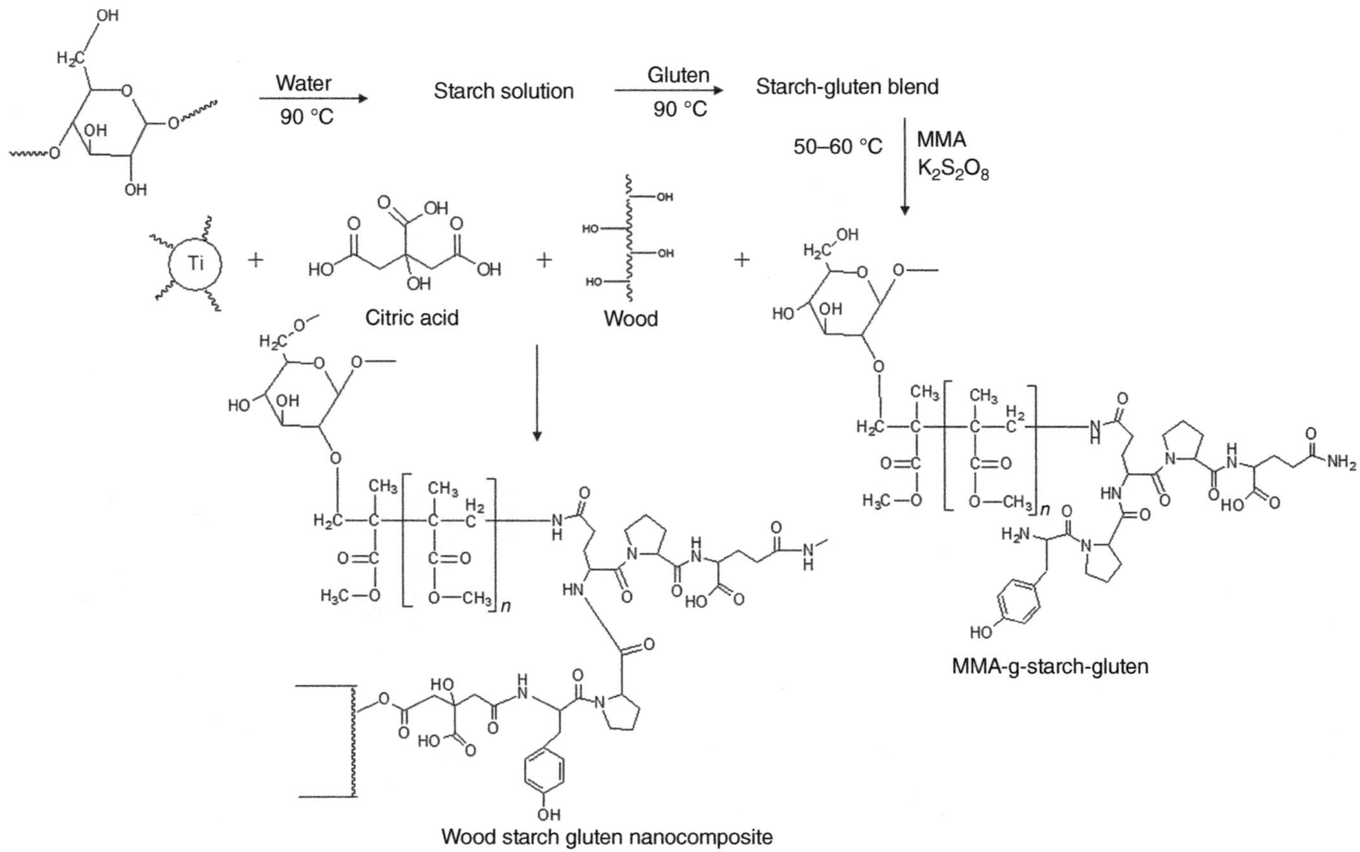

Figure 7.3 Grafting of starch gluten blend and crosslinked with citric acid. Source: Baishya et al. (2018)/with permission of Elsevier.

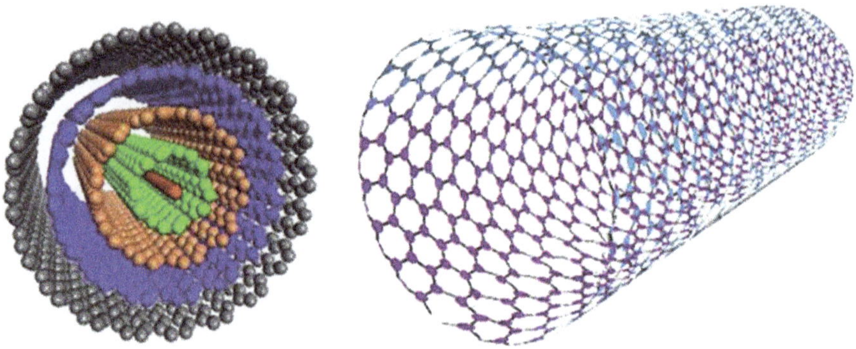

Figure 7.4 Structure of carbon nanotubes. Source: Hazarika et al. (2015)/Springer Nature.

7.2.3.4 Plasticizers

The brittle character of starch and starch-based composites restrict their applications in numerous areas. The properties of starch composites can also be improved by incorporating plasticizer. Incorporation of plasticizer helps to improve the properties of starch (Baishya and Maji 2014, 2016a). Plasticizers increase the flexibility and toughness of the starch-based composites by internal modification of the polymeric molecule. Plasticizer improves the flexibility and toughness of the starch-based composites by modifying the internal molecular structure of the polymer. Glycerol, sorbitol are most used plasticizers to modify the starch. Müller et al. (2014) have prepared the starch-based wood composites from starch plasticized with 36% glycerol which notably improved the tensile strength of the composites. De Carvalho et al. (2002) have prepared the composites from 30% glycerol plasticized corn starch containing 10–15% wood pulp fiber which exhibited tensility of 10 MPa. Chang et al. (2010) have prepared the glycerol-modified starch/cellulose composite by casting process. Muller et al. prepared wood composites from plasticized starch and the water adsorption properties, tensile strength, structure of the composite, and shrinkage were observed. The modified polymer was shown to have increase in strength and stiffness, whereas the shrinkage decreases at low level with 15–20 vol% wood content. The strong interfacial adhesion makes it less water absorbable maintaining the water uptake even in the presence of wood particle (Müller et al. 2014).

7.2.3.5 Use of Nanofillers

The use of nanotechnology in the field of composites to develop high-quality materials is one of the exciting techniques in recent days. Nanoscience and technology is widely used for manufacturing of multifunctional products with supreme quality (Candan and Akbulut 2013; Montanari et al. 2021). The nanoparticles can be of different dimensions like one dimension, two dimensions, three dimensions, and even spherical or polyhedral. Nanocomposites properties differ based on the preparation technique used and the nature of the nanofiller. Addition of nanoparticles into the polymeric matrix induces the chemical as well as physical characteristics of the composite as compared to the one incorporated with microparticle (Baishya and Maji 2017). Nanofillers like SiO_2, TiO_2, ZnO, graphene oxide, and carbon nanotube (CNT) are widely used with the polymeric matrix to improve properties like mechanical strength, thermal stability, flame resistance properties, water and chemical resistance properties of composites for advanced applications. Incorporating SiO_2 nanoparticles has improved the mechanical strength and thermal stability of polymer composites (Deka et al. 2014; Deka and Maji 2013). Wang et al. (2013) have also reported that SiO_2 nanofillers have improved the mechanical and the thermal properties of nanocomposites. Yang et al. (2005) in their work used ZnO, SiO_2, and Al_2O_3 as photo stabilizer with the polymer and reported the improvement in UV stability and the thermal stability of polymer nanocomposites. TiO_2 are used as photocatalyst to decompose organic pollutants in water and are also used extensively in industries like solar cells and catalysis (Etgar et al. 2012; Subramanian et al. 2004). Saraç et al. (2019) reported the beneficial use of nano-TiO_2, nano-SiO_2 on composites which shows improvement in the mechanical properties.

Graphene polymeric material has been used for many polymer matrixes, like epoxy, polystyrene, polypropylene, nylon in many applications. Graphene oxide gives outstanding properties because of its higher surface area. These graphene fillers make the material mechanically tougher. Graphene and TiO_2 are used extensively as fillers for the purpose of improving the functionality of polymers (Kavimani et al. 2020). CNTs are long, cylindrical shaped nanostructures (Hazarika et al. 2015; Ebbesen 1994). The length of the CNT varies from nanometers to micron to millimeters and radius can be in the size of nanometer (Dresselhaus et al. 2000). The awesome physical and the chemical properties of CNT nanocomposite have made many interested to work on the wood nanocomposite for the advancement in applications (Hazarika et al. 2015). Cao et al. prepared CNT nanocomposites from starch with MWCNTs as fillers and reported that with the presence of MWCNT it decreased water sensitivity and gives better mechanical properties to the nanocomposite (Hazarika et al. 2015; Cao et al. 2007).

7.3 Carboxymethyl Cellulose (CMC)

CMC is one of the most widely used anionic and water-soluble derivatives of cellulose (Pettignano et al. 2019). Etherification of cellulose results in the production of CMC which is used predominantly in the manufacturing of biodegradable products (Basta et al. 2016). In 1918, CMC was manufactured for the first time; however, the commercial manufacturing of CMC was approved in the early 1920s in Germany (Paria and Rai 2022). Its flexile characteristics include bio-adhesive nature, pH sensitivity, non-toxicity, hydrophilicity, thermal gelation, and viscosity due to which CMCs are given higher importance than the other derivatives of cellulose (Pettignano et al. 2019; Paria and Rai 2022). CMC is mainly obtained from its precursor, cellulose (Rahman et al. 2021). Cellulose is highly abundant in wood and other plant-based sources which contain naturally a high amount of cellulose, some of the examples being corn cobs, husk, and stalks, sugarcane bagasse, maize stalks, and banana pseudo-stems. Alternatively, many researchers were able to introduce some waste materials like waste textiles, knitted rags, cotton gin wastes, wastepaper, and waste cotton linters from textile industries as raw materials of cellulose for the production of CMC (Rahman et al. 2021). CMC is also present in its sodium salt form or Na-CMC, and can be prepared by carboxymethylation of cellulose obtained from banana stem, sago waste, paper sludge, and many other natural resources (Adinugraha and Marseno 2005; Pushpamalar et al. 2006). Cellulose undergoes etherification reaction with anionic carboxymethyl groups (CH_2COO^-) and produces CMC or its sodium salt which is capable of forming films with improved physico-chemical properties.

CMC when used in its raw form shows inadequate mechanical, thermal, and optical properties. The substitution of carboxymethyl groups into its structure induces the hydrophilic nature which is indeed an essential quality for its solvation in aqueous medium but also difficult to handle in some cases. The modifiable nature of CMC makes it easier to manufacture CMC-based products with enhanced physical and chemical properties. Various researches have been carried out for modification of CMC and CMC-based substances via grafting, blending, and crosslinking processes (Pettignano et al. 2019; Paria and Rai 2022). These modification processes proved to be advantageous for the future application of CMC-based materials in different fields. The application of CMC and CMC-based hybrid materials relies on the uniformity, purity, degree of substitution, and degree of polymerization which constitutes the basic criteria for the improved performance of the products like size of the particle, surface properties, thermal stability, mechanical strength, and viscosity (Rahman et al. 2021). Several wound-healing CMC-based materials like hydrogels, antimicrobial composites, films, 3D printing materials, and nanofibers have also gained importance in the recent years (Kanikireddy et al. 2020).

7.3.1 Structure and Synthesis

CMC is an anhydrous polysaccharide of glucose where the repeating units are connected by the – 1,4-glycosidic linkages and oriented in a linear fashion. At the molecular level, the key distinction between the structure of CMC and cellulose is that in the case of CMC anionic carboxymethyl groups are attached to the hydroxyl groups of the glucopyranose chain of cellulose (Paria and Rai 2022). Being the derivative of cellulose CMC possesses inherited structural properties such as (i) the glucose units in CMC are rigid; (ii) backbone is fully extended in nature owing to restricted rotation around the C—O—C bond between the glucose units because of strong steric hindrance; (iii) orientation of the bulkier groups (–OH, $-CH_2OH$, $-OCH_2COOH$) toward outside of the equatorial plane of the saccharic ring (Zhivkov 2013). The three hydroxyl groups present in the cellulose monomer unit can be substituted by $-CH_2COOH$ groups to reach the maximum value of degree of substitution to 3, but generally it reaches up to 2 (Zhivkov 2013).

Williamson ether synthesis is usually carried out to prepare CMC which involves two steps (Kanikireddy et al. 2020). At first, the cellulose obtained from various natural resources is treated with sodium hydroxide which results in the formation of alkali cellulose. The alkali cellulose is then allowed to react with monochloroacetic acid (Pettignano et al. 2019). In the final step, the anionic carboxymethyl groups are adjoined to the cellulose molecule with the replacement of the three hydroxyl groups with the varying degree of substitution ranging between 0.4 and 1.5 (Pettignano et al. 2019; Wong and Ramli 2014). The degree of substitution determines the availability of CMC with various molecular weights between the range of 90,000–2,000,00 g mol^{-1} (Kanikireddy et al. 2020).

The length of the CMC polymer is generally shorter as compared to cellulose, and it consists of some regions throughout the molecule where uneven distribution of carboxymethyl group is found. In lower concentrations CMCs show extended nature, while at higher concentrations overlapping and coiling of the molecules are found which is followed by formation of thermo-reversible gel when the concentration is higher (Mishra and Kumar 2012).

7.3.2 Advantages of CMC

The presence of innate properties like hydrophilicity, bio-adhesiveness, stabilizing power, odorless and non-toxic nature, viscosity, and pH sensitivity in CMC proved to be advantageous for its application in many fields, e.g., biomedicines, cosmetic appliances, drilling fluids, cleansing soaps, and toothpaste (Li et al. 2016).

The hydrophilic CH_2COO^- groups present in CMC increase the water absorption capacity and formation of complex hydrogen bonding occurs due to the uneven substitution of the carboxymethyl groups. Thermogravimetric tests of CMC molecule revealed that it is highly hydrophilic in nature as it contained water molecules even after drying at 50 °C in a vacuum oven for 24 hours (Javanbakht and Shaabani 2019). The presence of hydroxyl and carboxyl groups in the structure of CMC proved to be a good absorbent of heavy metal ions (Abdulsahib et al. 2020).

CMC being polyionic in nature shows bioadhesive character and easily gets bind to biological membranes as compared to other derivatives of cellulose and thus can be used as filling agent for the preparation of transdermic and transmucosal medicines (Paria and Rai 2022; Jiménez-Castellanos et al. 1994). It can be used as biopolymer carriers as it decreases the toxicity of therapeutic agents. CMC is biodegradable and it is broken down by the enzyme glycosidase inside the body. The 1,4 bond of the CMC is broken down to simple sugar units by glycosidase (Aboagye 2015). When present in the form of hydrogel which is an anionic polyelectrolyte, CMC exhibits pH sensitivity due to the presence of carboxylic groups (Qiu and Park 2001). The water uptake capacity of CMC is 5.11 g g^{-1} which makes it ideal for pharmaceutical and food additives. Low viscous CMC can be used as moisture adhesive agent in various food businesses, while the CMC having high viscosity are used in gelation purposes. The solution of polymeric CMC is viscous in aqueous solution due to its high solubility in water (Paria and Rai 2022).

The solubility of CMC is temperature independent and it is hydrated rapidly due to its hygroscopic nature. The rapid hydration of CMC creates lumps when introduced to water, which can be removed by high agitation or blending CMC with dry substances like sugar before dissolving in water (cellulose).

7.3.3 Drawbacks of CMC

In spite of having versatile physico-chemical properties, CMC in its raw form shows poor mechanical properties, acts as non-conductive polymer (Ampaiwong et al. 2019), and is a poor shale inhibitor (Mishra and Kumar 2012). Moreover, the single-electrode polymers of CMC are highly stiff, possess low electrochemical stability, and reduced elongation at break (Arefian et al. 2020). The swelling property of CMC-based adsorbents limits its application in various fields (Chen et al. 2019). CMC-based films have low water resistance and are soft in nature (He et al. 2020). The sodium salt of CMC is unstable in aqueous environment due to its high water solubility which restricts its medicinal use (Lee et al. 2015).

Viscosity also plays a crucial role for the application of CMC in different fields (Rahman et al. 2021). The high viscosity-type CMC is used in gelating purposes, while the CMC of low viscosity is used in moisture-binding applications (Theeuwen and Dijk 2006). Sodium bentonite acts as viscosity modifier when added to low viscosity-type CMC which helps in drilling mud system. Adding CMC induces the viscous nature of bentonite, and thus loss of fluid in drilling system can be reduced (Abu-Jdayil and Ghannam 2014). Although Na-CMC has excellent properties for its use in medicines, its water-soluble nature cannot be handled in wet processes (Lee et al. 2015).

7.3.4 Modification of CMC

CMC is given more preference than other cellulose derivatives because of its wide range of properties and its ability to undergo further modifications (Basta et al. 2016). The drawbacks of raw CMC and CMC-based materials can be eliminated by its surface modifications like grafting, blending, and introducing crosslinking agents. CMC backbone when grafted with synthetic polyacrylamide (PAM) forms good flocculating agent as the copolymerization of the two polymer increases shear stability (Dahlan et al. 2018).

Grafting of CMC backbone with acrylic acid and acrylamide modifies the surface and enhances the adsorption of dyes and helps to eliminate toxic trace metal ions (Radwan et al. 2018). CMC when blended with PVC forms a composite which is hydrophilic in nature and helps in producing biomaterials and in drug-delivery system (Arefian et al. 2020). The modified CMC with gelatin when crosslinked by

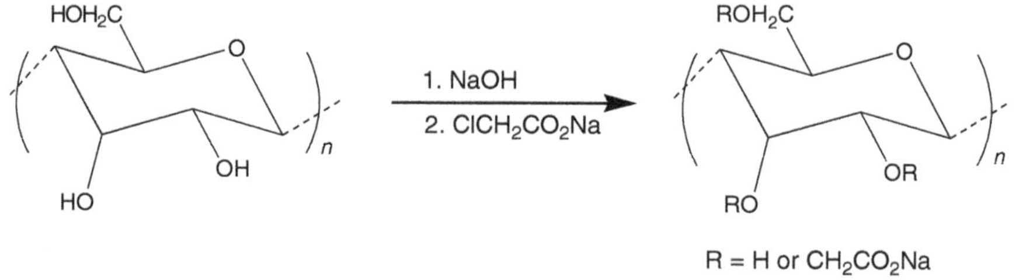

Figure 7.5 Synthesis of carboxymethyl cellulose. (Source: Iragavarapu et al. 2019.)

iron produces ibuprofen-encapsulated beads which are helpful for oral sustained delivery of drug, which is not suitable for CMC alone (Javanbakht and Shaabani 2019). Blending of CMC with polyvinylpyrrolidone polymer (PVP) and doped with SeO_2 results in the production of CMC/PVP nanocomposite which shows improved thermal, optical, and dielectric properties for its use in device applications (Gaabour 2020).

7.4 Properties of Water-Soluble Polymer-Based Composites

7.4.1 Mechanical Properties

The most important properties of the prepared water-soluble polymer-based composites are their high mechanical strength for their end uses. The mechanical properties of water-soluble polymer depend on the use of crosslinking agents, treatments of fibers, and the content of fiber (Hazarika et al. 2015). Also, the other parameters which influenced are the tensile strength, flexural strength and their modulus, the nature of crosslinkers, polymer–fiber ratio, effects of nanofillers, etc. The interactions among polymer, crosslinker, and reinforcing agent plays a pivotal role for improving the mechanical properties of the water-soluble polymer-based composites. Addition of nanofiller significantly enhances the interactions which improve the mechanical properties. The content of fiber in the polymeric matrix plays a significant role in enhancing the mechanical properties of a polymer. The modulus of a polymer is affected by the properties of the fiber content polymer.

Baishya and Maji (2014) studied the crosslinking of methyl-methacrylate-grafted starch with wood flour using three different crosslinkers NMA, glutaraldehyde (GA), and DMDHEU and compared their change in mechanical properties. The DMDHEU crosslinked composites have showed maximum improvement in mechanical strength. Incorporation of montmorillonite clay further improved the properties of starch-based composites (Baishya and Maji 2016a). The mechanical properties of the composites are depicted in Table 7.1. Fillers like mineral clay also improve the mechanical properties (Wilhelm et al. 2003). Liu et al. (2013) in their finding have shown that mechanical properties of the polymer improve by the reinforcing agents.

Table 7.1 Mechanical properties of wood starch composites.

Sample	Flexural strength (MPa)	Tensile strength (MPa)	Hardness (Shore D)
W/MMA-g-starch/GA	30.8	13.6	64.1 (±1)
W/MMA-g-starch/NMA	42.7	14.8	66.3 (±0.5)
W/MMA-g-starch/DMDHEU	44.7	15.6	72.0 (±1)
W/MMA-g-starch/DMDHEU/3 phr MMT	57.4	33.1	79.4 (±1)

Source: Adapted from Baishya and Maji (2014, 2016a).

The more the hydrophobic high degree of substitution, the superior is the mechanical properties (Duanmu et al. 2007b). Mali et al. (2005) tested the mechanical properties of starch by puncture test and tensile test. Modified thermoplastic starch with the inclusion of wood particles enhanced the stiffness and strength properties of the composite (Fekete et al. 2018).

CMC film developed from the H_2O_2-bleached palm bunch and bagasse with controlled amount of H_2O_2 incorporation shows strong mechanical properties, high solubility, and viscosity (Suriyatem et al. 2020). The films produced by combination of polyethylene glycol and sodium-based clay nanofillers induce the mechanical strength of the CMC film which suggests the applicability of these films for food packaging (de Melo Fiori et al. 2019). Several studies have also been made where incorporation of CMC has led to the improvement of other cellulose and starch-based films. Soy protein isolate when combined with CMC in the presence of an antioxidant like catechin induces the tensile strength as well as water solubility of the film. The stress–strain curve obtained from the uniaxial tensile strength tests validated the results (Han et al. 2015).

7.4.2 Thermal Resistance Properties

The most desired characteristics of the water-soluble polymer-based composites are their thermal stability. The thermal resistance characteristics of the composites are studied by DSC and TGA (LeCorre et al. 2012). The incorporation of nanoparticles like SiO_2, TiO_2, and CNT in the polymeric matrix improves thermal properties like initial decomposition temperature (T_i) and maximum pyrolysis temperature (T_m). Baishya et al. (2018) have reported that the addition of TiO_2 in the starch-based composites increased the thermal stability significantly compared to untreated one. The multi-walled carbon nanotubes (MWCNT) were used to prepare highly thermal-resistant starch-based natural fiber composites (Baishya and Maji 2016b). LeCorre et al. (2012) reported that starch with amylose content shows less influence on the thermal stability. Surface modifications of the composites are often carried out to improve the thermal properties (Rosa et al. 2009). The nature of crosslinkers also plays an important role on the thermal stability of starch-based composites. Baishya and Maji have reported that DMDHEU crosslinked wood starch composites showed better thermal resistivity than the NMA and glutaraldehyde (GA) crosslinked composites (Baishya and Maji 2014; Baishya et al. n.d.). Reddy and Yang (2010) have reported that starch films that are crosslinked have much higher strength and thermal stability than those that are not crosslinked.

For the facile application of the CMC-based substances thermal stability is often considered as a key factor. Incorporation of carbon dots into the hybrid film of CMC-Na salt and sodium alginate produces a fluorescent film which is sensitive to Ag^+ and Fe^{3+}. The produced film possesses low elongation at break (23.54%), higher tensile strength of 52.38 MPa, and good thermal stability (da Silva Junior et al. 2021). CMC when crosslinked with collagen (COL) nanofibers produced CMC/COL composites which has been able to exert low- and high-temperature stability better than native COL nanofibers. The hydrogen bonds between carboxyl or hydroxyl groups of CMC and amino or carboxyl groups of COL resulted in the crosslinking of CMC and COL nanofibers (Tian et al. 2021). Mallikajurna and Rajender have reported the metal ion incorporation effect into the CMC film. The nanocomposites of CMC with metals like Cu, Ag, Fe and In are found to provide enhanced thermal stability, decompose at higher temperatures, exhibit luminescence, and show antibacterial, antifungal properties (Nadagouda and Varma 2007).

7.4.3 Water Resistance and Dimensional Properties

The hydrophilicity of starch has been a major concern, and the main drawbacks of starch-based composites and thus limiting it to be applied directly for many applications (Duanmu et al. 2007b). Starches are/can be blended with various synthetic polymers to make better resistance to water. Water resistance property can be improved by converting the hydrophilic hydroxyl groups to a large number and less polar group. It can also be improved by adding crosslinking agents like glyoxal, which create network structures within the polymer. It has been shown that blending of starch with hydrophobic polymers, like latex or cassava foams, increases the water resistance of starch (Soykeabkaew et al. 2015). The grafting of starch with methyl methacrylate improved the water resistance properties as reported by Baishya and Maji (2014). The incorporation of nanoparticles like SiO_2, TiO_2, CNT, and graphene oxide also induces the water resistance properties of starch-based composites (Baishya and Maji 2017, 2018). Deng and Catchmark (2014) prepared an insoluble bio-composite by crosslinking potato starch and chitosan by microwave treatment and showed that the uptake behavior of the solution depends both on the pH and content of the chitosan. Introducing functionalized-MWCNT in the starch composites shows in the reduction of the water vapor permeability rate of the composites as the voids present inside the molecular structure of the composites were filled up by the MWCNT (Baishya and Maji 2016b).

Incorporation of cotton nanofibril obtained from cotton linter pulp with CMC forms an edible nanocomposite film which shows excellent water vapors resistance at 5 wt% along with high mechanical stability. The water vapor diffusion rate can be decreased by cellulose nanofibril incorporation, but at 10 wt%, permeability of water vapor is reduced due to agglomeration of nanofibrils (Oun and Rhim 2015). Cassava starch when modified to form CMC starch by etherification process at atmospheric pressure followed by the incorporation of waterborne polyurethane forms a polymeric film which shows better water resistivity due to the substitution of the –OH group of cassava starch with the ether CMC starch. Polyurethane induces the water-resistant property of the film as it acts as surface sizing material. It fills up the interspaces and forms a continuous layer (Rusman et al. 2017).

7.4.4 Chemical Resistance Properties

The chemical resistivity of water-soluble polymers is crucial for the enhanced quality and longevity of the desired polymer. Most of the starch composites are used in our daily life for food packing and medical applications (Gürler et al. 2021). There is a chance that these biocomposites may come in contact with moisture or water or any chemical like solvents there by degrading and losing its function. In recent times there have been many researches on starch biocomposites to improve the chemical resistance and so that they can be used without any problem (Benito-González et al. 2019; Wan et al. 2009). Though many articles are available, a few reports are available about the chemical resistivity of starch biocomposites. Starch biocomposites are excellent materials for applications, if their chemical resistivity properties are advanced.

The films prepared by CMC blending with polysaccharides are found to exhibit barrier against various lipids and gases, are resistant to microbial growth and therefore can be used for the application of food packaging (Varaprasad et al. 2011). Aulin et al. studied the O_2 permeability rate and oil-resistant nature of carboxymethylated microfibrillated cellulose films. The oxygen permeability of the film was found to be lower than other natural polymeric films made up of starch and whey proteins. Incorporation of fibrils reduced the surface porosity as investigated by scanning electron microscopy, which explains the excellent oil barrier property of the film (Aulin et al. 2010). Composite films made up of natural rubber and eucalyptus microcellulose fiber incorporated with Na-CMC are proved to be insoluble in solvents like toluene. Toluene being a non-polar solvent does not solubilize the polar cellulose matrix (Supanakorn et al. 2021).

7.5 Conclusion and Future Prospects

In recent years, researchers are more focused on the use of water-soluble polymers in the designing of new environmental friendly composites. The use of water as a solvent has multiple advantages, e.g., it reduces the use of chemically hazardous organic solvents responsible for long-lasting diversified affect in eco-system and environment, inheritably eco-friendly, easily accessible, and economically feasible. Use of naturally available sources of polymers for our everyday use is growing enormously due to their biocompatibility, easy synthesis, and benign nature. Natural polysaccharides like starch and CMC proved to be two most flexible water-soluble polymers for application in packaging, cosmetics, textiles, coating, adhesives, 3D printing, pharmaceuticals, etc. Starch is economically cheap, abundant in nature, easy to process, and does not require intensive purification; above all, they are eco-friendly making it more advantageous than other polymers. Starch in its native form can only be use for minimum purposes. One reason being that starch is sensitive to moisture and water, and hence loses its functionality and starts degrading once it come in contact. However, modified starch polymer can be used for advanced applications as they show more resistance to moisture, chemical, heat, micro-organism, etc., and also in their durability or long-time performance as their mechanical and thermal properties are improved. Similarly, CMC also possesses advantages like easy availability, biodegradation, and non-toxicity due to which CMC has been able to substitute synthetic polymers in recent times. Modified CMC products including CMC hydrogels, films, nanocomposites, and other bio-hybrids exhibit better properties which induces its application in food, cosmetics as well as in pharmaceutical industries. These two polymers show a great tendency to substitute synthetic polymer in the future.

References

Abdulsahib, W.K., Ganduh, S.H., Radia, N.D., and Jasim, L.S. (2020). New approach for sulfadiazine toxicity management using carboxymethyl cellulose grafted acrylamide hydrogel. *International Journal of Drug Delivery Technology* 10 (2): 259–264.

Abe, M.M., Martins, J.R., Sanvezzo, P.B. et al. (2021). Advantages and disadvantages of bioplastics production from starch and lignocellulosic components. *Polymers* 13 (15): 2484.

Aboagye, I.A. (2015). *Use of Exogenous Fibrolytic Enzymes to Improve the Nutritive Value of Preserved Forage for Ruminants.* Thompson Rivers University.

Abu-Jdayil, B. and Ghannam, M. (2014). The modification of rheological properties of sodium bentonite-water dispersions with low viscosity CMC polymer effect. *Energy Sources, Part A: Recovery, Utilization, and Environmental Effects* 36 (10): 1037–1048.

Adinugraha, M.P. and Marseno, D.W. (2005). Synthesis and characterization of sodium carboxymethylcellulose from cavendish banana pseudo stem (*Musa cavendishii* LAMBERT). *Carbohydrate Polymers* 62 (2): 164–169.

Ampaiwong, J., Rattanawaleedirojn, P., Saengkiettiyut, K. et al. (2019). Reduced graphene oxide/carboxymethyl cellulose nanocomposites: novel conductive films. *Journal of Nanoscience and Nanotechnology* 19 (6): 3544–3550.

Arefian, M., Hojjati, M., Tajzad, I. et al. (2020). A review of polyvinyl alcohol/carboxymethyl cellulose (PVA/CMC) composites for various applications. *Journal of Composites and Compounds* 2 (3): 69–76.

Aulin, C., Gällstedt, M., and Lindström, T. (2010). Oxygen and oil barrier properties of microfibrillated cellulose films and coatings. *Cellulose* 17 (3): 559–574.

Baishya, P. and Maji, T.K. (2014). Studies on effects of different cross-linkers on the properties of starch-based wood composites. *ACS Sustainable Chemistry & Engineering* 2 (7): 1760–1768.

Baishya, P. and Maji, T.K. (2016a). Studies on the physicochemical properties of modified starch-based wood nanocomposites. *Starch – Stärke* 68 (3–4): 249–257.

Baishya, P. and Maji, T.K. (2016b). Functionalization of MWCNT and their application in properties development of green wood nanocomposite. *Carbohydrate Polymers* 149: 332–339.

Baishya, P. and Maji, T.K. (2017). Enhancement in physicochemical properties of citric acid/nano SiO_2 treated sustainable wood-starch nanocomposites. *Cellulose* 24 (10): 4263–4274.

Baishya, P. and Maji, T.K. (2018). A comparative study on the properties of graphene oxide and activated carbon based sustainable wood starch composites. *International Journal of Biological Macromolecules* 115: 970–977.

Baishya, P., Mandal, M., Gogoi, P., and Maji, T.K.J. (2017). Natural polymer-based nanocomposites: a greener approach for the future. In: *Handbook of Composites from*

Renewable Materials, Nanocomposites: Science and Fundamentals, vol. 7 (ed. V.K. Thakur, M.K. Thakur, and M.R. Kessler), 433. Wiley.

Baishya, P., Nath, D., Begum, P. et al. (2018). Effects of wheat gluten protein on the properties of starch based sustainable wood polymer nanocomposites. *European Polymer Journal* 100: 137–145.

Baishya, P., Moon, M., Gogoi, P., and Maji, T.K. (2022). Natural polymer based nanocomposites a greener approach for future. In: *Handbook of Composites from Renewable Materials, Nanocomposites: Science and Fundamentals* (ed. V.K. Thakur, M.K. Thakur, and M.R. Kessler), 433–459. Wiley.

Basta, A.H., El-Saied, H., El-Deftar, M.M. et al. (2016). Properties of modified carboxymethyl cellulose and its use as bioactive compound. *Carbohydrate Polymers* 153: 641–651.

Benito-González, I., López-Rubio, A., and Martínez-Sanz, M. (2019). High-performance starch biocomposites with cellulose from waste biomass: film properties and retrogradation behaviour. *Carbohydrate Polymers* 216: 180–188.

Bien, F., Wiege, B., and Warwel, S. (2001). Hydrophobic modification of starch by alkali-catalyzed addition of 1,2-epoxyalkanes. *Starch – Stärke* 53 (11): 555–559.

Bikiaris, D., Prinos, J., Koutsopoulos, K. et al. (1998). LDPE/plasticized starch blends containing PE-g-MA copolymer as compatibilizer. *Polymer Degradation and Stability* 59 (1–3): 287–291.

Candan, Z. and Akbulut, T. (2013). Developing environmentally friendly wood composite panels by nanotechnology. *BioResources* 8 (3): 3590–3598.

Cao, X., Chen, Y., Chang, P.R., and Huneault, M.A. (2007). Preparation and properties of plasticized starch/multiwalled carbon nanotubes composites. *Journal of Applied Science* 106 (2): 1431–1437.

Carvalho, A.J.J.M. (2008). Starch: major sources, properties and applications as thermoplastic materials. In: *Monomers, Polymers and Composites from Renewable Resources* (ed. M.N. Belgacem and A. Gandini), 321–342. Elsevier.

Chang, P.R., Jian, R., Zheng, P. et al. (2010). Preparation and properties of glycerol plasticized-starch (GPS)/cellulose nanoparticle (CN) composites. *Carbohydrate Polymers* 79 (2): 301–305.

Chen, L., Wang, Y., Fei, P. et al. (2017). Enhancing the performance of starch-based wood adhesive by silane coupling agent (KH570). *International Journal of Biological Macromolecules* 104: 137–144.

Chen, Y., Long, Y., Li, Q. et al. (2019). Synthesis of high-performance sodium carboxymethyl cellulose-based adsorbent for effective removal of methylene blue and Pb(II). *International Journal of Biological Macromolecules* 126: 107–117.

Copeland, L., Blazek, J., Salman, H., and Tang, M.C. (2009). Form and functionality of starch. *Food Hydrocolloids* 23 (6): 1527–1534.

Cvjetko Bubalo, M., Vidović, S., Radojčić Redovniković, I., and Jokić, S. (2015). Green solvents for green technologies. *Journal of Chemical Technology & Biotechnology* 90 (9): 1631–1639.

Dahlan, N.A., Pushpamalar, J., Veeramachineni, A.K., and Muniyandy, S. (2018). Smart hydrogel of carboxymethyl cellulose grafted carboxymethyl polyvinyl alcohol and properties studied for future material applications. *Journal of Polymers and the Environment* 26 (5): 2061–2071.

De Carvalho, A.J.F., Curvelo, A.A.S., and Agnelli, J.A.M. (2002). Wood pulp reinforced thermoplastic starch composites. *International Journal of Polymeric Materials and Polymeric Biomaterials* 51 (7): 647–660.

Deka, B.K. and Maji, T.K. (2013). Effect of SiO_2 and nanoclay on the properties of wood polymer nanocomposite. *Polymer Bulletin* 70 (2): 403–417.

Deka, B.K., Baishya, P., and Maji, T.K. (2014). Synergistic effect of SiO_2, ZnO and nanoclay on mechanical and thermal properties of wood polymer nanocomposite. *Journal of Thermoplastic Composite Materials* 27 (4): 464–480.

Deng, Y. and Catchmark, J.M. (2014). Insoluble starch composite foams produced through microwave expansion. *Carbohydrate Polymers* 111: 864–869.

Djordjevic, S., Nikolic, L., Kovacevic, S. et al. (2013). Graft copolymerization of acrylic acid onto hydrolyzed potato starch using various initiators. *Periodica Polytechnica, Chemical Engineering* 57 (1–2): 55–61.

Dresselhaus, M.S., Dresselhaus, G., Eklund, P., and Rao, A. (2000). Carbon nanotubes. In: *The Physics of Fullerene-Based and Fullerene-Related Materials* (ed. W. Andreoni), 331–379. Springer.

Duanmu, J., Gamstedt, E.K., and Rosling, A. (2007a). Synthesis and preparation of crosslinked allylglycidyl ether-modified starch-wood fibre composites. *Starch – Stärke* 59 (11): 523–532.

Duanmu, J., Gamstedt, E.K., and Rosling, A. (2007b). Hygromechanical properties of composites of crosslinked allylglycidyl-ether modified starch reinforced by wood fibres. *Composites Science and Technology* 67 (15–16): 3090–3097.

Ebbesen, T.W. (1994). Carbon nanotubes. *Physics Today* 24 (1): 235–264.

Emmerich, L., Bollmus, S., and Militz, H. (2019). Wood modification with DMDHEU (1.3-dimethylol-4.5-dihydroxyethyleneurea) – state of the art, recent research activities and future perspectives. *Wood Material Science and Engineering* 14 (1): 3–18.

Erdmenger, T., Guerrero-Sanchez, C., Vitz, J. et al. (2010). Recent developments in the utilization of green solvents in polymer chemistry. *Chemical Society Reviews* 39 (8): 3317–3333.

Etgar, L., Gao, P., Xue, Z. et al. (2012). Mesoscopic $CH_3NH_3PbI_3/TiO_2$ heterojunction solar cells. *Journal of the American Chemical Society* 134 (42): 17396–17399.

Fekete, E., Kun, D., and Móczó, J. (2018). Thermoplastic starch/wood composites: effect of processing technology, interfacial interactions and particle characteristics. *Periodica Polytechnica, Chemical Engineering* 62 (2): 129–136.

Feldman, D. (2008). Polymer history. *Designed Monomers and Polymers* 11 (1): 1–15.

Freinkel, S. (2011). A brief history of plastic's conquest of the world. *Scientific American*. https://www.scientificamerican.com/article/a-brief-history-of-plastic-world-conquest/

Gaabour, L. (2020). Effect of selenium oxide nanofiller on the structural, thermal and dielectric properties of CMC/PVP nanocomposites. *Journal of Materials Research and Technology* 9 (3): 4319–4325.

Gadhave, R.V., Mahanwar, P.A., and Gadekar, P.T. (2017). Starch-based adhesives for wood/wood composite bonding.

George, J. and Sabapathi, S. (2015). Water soluble polymer-based nanocomposites containing cellulose nanocrystals. In: *Eco-friendly Polymer Nanocomposites* (ed. V.K. Thakur and M.K. Thakur), 259–293. Springer.

Geyer, R. (2020). Production, use, and fate of synthetic polymers. In: *Plastic Waste and Recycling*, 13–32. Elsevier.

Godavitarne, C., Robertson, A., Peters, J., and Rogers, B. (2017). Biodegradable materials. *Orthopaedics and Traumatology* 31 (5): 316–320.

Goel, C., Semwal, A.D., Khan, A. et al. (2020). Physical modification of starch: changes in glycemic index, starch fractions, physicochemical and functional properties of heat-moisture treated buckwheat starch. *Journal of Food Science and Technology – Mysore* 57 (8): 2941–2948.

Griffin, G.J.L. (1977). Biodegradable synthetic resin sheet material containing starch and a fatty material. Google Patents. US4016117A. Application filed: 25 November 1974. Published: 5 April 1977.

Gürler, N., Paşa, S., Erdoğan, Ö., and Cevik, O. (2021). Physicochemical properties for food packaging and toxicity behaviors against healthy cells of environmentally friendly biocompatible starch/citric acid/polyvinyl alcohol biocomposite films. *Starch – Stärke* 75: 2100074.

Han, J., Shin, S.-H., Park, K.-M., and Kim, K.M. (2015). Characterization of physical, mechanical, and antioxidant properties of soy protein-based bioplastic films containing carboxymethylcellulose and catechin. *Food Science and Biotechnology* 24 (3): 939–945.

Haroon, M., Wang, L., Yu, H. et al. (2016). Chemical modification of starch and its application as an adsorbent material. *RSC Advances* 6 (82): 78264–78285.

Hazarika, A. and Maji, T.K. (2013). Effect of different crosslinkers on properties of melamine formaldehyde-furfuryl alcohol copolymer/montmorillonite impregnated softwood (*Ficus hispida*). *Polymer Engineering & Science* 53 (7): 1394–1404.

Hazarika, A., Baishya, P., and Maji, T.K. (2015). Bio-based wood polymer nanocomposites: a sustainable high-performance material for future. In: *Eco-friendly Polymer Nanocomposites* (ed. V.K. Thakur and M. Kumari), 233–257. Springer 2015; pp 233–257.

He, B., Wang, W., Song, Y. et al. (2020). Structural and physical properties of carboxymethyl cellulose/gelatin films functionalized with antioxidant of bamboo leaves. *International Journal of Biological Macromolecules* 164: 1649–1656.

Hemamalini, T. and Dev, V.R.G. (2018). Comprehensive review on electrospinning of starch polymer for biomedical applications. *International Journal of Biological Macromolecules* 106: 712–718.

Iragavarapu, A.G., Dash, I., Jayabalan, R., and Balasubramanium, P. (2019). Optimization of etherification reactions for recycling of tea fungal biomass waste into carboxymethylcellulose. *Green Buildings and Sustainable Engineering: Proceedings of GBSE 2018.* Harald Drück, Radhakrishna G. Pillai, Manoj G. Tharian, Aysha Zeneeb Majeed. Springer, pp 337–346.

Javanbakht, S. and Shaabani, A. (2019). Carboxymethyl cellulose-based oral delivery systems. *International Journal of Biological Macromolecules* 133: 21–29.

Jiang, T., Duan, Q., Zhu, J. et al. (2020). Starch-based biodegradable materials: challenges and opportunities. *Advanced Industrial and Engineering Polymer Research* 3 (1): 8–18.

Jimenez, A., Fabra, M.J., Talens, P., and Chiralt, A. (2012). Edible and biodegradable starch films: a review. *Food and Bioprocess Technology* 5 (6): 2058–2076.

Jiménez-Castellanos, M.R., Zia, H., and Rhodes, C.T. (1994). Design and testing in vitro of a bioadhesive and floating drug delivery system for oral application. *International Journal of Pharmaceutics* 105 (1): 65–70.

Jyothi, A.N., Moorthy, S.N., and Rajasekharan, K.N. (2006). Effect of cross-linking with epichlorohydrin on the properties of cassava (*Manihot esculenta* Crantz) starch. *Starch – Stärke* 58 (6): 292–299.

Kadajji, V.G. and Betageri, G.V. (2011). Water soluble polymers for pharmaceutical applications. *Polymers* 3 (4): 1972–2009.

Kanikireddy, V., Varaprasad, K., Jayaramudu, T. et al. (2020). Carboxymethyl cellulose-based materials for infection control and wound healing: a review. *International Journal of Biological Macromolecules* 164: 963–975.

Kavimani, V., Prakash, K.S., Thankachan, T., and Udayakumar, R. (2020). Synergistic improvement of epoxy derived polymer composites reinforced with graphene oxide (GO) plus titanium di oxide (TiO_2). *Composites Part B: Engineering* 191: 107911.

Koelmans, A.A., Besseling, E., and Shim, W.J. (2015). Nanoplastics in the aquatic environment. Critical review. In: *Marine Anthropogenic Litter* (ed. M. Bergmann, L. Gutow, and M. Klages), 325–340. Springer Open.

LeCorre, D., Bras, J., and Dufresne, A. (2012). Influence of native starch's properties on starch nanocrystals thermal properties. *Carbohydrate Polymers* 87 (1): 658–666.

Lee, J.Y., Im, J.N., Kim, T.H. et al. (2015). Structure and liquid handling properties of water-insoluble carboxymethyl cellulose foam. *Fibers and Polymers* 16 (4): 726–734.

Li, C.-J. and Trost, B.M. (2008). Green chemistry for chemical synthesis. *Proceedings of the National Academy of Sciences* 105 (36): 13197–13202.

Li, M.-C., Mei, C., Xu, X. et al. (2016). Cationic surface modification of cellulose nanocrystals: toward tailoring dispersion and interface in carboxymethyl cellulose films. *Polymer* 107: 200–210.

Liu, R., Cao, J., Luo, S., and Wang, X. (2013). Effects of two types of clay on physical and mechanical properties of poly(lactic acid)/wood flour composites at various wood flour contents. *Journal of Applied Polymer Science* 127 (4): 2566–2573.

Lu, D., Xiao, C., and Xu, S. (2009). Starch-based completely biodegradable polymer materials. *Express Polymer Letters* 3 (6): 366–375.

Mali, S., Grossmann, M.V.E.; García, M. A. et al. (2002). Microstructural characterization of yam starch films. *Carbohydrate Polymers* 50 (4): 379–386.

Mali, S., Grossmann, M.V.E., García, M. A. et al. (2005). Mechanical and thermal properties of yam starch films. *Food Hydrocolloids* 19 (1): 157–164.

McCormick, C.L., Lowe, A.B., and Ayres, N. (2002). Water-soluble polymers. In: *Encyclopedia of Polymer Science and Technology* (ed. H.F. Mark), 452–521. Wiley.

Meimoun, J., Wiatz, V., Saint-Loup, R. et al. (2018). Modification of starch by graft copolymerization. *Starch – Stärke* 70 (1–2): 1600351.

de Melo Fiori, A.P.S., Camani, P.H., dos Santos Rosa, D., and Carastan, D.J. (2019). Combined effects of clay minerals and polyethylene glycol in the mechanical and water barrier properties of carboxymethylcellulose films. *Industrial Crops and Products* 140: 111644.

Mishra, V. and Kumar, R. (2012). Graft copolymerization of carboxymethylcellulose: an overview. *Trends in Carbohydrate Research* 4 (3): 1–17.

Moad, G. (2011). Chemical modification of starch by reactive extrusion. *Progress in Polymer Science* 36 (2): 218–237.

Molavi, H., Behfar, S., Shariati, M.A. et al. (2021). A review on biodegradable starch based film. *Journal of Microbiology, Biotechnology and Food Sciences* 2021: 456–461.

Montanari, C., Olsén, P., and Berglund, L.A. (2021). Sustainable wood nanotechnologies for wood composites processed by in-situ polymerization. *Frontiers in Chemistry* 9: 483.

Müller, P., Renner, K., Móczó, J. et al. (2014). Thermoplastic starch/wood composites: interfacial interactions and functional properties. *Carbohydrate Polymers* 102: 821–829.

Muller, J., González-Martínez, C., and Chiralt, A. (2017). Combination of poly(lactic) acid and starch for biodegradable food packaging. *Materials (Basel)* 10 (8): 952.

Muneer, F., Andersson, M., Koch, K. et al. (2015). Nanostructural morphology of plasticized wheat gluten and modified potato starch composites: relationship to mechanical and barrier properties. *Biomacromolecules* 16 (3): 695–705.

Nadagouda, M.N. and Varma, R.S. (2007). Synthesis of thermally stable carboxymethyl cellulose/metal biodegradable nanocomposites for potential biological applications. *Biomacromolecules* 8 (9): 2762–2767.

Oun, A.A. and Rhim, J.-W. (2015). Preparation and characterization of sodium carboxymethyl cellulose/cotton linter cellulose nanofibril composite films. *Carbohydrate Polymers* 127: 101–109.

Pareta, R. and Edirisinghe, M.J.C. (2006). A novel method for the preparation of starch films and coatings. *Carbohydrate Polymers* 63 (3): 425–431.

Paria, A. and Rai, V.K. (2022). The fate of carboxymethyl cellulose as a polymer of pharmaceutical importance. *Biological Sciences* 2 (2): 204–215.

Pettignano, A., Charlot, A., and Fleury, E. (2019). Carboxyl-functionalized derivatives of carboxymethyl cellulose: towards advanced biomedical applications. *Polymer Reviews* 59 (3): 510–560.

Pushpamalar, V., Langford, S.J., Ahmad, M., and Lim, Y.Y. (2006). Optimization of reaction conditions for preparing carboxymethyl cellulose from sago waste. *Carbohydrate Polymers* 64 (2): 312–318.

Qiu, Y. and Park, K. (2001). Environment-sensitive hydrogels for drug delivery. *Advanced Drug Delivery Reviews* 53 (3): 321–339.

Radwan, E.K., Kafafy, H., El-Wakeel, S.T. et al. (2018). Remediation of Cd(II) and reactive red 195 dye in wastewater by nanosized gels of grafted carboxymethyl cellulose. *Cellulose* 25 (11): 6645–6660.

Rahman, M.S., Hasan, M.S., Nitai, A.S. et al. (2021). Recent developments of carboxymethyl cellulose. *Polymers* 13 (8): 1345.

Ramesh Kumar, S., Shaiju, P., and O'Connor, K.E. (2020). Bio-based and biodegradable polymers-state-of-the-art, challenges and emerging trends. *Current Opinion in Green and Sustainable Chemistry* 21: 75–81.

Ratnayake, W.S. and Jackson, D.S. (2009). Starch gelatinization. *Advances in Food and Nutrition Research* 55: 221–268.

Reddy, N. and Yang, Y. (2010). Citric acid cross-linking of starch films. *Food Chemistry* 118 (3): 702–711.

Ribba, L., Garcia, N.L., D'Accorso, N., and Goyanes, S. (2017). Disadvantages of starch-based materials, feasible alternatives in order to overcome these limitations. In: *Starch-Based Materials in Food Packaging* (ed. S.E. Barbosa, M.A. García, L. Castillo, and O.V. Lopez), 37–76. Elsevier.

Rivas, B.L., Urbano, B.F., and Sánchez, J. (2018). Water-soluble and insoluble polymers, nanoparticles, nanocomposites and hybrids with ability to remove hazardous inorganic pollutants in water. *Frontiers in Chemistry* 6: 320.

da Rosa Zavareze, E., Pinto, V.Z., Klein, B. et al. (2012). Development of oxidised and heat-moisture treated potato starch film. *Food Chemistry* 132 (1): 344–350.

Rosa, M.F., Chiou, B.-s., Medeiros, E.S. et al. (2009). Effect of fiber treatments on tensile and thermal properties of starch/ethylene vinyl alcohol copolymers/coir biocomposites. *Bioresource Technology* 100 (21): 5196–5202.

Rusman, R., Majid, R.A., Abd Rahman, W.W., and Low, J.H. (2017). Carboxymethyl cassava starch/polyurethane dispersion blend as surface sizing agent. *Chemical Engineering Transactions* 56: 1171–1176.

Saraç, İ., Adin, H., and Temiz, Ş. (2019). Investigation of the effect of use of nano-Al_2O_3, nano-TiO_2 and nano-SiO_2 powders on strength of single lap joints bonded with epoxy adhesive. *Composites Part B: Engineering* 166: 472–482.

Shah, U., Naqash, F., Gani, A., and Masoodi, F. (2016). Art and science behind modified starch edible films and coatings: a review. *Comprehensive Reviews in Food Science and Food Safety* 15 (3): 568–580.

da Silva Junior, A.H., Macuvele, D.L.P., Riella, H.G. et al. (2021). Polymeric blends of carboxymethyl cellulose and sodium alginate containing functionalized carbon dots result in stable and efficient fluorescent films for silver and iron (III) sensing. *Journal of Environmental Chemical Engineering* 9 (4): 105728.

Soykeabkaew, N., Thanomsilp, C., and Suwantong, O. (2015). A review: starch-based composite foams. *Composites Part A Applied Science and Manufacturing* 78: 246–263.

Subramanian, V., Wolf, E.E., and Kamat, P.V. (2004). Catalysis with TiO_2/gold nanocomposites. Effect of metal particle size on the Fermi level equilibration. *Journal of the American Chemical Society* 126 (15): 4943–4950.

Supanakorn, G., Taokaew, S., and Phisalaphong, M. (2021). Ternary composite films of natural rubber, cellulose microfiber, and carboxymethyl cellulose for excellent mechanical properties, biodegradability and chemical resistance. *Cellulose* 28 (13): 8553–8566.

Suriyatem, R., Noikang, N., Kankam, T. et al. (2020). Physical properties of carboxymethyl cellulose from palm bunch and bagasse agricultural wastes: effect of delignification with hydrogen peroxide. *Polymers* 12 (7): 1505.

Thakur, R., Pristijono, P., Scarlett, C.J. et al. (2019). Starch-based films: major factors affecting their properties. *International Journal of Biological Macromolecules* 132: 1079–1089.

Theeuwen, C. and Dijk, B. (2006). Use of carboxymethyl cellulose (cmc) in fruit-based products. Google Patents.

Tian, Z., Wang, Y., and He, J. (2021). Nanofiber formation of self-crosslinking dialdehyde carboxymethyl cellulose/collagen composites. *Polymer Engineering & Science* 61 (7): 2117–2126.

Uliniuc, A., Popa, M., Drockenmuller, E. et al. (2013). Toward tunable amphiphilic copolymers via CuAAC click chemistry of oligocaprolactones onto starch backbone. *Carbohydrate Polymers* 96 (1): 259–269.

Varaprasad, K., Vimala, K., Ravindra, S. et al. (2011). Fabrication of silver nanocomposite films impregnated with curcumin for superior antibacterial applications. *Journal of Materials Science: Materials in Medicine* 22 (8): 1863–1872.

Wan, Y., Luo, H., He, F. et al. (2009). Mechanical, moisture absorption, and biodegradation behaviours of bacterial cellulose fibre-reinforced starch biocomposites. *Composites Science and Technology* 69 (7–8): 1212–1217.

Wang, X., Wang, L., Su, Q., and Zheng, J. (2013). Use of unmodified SiO_2 as nanofiller to improve mechanical properties of polymer-based nanocomposites. *Composites Science and Technology* 89: 52–60.

Whistler, R.L. and Daniel, J.R. (2000). Starch. In: *Kirk–Othmer Encyclopedia of Chemical Technology*.

Wilhelm, H.-M., Sierakowski, M.-R., Souza, G., and Wypych, F. (2003). Starch films reinforced with mineral clay. *Carbohydrate Polymers* 52 (2): 101–110.

Wong, T.W. and Ramli, N.A. (2014). Carboxymethylcellulose film for bacterial wound infection control and healing. *Carbohydrate Polymers* 112: 367–375.

Yang, R., Li, Y., and Yu, J. (2005). Photo-stabilization of linear low density polyethylene by inorganic nano-particles. *Polymer Degradation and Stability* 88 (2): 168–174.

Zeng, G.S., Hu, C., Zou, S. et al. (2019). BP neural network model for predicting the mechanical performance of a foamed wood-fiber reinforced thermoplastic starch composite. *Polymer Composites* 40 (10): 3923–3928.

Zhang, J.-F. and Sun, X. (2004). Mechanical properties of poly(lactic acid)/starch composites compatibilized by maleic anhydride. *Biomacromolecules* 5 (4): 1446–1451.

Zhivkov, A.M. (2013). Electric properties of carboxymethyl cellulose. In: *Cellulose-Fundamental Aspects* (ed. T.G.M. Van De Ven and L. Godbout), 1–31. InTech Open.

8

Future Roadmap of Organic Polymers

Chanchal Bhardwaj[1], Manisha Sharma[1], Pinki R. Agrawal[1,2], Chankit Kaushik[1], Rahul Sharma[1,2], and A.K. Sharma[1]

[1]Deenbandhu Chhotu Ram University of Science and Technology, Department of Chemistry, Murthal, Sonipat 131001, India
[2]AcSIR, CSIR-National Physical Laboratory, New Delhi 110012, India

8.1 Introduction

Porous organic polymers (POPs), a particular sort of porous substance, have pores that are connected by covalent bonds. The principal connecting factors among the frameworks are light elements like C, H, N, O, and B. This creates a structure with a low density, exceptional stability, and a high specific surface area. POPs are used in a wide range of industries, including gas adsorption, gas partition, catalysis, synthetic sensors, and energy storage. The concavities in the component molecules result in inefficient molecular packing, which gives rise to the porosity of the synthetic porous organic materials (Taylor et al. 2016). The manufactured porous organic materials achieve their porosity through ineffective molecular packing, which happens when the component molecules exhibit concavities (Taylor et al. 2016). POPs can be classified into the accompanying gatherings.

8.1.1 Hypercross-linked Polymers (HCPs)

In order to stop polymer chains from clustering tightly, HCPs belong to polymers created through intensive cross-linking (Wood et al. 2007). Scholars devote more attention to them because of their high specific surface area, simple response, and inexpensive preparation. HCPs are mostly synthesized using Friedel–Crafts chemistry (Wood et al. 2008) which proceeds swiftly to produce strong connections and results in a densely cross-linked network with high porosity. For example, through the Friedel–Crafts self-condensation of bischloromethyl (BCM) monomers such as dichloroxylene (DCX) and 4,40-dichlorobenzene, a number of hyper-cross-linked polymer networks have been created (Wood et al. 2008). HCPs are mostly made using the resulting techniques (Li et al. 2021):

The storage and adsorption of gases, catalysis, the elimination of aromatic chemicals from water, and other applications are all possible uses for HCPs.

Organic Polymers in Energy-Environmental Applications, First Edition.
Edited by Ramesh Oraon, Pardeep Singh, Sanchayita Rajkhowa, Sangita Agarwal, and Ravindra Pratap Singh.

Three techniques are employed to create HCPs: direct one-step polycondensation, post-cross-linking of polymers, and external cross-linkers. The third synthesis method development has increased the diversifiable of HCPs (Das et al. 2017).

- One-step self-condensation of small-molecule functionalized monomers after post-cross-linking functional-group-containing polymer precursors.
- By using brittle aromatic monomers weaved into diplomatic liaison officer's uniform.

In addition, the range of synthesis techniques, exceptional structural qualities, and simple synthesis process of HCPs make them viable adsorbents for volatile organic components (VOC) adsorption (Vallano and Remcho 2000).

8.1.2 Aromatic Frameworks with Pores (PAFs)

Over time, the aromatic frameworks with pores (PAFs) family has grown significantly. It is a brand-new kind of material composed of micropores throughout. The covalent bonds that bind the aromatic molecules that make up the open skeletal structures can serve as carriers for small-molecule visitors.

PAFs can quickly be functionalized to increase their utility. PAF-3 leads the way in this area in terms of PAFs' outstanding ability in the selective adsorption of greenhouse gases including carbon dioxide and methane. PAFs have had good results in other areas as well (Das et al. 2017).

PAFs originally attracted the attention of scientists in 2009 with the discovery of PAF-1, which has a stunning Brunauer-Emmett-Teller (BET) surface area of $5600\,m^2\,g^{-1}$. Materials with exceptional H_2 and CO_2 storage capacity and exceptional stability were generated using the Yamamoto coupling. With the help of this reaction, which provides a simple synthetic method for the homocoupling of monomers, large networks like PAFs conjugated microporous polymers (CMPs) and other structures can be created. PAFs lose their framework conjugation because the center atom of the tetrahedral monomers breaks the conjugation, but CMPs retain the conjugation that arose from the monomers (Ben et al. 2009). The design of PAF-1 was inspired by the crystalline structure of diamond, and it was logically hypothesized that maintaining the stability (comparable to diamond) in the structure of PAF-1 and allowing for sample exposure of the phenyl groups, which might help increase the internal surface area of PAF-1, would maintain the C—C covalent bonds of diamond with phenyl groups. PAF-1 was characterized using theoretical models, which led to the discovery of a previously unheard of BET surface area of $5640\,m^2\,g^{-1}$ and a density as low as $0.315\,g\,cm^{-3}$ (Ben et al. 2009).

PAFs can be viewed as pure benzene-ring-structured conjugated microporous organic polymers (Lu et al. 2021). Different building blocks and coupling reactions have been used to create PAFs with variable specific surface areas, pore volumes, and pore sizes. PAFs can tolerate extreme environments that are acidic, alkaline, and humid, which is crucial for their industrial uses. PAFs often have a hydrophobic structure because they contain benzene rings and covalent bonds between carbon atoms. PAFs' hydrophobicity aids in improving the ability of more aromatic chemical molecules, including benzene and toluene, to adsorb (Ben et al. 2009).

8.1.3 Covalent Organic Frameworks (COFs)

CMPs, the first microporous natural materials with tunable pore qualities, are built from a formed skeleton, and the whole organization is formed on the other hand with – formed or – joints. There are various CMPs materials that have formed ethynyl network models. Specialists' advantage in CMPs materials (Jiang et al. 2007) has developed because of their unbending designs with formation, high unambiguous surface regions, and excellent substance dependability. One of the interesting assortments of hugely formed macrocycles that have been utilized as building blocks to make high-porosity CMPs is porphyrins with 18-electron frameworks. The porosity of CMP networks is fundamental for adsorption, and earlier investigations exhibited that it was feasible to definitively control the CMP organizations' gap size by differing the length and size of the primary structure monomer (Xu et al. 2013).

CMPs are the major naturally occurring microporous materials with tunable pore characteristics. On the other hand, the entire organization is made up of conjugated or linked bonds. Several CMPs materials' conjugated ethynyl network architectures are obtained. Researchers' interest in CMPs materials has grown significantly due to their rigid, conjugated structures, large explicit surface areas, and exceptional synthetic security. Porphyrins with 18-electron frameworks are one of the fascinating types of large produced macrocycles that have been used as the building blocks to create high-porosity CMPs. A previous study discovered that altering the length and size of the primary structure monomer may precisely alter the opening size of the CMP organizations, which is important for adsorption given that CMPs networks should be permeable.

8.2 Polymers of Intrinsic Microporosity

Polymers of intrinsic microporosities (PIMs) are amorphous microporous polymers with stiff polymer chains. They resemble an interlocking network of aromatic rings with twisted or irregular locations. As a result, they are unable to pack efficiently. Some PIMs can be used to make membranes for gas separation because of their solubility in organic solvents and film-forming properties. Four years prior, they published the design of a porous polymer made of phthalocyanines connected to Spiro cyclic groups (Mckeown 2012, 2017). In 2002, a technique for synthesizing phthalocyanine-based network polymers was devised by a number of researchers using the phthalocyanine dye (Mckeown and Budd 2002). Metal ion templates help the process. Spirobisindane is a monomer that is created when 4,5-dichlorophthalonitrile reacts with 5,5′,6,6′-tetrahydroxy-3,3,3′,3′-tetramethyl-1,1′- (Mckeown and Budd 2006). In 2003, the group disclosed the synthesis of a PIM based on hexaazatrinaphthylene. This PIM may serve as both a catalyst and an adsorbent, and its surface area is equivalent to that of the PIMs mentioned above (Budd et al. 2003). Recent studies have focused on identifying applications for these porous polymers as well as enhancing their apparent qualities. One example is the electrospun reformation of PIM-1, which offers increased surface area in the form of polymer fibers rather than powder (Zhang et al. 2016). The most recent advancement in the tactical synthesis of PIMs is the mechanochemical synthesis approach, which provides benefits including fast synthesis and solvent-free reaction (particularly for PIMs with high molecular weight) (Yu et al. 2016) PIMs, in summary, differ significantly from other porous polymers. Despite having designed frames, they are microporous. The porosity of these materials is a property of the flexibility of the polymer chains rather than connected linkages within the substance itself. PIMs have large surface areas in addition to special properties that make these porous polymers liquid, increasing their industrial usage and allowing for the production of porous membranes.

In the adsorption of unstable natural atoms, POPs offer distinct advantages over inorganic or inorganic-natural mixture permeable adsorbents, including an engaging skeletal structure, a sizable explicit surface area, a flexible pore structure, and super hydrophobicity. These characteristics lead to their common use in a few applications.

8.2.1 Water Purification

- Covalent organic polymers (COPs) are structures made of permeable natural atoms that have covalent bonds between their monomers. Owing to their minimal cost, reusability, potent characteristics, and considerable gas particle selectivity, they have lately been identified as being brand-new, widely available, and functionalized adsorbents, making them an anticipated adsorbent for wastewater treatment, particularly for hazardous metals. Additionally, COPs remain remarkably stable throughout a variety of testing circumstances, including high temperatures and, in any case, boiling water. They have been used successfully for natural applications like groundwater remediation, carbon dioxide capture, and soluble take-up.
- Mines et al. have demonstrated composites of nano zero-valent iron impregnated within various COPs (water miscible [COP-1, COP-19], and immiscible [COP6, COP-60, and COP-61]) for decolorization of azo dye.
- Ko et al. created disulfide-connected polymer organizations (COP-63) and connected them as an adsorbent for the expulsion of weighty metal particles from storm water. They showed that COP-63 shows Cd^{2+} evacuation proficiency identical to that of incredibly permeable AC in spite of its moderate surface region. Extracellular polymeric substances (EPS) are intricate polymers using microbial totals as their framework. These macromolecules, which surround the microbial cell in a three-layered grid and have a high subatomic weight, include a mixture of polysaccharides, proteins, uronic acids, lipids, DNA, and compounds that resemble humic substances. Similar to natural wastewater treatment methods including membrane bioreactors (MBRs), activated sludge processes, and fixed-film processes, EPS plays a crucial role.
- Thiol-functionalized POP was made using simple free radical polymerization techniques. It is a cheap substance that is designed to capture mercury and then

clean up the environment. The resultant POP had absorption capabilities of 1216 mg g^{-1} for aqueous mercury removal and 630 mg g^{-1} for airborne mercury deletion. The substance exhibits quick kinetics and may reduce the mercury concentration from 5 ppm to 1 ppb in less than 10 minutes, which is below the drinking water limit set by the US Environmental Protection Agency (2 ppb) (Aguila et al. 2017).

- Dai and associates disclosed a mesoporous polymer for recovering uranium from seawater. Vinyl benzyl chloride (VBC) and divinyl benzene were used to create the copolymer supports using 2,2′-azobisisobutyronitrile (AIBN) and traditional free radical polymerization Divinyl benzene (DVB). The poly(DVB: VBC surface's) region and pore geographies were adjusted as part of the DVB: VBC percentage. The revolutionary polymerization atom transfer radical polymerization (ATRP) of acrylonitrile (AN), which was later converted into amidoxime by the use of hydroxylamine, was sparked by the chloride present in the poly(VBC-*co*-DVB). When tested in a saltwater solution that was coupled with uranyl, the resulting sorbents demonstrated an enormous retention of 80 mg U/g Ads (c. 6 ppm, pH 8). Five gallons of genuine seawater were sent in for 27 days, and the result was a take-up of 2 mg U/g Ads, exceeding the capability of the light-connected nonwoven polyethylene benchmark material's (0.75 mg U/g Ads) texture (Cheong et al. 1997; Sun et al. 2019).
  - A new family of hexaazatriphenylene-based conjugated triazine frameworks (CTFs) for efficient CO_2 collection was recently created by the Dai group using an in situ doping technique (Zhu et al. 2016).
  - Through the Friedel–Crafts self-condensation of 50 bischloromethyl monomers, including DCX and 4,4′-bis(chloromethyl)-1,10-biphenyl (BCMBP), Cooper and coworkers have investigated the methane storage capacity of a number of hypercross-linked polymers (Wood et al. 2008).
  - The El-Kaderi group developed a new POP known as BILP-101, which was produced utilizing a cost-effective Schiff-based polymerization method with simple, readily available monomers. BILP-101 has a greater capacity for CO_2 adsorption compared to other sorbents because of its high porosity and imidazole functionalities, and its all-organic structure improves compatibility with a polymeric matrix (Arab et al. 2014).

8.2.2 Catalysis

One huge area of POP application is drawing attention toward catalysis application. POPs can integrate an assortment of usefulness into their skeletons. POPs have synergist locales coordinated into them on account of the expansion of reactant destinations to the skeletons. POPs are a sophisticated type of nanoreactors that, because of their 3D open systems, enable the correlative usage of skeletons and pores (Arab et al. 2014). According to a few POP types, the synergist properties of POPs and their functionalized offspring are assessed in the section below. Permeable natural polymers without metals, chiral permeable natural polymers, permeable natural polymers supported by nanoparticles, and permeable natural polymers that are photoactive are just a few examples.

In their combined study, Farha, Hupp, and Nguyen demonstrated how to insert catalytically active Al(porphyrin) moieties into an all-organic porous network for use in the methanolysis breakdown of a nerve agent mimic (Arab et al. 2014).

- Cooper and coworkers have developed a coordinated metal-organic conjugated microporous polymer that catalyzes a reductive amination process. CMP-CpIr-3 is efficient for a range of alkyl, aromatic, and cyclic substrates and has a yield of 90% and high catalytic activity (Arab et al. 2014).
- The Coskun group revealed that a new family of nanoporous polymers containing sterically limited N-heterocyclic carbenes (NP-NHCs) had the greatest CO_2 capture fixation efficacy for materials assessed as carbene-based materials in the solid state, 97% at ambient temperature (Talapaneni et al. 2015).

The aforementioned polymeric adsorbents have a few benefits over other conventional adsorbents, such as synthetic ion-exchange resins and commercial activated carbons, including being more reasonable, suitable for removing a variety of (hydrophobic) impurities, high adsorption and rate limits, high selectivity for various concentrations, a large surface area, high strength, and ease of recovery.

8.2.3 Supercapacitors

- Recent research has shown that glass-like covalent natural structures (COFs) and randomly generated microporous polymers (CMPs) in permeable natural polymers (POPs) might provide attractive terminal materials for energy capacity frameworks like supercapacitors and battery-powered batteries. Permeable natural polymers (POPs) with nano-, micro-, or mesoporous underlying properties stand out for their true ability to be used in optoelectronics, catalysis, gas capacity, and partition, among other fields, due to the uprightness of their enormous available surface regions, magnificent warm soundness, and variety of functionalities. Indistinct formed miniature/mesoporous polymers (CMPs) and POPs made of covalent natural structures (COFs) have distinguished themselves by fusing the benefits of conductive polymers with the high unambiguous surface area of carbon materials, making them rather alluring candidates as dynamic materials.
- The 1,3,5-triformylphloroglucinol with phenylenediamine moieties reversible Schiff base reaction and irreversible enol-to-keto tautomerization were combined in 2012 to produce two chemically stable COFs. A BET surface area of 535 $m^2\ g^{-1}$ is present in the COFs that are produced (Segura et al. 2016).
- On to 3D graphene the author a noncovalent method to combine anthraquinone-based COFs COFDAAQ-BTA with the conjugated surface (Zhang et al. 2021; Mulzer and Shen et al. 2016) altered redox-active 2D COFs films by electropolymerizing 3,4-ethylenedioxythiophene (EDOT) inside COFs pores (Mulzer and Shen et al. 2016)

- By producing a series of aza-CMP utilizing an ion thermal condensation of 1,2,4,5-benzenetetramine with triquinoyl hydrate at 300–500 °C, Jiang and coworkers investigated the potential of aza-CMP for energy storage (Xu et al. 2013)
- Wei and colleagues grew TpPa-COF on MWCNTs under solvothermal conditions to create a unique core-shell COF/multi-walled (MWCNTs) nanocomposite (TpPa-COF@MWCNTs) (Liu et al. 2020).

8.2.4 Removal of Organic Pollutants from Water

Porous organic materials are crucial for applications like wastewater treatment because they have proven to be effective in identifying and removing pollutants and impurities from water and other liquids. The HCPs work as potent catalysts for the conversion of the adsorbed compounds when they are functionalized with gold nanoparticles. They are able to efficiently adsorb aromatic molecules from water (Yu et al. 2016). The bipyridinium cations display affinity for anionic organic dyes, confirming the COF's aptitude for the adsorption of anionic dyes from water. This is the first report of a 2D polycationic COF. The COF-thioether LZU8's functional groups facilitate the removal of Hg^{2+} from water (Luo et al. 2016). The presence of fluorine atoms and the significant specific surface area of the porous polymer are responsible for PFCMP-0's remarkable performance (Berton et al. 1999). PFCMP-0 has a great capacity for adsorption that makes it possible to remove different types of impurities from water. PAF-45 may efficiently remove pollutants from water because it has a positively charged surface and attracts negatively charged perfluorooctanesulfonate molecules.

8.2.5 Template Synthesis Methods of Porous Organic Polymers

- *Molecular imprinting*: In contrast to the casting of macroscopic objects, colloid crystal imprinting does not include any self-organization. Micellar aggregates are the most basic self-assembling form, and it is advised to use them before imprinting since it is the next degree of complexity (Cooper and Holmes 1999). To create porous polymer monoliths, ethylene trimethyl acrylate (TRM) was polymerized in supercritical scCO in the presence of AOT micelles that were swelled with water. No extra stabilizers were required to create a single-phase system prior to polymerization since the high monomer concentration of 50% prevented them from being needed. BET tests revealed that the radical polymerization-produced gels had a median pore width of 25 nm and a specific surface area of 260 $m^2\,g^{-1}$. Purified scCO without AOT micelles allowed polymerization, which resulted in substantially bigger pores with 100 nm diameters. Acrylamide and bisacrylamide were copolymerized in the continuous aqueous phase of a sodium dodecylsulfate-stabilized macroemulsion. Therefore, compared to micellar imprinting, the final polymer had much greater pore widths of 1 mm (Imhof and Pine 1998). By polymerizing water-miswollen reverse AOT micelles in the continuous phase, gels with closed pore topologies are created. Open porous polymers can be produced by polymerizing within additional self-organized media, such as bi-continuous microemulsions.

- *Colloid crystal templating*: Micellar engraving is a transcriptive templating strategy, much as sub-atomic engraving and colloid gem templating. On a mesoscopic level, colloid gems show gigantic order, which could prompt different captivating qualities, including especially high porosities at ideal mechanical strength and photonic band holes. Greatest pressing thickness is accomplished by means of the helpful crystallization process that makes colloidal gems. Colloid gems display unprecedented request at the mesoscopic level, which might bring about various amazing properties, like exceptionally enormous porosities at the ideal mechanical strength and photonic band holes, for instance. The helpful crystallization technique used to create colloidal precious stones takes into account the greatest pressing thickness. Utilizing extraction, disintegration, or calcination to eliminate the format after polymerization makes a permeable, consistently requested polymer. The accompanying models exhibit how the colloid gem-templating approach, which was first produced for inorganic materials, likewise applies to natural polymer gels (Xia et al. 2000; Velev and Lenhoff 2000). Latex crystals formed by injection and buildup between two solid plates were molded into porous polyurethane membranes. Following UV polymerization, toluene was used to selectively extract the polystyrene lattices.
- *Micellar imprinting*: A method for creating polymers having molecularly organized pores is called molecular imprinting (Wulff 1995). Thus, it is possible to create synthetic polymers with complementary holes and molecules. Therefore, exceptionally specific acknowledgment capacities similar to those seen in natural frameworks are to be anticipated (for example enzymatic catalysis). Regularly, a sub-atomic format (engrave particle) and additional solvents going about as porogenes are available while utilitarian and cross-connecting monomers are copolymerized. A few surveys on the production of microscopically engraved polymers (MIPs) and their utilization in separation (Vallano and Remcho 2000) and sensing (Takeuchi and Haginaka 1999) have been released. The templates used in the majority of molecular imprinting techniques are smaller molecules such as amino acids (Lin et al. 1998), sugars (Mulzer and Shen et al. 2016; Mayes et al. 1994), oligo-peptides (Klein et al. 1999), or steroids (Cheong et al. 1997).

8.2.6 Reaction Elaborated in the Chemical Synthesis of Porous Organic Polymers

All relevant information about these reactions is provided in Table 8.1.

8.3 Conclusion

Despite the fact that there has been critical headway in the improvement of different applications for permeable natural polymers, like gas stockpiling/partition, catalysis, squander water remediation, and energy stockpiling, there is still a lot

Table 8.1 Porous organic polymers' preparation and reaction.

Reaction	Reactants	Catalyst	Reaction conditions	Porous materials synthesized	Kinetically or thermodynamically controlled	Surface area of pop materials	References
Imide reaction	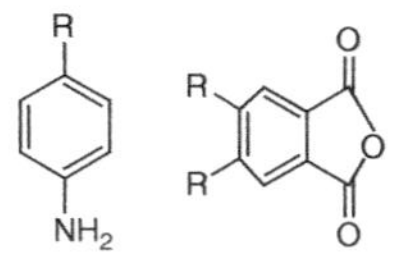	Not required	Anhydrous and inert conditions and heat and solvothermal conditions	PIM, COF	Thermodynamically controlled	551(p4) 2346(PI-COF3)	(Côté et al. 2005)
Amide reaction	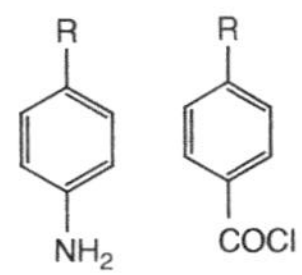		Anhydrous and inert condition and heat	PIM	Thermodynamically controlled	156 (P3)	Côté et al. (2005) and Fang et al. (2014)
Boroxine anhydride formation reaction	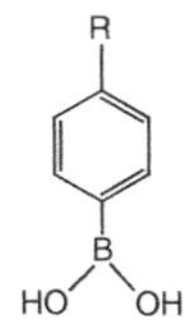	None	Solvothermal reaction	COF	Thermodynamically controlled	711(COF-1)	Côté et al. (2005)
Oxidative coupling reaction	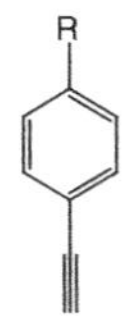	Transition metal compound	Anhydrous and inert environmental conditions	PAFs, CMPs	Kinetically controlled	827(PPN-1) 842(HCMP-1)	Jiang et al. (2002)
Phenazine ring fusion reaction	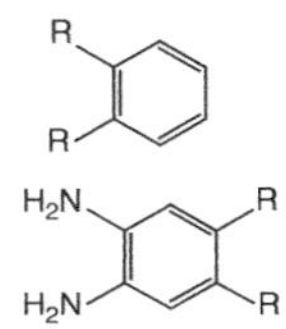	$AlCl_3$	Ionothermal reaction	CMPs	Kinetically controlled	1227(Aza-CMP@500)	Kou et al. (2011)

of work to be finished as far as pore structure customization and pore surface customization are concerned. Permeable natural polymers have the advantage of being entirely versatile concerning pieces, inside designs, and qualities, empowering their coherent plan for a specific application. The scope of highlights and applications for permeable natural polymers, which are dependent on their utilization for useful applications, will without a doubt be extended by future review, which will include specialists from a wide range of areas. Then, making permeable natural polymers with generally scattered dynamic destinations addresses a huge trouble. Subsequently, making permeable natural polymers with exceptionally scattered dynamic destinations and better than the ones that as of now exist is a huge test. Doing so will empower the making of surface locales with a known coordination circle, working with structure–action relationship studies. As enormous scope creation is a necessity for the vast majority potential applications, endeavors ought to be made to keep creating methods that permit the versatile readiness of permeable natural polymers using reasonable and environmentally harmless innovations.

References

Aguila, B., Sun, Q., Perman, J.A. et al. (2017). Efficient mercury capture using functionalized porous organic polymer. *Advanced Materials* 29 (31): 1–6. https://doi.org/10.1002/adma.201700665.

Arab, P., Rabbani, M.G., Sekizkardes, A.K. et al. (2014). Copper(I)-catalyzed synthesis of nanoporous azo-linked polymers: impact of textural properties on gas storage and selective carbon dioxide capture. *Chemistry of Materials* 26 (3): 1385–1392. https://doi.org/10.1021/cm403161e.

Ben, T., Ren, H., Shengqian, M. et al. (2009). Targeted synthesis of a porous aromatic framework with high stability and exceptionally high surface area. *Angewandte Chemie International Edition* 48 (50): 9457–9460. https://doi.org/10.1002/anie.200904637.

Berton, B., Fo, S., and Antonietti, M. (1999). Lyotropic mesophases of poly(ethylene oxide)-*b*-poly(butadiene) diblock copolymers and their cross-linking to generate ordered gels. *Macromolecules* 32 (18): 5803–5809. https://doi.org/10.1021/acscentsci.6b00220.

Budd, P.M., Ghanem, B., Msayib, K., and Mckeown, N.B. (2003). A nanoporous network polymer derived from hexaazatrinaphthylene with potential as an adsorbent and catalyst support. *Journal of Materials Chemistry* 13: 2721–2726. https://doi.org/10.1039/b303996j.

Cheong, S., Rachkov, A.E., Park, J.-K. et al. (1997). Synthesis and binding properties of a noncovalent. *Journal of Polymer Science* 36: 1725–1732.

Cooper, A.I. and Holmes, A.B. (1999). Synthesis of molded monolithic porous polymers using supercritical carbon dioxide as the porogenic solvent. *Advanced Materials* 11 (15): 1270–1274. https://doi.org/10.1002/(SICI)1521-4095(199910)11:15<1270::AID-ADMA1270>3.0.CO;2-E.

Côté, A.P., Benin, A.I., Ockwig, N.W. et al. (2005). Chemistry: porous, crystalline, covalent organic frameworks. *Science* 310 (5751): 1166–1170. https://doi.org/10.1126/science.1120411.

Das, S., Heasman, P., Ben, T., and Qiu, S. (2017). Porous organic materials: strategic design and structure-function correlation. *Chemical Reviews* 117 (3): 1515–1563. https://doi.org/10.1021/acs.chemrev.6b00439.

Fang, Q., Zhuang, Z., Gu, S. et al. (2014). Designed synthesis of large-pore crystalline polyimide covalent organic frameworks. *Nature Communications* 5 (1): 4503. https://doi.org/10.1038/ncomms5503.

Imhof, A. and Pine, D.J. (1998). Uniform macroporous ceramics and plastics by emulsion templating. *Advanced Materials* 10 (9): 697–700. https://doi.org/10.1002/(sici)1521-4095(199806)10:9<697::aid-adma697>3.3.co;2-d.

Jiang, J.X., Su, F., Niu, H. et al. (2002). Conjugated microporous poly(phenylene butadiynylene)s. *Chemical Communications* 8 (4): 486–488. https://doi.org/10.1039/b715563h.

Jiang, J.-X., Su, F., Trewin, A. et al. (2007). Conjugated microporous poly(aryleneethynylene) networks. *Angewandte Chemie* 119 (45): 8728–8732. https://doi.org/10.1002/ange.200701595.

Klein, J.U., Whitcombe, M.J., Mulholland, F., and Vulfson, E.N. (1999). Template-mediated synthesis of a polymeric receptor specific to amino acid sequences. *Angewandte Chemie International Edition* 38 (13/14): 2057–2060. https://doi.org/10.1002/(sici)1521-3773(19990712)38:13/14<2057::aid-anie2057>3.3.co;2-7.

Kou, Y., Xu, Y., Guo, Z., and Jiang, D. (2011). Supercapacitive energy storage and electric power supply using an aza-fused π-conjugated microporous framework. *Angewandte Chemie* 123 (37): 8912–8916. https://doi.org/10.1002/ange.201103493.

Li, M., Liu, J., Zhang, T. et al. (2021). 2D Redox-active covalent organic frameworks for supercapacitors: design, synthesis, and challenges. *Small* 17 (22): 1–16. https://doi.org/10.1002/smll.202005073.

Lin, J.M., Uchiyama, K., and Hobo, T. (1998). Enantiomeric resolution of dansyl amino acids by capillary electrochromatography based on molecular imprinting method. *Chromatographia* 47 (11–12): 625–629. https://doi.org/10.1007/BF02467444.

Liu, X., Liu, C.F., Lai, W.Y., and Huang, W. (2020). Porous organic polymers as promising electrode materials for energy storage devices. *Advanced Materials Technologies* 5 (9): 1–20. https://doi.org/10.1002/admt.202000154.

Lu, S., Liu, Q., Han, R. et al. (2021). Potential applications of porous organic polymers as adsorbent for the adsorption of volatile organic compounds. *Journal of Environmental Sciences (China)* 105: 184–203. https://doi.org/10.1016/j.jes.2021.01.007.

Luo, Q., Zhao, C., Liu, G., and Ren, H. (2016). A porous aromatic framework constructed from benzene rings has a high adsorption capacity for perfluorooctane sulfonate. *Scientific Reports* 6 (1): 20311: 1–9. https://doi.org/10.1038/srep20311.

Mayes, A.G., Andersson, L.I., and Mosbach, K. (1994). Sugar binding polymers showing high anomeric and epimeric discrimination obtained by noncovalent molecular imprinting. *Analytical Biochemistry* 222 (2): 483–488. https://doi.org/10.1006/abio.1994.1521.

Mckeown, N.B. (2012). Polymers of intrinsic microporosity. *International Scholarly Research Notices*. https://doi.org/10.5402/2012/5139

Mckeown, N.B. (2017). The synthesis of polymers of intrinsic microporosity (PIMs). *Science China Chemistry* 60 (8): 1023–1032. https://doi.org/10.1007/s11426-017-9058-x.

Mckeown, N.B. and Budd, P.M. (2002). Phthalocyanine-based nanoporous network polymers. *Chemical Communications* 44: 2780–2781.

Mckeown, N.B. and Budd, P.M. (2006). Polymers of intrinsic microporosity (PIMs): organic materials for membrane separations, heterogeneous catalysis and hydrogen storage. *Chemical Society Reviews* 35: 675–683. https://doi.org/10.1039/b600349d.

Mulzer, C.R., Shen, L. et al. (2016). Superior charge storage and power density of a conducting polymer-modified covalent organic framework. *ACS Central Science* 2 (9): 667–673. https://doi.org/10.1021/acscentsci.6b00220.

Segura, J.L., Mancheño, M.J., and Zamora, F. (2016). Covalent organic frameworks based on Schiff-base chemistry: synthesis, properties and potential applications. *Chemical Society Reviews* 45 (20): 5635–5671. https://doi.org/10.1039/c5cs00878f.

Sun, Q., Aguila, B., and Ma, S. (2019). Opportunities of porous organic polymers for radionuclide sequestration. *Trends in Chemistry* 1 (3): 292–303. https://doi.org/10.1016/j.trechm.2019.02.010.

Takeuchi, T. and Haginaka, J. (1999). Separation and sensing based on molecular recognition using molecularly imprinted polymers. *Journal of Chromatography. B, Biomedical Sciences and Applications* 728 (1): 1–20. https://doi.org/10.1016/S0378-4347(99)00057-2.

Talapaneni, S.N., Buyukcakir, O., Je, S.H. et al. (2015). Nanoporous polymers incorporating sterically confined N-heterocyclic carbenes for simultaneous CO_2 capture and conversion at ambient pressure. *Chemistry of Materials* 27 (19): 6818–6826. https://doi.org/10.1021/acs.chemmater.5b03104.

Taylor, R.G.D., Bezzu, C.G., Carta, M. et al. (2016). The synthesis of organic molecules of intrinsic microporosity designed to frustrate efficient molecular packing. *Chemistry – A European Journal* 22 (7): 2466–2472. https://doi.org/10.1002/chem.201504212.

Vallano, P.T. and Remcho, V.T. (2000). Highly selective separations by capillary electrochromatography: molecular imprint polymer sorbents. *Journal of Chromatography. A* 887 (1–2): 125–135. https://doi.org/10.1016/S0021-9673(99)01199-1.

Velev, O.D. and Lenhoff, A.M. (2000). Colloidal crystals as templates for porous materials. *Current Opinion in Colloid & Interface Science* 5 (1–2): 56–63. https://doi.org/10.1016/S1359-0294(00)00039-X.

Wood, C.D., Bien, T., Trewin, A. et al. (2007). Hydrogen storage in microporous hypercrosslinked organic polymer networks. *Chemistry of Materials* 19 (8): 2034–2048. https://doi.org/10.1021/cm070356a.

Wood, C.D., Tan, B., Trewin, A. et al. (2008). Microporous organic polymers for methane storage. *Advanced Materials* 20 (10): 1916–1921. https://doi.org/10.1002/adma.200702397.

Wulff, G. (1995). Molecular imprinting in cross-linked materials with the aid of molecular templates—a way towards artificial antibodies. *Angewandte Chemie International Edition in English* 34 (17): 1812–1832. https://doi.org/10.1002/anie.199518121.

Xia, Y., Gates, B., Yin, Y., and Lu, Y. (2000). Monodispersed colloidal spheres: old materials with new applications. *Advanced Materials* 12 (10): 693–713. https://doi.org/10.1002/(sici)1521-4095(200005)12:10<693::aid-adma693>3.3.co;2-a.

Xu, Y., Jin, S., Xu, H. et al. (2013). Conjugated microporous polymers: design, synthesis and application. *Chemical Society Reviews* 42 (20): 8012–8031. https://doi.org/10.1039/c3cs60160a.

Yu, S.-B., Lyu, H., Tian, J. et al. (2016). A polycationic covalent organic framework: a robust adsorbent for anionic dye pollutants. *Polymer Chemistry* 7 (20): 3392–3397. https://doi.org/10.1039/C6PY00281A.

Zhang, C., Li, P., and Cao, B. (2016). Electrospun polymer of intrinsic microporosity fibers and their use in the adsorption of contaminants from a nonaqueous system. *Journal of Applied Polymer Science* 133: 43475: 1–10. https://doi.org/10.1002/app.43475.

Zhang, B., Wang, W., Liang, L. et al. (2021). Prevailing conjugated porous polymers for electrochemical energy storage and conversion: lithium-ion batteries, supercapacitors and water-splitting. *Coordination Chemistry Reviews* 436: 213782. https://doi.org/10.1016/j.ccr.2021.213782.

Zhu, X., Tian, C., Veith, G.M. et al. (2016). In situ doping strategy for the preparation of conjugated triazine frameworks displaying efficient CO_2 capture performance. *Journal of the American Chemical Society* 138 (36): 11497–11500. https://doi.org/10.1021/jacs.6b07644.

9

Covalent–Organic Frameworks (COF): An Advanced Generation of Reticular Organic Polymers for Energy and Environmental Applications

Adhithya Ravi, Athira J. Ajith, and Abhijeet K. Chaudhari

Christ University, Department of Chemistry, Laboratory for Nano-Void Chemistry (NAVICHEM), Bangalore 560029, India

9.1 Introduction

A growing curiosity in renewable and clean energy, such as splitting of water for the production of hydrogen, cells and supercapacitors for the storage of energy, and fuel cells for the generation of emission-free electricity, is largely driven by the serious environmental problems brought on by increased emissions of greenhouse gases and pollutants (CO_x, NO_x, SO_x, and fine particulates) as well as the rising global energy demand.

The production of innovative materials is a crucial step for improving the equipment and procedures required for the generation, conversion, and storage of sustainable energy in the future.

Recently, a family of porous materials with fine-tuned morphologies has been created; these materials have been dubbed covalent organic frameworks (COFs).

The concept of COFs was introduced by Omar. M. Yaghi and team in 2005 (Côté et al. 2005). Porous organic frameworks known as COFs are created from lighter periodic elements such as C, H, O, N, and B. Strong covalent bonds such as borosilicate bonds are then used to form the overall frameworks, thus giving them excellent chemical stability (Wang et al. 2021a).

From then onward, the frameworks were tuned to fit the criteria for numerous applications such as:

- Energy storage
- Elimination and detection of heavy metals and various industrial pollutants
- Electrochemical applications
- Supercapacitors and Batteries
- Energy production.

Although research on COFs and their electrochemical applications is still in its preliminary stage, intrinsic properties like tuneable skeletons, large surface areas, and almost perfect crystallinity have made them exemplary candidates to advance electrochemical processes.

Organic Polymers in Energy-Environmental Applications, First Edition.
Edited by Ramesh Oraon, Pardeep Singh, Sanchayita Rajkhowa, Sangita Agarwal, and Ravindra Pratap Singh.

COFs are made up of lower mass elements including hydrogen, carbon, boron, nitrogen, and oxygen as opposed to metal-based nodes and organic structures. Furthermore, COFs can create both 2-dimensional and 3-dimensional constructs. COFs are made solely of covalent bonds, as opposed to MOFs.

Reversible covalent linkages exist in COFs. When COFs were still a relatively novel idea, it was expected that this property would lower the stability. However, after extensive study done specially to improve the stability, COFs are now much more stable thermally than MOFs, which were first believed to be much more stable than COFs. Often, MOFs are heavier than COFs. However, since only specific materials and conditions may be used to build COFs, the number of MOFs that have been produced significantly exceeds the number of COF structure moieties that have so far been constructed (Li et al. 2020a).

9.2 Synthesis

COFs are generally produced by solvothermal condensation. It is the first technique for synthesizing COFs and enables a proper construction of the COF with acceptable crystallinity and porosity. Usually, this synthesis is performed at appropriate temperatures (80–160 °C) for specific days in a sealed airtight container under an inert atmosphere. The monomers are added in presence of a catalyst under these conditions. Parameters like temperature and nature of the solvent significantly affect the quality of the product formed. On top of these, parameters such as type of catalyst used and time is also monitored.

Other synthesis methods include microwave-assisted synthesis, mechanical grinding, ionothermal synthesis, interfacial polymerization, and microfluidic synthesis.

9.3 COFs as Thin Films

In addition to the COF films being developed on conventional electrodes to enable adequate transport of e^- and control of structure, developing COF films is essential to building other vital components of various devices, such as the thin membranes that separate the electrodes.

The bottom-up strategy, which is frequently linked to functional coatings on selected surfaces, has been heavily utilized for the production of COF thin films (Wang et al. 2021a). This method relies heavily on controlling the film's growth orientation and avoiding random extension. Numerous oriented COF thin films have been synthesized using the bottom-up approach so far on a variety of interfaces.

Solvothermal synthesis can be defined as the process of coating the powder of the frameworks on any type of support by drenching suitable substrates in the reaction mixture, rinsing them with required solvents (organic), and then drying them under a stream of nitrogen gas.

Interfacial synthesis is a common approach of producing graphene nanosheets and polyamide nanofilms. In this approach, reaction between the monomers takes place at the point of contact. The confining of the COF development to the contact area results in the formation of said thin films. In order to create COF thin films by this method two interfaces have been used: the liq–air and the liq–liq (Dey et al. 2017; Feldblyum et al. 2015).

Bein and co-workers developed the mild process known as **room-temperature vapor-assisted conversion synthesis** in order to compensate for the instability of building blocks which are delicate to the harsh conditions utilized in solvothermal synthesis (Medina et al. 2015).

This excellent technique enables the creation of COF films using delicate precursors and on delicate substrates.

Although not as popular as the above methods, the top-down method, which is another way to make COFs, includes making free-standing, one- or two-layered COF thin films.

9.4 Polygon Skeletons

A special topology diagram that is based on reticular chemistry can be used to determine the structure of COFs. These are categorized as two-fold, three-fold, four-fold, and six-fold axes of symmetry. Thus, the final structures of the COFs are just combinations of the building blocks (Figure 9.1a,b).

Depending on which two building blocks are combined, the structure of the COF can be classified as: **2D COFs or 3D COFs** (see Figure 9.1). Depending upon the structural arrangements, COFs could exert extended conjugation in two different directions found in 3D COFs or weak chemical interactions like hydrogen bonding or van der Waal's interaction between two layers of 2D sheets arranged along one axis of the structural arrangement (Figure 9.1e). COFs offer several ways to design structural motifs leading to cationic COFs having anions seating within their pore or vice versa anionic COF having cation neutralizing the framework charge. They can have different functional guest species occluded inside the porosities playing major role in overall functioning of the structure. As COFs are purely organic building blocks, they can be easily modified synthetically even after their structural motif gets synthesized (Figure 9.1f)

9.5 Pore Engineering

COFs have a high degree of molecular flexibility, enabling atomically precise integration of their constituent parts into porous structures.

The crystalline nature of the structure and ordered monomer units dictate the porosity of these frameworks, which display surface areas of up to 4000 $m^2 g^{-1}$ (Medina et al. 2015). Chemical stability and its property to remain stable even at high temperatures (greater than 400 °C) is a result of the high strength of its bonds, which can

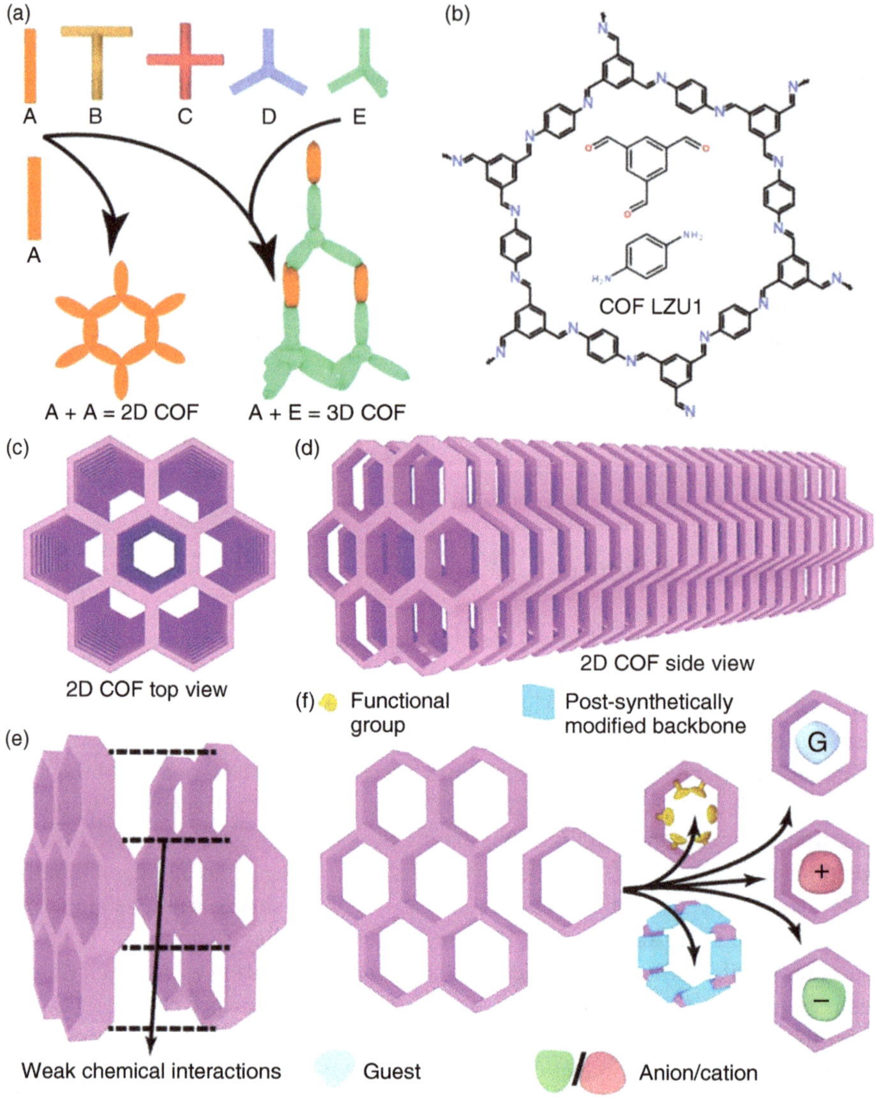

Figure 9.1 Scheme showing (a) different shapes of building blocks that make up different types of COFs like 2D or 3D (b) an example of COF LZU1 made from its constituent molecules (inside the pore of COF) (c) 2D COF top view and (d) side view (e) weak interactions observed between two 2D layers of COF (f) varieties of COFs like anionic COF, cationic COF, guest included COF, post synthetically modified COF.

endure challenging circumstances including hydrolysis, oxidation, reduction, and pH range variations. Additionally, its reduced density permits a superior gravimetric performance by virtue of the homogeneous composition and lightweight parts. In order to modify the structural and functional features of COFs, it is necessary to change the tuneable pore size of the pristine frameworks utilizing well-established molecular organic chemistry.

Nagai et al. published a method for pore engineering of COFs (Nagai et al. 2011). A synthetic phenylene block with an azide added to it was employed as the wall component in a 2-component condensation technique to fabricate frameworks with azide-integrated walls. The click reaction and a 3-component condensation system were combined to create a method that allows for control of the density and composition of groups within the porous sites. It has been established that employing this technology makes it incredibly advantageous to build various specialized surfaces in COF pores.

By using this strategy, several COFs with micro- or meso-scale tetragonal or hexagonal pores have been created. Despite the availability of numerous COF skeletons, regulating the pore surface is still challenging since it has proven difficult to develop a technique that allows for the systematic engineering of COF pore walls. COFs would have a unique interface by customizing the pore surface to meet the requirements of each application.

The importance of creating reticular frameworks with a variety of connections and their application in frameworks with different pore sizes, pore volumes, and dual pores have been extensively discussed in recent years in a number of theme viewpoints and reviews. Some highlighted linkages and pore geometry throughout the years are being discussed below.

- *BO linkage*: Yaghi and co-workers described the self-condensation of a boroxin building unit (COF-1, COF-102) and the same base building block was condensed with catechol for creating the primary class of boronic acid frameworks (COF-5, COF105) (Côté et al. 2005) hydrolyzed in moisture because of the B—O linkage susceptibility. A further point made by Dichtel and co. was the rapid dissociation of HHTP-DPB COF in aqueous solution (Spitler et al. 2011). Yaghi's group modified the framework created by the reaction of silanols with $RB(OH)_2$ in order to boost stability.
- *CN linkage*: Yaghi synthesized a framework employing dynamic imine linkage, which was more stable than boron-linked frameworks (Uribe-Romo et al. 2009). By reacting 1,3,5-triformylphloroglucinol, p-phenylenediamine, and 2,5-dimethyl-p-phenylenediamine, Banerjee and co-workers produced TpPa-1 and TpPa-2, which further increase the stability of imine-based frameworks (Chandra et al. 2013). Aldimine condensation led to stability and was then followed by irreversible tautomerism from enol-imine to ketoenamine. The crucial significance of C–N links has also reportedly been highlighted in recent years by COFs based on squaraine and viologen (Nagai et al. 2011).
- *CC linkage*: The Hecht and co-workers published the pioneer C–C single bond framework made by utilizing the Ullmann coupling reaction followed by Knoevenagel condensation for constructing framework (Grill et al. 2007). By combining a formyl-substituted monomer with a nitrile-functionalized precursor like 1,4-phenylenediactonirtrile, Jiang and colleagues were able to create a sp^2 carbon-conjugated framework (Jin et al. 2017). Yaghi's team recently reported the creation of a -C–C double bonded framework utilizing Aldol condensation of 2,4,6-trimethyl 1,3,5-triazine and 4,4'-biphenyldicarbaldehyde, which

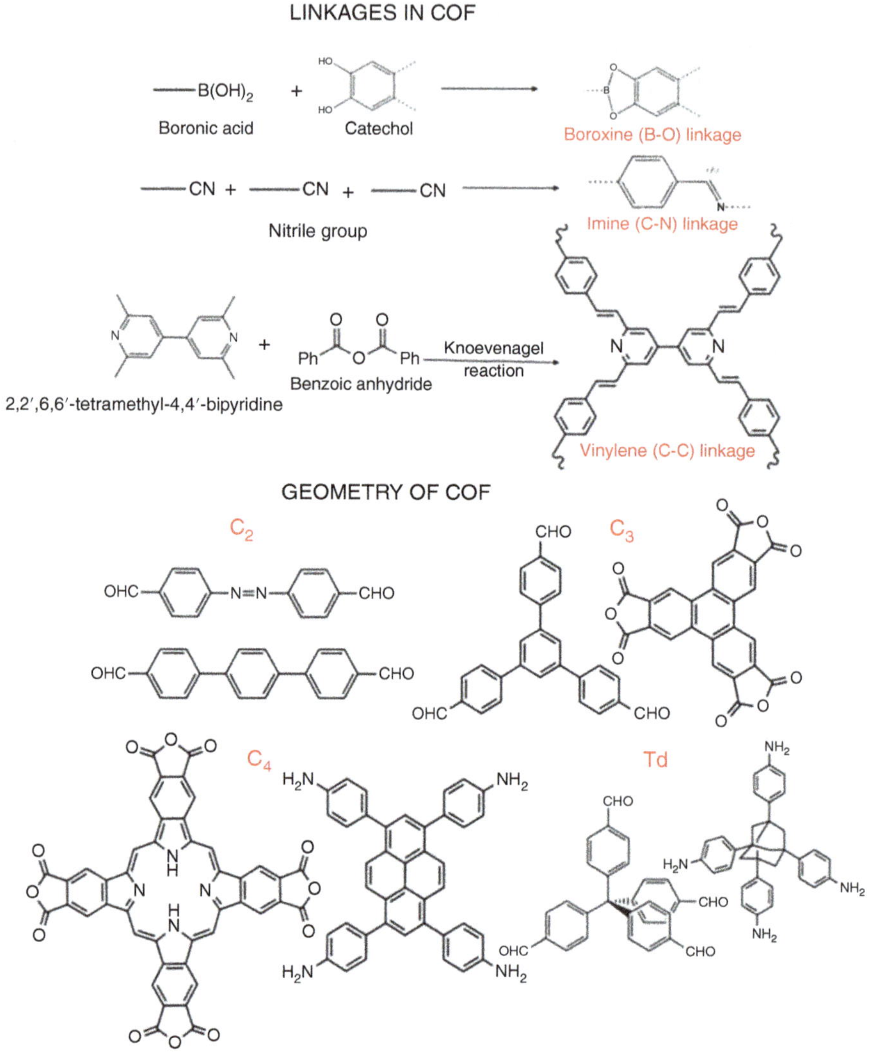

Figure 9.2 Schematic representation of B–O, C–N, and C–C linkage formation followed by the geometry of COF formed.

showed exceptional stability in Brønsted acid or base (Lyu et al. 2019) (refer to Figure 9.2).

In addition to the commonly used connections in COF building, azodioxide-, and hetero-linked frameworks were reported (Jackson et al. 2012; Beaudoin et al. 2013; Chen et al. 2015). Synthetic techniques such as solvothermal, ionothermal, microwave, mechanochemical, interfaces, and room temperature enhance properties like surface area of synthesized frameworks and crystallinity. In addition to the linkage used, these techniques are employed to construct the framework (Kuhn et al. 2008; Campbell et al. 2009; Dogru et al. 2011).

9.5.1 Pore Design

Porous material chemistry revert to the discoveries of complexes like β-Co(4-methylpyridine)$_4$(NCS)$_2$ and the first crystallographic coordination networks with metal–neutral ligand coupling published in 1959, both of which were the work of Nobel winner Alfred Werner (Werner 1893). In COFs, the reticular chemistry is implemented from two perspectives: first, the support structure of the framework, and second, the space that the structure takes up. The systematic pore network in COFs facilitates the inclusion of numerous guest molecules under non-covalent interaction; therefore, the geometrical structure of the building blocks is of utmost importance to translate the shape and size selective entry of molecules. As Jiang's group demonstrated the connection between the pore structure and creating units in a topological diagram, hexagonal, tetragonal, trigonal, and rhombic architectures have been used extensively (Werner 1893). As per the reports, COF-6 has the smallest pore (0.87 nm) (Côté et al. 2007) and PC-COF shows the largest pore (5.8 nm) (Yu et al. 2016).

In addition to this, HHTP-DPB COF39 (4.7 nm) (Spitler et al. 2011), BTP-COF64 (4.0 nm) (Dogru et al. 2011), DTP-ANDI COF75 (5.3 nm) (Jin et al. 2013) have the significant large pores synthesized so far. The Lavigne research team saw significant reductions in surface area and pore size to 1.6, 1.4, and 1.1 nm, respectively, after adding methyl, ethyl, and propyl functionalities to a framework with variable pore size (Tilford et al. 2008).

This suggests that size of substituents (–Me, –Et, –Pr) has a direct relationship with the decrease in pore size. Jiang and colleagues further emphasize the observed reduction in pore size compared to pure frameworks and alkynyl-, alkenyl-, and azide-decorated frameworks (Nagai et al. 2011). Using the D_{2h} symmetric 4,4′,4″,4‴-(ethane-1,1,2,2-tetrayl) tetraaniline building block, Zhao study group also reported a dual pore and three different types of pores (ETTA). Yaghi, Wang, and other research teams have also noted heterogeneous pore structure and variations in pore shape with changes in building block concentration (Dalapati et al. 2016; Zhu et al. 2015).

9.5.2 Increasing the Vulnerability of the Active Sites

The functional groups in COFs occupy inner surfaces of the long-range ordered porosities (see Figure 9.1f). The active sites present in those groups are buried deep inside the 1D pore since there is strong π-π interaction acting between the molecules. This causes a decrease in rate performance and reactivity toward alkali metal ions. Through pore engineering, the accessible exposure of active sites is increased thereby increasing the efficiency of COF.

9.5.3 Pore Surface Engineering for Environmental Application

Porous frameworks play a crucial part in the fight against pollution because of the adjustable pore size, extensive surface area, and by significantly increasing the adsorptive capacity, kinetics, and selectivity. The character and virtue of the host molecules are crucial; for example, the tiny pore size of amorphous hosts prevented the access to grafting groups, whereas the high pore size and other crucial characteristics of COFs resulted in easy capturing of heavy metal ions. More significantly, the COFs' inclusion of binding groups that have been evenly dispersed through pore surface engineering or de novo method places COFs at the forefront of this problem.

Although various binding molecules have been shown to bind uranium, amidoxime-adorned β-ketoenamine frameworks beat amorphous substances in terms of affinities, kinetics, and absorption abilities (Sun et al. 2018). The embellished framework was successful in lowering the water's uranium content (408 mg g^{-1}) to 0.1 parts per billion, which is substantially lower than the limitations specified by the UEPA (Sun et al. 2018).

Hg (Mercury) is a significant metal that has an impact on public health. To rid water of contamination, selective mercury collection is a pressing necessity. The fluorescent LZU8 COF was made by condensing the 1,3,5-triformylbenzene and the thioether-functionalized scaffold. The LZU8 COF was then uniformly and densely decorated with thioether functionalities in a 1-dimensional porous channel for greater affinity toward Hg ions. It also demonstrated a 300 mg g^{-1} uptake capacity, a detection limit of 25 ppb, and high recyclability. Jiang research team generated TAPB-BMTTPA-COF, which was designed to remove mercury from aqueous solutions (Ding et al. 2016). The framework that functionalized using thioether has been richly embellished with sulfur functionalities, resulting in an absorption capacity of 734 mg g^{-1}, almost two and a half times greater than LZU8 COF (Huang et al. 2017a).

Qiu and co-developed two distinct 3D-ionic-COF-1 and 3D-ionic-COF-2 via condensation of tetrakis(4-formylphenyl) CH_4 with diiridium bromide and ethidium bromide (Huang et al. 2017b). High levels of crystallinity, porosity, stability, and the ability to remove permanganate specifically were all displayed by both ionic frameworks. Signature materials like PVBTAH-ZIF-8 and LDHs are superseded by the measured kinetics with outstanding recyclability (Medina et al. 2015).

The Trabolsi research team initially reported a COF which was linked with viologen with good stability, porosity, and controlled shape, by employing the Zincke reaction (Das et al. 2017). In order to efficiently extract iodine from both solutions and vapors, 1,1′-bis(2,4-dinitrophenyl)-[4,4′-bipyridine]-1,1′-diium dihydrochloride and 1,3,5-tris(4-aminophenyl)benzene must be combined. The 5,6-bis(4-formylbenzyl)-1,3-dimethylbenzimidazolium bromide (BFBIm) is an ionic building block with a BET surface area of 1532 m^2 g^{-1} which was synthesized by the Jiang research group via dynamic imine linkage of PyTTA (Huang et al. 2017a).

The reverse AA stacking mode's decoration of benzimidazolium cations on each end of the porous sites produced the best absorption capacity of negatively charged methyl orange dye among various porous materials (553 mg g^{-1}), expected to be a result of the great electrostatic force of interactions (Huang et al. 2017b).

SA-COF was investigated as a size-dependent organic pollutant separation due to the existence of well-organized pore channels and reversible proton tautomerism (Ning et al. 2017). Size exceeds the host pore size, the framework's ability to bind a variety of dye molecules decreases with increasing molecular size and becomes impassable. Notably, the type of dye and how sensitive it is to pH affect the pore chemical environment and, thus, the selectivity of separation of charged molecules (Xia et al. 2022).

9.6 Thermal Stability

As COFs are covalently bonded frameworks, their natural stability is much greater than conventional polymers. This stability only increases when the strong aromatic units and crystalline nature of COFs are taken into consideration making them thermally stable up to 450 °C under an inert atmosphere (Song et al. 2014).

The strength of the chemical bonds determines its stability. Linkages are critical to the framework's stability. A stable link, however, limits the crystallinity of COF formation, reducing its reversibility.

Frameworks like CS-COF and CTFs are exceptionally stable due to their formation of highly stable ring-fused phenazine and triazine units, respectively (Evans et al. 2021).

9.7 Advantages Over Conventional Polymers

For a relatively recent class of porous materials or materials in general, COFs have proven to be outstanding alternatives to everything from catalysts to mitigating pollutants. COFs are superior to conventional polymers as they are more modifiable compared to said polymers. Conventional polymers are mostly found in amorphous conditions with partial crystalline nature. This relates directly to its physico-chemical behavior in the bulk form of material. For applications like energy harvesting, facile ionic/charge conduction, selective molecular separation material needs to possess peculiar characteristics like ordered molecular arrangements with tailored chemical struts having greater degree of flexibility in terms of pore size and chemical functionality. Needless to say, conventional polymers have been of great use in the scientific world as well as for daily life usage in the form of different tools and appliances. However, with the advent of smart generation of porous materials, COFs as an advanced class of polymers may serve better for chemically complex and challenging problems.

9.8 Backbone Modifications

9.8.1 Linkage Conversion

Reversibility plays a vital role in obtaining COFs as it helps in error correction and defect healing, in other words everything from the inherent properties (both physical and chemical) to the limitations of the COFs can be determined by just the linkages.

This reversibility sometimes limits some inherent properties of the COFs like crystallinity or even stability toward chemical solvents. For example, imine-based COFs (C=N) which are the most well-studied COFs as they are highly stable hydrothermally. Now although they are extremely stable, imine-based COFs can still go through hydrolysis under an acid medium or even exchange with amines as imines are reversible.

Yaghi and his colleagues came up with a technique to keep molecular cages made of dynamic covalent bonds stable. The amide-linked COFs are created using this method from the corresponding imine-COFs. The imine-based COFs were treated in the presence of dioxane with sodium chlorite (eleven equivalents per imine functionality), CH_3COOH (ten equivalents), & 2-methyl-but-2-ene (100 equivalents) (Waller et al. 2016).

As imine-based COFs undergo hydrolysis under acidic conditions, when subjected to an acid the COF loses its crystallinity and almost if not completely dissolves. Contrary to this, if the imine is converted into an amide linkage the COFs retains its crystallinity even under a strong acidic medium. Thus, the COF obtained from linkage conversion is more stable and shows great performance (Kandambeth et al. 2021).

9.8.2 Linker Changing

As mentioned above, dynamic covalent chemistry (DCC) is based on breaking and making of covalent bonds within the same reaction under thermodynamic conditions, i.e. a reversible chemical reaction. It is crucial for the production of COFs since it facilitates COF formation by enabling self-correction which directly increases the crystallinity of the COF. However, DCC is predominantly used in synthesis rather than post-synthesis in COFs. Similarly, in metal–organic frameworks or MOFs the method of linkage conversion has been used extensively to synthesize MOFs which are generally harder to prepare.

It is possible that this linker modification process can also be used with COFs because this reversible bond formation is analogous to both MOFs and COFs. Zhao and his colleagues demonstrated how this dynamic character can facilitate the exchange of COF linkers post-synthetically.

In order to show this, 2 distinct frameworks were synthesized and the subsequent linker change was investigated. Firstly, for the synthesis [1,1′:3′,1″-terphenyl]-3,3″, 5,5″-tetracarbaldehyde in which a core benzene and two aldehyde groups are introduced to the ends.

Secondly, 1,4-diaminobenzene and benzidine were chosen as the linkers to produce

TP-COF-DAB and ***TP-COF-BZ***.

Let the DAB-based COFs be termed as (A) and the BZ-based COF be termed as (B).

[B] is more thermally stable and shows higher crystallinity in comparison with [A].

In comparison with benzidine, 1,4-diaminobenzene is more active as it contains amine group in the para position (electron-donating group). Upon understanding this concept, Zhao and co-workers attempted to substitute benzidine units in [B] with 1,4-diaminobenzene. Thus, this transformation increases the crystallinity of [B] significantly.

9.9 Functional Group Changes

9.9.1 Inverse Vulcanization

Vulcanization is a process of heating rubber in presence of sulfur, an accelerator and an activator.

Pyun and co-workers developed a synthesis called inverse vulcanization primarily to synthesize a polymer with high sulfur content to be used as cathode materials in sulfur-based batteries like Li–S batteries (Simmonds et al. 2014; Chung et al. 2013). These lithium batteries, according to analysis, have one of the finest performances of any known COF-based sulfur cathodes, retaining 959 mA h g^{-1} after 100 cycles at 0.2C while having an initial capacity of 1400 mA h g^{-1}.

Sulfur loading and robust carbon-sulfur binding are made possible by this method's utilization of thermally driven radical polymerization between vinyl and sulfur groups. Thus, Xu, Chen, and associates have recently trapped sulfur inside the pores of the vinyl-functionalized COF-V utilizing the inverse vulcanization approach. By grinding, elemental sulfur and COF-V are combined evenly during the operation (Jiang et al. 2018). Molten sulfur is able to enter the pores with subsequent heat treatment at 155 °C while under vacuum in a sealed tube. In order to allow the ring-opening polymerization of sulfur-8 and complete co-polymerization between the poly-sulfur diradicals and vinyl groups, additional heat treatment at a greater temperature of 200 °C is required.

9.9.2 Thiol-ene Rx

The thiol-ene reaction, as the name suggests, comprises addition of a thiol to C=C under mild conditions. This method is especially useful in the field of coatings as it supports greater spatial control which starts UV initiation and is insensitive to oxygen.

In 2018, Ma and co-workers utilized this to post-synthetically modify COFs via thiol-ene reaction to successfully remove mercury (Sun et al. 2017). The COF was a vinyl-based COF which was synthesized by condensing 2,5 divinylterephthalaldehyde with 1,3,5-tris(4-aminophenyl) benzene. In addition to this, thiol-ene reaction was carried out with 1,2-ethanedithiol in the presence of azobisisobutyronitrile (AIBN) as a radical indicator. This material shows incredible potential in environmental remediation as mercury is extremely toxic and is one of the most prominent industrial pollutants.

Dong and co. added thiol-functionalized polysiloxane to an allyl-decorated COF that has previously supported Pd nanoparticles (Ma et al. 2020). An effective technique to create a self-standing and elastic membrane is through the interaction between the allyl-decorated COF with Pd nanoparticles and the thiol-functionalized polysiloxane. This membrane is consistently porous, strong, workable, homogeneous, and permeable to water. The dechlorination of chlorobenzene in water at ambient temperature may also be catalyzed using it (Waller et al. 2016).

9.10 COFs on Different Scales

As already stated, a fascinating benefit of COFs in electrochemical purposes is that, at least in theory, their structure and nanostructure are completely understood and can be controlled with high accuracy, but there are still a few issues with the microstructure that need to be resolved. It is challenging to transform the majority of COFs into bulk electrodes or films since they are often found as bulk powders that cannot be melted or dissolved.

The structural properties of COF materials on a macroscopic level, i.e. from a few hundred angstrom to centimeters, should also be taken into consideration in COF design in addition to a focus on microscopic length scales. The goal is to improve the properties of COFs by combining their chemical as well as crystalline structures with the ability to manufacture larger physical forms (thin films or coatings) with specified sizes, carefully defined morphologies, and/or hybrid components.

9.11 Terracotta Process

Because COFs are polycrystalline, it is extremely challenging to convert COF powders into the necessary physical forms without utilizing a binder or plasticizer. In recent times, a method known as the "terracotta process" has been used to shape the COFs into distinct forms and create hierarchical porosity in the COF matrix.

Since the reagent mixture maintains the properties of the fabric, these COFs can be processed into a variety of shapes without the addition of additives or binders, enabling the construction of numerous COF sculptures. Using baking soda throughout the initial phase of COF synthesis, a hierarchical porosity was added with a network of interconnected pores to the COF matrix.

In recent times, a method known as the "terracotta process" has been used to shape the COFs into distinct physical forms and create hierarchical porosity in the COF matrix. Grinding diamine linkers with p-toluenesulfonic acid (PTSA), water, and multitopic aldehydes, followed by brief heating, yielded significant amounts of ketoenamine COFs with improved porosity (Karak et al. 2017).

9.12 Pyrolysis of COFs

Carbonaceous electrode materials have been made using a wide range of organic precursors. Such precursors must be cheap, accessible, and plentiful for use in practical applications since they can increase the surface areas of carbon compounds or introduce heteroatoms into them. Since COFs barely meet these latter requirements, they don't seem to be the best precursors for carbon-based materials. However, there has recently been an increase in interest in crystalline porous materials as carbon precursors, partly because of the notion that their crystalline structure, which contains organized carbon sheets, may be a suitable area to direct the carbonization process (Zhang et al. 2016a).

Since the framework frequently breaks when carbonized, the widely held belief that the crystalline structure acts as a scaffold for the carbon elements is severely damaged. For the synthesis of porous carbonaceous materials from COFs, templates are often required. COF-derived carbon compounds outperform similar monomer-derived carbon compounds.

By simply carbonizing already-formed carbon-rich tubular and fiber-like polyphenylene precursors under argon, Feng et al. were able to create 1-D C fibers and tubes (Feng et al. 2009).

Improved electrodes constructed of carbon nanotubes and carbon nanofibers with areas up to nine hundred $m^2\ g^{-1}$ were tested. Xiang et al. created a graphitic porous carbon that was N-doped (Xiang et al. 2014) using 2D covalent organic polymer precursors, allowing for precise control over the heteroatom's placement. These materials demonstrated a strong relationship between the electrocatalytic activity and the properties of N-doped carbon.

9.13 COFs in Mitigation of Pollutants and Organic Dyes

COFs can analyze and eliminate various industrial pollutants [(heavy metals As, Cd, Cr, Cu etc.), radionuclides (UO_2^{2+} TcO_4^-, Nd^{3+}), organic waste (dyes, pharmaceutical waste, organo-phosphorous retardants)].

Due to their porous structure, COFs are fantastic candidates for eliminating dye molecules. Afshari and colleagues created a brand-new imine-linked COFA as a highly effective sorbent for eliminating hazardous dye (Afshari and Dinari 2020).

Heavy metals are significant environmental contaminants, and their toxicity is a concern that has grown in importance due to ecological, evolutionary, dietary, and environmental factors. The heavy metals are found in wastewater, all of which are hazardous to both human health and the environment. Toxic metals can cause a variety of ailments, such as cancer, mutation, and allergies. It is necessary to get rid of the organic contamination hydrazine, which is particularly dangerous for soil and water. Liang et al. developed redox-active COF to effectively recognize and eliminate hydrazine (Liang et al. 2020). The prepared COF showed astounding crystallinity and stability.

Cai et al. treated copper ions (Cu^{2+}) by preparing a Q-graphene-scaffolded COF which were used as adsorbents which showed a high ability to remove Cu^{2+} ions (Cai et al. 2019). Further, a C_3N_4 component was integrated into the COF backbone and the obtained COF can degrade dyes like methyl orange and organic compounds like phenol. The triazine moieties present in this COF act as photoactive sites with the cyclic ketone constituents and the conjugated structure acting as electron-withdrawing groups and electron-shift platform, respectively.

Zhao et al. mechanochemically synthesized TpMA to produce COFs with thread and ribbon-like morphology by further controlling the solvent and catalysts during the preparation (Kumar et al. 2021). This ribbon-like COF shows much greater efficiency when compared to its solvothermally synthesized counterparts.

9.14 COFs for Energy Applications

9.14.1 Hydrogen Evolution Reaction (HER)

Among all renewable energy alternatives, hydrogen (H_2) is one of the most promising. The most effective way to create high-purity H_2 via HER is through water electrolysis.

Adsorption, reduction, and desorption reactions are all involved in the multi-step electrocatalytic HER process:

$$(H^+ + e^- \rightarrow H_{\mathbf{ads}})$$

$$(H_{\mathbf{ads}} + H^+ + e^- \rightarrow H_2)$$

Or

$$(2H_{ads} \rightarrow H_2).$$

At present, platinum (Pt) is regarded as the top HER catalyst for fuel cells but due to scarcity and poor long-term stability COFs act as great catalysts (Hu et al. 2018).

9.14.2 Oxygen Evolution Reaction (OER)

OER is produced as a reaction at the anode during splitting and also during the charging process of metal–air cells. The procedure can be carried out in either an acidic or alkaline medium.

pH < 7: Acidic medium

$$2H_2O \rightarrow O_2 + 4H^+ + 4e^-$$

pH > 7: Basic medium

$$4OH^- \rightarrow O_2 + 2H_2O + 4e^-$$

Materials based on iridium (Ir) and ruthenium (Ru) are thought to be the main catalysts for OER (Kumar et al. 2021), but their limited storage and poor availability of these noble-metal based electrocatalysts hinder the commercial use. As a result, COFs are a fantastic substitute for Ru/Ir-catalysts (Zhang et al.

2021). Since COFs can be precisely built with only desired components, they offer a straightforward but effective method for analyzing the makeup of active sites.

9.14.3 Oxygen reduction Reaction (ORR)

The reaction:

$$O_2 + 4H^+ + 4e^- \rightarrow 2H_2O$$

plays a crucial part in storage and conversion of energy, such as fuel cells and metal–air batteries.

Continuously pumped into the cathode as the reactant to conduct ORR, oxygen from the surrounding air's oxygen content has a significant impact on the battery's voltage and energy density. Similar to HER, in fuel cells platinum is the ideal catalyst but due to scarcity and poor long-term stability carbon nanohybrids prepared by pyrolysis act as a great replacement to Pt catalysts (Li et al. 2020b).

9.14.4 Bifunctional OER/ORR Catalysts

High activity for both the oxygen evolution reaction (OER) and the oxygen reduction reaction (ORR) is displayed by bifunctional oxygen electrocatalysts.

Frameworks, which have ordered pores and high-precision functionalization, have been used in electrochemical energy and conversion systems due to their fast ion transport channels and precisely defined electrochemical active sites.

Even though the development of active sites in COFs toward ORR and OER is rarely studied, their electroactivities were hampered by the limited electron transit along the frameworks (Zhang et al. 2020).

9.14.5 Carbon Dioxide Reduction Reaction (CO_2RR)

The use of fossil fuels produces carbon dioxide (CO_2) emissions, which could be dangerous for the world's climate. Therefore, turning CO_2 into a useful carbon feedstock is a crucial objective for a sustainable economy.

CO_2 can be transformed into a variety of compounds depending on the number of migrating H^+ and e^-. Using electrocatalysis, CO_2 can be transformed into compounds of higher value, such as carbon monoxide (CO) and methanol (CH_3OH), or chemical energy can be converted into electricity (Hu et al. 2018).

Three fundamental stages for heterogeneous catalysis are:

1. Carbon dioxide is adsorbed on the sites that are active.
2. e^- transfer or migration of protons in the carbon-oxygen bonds can be formed, as well as carbon-hydrogen bonds.
3. Product formation by rearranging the configuration.

A few obstacles remain to attain perfect conversion of CO_2

- To produce CO_2 – an immense overpotential is required, due to its high-energy barrier
- Slow kinetics
- High cost.

Efficient electron transport is made possible by COFs having phthalocyanine (Pc) centers in the framework.

The first cobalt-COFs were created by Gao et al. Due to the improved Co-Pc-integration PBBAs of a storage function, the electrocatalytic reduction was conducted in the CO_2 gas phase, resulting in a nearly 100-fold increase in CO_2 concentration above the typical aqueous phase (Wang et al. 2018a).

The ultra-large surface area and easy accessibility of modifying surface functions by adding heteroatom doping are two features of the covalent triazine frameworks. Later, Huang, Cao, and colleagues also developed copper nanoparticles (Cu NPs) immobilized on the imidazolium-functionalized covalent triazine framework (Cu/ICTF) to produce C_2H_4 in electrochemical CO_2RR because Cu-based catalysts have frequently been used to produce multiple electron transfer C_2^+ products with relatively high selectivity (Mao et al. 2020; Yusran et al. 2019).

9.15 COFs in Batteries and Supercapacitors

In order to address the unsustainable depletion of fossil fuels and the accompanying energy problem, renewable energy sources, as well as ecologically benign energy technology, are necessary. Such technologies include solar energy and new electrochemical energy conversion and storage systems. The main focus of energy research has been the advancement of supporting infrastructure for converting and storing energy that improves transfer of mass and e^- mobility. COFs have showcased great potential in energy-related industries such as supercapacitors and batteries.

The sheet-like structure of anode materials allows ions to be accommodated between layers depending on how they are introduced. The shorter paths for diffusion of ions and greater accessibility active sites of nanosheets result in better rate performance in battery electrodes and supercapacitors (Wang et al. 2021a).

Post-synthetic modification: COF synthesis with high electrical conductivity as a first-stage is challenging. Post-synthetic alteration has made it possible to increase the conductivity of COFs. To improve the electrical conductivity, new functional groups or specific point alterations are made chemically following the synthesis of COF. The original structural framework of the COF is maintained throughout this process. That is, post-synthesis modification keeps the structural integrity of the pore surface while functional groups are added to or linked to it.

Yusran et al. functionalized azide-appended N_3-COF-5 series into RTrz-COF-5 series through triazole bond. They also synthesized hydroxyl-decorated square-like 2D porphyrin COFs by functionalization via ester bond (Kim and Choi 2019). This functionalization allows the COF to preserve its rigid structure with a well-organized functional group in the pores.

Surface doping: Another significant method for increasing conductivity is surface-doping modification of COF crystals. Organic linkers that are more adaptable and positively charged, and contain additional components can successfully coordinate with coordination functional groups. By establishing a continuous delocalized charge route, these organic linkers can improve energy matching using ligand orbital overlap. Particularly, azo ligand-containing compounds, which include nitrogen, have high electrical conductivity (Xia et al. 2022).

Bein and colleagues investigated the inherent conductivity of the two Wurster-COFs that formed. Dopants included iodine, a chemical dopant commonly used in COFs, antimony pentachloride (SbCl5), and 2,3,5,6-tetrafluoro-tetracyanoquinodimethane (Rotter et al. 2020).

In order to determine whether a comparable improvement in conductivity can also be seen, WTA was also doped along with F_4TCNQ in bulk. In this instance, the doping increased the conductivity [$4.91*10^{-6}$ Sm^{-1} to $1.51*10^{-5}$ Sm^{-1}] (Sugawara et al. 2022).

9.16 Batteries

The majority of the world's energy needs are still being met by conventional energy sources, whose extensive use raises a number of environmental concerns. Security in the environment and rising energy uses have been two of the biggest problems for human civilization. Since the last few decades, non-conventional energy sources, such as sun, wind, and sea tides, have been investigated for sustainable development. The intermittent nature of these renewable resources, however, limits their immediate use.

Scientists worldwide are attempting to create systems that effectively contain the power generated by renewable resources. When compared to conventional charge storage technologies, rechargeable metal ion batteries (MIBs) have improved redox reactivity, increased safety, and continuing device sustainability (Figure 9.3). Among these MIBs, Li-ion batteries (LIBs) are one of the most advanced and well-explored energy storage technologies. In addition to LIBs, numerous battery types, including Li–S batteries, sodium-ion (SIBs), are being studied for improved futures.

The four major components of a battery system are the electrolyte, cathode, anode, and separator. Which of these materials is employed to transmit charge carriers as electrodes or other electrode-like structures will determine the storage device's capacity values. In most situations, the active species is a single-phase substance or a composite material that directly alters electrodes. In batteries, between

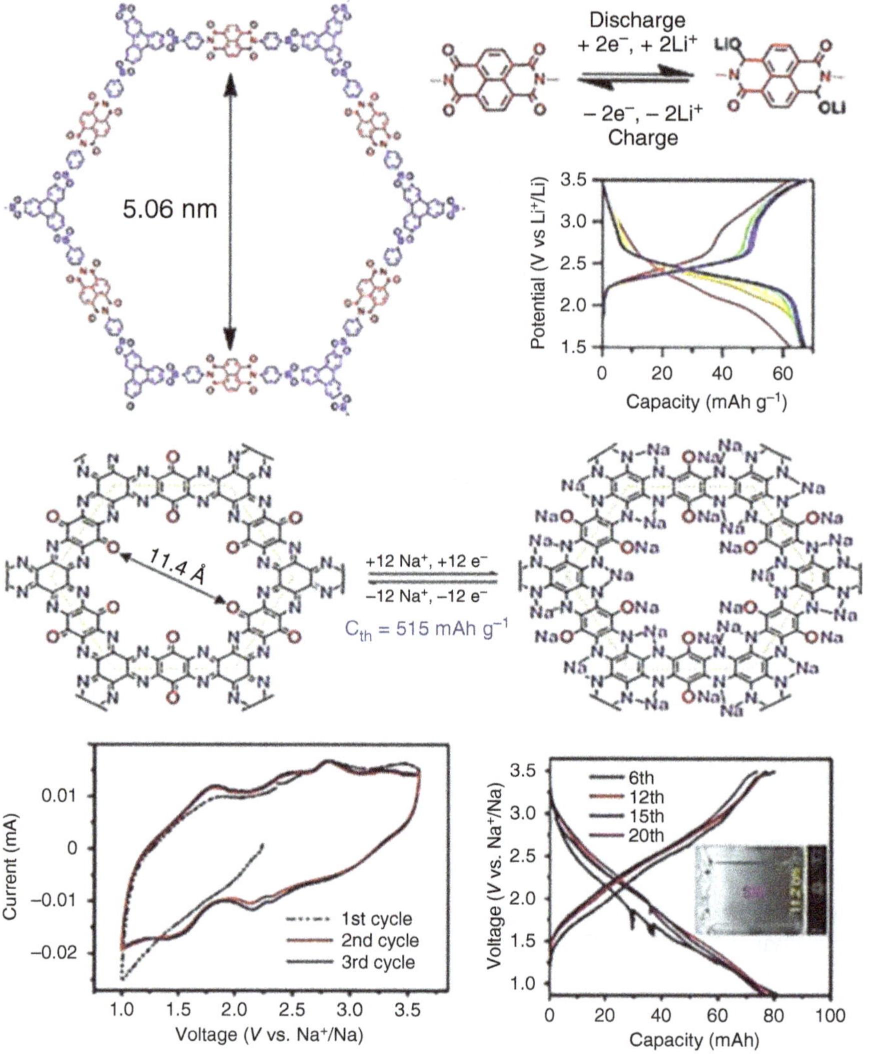

Figure 9.3 Structure of DTP-ANDI-COF with 2-e$^-$ redox step of an NDI backbone. Source: Adapted from Spitler et al. (2011). The structure-based redox process of TQBQ-COF, cyclic voltammetry graphs of TQBQ-COF scanned at the rate of 0.2 mV s^{-1}, and discharge curves of pouch-type Na//TQBQ-COF batteries under a current of 50 mA. Source: Adapted from Chen et al. (2015).

the cathode and the anode, a porous membrane serves as a separator to insulate the electrons while permitting ions to pass through, preventing short circuits. High ionic conductivity liquid or solid electrolytes are used in conjunction to improve ion transfer.

Because COFs are rectifying devices, rechargeable metal-COF batteries have been studied. The ability of COFs metal batteries to behave as electrodes depends on their reversible redox-active behavior in a charged state. Currently, liquid electrolytes pose safety issues, whereas solid electrolytes have restricted ion conductivity. The requirement for matching the supply and releasing energy whenever and wherever it is needed still cannot be satisfied by the capacity, rate performance, and lifetime of electrode materials. COFs are highly porous crystalline polymers that have a high ion storage capacity and can metrically change their inner surfaces with redox-active sites. Therefore, COFs could be utilized as battery electrode materials. COFs are also a preferred alternative in solid electrolytes and separators because of their open channels and electrical insulating properties.

9.16.1 COFs as Electrode Materials

(a) *Carbonyl-containing electrode materials*: As a sodium-ion battery anode, a novel vinyl polymer was developed and tested. It has a good electrochemical performance and terphenyl side chains. There was a significant improvement in electrochemical performance when compared to the vinyl polymer with disodium terephthalate. The discharge potential, in particular, dropped by roughly 0.12 to 0.27 V (vs. Na/Na+).
(b) *Nitrogen-containing electrode materials*: As a carbon-based electrode for Na-ion batteries, Kaskel's team discovered microporous (14 angstroms) bipolar frameworks having dynamic tri-azine rings and dormant benzene rings in the skeletons.
(c) *Lithium-sulfur batteries*: Wang and colleagues employed tetraaminobenzenequinone (TABQ) and dimethyl benzene anhydride (PMDA), two different types of active small molecules, to develop a COF. They then used the graphene solvent hot approach to in situ synthesize poly(imide benzoquinone) (PIBN), a two-dimensional microporous structure with an organic skeleton. As a result, PIBN-G exhibits excellent amplification and high reversibility in the electrochemical reaction (Chen et al. 2021).

9.16.2 Battery Additives and Functional Separators

A number of lithium-ion battery substitutes have been developed as a response of the limited supply and growing demand for Li capitals. The issue is slow intercalations of big ions (such as Na^+, K^+, or anions), which leads to poor capacity, confined cyclability, and primarily performance loss at a rapid rate when the majority of the material is utilized as electrodes. The COF is known as polypyrene sulfide (PPYS), and its use in the negative-ion storage was initially described by Wang et al. (2011). The novel theory also suggests that anions may be more easily transported by the electropositive COFs because of their potential to adsorb anions through electrostatic attractions.

This unique method acts as a proof-of-concept for PPYS cathodes by enhancing the intercalation kinetics of larger anions. Small amounts of COFs are added to

PPYS, significantly improving its capacity, rate capability, and cyclability (Yusran et al. 2020). Because of its electron-rich structure, the PPYS repeating unit can lose electrons, resulting in stable +1 charged states and the beginning of anions storage. To study the potential use of PPYS in anions storage, a half-cell is constructed with PPYS cathodes and Li metal as the counter electrode (Xia et al. 2022).

9.17 Supercapacitors

Rapid advancements in the electronic industries in recent years have increased the need for energy storage materials. The usage of coal, oil, natural gas, and other non-renewable resources seriously damages the environment, making the creation of new green energy storage technologies essential. Due to their excellent qualities of high charging and discharging effectiveness, extended cyclic life, and environmental friendliness, supercapacitors, as an electrochemical energy storage unit, aid a wide range of application possibilities in pocket-sized electronics, hybrid cars, etc.

Porous structural frameworks frequently have large area, which enhances the capability of the supercapacitor devices by speeding the movement of charge possessing ions in the electrolytes. Excellent supercapacitor performance and lengthy cycle lifetimes are provided by the crystalline porous MOF and COF structures. Additionally, their pore diameters are easily visible. As opposed to MOFs, which are made completely of heavy metallic components bound together by unstable covalent bonds, COFs showcase more stability and are not that heavy.

By possessing a discrete 1D channel that is aligned perpendicularly and an eclipsed layer stacking pattern, the typical 2D COF configuration resembles the structural core of graphene. Effective mass transfer and ion or charge mobility are made possible by this COF structural feature in an aligned diffusion tube. The development of EDLC (electric double-layer capacitors) electrode systems makes high surface area 2D COFs among the best materials for electrochemical energy storage applications. The samples below contrast modern COF supercapacitor capabilities. The molecular structure of COFs, including its shape, porosity, and so on, has a considerable impact on the efficiency and performance of supercapacitors (Kandambeth et al. 2021).

The effect of redox sites on the electrical and chemical functions of 2-dimensional COFs, a possible material for pseudo capacitors, has made substantial development in recent study. Nitrogen atoms have a significant impact on the transmission rate of e^- via the electrode material by increasing the electroactive surface area and making the system more wettable. By exfoliating a succession of three 2-dimensional mesoporous frameworks while maintaining the parent COFs' pore size, Yusran et al. demonstrated the impact of thin-layered 2D structure on electrochemical performance (Wei et al. 2018).

Micro-supercapacitors (MSCs) based on COFs materials may hold promise for next-generation miniaturized electronics by temporarily supplying the necessary power and energy. Flexible integrated assemblies can achieve significantly higher power densities in a smaller volume, which is usually sufficient for their purposes. Xu et al. used SWCNTs (single-walled carbon nanotubes) and a nanofibrous COF to create a flexible thin-film electrode for micro-supercapacitors (MSC) (An et al. 2021). The MSC, which is COF-based, provided an astonishing viewpoint on the growth of wearable & portable electronic systems.

Because they are stable and have porous properties that can be depended upon to provide COFs a high capacitance, SCs have a lot of potential for COFs. Despite the advances, it should be emphasized that COFs have a very low capacitance. Their entire organic makeup is to blame for this, which also affects their poor conductivity and slow charge transfer rate. Additionally, the bulk of electrode materials are currently made using COF powders, which has limited potential (Chen et al. 2020).

9.18 Electrochemical Sensors

Electrochemical sensors have garnered a lot of interest in the study of species with biological, environmental, and industrial importance. A working electrode, a counter electrode, and an electrolyte are generally used in electrochemical sensors. By observing how the working electrode interacts with the analytes, electric signals corresponding to the analyte concentrations can be determined. Only a small number of COFs have been used as electrochemical sensors, in contrast to their application in electrocatalysis and energy storage. COFs may be thought of as the optimal platform for adsorption/separation, detection, and sensing due to their low density and high surface areas. Furthermore, COFs' modular construction allows for the independent modification of both their pore topologies and functional groups. This enables the incorporation of electroactive sites and groups with the capacity to bind and detect particular species into COF channels.

Dynamic covalent chemistry is used to achieve periodicity in COF structures. When substantial crystallinity is observed in COFs, the newly generated covalent bonds are always guaranteed to be reversible. The bonds are logically implied to be easily reopenable by this, which is bad for stability (Zhao et al. 2021).

Generally, COFs are used to detect harmful organic compounds like hydrazine, catechol, nitrophenol, and hydroquinone (Liang et al. 2020). Further, metals like lead mercury also pose a threat to the environment as harms of elevated lead levels in the environment range from lower yield in crops to permanent damage to the neuro- and cardiovascular system. TAPB-DMTP-COF was synthesized by Kumar et al. (2021). The above COF acts as an excellent tool to remove lead by deposition.

Mercury (Hg) can affect cell function and proliferation by interfering with internal metabolic processes, which may also result in cell necrosis. Hg^{2+} is capable of making stable covalent bonds with functional groups containing elements such as oxygen, nitrogen, or sulfur. Because of its potent soft–soft interaction, mercury has a strong attraction for sulfur compounds. As a result, COFs and COF-based substances with high levels of O, N, or S have a positive impact on the detection and adsorption of Hg (Cukierman 2006).

9.19 Proton-Exchange Membrane Fuel Cells (PEMFC)

The only ion without its own electron shell is the proton. As a result, it has a considerable interaction with the environment's electron density. Proton-exchange membrane's crucial characteristic, proton conductivity, has a direct impact on how well a fuel cell functions. On the basis of the observed resistance, it is possible to determine the proton conductivity [ohm, cm^{-1}] according to

$$\sigma = l/A.1/R$$

where l is the electrode distance (in cm), R is the membrane resistance, A is the membrane's cross-sectional area (cm^2).

The discovery of protonic species dates back to 1806 (J.T. Von Grotthuss) (Litster and McLean 2004). In 1877, Bjerrum and Eigen worked on the proton conduction of ice; however, neither pure nor doped water was found to be a good conductor of protons (Cukierman 2006).

91 years later, in 1968 Nafion – A perflourinated sulfonated polymer was developed by Du Pont (Das et al. 2022). Nafion displayed high proton conductivity (10^{-1}–10^{-2} S cm^{-1}) while being extremely stable physically as well as chemically. Although Nafion displayed excellent properties, its commercial use was limited due to two major flaws:

9.19.1 Low-Temperature Conductivity

When the temperature exceeds 100 °C, conductivity rapidly drops as a result of the membrane's loss of the adsorbed water molecule.

9.19.2 Degradation

COFs overcome the poor hydrolytic stability and lesser tolerance to acidic guest doping of other porous materials which have been exploited in the field of proton conductivity throughout the years at higher operating temperatures.

9.20 Conclusion

Due to their many useful characteristics, such as high crystallinity, low density, tuneable porosity, wide surface area, and distinctive molecular architecture, the developing organic crystalline porous polymer materials called covalent organic frameworks (COFs) have been receiving more attention. Solvothermal condensation is a common method for creating COFs since it results in the formation of covalent bonds between lighter atoms including C, N, H, O, and B. COFs can be either 2-dimensional or 3-dimensional COFs, depending on the combination of the units and the type of weak interactions that result. Energy generation, energy storage, and pollution detection are just a few of the diverse applications of COFs. Through pore engineering, in which the pore surface is created by various linkages, such as B–O, C–N, and C–C linkages and functional groups, the structural and functional characteristics of COFs can be changed.

Metal atoms can also be added into them in order to improve the backbone properties. COFs play a vital role in environmental application by the mitigation of dyes and heavy metals. COFs like LZU8, TAPB-BMTTPA were developed through pore surface engineering to show high selectivity toward mercury. A Q-graphene-scaffold-COF was used as Cu^{2+} absorbent. COFs like imine-linked COFA are a highly efficient sorbent for hazardous dye elimination.

COFs have increased thermal stability compared to conventional polymers because of the strength of chemical bonds and linkages. In the energy production reactions like HER, OER, ORR, bifunctional OER/ORR and CO_2RR, COFs can act as good catalysts replacing the metals like Pt, Ir, and Ru. COFs are also an environment-friendly energy storage source used in batteries and supercapacitors after post-synthetic modification and surface doping. High surface area 2D COFs are the best for electrochemical energy storage applications like EDLC. They show high conductivity and excellent storage capacity, and thus "rechargeable metal-COF batteries" are a topic of interest for researchers across the globe.

In conclusion, unlike other crystalline porous materials, COFs are the promising reticular chemistry area for energy and environmental applications.

References

Afshari, M. and Dinari, M. (2020). Synthesis of new imine-linked covalent organic framework as high efficient absorbent and monitoring the removal of direct fast scarlet 4bs textile dye based on mobile phone colorimetric platform. *Journal of Hazardous Materials* 385: 121514. https://doi.org/10.1016/j.jhazmat.2019.121514.

An, Y., Tan, S., Liu, Y. et al. (2021). Designs and applications of multi-functional covalent organic frameworks in rechargeable batteries. *Energy Storage Materials* 41: 354–379. https://doi.org/10.1016/j.ensm.2021.06.010.

Beaudoin, D., Maris, T., and Wuest, J.D. (2013). Constructing monocrystalline covalent organic networks by polymerization. *Nature Chemistry* 5 (10): 830–834. https://doi.org/10.1038/nchem.1730.

Cai, Y., Jiang, Y., Feng, L. et al. (2019). Q-graphene-scaffolded covalent organic frameworks as fluorescent probes and sorbents for the fluorimetry and removal of copper ions. *Analytica Chimica Acta* https://doi.org/10.1016/j.aca.2018.12.054.

Campbell, N.L., Clowes, R., Ritchie, L.K., and Cooper, A.I. (2009). Rapid microwave synthesis and purification of porous covalent organic frameworks. *Chemistry of Materials* 21 (2): 204–206. https://doi.org/10.1021/cm802981m.

Chandra, S., Kandambeth, S., Biswal, B.P. et al. (2013). Chemically stable multilayered covalent organic nanosheets from covalent organic frameworks via mechanical delamination. *Journal of the American Chemical Society* 135 (47): 17853–17861. https://doi.org/10.1021/ja408121p.

Chen, X., Addicoat, M., Jin, E. et al. (2015). Designed synthesis of double-stage two-dimensional covalent organic frameworks. *Scientific Reports* 5 (1): https://doi.org/10.1038/srep14650.

Chen, S., Yuan, B., Liu, G., and Zhang, D. (2020). Electrochemical sensors based on covalent organic frameworks: a critical review. *Frontiers in Chemistry* 8: https://doi.org/10.3389/fchem.2020.601044.

Chen, Y., Kang, Y., Zhao, Y. et al. (2021). A review of lithium-ion battery safety concerns: the issues, strategies, and testing standards. *Journal of Energy Chemistry* 59: 83–99. https://doi.org/10.1016/j.jechem.2020.10.017.

Chung, W.J., Griebel, J.J., Kim, E.T. et al. (2013). The use of elemental sulfur as an alternative feedstock for polymeric materials. *Nature Chemistry* 5 (6): 518–524. https://doi.org/10.1038/nchem.1624.

Côté, A.P., Benin, A.I., Ockwig, N.W. et al. (2005). Porous, crystalline, covalent organic frameworks. *Science* 310 (5751): 1166–1170. https://doi.org/10.1126/science.1120411.

Côté, A.P., El-Kaderi, H.M., Furukawa, H. et al. (2007). Reticular synthesis of microporous and mesoporous 2D covalent organic frameworks. *Journal of the American Chemical Society* 129 (43): 12914–12915. https://doi.org/10.1021/ja0751781.

Cukierman, S. (2006). Et Tu, Grotthuss! and other unfinished stories. *Biochimica et Biophysica Acta (BBA) - Bioenergetics* 1757 (8): 876–885. https://doi.org/10.1016/j.bbabio.2005.12.001.

Dalapati, S., Jin, E., Addicoat, M. et al. (2016). Highly emissive covalent organic frameworks. *Journal of the American Chemical Society* 138 (18): 5797–5800. https://doi.org/10.1021/jacs.6b02700.

Das, G., Skorjanc, T., Sharma, S.K. et al. (2017). Viologen-based conjugated covalent organic networks via Zincke reaction. *Journal of the American Chemical Society* 139 (28): 9558–9565. https://doi.org/10.1021/jacs.7b02836.

Dey, K., Pal, M., Rout, K.C. et al. (2017). Selective molecular separation by interfacially crystallized covalent organic framework thin films. *Journal of the American Chemical Society* 139 (37): 13083–13091. https://doi.org/10.1021/jacs.7b06640.

Ding, S.-Y., Dong, M., Wang, Y.W. et al. (2016). Thioether-based fluorescent covalent organic framework for selective detection and facile removal of Mercury(ii). *Journal of the American Chemical Society* 138 (9): 3031–3037. https://doi.org/10.1021/jacs.5b10754.

Dogru, M., Sonnauer, A., Gavryushin, A. et al. (2011). A covalent organic framework with 4 nm open pores. *Chemical Communications* 47 (6): 1707. https://doi.org/10.1039/c0cc03792c.

Evans, A.M., Ryder, M.R., Ji, W. et al. (2021). Trends in the thermal stability of two-dimensional covalent organic frameworks. *Faraday Discussions* 225: 226–240. https://doi.org/10.1039/d0fd00054j.

Feldblyum, J.I., McCreery, C.H., Andrews, S.C. et al. (2015). Few-layer, large-area, 2D covalent organic framework semiconductor thin films. *Chemical Communications* 51 (73): 13894–13897. https://doi.org/10.1039/c5cc04679c.

Feng, X., Liang, Y., Zhi, L. et al. (2009). Synthesis of microporous carbon nanofibers and nanotubes from conjugated polymer network and evaluation in electrochemical capacitor. *Advanced Functional Materials* 19 (13): 2125–2129. https://doi.org/10.1002/adfm.200900264.

Grill, L., Dyer, M., Lafferentz, L. et al. (2007). Nano-architectures by covalent assembly of molecular building blocks. *Nature Nanotechnology* 2 (11): 687–691. https://doi.org/10.1038/nnano.2007.346.

Hu, C., Xiao, Y., Zou, Y., and Dai, L. (2018). Carbon-based metal-free electrocatalysis for energy conversion, energy storage, and environmental protection. *Electrochemical Energy Reviews* 1 (1): 84–112. https://doi.org/10.1007/s41918-018-0003-2.

Huang, N., Zhai, L., Xu, H., and Jiang, D. (2017a). Stable covalent organic frameworks for exceptional mercury removal from Aqueous Solutions. *Journal of the American Chemical Society* 139 (6): 2428–2434. https://doi.org/10.1021/jacs.6b12328.

Huang, N., Wang, P., Addicoat, M.A. et al. (2017b). Ionic covalent organic frameworks: design of a charged interface aligned on 1D channel walls and its unusual electrostatic functions. *Angewandte Chemie International Edition* 56 (18): 4982–4986. https://doi.org/10.1002/anie.201611542.

Jackson, K.T., Reich, T.E., and El-Kaderi, H.M. (2012). Targeted synthesis of a porous borazine-linked covalent organic framework. *Chemical Communications* 48 (70): 8823. https://doi.org/10.1039/c2cc33583b.

Jiang, Q., Li, Y., Zhao, X. et al. (2018). Inverse-vulcanization of vinyl functionalized covalent organic frameworks as efficient cathode materials for Li–S batteries. *Journal of Materials Chemistry A* 6 (37): 17977–17981. https://doi.org/10.1039/c8ta07008c.

Jin, S., Furukawa, K., Addicoat, M. et al. (2013). Large pore donor–acceptor covalent organic frameworks. *Chemical Science* 4 (12): 4505. https://doi.org/10.1039/c3sc52034j.

Jin, E., Asada, M., Xu, Q. et al. (2017). Two-dimensional sp_2 carbon–conjugated covalent organic frameworks. *Science* 357 (6352): 673–676. https://doi.org/10.1126/science.aan0202.

Kandambeth, S., Kale, V.S., Shekhah, O. et al. (2021). 2D covalent-organic framework electrodes for supercapacitors and rechargeable metal-ion batteries. *Advanced Energy Materials* 12 (4): 2100177. https://doi.org/10.1002/aenm.202100177.

Karak, S., Kandambeth, S., Biswal, B.P. et al. (2017). Constructing ultraporous covalent organic frameworks in seconds via an organic terracotta process. *Journal of the American Chemical Society* 139 (5): 1856–1862. https://doi.org/10.1021/jacs.6b08815.

Kim, S. and Choi, H.C. (2019). Light-promoted synthesis of highly-conjugated crystalline covalent organic framework. *Communications Chemistry* 2 (1): https://doi.org/10.1038/s42004-019-0162-z.

Kuhn, P., Antonietti, M., and Thomas, A. (2008). Porous, covalent triazine-based frameworks prepared by ionothermal synthesis. *Angewandte Chemie International Edition* 47 (18): 3450–3453. https://doi.org/10.1002/anie.200705710.

Li, J., Jing, X., Li, Q. et al. (2020a). Bulk COFs and COF nanosheets for electrochemical energy storage and conversion. *Chemical Society Reviews* 49 (11): 3565–3604. https://doi.org/10.1039/d0cs00017e.

Li, D., Li, C., Zhang, L. et al. (2020b). Metal-free thiophene-sulfur covalent organic frameworks: precise and controllable synthesis of catalytic active sites for oxygen reduction. *Journal of the American Chemical Society* 142 (18): 8104–8108. https://doi.org/10.1021/jacs.0c02225.

Liang, C., Lin, H., Wang, Q. et al. (2020). A redox-active covalent organic framework for the efficient detection and removal of hydrazine. *Journal of Hazardous Materials* 381: 120983. https://doi.org/10.1016/j.jhazmat.2019.120983.

Litster, S. and McLean, G. (2004). PEM fuel cell electrodes. *Journal of Power Sources* 130 (1–2): 61–76. https://doi.org/10.1016/j.jpowsour.2003.12.055.

Lyu, H., Diercks, C.S., Zhu, C. et al. (2019). Porous crystalline olefin-linked covalent organic frameworks. *Journal of the American Chemical Society* 141 (17): 6848–6852. https://doi.org/10.1021/jacs.9b02848.

Ma, H.-C., Chen, G.J., Huang, F., and Dong, Y.B. (2020). Homochiral covalent organic framework for catalytic asymmetric synthesis of a drug intermediate. *Journal of the American Chemical Society* 142 (29): 12574–12578. https://doi.org/10.1021/jacs.0c04722.

Mao, M.J., Zhang, M.D., Meng, D.L. et al. (2020). Imidazolium-functionalized cationic covalent triazine frameworks stabilized copper nanoparticles for enhanced CO_2 electroreduction. *ChemCatChem* 12 (13): 3530–3536. https://doi.org/10.1002/cctc.202000387.

Medina, D.D., Rotter, J.M., Hu, Y. et al. (2015). Room temperature synthesis of covalent–organic framework films through vapor-assisted conversion. *Journal of the American Chemical Society* 137 (3): 1016–1019. https://doi.org/10.1021/ja510895m.

Nagai, A., Guo, Z., Feng, X. et al. (2011). Pore surface engineering in covalent organic frameworks. *Nature Communications* 2 (1): https://doi.org/10.1038/ncomms1542.

Ning, G.-H., Chen, Z., Gao, Q. et al. (2017). Salicylideneanilines-based covalent organic frameworks as chemoselective molecular sieves. *Journal of the American Chemical Society* 139 (26): 8897–8904. https://doi.org/10.1021/jacs.7b02696.

Rotter, J.M., Guntermann, R., Auth, M. et al. (2020). Highly conducting Wurster-type twisted covalent organic frameworks. *Chemical Science* 11 (47): 12843–12853. https://doi.org/10.1039/d0sc03909h.

Simmonds, A.G., Griebel, J.J., Park, J. et al. (2014). Inverse vulcanization of elemental sulfur to prepare polymeric electrode materials for Li–S batteries. *ACS Macro Letters* 3 (3): 229–232. https://doi.org/10.1021/mz400649w.

Song, J.-R., Sun, J., Liu, J. et al. (2014). Thermally/hydrolytically stable covalent organic frameworks from a rigid macrocyclic host. *Chemical Communications* 50 (7): 788–791. https://doi.org/10.1039/c3cc47652a.

Spitler, E.L., Koo, B.T., Novotney, J.L. et al. (2011). A 2D covalent organic framework with 4.7-nm pores and insight into its interlayer stacking. *Journal of the American Chemical Society* 133 (48): 19416–19421. https://doi.org/10.1021/ja206242v.

Sugawara, K., Yabuuchi, N., and Miyasaka, M. (2022). Extended conjugated carbonyl-containing polymer as a negative electrode material for Na-Ion batteries. *Polymer Journal* 54 (9): 1111–1118. https://doi.org/10.1038/s41428-022-00658-4.

Sun, Q., Aguila, B., Perman, J. et al. (2017). Postsynthetically modified covalent organic frameworks for efficient and effective mercury removal. *Journal of the American Chemical Society* 139 (7): 2786–2793. https://doi.org/10.1021/jacs.6b12885.

Sun, Q., Aguila, B., Earl, L.D. et al. (2018). Covalent organic frameworks as a decorating platform for utilization and affinity enhancement of chelating sites for radionuclide sequestration. *Advanced Materials* 30 (20): 1705479. https://doi.org/10.1002/adma.201705479.

Tilford, R.W., Mugavero, S.J. III, Pellechia, P.J., and Lavigne, J.J. (2008). Tailoring microporosity in covalent organic frameworks. *Advanced Materials* 20 (14): 2741–2746. https://doi.org/10.1002/adma.200800030.

Uribe-Romo, F.J., Hunt, J.R., Furukawa, H. et al. (2009). A crystalline imine-linked 3-D porous covalent organic framework. *Journal of the American Chemical Society* 131 (13): 4570–4571. https://doi.org/10.1021/ja8096256.

Waller, P.J., Lyle, S.J., Osborn Popp, T.M. et al. (2016). Chemical conversion of linkages in covalent organic frameworks. *Journal of the American Chemical Society* 138 (48): 15519–15522. https://doi.org/10.1021/jacs.6b08377.

Wang, C., Dong, H., Hu, W. et al. (2011). Semiconducting π-conjugated systems in field-effect transistors: a material odyssey of organic electronics. *Chemical Reviews* 112 (4): 2208–2267. https://doi.org/10.1021/cr100380z.

Wang, Y., Chen, J., Wang, G. et al. (2018a). Perfluorinated covalent triazine framework derived hybrids for the highly selective electroconversion of carbon dioxide into methane. *Angewandte Chemie International Edition* 57 (40): 13120–13124. https://doi.org/10.1002/anie.201807173.

Wang, D.-G., Qiu, T., Guo, W. et al. (2021a). Covalent organic framework-based materials for energy applications. *Energy & Environmental Science* 14 (2): 688–728. https://doi.org/10.1039/d0ee02309d.

Wei, P.-F., Qi, M.Z., Wang, Z.P. et al. (2018). Benzoxazole-linked ultrastable covalent organic frameworks for photocatalysis. *Journal of the American Chemical Society* 140 (13): 4623–4631. https://doi.org/10.1021/jacs.8b00571.

Werner, A. (1893). Beitrag zur Konstitution Anorganischer verbindungen. *Zeitschrift fr anorganische Chemie* 3 (1): 267–330. https://doi.org/10.1002/zaac.18930030136.

Xia, J., Wang, R., Qian, C. et al. (2022). Supercapacitors of nanocrystalline covalent organic frameworks—a review. *Crystals* 12 (10): 1350. https://doi.org/10.3390/cryst12101350.

Xiang, Z., Cao, D., Huang, L. et al. (2014). Nitrogen-doped holey graphitic carbon from 2D covalent organic polymers for oxygen reduction. *Advanced Materials* 26 (20): 3315–3320. https://doi.org/10.1002/adma.201306328.

Yu, S.-B., Lyu, H., Tian, J. et al. (2016). A polycationic covalent organic framework: a robust adsorbent for anionic dye pollutants. *Polymer Chemistry* 7 (20): 3392–3397. https://doi.org/10.1039/c6py00281a.

Yusran, Y., Guan, X., Li, H. et al. (2019). Postsynthetic functionalization of covalent organic frameworks. *National Science Review* 7 (1): 170–190. https://doi.org/10.1093/nsr/nwz122.

Yusran, Y., Li, H., Guan, X. et al. (2020). Exfoliated mesoporous 2D covalent organic frameworks for high-rate electrochemical double-layer capacitors. *Advanced Materials* 32 (8): 1907289. https://doi.org/10.1002/adma.201907289.

Zhang, P., Zhang, J., and Dai, S. (2016a). Mesoporous carbon materials with functional compositions. *Chemistry – A European Journal* 23 (9): 1986–1998. https://doi.org/10.1002/chem.201602199.

Zhang, H., Zhu, M., Schmidt, O.G. et al. (2021). Covalent organic frameworks for efficient energy electrocatalysis: rational design and progress. *Advanced Energy and Sustainability Research* 2 (4): 2000090. https://doi.org/10.1002/aesr.202000090.

Zhao, X., Pachfule, P., and Thomas, A. (2021). Covalent organic frameworks (cofs) for electrochemical applications. *Chemical Society Reviews* 50 (12): 6871–6913. https://doi.org/10.1039/d0cs01569e.

Zhu, Y., Wan, S., Jin, Y., and Zhang, W. (2015). Desymmetrized vertex design for the synthesis of covalent organic frameworks with periodically heterogeneous pore structures. *Journal of the American Chemical Society* 137 (43): 13772–13775. https://doi.org/10.1021/jacs.5b09487.

10

A Multifunctional Polymer – POLYOX – and Its Uses as a Novel Drug-Delivery System

Preeta Bose, Dibya Das, Devlina Pal, Pintu K. De, and Himangshu S. Maji

JIS University, Department of Pharmaceutical Technology, Deshpriya Nagar, Agarpara, Kolkata, West Bengal 700109, India

10.1 Introduction

Oral dosage forms have historically been the most common and widely used kind of formulation. The majority of traditional oral pharmaceutical preparations, also known as instant-release preparations, are designed to start releasing the active ingredient as soon as the tablet or capsule is taken. This includes both tablets and capsules. When typical drug preparations are made, the pace at which the medication is released into the body is not particularly adjusted. With immediate-release medications, the onset of the pharmacodynamic effects and the absorption of the medication often occur quite rapidly. It is possible that the pharmacodynamic effect will be delayed as a result of the conversion of prodrugs in traditional oral preparations to the active treatment. This conversion might take place via hepatic or intestinal metabolism or chemical hydrolysis. In contrast, the absorption of medication from traditional oral formulations that contain poorly soluble medicines (lipid-loving medicines) may be prolonged due to steady dissolution in or selected absorption throughout the gastrointestinal tract, resulting in a similarly late inception time. This is because these formulations contain poorly soluble medications. In the pharmaceutical industry, medications with modified release are referred to by a number of names depending on their specific formulation. The pharmaceutical business is always developing fresh and cutting-edge methods of drug administration in order to modify the drug's release profile. This has a one-of-a-kind influence not only on the pharmacodynamic effect but also on the plasma drug conc vs time profile. The use of modified-release dosage forms rather than regular ones has a variety of advantages, including the following:

- Blood levels are less likely to vary when the rate of release is regulated, and peaks and troughs in blood levels are minimized when this is done.
- A reduction in the number of times daily that the drug has to be taken, which, given that extended-release medications typically include more than one serving, increased both convenience and adherence.

Organic Polymers in Energy-Environmental Applications, First Edition.
Edited by Ramesh Oraon, Pardeep Singh, Sanchayita Rajkhowa, Sangita Agarwal, and Ravindra Pratap Singh.

- Lessening of negative side effects.
- A general reduction in the expense of receiving medical care.

Polymers having a range of molecular weights are put to use in the process of modifying the release of drugs in dosage form. The utilization of polymers has resulted in the development of dosage forms that exhibit particular qualities. In the beginning, polymers were utilized for the purpose of solubilizing, stabilizing, and providing mechanical support for drugs in order to achieve a delayed release. However, during the course of time, the functionalities of polymers have progressed. Polymers were manufactured by synthetic processes in order to fulfil specific prerequisites or, to be more specific, to provide solutions to specific challenges that were associated with the production of drug-delivery systems. The first and most important parameters to consider are the source, kind of polymerization, chain growth polymerization, and degradability of the material. The method in which water and other polymers interact with one another. Numerous sectors, including those dealing with food and medicine, varnishes, clothes, paper, construction, gums, coating material, and water treatment, make extensive use of polymers that are soluble in water.

POLYOX, one of the types of polymers, is swiftly becoming more well known for its capacity to gently release medications over a long period of time. As a direct consequence of this, the present research provides a comprehensive analysis of POLYOX as a polymer for modifying drug release. It all depends on how a polymer and water interact with one another. Polyethylene oxide (PEO) is an example of a water-loving polymer. It is a homopolymer of ethylene oxide that does not include any ions. It's also known as POLYOX in some circles.

POLYOX, sometimes referred to simply as poly, is a water-soluble resin (ethylene oxide). They are crystalline powders that are hydrophilic, white, and free-flowing. The typical particle size of these powders is around 150 mm, and they are available in a wide variety of molecular weight grades ranging from 100,000 to 80,000. PEO is utilized for usage in the matrix materials. Because of the use of high-molecular-weight PEO, the dissolution of drugs that were both liquid and insoluble was successfully prevented in matrix tablets that were created by the use of direct compression. In addition to this, it acts as a polymer that is mucoadhesive. POLYOX water-soluble resins can be utilized in a variety of different applications, including pharmaceutical products, coating materials, TDDS, rapid-release dosage forms, and gastro-retentive dosage forms (Chandel & Kapoor 2013; Shah et al. 2014). They are capable of producing films and retaining water in their cells. In addition to being non-toxic, it has a high solubility in water (Kojima et al. 2008). POLYOX resins have the ability to create related compounds with a number of different substances, including polystyrene, poly-caprolactone, polyethylene, polypropylene, and polylactides. These related compounds can then result in distinctive blends and a variety of other useful formulation characteristics. Typical skeletal ($-O-CH_2-CH_2-$)nOH. They share the same chemical structure as PEG, although having molecular weights that are far higher. The usual number of oxyethylene groups is denoted by the letter n (British National Formulary 2009). The structure of the ethylene oxide monomer is that of an epoxide ring.

POLYOX is made as NF and other official-grade compounds, and it is also available in a variety of molecular weight classes. The different grades of POLYOX are given as:

10.1.1 Grades of POLYOX

POLYOX(tm) Water-Soluble Resin NF Products are WSR N, WSR N-80, WSR N-205, WSR-301, WSR-303, WSR N-750, WSR N-1105, WSR N-12K, WSR N-60K, and WSR Coagulant.

10.2 Advantages of Using POLYOX

Additionally, POLYOX has the following benefits to offer:

- The ability to formulate in a variety of ways is made possible by the presence of a broad molecular weight range.
- Versatility in applications including granulation and direct compression.
- Quick hydration and swelling, essential for the development of osmotic pump technology.
- Efficient generation of gel and hydration for use in applications involving hydrophilic matrices.
- In terms of bio adhesion, POLYOX films fared better than HPC films. This is due to the fact that POLYOX hydrates more quickly, absorbs a lot of water, and swells better than HPC does.
- POLYOX can be used in place of HPMC as the binding agent when direct compression is used. Under circumstances of testing that were both acidic and neutral, the swelling ratio of PEO was almost double as high as that of HPMC.
- PEO hydrogels' swelling behavior is pH independent. Despite the fact that Carbopol displays pH-dependent bioadhesion and swelling behavior.
- POLYOX has evolved into a reliable polymer that can be used for bio-adhesion in the GIT.
- Meets the needs of both the United States Food Chemicals Codex and the USP.

10.3 Physical and Chemical Constituents

POLYOX has a high binding efficiency for powders, giving it the name "high binding efficient." The capacity to cross-link POLYOX may result in the formation of extremely water-retentive gels when used in the process of cross-linking.

POLYOX releases a substantial quantity of lubricity into the system when it comes into contact with water. At various temperatures, POLYOX easily solubilizes and will become thick in a wide range of solvents which are organic, including halogenated hydrocarbons, alcohols, aromatic hydrocarbons, and esters. Due to this quality, POLYOX may be employed in a variety of applications. In general glycols, diols, aliphatic hydrocarbon solvents, and aliphatic ethers are insoluble in POLYOX.

Thermoplasticity: It allows POLYOX to be easily extruded, injection molded, or cast despite its status as a thermoplastic. This material is capable of being heat-sealed into sheets and films, and it can also be orientated to create high levels of strength. Films naturally possess the characteristics of being flexible, durable, and impervious to the majority of oils and greases. These resins may be used in conjunction with a wide variety of polymers, both natural and manufactured.

Wet tack: POLYOX have a high degree of wet tack and are thus suitable as wet adhesives due to this property. The residue, after it has dried, does not have a sticky texture. Reduced resistance to forward motion is achieved by very low quantities of heavier molecules. POLYOX may cut down on the turbulent frictional drag of the water by as much as 80% when it is dissolved in the water.

Viscosity: The viscosity of the solution quickly rises when the concentration of polyol water-soluble resin is raised. A POLYOX solution will have a pH between 8 and 10. At pH levels ranging from neutral to slightly alkaline, POLYOX solutions will exhibit the greatest degree of stability. When the pH is higher than 12, POLYOX resins will begin to precipitate. As ethers are susceptible to acidic hydrolysis, the polymer will deteriorate when the pH falls below 2. Viscoelasticity is produced in solutions when the adaptability of ether bonds is paired with the extraordinarily large molecular weight of POLYOX, which results in the molecules having an elastic behavior.

Glass transition temperature: For the class of products manufactured from poly(ethylene oxide), the glass transition temperature (T_g) ranges from −50 to −57 °C. In the context of this class of products, the molecular weight has little bearing on the T_g.

10.3.1 Pharmaceutical Property

Very quick hydration and the production of gels: POLYOX are the most water-soluble polymers that are utilized in pharmaceutical formulations that hydrate the quickest and form gels the quickest. They produce hydrogels very fast, which begin and control the release of active chemicals.

Emollient: The application of POLYOX to the skin and hair results in a feeling that is both smooth and velvety. Both thermoplastic processing and casting procedures may be used in order to shape POLYOX into flexible films. These films can then be used for a variety of applications.

Flocculent activity: Increased molecular weight types of POLYOX are able to efficiently adsorb onto a wide variety of colloidal materials and function as potent flocculating agents.

POLYOX are non-ionic and entirely aqueous soluble at all types of temperatures. They are also exceptionally efficient agents which can make other substances dense in both fresh and salty water. Water-soluble solutions are pseudo plastic. The flowability characteristics of the various grades of POLYOX are comparable to one another. POLYOX has a small amount of silica (about 1.5%), which contributes to its flowability. POLYOX is an excellent option for formulations requiring timed release due to its exceptional swelling capabilities, which, when combined with its carefully regulated rate of dissolution, make it a perfect alternative. POLYOX may be used to extrude components for devices that give a timed release of a variety of substances, including fragrance, colors, surfactants, and reagents, among other things.

Concerns relating to the risks: POLYOX demonstrates just the lowest possible order of toxicity in animal experiments, and this is the case regardless of the route of exposure. When administered orally, the maximum safe dose is equivalent to about $2\,g\,kg^{-1}$ of body weight. These gums do not irritate the skin in any way, nor do they sensitize the skin, and neither the dry powder nor the aqueous solutions of these resins cause irritation to the eyes.

Non-hazardous and biodegradable nature: POLYOX products are non-hazardous and have been granted FDA certification for a variety of uses in the food and medicine industries. Due to oxidation and aerobic biodegradation, aqueous solutions of POLYOX are capable of being degraded in the natural environment. Casting, extrusion, injection molding, and blown film are some of the techniques that work especially well with POLYOX because of the numerous exceptional qualities that it has.

10.3.2 Characteristics

Both Physical and Chemical High efficacy in binding: POLYOX has a high level of effectiveness when it comes to binding to powders. Capability to form crosslinks POLYOX possesses the capability to form crosslinks, which enables it to form very water-storing gels. POLYOX bestows a high power of lubricity upon its surroundings when it comes in close contact with water.

Organic solvent solubility and thickening: POLYOX are easily soluble in a wide range of organic solvents, including aromatic hydrocarbons, alcohols, halogenated hydrocarbons, ketones, and esters, at a variety of temperatures. Additionally, POLYOX may thicken in organic solvents. POLYOX is not made soluble by aliphatic hydrocarbon solvents, glycols, or aliphatic ethers in most cases.

Thermoplasticity is the property that allows POLYOX to be readily extruded, injection molded, or cast as a material. This material is available in the form of sheets and films, and it may both be heat-sealed and oriented in order to achieve high levels of strength. Films, by their very nature, are extremely durable, flexible, and resistant to the vast majority of oils. These resins are unanimous with a good range of chemical and natural polymers, and they perform very well with them. POLYOX is a kind of adhesive that is particularly useful for wet applications because of the high wet tack qualities it possesses. Stickiness is not seen in the dried residue. Low concentrations of molecules with a higher-molecular weight help to minimize friction. POLYOX, when dissolved in water, has the potential to lessen the turbulent frictional drag of the water by up to 80%.

Viscosity: The amount of POLYOX aqueous-soluble resin in the solution has a direct correlation to the rate at which the solution's viscosity increases. A solution of POLYOX has a pH that ranges between 8 and 10. The most effective and long-lasting POLYOX solutions will have a pH that falls anywhere between neutral and slightly alkaline. When the pH is greater than 12, POLYOX resins will begin to precipitate out. Since ethers hydrolyze into an acidic form when exposed to pH 2, the polymer will degrade at that value.

Viscoelasticity is achieved in solutions by combining the extremely high-molecular weight of POLYOX with the flexibility of ether bonds, which results in the formation of solutions that exhibit elastic behavior. The glass transition temperature (T_g) for the poly(ethylene oxide) product family ranges from −50 to −57 °C. The molecular will not matter to the T_g across the board for this class of goods. Pharmaceutical Qualitative Extremely Rapid Hydration and the Formation of a Gel: Polymers that are water-soluble and used in pharmaceutical systems, like POLYOX, have one of the fastest hydration rates. This is one of its advantages. They quickly produce hydrogels, which can both initiate and regulate the release of active compounds. After usage, POLYOX makes both the skin and the hair feel silkier and smoother. It is an emollient. POLYOX may be treated with thermoplastics or cast to generate flexible films, making it an ideal ingredient for use in film formers. Varieties of POLYOX with a high-molecular weight perform very well as flocculating agents because they adhere to a wide range of colloidal substances in an effective manner. Thickening capability (aqueous) – POLYOX are particularly effective thickening agents in both fresh and salty water since they are nonionic, completely soluble in water, and soluble at all temperatures. This makes them ideal for use in thickening aqueous solutions. Pseudoplastic aqueous solutions (i.e., shear thinning). In terms of flow properties, there is not much of a difference between the various grades of POLYOX.

POLYOX has a little amount of silica (1.5%), which helps the material flow more easily. POLYOX's unique swelling capabilities and sustained rate of dissolution make it a perfect component for time-release formulations. This property makes POLYOX the ideal element for time-release formulations. Extrusion techniques can be employed with POLYOX to create devices that, over time, gradually release a scent, color, surfactant, reagents, etc. Concerns about the safety POLYOX has been shown to have poor levels of toxicity in animal studies, regardless of the method of exposure. At oral dosages of around $2\,g\,kg^{-1}$ of body weight, which is the maximum amount that may be safely administered by mouth. The usage of these resins in any form, including dry powder, aqueous solutions, skin irritants, or sensitizers, does not result in eye discomfort under any circumstances. In addition to being safe and biodegradable, the FDA has given its blessing for the use of POLYOX in many types of applications including food and medicine. Aqueous solutions of POLYOX are ecologically biodegradable because they may be broken down by oxidation and aerobic biodegradation. As a result of their extensive range of distinguishing characteristics, POLYOX is ideally suited for a variety of manufacturing processes, including casting, extrusion, injection molding, and blown film production.

Below shows the various physical features of POLYOX.

- Having the look of powdery off-white.
- Crystals will begin to melt between 62 and 67 °C.

- A whiff of nitrate can be detected in the odor.
- 1.0% by weight of volatile substances (at 105 °C).
- One point 1% by weight of alkaline earth metals.
- 19–37 pounds per cubic foot is the bulk density of powder.
- 1% moisture content.
- Fusion heat – 33 calories per gram.
- Alternate forms of pH 8–10.

10.4 Release Mechanism

The POLYOX component is evenly distributed over the whole tablet. It hydrates and creates a gel coating on the exterior tablet surface when it comes into contact with water. The rate at which the active ingredient diffuses out of the gel layer and the rate at which the tablet is eroded both control the rate at which the active ingredient is distributed and the total dissolving rate. The molecular weight (MW) of the POLYOX employed in the formulation should be taken into account in order to achieve the correct release profile. In comparison to higher MW grades, lower-molecular-weight polymers often have a faster rate of release. The percentage of POLYOX in the pill should range from 20% to 90% in order to have a predictable and repeatable release profile. The drug release from high-molecular weight POLYOX tablets is controlled by swelling rather than by the breakdown of the polymer, which leads to abnormal release kinetics. The key mechanism controlling the release of drugs from low-molecular-weight POLYOX, however, is the swelling and erosion of the polymer, which encourages front synchronization and a consistent release rate. It has been demonstrated that the amount of medicine released from low-molecular-weight POLYOX tablets is unaffected by the drug's solubility or drug loading. There is no relationship between the molecular weight of POLYOX and the dissolving medium's pH or between the speed of stirring and the release of the medication. In drug-delivery systems that have been described for oral, mucosal, and transdermal administration, POLYOX has also been used as a component or carrier. Pharmaceutical Industry Application Areas POLYOX has previously achieved success with osmotic pump technologies, hydrophilic matrices, gastro-retentive dosage forms, and other drug-delivery systems such as transdermal and mucoadhesive technologies for sustained-release applications. Additionally, POLYOX has utilized transdermal and muco-adhesive technology. Currently, POLYOX is a popularly utilized, efficient component in the creation of instant-release dosage forms, such as films. Drug dosage formulations known as controlled-release drug-delivery systems are designed to release an active component at rates that are much slower than those of their comparable conventional dose forms. From a few milligrams to several hundred milligrams per hour, these speeds are possible. The purpose of the controlled release drug-delivery systems is to regulate the pace of drug administration, increase the therapeutic effect, and/or target drug delivery to a specific tissue. There are several approaches to achieve these aims Chowdary et al. (2009). High-molecular-weight linear PEO, or POLYOX, has been demonstrated to have a large amount of potential and is employed in controlled drug-delivery systems. The manufacture of matrix tablets based on POLYOX is a relatively simple procedure because of the remarkable compressibility of this polymer Yang et al. (1996). If POLYOX's polyether chains come into contact with water or stomach secretions, they may generate strong hydrogen bonds with it. This causes POLYOX to quickly hydrate and expand, creating hydrogels with properties that make them perfect for controlled drug-delivery systems. As a result, different release patterns may be achieved depending on the molecular weight of POLYOX and the drug's physicochemical characteristics. Maggi et al. (2000) developed hydrophilic matrix tablets using POLYOXWSRN 1105 (MW = 0.9106) and POLYOXWSRN 301 (MW = 4106), two distinct PEOs with two different molecular weights. A controlled release was established by doing this. The results indicate that the higher-molecular-weight PEO develops more quickly than the lower-molecular-weight PEO. The different degrees of success that the two polymeric products have had in controlling the delivery rate of the medicine that is water-soluble may be due to these differences in erosion behavior. The medication mebeverine hydrochloride was created as a sustained-release tablet Deshkar et al. (2013). By mixing the necessary excipients with Methocel K15M, K100M, and POLYOX WSR 303 at concentrations of about 11%, 21%, 31%, and 41% of the tablet weight, respectively. The tablet was then compressed using a method known as direct compression. Drug release may be postponed for up to 12 hours by Methocel K15M formulations (20% w/w), Methocel K100M formulations (10% w/w), and PEO-303 formulations (20% w/w). Verapamil-hydrochloride-containing controlled-release matrix tablets were created by Vidhyadhara et al. (2013) by the use of POLYOXWSR 303 and the direct compression method. The rate of drug release was seen to decrease as the amount of polymeric material in the matrix increased. The extended release of verapamil hydrochloride from matrix tablets was able to maintain the therapeutic action for up to 24 hours, according to the results of an in vivo pharmacokinetic trial. POLYOX is a binder that is used in direct compression tablet-binding processes. They usually exhibit improved flow and compaction properties in comparison to other binders, and the lubricity they provide facilitates the tableting process. Because of their large molecular weight, pills release verapamil hydrochloride slowly. Dimitrov and associates investigated POLYOX Dimitrov & Lambov (1999). The quantity of PEO detected in molecules controls how quickly the drug is released, which happens through a mechanism known as controlled diffusion. In all alkali mediums, PEO releases comparatively completely when hydrophilic polymers (Eudragit L, Eudisperthv, and Carbopol 934) with pH-dependent solubility (at concentrations ranging from 10% to 50%) are added. A relaxation-controlled diffusion on the matrix takes the role of a normal aberrant diffusion at lower drug doses. A study was conducted to examine the drug release from compressed tablets made from a powdered combination of the medication POLYOX and magnesium stearate by Kim et al. (1995). The anomalous release kinetics of the medication from high-molecular-weight PEO tablets (MW = 2106 and 4106) are regulated by swelling of the polymer rather than erosion of the polymer. However, the main factor controlling the drug release from the PEO with a low-molecular-weight ($MW = 0.9106$) is the swelling and erosion of the polymer, which leads to front synchronization and a consistent release rate. Neither the molecular weight of the PEO nor the drug release, nor the pH of the dissolving fluid nor the stirring rate, are correlated with one another. Adhering to the mucosa are bioadhesives: POLYOX's water solubility, hydrophilicity, high-molecular weight, hydrogen bonding activity, and strong biocompatibility make it a fantastic option for mucoadhesion. These are just a few of the many advantages it provides. These resins may mix with mucus and create a strong interpenetrating network because of their vast linear chain structure. The statistics indicate that the strongest adhesion appears to occur at molecular weights of 4,000,000 or more. Cappello et al. (1994) conducted study on monolithic devices to look at their release behavior and matrix properties. The intended outcome of changing the release mechanism has been achieved by combining PEOs with different molecular weights. Both the pace of dissolution and the expansion of the tablet's outer swelling layer have attracted a considerable lot of attention. The numerous various matrix parameters were connected to the numerous diverse drug-delivery behaviors that were found. Additionally, studies on the matrices' viscoelastic characteristics have been done. Regarding excellent release, viscoelastic, and mucoadhesive qualities, the two polymer fractions that were used in combination produced the greatest results. The weight ratio of the mixture was set at 50% (600,000 and 4,000,000 molecular weight). In order to buccally administer the medication carvedilol (CAR), which has a low solubility, Cappello et al. (2006) developed a tablet that used PEO as a bioadhesive sustained-release substrate and hydroxypropyl-beta-cyclodextrin (HPbetaCD) as a drug release modifier. The tablet was then encapsulated in an enteric coating. First, the CAR and HPbetaCD release properties of PEO tablets that included CAR/HPbetaCD binary systems with a variety of dissolving capabilities were examined and compared to PEO tablets that just contained CAR. In order to determine whether or not the buccal administration of CAR is a viable option, permeation tests were conducted with the mucosa of pigs that had been surgically removed. More CAR was able to penetrate from the PEO tablet when the tablets contained HPbetaCD, and the CAR/HPbetaCD freeze-dried system achieved the maximum value. Rama Bukka et al. (2012) used a 23 factorial design to make buccal mucoadhesive films of felodipine by casting PEO with either hydroxy propyl cellulose (HPC) or ethyl cellulose. This process created felodipine films that adhere to the mucosa of the buccal cavity. Both dichloromethane and ethanol were used as the solvent in this experiment (1 : 1 ratio). The films' ex vivo mucoadhesive effectiveness, in vitro residence time, drug release, and percentage swelling were all assessed. The non-Fickian, zero-order release theory of kinetic release was scrupulously followed in every formulation.

10.5 Ocular Drug Administration

In order to control the ocular delivery of ofloxacin (OFX) via powder compression, Di Colo et al. (2001) examined a unique usage of high-molecular-weight (400 kDa) linear PEO in gel-forming erodible inserts. This was done in order to control the ocular administration of ofloxacin via powder compression. Alterations were made to the erosion time scale by the use of PEO in conjunction with Eudragit L100 (EUD), 17% neutralized (EUDNa17), or 71% neutralized (EUDNa71). As soon as they were injected into the lower conjunctival sac of the rabbit eyes, the mucoadhesive gels produced by the inserts based on plain PEO, PEO-EUDNa17, were well tolerated by the animals. These inserts were built using PEO-EUDNa17, PEO-EUDNa71, or ordinary PEO. After some time had passed, the gels dispersed themselves throughout the corneal surface and disintegrated. The sequence in which PEO-EUDNa71, PEO, and PEO-EUDNa17 were used to describe the gel's residence time in the precorneal region is as follows:

10.5.1 POLYOX in Osmotic Pump Systems

Patel et al. built push–pull osmotic pumps (PPOP) using POLYOX in order to administer a model drug that is practically insoluble at a constant pace (zero order release) over the course of a considerable amount of time. In order to evaluate the effect that the material has, many grades of POLYOX were tested. The formulations for the pull layer used POLYOXWSR-80 NF or N-750 NF, while the push layer used POLYOXWSR-301 NF, Coagulant NF, or 303 NF. According to the findings of an investigation into the effects of utilizing various grades of POLYOX in the push layer ($2 > 70$), the grade of POLYOX utilized did not have a major impact on the release of the medication. Using a higher viscosity grade (N750) for the pull layer instead of the control formulation increased the lag time and delayed the drug release ($2 = 32$–34). This was in comparison to the formulation that served as the control. A developed an osmotic, oral, controlled-release capsule that releases pharmaceuticals at predetermined delivery rates (T80% = 6 or 14 hours) Waterman et al. (2011). This was accomplished in spite of drug characteristics (such as solubility) or drug loading. The osmotic capsule comprises both the active tablet and the push tablet. The active tablet is formed when the drug and PEO are mixed together. In addition, the body of the capsule is stuffed with a "push" tablet that is composed of sodium chloride, microcrystalline cellulose, and high-molecular-weight PEO. The results show that the capacity of the API to dissolve is unaffected by medication loadings ranging from 65 to 375 mg of the active component. The pH did not have any effect on the rate of dissolution anywhere in the biorelevant range (2–6.8).

10.6 Drug Delivery for the Gastroretentive System

Developed and optimized bisoprolol drug-delivery systems that consisted of a core tablet that contained the active ingredient and an erodible outer shell that included a gas-generating agent Jagdale et al. (2013). These systems were constituted of a bisoprolol drug-delivery system. POLYOXWSR205 and POLYOXWSR N12K were utilized throughout the process of press coating the optimized RRCT. In a 32-way full factorial design, the independent variables that were chosen to investigate were the amount of POLYOXWSR205 and POLYOXWSR N12K. The lag time, the rate at which medicine was released, and the swelling index were selected to be the dependent variables. The floating pulsatile release formulation F13 showed a lag time of four hours with more than 90% of the medication being released at level 0 (55 mg) for POLYOXWSR205 and level +1 (65 mg) for POLYOXWSR N12K. The release kinetics of the optimized formulation were the ones that offered the greatest match for the zero-order models. An in vivo investigation provides evidence that the burst effect can last for up to four hours, which suggests that the dose form has been perfected. In their study of sustained-release floating tablets containing cefuroxime axetil, Bomma & Veerabrahma (2013) utilized HPMC K4M and HPMC K100M polymers both independently and in conjunction with POLYOXWSR 303. Every formulation was able to support a medication release that lasted for 12 hours. After putting the dissolution profiles through their paces with a variety of kinetic release models, it was found that the drug release mechanism corresponded to Peppas' model. In vivo radiography investigations revealed that the pills were broken down and digested in the stomach over a period of 225–230 minutes. An asymmetric triple layer pill was presented by Yang et al. (1999) as a potential method for treating *Helicobacter* using the three medications tetracycline, metronidazole, and bismuth salt. PEO and hydroxypropyl methylcellulose were the two primary rate-controlling polymeric excipients that were utilized in this study. Tetracycline and metronidazole were injected into the core layer of the triple-layer matrix in order to provide controlled delivery. However, bismuth salt might be added to one of the outside layers of the matrix in order to achieve quick release. According to the findings, the continuous administration of metronidazole and tetracycline lasted for between six and eight hours. To distribute both water-soluble and water-insoluble compounds in the stomach, Mahalingam et al. (2009) produced compacts with specific bioadhesive polymers, fillers, and binders. Compacts made of polyvinylpyrrolidone (PVP) and PEO in ratios of 90 : 10, 75 : 25, and 60 : 40 were investigated for swelling, dissolving, bioadhesion, and in vitro gastric retention. In vitro gastric retention was also evaluated. Compacts with more PEO demonstrated increased swelling (110.99%) and bioadhesion (0.620,03 $N\,cm^{-2}$) under in vitro conditions. These compacts also retained their integrity and adherence to stomach mucosa for around nine hours. Compacts comprising 58.1% PVP, 39.9% PEO, and 2.1% of water-soluble or water-insoluble marker compounds were able to display gastroadhesive and retentive characteristics in vivo. It has been established that the combination of PEO and PVP results in the production of a bioadhesive dosage form of the non-disintegrating type that is suitable for use in gastroretentive applications.

10.7 Film with a Fast Turnaround Time

Particularly in the creation of fast-dissolving films, low-molecular-weight linear PEO (POLYOX) has demonstrated tremendous potential as a substance for rapid-release medication administration. Producing film from this material is a straightforward process since POLYOX possesses exceptional film-forming capabilities.

POLYOX has a high water absorption capacity; as a result, when its film comes into contact with aqueous medium, it immediately dissolves, leading to fast disintegration. This is because to POLYOX's high water absorption capacity. Numerous research have been conducted on low-molecular-weight POLYOX-based rapidly dissolving films for oral administration. These films have been shown to be effective. Using a variety of grades of POLYOX, including POLYOXN10, N80, N750, and N205, Chaudhary et al. (2012) were able to produce films that dissolve in a short amount of time. With the use of CCD design, the concentration of the plasticizer and the polymer in the film may be made to work better. It was agreed that the tensile strength, folding endurance, percentage of medication released at 10 minutes (Y10), and disintegration duration would be used as the dependent variables in this study. According to the findings, using 2% POLYOXN 750 and 15% PEG 400 resulted in a great film-forming property and 100% drug release in 15 minutes.

10.8 Extended Duration of Effect

The feasibility of utilizing a counter polymer in PEO/polyethylene glycol (PEG) polymeric matrices to achieve the prolonged release of a sizeable quantity of a very water-soluble medication was examined by Kiss et al. (2006). Their goal was to do this in a manner that was more controlled. Tablets known as CR-A have a PEO/PEG matrix in addition to four other drugs, each of which dissolves differently in water. In addition, the medicine diltiazem hydrochloride (DTZ), which is water-soluble, was included in the formulation of the cross-linked carboxyvinyl polymer (CVP)/PEO/PEG matrix tablets (CR-B), and the in vitro characteristics of these tablets were compared to those of the CR-A formulation. The diffusion of DTZ was effectively reduced, resulting in sustained release for 24 hours or more, as part of an effort to regulate the drug release of more water-soluble medications. This was accomplished with the use of a polymer that possessed a charge that was opposite to that of the drug.

Disadvantages of using POLYOX

- They can endow the finished product with lubricity and water solubility. Crazing and tensile failure, on the other hand, are possibilities if even a moderately consistent force is given to POLYOX materials.
- Crystallization may also occur with polymers, and it does so more easily and more rapidly with decreasing molecular weight because there is a correlation between the decreasing molecular weight and an increase in water absorption.
- Due to LMW POLYOX's strong water absorption, a glassy rubberized transition may be visible on SEM photomicrographs; as a result, it must be more noticeable. This is because it is a result of the high water absorption of LMW POLYOX.
- Both the dosage form and the pure substance are impacted by these structural changes that are carried on by the natural ageing process.

- There is a possibility that the mechanical qualities and drug release properties will shift over time, which may result in stability problems. Future issues must be avoided; thus, it's critical to keep an eye out for any potential structural changes (Kiss et al. 2006; Madhar and Saini 2011; Alcantor et al. 2010; Arunachalam et al. 2010; Sharma et al. 2011; Easterbrook et al. 2011; Tang, 2011).

10.9 Regulatory Aspects of POLYOX

Adherence to FDA and other regulatory agencies' regulations: The USP PEO NF monograph is satisfied by POLYOX Water-Soluble Resins, NF Grade. The American National Formulary, the International Codex Alimentarius, and the Food Chemicals Codex (NF) have all established standards for these goods. The majority of the countries that make up Europe have approved the usage of these components in their various medicinal products. It is predicted that the usage of this product will be permitted in Japan Shah et al. (2014).

10.10 The Consistency of POLYOX

The oxidative breakdown of POLYOX polymers can lead to chain cleavage and a decrease of viscosity during storage if it does place. All polymers developed by POLYOX for use in the pharmaceutical industry contain butylated hydroxyl toluene (BHT) as an antioxidant, at quantities ranging from 300 to 500 parts per million (ppm). However, by making the protons on carbon atoms more labile, oxygen significantly improves the process in the molecules. This is accomplished by the following mechanism: There are a lot of striking parallels between the degrading process of POLYOX and that of hydrocarbon chains (Tang 2011; Nisato and Baret 2010; Heiati and Weimann 2010; Vinod et al. 2012). A number of different types of metal ions, such as ferrous, cuprous, cupric, and silver, have the potential to speed accelerate the decomposition of POLYOX. In general, catalysts containing ions with a lower valence are more efficient. Increasing the polymer's stability may be accomplished in a number of ways, one of which is by applying an oxygen barrier film layer to the final dosage form. BHT and vitamin E are the two antioxidants that are utilized with POLYOX the vast majority of the time. Although BHT is already present in polymers designated for medical use of POLYOX in a concentration of approximately 301–501 ppm (0.03–0.05% w/w), it is recommended that an additional 101–1001 ppm (0.01–0.10%) be added to the formulation of POLYOX tablets. Vitamin E is utilized not only in preventative but also in curative applications due to its antioxidant properties. The typical range for application is between 501 and 1001 ppm. In order to guarantee that the POLYOX(tm) powder is properly mixed, Vitamin E is first diluted with a very little amount of isopropyl alcohol and then stirred into the powder. Industrial applications may call for the usage of sodium thiosulfate at a concentration of 0.1% weight-per-volume as a solid stabilizer. The processes that gain the most from sodium thiosulfate's capacity to scavenge chlorine are those that use municipal (treated) water as their source of water (Shah et al. 2014). POLYOX does not have a predetermined shelf life; nonetheless, it is recommended that the material be routinely retested in order to guarantee that it is suitable for any use. Batches of the material are retested in order to make certain that they adhere to the viscosity criteria.

In order to evaluate the resiliency of POLYOX, the following grades of the material are available (Yoon 2010; Zerbe and Paiement 2011; Sambasivam 2010; David et al. 2009; Gupta et al. 2011; Patel et al. 2012; Nghiem and Jackson 2009; Patel 2008; Kidane and Bhatt 2007):

- Six months for grades ranging from moderate-to-high molecular weight (WSR N all)
- Two years for grades ranging from moderate-to-low-molecular weight (WSR N-10 and N-80). Because of its safety, processing convenience (possibility of direct compression), and the potential to control drug release, a biocompatible eroding polymer called poly(ethylene oxide) is gaining popularity as a sustained-release and immediate platform.

10.11 Conclusion

PEO comes in a variety of molecular weights, and it is gaining popularity as a sustained-release and immediate platform because of its versatility. HMW POLYOX, which releases substances by swelling rather than erosion, is the root cause of anomalous release kinetics. LMW both erosion and the development of POLYOX hydrate take place at a faster pace. POLYOX has been used to make oral sustained-release tablets, gastroretentive tablets, extended release, buccal films, ocular inserts, colon-targeted tablets, osmotic tablets, and quick dissolving films based on the variable molecular weight and release rate that it generates. Polymers are currently being used extensively in the development of innovative drug-delivery strategies for therapies that are either highly or weakly soluble (Kumar et al. 2011; Saraf 2010; He et al. 2008; Rane and Prabhakar 2009; Godin and Touitou 2004; Youfang et al. 2009; Shimada 2012).

References

Alcantor, N., Williams, E.C., Toomey, R., Inventors; University of South Florida, FL, Assignee (2010). Niosome hydrogel drug delivery systems. US Patent 20100068264. Filed 20 November 2009.

Arunachalam, A., Karthikeyan, M., Vinay Kumar, D. et al. (2010). Transdermal drug delivery system: a review. *Current Pharma Research* 1 (1): 70–81.

Bomma, R. and Veerabrahma, K. (2013). Development of gastroretentive drug delivery system for cefuroxime axetil: in vitro and in vivo evaluation in human volunteers. *Pharmaceutical Development and Technology* 18 (5): 1230–1237. https://doi.org/10.3109/10837450.2012.660698.

British National Formulary, Issue 58, ISSN 0260-535X Pharmaceutical Press, 2009.

Bukka, R., Dwivedi, M., Nargund, L.V.G., and Prasam, K. (2012). Formulation and evaluation of felodipine buccal films containing polyethylene oxide. *International Journal of Research in Pharmaceutical and Biomedical Sciences* 3 (3): 1153–1158.

Cappello, B., Del Nobile, M.A., La Rotonda, M.I. et al. (1994). Water soluble drug delivery systems based on a nonbiological bioadhesive polymeric system. *Farmaco* 49 (12): 809–818.

Cappello, B., De Rosa, G., Giannini, L. et al. (2006). Cyclodextrin-containing poly(ethyleneoxide) tablets for the delivery of poorly soluble drugs: potential as buccal delivery system. *International Journal of Pharmaceutics* 319 (1–2): 63–70.

Chandel, P. and Kapoor, A. (2013). Polymer: a boon to controlled drug delivery system. *International Research Journal of Pharmacy* 4 (4): 28–34.

Chaudhary, S.A., Chaudhary, A.B., and Mehta, T.A. (2012). Formulation development and optimization of POLYOX based quick dissolving film of quetiapine. *Journal of Pharmacy & Bioallied Sciences* 4 (1): S19–S20.

Chowdary, K.P., Deepthi, K.S., and Rao, S. (2009). Mucoadhesive microcapsules of indomethacin: evaluation for controlled release and ulcerogenic activity. *International Journal of Pharmaceutical Sciences and Drug Research* 1 (2): 74–79.

David, A.E., Zhang, R., Park, Y.J. et al., Inventors (2009). Mucoadhesive vesicles for drug delivery. US Patent 20090232899. Issued 17 September 2009

Deshkar, S., Pawar, M., Shirsat, A., and Shirolkar, S. (2013). Development of sustained release tablet of Mebeverine hydrochloride. *Indian Journal of Pharmaceutical Education and Research* 4 (1): 64–69.

Di Colo, G., Burgalassi, S., and Chetoni, P. (2001). Gel-forming erodible inserts for ocular controlled delivery of ofloxacin. *International Journal of Pharmaceutics* 215 (1–2): 101–111.

Dimitrov, M. and Lambov, N. (1999). Study of verapamil hydrochloride release from compressed hydrophilic POLYOX-Wsr tablets. *International Journal of Pharmaceutics* 189: 105–111.

Easterbrook, T.J., Gosden, E., and Meyer, E., Inventors (2011). Transdermal drug delivery device. US Patent 20110190716. Filed 26 January 1988, Issued 19 September 1989.

Godin, B. and Touitou, E. (2004). Mechanism of bacitracin permeation enhancement through the skin & cellular membrane from an ethosomal carrier. *Journal of Controlled Release* 94 (2–3): 365–379.

Gupta, S., Singh, R.P., Sharma, R. et al. (2011). Osmotic pumps: a review. *International Journal of Comprehensive Pharmacy* 6 (1): 1–8.

He, Z.F., Liu, D.Y., Zeng, S., and Ye, J.T. (2008). Study on preparation of ampelopsin liposomes. *Journal Chinese Matria Medica* 33 (1): 27–30.

Heiati, H, Weimann, L., Inventor; Pharmapatch LLC, San Diego, CA, Assignee (2010). Multiple nozzle transdermal drug delivery system. US Patent 20100143448. Issued 28 July 2010.

Jagdale, S.C., Bari, N.A., Kuchekar, B.S., and Chabukswar, A.R. (2013). Optimization studies on compression coated floating-pulsatile drug delivery of bisoprolol. *BioMed Research International* 801769: 11 pages. http://dx.doi.org/10.1155/2013/801769.

Kidane A, Bhatt PP, Inventors (2007). Osmotic drug delivery system. US Patent 20070254032. Issued 1 November 2007.

Kim, C. (1995). Drug release from compressed hydrophilic POLYOX-WSR tablets. *Journal of Pharmaceutical Sciences* 84: 303–306.

Kiss, D., Karoly, S., and Marek, T. (2006). Tracking the physical aging of poly (ethylene oxide): a technical note. *AAPS PharmSciTech* 7 (4): E95–E98.

Kojima, H., Yoshihara, K., and Sawada, T. (2008). Extended release of a large amount of highly water-soluble diltiazem hydrochloride by utilizing counter polymer in polyethylene oxides (PEO)/polyethylene glycol (PEG) matrix tablets. *European Journal of Pharmaceutics and Biopharmaceutics* 70: 556–562.

Kumar, A., Sharma, P., and Banik, A. (2011). Microencapsulation as novel drug delivery system. *Internationale Pharmaceutica Sciencia* 1 (1): 1–7.

Madhar, N.V.S. and Saini, A. (2011). Niosomes: a novel drug delivery system. *International Journal of Research in Pharmacy and Chemistry* 1 (3): 498–511.

Maggi, L., Bruni, R., and Conte, U. (2000). High molecular weight polyethylene oxides (PEOs) as an alternative to HPMC in controlled release dosage forms. *International Journal of Pharmaceutics* 195: 229–238.

Mahalingam, R., Jasti, B., Birudaraj, R. et al. (2009). Evaluation of polyethylene oxide compacts as gastroretentive delivery systems. *AAPS PharmSciTech* 10 (1): 98–103.

Nghiem, T., Jackson, G., Inventors; Biovail Laboratories International S.R.L., Assignee (2009). Multiparticulate osmotic delivery system. US Patent 20090004281. Issued 1 January 2009.

Nisato, G. and Baret, J.C., Inventors; Koninklijke Philips Electronics N.V, NC, Assignee (2010). Transdermal drug delivery patch. US Patent 20100222751.Issued 10 June 2010.

Patel, H.B., Inventor; Reliant pharmaceuticals INC., Assignee (2008). Oral osmotic drug delivery system. US Patent 20080248114. Issued 9 October 2008.

Patel, H., Patel, U., Kadikar, H. et al. (2012). A review on osmotic drug delivery system. *International Research Journal of Pharmacy* 3 (4): 88–94.

Rane, S. and Prabhakar, B. (2009). Influence of liposome composition on paclitaxel entrapment & pH sensitivity of liposomes. *International Journal of Pharmaceutical Research & Technology* 1 (3): 914–917.

Sambasivam, M., Inventor; Convatec Technologies INC, NV, Assignee (2010). Ostomy devices with mucoadhesives. US Patent 2010010064. Issued 22 April 2010

Saraf, A.S. (2010). Applications of novel drug delivery system for herbal formulations. *Fitoterapia* 81: 680–689.

Shah, K.R., Chaudhary, S.A., and Mehta, T.A. (2014). Polyox (polyethylene oxide) multifunctional polymer in novel drug delivery system. *International Journal of Pharmaceutical Sciences and Drug Research* 6 (2): 95–101.

Sharma, B., Saroha, K., and Yadav, B. (2011). Sonophoresis: an advanced tool in transdermal drug delivery system. *International Journal of Current Pharmaceutical Research* 3 (3): 89–97.

Shimada, S. (2012). Composition comprising nanoparticle *Ginkgo biloba* extract with the effect of brain function activation. US Patent 8105637.Issued 31 January 2012.

Tang, J., Inventor (2011). Stabilised transdermal drug delivery system. US Patent 20110182949. Issued 28 July 2011.

Vidyadhara, S., Sasidhar, R.L.C., and Nagaraju, R. (2013). Design and development of polyethylene oxide based matrix tablets for verapamil hydrochloride. *Indian Journal of Pharmaceutical Sciences* 75 (2): 185–190.

Vinod, K.R., Reddy, R., Banji, D. et al. (2012). Critical review on mucoadhesive drug delivery systems. *Hygeia Journal for Drugs and Medicines* 6 (1): 7–28.

Waterman, K.C., Goeken, G., Konagurthu, S. et al. (2011). Osmotic capsules: a universal oral, controlled-release drug delivery dosage form. *Journal of Controlled Release* 152: 264–269.

Yang, L., Venkatesh, G., and Fassihi, R. (1996). Characterization of compressibility and compactibility of poly (ethylene oxide) polymers for modified release application by compaction simulator. *Journal of Pharmaceutical Sciences* 85: 1085–1090.

Yang, L., Eshraghi, J., and Fassihi, R. (1999). A new intragastric delivery system for the treatment of *Helicobacter pylori* associated gastric ulcer: in vitro evaluation. *Journal of Controlled Release* 57 (3): 215–222.

Yoon, H.J. and Jang, W.D. (2010). Polymeric supramolecular systems for drug delivery. *Journal of Materials Chemistry* 2: 211–222.

Youfang, C., Xianfu, L., Hyunjin, P., and Richard, G. (2009). Artemisnin nanoparticles. *Nanomedicine: Nanotechnology, Biology and Medicine* 5: 316–322.

Zerbe, H.G., and Paiement, N., Inventor (2011). Oral mucoadhesive dosage form. US Patent 20110028431.

11

Green Synthesis of Polymers and Its Application in Industry

Raja Chakraverty[1], Dibya Das[2], and Tatini Debnath[3]

[1] *Institute of Post Graduate Medical Education and Research, Department of Critical Care Medicine, 244, A.J.C. Bose Road, Kolkata 700 020, West Bengal, India*
[2] *JIS University, Department of Pharmaceutical Technology, 81,Nilgunj Road, Jagarata Pally, Deshpriya Nagar, Kolkata 700109, West Bengal, India*
[3] *Maulana Abul Kalam Azad University of Technology, Department of Pharmaceutical Technology, NH 12(Old NH 34), Simhat, Haringhata, Nadia 741249, West Bengal, India*

11.1 Introduction

The term "green" polymer synthesis is a newer approach in the field of polymer chemistry that revolves around the fact of biodegradable aspect to make the entire process ecologically sustainable. This approach has been encouraged by the scientists owing to the disadvantages incurred by the synthetic materials due to their chemical and physical degradation while used biomedically. Their biodegradability and biocompatibility with lower amount of toxicity is a major advantageous feature rendering them to be more convenient for use in human-health-related areas, agriculture, environment, etc. Plant-based renewable source of synthesis is gaining lots of attention worldwide so that the hazards due to the non-degradable chemicals induced into the environment could be minimized. To improve the standard of human health this research has provided right set of circumstances (Satyanarayana et al. 2009). Transition from the non-biodegradable to biodegradable materials is a greater achievement in itself which helps in the reduction of harmful chemicals to percolate into the environment. The composite materials of polymer include various applications that are used in transportation, packaging of food, pharmaceuticals, etc. Certain limitations of the polymers were overcome by use of intractable materials. Nanocomposites of polymers comprise nanoparticles and nanofillers distributed all over in the matrix of polymer matrix. Nanoparticles and nanofillers mostly have various shapes (fibers, spheroids, and platelets) where the size ranges from 1 to 50 nm. The nanoparticle implementation improvised the basic polymers properties in terms of permeability and mechanical (water and gases) properties with (elastic modulus, strength, and dimensional stability). Composites of polymers have opened new opportunities in the industries like rubber and thermoplastic having strengthened properties.

Organic Polymers in Energy-Environmental Applications, First Edition.
Edited by Ramesh Oraon, Pardeep Singh, Sanchayita Rajkhowa, Sangita Agarwal, and Ravindra Pratap Singh.

The intriguing fact about the polymers is their green variants having an environment sustainable side to it. Features of green polymer nanocomposites are their properties should encompass renewable and biodegradable genre since the inception of polymer formation on earth took place. Artificially and naturally in both the cases the polymers play important role. Polymers involve superstructures modern processing technologies and polymerization processes. In everyday life the involvement of developmental scientific technologies has a key role, and the difficulty level would have reached summit if polymers were not used which has one major role in pacemaker that aids life for those suffering from cardiovascular diseases. In earlier times chemically modified biopolymers were used. The abundant organic compounds produced by the organisms are sugar-based cellulose, organic biomass, and various other components. The incorporation of 20 amino acids under controlled condition in biotechnological labs helps in producing polypeptides thereby producing variety of synthetic proteins for instance insulin, which has both industrial and biomedical applications. On a large-scale basis synthetic polymers in the twentieth century are obtained and procured industrially. Indiscriminate use of fossils and natural source raw materials by the chemical industries expedites production of polymers thereby reducing manufacturing cost. Recycling of those polymer products which have completed their shelf life, therefore the by-products and the waste from polymers, has become an important source of energy and raw materials. Due to the advancement in polymer manufacturing processes, they are optimized to be insulating, conducting, stiff, rubbery or soft, biodegradable or stable, etc. Simply the most important polymer applications are mostly found to be in medical and food packaging, lightweight packaging, aerospace industries, automotive industries, modern architecture, high strength fibers, textile industries, microelectronics, electrical circuits, in paints, corrosion protection, drug-delivery system, dental filings, implants, and many more. Plastics are the most essential part of any consumer product.

11.2 Polymer

"Polymer" word originated from the Greek word "poly" which denotes "many" and "mer" means "parts." Polymers are long-chain molecules that comprise identical or similar building blocks covalently bonded. Polymers are macromolecules and mostly large molecules. Due to their respective characteristics polymer molecules are bonded accordingly in different manner. Rubber and polyesters are stretchable and flexible polymers, while glass and epoxies are resistant and tough.

Polymers have become a considerable part of everyday life, for example tires, electronic gadgets, and water bottles. Monomers are the single molecular units which combine and help in the formation of polymers. Consecutive units create repeating units. The repeating unit is the one that is enclosed in square brackets. But even though both of them contain identical atoms in the same relative position, the repeating unit's structure is a bit different from the monomers. Electrons must be rearranged in order for the monomer to become a polymer. The structural unit is the leftover from the monomer that was used to create the polymer. The polystyrene chain is the identical repeating unit with its structural unit since its polymer in the instance of polystyrene was created from a single monomer (styrene). Polyethylene, polyacrylonitrile, and polypropylene are more examples of this kind of polymer. Many factors may be used to characterize polymers. The classification depends on the polymer's origin, structure, mechanism of polymerization, preparative methods, or thermal behavior.

11.2.1 Naturally Occurring Polymer

Natural polymers are mainly found in nature (in animal and plants). These are nucleotides and nucleic acids, proteins, carbohydrates, etc. Essentially a polymer is made up of chemically joining multiple smaller molecules or substances to build bigger molecule. Since naturally occurring polymers are formed in the environment on its own, they are formed by either condensation or addition polymerization.

11.2.1.1 Natural Polymer Examples

In nature, there are several instances of natural polymers. Often, their structures are highly conspicuous. As compared to proteins or enzymes, which have rather complex structures, natural rubber, starch, and cellulose are some examples of polymers that originate from plant (natural). This is a list of some of them, along with a brief description:

Polypeptides and proteins: Proteins form the rudimentary class of organic polymers that creates up the majority of living things. Proteins are the most adaptable; also, they act as catalysts. Enzymes are the name for several proteins. The biochemical processes occurring inside are due to the presence of enzymes inside the cells in our bodies. Hemoglobin, a particular class of protein found in blood, transports oxygen from the lungs to the body's cells.

A naturally occurring polyamide is often a protein. This amide group in polymer forms the body's backbone chain.

Cellulose ($(C_6H_{10}O_5)_n$): It is an organic polysaccharide comprising chain (Scheller et al. 2007). In plants, cellulose, hemicelluloses, and pectin form the cell wall mostly (Nishiyama et al. 2002).

Applications:

- In the pharma industry, cellulose in microcrystalline form is readily used as diluent or binder in tablet manufacturing for direct compression and granulation process (Hon et al. 1996).
- *Starches*: The primary stored form of carbohydrates in plants, starch is mostly obtained from seeds and subterranean organs. Granules of starch can be found in nature (starch grains). A multitude of starches, including those from potato, rice, wheat, and maize (*Zea mays*), are approved for use in pharmaceuticals (*Solanum tuberosum*) (Kokate et al. 2003).

A carbohydrate known as starch or amylum is made up of many units of glucose units bonded together by glycosidic linkages (Te-Wierik et al. 1997).

11.2.2 Synthetic Polymer

They are artificially created by humans. Polymers are the amalgamation of repeating units known as monomers. Polyethylene is made up of ethene and ethylene monomeric units which are the simplest form of polymer and is also known as the high-density polyethylene-HDPE. Many of the polymeric materials have chain-like structures which resemble polyethylene. Synthetic polymers contain polyvinylchloride, polystyrene, etc.

11.2.3 Types of Synthetic Polymers with Examples

Different polymers have been created so far. Let us take a quick look at a few of the synthetic polymers that are utilized on a daily basis.

11.2.3.1 Nylon

Nylon, commonly known as polyamides, is a member of synthetic polymers. It was created on 28 February 1935, at the DuPont research facility by Wallace Carothers. A frequently used polymer is known as nylon. Its amide-based structure makes it more hydrophilic than other polymers. Contrary to the pure hydrocarbon polymers that are used to create the majority of plastics, nylon engages in hydrogen bonding with water.

11.2.3.2 Polyvinyl Chloride

Polyvinyl chloride is the most common plastic after polyethylene and polypropylene. Since its inexpensive unlike copper or iron they are used in buildings and constructions. PVC (polyvinyl chloride) replaces rubber in several places including clothing and electrical wire insulation.

11.2.3.3 Low-Density Polyethylene

Most common type of synthetic polymer is low-density polyethene, a polymer widely used in households. Ethylene is the basis of LDPE (low density polyethylene) which is the monomer.

11.2.3.4 Polypropylene

Polypropylene is a type of thermoplastic which is used in the plastics, reusable containers labeling, packaging, stationery, laboratory equipment, etc.

Other examples are neoprene, Teflon, polystyrene, high-density polythene, etc.

Synthetic polymers are formed by the following methods:

Polymerization is a chemical process by which monomeric units get added and linked to form a polymeric chain; hence, the process linked to form a polymer is called polymerization. The process of polymerization begins with monomers reacting to create a dimer, which then combines with some other monomer to create a trimer, and so on. Any chemical species present in the reaction medium may react with other molecules to generate dimers, trimers, or even larger molecules. Regardless of the situation, a network of connections is created in between consecutive units and the final polymer molecule, also known as a polymer chain. Oligomer's polymerization is with low-molecular-weight by-products like dimers, trimers, and tetramers. They often have unfavorable mechanical and thermal qualities. A substance often has to be highly polymerized before it can develop useful characteristics and be correctly referred to be a polymer (Zhou et al. 2020). There are two types:

1) Addition
2) Condensation

Addition Polymerization This is a process which involves coupling of monomeric units utilizing their multiple bonds known as addition polymerization. For instance, the formation of polyethylene from ethylene molecule is the most commonly observed chemical reaction process. In this type of reaction, the double bond opens up in each of ethylene molecules, and each of the two electrons previously present in this bond was used to form carbon-carbon new single bonds with the other two ethylene molecules. The formation of addition polymers takes place by repetitive addition of monomer molecules possessing triple or double bonds. For instance, the formation of $((C_3H_6)_n)$ polypropene and $((C_2H_4)_n)$ polythene is produced from (C_3H_6) propene and (C_2H_4) ethene, respectively.

Homopolymers: Homopolymers are created when one polymeric species is polymerized to create other polymers. As an illustration polythene $((C_2H_4)_n)$.

Copolymers: A copolymer is created when addition polymerization takes place between two distinct monomers, forming addition polymers as a result. For example, Buna-N, Buna-S, etc.

Condensation Polymerization The chemical reaction mechanism used to make and break polymers inside the cells is mostly the same in all cases.

11.2.4 Natural Polymer

Natural polymers are proteins, cellulose, wool, silk, DNA and RNA, chitosan, natural rubber, starch.

1) Polynucleotides are those types which is formed of large chain of nucleotides.
2) Polyamides are those types which are formed of large molecules of proteins.
3) Polysaccharides are those types which are formed from large chain of sugar molecules.

Need for biodegradable polymers:

1) Since polymer plays a significant part in everyday life and has various other uses and advantages in every sphere.
2) If the polymeric by products are deleterious which will account to a huge hazardous effect into the environment? Therefore, its biodegradable property will make it more eco-friendly.

3) Hence a large amount to waste from the polymers should be curbed to bring down the detrimental effects, so biodegradable polymers uses are nowadays encouraged.

11.2.5 The Strategic Green Synthesis and Their Application

11.2.5.1 Developing Polymer Raw Materials

Polymeric raw material production is also moving toward becoming more environmentally friendly. Several polymer chemical materials may be created during the chemical synthesis process in a single step using catalysis, and the conversion utilization rate can reach 100%. Also, this method modifies the conventional operation mode and avoids the usage of hazardous chemical catalysts (Li et al. 2018). For instance, biosynthetic technology will be used to entirely green, safely, and operationally produce adipic acid. Chlorine gas is added in the middle of the usual two-step process to produce propylene oxide.

This particular gas has a specific level of toxicity and will harm the environment. But today, both locally and globally, this manufacturing technique has been altered, and the catalytic oxidation method is used to fully utilize raw materials in the industrial reaction system without generating certain compounds that contaminate the environment. Several synthetic chemical compounds now being created are steadily reducing the amount of hazardous waste or emissions produced during material synthesis, moving in the direction of green ecology, and protecting the environment (Fu 2019).

11.3 The Difference Between Degradable and Biodegradable Polymer

11.3.1 Degradable Polymer

The term "degradable" means "that which is easily breakdown able." Mostly plastics are degradable type in nature. When an additive is administered to any petroleum-based plastic it becomes photodegradable becomes brittle. We can add an additive to normal, petroleum-based plastic that will make it become brittle. Sometimes when the plastics are subjected to oxidation it becomes easily breakable such is known as oxo degradable plastics. By involving this method the bulk amount seems to reduce in the process of degradable plastic manufacturing. It's a long process and takes several years to get reduced to small pieces by assimilation with the help of microorganisms and degrade. Hence whether a polymer is degradable or biodegradable should be verified quite consciously since plastic materials do not easily get degraded easily whatsoever back into the earth. In the environment, biodegradable polymers entirely decompose into CO_2, methane, water, biomass, and inorganic chemicals that microbes may readily mineralize through a thermochemical process. Both renewable and nonrenewable resources can be used to make a biodegradable polymer. Biobased polymers include organic carbon from sustainable feedstock, albeit they are not necessarily biodegradable. Furthermore, it is possible to make biobased polymers using components that are found in nature.

11.3.2 Biodegradable Polymer

In case of biodegradable polymers, the breakdown into the simpler forms is brought about by the microorganisms. The biodegradable plastics are broken into simpler forms and the carbon chains take part in the formation of organic molecules. Sometimes, they become a part of organic livings things; thus, it gets back to the nature taking part in the carbon cycle maintaining the ecological balance of the earth. Within a time frame only biodegradable plastics degrade. Sometimes with the help of microorganisms the petroleum-based plastics simply could not be decomposed easily, or otherwise it will take several years to degrade slowly. Recent insights into this area have focused upon oxo degradable approach which in turn will be more convenient. There are certain plastics which are formed from the renewable resources that are non-biodegradable. Even if they are degradable, they do not sublime into the earth and they cannot be decomposed by microorganisms; hence, there are significant differences between biodegradable and non-biodegradable plastics. Most frequently referred to as "biopolymers," biodegradable polymers are made primarily from different natural sources. The most often utilized biopolymers that have little to no impact on the environments expanding carbon footprint includes plastics like PLA (Polylactic acid), PHA (polyhydroxyalkanoate), and starch. It is crucial to know that biodegradability is a polymeric property which is nondependent upon its place of origin and may be changed by the molecular mechanism. As a result, certain polymers made from the petroleum feedstock can be broken by nature (Mohanty et al. 2005).

11.3.3 Biodegradable Polymers Classification (shown in Figure 11.1)

1) *Sugar-based biopolymers*: Poly lactic acid (PLA), mostly found, is adaptable and an exclusively studied sugar-based biopolymer with enormous promise for several uses, including the automobile industry. Lactic acid, a naturally occurring sugar found in milk and other foods including wheat, sugarcane, and potatoes, serves as the monomers primary raw material. Another type of biopolymer created by microbes is polyhydroxybutyrate or (PHB), which is used as an energy store molecule when the microorganisms access to nutrients is constrained. The bacterium produces PHB as a polymer by assimilating carbon from glucose.
2) *Starch-based biopolymers*: A biopolymer which is a form of carbohydrate known as starch is exclusively found in plants and stored in tissues. Although glucose makes up the majority of starch may be produced by applying heat or other chemical stimulants, it may be derived from many different foods such as corn, wheat, and tapioca.

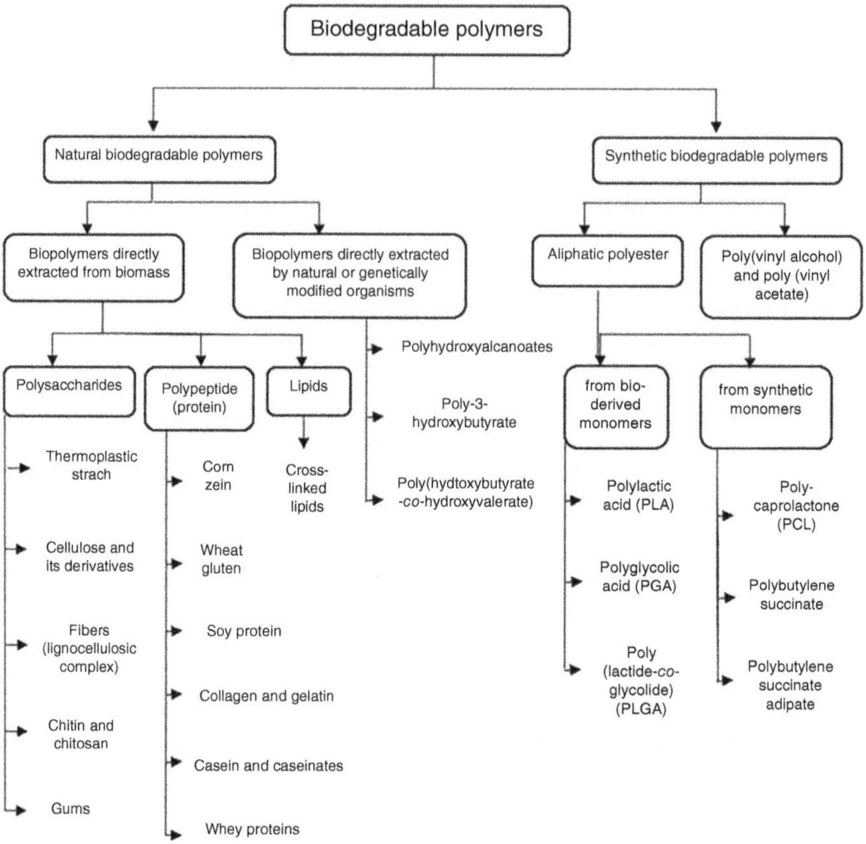

Figure 11.1 Schematic shows biodegradable polymers classification based on their origin and method of production. Source: Samir et al. (2022)/Springer Nature/CC BY 4.0.

3) *Cellulose-based biopolymers*: Cellulose is found abundantly and in surplus inside plant cell wall. The most abundant biopolymer found in the plant cell walls. Glucose is majorly found in cellulose. They are widely used as fillers and plasticizers in biopolymers.

11.3.4 Compostable Plastic

Compostable plastics means it biodegrades at a stipulated time under favorable conditions. Mostly bioplastic breakdown easily, but once it gets placed in a vacuum room it takes the prolonged years to breakdown. To eliminate plastics from the landfills newer biodegradable materials could be engineered. Since the name biodegradable means upon subjecting to degradation with the help of microorganisms it could be broken down to simpler forms. Natural polymers like cellulose and starch are the biopolymers that are obtained extracted from plants and again gets combined with the synthetic polymers to produce biodegradable polymers. Different properties of plastics could be obtained by varying uses of cellulose and starch. The uses are for instance toothbrush handles and adhesive packing. When it reaches the municipal waste collection site, they are subjected to degradation by the treatment with microorganisms to release carbon dioxide, water biomass which are useful nutrients for growth of plants. The sunlight plays the key role in the process of plant growth which produces a set of biodegradable polymers repeating the cycle consecutively.

Criteria of any material to be a compostable plastic:

1) Fragmentation and loss of transparency and disintegration, particularly in the final compost, is determined in a pilot decomposition test in which samples of the test substance are composted with biodegradable garbage for three months. The mass of the test material leftovers must be less than 10% of the original mass after this period (Sin and Bee 2019).
2) Biodegradability or the capacity of the biodegradable substance to break down into carbon dioxide when subjected to microbial activity. The standard specifies a minimum biodegradation threshold of 90%, which must be met in fewer than six months (Rudeekit et al. 2008).
3) Adverse impacts on the composting process are not seen.
4) The number of heavy metals must be less than the specified limit values, and the final compost cannot be adversely impacted (no reduction of agronomic value and no ecotoxicological effects on plant growth). The "Seedling" emblem may be used to identify plastics that have received EN13432 certification, making it easier for customers to decide whether to buy and dispose of a product. Products made of certified biodegradable polymers, including PLA, are disposed of at commercial composting facilities together with other organic waste. Contrary to what is commonly claimed in the gray literature, PLA, which is already manufactured at a rate of more than 100,000 t $year^{-1}$ and is a very versatile material that makes a great alternative with polystyrene and polypropylene in demanding applications, is in fact compostable and biodegradable.

For examples, a 2008 study conducted in Thailand on the deterioration of PLA(Polylactic acid) sheets in composting plants revealed that after eight days, the sheets were brittle and began to shatter into little pieces due to the relatively high temperature and humidity (50–60 °C and RH(Relative humidity 60%). As a result, water molecules may pass through the disordered polymer chains at a land composting facility's average temperature, which is higher than PLA's glass transition temperature and so facilitates the hydrolysis reaction. Even at home, PLA(polylactic acid) that is not home compostable becomes home compostable when combined with polycaprolactone (PCL). In other words, depending on the intended use of the commercially produced bioplastic items, PLA(polylactic acid)-based plastic blends may be created to be biodegradable at home or in the workplace and to biodegrade more or less quickly (Narancic et al. 2018).

As a result, food waste and biodegradable tableware and cutlery made of PLA(polylactic acid) or PHA that were used, for instance, at big events (sports, concerts, ceremonies, etc.) and in restaurants could be disposed of in the same stream. But waste will be converted using the principles of the circular economy

Table 11.1 Categories for green polymer chemistry.

Pathways	Examples
Green (catalysts)	Enzymes and biocatalyst.
Feedstock base (diverse)	Agricultural feedstock, building blocks (biobased) (ferulic acid, lignin, sugars, peptides, triglycerides). Fillers (natural) like carbon dioxide monomers.
Minimization of waste and degradable polymers	Renewable source agriculture-based materials. Polyamides and polyesters.
Polymer products and catalysts recycling	There are degradable polymers can possibly can be recycled. Reuse of immobilized enzyme.
Energy minimization and generation	Saving energy (e.g. jet cooking, microwave), biofuels.
Molecular design and activity (optimal)	Structured and designed polymers. Improved biocatalyst's improvement. Synthesis of proteins. Genetic engineering.
Benign solvents	Reactions without solvents or using ionic liquids and water.
Improved syntheses and processes	Reduction in toxicity, increasing efficiency, decreasing number of by-products.

Source: Adapted from Gross et al. (2001).

into valuable biogas, high-quality compost, and fertilizer that will enrich the land. The same composting facilities, which are common in countries like Germany or Italy, and areas where food and other organic waste is converted into compost and biogas, are were composting takes place (Table 11.1).

Biocatalysis requires utilization of the microorganisms, enzymes, microorganisms, and sometimes involvement of higher form of organism that facilitates and propels the biochemical reactions. In the absence of a solvent most of the biochemical reactions take place thereby inducing a lower temperature of around (60 °C or lower) resulting into production of reduced by products. The exceptional green method is the production of polymer and the polymerization processes. Recent studies in this field by the exemplary group of scientists (Gross et al. 2001; Kobayashi et al. 2001; Kobayashi and Makino 2009; Cheng and Gu 2012; and books Loos 2011; Cheng and Gross 2008; Cheng and Gross 2010; Cheng et al. 2015) are available. Biobased materials involve the use of agriculture-based and natural-based building blocks of polymers and their derivatives. Article reviews by Roach et al. (2007), Meier et al. (2007), Bhardwaj and Mohanty (2007), Raquez et al. (2010), Cheng (2012), Khemani and Scholz (2012), and books Mohanty et al. (2005), Sun (2005), Cheng et al. (2013), and Greene (2022) based upon this topic. The favorable use of these materials is supported by various factors:

1) Reduction of dependence upon petroleum raw materials.
2) Price stability between petroleum-based products and agriculture-based products.

3) Biodegradability, viability, and beneficial effects upon the environments of bio-based materials.
4) Opportunities for farmers from agricultural commodities Roland-Holst et al. (2013).

11.3.5 Pipeline to Green Polymers

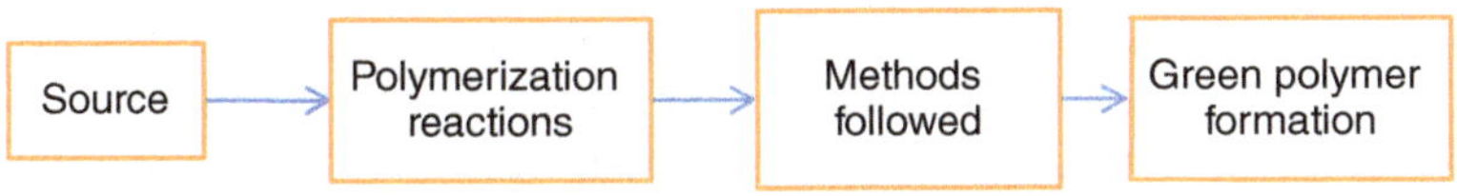

The flowchart mentioned above the source could be a monomeric unit or might be an agricultural raw material (e.g. triglyceride, starch, and cellulose). Polymerization is a process which occurs with or without involvement of a solvent or catalyst. It is an organic biocatalysts, acid or base, containing metal. The solvent could be organic, supercritical CO_2 aqueous, perfluorinated medium, ionic liquid, or none. The process of polymerization process involves extraction, separation, purification, waste removal with several other operations. The key considerations are processing, yield, efficiency, and compliance with environmental and energy usage. The ultimate product development and improvement aims toward compounding, formulating, stabilizing, packaging with evaluation.

11.3.6 Green Reactions

Broadly bio-catalysis and other conventional green efficient solvent-free media make it systematic. Based upon other studies as we all know bio-catalysis is an illustrative example of green reactions. For Puskas et al. (2015) and Barrera-Rivera et al. (2015) used lipases, Kadokawa (2015) phosphorylase, Côté and Skory (2015) glucan sucrose, and Nduko et al. (2015) lactate polymerizing enzyme for numerous polymerization and modification reactions. The treatment for wastewater an important point by Maloney et al. (2015) was use of oxidase and peroxidase. The non-biocatalytic catalysts applications were also disseminated in several chapters by MacDonald and Shaver (2015), Romain and Williams (2015), O'Leary Havelka and Gerhardt (2015), Hernández et al. (2015), Hong et al. (2015), Re et al. (2015), Cheng and Biswas (2015), and various others like Wu et al. (2019). The "evolutionary molecular engineering" is supposedly another new avenue to green synthesis reaction; hence Tada et al. (2015) demonstrated in vitro selection method for production of functional protein.

11.3.7 Green Processing

Owing to environmental hazards and huge energy consumption industries are trying to curb the chemical waste to minimize biohazards thereby propelling the yield. Newer process design and planning may expedite improvement with reduction in logistical difficulty. From agricultural point of view the biobased starting materials like proteins, carbohydrates, and other natural products are more advantageous and with almost minimal toxicity, so they may be considered for safe handling of reactants and products. Three papers in this book may be highlighted in this regard. Significant example is a paper by Toyama et al. (2015) which works on the novel cellulosic derivative synthesis with reduced solvent and energy consumption without affecting the properties of the product formed. The dimer acid epoxide preparation method followed of Li and Li (2015) does not involve organic solvents or toxic chemicals and is therefore more eco-friendly. Ideal eco-friendly and green polymers represent the polymers whose raw materials, polymerization process, processing, and final products are all harmless to the environment and human health. Our laboratory has been applying the principles of green chemistry and chemical engineering to all the areas of polymer production. On the basis of the use and transformation of renewable resources as well as the developments of eco-friendly polymerization, catalysis, processing, and polymeric materials, we are aiming to establish the new theories and technologies for the eco-friendly polymerization, processing, and modification of polymeric materials.

11.3.8 Greening of Synthetic Raw Materials

Raw materials created by polymers are used to create many living things. Certain materials, particularly medical ones, must not be hazardous when in use, be biodegradable, and be acceptable to the human immune system. As a result, environmentally friendly and secure standards for synthetic raw materials are necessary. Many successes have been realized both domestically and internationally recently.

Studies on polylactic acid network elastomers that could be biodegraded and were produced from green raw materials were conducted in a similar manner in Holland in 1988. They added lysine diisocyanate vinegar to the star-shaped prepolymer chain made of inositol, L-lactide, and other molecules. Because the ultimate breakdown products of LDI (lysine diisocyanate) chain extender include ethanol and lysine, LDI can be referred to as a "green" diisocyanate chain extender. These breakdown byproducts are completely biodegradable and non-toxic. In addition to being safe and favorable to the environment, this polymer's basic ingredient, inositol, is one of the vitamins required by the human body. In biomedicine, substances like lactic acid and 6-alkyl caproic acid are often used. They are also among the safest and "greenest" substances. This technique may be termed as "almost completely green." On this basis, Strey and other researchers performed more research in 1994 and used this green reagent to create LDI polylactic acid derivatives. Orthopedic fixing composites were made using high crystalline polyglycolic acid fiber as the reinforcing ingredient.

11.3.9 Green Catalyst

The utilization of safe catalysts may be overlooked in the research process for polylactic acid materials, despite the fact that conventional polymer raw materials and polymers are largely eco-friendly and non-toxic. To speed up the chemical process, stannous octoate, for instance, is frequently utilized as an intermediate catalyst. Yet, because it includes tin salt, this catalyst could be physiologically harmful and poisonous if ingested by humans. Biological enzymes can be used as reliable and safe catalysts in comparison (Wang and Ma 2019).

The insufficient availability of suitable biological enzymes for catalysis in various chemical processes is the bottleneck of employing biological enzymes for catalysis due to the restricted range of enzymes. The employment of various catalysts and compounds has an influence on the safety situation, even though some polyaddition processes among the present high-molecular polymers have an atomic utilization rate that may exceed 100%. The efficacy of these materials has to be investigated and evaluated, particularly in medical papers. It would be good to learn more about the creation of the green catalytic road.

11.3.10 Application of the Synthetic Polymer Materials

Synthetic polymer materials might be environmentally harmful; hence, the issue of eco-friendly and non-toxic waste disposal should be considered. To use these biodegradable polymers securely, the appropriate technique must be adopted. Synthetic materials made of biodegradable polymers can be put into landfills for disposal. Wastes made from non-biodegradable polymer materials are divided into recyclable and non-recyclable wastes. Materials made of recyclable polymers are categorized and processed to facilitate recycling and minimize resource waste. It is possible to burn incinerable polymer materials, and the heat energy created during rubbish incineration may also be used (Song and Li 2018).

1) While processing ecologically inert plastic waste, landfills should be avoided wherever feasible in favor of recycling and reusing environmentally inert polymer materials including PP, PE, PET, Nylon 66, PMMA, and PS, among others.
2) Polyolefins, such as PP and PE, burn safely and have high calorific values, which are comparable to fuel oil. In order to minimize heat energy pollution, the enormous heat energy produced by the burning of these polymer materials can instead be turned into electric energy or other kinds of energy. Nowadays, removing PVC and avoiding combining it with PP, PE, etc. is essential for the efficient conversion of municipal solid waste into energy and avoid to waste the energy due to difficulty in energy recovery.
3) Reasonable usage of PVC is required. Environmental issues arise throughout the production, processing, usage, and waste disposal of PVC, with the disposal of PVC waste being the most hazardous. PVC is processed using a variety of additives, and inappropriate handling will allow any harmful components to leak out. As a result, contact with food and medical items should be avoided as much as possible. While handling PVC trash, combustion should be avoided as much as possible since this polymer material will generate harmful compounds during incineration, which will seriously affect the environment. As quickly as feasible, PVC should be eliminated from packaging, toys, plastic film, and other application industries with short service lives. Additional study should be put into

the investigation of PVC manufacturing, processing, usage, and waste treatment in consideration of the advantages of minimizing environmental impact, wide applicability, cheap cost, non-flammability, good blood compatibility, etc.

References

Barrera-Rivera, K.A., Marcos-Fernández, Á., and Martínez Richa, A. (2015). Chemoenzymatic synthesis and characterization of polyester-urethanes bearing amino-acids moieties. In: *Green Polymer Chemistry: Biobased Materials and Biocatalysis* (ed. H.N. Cheng, R.A. Gross, and P.B. Smith), 27–40. American Chemical Society.

Bhardwaj, R. and Mohanty, A.K. (2007). Advances in the properties of polylactides based materials: a review. *Journal of Biobased Materials and Bioenergy* 1 (2): 191–209.

Cheng, T.C. (2012). *Pathogens of Invertebrates: Application in Biological Control and Transmission Mechanisms*, vol. 7. Springer Science & Business Media.

Cheng, H.N. and Biswas, A. (2015). Modifications of plant oils for value-added uses. In: *Green Polymer Chemistry: Biobased Materials and Biocatalysis* (ed. H.N. Cheng, R.A. Gross, and P.B. Smith), 235–247. American Chemical Society.

Cheng, H.N. and Gross, R.A. (ed.) (2008). *Polymer Biocatalysis and Biomaterials II*. American Chemical Society.

Cheng, H.N. and Gross, R.A. (2010). Green polymer chemistry: biocatalysis and biomaterials. In: *Green Polymer Chemistry: Biocatalysis and Biomaterials* (ed. H.N. Cheng, R.A. Gross, and P.B. Smith), 1–14. American Chemical Society.

Cheng, H.N. and Gu, Q.M. (2012). Enzyme-catalyzed modifications of polysaccharides and poly(ethylene glycol). *Polymers* 4 (2): 1311–1330.

Cheng, H.N., Smith, P.B., and Gross, R.A. (2013). Green polymer chemistry: a brief review. In: *Green Polymer Chemistry: Biocatalysis and Materials II*, ACS Symposium Series, vol. 1144, 1–12.

Cheng, H.N., Gross, R.A., and Smith, P.B. (2015). Green polymer chemistry: some recent developments and examples. In: *Green Polymer Chemistry: Biobased Materials and Biocatalysis*, ACS Symposium Series, vol. 1192, 1–13.

Côté, G.L. and Skory, C.D. (2015). Water-insoluble glucans from sucrose via glucansucrases. Factors influencing structures and yields. In: *Green Polymer Chemistry: Biobased Materials and Biocatalysis* (ed. H.N. Cheng, R.A. Gross, and P.B. Smith), 101–112. American Chemical Society.

Fu, X. (2019). Application analysis of organic chemistry in polymer material synthesis (in Chinese). *Chinese Information* 8: 111.

Greene, J.P. (2022). *Sustainable Plastics: Environmental Assessments of Biobased, Biodegradable, and Recycled Plastics*, 2e. John Wiley & Sons.

Gross, R.A., Kumar, A., and Kalra, B. (2001). Polymer synthesis by in vitro enzyme catalysis. *Chemical Reviews* 101 (7): 2097–2124.

Hernández, N., Yan, M., Williams, R.C., and Cochran, E. (2015). Thermoplastic elastomers from vegetable oils via reversible addition-fragmentation chain transfer polymerization. In: *Green Polymer Chemistry: Biobased Materials and Biocatalysis* (ed. H.N. Cheng, R.A. Gross, and P.B. Smith), 183–199. American Chemical Society.

Hon, D.N.-S. (1996). Cellulose and its derivatives: structures, reactions and medical uses. In: *Polysaccharides in Medicinal Applications* (ed. S. Dumitriu), 87–106. New York, NY: Marcel Dekker, Inc.

Hong, J., Hairabedian, D., Petrović, Z.S., and Myers, A. (2015). Polyacids from corn oil as curing agents for epoxy resins. In: *Green Polymer Chemistry: Biobased Materials and Biocatalysis* (ed. H.N. Cheng, R.A. Gross, and P.B. Smith), 223–233. American Chemical Society.

Kadokawa, J.I. (2015). Enzymatic synthesis of non-natural oligo-and polysaccharides by phosphorylase-catalyzed glycosylations using analogue substrates. In: *Green Polymer Chemistry: Biobased Materials and Biocatalysis* (ed. H.N. Cheng, R.A. Gross, and P.B. Smith), 87–99. American Chemical Society.

Khemani, K. and Scholz, C. (2012). Introduction and overview of degradable and renewable polymers and materials. In: *Degradable Polymers and Materials: Principles and Practice*, 2e, 3–10. American Chemical Society.

Kobayashi, S. and Makino, A. (2009). Enzymatic polymer synthesis: an opportunity for green polymer chemistry. *Chemical Reviews* 109 (11): 5288–5353.

Kobayashi, S., Uyama, H., and Kimura, S. (2001). Enzymatic polymerization. *Chemical Reviews* 101 (12): 3793–3818.

Kokate, C.K., Purohit, A.P., and Gokhale, S.B. (2003). *Pharmacognosy*, 22e, 133–166. Nirali Prakashan.

Li, H. (2018). Application prospect of polymer chemical materials (in Chinese). *China Chemical Trade* 3: 252.

Li, A. and Li, K. (2015). Development and characterization of pressure-sensitive adhesives from dimer acid and epoxides. In: *Green Polymer Chemistry: Biobased Materials and Biocatalysis* (ed. H.N. Cheng, R.A. Gross, and P.B. Smith), 411–429. American Chemical Society.

Loos, K. (ed.) (2011). *Biocatalysis in Polymer Chemistry*. Wiley.

MacDonald, J.P. and Shaver, M.P. (2015). Aluminumsalen and salan polymerization catalysts: from monomer scope to macrostructure control. In: *Green Polymer Chemistry: Biobased Materials and Biocatalysis* (ed. H.N. Cheng, R.A. Gross, and P.B. Smith), 147–167. American Chemical Society.

Maloney, A.J., Dong, C., Campbell, A.S., and Dinu, C.Z. (2015). Emerging enzyme-based technologies for wastewater treatment. In: *Green Polymer Chemistry: Biobased Materials and Biocatalysis* (ed. H.N. Cheng, R.A. Gross, and P.B. Smith), 69–85.

Meier, M.A., Metzger, J.O., and Schubert, U.S. (2007). Plant oil renewable resources as green alternatives in polymer science. *Chemical Society Reviews* 36 (11): 1788–1802.

Mohanty, A.K., Misra, M., and Drzal, L.T. (ed.) (2005). *Natural Fibers, Biopolymers, and Biocomposites*. CRC Press.

Narancic, T., Verstichel, S., Reddy Chaganti, S. et al. (2018). Biodegradable plastic blends create new possibilities for end-of-life management of plastics but they are not a panacea for plastic pollution. *Environmental Science & Technology* 52: 10441–10452.

Nduko, J.M., Sun, J., and Taguchi, S. (2015). Biosynthesis, properties, and biodegradation of lactate-based polymers. In: *Green Polymer Chemistry: Biobased Materials and Biocatalysis* (ed. H.N. Cheng, R.A. Gross, and P.B. Smith), 113–131.

Nishiyama, Y., Langan, P., and Chanzy, H. (2002). Crystal structure and hydrogen-bonding system in cellulose Iβ from synchrotron X-ray and neutron fiber diffraction. *Journal of the American Chemical Society* 124 (Suppl. 31): 9074–9082.

O'Leary Havelka, K. and Gerhardt, G.E. (2015). From biorefinery to performance technology: transforming renewable olefinic building blocks into lubricants and other high-value products. In: *Green Polymer Chemistry: Biobased Materials and Biocatalysis* (ed. H.N. Cheng, R.A. Gross, and P.B. Smith), 201–222. American Chemical Society.

Puskas, J.E., Castano, M., and Gergely, A.L. (2015). Green polymer chemistry: enzyme-catalyzed polymer functionalization. In: *Green Polymer Chemistry: Biobased Materials and Biocatalysis* (ed. H.N. Cheng, R.A. Gross, and P.B. Smith), 17–25. American Chemical Society.

Raquez, J.M., Deléglise, M., Lacrampe, M.-F., and Krawczak, P. (2010). Thermosetting (bio) materials derived from renewable resources: a critical review. *Progress in Polymer Science* 35: 487–509.

Re, G.L., Dubois, P., and Raquez, J.M. (2015). In situ metal-free synthesis of polylactide enantiomers grafted from nanoclays of high thermostability. In: *Green Polymer Chemistry: Biobased Materials and Biocatalysis* (ed. H.N. Cheng, R.A. Gross, and P.B. Smith), 287–303. American Chemical Society.

Roach, P., Eglin, D., Rohde, K., and Perry, C.C. (2007). Modern biomaterials: a review—bulk properties and implications of surface modifications. *Journal of Materials Science: Materials in Medicine* 18 (7): 1263–1277.

Roland-Holst, D., Triolo, R., Heft-Neal, S. and Bayrami, B., 2013. Bioplastics in California.

Romain, C. and Williams, C.K. (2015). Combining sustainable polymerization routes for the preparation of polyesters, polycarbonates, and copolymers. In: *Green Polymer Chemistry: Biobased Materials and Biocatalysis* (ed. H.N. Cheng, R.A. Gross, and P.B. Smith), 135–146. American Chemical Society.

Rudeekit, Y., Numnoi, J., Tajan, M. et al. (2008). Determining biodegradability of polylactic acid under different environments. *Journal of Metals, Materials and Minerals* 18: 83–87.

Samir, A., Ashour, F.H., Hakim, A.A., and Bassyouni, M. (2022). Recent advances in biodegradable polymers for sustainable applications. *Npj Materials Degradation* 6 (1): 68.

Satyanarayana, K.G., Arizaga, G.G., and Wypych, F. (2009). Biodegradable composites based on lignocellulosic fibers—an overview. *Progress in Polymer Science* 34 (9): 982–1021.

Scheller, H.V., Jensen, J.K., Sørensen, S.O. et al. (2007). Biosynthesis of pectin. *Physiologia Plantarum* 129: 283–295.

Sin, L.T. and Bee, S.T. (2019). Degradation and stability of poly(lactic acid). In: *Poly Lactic Acid: A Practical Guide for the Processing, Manufacturing, and Applications of PLA*, 2e, 97–344. Amsterdam: Elsevier.

Song, R. and Li, R. (2018). Application of new chemical materials from the perspective of environmental protection (in Chinese). *China Metal Bulletin* 12: 118–120.

Sun, X.S. (2005). Overview of plant polymers: resources, demands, and sustainability. In: *Bio-Based Polymers and Composites*, 1e, 1–14.

Tada, S., Uzawa, T., and Ito, Y. (2015). Creation of functional peptides by evolutionary engineering with bioorthogonal incorporation of artificial components. In: *Green Polymer Chemistry: Biobased Materials and Biocatalysis* (ed. H.N. Cheng, R.A. Gross, and P.B. Smith), 169–180. American Chemical Society.

Te-Wierik, G.H., Eissens, A.C., Bergsma, J. et al. (1997). A new generation starch product as excipient in pharmaceutical tablets, III: parameters affecting controlled drug release from tablets based on high surface area retrograded pregelatinized potato starch. *International Journal of Pharmaceutics* 157: 181–187.

Toyama, K., Soyama, M., Tanaka, S., and Iji, M. (2015). Development of cardanol-bonded cellulose thermoplastics: high productivity achieved in two-step heterogeneous process. *Cellulose* 22 (3): 1625–1639.

Wang, M. and Ma, Y. (2019). Green strategy in the synthesis and application of polymer materials (in Chinese). *China Chemical Trade* 13: 115.

Wu, X., Wang, S., Fu, B. et al. (2019). Socio-ecological changes on the Loess Plateau of China after Grain to Green Program. *Science of the Total Environment* 678: 565–573.

Zhou, Y.-N., Li, J.-J., Yi-Yang, W., and Luo, Z.-H. (2020). Role of external field in polymerization: mechanism and kinetics. *Chemical Reviews* 120 (5): 2950–3048. https://doi.org/10.1021/acs.chemrev.9b00744.

12

Organic Polymers and Their Role in Pharmaceutical and Chemical Industries

Raja Chakraverty[1], Dibya Das[2], and Tatini Debnath[3]

[1] *Institute of Post Graduate Medical Education and Research, Critical Care Medicine, 244, A.J.C. Bose Road, Kolkata 700 020, West Bengal, India*
[2] *JIS University, Department of Pharmaceutical Technology, 81, Nilgunj Rd, Jagarata Pally, Deshpriya Nagar, Kolkata 700109, West Bengal, India*
[3] *Maulana Abul Kalam Azad University of Technology, Department of Pharmaceutical Technology, NH-12 (Old NH-34) Simhat Haringhata, Nadia 741249, West Bengal, India*

12.1 Natural Polymers: Inorganic and Organic

Organic polymers play a significant role in life processes pertaining to formation of basic structure, for instance a plant's structural parts are made up of polymers and they are of different types of resins, lignin, cellulose, etc. Cellulose is polysaccharides composed of complex sugar molecules, while lignin which is a complex three-dimensional polymeric network forms the very basis. Wood resins and rubber are simple polymers of hydrocarbon isoprene. Other important natural polymers as biomolecules include the nucleotides, proteins, amino acids, nucleic acids – nitrogen-containing complex molecules composed of sugars, bases, and phosphoric acid. The coding of genetic information is stored inside these nucleic acids inside the cell structure. Another important natural polymer which forms the food energy source present in plants includes starches which are composed of glucose biomolecules

12.2 Synthetic Organic Polymers

(Synthetic) polymers are artificial polymers. Synthetic polymers are mainly categorized into four types: fibers, elastomers, thermoplastics, and thermosets. Thermoplastics are a type of polymers; they are malleable at a particular temperature and hence upon cooling they solidify and easily moldable. On the other hand, thermosets are rigid and once shaped they are unchangeable; therefore, they are mostly used as adhesives. Elastomer is a flexible polymer, for example rubber. Synthetic fibers are obtained from plant and animal origin by constant improvement of polymers.

Organic Polymers in Energy-Environmental Applications, First Edition.
Edited by Ramesh Oraon, Pardeep Singh, Sanchayita Rajkhowa, Sangita Agarwal, and Ravindra Pratap Singh.

Polyacrylates form the very basic structure for synthetic polymers like polystyrene and polythene. They are composed of carbon-carbon bonds; on the other hand, hetero chain polymers for example, polycarbonates, polysulfides, polyesters, polyamides, and polyurethanes, having other intrinsic elements (e.g. nitrogen, sulfur, oxygen), are included along the foundation structure. In the presence of non-covalent bonding the coordination polymers contain an extensive range of metals in their backbone structure. On account of variations in their side chain and main chains a vast range of synthetic polymers are available.

12.3 Synthetic Polymers in Everyday Use

There are certain regular synthetic polymers used in household include nylons. Teflon and polyvinyl chlorides are mostly used in pipes and in non-stick cookware, respectively. Whereas common PET (polyethylene terephthalate) bottles are usually synthetic polymer manufactured from (polyethylene terephthalate). Numerous plastic covers and kits are predominantly made of polythene, and Buna rubbers are used in the manufacture of tires. Due to environmental concerns non-biodegradable synthetic polymers pose hazards; therefore, alternative methods of obtaining polymers are more encouraged, for instance bioplastics even if they are expensive. Since the polymers are made up of hydrocarbon, they have a tendency of hydrophobicity which means they do not easily absorb water; otherwise, it would have been a misfit phenomenon.

12.4 Synthetic Polymers Types

12.4.1 Polyethylene (Low Density)

Polyethylene (low density) is a common source of organic polymers (synthetic in nature) often used in everyday households activities. Low-density polyethylene is a common thermoplastic formed out of ethylene monomer. The first polymer was produced in the year 1933 by Imperial Chemical Industries using the technique of polymerization of free radical under high pressure. LDPE (low density polyethylene) is usually recycled and depicted by number 4 owing to its recycling symbol. Despite other modern replaceable polymers in market, LDPE has retained its own importance.

12.4.2 (Polyethylene of High Density)

High-density polyethylene is obtained from petroleum which is also a polyethylene (thermoplastic). Approximately 1.75 kg of petroleum (based upon raw materials and energy) is required to manufacture a kilogram of HDPE (high-density polyethylene).

12.4.3 (Polypropylene)

Polypropylene (PP), which is usually called polypropene, is also a thermoplastic polymer widely used in various labeling and packaging, stationery, plastic parts, textiles, and sometimes in various types of reusable containers. It is also used to make automobile equipment, laboratory equipment, polymer banknotes, loudspeakers, etc. Additionally, a new set is made up from monomer of propylene which is quite durable and non-reactive to acids bases and chemical solvents.

12.4.4 (Polyvinyl Chloride)

Polyvinyl chloride (PVC) is considered third widely obtained plastic material next to polypropylene and polyethylene. PVC forms the core of foundations in construction due to its cost effectiveness and durability other than the conventional alternative like copper and iron. With the addition of plasticizers, it can reform into more soft and flexible structures phthalates being the most popular. Polyvinyl chloride has a usage in upholstery and clothing, insulation inside the electrical cable, inflatable products with wide applications where it can be used as alternative to rubber.

12.4.5 Polystyrene

Polystyrene (PS), a polymer aromatic in nature, is obtained from styrene monomer which is a liquid form of petrochemical. Popularly known plastic, it is a solid in nature and used in colorless plastic model, DVD, CD cases, disposable cutlery, and smoke detector inside multi-storied housings. Foamed form of polystyrene encompasses insulation, drinking cups, and packing materials. The biodegradation process of this polymer is slower and is quite controversial, and often found littered along seashore.

12.4.6 Nylon

It is a synthetic polymer popularly belonging to a polyamide and was produced by Wallace Carothers at DuPont's research facility. Nylon is one of the regularly and usually used polymers. The presence of amide in the main structure makes nylon hydrophilic in nature. That is the reason why nylon clothing absorbs water because nylon material forms a unique hydrogen bonding with water in contrast to the other hydrocarbon polymers.

12.4.7 Teflon

Teflon is synthetic in nature (fluoropolymer) of tetrafluoroethylene, and has various applications. It has high-molecular structure compound, solid in nature, consists totally of fluorine and carbon. It is non-hydrophobic in nature; therefore, water-containing substances interact with Teflon. Teflon is mostly used in manufacturing cookware (non-stick) coated pans and other cookware due to its less friction with other compounds. Due to its inactive nature use of PTFE it is mostly found in making containers and pipework where corrosive and reactive chemicals would not create any untoward impact. Often used as a lubricant, Teflon diminishes abrasion wear and tear, lowers energy exhaust of machinery. Teflon is also used by NASA in space project peripherals.

12.4.8 Thermoplastic (Polyurethane)

Thermoplastic polyurethane belongs to the family of compounds polyurethane plastic. Polyurethane has many utilities which includes transparency, elasticity, and resistance to abrasion, grease, and oil. Hydrophilic nature of thermoplastic polyurethane (TPU) attributes and propels it to reaction involving water. Technically, it is an elastomer which is thermoplastic in nature comprising long-chain fragmented components of soft and hard segments of copolymer.

12.5 Addition Reactions

In addition reactions, combination of two molecules creates a bigger molecule. In organic chemistry, an addition reaction means where two or more molecules combine together to form a bigger molecule. Addition reactions form chemical compounds due to multiple bonds between carbon-carbon double bonds (form alkenes) or with triple bonds (form alkynes). Molecules having carbon-carbon double bonds, like carbonyl (C=OC=O) groups or imine (C=NC=N) groups, are subjected to addition reaction. An addition reaction is basically the inverse of an elimination reaction.

Two most important types of polar addition reactions are nucleophilic addition and electrophilic addition with the presence of two (non-polar) addition reactions also present in free radical addition pertaining to other cycloadditions. In case of addition–elimination reaction, mostly an addition reaction is followed by an elimination reaction; consequently, this chemical reaction helps in the addition to carbonyl compounds in nucleophilic acyl substitution process. The best example of addition elimination reaction is hydrolysis of nitriles into respective carboxylic acids.

12.5.1 Electrophilic Addition

Most alkene chemical reaction goes with electrophilic addition reaction. For instance, in the Prins reaction a carbonyl group is an electrophile. In halogenation reaction, addition of elementary chlorine or bromine to alkenes yields dichloro- and dibromo-alkanes, respectively.

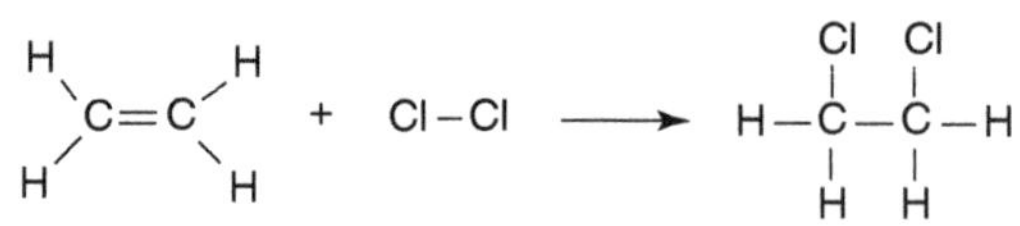

In above addition reaction of chlorine to ethane both chlorine and ethene they form a single molecule of 1,2-dichloroethane.

12.5.2 Nucleophilic Form (an Addition Reaction)

An electron pair in nucleophilic addition reaction is donated by a nucleophile to the electrophile (in the double bond single atoms are present). It is the halogen in hydro halogenation reaction. The electron pair gets donated by the halogen to double-bond carbons. This is followed by addition of hydrogen to the another carbon which was previously double bonded. When hydrohalic acids like HCl or HBr are added to alkenes they yield the corresponding haloalkanes, respectively. Chemical reaction is given as:

$$CH_3CH = CH_2 + HBr \rightarrow CH_3 - CHBr - CH_3CH_3CH$$
$$= CH_2 + HBr \rightarrow CH_3 - CHBr - CH_3$$

12.5.3 Free Radical

In case of free radical additions of any halogens to alkanes or any alkenes a radical halogen attacks an alkane in order to produce a radical; therefore, radical form of the alkane is formed. Similarly, the radical alkane has the ability to attack another compound generating a series of radicals which continue to attack eventually and this process continues.

Polymerization of alkenes is quite feasible and is a significant reaction which helps in the production of polymers in larger scale of high-throughput industrial value like plastics polypropylene and polyethylene. Polymerization proceeds either with anionic mechanism or with free radical generation mechanism, respectively.

12.5.4 Condensation Reactions

In a condensation reaction the chemical reaction occurs where one new molecule is created with a loss of another molecule (usually water is lost). Releasing a small molecule by combination of two molecules occurs here. When water is lost in the process of condensation reaction it is usually called dehydration reaction. The other molecules lost in the process could be chloride, methanol, acetic acid, and hydrogen chloride. When the chemical reaction takes place between the two separate molecules then the reaction is intermolecular. A common example of such chemical reaction is when two amino acids combine to form a peptide. Hydrolysis inverse is condensation reaction where two entity splits to give rise to hydroxide and hydrogen ions formation.

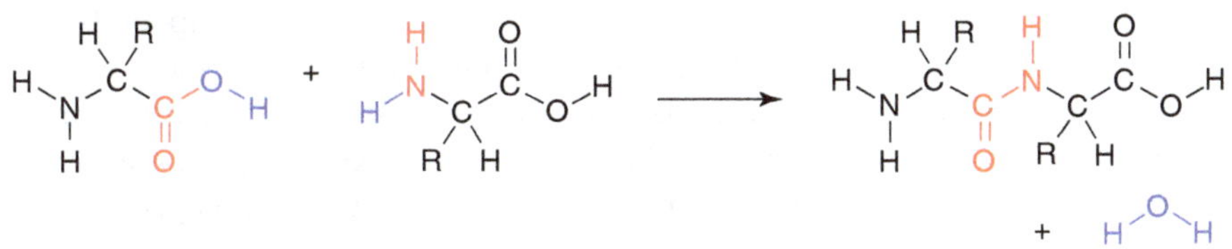

In peptide bond, for example, the condensation with dehydration of two particular amino acids forms a peptide bond in red, while water molecule is eliminated (blue).

Condensation reaction occurs in between different parts of the same molecule, which itself is known as intramolecular reaction therefore leading to ring formation. For instance, in Dieckmann condensation from a single diester molecule two ester groups react with each other to lose one small molecule of alcohol to form a product called β-ketoester. Condensation reactions follow an aldol condensation mechanism or a nucleophilic acyl substitution. Other condensations reactions are triggered by radical conditions like acyloin condensation.

12.5.5 Condensation Polymerization Reactions

The polymerization condensation reaction forms a continuous series of step occurs where monomer gets added consecutively to form longer chains. Hence such reaction is known as "condensation polymerization or step growth polymerization" which occurs through similar processes like nylon or polyesters. Nylon is a repeating unit linked by amide bonds which is a silky material often known as polyamide. It is a silky material most often used in clothes formed by consecutive repeating units linked by amide bonds, mostly attributed as polyamide. This reaction is known as homopolymerization or copolymerization having a single monomer A-B with two different end groups or co-monomers A-A and B-B. Small molecules generally get eliminated in condensation steps in contrast to poly addition reactions.

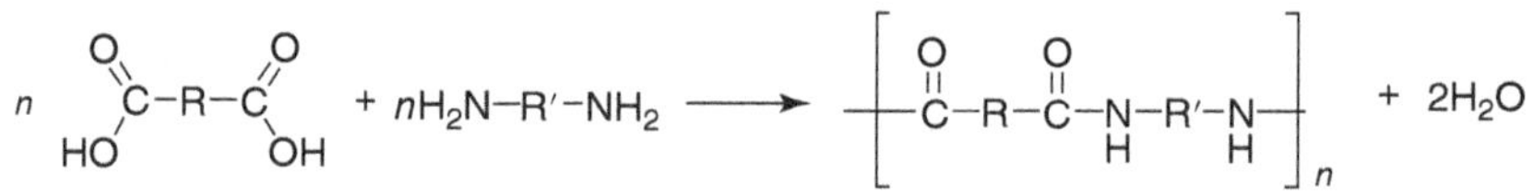

Condensation polymers require heat and are quite slower than addition polymers with very lower molecular weight. Many important polymers are formed by such reactions, for example polyester, various epoxies, and nylon; laboratory chemicals like polyphosphates and silicates; various biological modifications like terpene synthesis, polyketide synthesis, glycosylation, and phosphorylation are condensation polymerization.

12.6 Polymerization Method

12.6.1 Homogeneous Polymerization

Often the question arises on how polymers are made up from the monomers. Chemical reactions between homogeneous and heterogeneous compounds in a large scale lead to the formation of polymers. The previous system includes solution and bulk polymerizations, whereas the former encompasses dispersed system, for example emulsions, suspension with their reverse-phase counterpart, are inverse suspensions and emulsions. The reason behind the occurrence of homogeneous and bulk polymerization is because no other materials are used except monomer which acts as an initiator. When the monomer is water soluble a formation theoretically can be made of a linear water-soluble polymer. On the other hand, only oil-soluble monomers make oil-soluble polymers. Astonishingly, sometimes water-swellable polymer is prepared from olefinic water-soluble polymer. It causes excessive exothermic chemical reactions producing heat, resulting in hydrogen detachment from the polymer backbone, which propels cross-linking chemical reactions at the incomplete site. The intervention of the cross-linked polymer without using any chemical cross-linker is popularly known as popcorn polymer and the reaction is called "popcorn polymerization."Crospovidone in a solid-dose formulation that forms a super disintegrant and is a cross-linked polymer of vinyl pyrrolidone which is obtained from popcorn polymerization. Under certain circumstances when the monomer is temperature sensitive popcorn polymerization occurs without the interference of an initiator. The monomer acrylic melting point of monomer acrylic acid is glacial around 13 °C. While storing the monomer at freezing temperature polymerization stabilizer gets unequally distributed between the liquid and defrost phases. This results in generation of tremendous amount of heat leading to poor protection of the polymer. On account of the exothermic heat production alternatively the chemical reaction can be conducted in solution. Owing to the monomer stability organic solvents or water could be used as solvent or a diluent. An oil-soluble or a water-soluble polymer is obtained if monomers are water-soluble or oil-soluble, respectively. In between the neighboring monomer molecules, the solvents or the diluents reside and they decrease the interactions, respectively. Therefore, lesser amount of heat gets generated and less exothermic reactions take place with sustained effects. Polymers are usually soluble in their respective solvents, but they swell if any crosslinker is incorporated during polymerization. The cross-linker can be either oil-soluble or water-soluble. Swellability of a polymer solely depends upon simultaneous interchange ability between oil-soluble and water-soluble crosslinkers.

12.6.2 Dispersion Polymerization

This type of polymerization occurs mostly in suspensions, inverse emulsions, emulsions, and inverse suspension. This type of polymerization involves miscibility of two contradictory phases like oil and water dispersed into each other. It is divided into two phases: one is minor (dispersed) another one is critical (continuous). The active component (monomer) can be either oil-soluble or water-soluble. In a dispersed system polymerization is carried out, and the monomer gets dispersed into the continuous phase using a surfactant. Based upon the nature of the surfactant the continuous phase is selected. For the dispersion polymerization to be compatible surfactant should be in continuous-phase soluble one. Hence if the continuous phase is water, then the surfactant is hydrophilic in nature. Similarly, if the continuous phase is oil, then the surfactant nature should by more lipophilicity oriented for miscibility. Two basic factors govern the dispersion system: one is surface active agents' concentration and the other is surface tension parameter. Polymer particle size is determined by the surfactant concentrations. The system will be an inverse suspension or suspension with particle sizes ranging from 0.2 to 0.8 mm below CMC (critical micelle conc.). Above the critical micelle concentration (CMC), sizes ranging from10 to 100 μm particles are formed above the critical micelle concentration.

Sufficient surfactant is required in order to make nanosized particle. In pharmaceutical industry inverse nano emulsion is rarely used due to their stability issue of the system.

12.7 Polymers and Their Uses in Pharmaceutical and Chemical Industry

In a conventional process of tablet manufacturing, mostly polymers are used as binders to bind the excipients. Modern techniques have evolved to such extent that polymers are used drug delivery, taste masking, controlled release of drugs, increased bioavailability, and so on. Not only in solid dosage forms but in liquid dosage forms also polymers play an important role. Polymers help in maintaining viscosity, stabilize the suspension, and help in the process of granulations for the solid dosage forms. The enormous application of polymers is usually found in controlled drug release. In the biomedical field polymers are used as implants for long-lasting activities. Customized use of polymers is carried out in the pharmaceutical as well as chemical industries. Generally, the desired properties of polymers are barrier protection, adhesion, gelling, thickening, film forming, pH-dependent solubility, etc.

Polymers are broadly of two types: water-soluble and water-insoluble ethers with methyl and hydroxypropyl substitutions are generally water soluble; on the other hand, cellulose butyrate or phthalate are organic soluble. Hydrocolloid gums are used when water solubility is desired. Synthetic water-soluble polymers are also used extensively, for example polyvinyl pyrrolidine, polyethylene glycol, polyethylene glycol vinyl, polyacrylate or polymethacrylate esters, etc. Water-soluble synthetic polymers are polyacrylic, for example in cosmetic, pharmaceuticals, cationic drug immobilization, and carbopol polymer base. Polyethylene oxide is flocculant, coagulant, has high molecular weight upto few millions. Polyvinyl pyrrolidine, used in the manufacture of betadine, has less toxicity, plasma replacement process.

12.8 Polymeric Hydrogels

Polymeric hydrogels consist of self-supporting, water-swollen three-dimensional (3D) structures and possess viscoelastic network (Omidian et al. 2005). These materials have the capability of imbibing water or biological fluids up to several thousand % and swell readily without dissolving due to the presence of hydrophilic functional groups (carboxyl, amide, amino, and hydroxyl) in their polymeric chains (Titow et al. 2012). Hydrogels have existed for more than half a century, providing one of the earliest reports in the literature “by Wichterle and Lim” in 1954–1960, which described a flexible crosslinked structure of poly-hydroxyethyl methacrylate (pHEMA)-based hydrogel (Hovgaard and Brondsted 1996). Since then, hydrogel technology has diversified at a large scale, exhibiting good scope for their applications in the field of biomedical sciences (Basit 2005). Thus, the term hydrogel is self-explanatory in terms of their structure, properties, and applications. Hydrogels have fascinated materials scientists and biomedical researchers in terms of their formulations and applications (Omidian et al. 2006; Vandamme et al. 2002). The swelling properties of hydrogels can be attributed to their higher thermodynamical affinity toward solvents (Chan et al. 2002), which can be adjusted as per the requirement through the introduction of various crosslinkers via covalent bonds or physical interactions through intramolecular and intermolecular attractions (Dusseault et al. 2006). In addition, these hydrogels show a good capillary action, osmotic pressure, and water diffusion. Biopolymer (such as chitosan, starch, guar gum, alginate, etc.)-based hydrogels show desirable biodegradability and biocompatibility. For example, Vashist et al. (2012) developed biocompatible interpenetrating network using chitosan, methyl methacrylate (MMA), and oleo polyol for drug-delivery system. Later, they also reported chitosan, acrylonitrile, oleo polyol, and MWCNT-based Page 2 nanocomposite hydrogel films, which exhibits good mechanical strength, biodegradability, and biocompatibility. Noori et al. (2018) reported clay-dispersed poly(vinyl alcohol) –chitosan-honey based novel nanocomposite hydrogels for wound dressing applications. These hydrogels are biocompatible and show antibacterial activity. The concept of smart hydrogels is associated with their response toward external stimuli including temperature, pH, ionic concentration, light, magnetic fields, electrical fields, and chemicals. These features result in an enormous potential for their various advanced technological applications, for instance, the controlled drug-delivery application, which has been an area of active research. Many novel hydrogel-based delivery systems have been fabricated to fulfill the ever-increasing demands of the pharmaceutical industries. Hydrogels are classified based on their constituents, physical and chemical structures, various polymeric forms (like homo and co-polymers), blend, interpenetrating network (IPN), and composites. Their advantages, drawbacks, and applications are further discussed in proceeding chapters.

12.9 Conclusion and Future Scope

Polymorphism of the molecules can be defined as molecules having the same structure and varying in the lattice arrangement of crystal solids. Most of the drug substances are known to exhibit polymorphism. Various polymorphs exhibit chemical and physical properties, for example chemical reactivity, melting point, and solubility. The solvates/hydrates of the molecules are obtained quite often when the drug substances encapsulate solvent molecules during crystallization. These solvates/hydrates are referred to as pseudo polymorphs. The main attributes that vary due to different polymorphic forms, which need to be controlled for the pharmaceutical preparations, are solubility, thermal and kinetic stability of the molecules. These can influence pharmaceutical properties like dissolution rate and bioavailability. Thereby choosing a polymorphic form of a drug substance that has a desirable solubility, stability, and also circumventing the infringing of patents is challenging for the product development scientist. A few examples of pharmaceutical drug substances are Cefdinir, Finasteride, Sertaline HCl, Lamivudine, and Nateglinide which exhibit the polymorphism. Hence a suitable analytical technique should be adapted to identify/evaluate the different polymorphic forms to get a required polymorphic form of the drug substance. Different analytical techniques like FT-IR spectroscopy and X-Ray powder diffraction, FT-Raman spectroscopy, and thermal analysis (DSC/TGA and hot-stage microscope) have been used to evaluate the different polymorphic forms of the drug substances.

All the techniques mentioned herein are complementary to each other and to establish the identity of polymorphic forms, each technique has to be used within its scope and depends upon the nature of the molecule. Using the techniques mentioned herein and the proper interpretation of the test results, the identity of different polymorphic forms of a drug substance can be evaluated and undesirable polymorphic forms can be quantified. Nowadays Raman spectroscopy has gained popularity and is among the mostly used technique since past three decades and with an unbelievable 103 applications in the analysis of biological, pharmaceuticals, and environmental samples. This is the most important aspect of the deficient Raman scattering property of water that makes one to choose it over IR spectroscopy for analyzing the APIs in the formulation whenever the drug concentration is low. Different analytical techniques like X-ray powder diffraction (XRPD), FT-IR spectroscopy, FT-Raman spectroscopy, and thermal analysis (DSC, TGA, and hot-stage microscope) have been used to evaluate the different polymorphic forms of the drug substances. X-Ray powder diffraction, both powder diffraction (XRPD) and single-crystal XRD and X-Ray, were employed for the polymorphism detection; however, single-crystal XRD demonstrates the crystal structure of certain form and the XRPD can demonstrate both qualitative and quantitative information. It is not possible to study the purity of the powder sample using single-crystal XRD which is very important for the pharmaceutical industry.

In addition, XRPD is unable to demonstrate information at molecular level and also cannot differentiate the isomorphous solvates, for instance neotame, erythromycin and aspartame etc. In Antonio et al. (2014) finasteride polymorphic forms (forms I and II) were identified and investigated in mixtures using the analytical method Rietveld method (RM) and X-ray powder diffraction (XRPD) data. With anhydrous polymorphs of finasteride drug, the binary mixture samples were prepared and analyzed and were assessed in four diffractometers with various experimental conditions. For drug analysis XRPD is usually used and equipment represent the foremost resolutions. By subjecting to heat treatment batch of form I was obtained in pure state and later pure form II was obtained.

Generally, form A is usually found in a polymorphic mixture with form B, and hence authors 104 studied and characterized using X-ray powder diffraction (XRPD), Fourier transform infrared spectroscopy (FT-IR), and thermal analysis. It was found that XPRD method had a broad quantitative range (0–100%), good linear relationship (R^2–0.999), excellent repeatability and precision, low limits of detection (0.15% (w/w)), and quantification (0.5% (w/w)). Synchrotron X-Ray powder diffraction (Synchrotron XRPD) was used by Egusa et al. (2017) to determine the polymorphic differences of pure substance even if the substance is in low concentration in the formulation. Spiriva® inhalation powder (Boehringer Ingelheim) was used that contains nearly 0.4% (w/w) of the active pharmaceutical ingredient tiotropium bromide monohydrate in a matrix of α-lactose monohydrate as excipient. After investigation using XRPD and Synchrotron XRPD, it was known that the detection limit was around 2–5% (w/w) for tiotropium bromide by the laboratory XRPD and hence, detection of 0.4% (w/w) tiotropium bromide was impossible.

Hence, X-Ray powder diffraction is conventional, popular, and most used technique in the pharmaceutical industry to distinguish the different polymorphic forms. X-Ray powder diffraction has some limitations like preferred orientation effect, more sample quantity for analysis, 105 high limits of detection and quantification and same diffraction pattern for different hydrate forms having same unit cells. FT-IR spectroscopy is a most lucid analytical technique to identify the different polymorphic forms as it uses the spectral vibration of the molecules. It could not be used always because of its limitations like the sample preparation involved and the resolution of peaks. Lin et al. (2010) and Paul et al. (2018) studied the gabapentin polymorphic transformation involving hot-stage methods of FTIR microspectroscopy and well demonstrated the one-step polymeric interconversion induced thermally of gabapentin polymorphic forms (polymorphs I, II, III, and IV) solids in the three-step temperature ranges. FT-IR spectra for the polymorph forms I and II of finasteride at room temperature were recorded. After the background spectrum got recorded, the samples were placed on the particular crystal and data were recorded 32 scans for each sample with a respective resolution of 4 cm^{-1} were observed. The results were well correlated and the IR spectra showed characteristic bands, respectively. Paczkowska et al. (2018) studied the benzocaine's physicochemical properties and reported their analysis of the polymorphic form utilizing ($C_9H_{11}NO_2$) X-ray powder diffraction (PXRD) mediated by FT-IR spectroscopy combined with DFT calculations. The polymorphs were assembled using cryogenic grinding, ball milling, and micro milling. The polymorphic forms of benzocaine bands vibrate and lie between 500 and 2000 cm^{-1}.

In comparison to forms I and II, polymeric form III (FT-IR spectra) exhibits two bands with intense activity falling in the range 1280–1300 cm^{-1} pertaining to C–O, C–C, and C=O stretching vibrations, C–H wagging vibration at 773 cm^{-1} shows immense low intensity analogous to out-of-plane bending vibration of the O–C=O bond. Similarly, in our published research work for FT-IR spectra of lamivudine, similar peaks pattern was observed for both forms I and II and mixture containing 4% (w/w) the latter, and at the same time no characteristics peaks of form II were observed in the IR spectrum under (106) pattern. Hence, identification of form II becomes difficult due to low limit of detection mechanism by FTIR spectroscopy in 7.3 FT-Raman spectroscopy FT-Raman spectroscopy is emerging as a powerful technique to identify/quantify and distinguish the different polymorphic forms. Using laser light excitation method chemical compounds are characterized; therefore, elastically scattered lights are produced (Rayleigh) and on the other hand in elastically scattered lights are produced (Raman). The molecular bonding mechanism is provided and obtained by the spectrum regarding the sample. The sample required for the analysis of FT-Raman spectroscopy is less in quantity and reusable. The beta form of mannitol, being a polymorphic excipient, usually renders a major part in pharmaceutical products, although the alpha and delta are considerable contaminants.

Roberts et al. (2002) prepared binary mixtures of beta and delta mannitol using FT-Raman spectroscopy to assess the concentration of the beta form. The spectral results of IR indicated that 2% low levels of beta form were estimated using this innovative, harmless approach. Dandeu et al. (2006) well demonstrated the Raman spectroscopy technique as a dynamic tool for the assessable determination of polymorphic mixtures of $CaCO_3$ composition, because it exerts excellent spectral specificity to each crystal structure. Further it is well documented that Raman spectroscopy has gained much attention among various spectroscopic techniques, and is an extensively used method for discerning the polymorphic structures owing to non-invasiveness, for lesser quantity of sample requirement, and respective specificity. Hence this method involved here helps to recognize numerous physical forms like tablet, powder, cream, emollient etc. although there are exceptions where this procedure does not cause Raman scattering in case of some medicines. This methodology impelled, therefore, polymorphism of Finasteride and Lamivudine drugs were experimentally studied and compared using Fourier Transform-Raman spectroscopy in order to exemplify the beneficial aspect of this technique over the methods. A characteristic peak is observed at 697 and 107 $cm^{-1}$of lamivudine using Raman spectroscopic method of analysis, and no significant peaks were detected at 463 and 798 cm^{-1} which establishes the fact that absence of lamivudine form II at those peaks.

At certain point 6% (w/w) drug mixture was examined at various time intervals, for instance 5-, 10-, 15-, and 30-minutes using grinding method to study the observable effect of grinding on polymorphism and exhibits same peak strength at 798 cm^{-1} showing Raman shift which suggests the fact that grinding methods do not necessarily involve polymorphic conversion either for the mixture containing 6% (w/w) of form II in form I or polymorphic forms as well. Hence no observable differences were concluded by the grinding process for the polymorphic ground and the mixtures which were prepared using mortar and pestle. Following this process again form I mixture of components comprising this concentration 4% (w/w) of form II was studied further using FT-Raman spectra. Various patterns were observed at 798 cm^{-1} (Raman Band) attributing to the form II in polymorphic mixture containing 4% (w/w) of form II in form I. Distinguishable Raman shifts were observed between the two polymorphic forms of finasteride. When two polymorphs of finasteride (Antonio et al. 2014) were studied using FT-Raman analysis the results exhibited due to the Raman shift were visible and distinct between the two forms. The finasteride forms I, II tablets and placebo were recorded with Raman spectral analysis. The exceptional difference of the two polymorph forms showed striking difference in the Raman shift region between 1750 and 1550 cm^{-1} where placebo interference was not noticed in this region, and hence this region is profoundly appropriate for discerning and detection of the polymorphic forms of finasteride in tablets. The form I has a characteristic peak found at 1670 and 1695 cm^{-1}, while form II polymorph peaks were detected at 1656 and 1677 cm^{-1}.

The Raman spectral peak was detected at 1598 cm^{-1} which was usual for finasteride polymorphic forms, and the major characteristic peak of the polymorphic form I was noted at 1670 cm^{-1}, while the minor characteristic peak was found at 1695 cm^{-1}. The vital peak of polymorph forms I in form II tablets is used for detection of the same. The minor form II of polymorph was found at 1677 cm^{-1}, while the major peak was observed at 1656 cm^{-1}; concomitantly, the major peak was observed of polymorphic form II in form I tablets. In addition to the above techniques, differential scanning calorimeter (DSC) is useful to identify the different polymorphic forms as it measures the heat flow and temperature of transition of sample under the controlled temperature. We have studied 108 the differential scanning calorimeter for finasteride drug and the polymorphic form I exhibits at around 233 °C a solid-solid transition and ultimately melts at 258 °C. Therefore, the polymorph form II melts at around 258 °C. Hence the DSC values of the two forms were well equated with their published values. Thermo gravimetric analysis (TGA) is useful to identify the different hydrate forms, the loosely bound water, and the decomposition of the molecules as it measures the weight loss with respect to temperature. Hot-stage microscope is also used to find out the thermal behavior of the molecules. It can be used to confirm the thermal events like melting, evaporation, and decomposition that were obtained from DSC/TGA studies. All the techniques mentioned herein are complementary to each other and to establish the identity of polymorphic forms, each technique has to be used within its scope and depends upon the nature of the molecule. The most significant aspect is using FT-Raman spectroscopic method to detect the polymorphism in the final dosage form and is highly suggested due to their advantageous nature; they are of very less use of analytical sample, lucid sample preparation method, universally accepted sharp spectral method (Raman), and no sampling problems were observed in Raman spectroscopy unlike the cumbersome process involved in X-ray powder diffraction. It is evident that polymorphism in various pharmaceutical ingredients using FT-Raman spectroscopic methods in lower concentration level was investigated and correlated accordingly.

12.10 Future Scope

Future studies will corroborate further experiments in the analysis of the specific polymorphs which will directly help in the drug development thereby enhancing the efficiency in the manufacturing process by endorsing robust quality and stability of drug products.

References

Antonio, S.G., Paiva-Santos, C.O., and Bezzon, V.D. (2014). Limits of visual detection for finasteride polymorphs in prepared binary mixtures: analysis by X-ray powder diffraction. *Journal of Pharmaceutical Sciences* 103 (11): 3567–3575.

Basit, A.W. (2005). Advances in colonic drug delivery. *Drugs* 65: 1991–2007.

Chan, L.W., Lee, H.Y., and Heng, P.W.S. (2002). Production of alginate microspheres by internal gelation using an emulsification method. *International Journal of Pharmaceutics* 242 (1–2): 259–262.

Dandeu, A., Humbert, B., Carteret, C. et al. (2006). Raman spectroscopy–a powerful tool for the quantitative determination of the composition of polymorph mixtures: application to $CaCO_3$ polymorph mixtures. *Chemical Engineering & Technology: Industrial Chemistry-Plant Equipment-Process Engineering-Biotechnology* 29 (2): 221–225.

Dusseault, J., Tam, S.K., Ménard, M. et al. (2006). Evaluation of alginate purification methods: effect on polyphenol, endotoxin, and protein contamination. *Journal of Biomedical Materials Research Part A: An Official Journal of the Society for Biomaterials, The Japanese Society for Biomaterials, and The Australian Society for Biomaterials and the Korean Society for Biomaterials* 76 (2): 243–251.

Egusa, K., Okazaki, F., Schiewe, J. et al. (2017). Identification of polymorphic forms of active pharmaceutical ingredient in low-concentration dry powder formulations by synchrotron X-Ray powder diffraction. *Drugs in R&D* 17: 413–418.

Hovgaard, L. and Brondsted, H. (1996). Current applications of polysaccharides in colon targeting. *Critical Reviews™ in Therapeutic Drug Carrier Systems* 13 (3–4): 185–223.

Lin, Y.H., Chen, H.S., Lin, H.C. et al. (2010). Polarizer-free and fast response microlens arrays using polymer-stabilized blue phase liquid crystals. *Applied Physics Letters* 96 (11).

Noori, S., Kokabi, M., and Hassan, Z.M. (2018). Poly(vinyl alcohol)/chitosan/honey/clay responsive nanocomposite hydrogel wound dressing. *Journal of Applied Polymer Science* 135 (21): 46311.

Omidian, H., Qiu, Y., Yang, S. (2005). Hydrogels having enhanced elasticity and mechanical strength properties. US Patent 6,960,617, issued 1 November 2005.

Omidian, H., Rocca, J.G., and Park, K. (2006). Elastic, superporous hydrogel hybrids of polyacrylamide and sodium alginate. *Macromolecular Bioscience* 6 (9): 703–710.

Paczkowska, M., Wiergowska, G., Miklaszewski, A. et al. (2018). The analysis of the physicochemical properties of benzocaine polymorphs. *Molecules* 23 (7): 1737.

Paul, E.D., Garba, Z.N., and James, D.O. (2018). Chitosan: a natural polymer for health. *ATBU Journal of Science, Technology and Education* 6 (4): 183–193.

Roberts, S.N.C., Williams, A.C., Grimsey, I.M., and Booth, S.W. (2002). Quantitative analysis of mannitol polymorphs. FT-Raman spectroscopy. *Journal of Pharmaceutical and Biomedical Analysis* 28 (6): 1135–1147.

Titow, W.V. (2012). *PVC Plastics: Properties, Processing, and Applications*. Springer Science & Business Media.

Vandamme, T.F., Lenourry, A., Charrueau, C., and Chaumeil, J.C. (2002). The use of polysaccharides to target drugs to the colon. *Carbohydrate Polymers* 48 (3): 219–231.

Vashist, A., Gupta, Y.K., and Ahmad, S. (2012). Interpenetrating biopolymer network based hydrogels for an effective drug delivery system. *Carbohydrate Polymers* 87 (2): 1433–1439.

13

Current Trends in Organic Polymers and Nutraceutical Delivery

Sudeepto Biswas[1], Senjuti Bhattacharjee[2], Sukanta Roy[3], Dibya Das[4], Sourav Das[3], Panchali Dasgupta[2], Himangshu S. Maji[4], Subhasish Mondal[3], and Anirbandeep Bose[5]

[1] *West Bengal University Of Animal And Fishery Science, IGNOU, Kolkata, West Bengal 700 037, India*
[2] *Child Fund India, 22, Museum Road, Bengaluru, Karnataka 560001, India*
[3] *The Neotia University, School of Pharmacy, Sarisha, West Bengal 743368, India*
[4] *JIS University, Department of Pharmaceutical Technology, Deshpriya Nagar, Agarpara, Kolkata, West Bengal 700109, India*
[5] *Adamas University, School of Pharmaceutical Technology, Kolkata, West Bengal 700126, India*

13.1 Introduction

13.1.1 Nutraceutical and Its Global Progression

The Rig Veda says, "For a knowledgeable healer, the herbs rally together akin to an army of kings" (Saini 2016). There were many historical civilizations, such as ancient Egyptian, Greek, Roman, and others that used herbal products and plants in treating and preventing diseases.

In 1989 "The term 'nutraceutical' was coined by the Foundation for Innovation, to describe The realm of biomedical investigation that combines elements of nutrition and pharmaceuticals." This educational foundation, based in New York, was established to promote discoveries in medicine (Sharma 2017).

A nutraceutical refers to a substance that can be categorized in the form of food or as a constituent of food. It is known for its medical or health advantages, such as preventing and treating illnesses (van Boekel 2022).

The approximately calculated value of the worldwide nutraceuticals market was determined at USD 454.55 billion in 2021 and is expected to reach USD 493.06 billion in 2022 (Grand View Research 2021). It is projected that the nutraceuticals market will experience significant growth over the period from 2021 to 2030. This growth is anticipated to be at a compound annual growth rate (CAGR) of 9.0%. By the year 2030, the market is expected to achieve a value of approximately USD 991.09 billion. Functional food dominates the nutraceuticals market, representing 44.89% of its total value in 2021. It is projected to grow significantly at a robust CAGR of 9.3% until 2030.

Organic Polymers in Energy-Environmental Applications, First Edition.
Edited by Ramesh Oraon, Pardeep Singh, Sanchayita Rajkhowa, Sangita Agarwal, and Ravindra Pratap Singh.

Not only Cargill but also Incorporated as well as Nestle S. A are the prominent players available in the market. Along with them, Company of Archer Daniels Midland as well as Amway Corporation are also present in the market.

The growth of the nutraceutical market is primarily fueled by several key factors. One of these factors is the positive perception and outlook toward medical nutrition. Additionally, the rising utilization of nutraceuticals for addressing cardiovascular disorders and malnutrition also contributes to market expansion. Sports drinks are becoming increasingly popular among athletes and other individuals who are physically active and engaged in strenuous activities. The sports drinks market is propelled by the millennial generation, who possess significant purchasing power, a strong inclination toward health products, a keen interest in sports, and an increasing preference for fitness-related activities.

There is an anticipated substantial growth in the functional foods category of 11.4% between 2021 and 2030. This is primarily due to an increasing desire for healthy food choices and a shift in preference toward solutions for managing weight. The functional food sector is predicted to see rapid growth, driven by its potential to enhance immune system and serve as a preventative health supplement (Grand View Research 2021).

13.1.2 Classification of Nutraceuticals

Nutraceuticals can be classified in various ways, based on their source, mode of action, and chemical nature (Zaki 2014). For instance, natural food products are a common source of nutraceuticals, whereas pharmacological mechanisms are employed to categorize nutraceuticals according to their mode of action. Additionally, nutraceuticals may be classified as water-soluble or lipid-soluble, depending on their chemical nature. More recently, a further classification has been proposed to further characterize nutraceuticals illustrated in Figure 13.1 (Sharma 2017; Helal et al. 2019; Kalia 2005; Kokate et al. 2002).

13.1.3 Nutraceutical and Its Challenges in Delivery

Nutraceuticals possess a multitude of therapeutic benefits and play a significant role in the prevention or postponement of various age-related conditions. These conditions include osteoporosis, cancer, metabolic disorders, heart-related disorders, cataracts, and neurological disorders. Notably, nutraceuticals exhibit exceptional anti-fatigue properties, further enhancing their effectiveness in combating these ailments (González-Sarrías et al. 2013). Creating effective delivery methods for additives and dietary supplements is a complex task with various obstacles. Properly delivering specific types of food components necessitates the development of solutions that relies on a thorough understanding of the essential properties of the active core, enabling the formulation of comprehensive and effective solutions. This involves examining its interactions with the surrounding environment and its role within food throughout the stages of processing, storage, consumption. Additionally, it entails assessing how the delivery format, such as

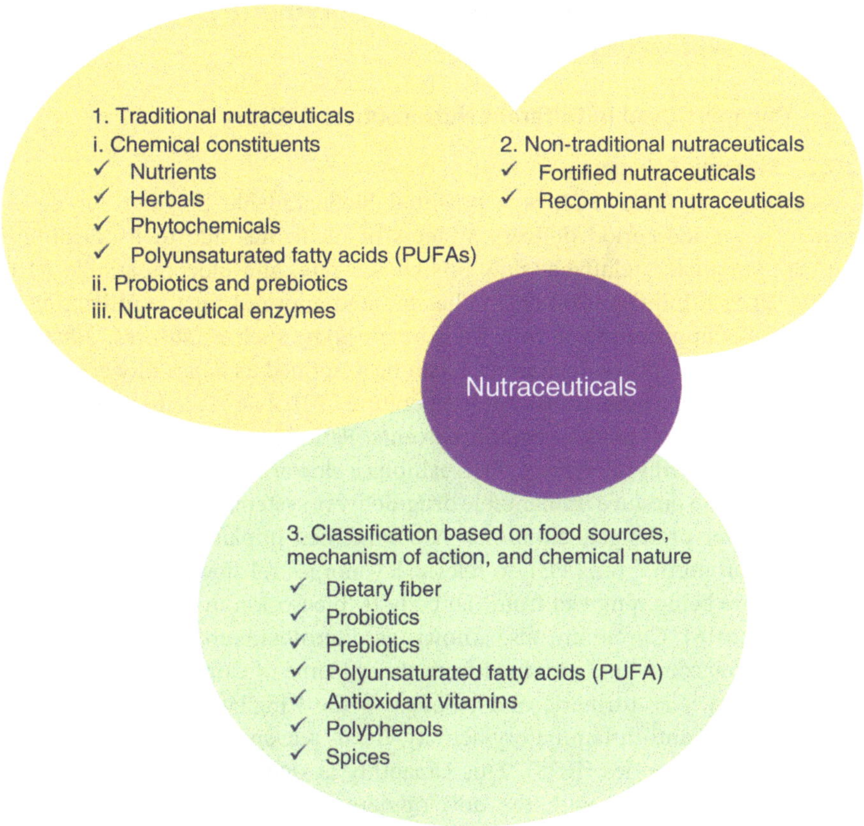

Figure 13.1 Classification of nutraceuticals.

powder, solution, or emulsion, affects its functionality (Augustin and Sanguansri 2012). Most nutraceutical products are taken orally; however, there are apprehensions regarding the development, absorption into the body, and targeted delivery of these substances. The primary challenges to delivering nutraceuticals orally are their unstable nature, poor absorption, and the difficulty of targeting them to the desired location. To overcome these obstacles, researchers have developed a variety of broad strategies tailored to each specific nutraceutical, and the most frequently used technique involves enhancing their solubility. Recently, nanosizing techniques have been employed to improve nutraceutical delivery, showing encouraging results in protecting and stabilizing these substances on the shelf and in the body, increasing their solubility and ability to dissolve, improving their permeability in the intestines, and targeting them to intracellular and subcellular organelles while also prolonging their circulation half-life. The primary focus of this chapter is on the progress and obstacles in orally delivering nutraceuticals, emphasizing the potential of polymeric strategies to overcome these challenges (Zaki 2014).

13.2 Current Advancements in the Polymeric Delivery System for Nutraceuticals

13.2.1 Polymers Used in Different Nutraceutical Delivery

13.2.1.1 Synthetic Polymers

Poly(e-caprolactone) (PCL) PCL is a potential biodegradable polymer for development of extended-period delivery systems of medicines due to its flexibility, production simplicity, relatively slow breakdown rate, and biocompatibility. The Food and Drug Administration (FDA) has granted approval for the utilization of PCL in various applications within the human body, such as sutures, adhesion barriers, and drug-delivery devices. PCL can be described as a partially crystalline polyester possessing a glass transition temperature (T_g) of 54 °C, a melting point of 55–60 °C, and high solubility in organic solvents (Patlolla et al. 2010). In particular, when compared to other polymers, PCL exhibits a slower rate of biodegradation, it has the potential to produce implantable drug-delivery systems capable of delivering medications over prolonged durations. A commercial implantable contraceptive pill called Capronorm™ uses PCL to release levonorgestrel slowly over the course of a year before being removed from the body by bioerosion in two to three years (Sinha et al. 2016). Curcumin, also known as diferuloylmethane, is a naturally occurring yellow compound obtained from the rhizome of Curcuma longa, which is typically known as turmeric and belongs to the Zingiberaceae plant family. Curcumin has an anti-inflammatory activity, it can scavenge free radicals as well as reactive oxygen species (ROS). This capability is significant as these harmful substances are associated with the development of various diseases, including cardiovascular diseases (CVD), neurological disorders, and diabetes. Bioavailability of curcumin is low due to its lipophilic nature, extensive metabolism, low water solubility, and chemical instability. Using nanocarriers could be a way to overcome the aforementioned difficulties. Nanocarriers might increase the bioavailability and therapeutic potency of curcumin. Resveratrol is an inherent polyphenolic compound that occurs naturally, is prominently present in grapes, peanuts, and red wine (Nair et al. 2010). These are some of the benefits: it can prevent or fight cancer, kill or prevent the growth of microbes, protect the nervous system, slow down the aging process, reduce inflammation, shield the heart from damage, and regulate blood sugar levels (Salehi et al. 2018). In vivo, its efficacy is compromised due to unfavorable biopharmaceutical attributes like limited solubility, rapid metabolism, and chemical instability, rendering it ineffective (Zupančič et al. 2015). PCL has its advantageous properties, including biodegradability, a slow rate of degradation, excellent drug permeability, and low toxicity, it is a viable material for designing nanocarrier-based systems. PCL is a polymer that possesses both biodegradable and biocompatible properties that has also shown sustained drug release kinetics and is used to produce nanofibers (Zupančič et al. 2015). PCL exhibits improved cellular absorption, regulated payload release, and greater stability in vivo. PCL seems to be an appropriate option for encapsulation purposes of lipophilic compounds like curcumin due to its increased hydrophobic character (Tinca et al. 2020), and

Resveratrol is also incorporated into PCL. Due to their low gingival fluid flow and gradual release, the resveratrol-loaded PCL nanofibers will supply resveratrol for treating periodontal disease for a longer period of time (Zupančič et al. 2015).

Poly(lactide) (PLA) Drug-delivery systems frequently use PLA, a synthetic biodegradable aliphatic polyester. PLA is a non-toxic, thermoplastic polymer with exceptional strength, modulus, with no harmful side effects. a study by Luzuriaga et al. (2018) developed microneedles (MNs) for transdermal drug delivery with biodegradable PLA-based.

A dietary flavonoid called quercetin, is found in many fruits, green leafy vegetables, seeds, nuts, buckwheat, flowers, barks, olive oil, broccoli, apples, and berries. It possesses anti-inflammatory properties, helps lower blood pressure, promotes vasodilation, aids in weight management, reduces high cholesterol levels, and exhibits anti-atherosclerotic effects. The substance doesn't dissolve well in hot water, but it dissolves fairly well in alcohol and fats. However, it doesn't dissolve at all in cold water. Before reaching the systemic circulation, quercetin exhibits low permeability, instability, and significant first-pass metabolism as a result of these characteristics. So, encapsulating this chemical on suitable nanocarriers is one of the successful solutions to these issues. PLA is a popular biodegradable polymeric nanoparticle used to encapsulate a variety of medicinal medicines. Biodegradability, biocompatibility, high hydrophobicity, great mechanical strength, and gradual drug release are just a few of the many qualities that PLA possesses. In order to improve the antioxidant quercetin's poor water solubility and stability, the solvent evaporation method has been utilized to encapsulate it within PLA nanoparticles. A study shows that quercetin-loaded PLA nanoparticles are an excellent prospect for the ongoing development of nanomedicines due to their high encapsulation efficiency, tiny size, and steady release (Tefas et al. 2015).

Poly(lactide-co-glycolide) (PLGA) PLGA is a copolymer composed of lactide and glycolide, which is both biocompatible and biodegradable. This polymer has obtained FDA approval for its application in the pharmaceutical field. It finds extensive utilization in the development of drug-delivery systems and is the most commonly used polymer for this purpose. The composition of the copolymer determines the shape that PLGA takes, and it is made by copolymerizing lactic and glycolic acid monomers, which are both aliphatic polyesters (Makadia and Siegel 2011).

PLGA has proven to be an effective drug-delivery medium because of its good biocompatibility, physiological degradability, and controlled release.

A formulation of biodegradable nanoparticles comprising poly(lactide-*co*-glycolide) (PLGA) and polyethylene glycol (PEG) stabilizer was employed to encapsulate curcumin with an exceptional bioavailability enhancement, achieving an efficiency of 97.5%. A study found that in vitro, encapsulated curcumin showed a significantly higher cellular absorption, suppressed the proliferation of different types of cancer cells, suppressing NF-kB activation than curcumin (Punfa et al. 2012).

One of the most significant plant flavanols is quercetin, which has a variety of pharmacological and biological applications. Quercetin's notable biological properties are hindered by its poor water solubility, limiting its efficacy in treating various disorders. Due to quercetin's extreme hydrophobicity and low cell absorption and retention, its use has been restricted. Alternative methods, including developing nanoformulations or combining phytochemicals with nanoparticles, can enhance the water solubility, bioavailability, and targeted delivery of phytochemicals to the tumor location, which improves their therapeutic potential. A study demonstrated that PLGA-quercetin nanoparticles have greater efficacy than quercetin on cervical and breast cancer cells.

Thymoquinone, a lipophilic phytochemical with a benzoquinone base, exhibits remarkable anti-inflammatory and anti-tumor characteristics. Due to its water insolubility, its use is restricted. In addition, excessive doses of thymoquinone have been given to rats, which have led to hypoactivity and breathing difficulties linked to decrease glutathione levels in the liver and kidney (Ganea et al. 2010). Similarly, Resveratrol is also more stable, water soluble when it is encapsulated in PLGA nanoparticle. It reduces lipogenesis and hepatocellular proliferation, and promotes lipolysis. To treat non-alcoholic fatty liver disease (NAFLD), RSV-PLGA-NPs can considerably improve stability, solubility, and bioactivity.

Daidzein (4′,7-dihydroxylisoflavone) is an isoflavone that is not water soluble and poor lipophilic compound, used for hypertension, coronary heart disease, and menopausal syndrome. Daidzein has a very low oral bioavailability, which reduces its therapeutic impact. Incorporation into PLGA nanoparticle daidzein, showed improved oral bioavailability (Ma et al. 2012).

Polyvinyl pyrrolidone (PVP) PVP, which is also referred to as polyvidone or povidone, is a biodegradable polymer that originates from *N*-vinylpyrrolidone monomer. This water-soluble polymer is hydrophilic and has remarkable solubility in polar solvents. PVP possesses binding properties and can stabilize suspensions and emulsions. As far as the system of delivery of drugs is concerned, the application of PVP has been found to be crucial, especially in pharmaceutical as well as biomedical sectors, where it has been employed for oral, topical, transdermal, and ocular administration. In addition, PVP has proven useful in gene delivery.

β-carotene is a molecule that lacks polarity and has a long chain of double bonds and unsaturated structures, which makes it highly vulnerable to damage from light and heat. To prevent its degradation, researchers have explored the use of electrospinning technology to encapsulate β-carotene. In order to create composite nanofibers, involving PVP as well as β-carotene, researchers have developed a finer method. These composite nanofibers are exceedingly known for their antioxidant properties. This electrospinning-based encapsulation method has the potential to protect β-carotene and preserve its beneficial properties (Kurakula and Rao 2020).

Hydroxypropyl Methylcellulose (HPMC) Those derivatives of cellulose ethers in which one or greater number of the hydroxypropyl groups available in the cellulose ring have been substituted with hydroxyl groups constitutes hydroxypropyl methylcellulose or HPMC. Besides being a hydrophilic (water soluble) polymer, HPMC is not only biodegradable but also biocompatible.

The release kinetics of pharmaceuticals are impacted by their high swellability.

The polymethoxylated flavone (PMF) nobiletin (3′,4′,5,6,7 and 8-hexamethoxy-flavone) is multihydroxylated polyphenolic molecule. They possess numerous prospective advantages in health, starting from their anti-carcinogenic along to their anti-atherosclerotic activities. PMFs are also known to possess several anti-inflammatory actions as well. When utilized as nutraceuticals in various foods, PMFs have several issues owing to their greater melting points, inadequate solubility in water, and lesser bioavailability.

HPMC contains polar and non-polar groups on its structure, allowing it to form both hydrogen and hydrophobic bonds, both within its own molecule and between other molecules. When HPMC is dissolved, its intramolecular bonds are broken, which can lead to the creation of new hydrogen and hydrophobic bonds with drugs, causing a delay in their precipitation. The consequence of HPMC on the storage and discharge of nobiletin, a hydrophobic nutraceutical, which is confined within a delivery system based on nanoemulsion, has been explored in a recent study. In order to stabilize nobiletin's supersaturated state, HPMC and thermal treatment were included thereby postponing its crystallization as well as the precipitation which is obtained from the nanoemulsions (Chen et al. 2015).

Ethyl Cellulose (EC) Additive manufacturing pharmaceuticals have been found to utilize thermoplastic polymer, like ethyl cellulose (EC), insoluble in water and derived from cellulose and contains some hydroxyl (–OH) functional groups in place of ethoxy groups (–O–CH_2–CH_3). The T_g of EC is roughly 130 °C. Depending on the molecular weight of the polymer, it possesses a melting point of approximately 180 °C. Apart from having great mechanical strength, EC has immense film-forming capacity along with flexibility as well as transparency.

A natural vitamin A derivative known as retinoic acid (RA) is regarded as a chemopreventive and/or chemotherapeutic drug. Patients who develop RA resistance compromise the effectiveness of therapy. Additionally, it has been reported that due to ongoing, recurrent oral RA administration, the concentration of this retinoid in patients' plasma has decreased further and their blood RA half-life has significantly decreased. To improve the treatment of acute promyelocytic leukemia, it is crucial to include RA into suitable carrier-based modes of delivery for its slow and continuous release. EC and PEG, two synthetic polymers, were taken into consideration while formulating microparticle compositions for RA oral administration and regulated gastrointestinal absorption.

EC-based nanoparticle was formulated to encapsulate gamma oryzanol (GO), an antioxidant which is not soluble in water. Without altering the quality of products, it has been applied several times to fortify liquid food products. On delivering through 3D-hydrogels and core–shell nanoparticles, the bioavailability of two powerfully synergistic nutraceuticals, i.e. resveratrol as well as curcumin, increases considerably.

The co-axial electrospraying method was used to create core–shell nanoparticles that included both bioactive in a zein core that was encased in a shell made of either PEG or EC (Leena et al. 2021).

13.2.1.2 Natural Polysaccharides

The attributes of natural polysaccharides such as biocompatibility, biodegradability, low immunogenicity, renewability, and ease of modification make them highly desirable as potential options for innovative drug-delivery systems. Polysaccharide-based nanocarriers are versatile tools with broader applications in different domains such as therapeutics, diagnostics along with safeguarding/delivering the bioactive substances or medications. The emphasis of research ought to be on the advancement of polysaccharide nanoparticles and the encouragement of their utilization in numerous areas, with a particular emphasis on the field of biomedicine (Qiu et al. 2022).

Chitosan Numerous polymers are found in the cell walls of fungi. Among those, chitin is a polymer which is found in the cellular walls of fungi as well as the shells, mainly the outer ones in case of numerous arthropods, such as crustaceans and insects. Its structure consists of a linear chain with units of 2-acetylamino-2-deoxy-β-D-glucopyranose interconnected to each other with the help of β-1,4 linkages. Chitosan, which is a natural cationic polysaccharide that can be easily obtained in large quantities from chitin, is not only non-toxic but also possesses biocompatible properties and are enzymatically biodegradable. Owing to the presence of the amino groups, they can be utilized as a bioadhesive, attaching to those mucosal cells which are negatively charged in order to increase the retention as well as their transit into cells. Additionally, chitosan has relatively low immunogenicity. In addition to promoting epithelial tight junction opening and enlarging paracellular routes, chitosan's positive charge facilitates this transport and aids in the delivery of medicines, particularly hydrophilic and high-molecular-weight compounds (Elieh-Ali-Komi and Hamblin 2016; Chinnaiyan 2019).

Hyaluronic Acid D-glucuronic acid along with the units of *N*-acetyl-D-glucosamine are the manufacturing blocks of hyaluronic acid, and is held strongly together by beta-1,4 and beta-1,3 glycosidic linkages. Typically, this molecule has a high molecular weight and tends to be negatively charged in the body. It has been found to be effective in treating both acute and chronic forms of gingivitis and periodontitis. Numerous research studies have revealed several beneficial properties of hyaluronic acid, such as its anti-inflammatory property along with its anti-edematous as well as anti-bacterial property thereby helping to combat microorganisms found in plaque located beneath the gums (Alipoor et al. 2022). Extensive research has been conducted on the possibility of hyaluronic acid being a carrier of the drug mainly in the case of dental care by developing micro- and nanoparticles loaded with drugs. Therefore, hyaluronic acid has emerged as a promising option for delivering drugs in dental treatment (Fallacara et al. 2018).

Dextran The polysaccharide known as dextran is composed of D-glucose, where it is linked by a-(1->6) bonds, as well as branches of D-glucose linked by a-(1->4), a-(1->3), or a-(1->2) bonds.

It is produced in the presence of sucrose by a type of bacteria or its enzymes, called the lactic acid bacteria. Dextran nanocarriers have the potential to enhance the dissolution capacity of various drugs and also improve the GI membrane permeability. Dextran, which can only be broken down by the enzyme dextranase found mainly in the large intestine, liver, spleen, or kidney, is more resistant to degradation by enzymes such as salivary amylase or malt amylase than other kinds of polysaccharides. Owing to this, the delivery systems of drugs which are based on dextrans provide a promising approach to protect drugs from degradation by stomach acids or small intestinal enzymes, improving their oral bioavailability and increasing the likelihood that they will be absorbed by the intestinal epithelium (Hu et al. 2021).

Cellulose Cellulose, a common biopolymer present in a wide range of plants, bacteria, algae, and fungi and connected together by -1,4 glycosidic bonds to form a chain of D-glucopyranosyl units, serves as a crucial structural component. It is frequently taken from the main cell walls of these species because it is the most prevalent biopolymer on the globe (Lynd et al. 2002). It is well known that chitosan has remarkable physical, mechanical, and chemical qualities in the field of medicine, as well as high biocompatibility, great biodegradability, lower toxicity, and is generally recognized as safe (GRAS). This makes it a viable option for effectively trapping and delivering drugs.

For the nanoencapsulation of drugs, numerous modified celluloses such as cellulose nanocrystals (CNCs), nanofibrillar cellulose (NFC) as well as carboxymethyl cellulose (CMC) are available. Different disease targets can benefit from the use of these celluloses. Not only one-dimensional nanofibers microparticles, two-dimensional films, been made of nanocellulose but three-dimensional hydrogels as well as aerogels have all been made using nanocellulose for several different purposes. These structures can be employed for the delivery of drug, curing of wounds, and also in tissue engineering. Materials, which are created from nanocellulose, have garnered considerable interest because they are easy to chemically modify, have a large surface area, are biocompatible, and can naturally degrade (Liu et al. 2015).

Alginic Acid Alginate is a type of natural, water-soluble polysaccharide that is obtained from the cellular walls of several varieties of brown algae. This polymer consists of blocks of mannuronic and guluronic acids, which can be arranged in a homogeneous or heterogeneous pattern. When in interaction with specific bi- as well as trivalent cations (such as Al^{3+}, Ba^{2+}, Sr^{2+}, or Ca^{2+}), the carboxylate groups present on the blocks of guluronate on the backbone of Alginate polymer, cross-link with each other and form gels. Additionally, gels can be formed through hydrogen bonding in acidic environments. Alginate is biodegradable, non-harmful, and biocompatible, making it suitable for the utilization in drug-delivery vehicles and numerous other oral insulin nanoparticles. The polymers, being flexible, can be molded into biomaterial forms, like fibers, films, sponges, and injectable gels, as well as nano, micro, and bead particles. Alginate has gained popularity recently as a material for nanomedicine delivery systems, either on its own or in combination with other polysaccharides including chitosan, pectin, along with carrageenan to treat a variety of ailments (Abka-Khajouei et al. 2022).

Pectin Natural polysaccharide pectin can be described as hydrophilic, anionic, and linear in nature. It is composed of D-galacturonic acid connected to – (1-4) linkages of glycosidic bonds (Chinnaiyan 2019). An important component that affects the properties of pectin is the degree of methylation (DM). According to DM and the number of methoxyl groups, high methoxyl pectin (HMP), having more than 50% methylated carboxyl groups, along with low methoxyl pectin (LMP), which has lesser methylation, are typically divided into two primary categories. Pectin has been thought of being a potential carrier in the process of nanoencapsulation. It has the property to prevent pharmaceuticals from proteolysis by enzymes and increases the capacity of nanoparticles thereby helping them to pass through the upper gastrointestinal system intact by acting as a protease inhibitor. In the intestine, pectinolytic enzymes from the colonic microflora break it down and absorb it as a colon-specific medication delivery. These features make pectin essential for the creation of a colon-specific controlled delivery system.

13.2.2 Different Nutraceutical Delivery Systems

13.2.2.1 Liposomes

Phospholipids are the building blocks of liposomes. They are amphiphilic molecules with a hydrophilic head along with a hydrophobic tail. In the phospholipid molecule, two fatty acid chains, each having 10–24 carbon atoms and 0–6 double bonds, make up its hydrophobic region. The majority of the hydrophilic component is composed of phosphoric acid bound to a water-soluble molecule. They position themselves in such a manner so that the polar head group faces the aqueous region and the fatty acid groups face each other when dispersed in aqueous media, thereby resulting in spherical and vesicle-like structures called the liposomes (Cheng et al. 2022; Kalepu et al. 2013).

According to the number as well as the size of the bilayers, liposomes are categorized (shown in Figure 13.2). On the basis of their size and structures, liposomes can be classified as multilamellar vesicles (MLV) as well as large and small unilamellar vesicles (LUV and SUV).

According to their chemical composition, they can also be categorized as conventional liposomes (CL), pH-sensitive, cationic, immune, and long-circulating liposomes (LCL) (Sandeep 2013). Quercetin may be able to delay the progress of several carcinogenic cells, including the cells of many organs like colorectal, prostate as well as liver. The pancreatic, breast, kidney, and lung cells along with the ovarian cancer cells are also delayed via altering a number of cellular processes. Before reaching the systemic circulation, quercetin exhibits low permeability, instability, and significant first-pass metabolism as a result of these characteristics. So, encapsulating this chemical on appropriate nanocarriers is a highly fruitful

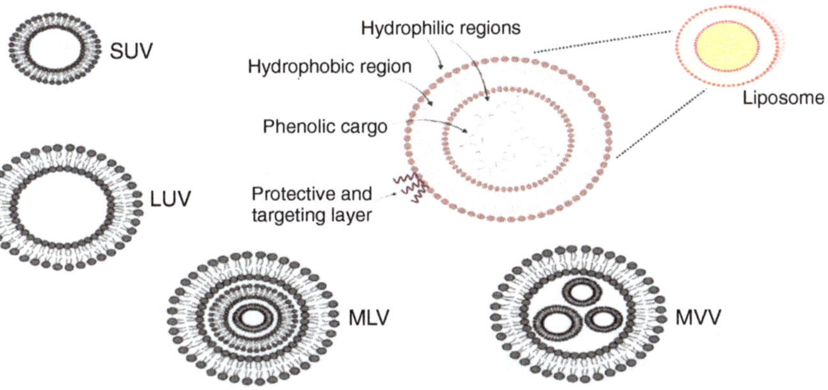

Figure 13.2 Different types of liposomes depending on their vesicular arrangement.

solution to solve these issues. In vitro, quercetin-containing liposomes in ovarian cancer models caused cell-cycle arrest, triggered apoptosis, and decreased cell growth (Vafadar et al. 2020).

Curcumin: When the pH of the solution exceeds 5, curcumin becomes unstable, and as the pH rises, its rate of disintegration accelerates significantly. The effects of curcumin-encapsulated liposomes on cancer cells are more pronounced and have a proapoptotic and growth-inhibitory effect. Numerous malignancies have been examined using curcumin liposomes (such as cancer of the lung, cervix, breast along with liver cancer, osteosarcoma). Curcumin's anticancer and pharmacological properties can be enhanced by liposomes. By enhancing the pharmacokinetic as well as the pharmacodynamic outcomes, liposomes lower the doses needed to end the tumor.

Like other polyphenols, Resveratrol possesses lower bioavailability as a result of poor absorption, fast metabolism of glucuronidated and sulfated metabolites, and subsequent excretion of these substances. Many studies showed the viability and effectiveness of using resveratrol-loaded liposomes to specifically target cancer cells. On tumor growth and volume, resveratrol-encapsulated liposomes had a potent inhibitory effect.

Similar to other forms of polyphenols, anthocyanin exhibits anti-inflammatory and anticancer properties on a range of cancer cell lines. Anthocyanins are affected by environmental elements such as pH, temperature along with light, oxygen as well as enzymes. Additionally, their bioavailability is low, thereby lowering their effectiveness. A study examined the effects of liposome-encapsulated anthocyanin from *Hibiscus sabdariffa* on the A375 melanoma cell line to improve their stability. This method thus markedly reduced the amount of melanin in the melanocytes of the cell line.

The most frequent and physiologically active catechin in green tea is epigallocatechin-3-gallate (EGCG), which is an ester of epigallocatechin as well as gallic acid. The chemopreventive effect of EGCG is well supported by research on cancers of the breast, duodenum, and prostate. It also supports cancers of the colon, skin cancer, lung along with cervical and liver cancers. Individuals rarely drink much green tea in daily life, and this substance has a poor bioavailability. Only 0.2–2% of the initial amount of ingested EGCG remains in the body. Majority of the absorption takes place mainly in the small intestine, and then it is broken down with the help of several bacteria residing in the large intestine. Integrating EGCG into liposomes is one way to enhance EGCG's effects, better stability, prolonged release, and higher intracellular concentration of EGCG when it was encapsulated in liposomes, it can also limit enzymatic breakdown (Enaru et al. 2021).

13.2.2.2 Solid-Lipid Nanoparticle

The solid-lipid nanoparticles (SLNs), which are composed of a lipid matrix and have ultimate particle sizes ranging from 50 to 1000 nm, were introduced in the early 1990s. They were created to address issues with conventional unstable colloidal systems like polymeric particles, emulsions, and lipospheres. With advantages like excellent tolerability, regulated medication release and drug targeting, the ability to incorporate hydrophilic and lipophilic drugs. Because of the reduction in particle size and the use of harmless materials, SLNs are one of the most commonly used systems today.

The protection of phytochemical compounds during digestion with enhanced absorption in the gastrointestinal tract (GIT) and precise transport to the targeted sites are all possible benefits of using SLNs to enhance bioavailability.

Many drugs' high first-pass metabolism is the primary cause of their low oral bioavailability. According to studies, SLNs may also demonstrate bio-adhesion to the GIT wall, lengthening their stay there, boosting bioavailability, and diminishing or eliminating irregular absorption.

Recent experimental bioavailability studies on SLNs showed promising outcomes. To improve the bioavailability of phytonutrients such as a substance known as quercetin, curcumin, resveratrol, and puerarin solid lipid nanoparticles have been employed as drug-delivery medium. Curcumin is poorly soluble in water, and is less bioavailable, it is rapidly broken down by hepatic enzymes system (cytochrome P450) in individuals. Curcumin can be encapsulated into SLNs, nanoemulsions, and PLGA nanoparticles, increasing its oral bioavailability by more than twofold. Male Wistar rats were given both free and nanoresveratrol orally (daily for 14 days) in the form of $5\,mg\,kg^{-1}$ body weight, and nanoresveratrol considerably elevated the rat tissue resveratrol concentrations by more than double, when compared to free resveratrol. Since puerarin has a poor solubility and a shorter half-life, its use as a cardioprotective agent is severely constrained. Oral absorption of puerarin has yet not been formulated, and repeated, high-dose intravenous administration may cause severe, acute adverse effects. Puerarin-loaded SLNs had a 300% improvement in oral bioavailability following a single intragastric dose. The research listed above strongly supports the idea that SLNs can be effectively employed as drug carriers, notably for phenolic chemicals, overcoming the disadvantages associated with oral bioavailability problems.

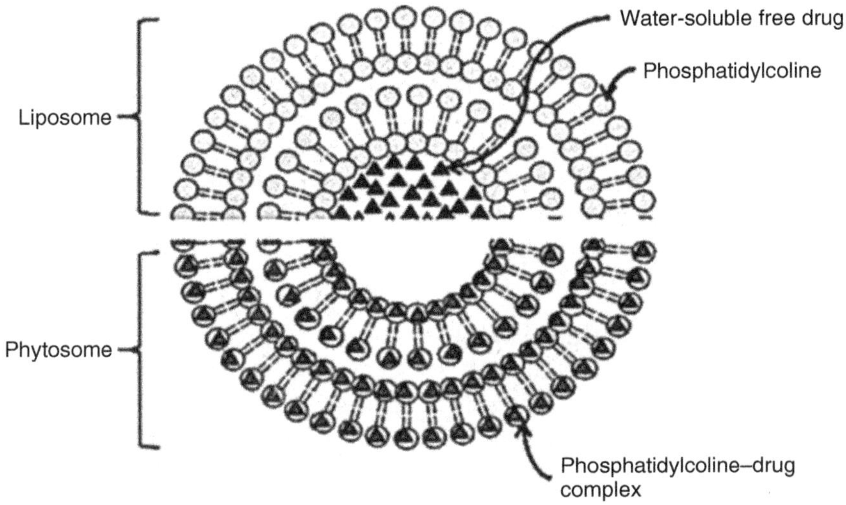

Figure 13.3 Structural difference in outer layer arrangements between the liposome and phytosome. Source: Adapted from Kumar et al. (2009).

13.2.2.3 Phytophospholipid

The term "phyto" refers to a plant, whereas "some" refers to a cell-like structure. The phytosomes method creates a tiny cell that prevents gastric fluids and intestinal microbes from destroying the plant extract or active ingredient. This is because phosphatidylcholine has gastro-protective properties. The primary distinction between liposomes and phytosomes is the fact that the active substance in liposomes is dissolved in the media contained in the cavity or folds that make up the membrane. In phytosomes, on the other hand, the therapeutic ingredient is chemically bonded to the polar head of the phospholipid (shown in Figure 13.3) (Karimi et al. 2015).

A research was conducted on silybin phytosome. This included 232 patients suffering from chronic hepatitis caused by a virus, alcohol, or narcotics. These patients were provided with silybin phytosome at a dose of 120 mg twice or three times each day for up to 120 days. The study found that the liver function of treated patients recovered to normal faster than the control group. Grape seed phytosomes, which are composed of oligomeric polyphenols complexed with phospholipids of varying molecular sizes, were prepared and found to increase the total antioxidant capacity, activate physiological antioxidant defenses in plasma, and beyond their strong antioxidant potency, they considerably protect the cardiovascular system and other organs via a network of mechanisms.

A study reported that the complex's antioxidant activity of curcumin phytosome was noticeably greater than that of pure curcumin at all tested dose levels. On the other hand, the naringenin phytosome provided stronger free radical scavenging activity than the free molecule with a longer half-life (Kumar et al. 2009).

13.2.2.4 Biphasic System

The biphasic drug-delivery system is a novel drug-delivery device for oral administration with two layers: a fast-release layer and a sustained-release layer. To create formulations for controlled and sustained release, the layered tablet concept is used, and it is intended to release the medication at two different rates. Traditional sustained-release dosage forms do not allow for a quick start to action following oral administration and delayed drug release. Due to the rapid rise in drug plasma concentration caused by the quick release of the drug from the fast layer, this layer tablet has a pharmacokinetic advantage over conventional dosage forms. This is because the sustained-release layer continues to release the drug after the quick-release layer (Preetha et al. 2015).

13.2.2.5 Self-emulsifying Delivery Systems

Pouton observed SEDS for the initial time back in 1982. He discovered the bright side of these delivery methods which include better solubility, greater bioavailability, increased absorption by the system of lymphocytes, and increased P-gp efflux. Basically, these are isotropic combinations of emulsifiers, lipids, and biologically active substances that may or may not also contain cosolvents or co-emulsifiers. These were basically classified depending upon their dispersion with a diameter of droplets up to a couple of microns, and SEDS have been broadly categorized. If the droplet size of the distribution is between 100 and 250 nm, the SEDS are commonly referred to as SMEDS, and if they are below 100 nm, they are alluded to as SNEDS.

Self-emulsifying delivery systems (SEDS), which are similar to many other lipid-based delivery systems, are one such innovative successfully implemented technology evaluated for oral delivery. To ensure their optimal distribution, a range of nutraceuticals have been created in the form of many different SEDS using diverse blends of excipients. Taking the self-emulsifying vitamin E formulation only once led to a faster and more significant absorption than the usual vitamin E formulation, which is marketed as soft gelatin capsules. The antioxidant coenzyme Q10, which has poor solubility and bioavailability, is utilized for the management of heart conditions, such as angina pectoris. Its uses are therefore constrained. Coenzyme Q10 was delivered using SEDS, and an optimized formulation made up of Labrasol (50%) and Lauroglycol (10%) was developed in addition to Myvacet 9-45 (40%). In comparison to the standard formulation, this formulation's increased bioavailability indicates that SEDS can be used to successfully improve the oral absorption of medicines that are lipophilic (Kommuru et al. 2001).

Pueraria lobata isoflavone was developed as SMEDS to make it easier to dissolve and bioavailable. When using SMEDS to dissolve the bioactive compound in vitro, a threefold increase in solubility was seen compared to when using tablets. SMEDS of puerarin, a molecule of protein having advantageous effects on the heart and against diabetes, were designed to accomplish their oral formulation. After oral dosing, puerarin was found to have a 24.8% absolute bioavailability in beagle dogs. This demonstrated SEDS's efficacy in successfully delivering this protein molecule via oral route. Another study recorded, an in vivo investigation, through passive diffusion across lipid membranes, curcumin from SMEDS was absorbed (Zhongfa et al. 2012).

13.3 Scope of Developing New Polymeric Nutraceutical Delivery

13.3.1 Enhanced Medication Delivery Method Brought About by the Presence of Chitosan on Nanoparticle Surfaces

Chitosan possesses the capacity to foster mucoadhesion, open tight junctions, and induce positive or negative interactions. Due to these properties, researchers have tried coating lipid as well as polymeric-based nanoparticles with chitosans in order to enhance drug delivery and nutraceuticals. When the nanoparticles are coated with chitosan, an improvement is seen in their physiochemical stability. Chitosan also enhances the interactions between cells and tissues, controlled release of drugs, and bioavailability of and efficiency of active substances (Elkomy et al. 2022). This can provide very promising results in the upcoming future and will be effective in ocular, oral, nasal, mucosal, pulmonary, and gene drug delivery (shown in Figure 13.4) (Garg et al. 2019).

13.3.2 Addition of Protein Nanohydrogels in the Drug-Delivery System

A distinctive three-dimensional interconnected network of polymers is seen in hydrogels, allowing them to absorb more water and bodily fluids without dissolving. Protein nanohydrogels function as a storage space for drugs and other bioactive substances. Bioactive chemical release patterns can be modified by manipulating various factors such as the type and concentration of protein, interaction with other

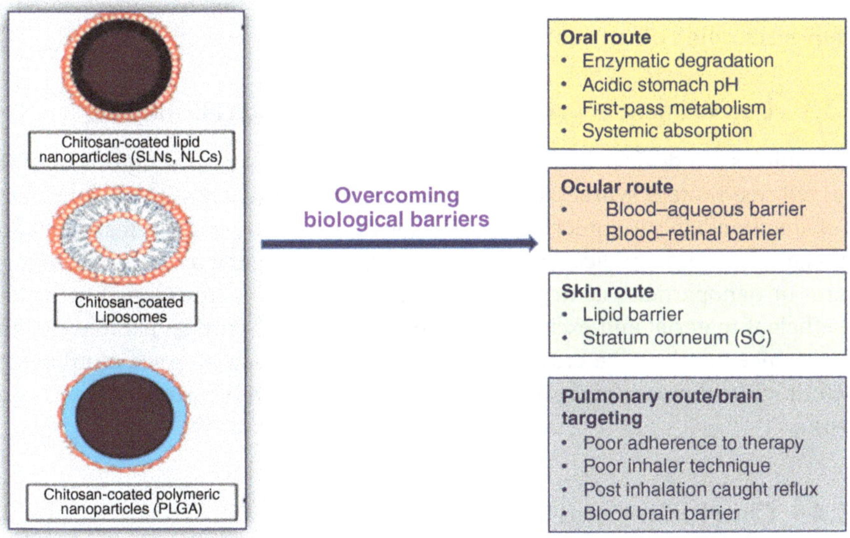

Figure 13.4 The figure shows the presence of chitosan on the surface of nanoparticles overcomes the biological barriers. Source: Adapted from Elkomy et al. (2022).

polymer systems, pH, ionic concentration, and other physicochemical influences from outside (Chander et al. 2021).

Nanogels have been utilized in the management of various illnesses, including carcinoma. In an investigation, Where recombinant murine interleukin-12 (IL-12) was subcutaneously injected into animals with subcutaneous fibrosarcoma combined with the nanogel of cholesteryl pullulan. The incorporation of active substances such as cells, peptides, biomolecules, and drugs into protein hydrogels shields them from the hostile environment. The use of nanogels that are loaded with active substances can alter the thixotropic property of the body at physiological pH and body temperature, leading to a prolonged increase in IL-12 levels and the inhibition of tumor growth. Protein hydrogels have numerous potential applications and should be investigated in various fields, including the delivery of bioactive compounds (Chander et al. 2021; Nandi et al. 2020).

13.3.3 Supersaturating Drug-Delivery System

A supersaturating medication delivery system is a promising idea for improving oral bioavailability. To achieve intraluminal concentrations above the drug's saturation solubility, the drug is in a high-energy form in SDSS (Brouwers et al. 2009).

To amplify the capacity of less water-soluble medication to be absorbed orally, one can make use of supersaturated drug-delivery systems (SDDSs). Amorphous solid dispersions (ASDs), nanoparticulate systems, and lipid-based delivery systems (PWSDs) are a few examples of these systems. When a drug is supersaturated for a significant duration, its concentration in the gastrointestinal tract remains higher than its equilibrium solubility, permitting absorption through the intestinal barrier. The "spring" and "parachute" phenomena support SDDSs that have reached a state of super saturation (Sharma et al. 2022).

13.3.4 Development of Drug-Delivery System Through Microfluidic

Microfluidics are devices that are utilized for the generation of drug carrier particles, cell-free protein synthesis, and drug screening approaches. It has attracted a lot of attention lately. In order to create viable drug-delivery systems for micro- and nanoparticles, microfluidics has a lot of promise. Utilizing microfluidics to produce micro- or nanoparticles as drug-delivery vehicles has numerous benefits, including efficient material and excipient use, fine control over particle parameters, and performance metrics. The creation of Janus particles is one instance of utilizing a microfluidic technology to create a smart medication delivery system (Damiati et al. 2018).

13.4 Conclusion and Future Prospects

The use of polymeric materials for nutraceutical delivery is still in its infancy, but the potential for these materials to improve the efficacy of nutraceuticals is great.

Polymeric materials have emerged as a promising class of drug-delivery vehicles due to their ability to protect and deliver drugs to specific sites in the body. Nutraceuticals, which are often poorly absorbed and have poor bioavailability, are a particularly attractive application for polymeric drug delivery. There are many different types of polymers that can be used to deliver nutraceuticals, and each has its own advantages and disadvantages. The ideal delivery system would be one that is safe, effective, and affordable. In addition, polymeric materials can be designed to target specific organs or tissues, and this is essential to the administration of nutraceuticals with a variety of therapeutic advantages.

However, there is still much research to be done in this area before such a system can be developed. In the meantime, polymeric nutraceutical delivery systems show promise as a way to improve the health of people around the world.

References

Abka-Khajouei, R., Tounsi, L., Shahabi, N. et al. (2022). Structures, properties and applications of alginates. *Marine Drugs* 20 (6): 364.

Alipoor, R., Ayan, M., Hamblin, M.R. et al. (2022). Hyaluronic acid-based nanomaterials as a new approach to the treatment and prevention of bacterial infections. *Frontiers in Bioengineering and Biotechnology* 10: 913912.

Augustin, A.M. and Sanguansri, L. (2012). *Challenges in Developing Delivery Systems for Food Additives, Nutraceuticals and Dietary Supplements*. Cambridge: Woodhead Publishing.

van Boekel, M.A.J.S. (2022). Nutraceuticals: possible future ingredients and food safety aspects. In: *Ensuring Global Food Safety*, 2e (ed. C.E. Boisrobert, A. Stjepanovic, S. Oh, and H.L.M. Lelieveld), 379–382. Elsevier.

Brouwers, J., Brewster, M.E., and Augustijns, P. (2009). Supersaturating drug delivery systems: the answer to solubility-limited oral bioavailability? *Journal of Pharmaceutical Sciences* 98 (8): 2549–2572.

Chander, S., Kulkarni, G.T., Dhiman, N., and Kharkwal, H. (2021). Protein-based nanohydrogels for bioactive delivery. *Frontiers in Chemistry* 9: 573748.

Chen, H., An, Y., Yan, X. et al. (2015). Designing selfnanoemulsifying delivery systems to enhance bioaccessibility of hydrophobic bioactives (nobiletin): influence of hydroxypropyl methylcellulose and thermal processing. *Food Hydrocolloids* 51: 395–404.

Cheng, X., Yan, H., Pang, S. et al. (2022). Liposomes as multifunctional nano-carriers for medicinal natural products. *Frontiers in Chemistry* 10: 963004.

Chinnaiyan, S.K., Deivasigamani, K., Gadela, V.R (2019). Combined synergetic potential of metformin loaded pectin-chitosan biohybrids nanoparticle for NIDDM. *Int J Biol Macromol* 125: 278–289. doi:10.1016/j.ijbiomac.2018.z2.009

Damiati, S., Kompella, U.B., Damiati, S.A., and Kodzius, R. (2018). Microfluidic devices for drug delivery systems and drug screening. *Genes* 9 (2): 103.

Elieh-Ali-Komi, D. and Hamblin, M.R. (2016). Chitin and chitosan: production and application of versatile biomedical nanomaterials. *International Journal of Advanced Research* 4 (3): 411–427.

Elkomy, M.H., Ali, A.A., and Eid, H.M. (2022). Chitosan on the surface of nanoparticles for enhanced drug delivery: a comprehensive review. *Journal of Controlled Release: Official Journal of the Controlled Release Society* 351: 923–940.

Enaru, B., Socaci, S., Farcas, A. et al. (2021). Novel delivery systems of polyphenols and their potential health benefits. *Pharmaceuticals (Basel, Switzerland)* 14 (10): 946.

Fallacara, A., Baldini, E., Manfredini, S., and Vertuani, S. (2018). Hyaluronic acid in the third millennium. *Polymers* 10 (7): 701.

Ganea, M.G., Fakayode, O.S., Losso, J.N. et al. (2010). Delivery of phytochemical thymoquinone using molecular micelle modified poly(D, L lactide-co-glycolide) (PLGA) nanoparticles. *Nanotechnology* 21 (28): 285104.

Garg, U., Chauhan, S., Nagaich, U., and Jain, N. (2019). Current advances in chitosan nanoparticles based drug delivery and targeting. *Advanced Pharmaceutical Bulletin* 9 (2): 195–204.

González-Sarrías, A., Larrosa, M., García-Conesa, M.T. et al. (2013). Nutraceuticals for older people: facts, fictions and gaps in knowledge. *Maturitas* 75 (4): 313–334.

Grand View Research. (2021). Transdermal drug delivery system market size report, 2030. https://www.grandviewresearch.com/industry-analysis/transdermal-drug-delivery-systems-industry (accessed 1 July 2023).

Helal, N.A., Eassa, H.A., Amer, A.M. et al. (2019). Nutraceuticals' novel formulations: the good, the bad, the unknown and patents involved. *Recent Patents on Drug Delivery & Formulation* 13 (2): 105–156.

Hu, Q., Lu, Y., and Luo, Y. (2021). Recent advances in dextran-based drug delivery systems: from fabrication strategies to applications. *Carbohydrate Polymers* 264: 117999.

Kalepu, S., Kumar, K.T.S., Betha, S., and Verma, M.M. (2013). Liposomal drug delivery system - a comprehensive review. *International Journal of Drug Development & Research* 5: 62–75.

Kalia, A.N. (2005). *Textbook of Industrial Pharmacognosy*. New Delhi: CBS Publishers & Distributors.

Karimi, N., Ghanbarzadeh, B., Hamishehkar, H. et al. (2015). Phytosome as novel delivery system for nutraceutical materials. *International Journal of Current Microbiology and Applied Sciences* 4 (6): 152–159.

Kokate, C.K., Purohit, A.P., and Gokhale, S.B. (2002). *Nutraceutical and Cosmaceutical*. Pune: Nirali Prakashan.

Kommuru, T.R., Gurley, B., Khan, M.A., and Reddy, I.K. (2001). Self-emulsifying drug delivery systems (SEDDS) of coenzyme Q10: formulation development and bioavailability assessment. *International Journal of Pharmaceutics* 212 (2): 233–246.

Kumar, P., Yadav, S.S., Agarwal, A., and Kumar, N. (2009). Phytosomes: a noval phyto-phospholipid carriers: an overview. *International Journal of Pharmaceutical Research and Development* 2 (6): 1–7.

Kurakula, M. and Rao, G.S.N.K. (2020). Pharmaceutical assessment of polyvinylpyrrolidone (PVP): as excipient from conventional to controlled delivery systems with a spotlight on COVID-19 inhibition. *Journal of Drug Delivery Science and Technology* 60: 102046.

Leena, M.M., Anukiruthika, T., Moses, J.A., and Chinnaswamy, A. (2021). Co-delivery of curcumin and resveratrol through electrosprayed core-shell nanoparticles in 3D printed constructs. *Food Hydrocolloids* 124: 107200.

Liu, D.Y., Sui, G.X., and Bhattacharyya, D. (2015). Properties and characterization of electrically conductive nanocellulose-based composite films. In: *Fillers and Reinforcements for Advanced Nanocomposites* (ed. Y. Dong, R. Umer, and A.K.-T. Lau), 3–25. Woodhead Publishing.

Luzuriaga, M., Berry, D., Reagan, J. et al. (2018). Biodegradable 3D printed polymer microneedles for transdermal drug delivery. *Lab on a Chip* 18: 1223–1230.

Lynd, L.R., Weimer, P.J., van Zyl, W.H., and Pretorius, I.S. (2002). Microbial cellulose utilization: fundamentals and biotechnology. *Microbiology and Molecular Biology Reviews: MMBR* 66 (3): 506–577.

Ma, Y., Zhao, X., Li, J., and Shen, Q. (2012). The comparison of different daidzein-PLGA nanoparticles in increasing its oral bioavailability. *International Journal of Nanomedicine* 7: 559–570.

Makadia, H.K. and Siegel, S.J. (2011). Poly lactic-co-glycolic acid (PLGA) as biodegradable controlled drug delivery carrier. *Polymers* 3 (3): 1377–1397.

Nair, H.B., Sung, B., Yadav, V.R. et al. (2010). Delivery of antiinflammatory nutraceuticals by nanoparticles for the prevention and treatment of cancer. *Biochemical Pharmacology* 80 (12): 1833–1843.

Nandi, A.K., Mondal, S., and Das, S. (2020). A review on recent advances in polymer and peptide hydrogels. *Soft Matter* 16: 1404–1454.

Patlolla, A., Collins, G., and Arinzeh, T.L. (2010). Solvent-dependent properties of electrospun fibrous composites for bone tissue regeneration. *Acta Biomaterialia* 6 (1): 90–101.

Preetha, P., Rao, A.S., and Pushpalatha, P. (2015). Biphasic drug delivery in controlled release formulations – a review. *International Journal of Pharmacy and Technology* 6 (4): 3046–3060.

Punfa, W., Yodkeeree, S., Pitchakarn, P. et al. (2012). Enhancement of cellular uptake and cytotoxicity of curcumin-loaded PLGA nanoparticles by conjugation with anti-P-glycoprotein in drug resistance cancer cells. *Acta Pharmacologica Sinica* 33: 823–831.

Qiu, A., Wang, Y., Zhang, G., and Wang, H. (2022). Natural polysaccharide-based nanodrug delivery systems for treatment of diabetes. *Polymers* 14 (15): 3217.

Saini, A. (2016). Physicians of ancient India. *Journal of Family Medicine and Primary Care* 5 (2): 254–258.

Salehi, B., Mishra, A.P., Nigam, M. et al. (2018). Resveratrol: a double-edged sword in health benefits. *Biomedicines* 6 (3): 91.

Sandeep, K., Sunil, K.T., Sudheer, B., Mohan, M. (2013). Liposomal drug delivery system - A Comprehensive Review. *International Journal of Drug Development & Research* 5: 62–75.

Sharma, R. (2017). Role of nutraceuticals in health care: a review. *International Journal of Green Pharmacy* 11 (3): S385–S394.

Sharma, A., Arora, K., Mohapatra, H. et al. (2022). Supersaturation-based drug delivery systems: strategy for bioavailability enhancement of poorly water-soluble drugs. *Molecules (Basel, Switzerland)* 27 (9): 2969.

Sinha, V., Sharma, S., Walia, S. et al. (2016). Current polyesteric systems for advanced drug delivery. In: *Nanoarchitectonics for Smart Delivery and Drug Targeting*, 143–168. *Elsevier*.

Tefas, L.R., Tomuţă, I., Achim, M., and Vlase, L. (2015). Development and optimization of quercetin-loaded PLGA nanoparticles by experimental design. *Clujul Medical* 88 (2): 214–223.

Tinca, A.L., Lurian, S., Rus, L.M. et al. (2020). Development and characterization of curcumin-loaded polycaprolactone nanoparticles by design of experiments. *Pharmaceutical Practice* 13: 127–138.

Vafadar, A., Shabaninejad, Z., Movahedpour, A. et al. (2020). Quercetin and cancer: new insights into its therapeutic effects on ovarian cancer cells. *Cell & Bioscience* 10: 32.

Zaki, N.M. (2014). Progress and problems in nutraceuticals delivery. *Journal of Bioequivalence & Bioavailability* 6: 75–77.

Zhongfa, L., Chiu, M., Wang, J. et al. (2012). Enhancement of curcumin oral absorption and pharmacokinetics of curcuminoids and curcumin metabolites in mice. *Cancer Chemotherapy and Pharmacology* 69 (3): 679–689.

Zupančič, S., Baumgartner, S., Lavrič, Z. et al. (2015). Local delivery of resveratrol using polycaprolactone nanofibers for treatment of periodontal disease. *Journal of Drug Delivery Science and Technology* 30: 408–416.

14

Conducting Organic Polymers Used in Biosensors for Diagnostic and Pharmaceutical Applications

Ritambhara Dash[1], Aritri Das[2], and Arnab S. Bhattacharyya[1,3]

[1] *Central University of Jharkhand, Department of Metallurgical and Materials Engineering, Cheri-Manatu, Ranchi, Jharkhand, 835205, India*
[2] *University of North Bengal, Raja Rammohunpur, Raja Rammohunpur, Darjeeling, West Bengal, 734013, India*
[3] *Central University of Jharkhand, Centre of Excellence in Green and Efficient Energy Technology (CoE-GEET), Cheri-Manatu, Ranchi, Jharkhand, 835205, India*

14.1 Introduction

Conducting polymers (CPs) are becoming popular as there is an increasing need for materials that are high performing, lightweight, and inexpensive for electrical and electronic product components and devices. CPs are less expensive to produce, lighter than other materials used in current technologies, and take up less space. They are also adaptable and indestructible. These qualities make them ideal for use in satellites, spacecraft, and robotic or human exploration. CPs are employed in molecular electronics, solid-state batteries, light-emitting diodes, and rechargeable batteries. In rechargeable aluminum batteries that operate at room temperature. CPs like *polypyrrole* (PPy) and *polythiophene* (PTh) serve as active components for the positive-electrode lithium-ion batteries as hybrid composites (Mishra 2018). Due to their lightweight nature, which is crucial for aerospace applications, corrosion resistance, ease of processing, and tunable conductivities, conducting polymers, especially polyaniline (PANI) and PPy, are useful for various application including electromagnetic Interference (EMI) shielding as well. There are numerous possible uses for CPs in the microelectronics sector also, both in the pristine conducting state and doped conducting state (Hudak 2013; Angelopoulos 2001; Sengodu and Deshmukh 2015; Wu et al. 2017; Wang and Jing 2005). CP nanostructures have also drawn more and more attention in both fundamental research and diverse application sectors. They exhibit novel features related to their nanoscale size and have significantly increased device performance. The conducting polymer nanostructures have emerged as a fast-expanding topic of research (Huang et al. 2003; Tseng et al. 2005; Long et al. 2005, 2010; Pan et al. 2006; Forzani et al. 2004; Chiou et al. 2007; Verma et al. 2023).

Organic Polymers in Energy-Environmental Applications, First Edition.
Edited by Ramesh Oraon, Pardeep Singh, Sanchayita Rajkhowa, Sangita Agarwal, and Ravindra Pratap Singh.

14.2 Processibility and Sensitivity Issues

Most conductive polymers undergo oxidative doping. Due to their salt-like nature (polymer salt), these materials are less soluble in organic solvents and water, which reduces their processability. The charged organic backbone frequently becomes unstable in the presence of air moisture. Organic conductors can be more expensive than metals and need a multi-step synthesis process. Many polymers' poor processability necessitates the addition of solubilizing agents or substituents, which can make the synthesis even more difficult. The non-processibility of the solvent of conducting polymers is one of their main issues. Most CPs degrade before melting because they are insoluble in ordinary solvents. Additionally, they are unstable in the environment and have limited mechanical strength. These issues have been successfully resolved by numerous researchers, and they are now creating polymer composites and blends that can demonstrate improved stability and mechanical strength. The sensitivity of the polymers changes in fabrication, effects that depend on time and temperature, and nonlinearity of their associated properties present difficulties. CPs must be able to carry out their tasks in dynamic settings with a wide variety of temperatures and frequencies (MacDiarmid and Epstein 1994; Waware and Umare 2003; Ismail et al. 2008; Pant et al. 2006; Mzenda et al. 2002).

14.3 Side Chain and π-Electron Backbone

CPs in their pristine state are intrinsic semiconductors whose band gap is influenced by both the type of substituents connected to the main chain and the chemical makeup of the conjugated backbone. Thus, depending on the functionalization, the optical and electrical properties can be changed. The side chains are significant in the case of CPs as they are the sites where dopant ions get attached affecting the optical, mechanical, and electronic properties. Bromine-doped poly(acetylene) has been found to enhance conductivity by many folds (Inal et al. 2018; Chen et al. 2021; Wu et al. 2019). Being lightweight, flexible, and having ease of production, portable devices can be made of CPs. The alternate single and double bonds provide mechanical stability to the structure. The microstructure and surface morphology of CP used in the form of electrodes can be modified to increase conductivity and signal sensitivity (Russo et al. 2021; Tao et al. 2020). CPs are found in crystalline and partially amorphous states. As the delocalization of the π-electrons depends on the disorder of the system, both localized and delocalized conditions of the π-electrons exist in CPs. The π-electron backbone of conducting polymers is responsible for their electrical conductivity, due to which CPs are widely used in sensing as well as in bioelectronics. They provide routes for electrons or holes to migrate along a polymer chain and jump from chain to chain. Materials with directional conductivity are produced by the quasi-infinite system, which spans numerous recurrent monomer units (Namsheer and Rout 2021; Tadesse et al. 2024).

The conjugated polymer's extended system is extremely vulnerable to chemical or electrical oxidation or reduction. Controlling this oxidation and reduction allows for extremely precise systematic control of the optoelectronic properties of the polymer, which are altered by these processes. The polymer may either lose an electron from one of the bands during the oxidation process or localize the charge during a brief period of the chain. The polymer expands some energy as a result of the local distortion brought on by the localization of the charge due to a change in geometry. The creation of this local geometry, however, lowers the polymer chain's ionization energy and raises its electron affinity, making it more suited to accept the newly produced charges (Le et al. 2017).

14.4 Polarons, Bipolarons, and Solitons

When PPy is oxidized, *polarons and bipolarons* are formed. Polarons are negatively charged particles that are formed due to the interaction of excess electrons with surrounding ionic vibrations. They are heavier than electrons. Two simultaneous polarons are called a *bipolaron* and mainly occurs in CPs. As the ground state in such a case is twofold degenerate, the charged cations are not connected to each other and can freely dissociate along the chain. As a result, the charged defects are independent of one another and can form domain walls separating two phases with opposite orientations and identical energy. These are known as solitons which are quasi-one-dimensional quantized wave packets that have only one crest and can travel long distances. The generation of *Solitons* leads to the formation of new localized electronic states in the energy gap. At high doping levels, charged Solitons interact with one another to produce a *Soliton* band, which can eventually merge with the band edges to achieve real metallic conductivity. Solitons in polyaniline are delocalized over around 12 hydrocarbon units, with the highest charge density near to the dopant counterion (Karlsson 2014 and Heeger 2001).

An electron is removed from the backbone system during oxidative doping of PPy, resulting in a free radical and a spinless positive charge. The charge and radical's local resonance couples the radical and the cation. In this scenario, a sequence of quinoid-like rings is utilized. This causes a greater energy distortion than the rest of the chain. These flaws require a large amount of energy to generate and separate. The number of quinoid-like rings that can connect these two bound species is reduced as a result. The lattice distortion in PPy is thought to extend over four pyrrole rings. A combination of a radical (anion or cation) and charged state, gives rise to Polaron. As a result, a new localized electronic state forms in the gap, with the lower energy states inhabited by a single unpaired electron.

The polaron state of PPy is positioned symmetrically at approximately 0.5 eV from the band boundaries. Further oxidation removes the free radical from the polaron, resulting in the development of a new spinless defect called a bipolaron. This requires less energy than creating two distinct polarons (Samwang et al. 2024). Two polarons can combine to generate a bipolaron at higher doping levels. As a result, at increasing doping levels, polarons are replaced by bipolarons (Celiberti et al. 2024). The bipolarons in PPy are symmetrically arranged, with a band gap of 0.75 eV. Repeated doping eventually leads to the formation of continuous bipolaron

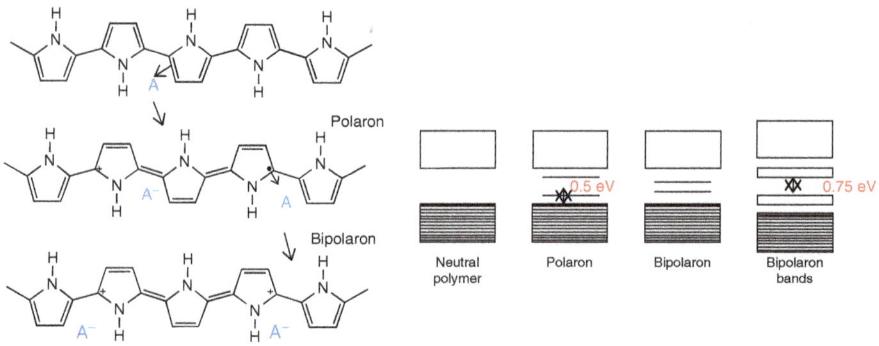

Figure 14.1 Conduction in polypyrrole (PPy). Source: Reproduced under CC-BY License (Mishra 2018).

bands. As additional bipolarons emerge at the expense of the band edges, their band gap widens. In a heavily doped polymer, the upper and lower bipolaron bands may overlap with the conduction and valence bands, resulting in partially filled bands and metallic-like conductivity (Figure 14.1) (Mishra 2018).

14.5 Doping in CPs

An organic polymer that is either an insulator or a semiconductor with low conductivity is changed into CP through the doping process. The electronic, electrical, magnetic, optical, and structural characteristics of the polymer are drastically altered by doping. The breakdown of the polymer backbone is minimal or nonexistent when doping is reversed to generate the original polymer (Salzmann et al. 2016). The first doping agents used were halogens (Br_2 and I_2), $FeCl_3$, and arsenic pentafluoride (AsF_5) for doping polyacetylene. Other popular doping agents involve strongly or mildly oxidizing acids like $HClO_4$ and H_2SO_4 (Mishra 2018).

Doping and undoping operations, which involve dopant counter-ions that stabilize the doped state, can be performed chemically or electrochemically. It is also possible to perform transitory doping using no dopant ions. A conductivity anywhere between that of the non-doped (insulating or semiconducting) and that of the completely doped (very conducting) version of the polymer can be easily achieved by controllably altering the doping level (Le et al. 2017). Conducting blends comprising a (doped) conducting polymer and a normal polymer (insulator) can be created, and their conductivity can be changed by adjusting the relative quantities of each polymer. This allows for the best qualities of each type of polymer to be optimized.

The poly(*p*-phenylene sulfide, PPS) was the first nonrigid, partially carbon-backbone connected polymer that became highly conducting after doping. PPS has been reported to have an orthorhombic unit cell with the space group *Pbcn* in its crystal structure. Diffraction investigations revealed that the adjacent phenyl rings were nearly perpendicular. The bandwidth is only 0.2 eV with a high ionization potential of 6.3 eV (Rabolt et al. 1980; Chance et al. 1980; Tabor et al. 1971; Edmonds and Hill 1967). The p-type doping involving electron donation from CP and n-type doping involving electron capture can be used to control the electrical conductivity of CPs. CPs like polyacetylene (PA) and polypyrrole (PPy) have been used to this effect in many electronic devices. Different polymerization methods can be used to create PPy with various morphologies useful for tissue engineering (Albaaj et al. 2017; Chauhan et al. 2017; Khan et al. 2020). The efficiency of the CPs is enhanced by blending them with nanomaterials like graphene and iridium oxide to form composites (Yang et al. 2014; Hassanein et al. 2017; Mao et al. 2018; Noh et al. 2014; Prajapati and Kandasubramanian 2019; Wongkaew et al. 2019; Zare et al. 2020; Erkmen et al. 2020).

14.6 CPs for Biosensing

CPs are excellent for use in sensors due to their chemical characteristics. CPs are widely used in gas sensors because they provide excellent design flexibility. They can form specific layers in which the transduction mechanism's primary physical parameter is changed as a result of the interaction between the analyte gas and the conducting matrix (Janata and Josowicz 2003; McQuade et al. 2000; Gerard et al. 2002). Biosensors are devices that are required for quick and accurate medical diagnosis of a living body. The sophistication of biosensors is increasing day by day as technology and new materials advance. Biosensors are one of the most important applications of conjugated conductive polymers (CPs) (Wu et al. 2020; Prabakaran et al. 2021; Huang et al. 2018; Park et al. 2016). CPs are used in the sensing of biomolecules like glucose, dopamine, DNA, L-tryptophan, drug detection in foods, urinary progesterone, as nuclei in human brain organoids, nitrofurantoin, uric acid, ascorbic acid, micro-RNA (Chen and Luo 2015; Luo et al. 2017; Lee et al. 2020; Aydogdu et al. 2014; Yang et al. 2020; Wang et al. 2019). They are used in detecting tumor cells and also food spoilage by being used in a wireless sensor (Xu et al. 2019; Ma et al. 2018). The following section discusses some of the significant applications of CPs in biosensing and pharmaceutical applications.

The CPs used in biosensing are poly(acetylene), PEDOT, poly(thiophene), PPV, poly(pyrrole), and PANI (Figure 14.2). Conduction occurs due to delocalized electrons in the long-conjugated system in the molecular skeleton. They are used in diagnostics to improve the speed, sensitivity, and versatility of biosensors for measuring vital analytes (Hsiao et al. 2014; Das and Sarkar 2016).

Some of the well-known CPs in use are polythiophene (PTh), poly sulfur nitride (PSN_x), poly(3,4-ethylenedioxythiophene) (PEDOT), and their derivatives. Both P_{Th} and PEDOT are used in organic electronics like organic light-emitting devices and field effect transistors as they do not cause any health hazards. PEDOT further solves the insoluble problem of other CPs by incorporating a water-soluble polyelectrolyte, poly(styrene sulfonic acid) (PSS), into the film without sacrificing the electronic conductivity. PSN_x has a metallic gold appearance, has conductivity comparable to that of metal at room temperature, and shows superconductivity at 0.3 K. The

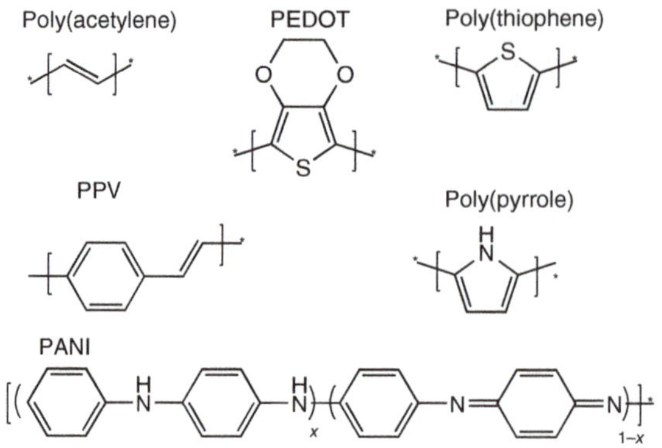

Figure 14.2 Structure of conducting polymers used in biosensing.

conductivity in PSN_x is more dominant along the molecular chain than perpendicular to it. Several CPs have found widespread application in organic electronics. The flexible mechanical properties of the present CPs being biocompatible are particularly advantageous for biomedical applications (Regasa et al. 2020; Singh et al. 2019; Lu et al. 2021; Namsheer and Rout 2021; Guo et al. 2019).

14.7 Oxidation and Charge Transfer

Each of a film's constitutive polymer chains can be oxidized by extracting or injecting n consecutive electrons generating p or n doping, respectively. The counter ions and solvent are exchanged with the electrolyte to maintain the charge and osmotic balance. The materials are transformed into reactive polyelectrolyte gels with variable polymer-ion compositions. The distribution of double bonds in polymer chains changes during these reactions from s bonds and free rotation among consecutive monomeric units to conjugated bonds with a quinoid structure along several monomeric units (Figure 14.3) (Yu et al. 2016; Toribio et al. 2013).

14.8 Color Change in PDA Polymers

PDA-based polymers undergo reversible temperature-induced color changes, at a temperature of 75 °C, extended conjugation causes an absorption at 640 nm giving a blue color that gets shifted to purple at 585 nm. The twisting of the polymer backbone and changes in the degree of conjugation in the system produces a red

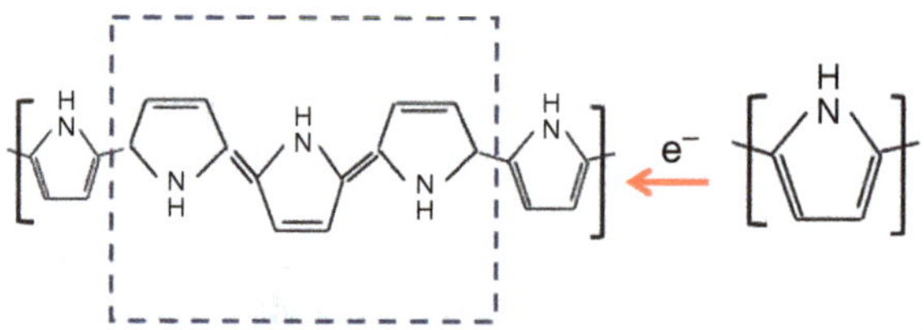

Figure 14.3 Change of the double bond distribution during electrochemical reactions in polypyrrole.

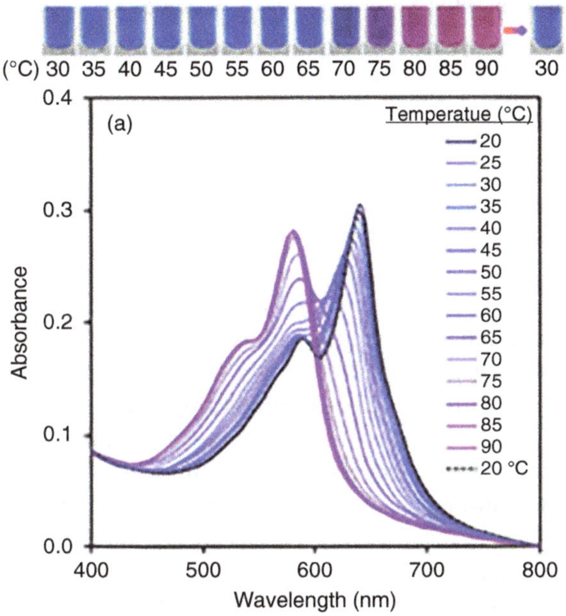

Figure 14.4 Absorption spectra of PDA (8, 9)/PVP10 nanocomposites measured upon increasing temperature. Source: Adapted from Lu et al. (2021).

color (Figure 14.4). These properties can be used for temperature monitoring, biosensors, building materials, and intelligent screens and have a variety of applications, including biomedicine, optical storage, sensing, building materials, and anti-counterfeiting (Kamphan et al. 2017; Wu et al. 2020; Prabakaran et al. 2021; Huang et al. 2018; Lu et al. 2021).

14.9 Ionic Detection

In potentiometric ion-selective electrodes (ISE), CP is used as a sensor. The first attempt used poly(4,4-diaminodiphenyl ether) pH sensors. The ion-exchange characteristics of polymers that have been suitably doped can be used to detect a specific ion in a solution. The substituent and size of the acid anion utilized as a dopant during electro-polymerization affect the pH sensitivity of polyaniline and its derivatives. Tetraphenylborate-doped polyaniline produced in dry acetonitrile can be employed as a pH potentiometric sensor in both aqueous and non-aqueous conditions. The pH sensitivity of a poly(1-aminoanthracene) film electrodeposited on a platinum electrode was also reported without interference from alkali metal ions. The polymer thickness was found to be important in potentiometric pH responses for PPy-modified glassy carbon electrodes. By using the template effect of a non-conducting polymer PVC cast over the graphite surface before PPy deposition, the time response of the pH sensitivity of the doped PPy electrode was dramatically improved (Trojanowicz 2003; Ohnuki et al. 1983; Lindfors and Ivaska 2002; Pandey and Singh 2001; Faria and Bulhões 1998; Shiu et al. 1999; Masalles et al. 2002; Michalska et al. 1997).

14.9.1 Cation Detection

Doping CPs with immobile anions results in cationic potentiometric sensitivity due to the presence of mobile metal ions in the polymer layer. The addition of a cation-exchanging, ion-conductive *Nafion* proton-exchange membrane (PEM) layer on top of PPy increases the cationic response slope by a factor of two. *Nafion* membranes have applications in hydrogen fuel cells as well as potentiometric sensitivity to cations was also discovered in oxidized PPy films, which demonstrated significantly lower redox interference than PPy electrodes. A calcium potentiometric sensor made of PPy doped with the calcium-complexing ligand *Calcion* was successfully developed. It can be utilized as an indication for potentiometric calcium titrations, particularly mixed solvent titrations when traditional PVC-based electrodes cannot be employed. The adsorption of calcion ions on both the substrate and the film aids in the deposition of the film. Calcion adsorption on PPy film is much stronger than on the electrode substrate, and the redox process of calcion is more reversible. Membranes prepared by electro-polymerizing a pyrrole dibenzo-18-crown-6monomer with pyrrole showed cationic sensitivity to divalent cations. As an internal sensor, a nitrate-selective PPy electrode was used to create a NO_x gas sensor (Michalska et al. 2002; Ersöz et al. 2002; Blaz et al. 2000; Gandasasmita et al. 2002; Zhu et al. 2022). The enhanced response stability of CP-based ion-selective electrodes was attributed to the capacity of PPy to transfer both ionic and electronic charges. PPy film was used as an electric connector support to create ion-selective electrodes sensitive to potassium ions, as well as sulfate ions and anionic and cationic surfactants. To make a solid contact calcium-sensitive electrode, a PVC layer containing an ionophore was cast on a PPy film saturated in calcium hydroxide solution (Oyama and Hirokawa 1987; Cadogan et al. 1992; Bobacka 1999; Gyurcsányi et al. 1998; Pandey and Prakash 1998; Berthold et al. 2001; Kovács et al. 2001). PANI was used to fabricate potentiometric calcium sensors with various ionophores. The chemical sensing of Cr^{+6} was performed using a polyaniline-polystyrene composite-modified electrode (Bobacka et al. 1995; Lindfors and Ivaska 2000, 2001; Alizadeh and Mahmodian 2000; Yang and Huang 2001; Dominguez-Ballart et al. 2000).

14.9.2 Anion Detection

Anions like Fl^- and Cl^-, although beneficial for human health, become detrimental when present in excess amounts. PVC membrane containing dodecyl sulfate-doped PPy was used as an ionophore and demonstrated good selectivity over diverse inorganic anions (Diniz et al. 1999; Wang et al. 2014). A diketopyrrolo[3,4-c] pyrrole (DPP)-based polymer with NH units on the polymer backbone has been reported to be a good anion detector. The functional group NH reacts with the fluoride anions in the organic solvents, resulting in a color change from red to blue. These CPs not only detect the anions but also filter them (Wu et al. 2020; Prabakaran et al. 2021; Huang et al. 2018). They are used to detect cyanide ions (CN^-), found in industrial and biological wastes in the water, by being used in fluorescence and colorimetric chemo-sensors (Kumar et al. 2019).

14.10 Conductometry

Conductometric biosensors measure changes in a biological component's conductance induced by the interaction of two metal electrodes. Conductometric detection relies on a process that can change the concentration of ionic species. This mechanism causes changes in electrical conductivity or current flow. How changes in ionic composition during biorecognition affect conductance between electrodes is investigated. They possess speed, usability, specificity, sensitivity, near real-time detection, and demand for tiny sample amounts. Biosensors for glucose, urea-neutral lipid/lipase, and hemoglobin/pepsin estimation detect changes in electronic conductivity caused by changes in the redox potential and/or pH. Polyaniline Langmuir Blodgett films have been used as a glucose biosensor. The dielectric spectroscopic measurements of the PPy/glucose biosensor were also taken into consideration. For penicillin, as well as glucose, urea, lipids, and hemoglobin, conductivity biosensors based on conducting polymers were developed. Microelectronic devices with millimolar sensitivity to urea concentration have been developed for measurements of conductance, capacitance, and admittance using urease (Wu et al. 2020; Prabakaran et al. 2021; Huang et al. 2018).

The bio-recognition event changes the concentration of ionic species during conductometric detection, resulting in a quantifiable change in electrical conductivity or conductance. Conductometric biosensors are tiny two-electrode devices that allow current to flow when an alternating current voltage is applied across the electrodes. Because of its short detection time, sensitivity, and ease of use, this device is most likely to be employed on-site for pathogen detection. It can detect pathogens such as *Bacillus cereus* and *Escherichia coli* present in food. The link between conductance and bioreceptor approaches serves as the foundation for establishing conductometric detection employing a CP. The CP converts biological impulses into electrical signals by acting as an electrochemical transducer. Polyaniline, polythiophene, polyacetylene, and PPy are examples of CPs used for this (Yahaya et al. 2019; Tsai et al. 2021).

14.11 Enzyme Entrapment

Conducting polymers serve as a suitable matrix for enzyme entrapment. A conductometric biosensor using immobilized *Chlorella vulgaris* microalgae as bio-receptors has been used as a bi-enzymatic biosensor. The incorporation of enzymes into electro-depositable conducting polymeric films allows for the localization of biologically active molecules on electrodes of any size or geometry, making it particularly suitable for the fabrication of multi-analyte micro-amperometric biosensors. The available chemical structure of CPs is highly flexible and can be modified as needed. The loss of enzymes is one of the problems associated with the operation of the enzyme system. Enzyme stability and biosensor shelf life are maintained. Several immobilization procedures have been developed to address these issues. Immobilization of biological molecules on transducers is a critical step in the development of biosensors, and conducting polymer-made transducers are widely used for this purpose (Promsuwan et al. 2020; Lian et al. 2016; Kucherenko et al. 2019).

14.12 DNA Sensing

DNA biosensors are used in the electrochemical detection of amino acids, pharmaceuticals, proteins, and other biomaterials contained in urine and blood. They provide quick sensing, have good storage as well as cyclic stability, and can be made at lower expenses, DNA biosensors are regarded as general platforms using CPs as substrates for biosensor applications. These are mainly DNA hybridization-based biosensors that use enzyme-free nucleic acid amplification strategies to enhance their responses. The recognition events between aptamers and analytes cause a change in the surface morphology compared to pristine electrodes, which may alter the path of electron transport (Rashid and Yusof 2017; Yang et al. 2017; Bizid et al. 2017; Hui et al. 2017; Kulikova et al. 2018; Karimi-Maleh et al. 2015; Bizid et al. 2016). The biochemical signals are converted into an electrical readout via transducing mechanisms. A transducer surface is an electrode where CPs can be used to capture targets with selectivity by immobilizing single-stranded probe oligonucleotides into a self-assembled monolayer. The physicochemical parameters of the electrode surface have a considerable impact on the performance of DNA biosensors properties of these electrochemical biosensors (Williamson et al. 2023). The use of PPy has caused improved performance in detecting ds-DNA and *Idarubicin* (Foroughi and Jahani 2022).

14.13 Hydrogel-Based Biosensors

Hydrogel biosensors are multifunctional with high flexibility and biocompatibility. CP-based hydrogels have been widely used in several biomedical applications, including medication delivery, wound healing by means of tissue regeneration, and molecular diagnostics (Herrmann and Haag 2021). Wichterle and Lim created the first hydrophilic gel (hydrogel) in 1960, the objective was to create a plastic alternative to human tissue (Wichterle and Lím 1960). Hydrogels are made of biocompatible polymers with reasonable mechanical characteristics. CPs can be bound with hydrogels utilizing electro-polymerization and hydrogel coating to make hydrogels conductive. PEDOT, PANI, and PPy are mainly used in hydrogels. The conductive hydrogel is implantable in live creatures and functions as a flexible electrode to detect cell impulses from the brain (Ahmed 2015; Feig et al. 2018; Tavakoli and Tang 2017; Qu et al. 2014; Goh et al. 2017). *Sulfonated lignin-conductive* polymer nanoparticles having hydrophilic properties have been employed for hydrogel bioelectronics (Gan et al. 2020).

14.14 Urea and Melamine Detection

Urea and melamine in milk, if not in property quantity, may cause kidney diseases. To detect urea and melamine, biosensors based on CP composite materials have been developed for their detection using electrochemical sensing platforms. To produce a urea sensor, PANI-based hydrogels were coated with the enzyme urease, which was immobilized on the working electrode. The hydrogel supplied an extended surface area, while the PANI increased the sensitivity. When these CP-based nanocomposites are utilized as electrode materials, the signals of biosensors are improved, and enzyme signaling and activity are preserved (Azad and Ahmed 2016).

14.15 Summary

Conducting polymers (CPs) have significant use in biosensing. They have the advantages of being lightweight, flexible, and can be produced in large quantities. CPs like poly(acetylene), PEDOT, poly(thiophene), PPV, poly(pyrrole), and PANI are the ones most used for biosensing. The sensing is done by a change in conduction and in some cases change in color occurring in CPs. Detection of anions, milk adulteration urea, melamine, enzyme entrapment, and DNA sensing are some of the major applications of CP-based biosensors. Their sensing properties can be further enhanced by doping and making nanocomposites.

References

Ahmed, E.M. (2015). Hydrogel: preparation, characterization, and applications: a review. *Journal of Advanced Research* 6: 105–121.

Albaaj, A., Foucras, G., and Raboisson, D. (2017). Changes in milk urea around insemination are negatively associated with conception success in dairy cows. *Journal of Dairy Science* 100: 3257–3265.

Alizadeh, N. and Mahmodian, M. (2000). A new dodecylsulfate ion-selective sensor based on electrochemically prepared polypyrrole and PVC. *Electroanalysis* 12: 509–512.

Angelopoulos, M. (2001). Conducting polymers: concepts and applications. *Organic Electronics* 45: 57.

Aydogdu, G., Gunendi, G., Zeybek, D.K. et al. (2014). A novel electrochemical DNA biosensor based on poly-(5-amino-2-mercapto-1, 3, 4-thiadiazole) modified glassy carbon electrode for the determination of nitrofurantoin. *Sensors and Actuators B: Chemical* 197: 211–219.

Azad, T. and Ahmed, S. (2016). Common milk adulteration and their detection techniques. *International Journal of Food Contamination* 3: 1–9.

Berthold, M., Jahn, H., and Kaden, H. (2001). Festkontaktierte potentiometrische Elektroden für Sulfatbestimmungen in wässrigen Lösungen. *Acta Hydrochimica et Hydrobiologica* 29: 219–226.

Bizid, S., Mlika, R., Haj Said, A. et al. (2016). Investigations of poly(p-phenylene) modified with ferrocene and their application in electrochemical DNA sensing. *Sensors and Actuators B: Chemical* 226: 370–380.

Bizid, S., Blili, S., Mlika, R. et al. (2017). Direct electrochemical DNA sensor based on a new redox oligomer modified with ferrocene and carboxylic acid: application to the detection of *Mycobacterium tuberculosis* mutant strain. *Analytica Chimica Acta* 994: 10–18.

Blaz, T., Migdalski, J., and Lewenstam, A. (2000). Polypyrrole-calcion film as a membrane and solid-contact in an indicator electrode for potentiometric titrations. *Talanta* 52: 319–328.

Bobacka, J. (1999). Potential stability of all-solid-state ion-selective electrodes using conducting polymers as ion-to-electron transducers. *Journal of Analytical Chemistry* 71: 4932–4937.

Bobacka, J., Lindfors, T., McCarrick, M. et al. (1995). Single-piece all-solid-state ion-selective electrode. *Analytical Chemistry* 67: 3819–3823.

Cadogan, A., Gao, Z., Lewenstam, A. et al. (1992). All-solid-state sodium-selective electrode based on a calixarene ionophore in a poly(vinyl chloride) membrane with a polypyrrole solid contact. *Analytical Chemistry* 64: 2496–2501.

Celiberti, L., Fiore Mosca, D., Allodi, G., et al. (2024). Spin-orbital Jahn-Teller bipolarons. *Nat Commun* 15: 2429. https://doi.org/10.1038/s41467-024-46621-0.

Chance, R.R., Shacklette, L.W., Miller, G.G. et al. (1980). Highly conducting charge-transfer complexes of a processible polymer: poly(*p*-phenylene sulphide). *Journal of the Chemical Society, Chemical Communications* 348–349.

Chauhan, N., Chawla, S., Pundir, C.S., and Jain, U. (2017). An electrochemical sensor for detection of neurotransmitter-acetylcholine using metal nanoparticles, 2D material and conducting polymer modified electrode. *Biosensors & Bioelectronics* 89: 377–383.

Chen, C.H. and Luo, S.C. (2015). Tuning surface charge and morphology for the efficient detection of dopamine under the interferences of uric acid, ascorbic acid, and protein adsorption. *ACS Applied Materials & Interfaces* 7: 21931–21938.

Chen, C.A., Yuan, H., Chen, C.W. et al. (2021). An electricity- and instrument-free infectious disease sensor based on a 3D origami paper-based analytical device. *Lab on a Chip* 21: 1908–1915.

Chiou, N.R., Lui, C.M., Guan, J.J. et al. (2007). Growth and alignment of polyaniline nanofibres with superhydrophobic, superhydrophilic and other properties. *Nature Nanotechnology* 2: 354–357.

Das, J. and Sarkar, P. (2016). Enzymatic electrochemical biosensor for urea with a polyaniline grafted conducting hydrogel composite modified electrode. *RSC Advances* 6: 92520–92533.

Diniz, F.B., de Freitas, K.C.S., and de Azevedo, W.M. (1999). Salinity measurements with polyaniline matrix coated wire electrodes. *Electrochemistry Communications* 1: 271–273.

Dominguez-Ballart, M., Bachas, L.G., and Salvado, V. (2000). Potentiometric chloride-selective electrodes based on polypyrrole: preparation and membrane characterization by SEM and EDX. *Quimica Analitica* 19: 49–56.

Edmonds, J.T. Jr., and Hill, H.W. Jr., (1967). Production of Polymers from Aromatic Compounds. US Patent 3, 354,129.

Erkmen, C., Kurbanoglu, S., and Uslu, B. (2020). Fabrication of poly(3,4-ethylenedioxyt hiophene)-iridium oxide nanocomposite based tyrosinase biosensor for the dual detection of catechol and azinphos methyl. *Sensors and Actuators B Chemical* 316: 128121.

Ersöz, A., Gavalas, V.G., and Bachas, L.G. (2002). Potentiometric behavior of electrodes based on overoxidized polypyrrole films. *Analytical and Bioanalytical Chemistry* 372: 786–790.

Faria, R.C. and Bulhões, L.O.S. (1998). Hydrogen ion selective electrode based on poly(1-aminoanthracene) film. *Analytica Chimica Acta* 377: 21–27.

Feig, V.R., Tran, H., Lee, M., and Bao, Z. (2018). Mechanically tunable conductive interpenetrating network hydrogels that mimic the elastic moduli of biological tissue. *Nature Communications* 9: 2740.

Foroughi, M.M. and Jahani, S. (2022). Investigation of a high-sensitive electrochemical DNA biosensor for determination of Idarubicin and studies of DNA-binding properties. *Microchemical Journal* 179: 107546.

Forzani, E.S., Zhang, H.Q., Nagahara, L.A. et al. (2004). A conducting polymer nanojunction sensor for glucose detection. *Nano Letters* 4: 1785–1788.

Gan, D., Shuai, T., Wang, X. et al. (2020). Mussel-inspired redox-active and hydrophilic conductive polymer nanoparticles for adhesive hydrogel bioelectronics. *Nano-Micro Letters* 12: 169.

Gandasasmita, S., Persin, M., Cretin, M., and Sarrazin, J. (2002). Elaboration and characterization of new membranes based on the electropolymerization of mixtures of pyrrole and 4,4′-bis[N-(1-oxo) hexylpyrrole] dibenzo-18-crown-6 monomers. *Desalination* 148: 5–9.

Gerard, M., Chaubey, A., and Malhotra, B.D. (2002). Application of conducting polymers to biosensors. *Biosensors & Bioelectronics* 17: 345–359.

Goh, K.B., Li, H., and Lam, K.Y. (2017). Development of a multiphysics model to characterize the responsive behavior of urea-sensitive hydrogel as biosensor. *Biosensors & Bioelectronics* 91: 673–679.

Guo, B., Ma, Z., Pan, L., and Shi, Y. (2019). Properties of conductive polymer hydrogels and their application in sensors. *Journal of Polymer Science Part B: Polymer Physics* 57: 1606–1621.

Gyurcsányi, R.E., Nybäck, A.S., Tóth, K. et al. (1998). Novel polypyrrole based all-solid-state potassium-selective microelectrodes. *Analyst* 123: 1339–1344.

Hassanein, A., Salahuddin, N., Matsuda, A. et al. (2017). Fabrication of biosensor based on Chitosan-ZnO/polypyrrole nanocomposite modified carbon paste electrode for electroanalytical application. *Materials Science and Engineering C: Materials for Biological Applications* 80: 494–501.

Heeger, A.J. (2001). *J. Phys. Chem. B* 105: 8475–8491.

Herrmann, A. and Haag, R. (2021). Hydrogels and their role in biosensing applications. *Advanced Healthcare Materials* 10: 2100062.

Hsiao, Y.S., Luo, S.C., Hou, S. et al. (2014). 3D bioelectronic interface: capturing circulating tumor cells onto conducting polymer-based micro/nanorod arrays with chemical and topographical control. *Small* 10: 3012–3017.

Huang, J.X., Virji, S., Weiller, B.H., and Kaner, R.B. (2003). Polyaniline nanofibers: facile synthesis and chemical sensors. *Journal of the American Chemical Society* 125: 314–315.

Huang, P.-C., Shen, M.-Y., Yu, H.-h. et al. (2018). Surface engineering of phenylboronic acid-functionalized poly(3, 4-ethylenedioxythiophene) for fast responsive and sensitive glucose monitoring. *ACS Applied Bio Materials* 1: 160–167.

Hudak, N.S. (2013). Rechargeable aluminum batteries with conducting polymers as positive electrodes. SANDIA REPORT SAND2014-0068.

Hui, N., Sun, X.T., Niu, S.Y., and Luo, X.L. (2017). PEGylated polyaniline nanofibers: antifouling and conducting biomaterial for electrochemical DNA sensing. *ACS Applied Materials & Interfaces* 9: 2914–2923.

Inal, S., Rivnay, J., Suiu, A.O. et al. (2018). Conjugated polymers in bioelectronics. *Accounts of Chemical Research* 51: 1368–1376.

Ismail, Y.A., Ahmad, A., and Mohammed, F. (2008). Synthesis, electrical, electronic and charge transport properties of poly(aniline-co-p-toluidine). *Journal of Macromolecular Science Part A Pure and Applied Chemistry* 45: 650–657.

Janata, J. and Josowicz, M. (2003). Conducting polymers in electronic chemical sensors. *Nature Materials* 2: 19–24.

Kamphan, A., Khanantong, C., Traiphol, N., and Traiphol, R. (2017). *J. Ind. Eng. Chem.* 46: 130–138.

Karimi-Maleh, H., Tahernejad-Javazmi, F., Atar, N. et al. (2015). A Novel DNA biosensor based on a pencil graphite electrode modified with polypyrrole/functionalized multiwalled carbon nanotubes for determination of 6-mercaptopurine anticancer drug. *Industrial and Engineering Chemistry Research* 54: 3634–3639.

Karlsson, C. (2014). Conducting Redox Polymers for Electrical Energy Storage. Backbone-Substituent Interactions in Quinone Polypyrrole Model Systems. *Digital Comprehensive Summaries of Uppsala Dissertations from the Faculty of Science and Technology* 1177, 72–43 pp. Uppsala: Acta Universitatis Upsaliensis. ISBN 978-91-554-9033-1.

Khan, M.Z.H., Ahommed, M.S., and Daizy, M. (2020). Detection of xanthine in food samples with an electrochemical biosensor based on PEDOT:PSS and functionalized gold nanoparticles. *RSC Advances* 10: 36147–36154.

Kovács, B., Csóka, B., Nagy, G., and Ivaska, A. (2001). All-solid-state surfactant sensing electrode using conductive polymer as internal electric contact. *Analytica Chimica Acta* 437: 67–76.

Kucherenko, I.S., Soldatkin, O.O., Kucherenko, D.Y. et al. (2019). Advances in nanomaterial application in enzyme-based electrochemical biosensors: a review. *Nanoscale Advances* 1: 4560–4577.

Kulikova, T.N., Porfireva, A.V., Shamagsumova, R.V., and Evtugyn, G.A. (2018). Voltammetric sensor with replaceable polyaniline-DNA layer for doxorubicin determination. *Electroanalysis* 30: 2284–2292.

Kumar, A., Sharm, T.K., Malinee, M., and Dhiman, A. (2019). Aptamer-mediated Nanobiosensing for health monitoring. In: *Advanced Biosensors for Health Care Applications* (ed. K.R. Inamuddin, A. Mohammad, and A.M. Asiri), 227–248. Elsevier.

Le, T.-H., Kim, Y., Yoon, H. (2017). Electrical and Electrochemical Properties of Conducting Polymers. *Polymers* 9: 150–8036. https://doi.org/10.3390/polym9040150.

Lee, M.-H., Liu, K.T., Thomas, J.L. et al. (2020). Peptide-imprinted poly(hydroxymethyl 3, 4-ethylenedioxythiophene) nanotubes for detection of α synuclein in human brain organoids. *ACS Applied Nano Materials* 3: 8027–8036.

Lian, M., Chen, X., Lu, Y., and Yang, W. (2016). Self-assembled peptide hydrogel as a smart biointerface for enzyme-based electrochemical biosensing and cell monitoring. *ACS Applied Materials & Interfaces* 8: 25036–25042.

Lindfors, T. and Ivaska, A. (2000). All-solid-state calcium-selective electrode prepared of soluble electrically conducting polyaniline and di (2-ethylhexyl) phosphate with tetraoctylammonium chloride as cationic additive. *Analytica Chimica Acta* 404: 111–119.

Lindfors, T. and Ivaska, A. (2001). Calcium-selective electrode based on polyaniline functionalized with bis[4-(1,1,3,3-tetramethylbutyl) phenyl] phosphate. *Analytica Chimica Acta* 437: 171–182.

Lindfors, T. and Ivaska, A. (2002). pH sensitivity of polyaniline and its substituted derivatives. *Journal of Electroanalytical Chemistry* 531: 43–52.

Long, Y.Z., Zhang, L.J., Chen, Z.J. et al. (2005). Electronic transport in single polyaniline and polypyrrole microtubes. *Physical Review B* 71 (16): 165412.

Long, Y.Z., Duvail, J.L., Li, M.M. et al. (2010). Electrical conductivity studies on individual conjugated polymer nanowires: two-probe and four-probe results. *Nanoscale Research Letters* 5: 237–242.

Lu, H., Li, X., and Lei, Q. (2021). Conjugated conductive polymer materials and its applications: a mini-review. *Frontiers in Chemistry* 9: 732132.

Luo, S.-C., Thomas, J.L., Guo, H.-Z. et al. (2017). Electrosynthesis of nanostructured, imprinted poly(hydroxymethyl 3,4-ethylenedioxythiophene) for the ultrasensitive electrochemical detection of urinary progesterone. *Chemistry Select* 2: 7935–7939.

Ma, Z., Chen, P., Cheng, W. et al. (2018). Highly sensitive, printable nanostructured conductive polymer wireless sensor for food spoilage detection. *Nano Letters* 18: 4570–4575.

MacDiarmid, A.G. and Epstein, A.J. (1994). The concept of secondary doping as applied to polyaniline. *Synthetic Metals* 65: 103–116.

Mao, Y., Guo, L., Ning, X. et al. (2018). The signal amplification in electrochemical detection of chloramphenicol using sulfonated polyaniline-chitosan composite as redox capacitor. *Electroanalysis* 30: 2085–2093.

Masalles, C., Borros, S., Vinas, C., and Teixidor, F. (2002). Simple PVC–PPy electrode for pH measurement and titrations. *Analytical and Bioanalytical Chemistry* 372: 513–518.

McQuade, D.T., Pullen, A.E., and Swager, T.M. (2000). Conjugated polymer-based chemical sensors. *Chemical Reviews* 100: 2537–2574.

Michalska, A., Ivaska, A., and Lewenstam, A. (1997). Modeling potentiometric sensitivity of conducting polymers. *Analytical Chemistry* 69: 4060–4064.

Michalska, A., Walkiewicz, S., Maksymiuk, K., and Hall, E.A.H. (2002). Potentiometric responses of poly(pyrrole) films surface modified by Nafion. *Electroanalysis* 14: 1236–1244.

Mishra, A.K. (2018). Conducting polymers: concepts and applications. *Journal of Atomic, Molecular, Condensed Matter and Nano Physics* 5 (2): 159–193.

Mzenda, V.M., Goodman, S.A., and Auret, F.D. (2002). Conduction models in polyaniline: the effect of temperature on the current–voltage properties of polyaniline over the temperature range 30< T (K)< 300. *Synthetic Metals* 127: 285–289.

Namsheer, K. and Rout, C.S. (2021). Conducting polymers: a comprehensive review on recent advances in synthesis, properties and applications. *RSC Advances* 11: 5659–5697.

Noh, H.B., Revin, S.B., and Shim, Y.B. (2014). Voltammetric analysis of anti-arthritis drug, ascorbic acid, tyrosine, and uric acid using a graphene decorated-functionalized conductive polymer electrode. *Electrochimica Acta* 139: 315–322.

Ohnuki, Y., Matsuda, H., Ohsaka, T., and Ogama, N. (1983). Permselectivity of films prepared by electrochemical oxidation of phenol and amino-aromatic compounds. *Journal of Electroanalytical Chemistry and Interfacial Electrochemistry* 158: 55–67.

Otero, T.F. and Martinez, J.G. (2013). *J. Mater. Chem. B* 1: 26.

Oyama, N. and Hirokawa, T. (1987). Hydrogen ion selective microelectrode prepared by modifying an electrode with polymers. *Analytical Chemistry* 59: 258–262.

Pan, L.J., Pu, L., Shi, Y. et al. (2006). Hydrothermal synthesis of polyaniline mesostructures. *Advanced Functional Materials* 16: 1279–1288.

Pandey, P.C. and Prakash, R. (1998). Polyindole modified potassium ion-sensor using dibenzo-18-crown-6 mediated PVC matrix membrane. *Sensors and Actuators B: Chemical* 46: 61–65.

Pandey, P.C. and Singh, G. (2001). Tetraphenylborate doped polyaniline based novel pH sensor and solid-state urea biosensor. *Talanta* 55: 773–782.

Pant, H.C., Patra, M.K., Negi, S.C. et al. (2006). Studies on conductivity and dielectric properties of polyaniline-zinc sulphide composites. *Bulletin of Materials Science* 29: 379–384.

Park, S.J., Lee, S.H., Yang, H. et al. (2016). Human dopamine receptor-conjugated multidimensional conducting polymer nanofiber membrane for dopamine detection. *ACS Applied Materials & Interfaces* 8: 28897–28903.

Prabakaran, K., Jandas, P.J., Luo, J. et al. (2021). Molecularly imprinted poly(methacrylic acid) based QCM biosensor for selective determination of L-tryptophan. *Colloids and Surfaces A: Physicochemical and Engineering Aspects* 611: 125859.

Prajapati, D.G. and Kandasubramanian, B. (2019). Progress in the development of intrinsically conducting polymer composites as biosensors. *Macromolecular Chemistry and Physics* 220: 1800561.

Promsuwan, K., Meng, L., Suklim, P. et al. (2020). Bio-PEDOT: modulating carboxyl moieties in poly(3,4-ethylenedioxythiophene) for enzyme-coupled bioelectronic interfaces. *ACS Applied Materials & Interfaces* 12: 39841–39849.

Qu, F., Zhang, Y., Rasooly, A., and Yang, M. (2014). Electrochemical biosensing platform using hydrogel prepared from ferrocene modified amino acid as highly efficient immobilization matrix. *Analytical Chemistry* 86: 973–976.

Rabolt, J.F., Clarke, T.C. et al. (1980). Organic metals: poly-(*p*-phenylene sulphide) hexafluoroarsenate. *Journal of the Chemical Society, Chemical Communications* 8: 347–348.

Rashid, J.I.A. and Yusof, N.A. (2017). The strategies of DNA immobilization and hybridization detection mechanism in the construction of electrochemical DNA sensor: a review. *Sensing and Bio-Sensing Research* 16: 19–31.

Regasa, M.B., Refera Soreta, T., Femi, O.E., and Ramamurthy, P.C. (2020). Development of molecularly imprinted conducting polymer composite film-based electrochemical sensor for melamine detection in infant formula. *ACS Omega* 5: 4094–4099.

Russo, M.J., Han, M., Desroches, P.E. et al. (2021). Antifouling strategies for electrochemical biosensing: mechanisms and performance toward point of care based diagnostic applications. *ACS Sensors* 6: 1482–1507.

Samwang, T., Watanabe, N.M., Okamoto, Y., and Umakoshi, H. (2024). Exploring the Influence of Morphology on Bipolaron–Polaron Ratios and Conductivity in Polypyrrole in the Presence of Surfactants. *Molecules* 29: 1197. https://doi.org/10.3390/molecules29061197.

Salzmann, I., et al. (2016). *Acc. Chem. Res.* 49: 3, 370–378. https://doi.org/10.1021/acs.accounts.5b00438.

Sengodu, P. and Deshmukh, A.D. (2015). Conducting polymers and their inorganic composites for advanced Li-ion batteries: a review. *RSC Advances* 5: 42109–42130.

Shiu, K.K., Song, F.Y., and Lau, K.W. (1999). Effects of polymer thickness on the potentiometric pH responses of polypyrrole modified glassy carbon electrodes. *Journal of Electroanalytical Chemistry* 476: 109.

Singh, A.K., Singh, M., and Verma, N. (2019). Electrochemical preparation of Fe_3O_4/MWCNT-polyaniline nanocomposite film for development of urea biosensor and its application in milk sample. *Journal of Food Measurement and Characterization* 14: 163–175.

Tabor, B.J., Magre, E.P., and Boon, J. (1971). The crystal structure of poly-p-phenylene sulphide. *European Polymer Journal* 7: 1127–1133.

Tadesse, M.G., Ahmmed, A.S., Lübben, J.F. (2024). Review on Conductive Polymer Composites for Supercapacitor Applications. *J. Compos. Sci.* 8: 53. https://doi.org/10.3390/jcs8020053.

Tao, X., Peng, Y., and Liu, J. (2020). Nanomaterial-based fluorescent biosensors for veterinary drug detection in foods. *Journal of Food and Drug Analysis* 28: 576–595.

Tavakoli, J. and Tang, Y. (2017). Hydrogel based sensors for biomedical applications: an updated review. *Polymers* 9: 364.

Trojanowicz, M. (2003). Application of conducting polymers in chemical analysis. *Microchimica Acta* 143: 75–91.

Toribio, F. Otero and Jose, G. Martinez (2013). *J. Mater. Chem. B* 1: 26.

Tsai, M.H., Lin, Y.K., and Luo, S.C. (2021). Conducting polymer-based sensors for food and drug analysis. *Journal of Food and Drug Analysis* 29: 544–558.

Tseng, R.J., Huang, J.X., Ouyang, J. et al. (2005). Polyaniline nanofiber/gold nanoparticle nonvolatile memory. *Nano Letters* 5: 1077–1080.

Verma, A., Gupta, R., Verma, A.S., and Kumar, T. (2023). A Review of Composite Conducting Polymer-Based Sensors for Detection of Industrial Waste Gases. *Sens. Actuators Rep.* 5: 100143. https://doi.org/10.1016/j.snr.2023.100143.

Wang, Y. and Jing, X. (2005). Intrinsically conducting polymers for electromagnetic interference shielding. *Polymers for Advanced Technologies* 16: 344–351.

Wang, F., Wang, L., Chen, X., and Yoon, J. (2014). Recent progress in the development of fluorometric and colorimetric chemosensors for detection of cyanide ions. *Chemical Society Reviews* 43: 4312–4324.

Wang, M., Cui, M., Liu, W., and Liu, X. (2019). Highly dispersed conductive polypyrrole hydrogels as sensitive sensor for simultaneous determination of ascorbic acid, dopamine and uric acid. *Journal of Electroanalytical Chemistry* 832: 174–181.

Waware, U.S. and Umare, S.S. (2003). Chemical synthesis, spectral characterization and electrical properties of poly(aniline-co-m-chloroaniline). *Reactive and Functional Polymers* 65: 343–350.

Wichterle, O. and Lím, D. (1960). Hydrophilic gels for biological use. *Nature* 185: 117.

Williamson, P., Piskunen, P., Ijäs, H. et al. (2023). Signal amplification in electrochemical DNA biosensors using target-capturing DNA origami tiles. *ACS Sensors* 8 (4): 1471–1480.

Wongkaew, N., Simsek, M., Griesche, C., and Baeumner, A.J. (2019). Functional nanomaterials and nanostructures enhancing electrochemical biosensors and lab-on-a-chip performances: recent progress, applications, and future perspective. *Chemical Reviews* 119: 120–194.

Wu, Y., Wang, Z., Liu, X. et al. (2017). Ultralight graphene foam/conductive polymer composites for exceptional electromagnetic interference shielding. *ACS Applied Materials & Interfaces* 9 (10): 9059–9069.

Wu, J.G., Chen, J.H., Liu, K.T., and Luo, S.C. (2019). Engineering antifouling conducting polymers for modern biomedical applications. *ACS Applied Materials & Interfaces* 11: 21294–21307.

Wu, J.G., Wei, S.C., and Luo, S.C. (2020). In situ probing unusual protein adsorption behavior on electrified zwitterionic conducting polymers. *Advanced Materials Interfaces* 7: 2000470.

Xu, L.H., Li, J.J., Zeng, H.B. et al. (2019). ATMP-induced three-dimensional conductive polymer hydrogel scaffold for a novel enhanced solid-state electrochemiluminescence biosensor. *Biosensors & Bioelectronics* 143: 111601.

Yahaya, M.L., Noordin, R., Razak, K.A. et al. (2019). Advanced nanoparticle-based biosensors for diagnosing foodborne pathogens. In: *Advanced Biosensors for Health Care Applications* (ed. A. Inamuddin Mohammad, R. Khan, and A.M. Asiri), 1–43. Elsevier.

Yang, Y.J. and Huang, H. (2001). A polyaniline-modified electrode-based FIA system for sub-ppb-level chromium (VI) analysis. *Journal of Analytical Chemistry* 73: 1377–1381.

Yang, Y., Asiri, A.M., Du, D., and Lin, Y. (2014). Acetylcholinesterase biosensor based on a gold nanoparticle-polypyrrole-reduced graphene oxide nanocomposite modified electrode for the amperometric detection of organophosphorus pesticides. *Analyst* 139: 3055–3060.

Yang, Z.H., Zhuo, Y., Yuan, R., and Chai, Y.Q. (2017). Amplified impedimetric aptasensor combining target-induced DNA hydrogel formation with pH-stimulated signal amplification for the heparanase assay. *Nanoscale* 9: 2556–2562.

Yang, L., Wang, H., Lu, H., and Hui, N. (2020). Phytic acid functionalized antifouling conducting polymer hydrogel for electrochemical detection of microRNA. *Analytica Chimica Acta* 1124: 104–112.

Yu, Z., Li, H., Zhang, X. et al. (2016). Facile synthesis of $NiCo_2O_4$@polyaniline core–shell nanocomposite for sensitive determination of glucose. *Biosensors & Bioelectronics* 75: 161–165.

Zare, E.N., Makvandi, P., Ashtari, B. et al. (2020). Progress in conductive polyaniline-based nanocomposites for biomedical applications: a review. *Journal of Medicinal Chemistry* 63: 1–22.

Zhu, L., Li, Y., Liu, J. et al. (2022). Recent developments in high-performance Nafion membranes for hydrogen fuel cells applications. *Petroleum Science* 19 (3): 1371–1381.

15

Organic-Polymer-Based Photodetectors: Mechanism and Device Fabrication

Nasrin Sultana[1], *Indranee Hazarika*[2], *and Bedanta Gogoi*[2,3]

[1] *Institute of Advanced Study in Science and Technology, Physical Sciences Division, Paschim Boragaon, Guwahati, Assam 781035, India*
[2] *Gauhati University, Department of Chemistry, Gopinath Bordoloi Nagar, Jalukbari, Guwahati, Assam 781014, India*
[3] *Tezpur University, Department of Chemical Sciences, Napaam, Assam 784028, India*

15.1 Introduction

Optoelectronics is one of the most exciting areas in electronics as well as in material science. In the last 20–30 years, there have been dramatic developments in this field (Predeep 2011), and it plays a vital role in almost every aspect, starting from day-to-day life to national security. Optoelectronics is a very broad term that bridges optics and electronics and the development of hardware devices that convert electrical energy into light and vice versa. Optoelectronics is not only limited to a single area, but it also has plenty of applications such as military services, telecommunications, and medical equipment. For convenience, we can broadly classify optoelectronic devices into two categories as shown in Figure 15.1.

Herein, an attempt has been made to emphasize the importance of photodetectors in optoelectronics that finds many real-world applications (Chee 2018; Miao and Zhang 2019; Wang and Kim 2017). Advancements in photodetectors play a pivotal role for their wide range of applications such as in remote sensing, national security, detecting harmful radiations, and health monitoring (Chen et al. 2020b; Gogoi and Sen Sarma 2015; Lee et al. 2010). Photodetectors range from simple devices to receivers on VCR and TV remote controls, to the CCD in a video camera, to photodiodes in a fiber optic connection, and to boundless arrays used by astronomers to detect radiation.

15.1.1 Types of Photodetectors

Photodetectors are broadly divided into two categories

i. Thermal photodetectors and
ii. Photon photodetectors

Organic Polymers in Energy-Environmental Applications, First Edition.
Edited by Ramesh Oraon, Pardeep Singh, Sanchayita Rajkhowa, Sangita Agarwal, and Ravindra Pratap Singh.

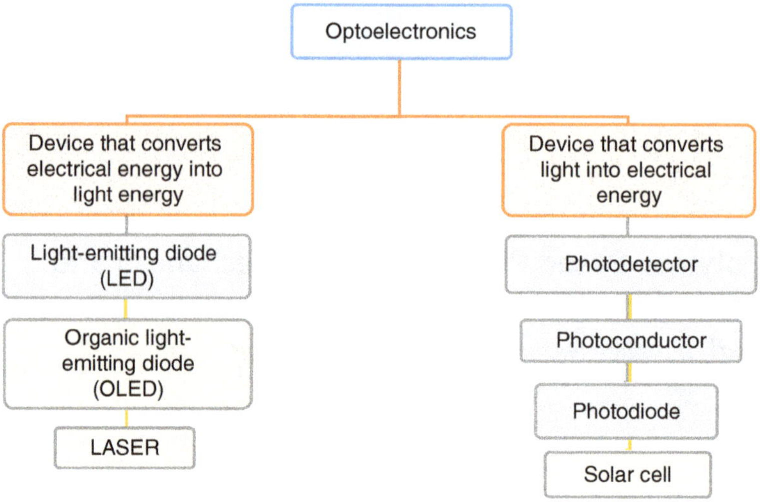

Figure 15.1 Classification of optoelectronics.

Thermal photodetectors generally absorb infra-red (IR) radiations, and this radiation increases the temperature of the material used in the photodetector. Further, some physical properties of the material may change to generate an electrical output. For example, the temperature changes are converted to changes in voltage by the use of a thermocouple, the Seebeck effect in a thermopile detector, changes in electrical polarization by the use of pyroelectric materials in pyroelectric detectors, and changes in resistance by the use of a barometer in a bolometer detector. The photonic nature of the incident radiations has no bearing on the signal. Therefore, thermal effects are wavelength-independent. Thermal photodetectors typically have poor sensitivity and response, but photon photodetectors, another type of photodetector, are more effective than thermal detectors in terms of sensitivity and response time.

However, after absorbing electromagnetic radiation, photon photodetectors change the distribution of electronic energy within the substance to produce an electrical signal. In addition, photon photodetectors depend upon a selective wavelength of the radiation. It requires cryogenic cooling for preventing the formation of carriers by thermal energy. Photon photodetectors can be further subdivided into different categories, but three of them are very common such as photoconductor, photodiode, and phototransistor (Miao and Zhang 2019).

15.1.2 Mechanism involved in the Photodetection

In photodetection, photoactive materials, semiconducting materials create an electron–hole pair (exciton) under the influence of suitable light energy. That is, at the point of a semiconductor exposed to photons with energies greater than or equal to its bandgap, then electrons are ejected from the valence band to the conduction band. Under the influence of an external applied electric field, ejected electrons can move freely for a long distance. Besides, the positively charged holes move from one site to another in the valence band that also contribute to electrical conduction (Omnes 2009).

Instead of the electron–hole pair mechanism, there are some materials that have their donor–acceptor properties, and via the donor–acceptor mechanism, they can also produce photocurrent in the photodetector.

15.1.3 Materials

Researchers around the world have been developing many organic, inorganic, hybrid materials, and nanomaterials for fabricating photodetectors. Such materials are essentially semiconducting materials. The vast majority of inorganic semiconducting materials are utilized as photoactive materials in photodetectors because of their better photophysical properties and durability. Photodetectors mainly based on p-n heterojunctions and some examples of n-type inorganic semiconducting materials are TiO_2 (Chahrour et al. 2019), ZnO (Gedamu et al. 2014), SnO_2 (Y. Li et al. 2018), Zn_2SnO_4 (Dong et al. 2018), Zn_2GaO_4 (Huang et al. 2020), Zn_2GeO_4 (Han et al. 2019), ZnS (Chen et al. 2020a), $Ga_2In_4S_9$ (Wang et al. 2019a) and p-type are GaN (Wang et al. 2019b), SiC (Liang et al. 2020), NiO (Zhang et al. 2020) metal oxides. The synthesis of such materials is not straightforward in spite of their excellent photophysical properties. Additionally, majority of the inorganic materials are not flexible and also quite expensive. Such disadvantages could be addressed by developing organic semiconductors (Miao and Zhang 2019). Organic semiconductors that include small molecules, polymers, fullerenes, and their derivatives are the main materials for organic photodetectors. Among them, small molecules and polymer materials are often used as single-component films as well as donor materials for bulk heterojunction (BHJ). Organic polymers can be easily tuned by adjusting their conjugate chain according to different applications. In this chapter, we will discuss organic-polymer-based photodetectors.

15.1.4 Photodetector Parameters

A brief discussion about the important parameters used for the assessment of photodetectors is given as follows:

15.1.4.1 Responsivity

Responsivity is the detector output signal in the form of current (I_{out}) or voltage (V_{out}) response to the input optical power (P_{in}) (Chow and Someya 2020). Mathematically it can be expressed as

$$\text{Current responsivity } R_i = \frac{I_{out}}{P_{in}} \qquad \text{unit} = \text{A W}^{-1}$$

$$\text{Voltage responsivity } R_v = \frac{V_{out}}{P_{in}} \qquad \text{unit} = \text{V W}^{-1}$$

15.1.4.2 Noise-Equivalent Power (NEP)

The input optical power that results in an output signal (voltage or current), identical to the device's noise, is known as the noise-equivalent power (NEP) of a photodetector. Since noise varies with bandwidth (Δf) of the circuit, the bandwidth must be restricted (Chow and Someya 2020). Mathematically:

$$\text{NEP} = \frac{HAV_\text{N}}{V_\text{S}(\Delta f)^{1/2}}$$

where

H = Irradiation on the detector of area A
V_N = Root means square noise voltage within (Δf)
V_S = Root means square signal voltage

15.1.4.3 Detectivity

Although responsively provides some information about output vs. input signals, it does not provide any information regarding the detector's sensitivity. Sensitivity is defined as the minimum input optical power that a photodetector can detect. This power is the NEP (Chow and Someya 2020).

$$\text{Detectivity, } D = \frac{1}{\text{NEP}} \qquad \text{unit} = \text{W}^{-1}$$

However, D cannot be used for comparing two detectors because it has no idea about the detector area. In order to solve this problem, the term specific detectivity is introduced, which can be defined as the square root of the detector area (A) per unit value of NEP (Chow and Someya 2020).

$$\text{Specific detectivity, } D^* = \frac{A^{1/2}}{\text{NEP}} \qquad \text{unit} = \text{cm W}^{-1}$$

15.1.4.4 Quantum Efficiency

The ratio of the number of electrons that are gathered to the number of photons that are incident is known as the quantum efficiency (Q) of a photodetector (Chow and Someya 2020). For example, in the case of photoemissive detectors, Q is equal to number of electrons emitted to the number of photon incidents. Similarly, for a p-n junction-based semiconductor device, Q denotes the number of excitons produced per unit incident photon. Therefore, Q is the fundamental way of expressing the credibility of the incident optical power to generate an output signal. The quantum efficiency is related to the responsivity (R_d) by the following relationship:

$$Q = 100 \times R_\text{d} \times \left(\frac{1.2395}{\lambda}\right)$$

where R_d = Responsivity, λ = Wavelength of light.

15.1.4.5 Detector Response Time

It requires a definite amount of time to produce photo current when light source turned instantly and illuminate's detectors. Again, it takes a moment to get back to the origin value for the photocurrent.

15.1.4.6 Spectral Response

Spectral response refers to the changes in the detector performance (responsivity or detectivity) with wavelength. The type of detector and the type of substance affect the precise shape of the spectral response curve.

Out of the many types of detectors available UV, Vis, and IR have the maximum spectral response curve. Again, the manufacturer also has a pronounced effect on the shape of the spectral response curve (Yang and Ma 2019).

15.2 Organic-Polymer-Based Photodetector

The development of organic polymers and their conjugates with other materials is a potential alternative to inorganic-chemical-based photodetectors. Polymer-based semiconductors can be designed to achieve their desired functions (Forrest 2004). Furthermore, polymer semiconductors have the potential for large-area printing and also, these are solution processable that simplifies the device fabrication process (Sekitani et al. 2008). The flexibility of the polymer materials is an important parameter to fabricate flexible devices and also possesses a broad spectral range beyond visible light (White et al. 2013). The molecular structure of some of the polymer and conjugated polymer used in the organic photodetector for different applications are mentioned in Figure 15.2.

15.2.1 Working Principle of Organic Photodetector

The working mechanism of organic-polymer-based photodetectors includes the general steps to convert the optical signal into the electrical signal:

1. Generation of electron-hole pairs, i.e. exciton
2. Production of charge transfer exciton

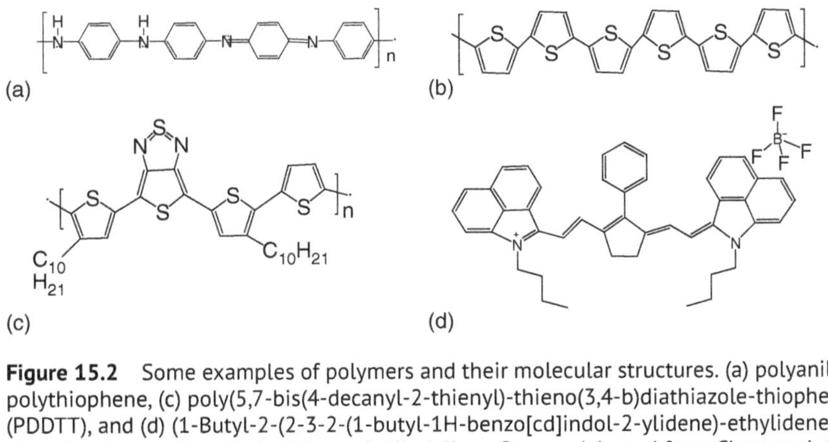

Figure 15.2 Some examples of polymers and their molecular structures. (a) polyaniline, (b) polythiophene, (c) poly(5,7-bis(4-decanyl-2-thienyl)-thieno(3,4-b)diathiazole-thiophene-2,5) (PDDTT), and (d) (1-Butyl-2-(2-3-2-(1-butyl-1H-benzo[cd]indol-2-ylidene)-ethylidene-2-phenylcyclopent-1-enyl-vinyl)-benzo[cd]indolium. Source: Adapted from Chen et al. (2016); Gong et al. (2009); Schilinsky et al. (2002); Young et al. (2016).

3. Separation of charge transfer exciton
4. Collection of free charges

In the case of organic-polymer-based photodetectors, the photogeneration of coulombically bound charge carriers, i.e. electron–hole pairs known as exciton is a key mechanism for detecting light. Literature reveals that organic materials typically have exciton binding energies between 0.3 and 0.5 eV. The efficient separation of charges creates a large photocurrent that can be measured and used to detect light. Therefore, this state can be generated by light absorption in a semiconductor material. For the diffusion process to effectively separate electron–hole pairs in organic materials the formation of a heterojunction between an electron-donating (donor) and an electron-accepting (acceptor) substance is crucial. After that, the excitons are spread over the interface between donor and acceptor for their dissociation into free charges. Therefore, to enhance the exciton dissociation, the exciton diffusion length is an essential factor. Unfortunately, for most organic materials the exciton diffusion length is typically short, i.e. in the range of 5–20 nm. This challenge is overcome by using materials like C_{60} that have a long exciton diffusion length. In general, organic photodetectors use planar heterojunction (PHJ) or bulk heterojunction (BHJ). The BHJ structure is used by most of the devices to improve the device performance by reducing the distance that is required to reach the interface for the excitons and generate a large number of free charges, since in organic semiconductors, excitons do not spontaneously dissociate into free electron–hole pairs. The external electric field is used to assist the exciton separation process. After the successful extraction of the charges at the electrodes, finally a photocurrent is created without undergoing recombination, which leads to the loss of the photocurrent. Improving the device designs can optimize the performance of organic photodetectors for a wide range of applications with respect to spectral response (Chow and Someya 2020).

15.2.2 Basic Device Architecture of Organic Photodetector

There are mainly three types of organic photodetectors such as photoconductors, photodiodes, and phototransistors.

One of the most fundamental kinds of photodetectors is the photoconductor. In these devices, the semiconducting material is placed in between the two ohmic contact electrodes as shown in Figure 15.3a. Electron–hole pairs (EHPs) are created as a result of the incident photons being absorbed by the exposed semiconductor region to increase the conductivity of the photodetector. Photoconductor exhibits intrinsic photocurrent amplification. It is possible to increase the sensitivity of photodetectors through photoconductive gain, in which the quantity of charge carriers that passes through the external circuit in reaction to an optical signal can be greater than the quantity of EHPs produced by photon absorption (He et al. 2015; Miao and Zhang 2019). Photoconductive detectors are often used for the detection of IR radiation. The attainable noise performance of the photoconductors is constrained in the case of small bandgap materials by both the thermal generation

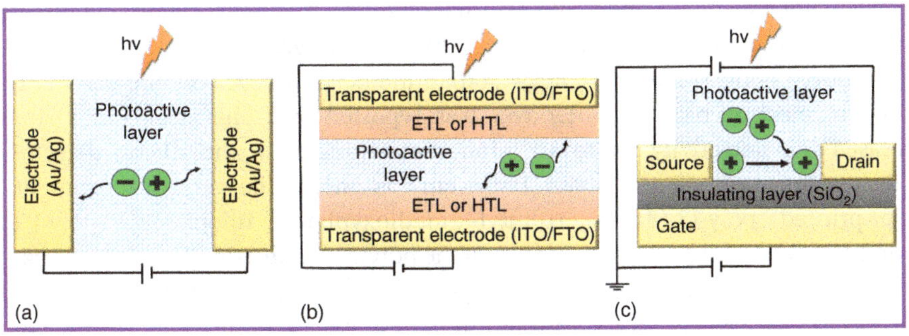

Figure 15.3 Device architectures of (a) photoconductor, (b) photodiode, and (c) phototransistor. Here, ETL, HTL, ITO, FTO are electron transport layer, hole transport layer, indium tin oxide, fluorinated tin oxide, respectively.

rate and the non-zero dark conductivity. Particularly for low-noise and high-speed applications, there is another device that can reduce the performance deficiencies of photoconductive devices and the semiconductor junction photodiode as shown in Figure 15.3b. In general, photodiodes are intrinsic photodetectors that sense light by utilizing EHPs, produced in the p-n junction's depletion region. Both p- and n-type dopants are available for p-n photodiodes. For example, the production of the detectors uses silicon, GaAs/AlGaAs, and InGaAs/InAlAs materials. By adding a lightly doped region in between the p- and n-type regions, which creates a p-i-n structure, the structure of the p-n junction is improved in order to combat the low quantum efficiency and slow nature caused by the tiny depletion width. Therefore, it leads to larger detection bandwidths that improve the quantum efficiency as well as the junction potential is decreased. Generally, photodiodes are used for UV photodetection along with a variety of electronic applications (Nasiri and Tricoli 2019). Phototransistors (Figure 15.3c), with a three-terminal configuration, operate under a normal field-effect transistors (FET) mode where the drain current is flowing in the accumulated channel by using the gate voltage at a given source to drain bias. Phototransistors could have a high external quantum efficiency (EQE) that results in high photocurrent with high responsivity (Wang and Kim 2017).

15.2.3 Advanced Organic-Polymer-Based Photodetectors

Numerous studies have been published in recent years about innovative methods and electro-optical processes that result in organic photodetectors with high-performance characteristics like low noise, high speed, high sensitivity, and selective spectral range (Eggenhuisen et al. 2015; Hazarika et al. 2021; Pierre et al. 2015; Tedde et al. 2009). Various methods and polymers to enhance their mechanical and photodetector characteristics are outlined here. A photodetector's detectivity can be increased by having a low background current. Various strategies have been devised to lower the noise current. Using charge-blocking interlayers between the active layer and the electrodes is one method that can effectively reduce noise current without affecting the photodetector's responsiveness and external quantum efficiency (EQE), claims the study (Saracco et al. 2013; Valouch et al. 2012). Polystyrene-N, N-diphenyl-N, N-bis(4-n-butylphenyl)-(1,10-biphenyl)-4,4-diamine-perfluorocyclobutane (PS-TPD-PFCB) an electron-blocking interlayer was reported by Gong et al. (2009) that increased the detectivity up to 10^{13} Jones matching with the silicon detectors by decreasing dark current about 1 nM cm^{-2} is one of the very first examples. Inkjet-printed poly-[3-(3,5-di-tert-butyl-4-methoxyphenyl)-thiophene] (poly-PT) reported by Grimoldi et al. (2016), is one of the polymeric electron-blocking interlayers that have since been described. The poly(9,9-bis(30-(*N,N*-dimethylamino)propyl) -2,7-uorene)-alt-2,7-(9,9-diocty-luorene) reported by Li et al. (2014), and the alcohol-soluble variety. Even though adding charge-blocking interlayers can effectively increase the detectivity, the fabrication process is complex because the blocking interlayers' processing solvents must be orthogonal. Consequently, Poly(3-hexylthiophene) (P3HT), a conjugated polymer, was reported by Xiong et al. (2017), to decrease noise current without the need for processing solvents through the use of transfer printing. It is a dry fabrication technique, so an orthogonal solvent is not needed during the fabrication process. After depositing the film on the substrate onto the target, a medium substrate like PDMS is used to transfer-print the substrate. As a consequence, the photodiode's detectivity was increased up to 10^{12} Jones, and it can now detect extremely weak signals of $pW\,cm^{-2}$. Recently, a novel method for the reduction of noise current in organic BHJ without involving additional charge-blocking interlayers was developed in an effort to streamline the earlier processes. Layer-by-layer (LBL) solution processing was described by Zhong et al. (2019). This design is comparable to the planar heterojunction, where the BHJ morphology is created by a partial mixing of the donor and acceptor layers. They disclosed a high-performance, all-polymer PTzBI-Ph: N2200 photodiode with a high detectivity of 5.68×10^{12}.

Minor carrier trap sites assist the organic-polymer-based photodetector to improve the responsivity and sensitivity of the photodetector. Moreover, the majority of carrier trap sites can efficiently reduce the noise current and hence the detection performance is increased. On this concept, Shou and co-workers introduced a spatially separated electron/hole double trap mode for the phototransistors (Ji et al. 2022). For trapping sites for electrons/holes, they used $PC_{61}BM$ and metallic Ag nanoparticles, respectively, and achieved spatial separation by using a polymethyl methacrylate (PMMA) layer. It gives a high photoresponsivity of $5.26 \times 10^3\,A\,W^{-1}$ and ultra-high detectivity rates of about 10^{16} Jones. A high gain is required for the fast detection of very weak optical signals to employ in a wide range of applications such as telecommunications, signal monitoring, and surveillance. Without using external circuitry like amplifiers, Hiramoto et al. (1994) introduced an organic perylene pigment film to achieve photoconductive gain via the trapping of holes for the development of a high-performance photodetector. Recently, Zhou et al. (2016) reported PDPP3T: PCBM as an interfacial layer with ZnO nanoparticles to achieve a high photoconductive gain in a photodiode. In addition to the standard phototransistors and photoconductors, at a low voltage of 1 V and also at room temperature, Nie et al. (2017) recently developed an organic BHJ photodiode with

PBDTT-DPP: PCBM that gives an extraordinary detectivity about 10^{14} Jones with high gain (~30 A W^{-1} or ~5000% EQE). The fabrication process for this device is very simple and on the basis of this design in the recent future, the organic photodetector may lead to even better photodetection performance for use in different applications.

A wide variety of material systems are available for organic photodetectors to achieve broadband detection range, or the full visible and near-infrared (NIR) range. But with these materials, achieving a narrowband sensing range is very difficult (Jansen-van Vuuren et al. 2016; Liu et al. 2018a). Narrowband detection is crucial for many uses, including communications, optical monitoring, color imaging, and biomedical imaging (Liu et al. 2018b; Park et al. 2018). By sensitizing the BHJ layer with PbS quantum dots, Rauch and his colleagues were able to create the first high-performance NIR organic/inorganic hybrid photodiode that extends absorption up to 1.8 m (Rauch et al. 2009). Following that, Armin et al. (2015) reported a high-performance organic photodiode for narrowband visible or NIR detection that was built on the charge collection narrowing (CCN) concept. The CCN idea makes use of the incident photons' wavelength-dependent absorption coefficients inside the thicker photoactive layer. Armin and his colleagues used PCDTBT:PCBM and DPP-DTT:PCBM blends in dense BHJ films of about 2 m. The optimized narrowband organic photodiodes achieved a maximum EQE of about 30% and high detectivity of about 10^{12} Jones by considerably lowering the noise current. This was made possible by the use of thick active interlayer film. Since then, a plethora of investigations have documented organic BHJ-based narrowband CCN photodiodes (Zhong et al. 2017). Shen et al. (2016) produced an organic photodiode by fusing the CCN idea for narrowband detection with the photoconductive gain with high EQE in order to get around the low EQE limitation of CCN organic photodiodes. In order to trigger the photoconductive gain by trapping holes through which EQE greatly increases by tenfold from before, they added PbS quantum dots into a thick BHJ layer.

Hence, we can conclude that organic polymers are promising candidates for high-performance photodetectors from the recent advances in synthetic techniques.

15.2.4 Perovskite/Polymer Conjugate Photodetectors

Spectrally effective photodetectors are used in many applications such as optical communication, fluorescence microscopy, national security, combating climate change, early missile detection, environment monitoring, and advanced communications (Chen et al. 2020c). Si, SiC p-n junction, or GaN is the photodetector that has been utilized the most frequently out of all the commercially accessible photodetectors. These commercially available photodetectors are expensive and only detect up to $<5 \times 10^{13}$ Jones of light (Guo et al. 2012; Shao et al. 2015; Chen et al. 2006). The photomultiplier tube is one of the most used photodetectors (PMT). They have high detectivity in both UV and visible regions of about $\sim 3 \times 10^{14}$ Jones. However, because of their size, poor UV-to-visible rejection ratio, and high operating voltage, their application is practically limited. These restrictions prompted researchers to create materials and equipment that are innovative and easily resolve the aforementioned problem. Light-emitting diodes (LEDs), solar cells with high power conversion efficiencies above 20%, and highly sensitive photodetectors are just a few of the high-performance devices that have recently been made possible by significant advancements in organic and inorganic lead halide perovskite materials (Shao et al. 2020). Their high carrier mobility, high optical absorption coefficient, and lengthy carrier recombination lifetime are what give them their fascinating performance. For photodetector applications, hybrid lead halide perovskite-based material systems are a good candidate because they can be produced at a low cost while still being able to achieve excellent detectivity (10^{13}–10^{14} Jones), a large linear dynamic range (over 100 db), and quick response times (in the order of microseconds) (Dou et al. 2014; Fang et al. 2015). However, these spectrally selective photodetectors involved the detection of a very narrow wavelength. The residual part should be removed by using a narrow-band photodetector (NPD). The majority of photoactive materials used nowadays absorb photons in a broad spectral range; hence, the standard NPD is typically created by fusing optical filters with photosensitive materials, which increases the cost of the developed photodetectors (Qin et al. 2020). A typical NPD photodetector device is shown in Figure 15.4. Evidently, it has a complex architecture and requires hole-transporting material (PDOT:PSS) and ETL (Bathocuproine-BCP) such as conductive, semiconducting polymer (PTB7), and C60.

The EQE of these NPDs is, however, comparatively low because a large portion of the carriers is likely trapped in the thick active layer before being collected by the electrode due to the low carrier mobility ($\sim 6 \times 10^{-4}$ cm^2 V^{-1} S^{-1}) of the organic materials (Liu et al. 2014; Xue et al. 2018). Armin et al. (2015) developed a new class of NPDs using 2 μm thick PCDTBT:PC70BM bulk heterojunction (BHJ) films without

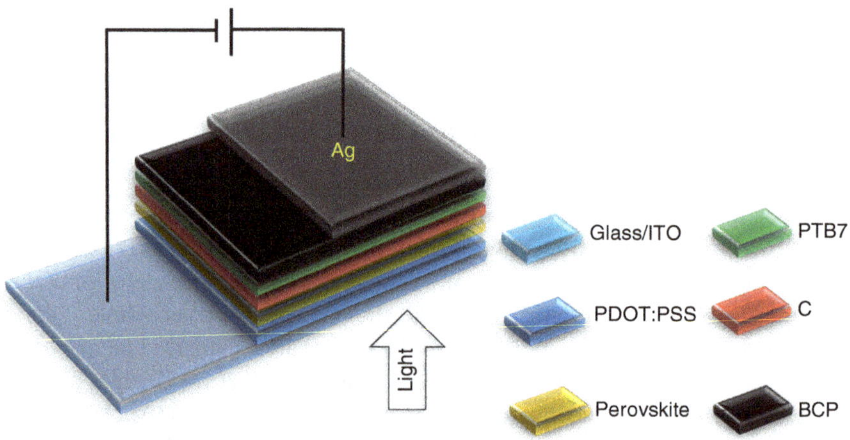

Figure 15.4 Schematic representation of the NPD device structure using synergetic layers of perovskite and polymer.

band pass optical filters (Wang et al. 2019b). It achieves CCN by making use of the variations in attenuation length of the light of various wavelengths.

Kim and co-workers used highly flexible and effective thin electrospun perovskite nanofiber materials to construct a photodetector. The pristine perovskite and polymer/perovskite composite films differ from the poly(vinylpyrrolidone) (PVP)/$MAPbI_3$ composite nanofibers in their distinctive moieties. The core and shell portions of the nanofibers, respectively, both concurrently displayed swollen nanoplates and embedded nanoparticles. Both the nanoparticles and nanoplates showed two wavelengths in PL emission spectra. The flexibility of the fibers has been evaluated by depositing the nanofibers in a rubbery substrate which can endure 15% of strain and started to fall apart at about 20% strain. Stress distribution in such materials can be determined using finite element analysis. Under green and red light irradiation, the PDs maintained 83.8% of their initial responsivity and detectivity after 15% strain. Additionally, under continuous green light irradiation, remarkable photostability for 720 hours at 35% relative humidity, as well as outstanding mechanical durability for 500 stretching cycles at 15% strain were observed (Kim et al. 2022).

Mathur et al. (2021), in their work, have developed a stretchable photodetector using the $MAPbI_3$ perovskite and polystyrene (PS-$MAPbI_3$). After 10,000 stretching cycles, 52% of its initial photocurrent value was obtained. However, in comparison to pristine $MAPbI_{3,}$ the photocurrent after 5 cycles decreases up to 80% and decreases up to 22% after 10,000 cycles. The stretchability in the detector device can be achieved by acquainting a wrinkled structure to the device. It is accomplished by putting the device on a stretched flexible substrate followed by releasing the tension to create a predictable wrinkled pattern. After only five stretching cycles, the perfect $MAPbI_3$ device exhibits considerable cracking, which significantly reduces performance. The PS-$MAPbI_3$, in contrast, exhibits a much lower crack density and smaller crack size which leads to the stretching that sustained device performance. The introduction of polystyrene to the $MAPbI_3$ moieties reduces the film modulus by ~38%. Apart from this, they also lead to a three-layer structure, which can resist the migration of ions across the grain boundaries and successively act as a shield against the ambient environment. This work demonstrates the use of room-temperature solution-processed platinum-gold (Pt-Au) nanoparticle chains to address the equally significant issue of fabricating economical electrodes that can endure considerable stretching, bending, and twisting. In comparison to the typically employed Ag and Cu nanowire electrodes in flexible electronics, the mesh of metallic Pt–Au nanochains offers confrontation against good conductivity and oxidation. Stretchable photodetector assemblies made of Pt-Au nanochain electrodes beneath a PS-$MAPbI_3$ film at a concentration of 1 wt% show high light switching ratios on the order of ~103 and can withstand high lateral strains of up to 100%.

15.2.5 Polymer–Metal Oxide-Based Photodetector

There are numerous applications of ultraviolet photodetectors in the field of missile early warning systems, medical diagnosis, ultraviolet astronomy, communication, and other emerging fields because of their potential properties. The most widely used photovoltaics for such applications include Schottky junction type (Iqbal et al. 2020), metal–semiconductor–metal type (M-S-M) (Lv et al. 2013), and p-n junction type (Chen et al. 2020c; Sharma et al. 2017). Due to the modest structure and relaxed construction, p-n junction has garnered a lot of attention. The fundamental benefit of p-n junctions is that they can function normally even in challenging circumstances. Both 0 bias voltage and external bias voltage are compatible with their operation. These junctions are the combination of both p- and n-type materials with opposite charges and potential differences. This facilitates the parting of photo generated holes and electrons. This resulted in a decrease in dark current and an increase in responsiveness, enabling operation even without an external power source. However, the main challenge of such UV photodetectors is to improve their performance (Hu et al. 2017; Ouyang et al. 2021). Nowadays researchers are trying to develop a photodetector based on p-n heterojunctions with high performance and selectivity.

The choice of material in the development of photodetectors is a crucial part. Semiconducting materials in this regard play an important role. They have been used widely because of their wide bandgap. Materials like TiO_2, ZnO, SnO_2, and NiO are the most commonly used material that has wide bandgap. Among all, TiO_2 gained a lot of attention due to its good physical and chemical properties as well as they are very less toxic, cost-effective, and stable to acidic and alkaline corrosion. They are mostly used in fields like solar cells, electrochemistry, photodetectors, and water splitting. Among the different morphologies, 1D nanomaterials of TiO_2 have a large surface area, strong electron conductivity, and a high aspect ratio, they perform the best among the various morphologies of TiO_2. One-dimensional TiO_2 can appear in a variety of morphologies, including ribbons, rods, flakes, and tubes, but nanorods are the best option for the creation of photodetectors (Che et al. 2021). Moreover, among all the semiconducting materials, organic semiconductors are comparatively easy to process and have good flexibility. As organic molecules contain different functional groups, modifying monomer molecules is relatively easy. However, carrier mobility in the case of organic semiconductors is very low. Therefore, a combination of both organic and inorganic semiconducting materials can overcome these shortcomings besides leading to some distinctive advantages. The common organic materials which have been used so far are poly-3-hexylthiophene-2,5-diyl (PFH), polyaniline (PANI), poly(3,4-ethylenedioxythiophene), poly(3,4-propylenedioxythiophene), and 3,4-ethylenedfcioxythiophene: polystyrene sulfonate (PEDOT: PSS). Among all the most widely used ones are the PEDOT-based ones as they are conducting in nature, can undergo reversible electrochemical doping and de-doping capability, and have high stability (Madhuvilakku et al. 2020). Researchers are trying to modify the 1D TiO_2 in order to achieve good physical and chemical properties. Zhang et al. (2015) in their work synthesized a modified TiO_2 nanorod by coating them with poly-3-hexylthiophene-2,5-diyl (PFH) spin-casting method. The developed device shows low dark current as well as good responsivity. Wang et al. (2015) in their work developed a heterojunction of polyaniline which is a p-type polymer and TiO_2 nanorod. Under UV irradiation of 365 nm at bias 0 V, this heterojunction displays a high photocurrent of around 3.5×10^{-4}. Apart from this, Tsay et al. (2021) fabricated a photodetector based on p-n junction based on GaN-ZnO nanowires. They have carried out a systematic study of the synthesized materials and checked their light response performance at different voltages such as 1, 2, and 0 V bias which corresponds to high sensitivity of about 8.48×10^4 under 0 V. In another work Song et al. (2021) prepared heterojunction of Ti_3C_2T-GaN. This heterojunction shows a high responsivity at 0 V bias. Che et al. (2021), in their work, developed a TiO_2 nanorod array (NRA) and PEDOT-type polymer composites using simple electrochemical polymerization. The TiO_2 NRAs were allowed to grow on the FTO substrate by a hydrothermal method where TiO_2 NRAs/FTO worked as a working electrode. The S-atoms present in the monomers can interact with the oxygen vacancies present in the TiO_2 NRAs to facilitate the forming of the polymeric films. TiO_2 NRAs in their respective composite act as a channel for light scattering and electron transmission, while the polymer acts as the hole transport layer. This combination constrains the recombination of photogenerated electrons and holes, which can improve the performance of the photodetector.

Tran and co-workers in their work developed a photodetector using ZnO as a base material. However, low responsivity, complicated fabrication, and slow speed limit their performance as a suitable photodetector. They have developed a p-n junction of graphene quantum dot (GQD) decorated on ZnO and poly(*N,N'*-bis-4-butylphenyl-*N,N'*-bisphenyl) benzidine (Poly-TPD). In their work firstly they have reduced the roughness and size of the ZnO using graphene quantum dot. This led to increase in the transport of photogenerated carriers as well as the charge lifetime of ZnO. They have checked the efficacy of the synthesized films using an appropriate GQD concentration and checked the efficiency of various ZnO:GQD devices based on charge transfer mechanism. Therefore, in comparison to the other photodiodes, the one based on ZnO:GQD/Poly-TPD exhibits the most balanced energy structure, improved mobility, and reduced charge recombination. These ZnO:GQD/Poly-TPD photodetectors display remarkable performance under 365 nm irradiation, have good stability, and can endure for around 10 weeks before experiencing severe photocurrent degeneration (Tran et al. 2021).

15.2.6 Polymer–Graphene Oxide (GO)/Reduced Graphene Oxide (rGO) Conjugate Photodetectors

Among all the 2D materials graphene can be considered as one of the ideal materials based on their both physical and chemical properties. It is a mono-elemental material composed of sp^2 hybridized C-atoms in a honeycomb arrangement. The remarkable properties make them suitable for a variety of applications such as photodetectors, super capacitors, modulators, and solar cells. Due to their excellent physical, chemical, optical, electrical, and electrochemical behavior, composites of graphene can be regarded as one of the most efficient materials. For instance, anisotropic polymer/graphene nanocomposites provide new ways to alter characteristics by changing their properties; the elements that are impacted include flaws, lateral size, and layer number. By creating conductive networks for carriers, it is possible to significantly enhance the electrical properties of polymer/graphene. By converting light energy to phonon energy as the composite absorbs the incident light, it is also possible to change the optical properties of polymer/graphene composites. However, the fabrication of GO is a bit difficult and destruction can occur during its transfer which can limit mass production. To overcome such difficulties, researchers are trying to use reduced graphene oxide (rGO) which can be synthesized using some easy and economical procedures. The devices based on rGO and rGO-composite possess high efficiency in different fields due to their large surface area, easy synthesis method, and cost effectiveness. This can be used for different fields such as sensing, green energy, and energy storage devices. The main advantage of rGO is that it has different reduction levels, which can be achieved by varying the annealing temperature. The performances of optoelectronic devices such as high selectivity, responsivity, detectivity, prolonged carrier lifetime, and light absorption can be improved by structure and surface engineering. Just like graphene composites, the use of rGO can also be beneficial for different applications. The incorporation of rGO flakes in polymer matrices like polyvinyl alcohol (PVA) can enhance and modify the desirable properties of the rGO-polymer-based films. Apart from such exceptional properties, the devices based on such membranes are hardly reported although both covalent and non-covalent approaches ensure the excellent dispersion of rGO flakes. Shih et al. (2022), in their work, developed a new method for the fabrication of low-power photodetectors of rGO-PVA films. The surface modification of such membranes enhances the performance and light response of such photodetectors which indicates the decrease of aggregation level as well as defective edges of rGO flakes. By controlling the percentage of PVA in rGO, one can enhance the detectivity as well as selectivity of the synthesized photodetectors. They have done surface engineering of rGO/PVA (polyvinyl alcohol) films to enhance the performance of photodetectors. Even when the photodetector is operated at a low bias of 0.3 V, the PVA/rGO photodetector's on/off ratio is 3.5 times better than that of the rGO photodetector, and its detectivity is 53% better.

A photodetector which can convert light into electrical signals has potential efficacy in different areas like night vision, weather surveillance, missile guidance, imaging, sensing, communication, and spectroscopy. Photodetectors are of two categories namely, thermal and photonic detectors.

(i) *Thermal or photothermoelectric (PTE) detectors*: This type of detector involves thermoelectric and photothermal conversion processes. They are responsive at ultra-broad bands without any cooling unit as well as external bias. However, the efficiency of such detectors mainly depends on the Seebeck coefficient of the thermoelectric materials and the difference in temperatures of the materials (Xiong et al. 2021).
(ii) *Photonic detectors*: These are based on the photoconductive and photoelectric effects of semiconductors. In the case of conventional semiconductors, only the photons which have energy $>h\upsilon$ can be detected. The main advantage of such devices is that they are highly sensitive and have narrow responsive bandwidths.

However, they need a cryogenic cooling unit as the dark current of such semiconductors is very high at room temperature. This will lead to an increase in the size and weight of the detector and limits their practical usability. So scientists are in search of small detectors without any cooling unit (Xiong et al. 2021).

Due to their intriguing properties, semiconductors are the preferred material for the creation of various optical and electrical devices. The ability to remove both inhomogeneous properties and crystal defects is a major benefit of small-sized crystal materials. Tiny materials are straightforward to analyze, and excellent material production makes the optoelectronics applications for these materials clear. They trap and scatter incident light, for example, and have a greater surface-to-volume ratio than their bulk counterparts (Graniel et al. 2018; Yildiz et al. 2016; Sharstniou et al. 2017; Vickers 2017).

In order to build optoelectronics, semiconducting devices must have a broad band gap in order to detect a variety of wavelengths and energies. This requirement can be met by two different organic, inorganic, or hybrid materials, as opposed to some expensive or underwhelming inorganic materials. Heterojunctions created specifically with nanoscale materials can therefore provide the opportunity to detect signals of different energies. This desire for electro-optical devices may be met by inorganic hybrid heterojunctions and rGO-PVP (polyvinyl pyrrolidone) fibers. Despite the potential for widespread usage of graphene-based materials due to their intriguing features, photodetectors produced from them have larger dark currents and higher power consumption.

Graphene or rGO can be combined with the appropriate nanostructured material to create composites to overcome this drawback. rGO has numerous advantages over graphene, including the ability to combine with other materials to create different composites, which makes it more advantageous in many ways. It has been extensively employed in numerous fields, including electrical devices and biosensors, due to its exceptional qualities, including ease of processing, scalable production at low cost, flexibility, and high potential (Kuilla et al. 2010; Herranz et al. 2010; Shao et al. 2012; Kim et al. 2012). Apart from their advantages, it has some drawbacks such as poor solubility for the formation of films on a particular substrate. To overcome such disadvantages, rGO/polymer is recommended which exhibits higher dispersion stability. PVP works well as a polymer and has excellent film-forming properties. It is also readily soluble in water. To prepare an appropriate viscous solution, it is therefore needed a sufficient amount of organic ingredients at which the uniform dispersion of rGO can be obtained. Additionally, PVP has frequently been used to improve surface capabilities of rGO and PVP as a stabilizer encourages the uniform distribution of the rGO because of the potent – interactions between the molecules of graphene and PVP. It's evident that for the deposition of rGO onto the Si substrate utilizing the electrospinning technique a suitably viscous solution of material must be given (Khalili and Chenari 2020). As a result of PVP's excellent film formation, superior compatibility, and big area, it has frequently been used in the fabrication of nanocomposite polymer films. For this reason, the electrospinning approach for the deposition of PVP-rGO film on Si substrate for device applications was developed. When used as an interfacial layer in electronic devices to enhance the electrical properties of metal–semiconductor (MS) Schottky diodes, PVP, which is an organic substance, offers numerous properties like ease of processing, cost-effectiveness, high water solubility, and flexibility (Alptekin and Altındal 2019).

Xiong et al. (2021), in their work, constructed a photodetector with broadband by coating a uniform photothermal layer on a thermistor. There is a conversion of light energy to heat due to the photothermal effect, which is then reflected by the voltage variation of the thermistor. They have synthesized a photothermal layer with high efficacy via encapsulating rGO film into a PDMS (polydimethylsiloxane) layer. The photodetectors based on rGO/PDMS exhibit high responsivity, broad spectral range (400–1300 nm), and good response. The simulation of sunlight at 0.8 kW m^{-2} shows responsivity up to 176.9 V W^{-1} with a response-recovery time of about ~80 s. At 473, 532, and 808 nm a laser emission can also be detected with high responsivities of about 15.6, 14.6, and 19.4 V W^{-1}, respectively. These rGO/PDMS films are not only considered as high-performing photothermal film but also create an effective plan for broadband photodetectors.

Apart from the advantages of rGO-based materials for the development of photodetectors, there are still some defects and aggregation in them which can effectively limit their performances in photodetectors. To overcome such difficulties, one has to control the composition of rGO, which can be considered as one of the vital factors for the improvement of their performances in different fields. In this regard, the interconnection of rGO and polymers can effectively modify the morphologies which can improve the enactment of devices.

15.3 Conclusion

In conclusion, it can be stated that the development of photodetectors from various organic and inorganic materials is a very important domain of research due to its wide range of practical applications. In this regard, designing new organic polymers plays a vital role as it offers the versatility of use in fabricating photodetector device architectures. Therefore, researchers around the globe have been working in developing polymers and polymer composites with superior photophysical and mechanical properties. A discussion about the available polymers used in photodetector fabrication is demonstrated in the book chapter. However, it is realized that there is still a wide scope for improvement and many other new things to be explored. For instance, synthesis of polymers is generally time consuming process and requires chemicals in substantial amount. It may lead to the accidental release of chemicals to the environment. To be able to solve that, alternative chemicals and greener synthesis routes can be employed for the synthesis of polymers. Another way to minimize chemical pollution is to use polymeric materials that are abundantly available in nature which will not only provide the necessary photophysical properties required for fabricating photodetector but also lower the cost of production.

References

Alptekin, S. and Altındal, Ş. (2019). A comparative study on current/capacitance: voltage characteristics of Au/n-Si (MS) structures with and without PVP interlayer. *Journal of Materials Science: Materials in Electronics* 30: 6491–6499.

Armin, A., Jansen-van Vuuren, R.D., Kopidakis, N. et al. (2015). Narrowband light detection via internal quantum efficiency manipulation of organic photodiodes. *Nature Communications* 6: 6343.

Chahrour, K.M., Yam, F., and Abdalrheem, R. (2019). High-performance UV photodetector of anodic rutile TiO_2 nanotube arrays. *Materials Letters* 248: 161–164.

Che, Y., Zhang, H., Abdiryim, T. et al. (2021). Ultraviolet photodetectors based on TiO_2 nanorod arrays/PEDOT-type conducting polymers. *Optical Materials* 122: 111805.

Chee, K.W. (2018). Introductory chapter: photodetectors. In: *Advances in Photodetectors-Research and Applications* (ed. K. Chee), 3–8. IntechOpen.

Chen, M.C., Sheu, J.K., Lee, M.L. et al. (2006). Improved performance of planar GaN-based p-i-n photodetectors with Mg-implanted isolation ring. *Applied Physics Letters* 89: 183509.

Chen, H., Yu, P., Zhang, Z. et al. (2016). Ultrasensitive self-powered solar-blind deep-ultraviolet photodetector based on all-solid-state polyaniline/MgZnO bilayer. *Small* 12: 5809–5816.

Chen, D., Xin, Y., Lu, B. et al. (2020a). Self-powered ultraviolet photovoltaic photodetector based on graphene/ZnO heterostructure. *Applied Surface Science* 529: 147087.

Chen, X., Sun, C., Liu, Y. et al. (2020b). All-inorganic perovskite quantum dots $CsPbX_3$ (Br/I) for highly sensitive and selective detection of explosive picric acid. *Chemical Engineering Journal* 379: 122360.

Chen, J., Ouyang, W., Yang, W. et al. (2020c). Recent progress of heterojunction ultraviolet photodetectors: materials, integrations, and applications. *Advanced Functional Materials* 30: 1909909.

Chow, P.C. and Someya, T. (2020). Organic photodetectors for next-generation wearable electronics. *Advanced Materials* 32: 1902045.

Dong, Y., Wang, S., Zou, Y. et al. (2018). Zinc stannate nanocrystal–based ultrarapid-response UV photodetectors. *Advanced Materials Technologies* 3: 1800085.

Dou, L., Yang, Y.M., You, J. et al. (2014). Solution-processed hybrid perovskite photodetectors with high detectivity. *Nature Communications* 5: 1–6.

Eggenhuisen, T., Galagan, Y., Biezemans, A. et al. (2015). High efficiency, fully inkjet printed organic solar cells with freedom of design. *Journal of Materials Chemistry A* 3: 7255–7262.

Fang, Y., Dong, Q., Shao, Y. et al. (2015). Highly narrowband perovskite single-crystal photodetectors enabled by surface-charge recombination. *Nature Photonics* 9: 679–686.

Forrest, S.R. (2004). The path to ubiquitous and low-cost organic electronic appliances on plastic. *Nature* 428: 911–918.

Gedamu, D., Paulowicz, I., Kaps, S. et al. (2014). Rapid fabrication technique for interpenetrated ZnO nanotetrapod networks for fast UV sensors. *Advanced Materials* 26: 1541–1550.

Gogoi, B. and Sen Sarma, N. (2015). Curcumin–cysteine and curcumin–tryptophan conjugate as fluorescence turn on sensors for picric acid in aqueous media. *ACS Applied Materials & Interfaces* 7: 11195–11202.

Gong, X., Tong, M., Xia, Y. et al. (2009). High-detectivity polymer photodetectors with spectral response from 300 nm to 1450 nm. *Science* 325: 1665–1667.

Graniel, O., Fedorenko, V., Viter, R. et al. (2018). Optical properties of ZnO deposited by atomic layer deposition (ALD) on Si nanowires. *Materials Science and Engineering: B* 236: 139–146.

Grimoldi, A., Colella, L., La Monaca, L. et al. (2016). Inkjet printed polymeric electron blocking and surface energy modifying layer for low dark current organic photodetectors. *Organic Electronics* 36: 29–34.

Guo, F., Yang, B., Yuan, Y. et al. (2012). A nanocomposite ultraviolet photodetector based on interfacial trap-controlled charge injection. *Nature Nanotechnology* 7: 798–802.

Han, X., Feng, S., Zhao, Y. et al. (2019). Synthesis of ternary oxide Zn_2 GeO_4 nanowire networks and their deep ultraviolet detection properties. *RSC Advances* 9: 1394–1402.

Hazarika, I., Goswami, K.J., Hussain, A.A. et al. (2021). Silk fibroin protein as dual mode picric acid sensor and UV photoactive material. *Journal of Materials Science* 56: 18959–18975.

He, X., Léonard, F., and Kono, J. (2015). Uncooled carbon nanotube photodetectors. *Advanced Optical Materials* 3: 989–1011.

Herranz, D., Muñoz-Martin, M., Cañamero, M. et al. (2010). Sirt1 improves healthy ageing and protects from metabolic syndrome-associated cancer. *Nature Communications* 1: 1–8.

Hiramoto, M., Imahigashi, T., and Yokoyama, M. (1994). Photocurrent multiplication in organic pigment films. *Applied Physics Letters* 64: 187–189.

Hu, K., Teng, F., Zheng, L. et al. (2017). Binary response Se/ZnO p-n heterojunction UV photodetector with high on/off ratio and fast speed. *Laser & Photonics Reviews* 11: 1600257.

Huang, P.H., Shen, Y.C., Tung, C.Y. et al. (2020). Energy-saving $ZnGa_2O_4$ phototransistor improved by thermal annealing. *ACS Applied Electronic Materials* 2: 3515–3521.

Iqbal, M.Z., Faisal, M.M., Afzal, A.M. et al. (2020). Ultraviolet-light-driven current modulation of Au/WS_2/Gr Schottky barrier. *Physica E: Low-Dimensional Systems and Nanostructures* 117: 113837.

Jansen-van Vuuren, R.D., Armin, A., Pandey, A.K. et al. (2016). Organic photodiodes: the future of full color detection and image sensing. *Advanced Materials* 28: 4766–4802.

Ji, Z., Zhao, W., Xiang, L. et al. (2022). Hierarchical heterojunction enhanced photodoping of polymeric semiconductor for photodetection and photothermoelectric applications. *ACS Materials Letters* 4: 815–822.

Khalili, S. and Chenari, H.M. (2020). Successful electrospinning fabrication of ZrO_2 nanofibers: a detailed physical–chemical characterization study. *Journal of Alloys and Compounds* 828: 154414.

Kim, D., Yang, S.J., Kim, Y.S. et al. (2012). Simple and cost-effective reduction of graphite oxide by sulfuric acid. *Carbon* 50: 3229–3232.

Kim, H.J., Oh, H., Kim, T. et al. (2022). Stretchable photodetectors based on electrospun polymer/perovskite composite nanofibers. *ACS Applied Nano Materials* 5: 1308–1316.

Kuilla, T., Bhadra, S., Yao, D. et al. (2010). Recent advances in graphene based polymer composites. *Progress in Polymer Science* 35: 1350–1375.

Lee, Y.H., Liu, H., Lee, J.Y. et al. (2010). Dipyrenylcalix [4] arene—a fluorescence-based chemosensor for trinitroaromatic explosives. *Chemistry–A European Journal* 16: 5895–5901.

Li, L., Huang, Y., Peng, J. et al. (2014). Highly responsive organic near-infrared photodetectors based on a porphyrin small molecule. *Journal of Materials Chemistry C* 2: 1372–1375.

Li, Y., Huang, W., Liu, H. et al. (2018). UV photodetector based on polycrystalline SnO_2 nanotubes by electrospinning with enhanced performance. *Journal of Nanoparticle Research* 20: 334.

Liang, S., Dai, Y., Wang, G. et al. (2020). Room-temperature fabrication of SiC microwire photodetectors on rigid and flexible substrates via femtosecond laser direct writing. *Nanoscale* 12: 23200–23205.

Liu, Y., Zhao, J., Li, Z. et al. (2014). Aggregation and morphology control enables multiple cases of high-efficiency polymer solar cells. *Nature Communications* 5: 1–8.

Liu, X., Lin, Y., Liao, Y. et al. (2018a). Recent advances in organic near-infrared photodiodes. *Journal of Materials Chemistry C* 6: 3499–3513.

Liu, J.Q., Gao, Y., Wu, G.A. et al. (2018b). Silicon/perovskite core–shell heterojunctions with light-trapping effect for sensitive self-driven near-infrared photodetectors. *ACS Applied Materials and Interfaces* 10: 27850–27857.

Lv, K., Zhang, M., Liu, C. et al. (2013). TiO_2 ultraviolet detector based on $LaAlO_3$ substrate with low dark current. *Journal of Alloys and Compounds* 580: 614–617.

Madhuvilakku, R., Alagar, S., Mariappan, R., and Piraman, S. (2020). Glassy carbon electrodes modified with reduced graphene oxide-MoS_2-poly (3, 4-ethylene dioxythiophene) nanocomposites for the non-enzymatic detection of nitrite in water and milk. *Analytica Chimica Acta* 1093: 93–105.

Mathur, A., Fan, H., and Maheshwari, V. (2021). Soft polymer–organolead halide perovskite films for highly stretchable and durable photodetectors with Pt–Au nanochain-based electrodes. *ACS Applied Materials & Interfaces* 13: 58956–58965.

Miao, J. and Zhang, F. (2019). Recent progress on photomultiplication type organic photodetectors. *Laser & Photonics Reviews* 13: 1800204.

Nasiri, N. and Tricoli, A. (2019). Nanomaterials-based UV photodetectors. In: *Industrial Applications of Nanomaterials* (ed. S. Thomas, Y. Grohens, and Y.B. Pottathara), 123–149. Elsevier.

Nie, R., Deng, X., Feng, L. et al. (2017). Highly sensitive and broadband organic photodetectors with fast speed gain and large linear dynamic range at low forward bias. *Small* 13: 1603260.

Omnes, F. (2009). Introduction to semiconductor photodetectors. In: *Optoelectronic Sensors* (ed. D. Decoster and J. Harari), 1–14. Wiley.

Ouyang, W., Chen, J., Shi, Z., and Fang, X. (2021). Self-powered UV photodetectors based on ZnO nanomaterials. *Applied Physics Reviews* 8: 031315.

Park, S., Fukuda, K., Wang, M. et al. (2018). Ultraflexible near-infrared organic photodetectors for conformal photoplethysmogram sensors. *Advanced Materials* 30: 1802359.

Pierre, A., Deckman, I., Lechêne, P.B., and Arias, A.C. (2015). High detectivity all-printed organic photodiodes. *Advanced Materials* 27: 6411–6417.

Predeep, P. (2011). *Optoelectronics: Devices and Applications*. BoD–Books on Demand.

Qin, Z., Song, D., Xu, Z. et al. (2020). Filterless narrowband photodetectors employing perovskite/polymer synergetic layers with tunable spectral response. *Organic Electronics* 76: 105417.

Rauch, T., Böberl, M., Tedde, S.F. et al. (2009). Near-infrared imaging with quantum-dot-sensitized organic photodiodes. *Nature Photonics* 3: 332–336.

Saracco, E., Bouthinon, B., Verilhac, J.M. et al. (2013). Work function tuning for high-performance solution-processed organic photodetectors with inverted structure. *Advanced Materials* 25: 6534–6538.

Schilinsky, P., Waldauf, C., and Brabec, C.J. (2002). Recombination and loss analysis in polythiophene based bulk heterojunction photodetectors. *Applied Physics Letters* 81: 3885–3887.

Sekitani, T., Noguchi, Y., Zschieschang, U. et al. (2008). Organic transistors manufactured using inkjet technology with subfemtoliter accuracy. *Proceedings of the National Academy of Sciences* 105: 4976–4980.

Shao, G., Lu, Y., Wu, F. et al. (2012). Graphene oxide: the mechanisms of oxidation and exfoliation. *Journal of Materials Science* 47: 4400–4409.

Shao, D., Gao, J., Chow, P. et al. (2015). Organic–inorganic heterointerfaces for ultrasensitive detection of ultraviolet light. *Nano Letters* 15: 3787–3792.

Shao, D., Zhu, W., Liu, X. et al. (2020). Ultrasensitive UV photodetector based on interfacial charge-controlled inorganic perovskite–polymer hybrid structure. *ACS Applied Materials & Interfaces* 12: 43106–43114.

Sharma, S., Khosla, R., Deva, D. et al. (2017). Fluorine-chlorine co-doped TiO_2/CSA doped polyaniline based high performance inorganic/organic hybrid heterostructure for UV photodetection applications. *Sensors and Actuators A: Physical* 261: 94–102.

Sharstniou, A., Niauzorau, S., Chubenko, E. et al. (2017). Morphology dependent optical properties of ZnO/SiNWs nanocomposites. *MRS Advances* 2: 3667–3672.

Shen, L., Zhang, Y., Bai, Y. et al. (2016). A filterless, visible-blind, narrow-band, and near-infrared photodetector with a gain. *Nanoscale* 8: 12990–12997.

Shih, Y.S., Li, W.C., Shen, J.H. et al. (2022). Low-power photodetectors based on PVA-modified reduced graphene oxide hybrid solutions. *Macromolecular Rapid Communications* 43: 2100854.

Song, W., Chen, J., Li, Z., and Fang, X. (2021). Self-powered MXene/GaN van der Waals heterojunction ultraviolet photodiodes with superhigh efficiency and stable current outputs. *Advanced Materials* 33: 2101059.

Tedde, S.F., Kern, J., Sterzl, T. et al. (2009). Fully spray coated organic photodiodes. *Nano Letters* 9: 980–983.

Tran, M.H., Park, T., and Hur, J. (2021). Solution-processed ZnO: graphene quantum dot/poly-TPD heterojunction for high-performance UV photodetectors. *Applied Surface Science* 539: 148222.

Tsay, C.Y., Hsiao, I.P., Chang, F.Y., and Hsu, C.L. (2021). Improving the photoelectrical characteristics of self-powered p-GaN film/n-ZnO nanowires heterojunction ultraviolet photodetectors through gallium and indium co-doping. *Materials Science in Semiconductor Processing* 121: 105295.

Valouch, S., Hönes, C., Kettlitz, S.W. et al. (2012). Solution processed small molecule organic interfacial layers for low dark current polymer photodiodes. *Organic Electronics* 13: 2727–2732.

Vickers, N.J. (2017). Animal communication: when I'm calling you, will you answer too? *Current Biology* 27: R713–R715.

Wang, H. and Kim, D.H. (2017). Perovskite-based photodetectors: materials and devices. *Chemical Society Reviews* 46: 5204–5236.

Wang, H., Yi, G., Zu, X. et al. (2015). A highly sensitive and self-powered ultraviolet photodetector composed of titanium dioxide nanorods and polyaniline nanowires. *Materials Letters* 138: 204–207.

Wang, F., Gao, T., Zhang, Q. et al. (2019a). Liquid-alloy-assisted growth of 2D ternary Ga2In4S9 toward high-performance UV photodetection. *Advanced Materials* 31: 1806306.

Wang, H., Wang, X., Luo, X. et al. (2019b). Optimization of all figure-of-merits in well-aligned GaN microwire array based Schottky UV photodetectors by Si doping. *ACS Photonics* 6: 1972–1980.

White, M.S., Kaltenbrunner, M., Głowacki, E.D. et al. (2013). Ultrathin, highly flexible and stretchable PLEDs. *Nature Photonics* 7: 811–816.

Xiong, S., Li, L., Qin, F. et al. (2017). Universal strategy to reduce noise current for sensitive organic photodetectors. *ACS Applied Materials & Interfaces* 9: 9176–9183.

Xiong, M., Shan, X., Liu, C. et al. (2021). Broadband photodetectors based on enhanced photothermal effect of polymer encapsulated graphene film. *Applied Surface Science Advances* 3: 100050.

Xue, J., Zhu, Z., Xu, X. et al. (2018). Narrowband perovskite photodetector-based image array for potential application in artificial vision. *Nano Letters* 18: 7628–7634.

Yang, D. and Ma, D. (2019). Development of organic semiconductor photodetectors: from mechanism to applications. *Advanced Optical Materials* 7: 1800522.

Yildiz, A., Uzun, S., Serin, N., and Serin, T. (2016). Influence of grain boundaries on the figure of merit of undoped and Al, In, Sn doped ZnO thin films for photovoltaic applications. *Scripta Materialia* 113: 23–26.

Young, M., Suddard-Bangsund, J., Patrick, T.J. et al. (2016). Organic heptamethine salts for photovoltaics and detectors with near-infrared photoresponse up to 1600 nm. *Advanced Optical Materials* 4: 1028–1033.

Zhang, M., Li, D., Zhou, J. et al. (2015). Ultraviolet detector based on TiO_2 nanowire array–polymer hybrids with low dark current. *Journal of Alloys and Compounds* 618: 233–235.

Zhang, Y., Ji, T., Zhu, J. et al. (2020). A high performance self-powered heterojunction photodetector based on NiO nanosheets on an n-Si (100) modified substrate. *Materials Letters* 285: 128995.

Zhong, Y., Sisto, T.J., Zhang, B. et al. (2017). Helical nanoribbons for ultra-narrowband photodetectors. *Journal of the American Chemical Society* 139: 5644–5647.

Zhong, Z., Bu, L., Zhu, P. et al. (2019). Dark current reduction strategy via a layer-by-layer solution process for a high-performance all-polymer photodetector. *ACS Applied Materials & Interfaces* 11: 8350–8356.

Zhou, X., Yang, D., Ma, D. et al. (2016). Ultrahigh gain polymer photodetectors with spectral response from UV to near-infrared using ZnO nanoparticles as anode interfacial layer. *Advanced Functional Materials* 26: 6619–6626.

16

Organic-Solvent-Resistant Polymeric Membranes for Emerging Application in Separation Science

Chinmoy Bhuyan[1,2], *Prarthana Bora*[1,2], *Parashmoni Rajguru*[1,2], *Achyut Konwar*[1], *and Swapnali Hazarika*[1,2]

[1] *CSIR-North East Institute of Science and Technology, Chemical Engineering Group and Centre for Petroleum Research, Jorhat, Assam, 785006, India*
[2] *Academy of Scientific and Innovative Research (AcSIR), Ghaziabad, Uttar Pradesh, 201002, India*

16.1 Introduction

In various industries like chemical and petrochemical, separation and purification processes have vital applications like purification of raw materials and removal of unwanted substances (Nunes et al. 2020). Besides these, separation technology has also great role in the treatment of pollutants generated from these industries (Li 2019). Membrane technology has been proven superior among the separation and purification processes in terms of efficient and easy separation. It works under a very simple permeation procedure which is easy to install and operate, and can reduce the cost by consuming minimum energy (Lee et al. 2011). Membrane technology has several other important advantages, e.g. high separation efficiency, ease of scaling, and avoidance of generating harmful by-products (Adler et al. 2000). A typical membrane separation technique has been depicted in Figure 16.1. Organic solvents are widely used in the petrochemical and pharmaceutical industries which need to be separated or purified at some point (Zeng et al. 2017). However, the corrosive nature of the organic solvents for various separating media makes organic solvent separation very difficult in terms of the selectivity and durability of the separating media. Polymers and their derivatives play an important role during separation and purification process and act as a vital component in the separating media, but most of them are completely or partially soluble in organic solvents, which hamper the selectivity and permeability as well as the durability of the separating media. Thus, it is quite essential to develop organic-solvent-resistant polymers having strong capability to resist degradation from the organic solvents (Marchetti et al. 2014).

In recent days, membrane technology has been facing lots of challenges due to rapid industrialization and high requirement in medical technologies. Separation of organic solvents is a challenge in order to minimize their severe impacts on environment as well as ecosystems. Organic solvents are commonly used in the pharmaceutical industry to extract high-value products. However, the processes for

Organic Polymers in Energy-Environmental Applications, First Edition.
Edited by Ramesh Oraon, Pardeep Singh, Sanchayita Rajkhowa, Sangita Agarwal, and Ravindra Pratap Singh.

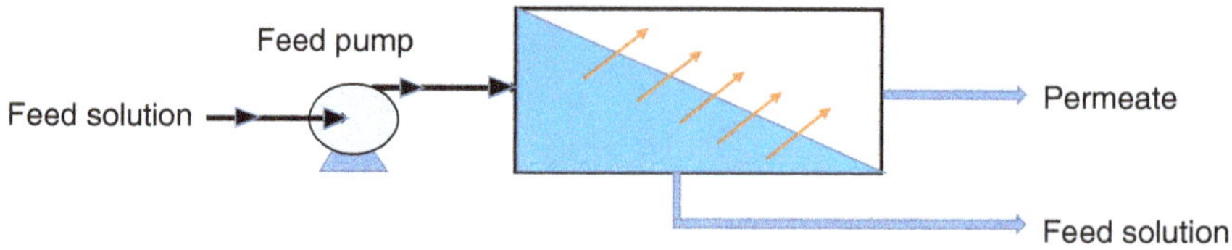

Figure 16.1 A typical membrane separation technique (flow of permeate through membrane is shown with arrows).

recycling, discarding, and reusing organic solvents are very expensive. To mitigate these day-to-day problems related to management of organic solvents, development of organic-solvent-resistant (OSR) membranes is becoming an important alternative.

OSR membranes can be efficiently used in nanofiltration, reverse osmosis, forward osmosis, ultrafiltration as well as pervaporation of organic solvents. Also, they can be extensively used in solvent recovery, separation and purification of solutes in food industry, pharmaceutical, and other chemical industries. However, designing an excellent OSR membrane is still very challenging (Marchetti et al. 2014).

Solvent-resistant nanofiltration based on membrane materials was reviewed by Vankelecom and their group in 2008 (Vandezande et al. 2008). Recent developments in OSR membrane and various mechanisms behind it were reviewed by Marchetti and their group (Marchetti et al. 2014). The sustainability issue related to the organic solvent nanofiltration was accessed by Szekely et al. (2014). Bye et al. in another review reported about the materials that may have potential application as an organic-solvent-resistant material. Development of cellulose-acetate-based membranes was the first attempt to separate organic solvents (xylene from ethanol-xylene mixture), and after that it was used as the base polymer in polymeric membranes for the use in organic separation. Cross-linked Lenzing P84 polyimide-based membranes are also very much effective for organic separation. Other polymers which have been extensively used in organic solvent separation and purifications even in harsh conditions also are polyether ether ketone (PEEK), polyacrylates, perfluorinated polymers, etc. (Galizia and Bye 2018). In this chapter, we have briefly discussed OSR membranes based on organic polymers and their applications in separation science.

16.2 Importance of Organic Polymers in Separation Science

Membrane, a separation barrier, has played an important role in separation science and technology, which is used in various industries (Geens et al. 2007). Membranes have been extensively used for separation and purification purposes of chemicals, biochemicals, water treatment, recovery of valuable components from liquids, gas streams, etc. (Gould et al. 2001). Separation processes account for 50–90% of capital investment in most chemical industries. Separating the value-added product from the organic solvent is a difficult task. To continue, membrane separation technology is preferred because of its ability to separate concentrated product without the use of high temperatures. Aside from these, the food and chemical industries make extensive use of organic solvent purification, recovery, and separation (Ren et al. 2021). The main challenge with membrane-based separation in contact with organic solvents is the membrane's stability and durability. When the membrane is in contact with a solvent for an extended period of time, it fouls. This fouling arises because of the possible interactions between the polymer of the membrane and the feed liquid. Plasticization and swelling cause pore blockage, which is referred to as fouling (Ulbricht 2020). Most of the organic solvents can cause swelling in polymeric membranes which ultimately lead to pore blockage. Common organic solvents that are responsible for membrane fouling and pore blockage during organic separations include acetone, chloroform, toluene, benzene, tetrachloroethane, toluene, diethyl ether, isopropanol, tetrahydrofuran (THF), methanol, and 1-octene. Petroleum industry wastewater also contains lots of organic solvents like n-hexane and cresol which also affect the membrane morphology and hence its performance (Marchetti et al. 2014; Al-Khalid and El-Naas 2018). To meet this type of complication and to attain the highest flux as well as selectivity, membrane material should be selected very carefully. Thus, the manufacture of OSR membranes to avoid fouling has become a great matter of concern in order to make membrane technology an irreplaceable one.

16.3 OSR Membranes: Materials and Classifications

Recently, membrane material development with strong resistivity toward organic solvents has become a growing topic of research. Such materials are specialty polymers, metal organic frameworks (MOF), covalent organic frameworks (COF), polymers of intrinsic microporosity (PIMs), carbon-based materials, and conjugated polymers. These can offer a solution to OSR membrane fouling. Based on the material type, membranes can be classified into different types, such as inorganic membranes, polymeric membranes, etc.

16.3.1 Inorganic Membrane Materials

Ceramic membrane can be prepared by using inorganic ceramic materials. In this type of membrane, alumina, titanium oxide, silicon oxide, zirconia, etc. are reported to be taken and coated on the membrane surface, followed by high-temperature treatment (Namvar-Mahboub and Pakizeh 2013). The structure of commercial membranes consists of three layers: a porous support layer, an interlayer, and a separative layer. Chemical stability: resistivity toward acids/bases/organic solvents, anti-microbial activity, mechanical strength and thermal stability are the main advantages associated with a ceramic membrane, which make it more suitable than a polymeric membrane. However, in comparison to polymeric materials, they are more brittle, hence difficult to upscale (Omar et al. 2022). It is seen from the literature that less variation in materials is possible for ceramic membranes as compared to polymeric materials. Therefore, hybrid composite membranes containing both polymers and inorganic materials are of great importance, especially for designing OSR membranes (Ren et al. 2021).

16.3.2 Organic Membrane Materials

The reported polymers for fabrication of (OSR) membranes include polyacrylonitrile (PAN) (Wu et al. 2012; Liu et al. 2020), polyimide (PI) (Omar et al. 2022), polypropylene (PP) (Li et al. 2018), polysulfone (PSf) (Kosaraju and Sirkar 2008) and polyaniline (PA) (Peyravi et al. 2012) etc. Using an interfacial polymerization reaction, a thin, dense film can be obtained on the porous supporting structure of the membrane (He et al. 2021). This may decrease the flux of the membrane, but the selectivity increases extensively. The addition of different additives and pore-inducing agents may also have a huge impact in selectivity and permeability of the membrane (Yang et al. 2021; Lalia et al. 2013; Buonomenna and Bae 2015).

Due to the presence of high number of polar groups with high biodegradability and biocompatibility, polyvinyl alcohols (PVA) are widely used in designing (OSR) membranes (Teodorescu et al. 2018; Gimenes et al. 2007). Polydimethylsiloxane (PDMS) is used as a membrane material, especially for the pervaporation separation approach of organic solvents (Si et al. 2022; Li et al. 2013). Polyacrylonitrile (PAN) membranes incorporated with FeOOH (Zhang et al. 2019), modified by tannic acid (Tham and Chung 2020), modified by chitosan (Musale and Kumar 2000), and composites with hyperbranched polyester (Wei et al. 2012) were also used for the fabrication of membranes with solvent-resistant properties. A natural polysaccharide, chitosan, blended with cellulose (Thakur and Voicu 2016), composite of chitosan with 2,5-furandicarboxaldehyde on a porous support of polyethylene terephthalate (PET) is used as thin-film composite (TFC) membranes (Park et al. 2021). Other polymers used as membrane materials for the fabrication of OSR membranes include polyvinyl pyrrolidone (PVP) (George et al. 2022), polyethylene (Heidari and Mahdavi 2021), Teflon (Shi and Chung 2020), polybenzimidazoles (PBI) (Tashvigh and Chung 2019), polyvinylidene fluoride (PVDF) (Zhang et al. 2020), and poly(ether sulfone) (PES) (George et al. 2022). A study reported on a PEEK membrane fabricated using the phase inversion method, controlling its pore size using heat treatment, and the effect of these pore sizes on the removal efficiency of the membrane toward MB (methylene blue), RB (rose-Bengal-sodium salt), and BSA (bovine serum albumin). In organic solvents and even in strong acids, the membrane shows very good reproducibility (Cao et al. 2020).

16.3.3 Conjugated Microporous Polymers (CMPs): Trending Novel Membrane Materials

CMPs are special category of porous polymeric materials that are formed by combining an extended π-conjugation with a permanently microporous skeleton.

Organic building blocks like CMPs exhibit some important characteristics like high specific surface area, low molecular density, and strong resistance to organic solvents, which make them superior as an ideal organic polymer to be used in organic-solvent separation and purification. Organic building blocks are easy to functionalize as compared to other novel membrane materials like activated carbon and graphite. CMPs are microporous materials that have an extended conjugate system and are connected to each other by multiple C—C bonds or aromatic systems (Cooper 2009). Copper created the first CMPs in 2009 by combining aryl halides and two-dimensional aryl alkylde. Due to their exceptional advantages like intrinsic three-dimensional porous microstructure, considerable specific surface area, varied synthetic methods, and superior stability, CMPs have gained much attention in the field of separation and purification (Jiang et al. 2007). In recent years, CMPs have found extensive application in environmental remediation, energy sources, and catalysis. CMPs can be used in harsh conditions of temperature, acidity, basicity, and organic solvent due to their prominent stability. Moreover, CMPs have been showing potential in pollutant recovery as well as organic solvent purification and separation as they have insoluble properties and a highly extended conjugated structure (Xu et al. 2013). The design of a superhydrophobic surface for easy absorption of oil and non-polar organic compounds was done by Lee et al. where they used superhydrophobic open-pore structured CMPs. Lipophilic material was created in that study by combining hydrophobic CMPs and hydrophilic sponge (Lee and Cooper 2020).

16.4 Types of Membrane Used for the Treatment of Organic Solvent

Organic solvents with high purity have been extensively used in chemicals, petrochemicals, and pharmaceutical industries during synthesis of organic materials, their separation, and purification. Due to strict protocols for environmental safety, organic solvents can't be reused or directly disposed. Conventional methods like distillation or evaporation often cost a lot, and energy consumption in these processes is very high. Pervaporation (PV), organic solvent nanofiltration (OSN), organic solvent reverse osmosis (OSRO), organic solvent forward osmosis (OSFO), and organic solvent ultrafiltration (OSU) are some advanced membrane-based separation processes which have been used in the separation of organic solvents.

16.4.1 Pervaporation (PV)

PV refers to a molecular-level liquid separation technology which proceeds through two mechanisms; evaporation with selective permeation (Yun et al. 2017). By this membrane-based separation technique a mixture of solvents can be separated on the basis of difference in their diffusion properties (Rdzanek et al. 2018). In PV partial pressure of the components that is to be separated is increased on the feed side whereas in the permeate side, a low partial pressure is maintained by using a vacuum pump (Paredes et al. 2017). The membrane selectively allows the passes of components under the pressure difference and the components condense on the downstream side. Separation of isomeric mixtures can also be easily done by PV method, which cannot be done by the conventional methods (Santoro et al. 2017). PV membranes should be improved in order to make them organic solvent resistant. Low permeability is a limitation in PV membrane technique, and it is only suitable for mixtures where component content is low. After a lots of research work related to PV many mass transfer models were proposed in order to explain PV, which includes pore flow model, the irreversible thermodynamic model, and the solution diffusion model. In designing a suitable selective membrane, the mass transfer model plays an important role (Van Hecke and De Wever 2017; Cheng et al. 2012).

16.4.2 Organic Solvent Nanofiltration (OSN)

Nanofiltration is the most widely used membrane separation technique used in various industries during separation of solvents. These types of membranes have pore size in the range of 0.5–2 nm and are effective in separating molecules of 200–1000 Da molecular weight (Marchetti and Livingston 2015; Tsarkov et al. 2012). OSN membranes has been gradually improved to meet the requirements like organic solvent and pollution resistance, making it suitable for solvent recovery as well as developing solvent exchange behavior in chemical synthesis (Xue et al. 2012; Cheng et al. 2012). OSN which is a green separation technique, is an efficient way in which molecular-level separation of organic molecules can be achieved in a single operating pressure, is also termed as solvent-resistant nanofiltration (SRNF) (Sourirajan 1964). Saurirajan, in 1964, used a cellulose acetate membrane for the separation of short-chain aliphatic hydrocarbons, and he is first to apply membrane technology for the separation of non-aqueous systems (Loeb and Manjikian 1965). Loeb et al. were the first to propose pressure-driven mechanism in membrane separation for OSN (Marchetti et al. 2014). Many patents have been published by various research groups as well as commercial laboratories for organic solvent nanofiltration since 1980. The first commercial membrane was developed by Grace, Davison, Koch, Solsep, and GMT in the mid-1990s (Liu et al. 2020). OSN have industrially been highly recognized for their large-scale application in the dewaxing process in the lubricating oil. However, OSN membrane is a new and challenging topic as many practical problems have always occurred at the time of its use which is needed to be solved with proper methodology (Szekely et al. 2014).

16.4.3 Organic Solvent Reverse Osmosis (OSRO)

Reverse osmosis (RO) is a completely inverse procedure to that of osmosis. In RO, water or organic solvent overcomes the osmotic pressure, and there is a flow from high concentration to the low concentration side which is separated by membrane (Li et al. 2016). Pore of the membrane traps the solute particles. RO generally works on the mechanism of dissolution–diffusion, which involves separation on the basis of difference in dissolution rate and diffusion rate of solvent and solutes (Choudhury et al. 2018; Baroña et al. 2012). RO mostly used in the desalination of salty sea water to produce drinking water in those areas. Nowadays, application of OSRO is also gaining importance in the purification and separation of organic solvents (Li et al. 2016; Song et al. 2018).

16.4.4 Organic Solvent Forward Osmosis (OSFO)

It is a pressure-driven membrane separation technique which works on the mechanism of difference in the osmotic pressure in both sides of the membrane. Diffusion of solvents occurs from the concentrated feed side to the dilute permeate side (Cath et al. 2006). Smaller flux value is observed than that measured by theoretical calculation which is due to the concentration polarization effect (Jung et al. 2011). Also, due to external concentration polarization the difference in osmatic pressure and flux declines (McCutcheon and Elimelech 2006). This problem can be remediated by increasing the surface velocity in the membrane and by increasing the cross-flow (Bai et al. 2011). However, it is difficult to improve the internal concentration gradient generated in porous layer due to its intrinsic properties. Bai et al. observed that by reducing the membrane thickness and increasing the porosity in the membrane the membrane performance could be greatly improved.

16.4.5 Organic Solvent Ultrafiltration (OSU)

It is also a pressure-driven membrane separation technology which works under the size exclusion principle. In OSU permeation of the small solute molecules occurs while large solute molecules are rejected. The pore size range of OSU membranes is between 1 and 100 nm. Molecular cut-off is in the range 300 and 500,000 Da (Liu et al. 2020). Though Ultrafiltration (UF) membrane technology is mature enough, simple, and easy to handle there are various shortcomings like swelling in organic solvents which limits its applicability in organic solvent purification and separation. Many studies have proved that organic polymers like polyimide, polybenzimidazole, and polyurea are resistant to organic solvents (Polotskaya et al. 2009). Moreover, polyester-based polymers have also shown very good organic-solvent-resistant property (Liu et al. 2020; Yeo et al. 2020; Baker 2012). Development of novel material impregnated organic-polymer-based membranes is a current trend in the field of separation and purification of organic solvents.

16.5 Modifications in Designing Membrane for Organic Solvent Purification

A membrane which is organic solvent resistant should be stable mechanically, chemically, and thermally. One of the drawbacks of polymeric membranes is that they often undergo fouling, and hence flux decline occurs. Aging and compaction are other problems with these types of membranes. However, in terms

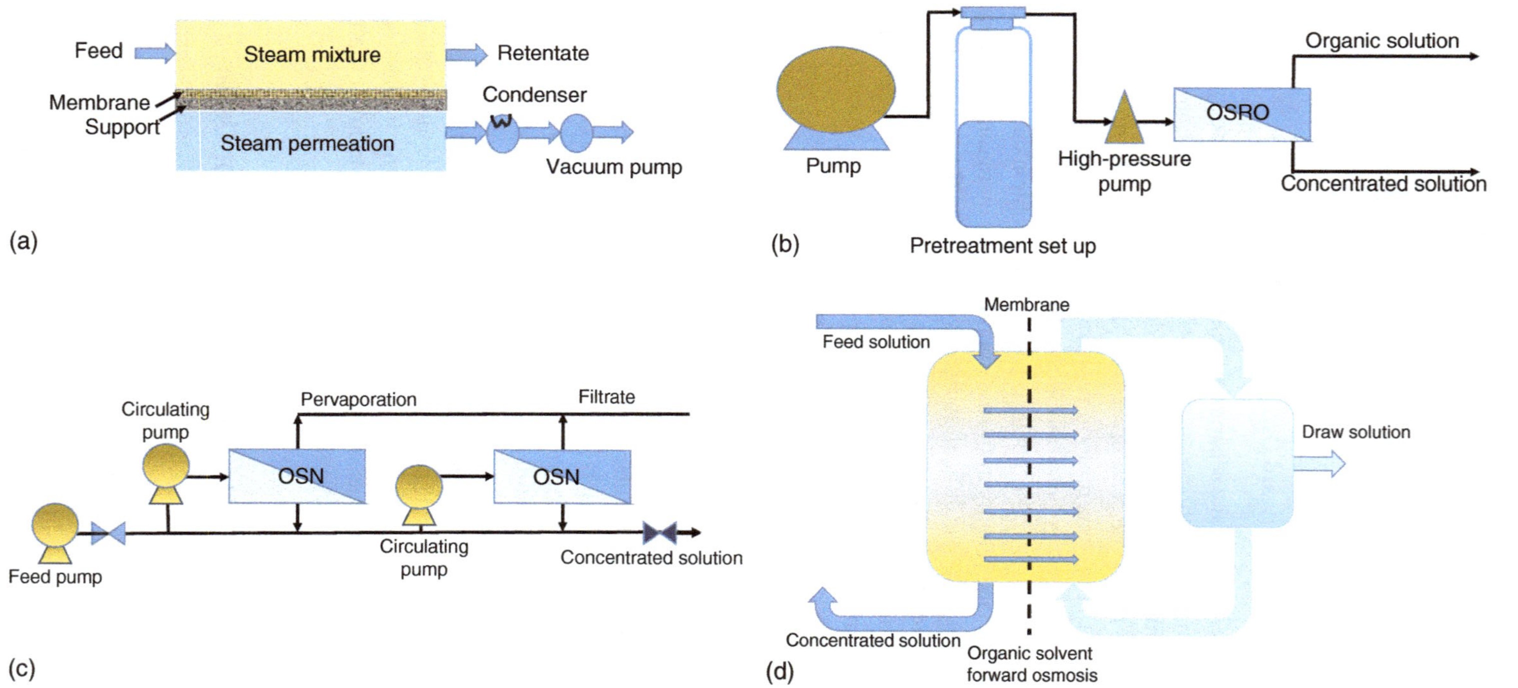

Figure 16.2 Schematic representation of (a) pervaporation, (b) organic solvent forward osmosis, (c) organic solvent nanofiltration, (d) organic solvent reverse osmosis, (e) Organic solvent ultrafiltration.

of mechanical, chemical, and thermal strength, ceramic membranes perform better, and also aging and compaction are not observed in case of ceramic membranes. Scaling up in this type of membrane is very difficult as it is more brittle than that of the polymeric membranes. Ceramic membranes on the other hand have no pores, and risk of leaching is also there. Very low operating pressure is needed in case of ceramic membranes. But as compared to polymers there are less variety of materials in case of ceramic membrane; also for separation of apolar solvents ceramic membrane is not suitable as metal hydroxides are intrinsically hydrophilic. Modification in polymeric membrane for more efficient separation organic solvents has always been topic of interest for various research groups. Mixed matrix membrane (MMM) is one such type of membrane in which properties of both inorganic and polymeric membranes can be tailored in order to achieve excellent performance in organic solvent purification and separation. These types of membrane have been also proven to be less prone to fouling and aging. Sometimes non-woven supports are also used in order to provide mechanical strength to the membranes. For OSN, OSFO, OSRO, and OSU, the non-woven support should also be organic solvent resistant and should have properties that are similar to the polymeric membrane in order to evade formation of crinkles that often lead to failure in membrane separation. For organic solvent separation and purification, the membranes should be stable to the solvent systems and should be competent to preserve their individualities. The exclusively used polymeric membranes are (i) Integrally skinned asymmetric (ISA) membranes and (ii) thin-film composite (TFC) membrane. Schematic representation of these type of membranes has been given in Figures 16.2 and 16.3a,b.

16.5.1 Integrally Skinned Asymmetric (ISA) Membranes

ISA membranes possess a skin layer on top of a more porous sublayer with the same composition. The top skin layer is formed simultaneously with the underlaying support layer. The final membrane performance is greatly affected by the texture of the top skinned layer. Loeb and Sourirajan used a simple phase inversion method to create the first ISA membrane. In phase inversion, a casting solution is precipitated by immersing it in a non-solvent bath. Due to high water absorption and solvent loss, the casted film precipitates rapidly from the top surface down (Park et al. 2000). The casting solution, which was originally a one-phase system, is precipitated into a two-phase system: one is the polymer-rich phase, and the

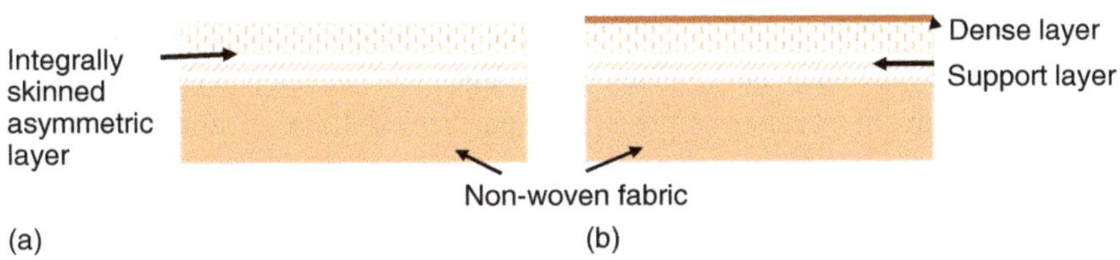

Figure 16.3 Membrane morphology for (a) integrally skinned asymmetric (ISA) membranes and (b) thin-film composite (TFC) membranes.

other is the polymer-poor phase, which forms the pores in the membrane. The membrane performance is greatly affected by the thermodynamic properties of the polymers as well as the kinetics of the solvent–nonsolvent exchange process (Vankelecom et al. 2005). A number of chemically as well as thermally stable polymeric materials have been used for the preparation of phase-inversion-based organic-solvent-resistant ISA membranes (Gevers et al. 2006; Cuperus 2005; Livingston et al. 2013). These include PAN, PI as nanofiltration or UF supports, and PSf, polyaniline, PBI, and PP as UF supports (Sairam et al. 2010; Li et al. 2008; Ali et al. 2020).

16.5.2 Thin-Film Composite (TFC) Membranes

TFC membranes are fabricated by the formation of a thin film over a porous membrane support and are done by the interfacial polymerization method, in which a polymerization reaction occurs at the interface of two immiscible phases that contain reactive monomers. In the TFC membrane, the overall membrane performance can be maximized by optimizing the performance of the barrier layer and the porous membrane support. PSf, PES, PAN, PVDF, PP, PI, and PBI are commonly used polymers that have been used as common UF supports in fabricating TFC membrane. The choice of the polymeric support is to be such that it will provide mechanical and thermal stability and should be able to form defect free top layers.

A novel TFC membrane was prepared by Ali, where he used bridged-bicyclic triptycene tetra-acyl chloride (Trip) and m-phenylenediamine (MPD). The resulting membrane was reported to be finely tuned microporous. Formation of cross-linked microporous morphology and free volume in the enhanced sub-microporous region improves the organic solvent separation efficiency of the developed membrane (Li et al. 2020). In another study by Li et al., they used sulfothiacalix[4]arene and sulfocalix[4]arene, which are hollow cup-shaped macrocyclic molecules, with an aqueous MPD solution for the fabrication of a polyamide layer with trimesoyl chloride (TMC). The formation of an ionic bond has a significant impact on the formation of molecular sieving, which enhances the organic-solvent-resistant property of the developed membrane (Li et al. 2020). An organic-solvent-resistant TFC membrane was also developed by Li et al., where they used aramid nanofibers, which provide a three-dimensionally interconnected hydrogel middle layer for efficient permeability for methanol and acetone permeance (Li et al. 2019). Goh et al. also developed a hollow fiber OSR membrane, employing a non-solvent-induced phase separation technique for membrane fabrication (Goh et al. 2020). Ag/graphene oxide quantum dots-based composite membrane was developed by Li et al., where organic solvent reverse osmosis was performed (Soroko and Livingston 2009).

Cross-linking with hexamethyl diamine gives rise to the substrate named polyimide, which is responsible for tightening the microstructure. Interfacial polymerization between polyethylene imine (PEI), piperazine (PIP), and TMC gives rise to the polyamide (PA) thin film on the inner surface of the polyimide substrate. This combination of PEI and PA results in a loosely selective layer that facilitates the enhancement of the flux of the organic solvent in the separation process.

16.5.3 Mixed Matrix Membranes (MMMs)

A hybrid membrane system with tuning properties can be developed by combining organic and inorganic polymers, which allow the formation of new membrane types with properties of both inorganic and organic materials. These new types of membranes are often termed "mixed matrix membranes". The addition of nanostructures, such as carbon nanomaterials, nanofibers, zeolite nanoparticles, clay, layered double hydroxides, and fullerenes, in the polymer matrix has been extensively studied in the field of membrane separation (Vanherck et al. 2012) as well as in organic solvent nanofiltration. Membranes that comprise nanoparticles in a polymer matrix are termed "nanocomposite membranes." The use of nanoparticles in membrane matrices can be done in three different ways: (i) by incorporating nanoparticles into the membrane matrix prior to casting via phase inversion; (ii) by imbuing nanoparticles onto the membrane surface; and (iii) by using nanoparticles to fill the pores of the membrane. After incorporation of nanoparticles, the overall properties of the membrane, like mechanical, thermal, and surface wettability, can be greatly changed (Vanherck et al. 2012). Various MMMs have been developed by impregnation of nanomaterials in the polymer matrix. Common nanomaterials used for the development of MMM are TiO_2 nanoparticles, transition-metal oxide nanoparticles, cellulose nanocrystal, carbon nanomaterials, etc. (Marchetti et al. 2014). Soroko and Livingston incorporated TiO_2 in polyimide polymer matrix. Addition of TiO_2 decreases the macrovoids in the membrane, and an increase in hydrophilicity of the membrane was observed (Vanherck et al. 2011). Among common nanoparticles gold nanoparticles have been extensively used for fabricating mixed matrix membrane for organic solvent nanofiltration (Li et al. 2013; Vanherck et al. 2011; Li et al. 2019). Vanherck and their group have developed a mixed matrix membrane where gold nanoparticles (NPs) were impregnated over a cellulose acetate polymer matrix. They successfully used the membrane for ethanol and isopropanol permeation. Solvent flux as well as the membrane stability were increased significantly after impregnation of the gold nanoparticle (Vanherck et al. 2011). In another work, a novel mixed matrix membrane was designed where gold NPs were added to a polymer solution of PI. Addition of gold NPs over the membrane matrix decreases the macrovoids in the membrane; also, the flux of isopropanol was increased up to 168% (Homayoonfal et al. 2013).

16.6 Application of OSR Polymeric Membranes in Different Industrial Processes

OSR membranes have various applications in the pharmaceutical industry, mostly for drug purification, peptide synthesis, and solvent recovery. Membrane-based separation aids in drug structure, purity, and product quality retention (Li et al. 2018). Furthermore, the membrane's energy efficiency and ease of handling aid in the removal of trace organic pollutants and salt. When titanium carbide nanosheets (2–4 μm) incorporated into MXene membranes are used for antibiotic separation from aqueous and ethanol-based solutions, a successful separation performance has been achieved. Again, for selective concentration of ibuprofen, an Au and polyaniline-based nanocomposite membrane has been used and found useful. Since a large quantity of solvents such as ethanol, acetone, methanol, ethyl acetate, and isopropanol are needed in herb production, OSR membrane can play a very important role in these processes. A ceramic membrane of varied pore sizes is also found effective in the removal of impurities from Shuanghuanglian water extract. Again, nanofiltration membranes can be used for the concentration of Angelica sinensis extract (Fu et al. 2020). The development of a PI based membrane has been reported (Marchetti et al. 2014), where the membrane is used for concentration of the spramycin antibiotic. In comparison to the conventional extraction of this antibiotic from bacterial broth using butyl acetate, the membrane-based treatment shows efficient separation with long durability. The membrane shows solvent resistance, and about 99% rejection is achieved. Also, membrane application in pharmaceuticals can also help in the recovery and reuse of solvents. In terms of energy efficiency and in terms of solvent recovery, OSN has been shown to be less intense than distillation. It is common practice in the pharmaceutical industry to synthesize important chemicals in organic media, which involves many steps along the synthetic route. To overcome the solvent resistivity, two nanofiltration membranes, viz., MPF-50 and MPF-60 (Sheth et al. 2003), are reported to have been prepared, and their application has been successfully done in solvent exchange with methanol. OSN membrane technology can also be used in the pharmaceutical industry for purification. In order to get better purification performance than conventional processes, membranes or membrane processes should be highly selective and show better separation. Since the purification step is a very important step in pharmaceuticals and costs almost half of the capital investment, improvisation in this step should be given more priority. The sustainability of membrane-based API purification when performed in organic solvents is a great matter of concern. Membranes outperform traditional purification processes such as recrystallization and chromatographic techniques in terms of energy, continuous processing, large scaling up, and so on. Thus, to improve OSN membranes, fabrication is done with molecular imprinting, which can be used for three-way ternary system separation (Zhu et al. 2010). However, membrane technology has been developing day by day to offer high performance at low energy costs.

16.6.1 Separation of Organic Solvent from Solvent Mixtures

Organic solvent separation from its mixture with water or other solvents is very common practice in chemical and petroleum industries. The problem arises in the separation of azeotropes, near azeotropes as well as for the substances that are thermally unstable. For organic solvent separation, PV is commonly used which is a liquid separation technology at molecular level and involves evaporation and membrane permeability. An OSRO membrane of polyketone support and grafted with polyamide layer is prepared out to apply in organic liquid mixture separation. For solvents toluene, pentane, hexane, and heptane, separation factor obtained with this membrane is 8.4, 11.1, 14.9, and 38.0, respectively (Liu et al. 2020). In fine chemicals, petrochemicals, and pharmaceuticals, anhydrous organic solvents are insoluble and hence dehydration of such solvents is very important. PV has been getting increased interest recently in dehydration of such solvents. Hydroxyl-polyimide-containing polybenzoxazole membrane can be prepared via in situ thermally induced transformation, and this membrane shows a separation factor of 2000 and permeation of 2000 $gm^{-2} h^{-1}$ in isopropyl alcohol dehydration (Xu et al. 2016). Cross-linking can be introduced into membrane. One such example is the cross-linking between polybenzimidazole-based membrane surface with polystyrene sulfonic acid chains (Han et al. 2015). This membrane is found to be stable as well as have rich hydrophilicity and is utilized for 80% tetrahydrofuran dehydration.

16.6.2 Food Processing

In food processing, OSR membrane have applications in natural ingredients purification, grain and oil processing, etc. In fruit and vegetable juice processing, membrane separation techniques such as nanofiltration and PV can be used with retention of texture as well as freshness of the juice (Conidi et al. 2020). Membrane can also be used for juice concentration via osmotic distillation with quality retention (Rehman et al. 2019). Membrane shows efficient sunflower pectin extraction at a pilot scale using ultrafiltration and microfiltration and results in greater than 90% pectin concentration with high yield (13.3%) (Muñoz-Almagro et al. 2020). Polypropylene-based and hybrid ZIF-8/PDMS membranes are being used for concentration of apple juice (Julian et al. 2020) and for separation of aromatic volatile component from blackberry juice, respectively (Rao et al. 2020). In dairy production also membrane technology can be used. Milk fat globule isolation (Hansen et al. 2020), milk phospholipid extraction and concentration are some examples of works done with membrane. From soybean milk, peptides with antioxidant and antibacterial properties can be prepared with the help of membrane cross-flow separation-assisted membrane bioreactor (Ren et al. 2021). In manufacturing of beer, wine, and low-alcohol wine production membrane can be used. PV technique is useful in beer industries for volatile compound separation (Paz et al. 2017). Recently, attention is focused on healthy drinking with non- or low-alcoholic drinks.

16.6.3 Wastewater Treatment

Water sources have become more and more polluted day by day, which results in the deterioration of ecological balance. To maintain the balance, immediate action must be taken to address urban sewage. For the treatment of municipal waste water, solvent-resistant membranes have been used and are still progressing. Nanofiltration membranes can be used for lupanine removal from wastewater via solvent extraction (Esteves et al. 2020). Membranes can be used for the separation of oily emulsion from waste water against oily phases such as hexane, petroleum ether, hexadecane, and soybean oil. An ultrathin membrane of polydopamine on a porous substrate of polyketone has been prepared and found useful for this purpose (Zhang et al. 2019). Another example of organic wastewater treatment via degradation is a three-channel Fe_3O_4/PVDF hollow fiber membrane with catalytic activity. Here, the membrane is able to remove methylene blue and also has antifouling characteristics due to the Fenton reaction (Huang et al. 2020). An innovative technology called the self-forming dynamic membrane bioreactor (SFDMBR) can be an attractive alternative for wastewater treatment that includes biological treatment in addition to membrane separation. This system can remove COD and ammonia up to 70% and 90%, respectively, and nitrogen as well. The passing of an ozone/air mixture can help in the cleaning of membrane bioreactors as well as waste water. Membrane bioreactors (MBRs) in conjunction with membrane biofilm reactors are also being researched for use in organics and nitrogen-containing compounds (Hasar 2009). This investigation shows the successful oxidation of COD and removal of nitrogen. The MBR system is very advantageous in terms of low sludge production, small floor space, and simple processing, which make it environmentally benign and cost effective.

16.7 Commercial OSR membrane

Due to easily tunable microstructures and ease of processing, polymeric membranes can be used for commercial purposes as OSR membranes. The commercial membranes available with organic solvent resistance include PI: StarMem 122, PI: StarMem 240, PI: DuraMem 300, PI: DuraMem 500, PI: DuraMem 900, PI: PuraMem 280, PI: DuraMem 150, PI: PuraMem 600, PI: PuraMem S600, PI: PuraMem 420, Nanopros S-3011, PI: DuraMem 150, PA: NF-90, PA: NF-270, PA: BW30XLE, PDMS: MPF-34, PDMS: MPF-44, PDMS: MPF-50, PDMS: PERVAP 4060, and PDMS: MPF-60 (Tandel et al. 2021).

Different works using these commercial membranes and their potential toward solvent resistance have been reported. The recovery of solvent from pharmaceutical mixtures was reported using DuraMem 150 and StarMem 122. In the study, the solvents in which the separation performances were carried out included methanol, ethanol, 2-propanol, ethyl acetate, and ethanol-ethyl acetate, whereas the pharmaceuticals included alprazolam, imatinib mesylate, donepezil HCl, riluzole, atenolol, disperse orange 3, and disperse red 19. DuraMem 150 showed better performance as compared to StarMem 122. A high removal efficiency of 99% was found for Disperse Red 19, and 95% rejection was found for Disperse Orange 3 mixtures in the case of DuraMem 150. DuraMem 150 membranes were more stable in organic solvents as compared to StarMem 122, as StarMem 122 possesses high swelling tendency as compared to DuraMem 150. Except for the ethyl acetate solution containing donepezil, DuraMem 150 shows >90% rejection for all the mentioned pharmaceutical mixtures (Darvishmanesh et al. 2011). Performance of PI-StarMem 122 was observed against Sudan 408 in the presence of solvents like ethanol, methanol, acetone, methyl-ethyl ketone, isopropyl alcohol (IPA), and toluene (Darvishmanesh et al. 2010). PI: starMem 240 works well as a solvent against carbon tetrachloride (CCl_4), IPA, hexane, and ethanol, and performs best in the separation of methylene blue when IPA is used. But when the used solvent is ethanol, the membrane gives very low rejection efficiency. The rejection of the membrane may be affected by the solvent treatment, which was done before the performance analysis (Zhang et al. 2017). PDMS: MPF-44 membranes showed 93% removal of Erythrosine B in the presence of methanol as a solvent (Volkov et al. 2009) and good removal of methylene blue in presence of ethanol, IPA, hexane and CCl_4 as solvents, respectively (Zhang et al. 2017). 97% removal of Erythrosine B was observed for PDMS: MPF-50 in presence of methanol as a solvent. Again, 98% removal of tetraoctyl-ammonium bromide was observed for PDMS: MPF-60 using methanol as a solvent (Morshed et al. 2020). In presence of THF, PI: DuraMem 150 showed very high rejection of polystyrene oligomers (approx. 99%) (Solomon et al. 2012). PI: DuraMem 500, PI: DuraMem 900, PI: PuraMem 280 & PI: PuraMem 380 membranes show removal of methyl vanillate when ethyl acetate is used as a solvent (Werhan et al. 2012). PuraMem 280 is also effective when toluene is used as a solvent and shows good rejection efficiency against PS oligomers (Siddique et al. 2013). In another study of the performance of commercial membranes such as DuraMem 150, DuraMem 300, DuraMem 500, PuraMem 280, and PuraMem 600 membranes gave good performance by showing rejection of glyceryl trilinolate up to 100%, 100%, 85.8% (approx.), 96.8% (approx.), and 88.5% (approx.), respectively, when acetone is used as a solvent (Shi et al. 2019). DuraMem 150 membranes are effective against atenolol, removing it with 97.5% efficiency in methanol (Shi et al. 2019). DuraMem 500 and PuraMem S600 membranes were used for the removal of palladium (II) acetate in the solvents IPA, methanol, and toluene, and it is seen that the effect of the solvent was more prominent in the case of DuraMem 500. On the other hand, the rejection efficiency of the PuraMem S600 membrane was mainly governed by the ligand effect (the ligands with which the palladium (II) acetate combined to form a stable catalyst) (Shen et al. 2020). DuraMem 500 and PuraMem 420 showed good separation performance with rejection efficiencies of 94.6% and 98%, respectively, against acid fuchsin when the solvent used was acetonitrile (Joseph et al. 2017).

Park et al. reported a polyethylene-incorporated polyamide-based TFC membrane having better tensile strength, chemical stability, and NaCl rejection performance as compared to commercial SWC4+ (RO) membranes (Park et al. 2018). The commercial OSR membranes of the MPF series also show very effective resistance toward organic solvents like acetone, hexanol, cyclohexanol, 3- propanol, methanol, 3-butanone, and other lower alcohols (Ren et al. 2021). Due to the inexpensiveness of the commercially available TFC membranes based on PA, they are also used in organic-solvent-resistant applications but are not as efficient as reported in the literature. Even commercially available organic-solvent-resistant membranes are only effective over a limited temperature range, and it is difficult for a membrane to be resistant to a wide range of organic solvents (Tandel et al. 2021).

16.8 Current Status of OSR Membrane

A high volume of organic solvents is used in the pharmaceutical, chemical, and other related industries. However, the separation and purification processes of these solvents are energy intensive, and the reusability rate of these solvents is very low (resulting in massive waste generation). Being an energy-efficient process with an easy operating procedure, OSR membrane is a greener approach for separation and purification of organic solvents as compared to other existing separation methods (Ren et al. 2021; Joshi and Adhikari 2019; Hermans et al. 2015).

The petrochemical industry's waste water contains varieties of inorganic elements as well as hydrocarbons in dissolved or non-dissolved form. Normal polymeric membranes get swollen easily and cannot tolerate harsh chemical environments. Therefore, in this case, specially designed OSR membrane are highly needed to recover oil products, produce quality water, and maintain the high reproducibility and lifespan of the membrane (Bhuyan et al. 2023; Lim et al. 2017). Similarly, OSR membrane has the potential for dye separation in the dye industry, processing in the food industry, solvent recovery, and the purification of drugs in the pharmaceutical industry (Baggio et al. 2021).

PET, being a low-cost material, is widely used for packaging application. The value addition to the plastic bottles which are composed of PET can be reused for separation and purification application. A chitosan-modified recycled PET-based membrane for effective oil–water separation (>95% rejection) (Park et al. 2021). A TFC membrane with an ultrathin layer of very high selectivity was obtained by interfacial polymerization of TA (tannic acid) and priamine on the porous support of recycled PET using p-Cymene (less toxic as compared to most of the organic solvents) as organic phase (Liu et al. 2019). A composite PET/PTFE (polytetrafluoroethylene) membrane combined with dopamine was reported by Liu et al. for more than 99% separation of oil and water with good recyclability (Lindman et al. 2010). Being a widely available natural amphiphilic polymer, cellulose can also be used for fabrication of hydrophilic membrane applicable for oil water separation (Park and Barnett 2001). As compared to ultrafiltration, nanofiltration membrane possesses pores of smaller diameter hence are more selective and can give better separation of oil–water emulsion even at high pressure to provide quality water (Junaidi et al. 2021). Different new-generation carbon-based nanomaterials/nanocomposites such as GO (graphene oxide) and CNT (carbon nanotubes) were also been reported for effective separation of oil and water (Gu et al. 2014; Saththasivam et al. 2018).

Common polymeric materials used in the fabrication of OSR membrane tend to swell in the presence of organic solvents (Baggio et al. 2021). Therefore, they show poor organic solvent resistance and tend to get degraded in organic solvents. To overcome the limitations, a variety of hybrid as well as novel or specialized membrane materials, as well as a membrane fabrication approach, should be used. For high separation efficiency, researchers have reported cellulose-based, intrinsically structured-material-based, carbon-based, and other low-cost material-based OSR membrane (Park et al. 2018). Various studies have been conducted to improve the separation performance and mechanical strength of the OSR membrane with improved resistance to a wide range of organic solvents.

16.9 Conclusion

Industrial processes use large amounts of solvents during the production, separation, and purification of organic compounds. When disposed of in natural systems, these organic solvents can have a high risk of damaging the sustainability of the ecosystem. Conventional methods for separation and purification of organic solvents often fail due to various issues like corrosion during separation, swelling of separating materials, and low selectivity in separation. Membrane separation, despite being known as the most simple and efficient method for organic solvent separation and purification, faces numerous challenges such as fouling, swelling caused by organic solvents, and corrosion caused by organic solvent action. Modification of the polymeric membranes to make them superior for organic separation in terms of flux and separation efficiency as well as resistant to organic solvent systems is a current challenge for various research groups. Inorganic membranes are one alternative that is resistant to organic solvents. However, the scalability of inorganic membrane-based separation is still a challenging task. On the other hand, some organic polymers have solvent-resistant properties, whereas others can be modified by means of hybridizing inorganic and organic membrane systems or by chemical cross-linking. It is necessary to reduce the free volume, inhibiting the inflammation in the membrane system and improving the solvent-resistant property of the polymeric membrane so as to make the membrane efficient for organic solvent separation and purification. Designing an excellent membrane structure for mitigating the day-to-day challenges and avoiding the traditional problems in the separation of organic solvents has received more attention. This new research area and its corresponding applications will definitely advance the evolution of membrane technology.

References

Adler, S., Beaver, E., Bryan, P. et al. (2000). *Vision 2020: 2000 Separations Roadmap*. Washington, DC: EERE Publication and Product Library.

Ali, Z., Ghanem, B.S., Wang, Y. et al. (2020). Finely tuned submicroporous thin-film molecular sieve membranes for highly efficient fluid separations. *Advanced Materials* 32 (22): 2001132.

Al-Khalid, T. and El-Naas, M.H. (2018). Organic contaminants in refinery wastewater: characterization and novel approaches for biotreatment. In: *Recent Insights in Petroleum Science and Engineering*, vol. 371 (ed. M. Zoveidavianpoor). IntechOpen.

Baggio, A., Doan, H.N., Vo, P.P. et al. (2021). Chitosan-functionalized recycled polyethylene terephthalate nanofibrous membrane for sustainable on-demand oil-water separation. *Global Challenges* 5 (4): 2000107.

Bai, H., Liu, Z., and Sun, D.D. (2011). Highly water soluble and recovered dextran coated Fe_3O_4 magnetic nanoparticles for brackish water desalination. *Separation and Purification Technology* 81 (3): 392–399.

Baker, R.B. (2012). *Membrane Technology and Applications*, 3e. Chichester, England: Wiley.

Baroña, G.N.B., Lim, J., and Jung, B. (2012). High performance thin film composite polyamide reverse osmosis membrane prepared via m-phenylenediamine and 2, 2′-benzidinedisulfonic acid. *Desalination* 291: 69–77.

Bhuyan, C., Konwar, A., Bora, P. et al. (2023). Cellulose nanofiber-poly(ethylene terephthalate) nanocomposite membrane from waste materials for treatment of petroleum industry wastewater. *Journal of Hazardous Materials* 442: 129955.

Buonomenna, M.G. and Bae, J. (2015). Organic solvent nanofiltration in pharmaceutical industry. *Separation & Purification Reviews* 44 (2): 157–182.

Cao, N., Sun, Y., Wang, J. et al. (2020). Strong acid- and solvent-resistant polyether ether ketone separation membranes with adjustable pores. *Chemical Engineering Journal* 386: 124086.

Cath, T.Y., Childress, A.E., and Elimelech, M. (2006). Forward osmosis: principles, applications, and recent developments. *Journal of Membrane Science* 281 (1–2): 70–87.

Cheng, S., Oatley, D.L., Williams, P.M., and Wright, C.J. (2012). Characterisation and application of a novel positively charged nanofiltration membrane for the treatment of textile industry wastewaters. *Water Research* 46 (1): 33–42.

Choudhury, R.R., Gohil, J.M., Mohanty, S., and Nayak, S.K. (2018). Antifouling, fouling release and antimicrobial materials for surface modification of reverse osmosis and nanofiltration membranes. *Journal of Materials Chemistry A* 6 (2): 313–333.

Conidi, C., Castro-Muñoz, R., and Cassano, A. (2020). Membrane-based operations in the fruit juice processing industry: a review. *Beverages* 6 (1): 18.

Cooper, A.I. (2009). Conjugated microporous polymers. *Advanced Materials* 21 (12): 1291–1295.

Cuperus, F.P. (2005). Recovery of organic solvents and valuable components by membrane separation. *Chemie Ingenieur Technik* 77 (8): 1000–1001.

Darvishmanesh, S., Degrève, J., and Van der Bruggen, B. (2010). Mechanisms of solute rejection in solvent resistant nanofiltration: the effect of solvent on solute rejection. *Physical Chemistry Chemical Physics* 12 (40): 13333–13342.

Darvishmanesh, S., Firoozpour, L., Vanneste, J. et al. (2011). Performance of solvent resistant nanofiltration membranes for purification of residual solvent in the pharmaceutical industry: experiments and simulation. *Green Chemistry* 13 (12): 3476–3483.

Esteves, T., Mota, A.T., Barbeitos, C. et al. (2020). A study on lupin beans process wastewater nanofiltration treatment and lupanine recovery. *Journal of Cleaner Production* 277: 123349.

Fu, Q., Ke, Y., Jiang, D., and Jin, Y. (2020). Chemical separation and characterization of complex samples with herbal medicine. *TrAC Trends in Analytical Chemistry* 124: 115775.

Galizia, M. and Bye, K.P. (2018). Advances in organic solvent nanofiltration rely on physical chemistry and polymer chemistry. *Frontiers in Chemistry* 6: 511.

Geens, J., De Witte, B., and Van der Bruggen, B. (2007). Removal of API's (active pharmaceutical ingredients) from organic solvents by nanofiltration. *Separation Science and Technology* 42 (11): 2435–2449.

George, J., Purushothaman, M., Singh, I. et al. (2022). Performance study of fouling resistant novel ultrafiltration membranes based on the blends of poly (ether ether sulfone)/poly (vinyl pyrrolidone)/nano-titania for the separation of humic acid, dyes and biological macromolecular proteins from aqueous solutions. *Journal of Hazardous Materials* 424: 127467.

Gevers, L.E., Aldea, S., Vankelecom, I.F., and Jacobs, P.A. (2006). Optimisation of a lab-scale method for preparation of composite membranes with a filled dense top-layer. *Journal of Membrane Science* 281 (1–2): 741–746.

Gimenes, M.L., Liu, L., and Feng, X. (2007). Sericin/poly (vinyl alcohol) blend membranes for pervaporation separation of ethanol/water mixtures. *Journal of Membrane Science* 295 (1–2): 71–79.

Goh, K.S., Chong, J.Y., Chen, Y. et al. (2020). Thin-film composite hollow fibre membrane for low pressure organic solvent nanofiltration. *Journal of Membrane Science* 597: 117760.

Gould, R.M., White, L.S., and Wildemuth, C.R. (2001). Membrane separation in solvent lube dewaxing. *Environmental Progress* 20 (1): 12–16.

Gu, J., Xiao, P., Chen, J. et al. (2014). Janus polymer/carbon nanotube hybrid membranes for oil/water separation. *ACS Applied Materials & Interfaces* 6 (18): 16204–16209.

Han, Y.J., Su, W.C., Lai, J.Y., and Liu, Y.L. (2015). Hydrophilically surface-modified and crosslinked polybenzimidazole membranes for pervaporation dehydration on tetrahydrofuran aqueous solutions. *Journal of Membrane Science* 475: 496–503.

Hansen, S.F., Hogan, S.A., Tobin, J. et al. (2020). Microfiltration of raw milk for production of high-purity milk fat globule membrane material. *Journal of Food Engineering* 276: 109887.

Hasar, H. (2009). Simultaneous removal of organic matter and nitrogen compounds by combining a membrane bioreactor and a membrane biofilm reactor. *Bioresource Technology* 100 (10): 2699–2705.

He, P., Zhao, S., Mao, C. et al. (2021). In-situ growth of double-layered polyaniline composite membrane for organic solvent nanofiltration. *Chemical Engineering Journal* 420: 129338.

Heidari, A.A. and Mahdavi, H. (2021). Thin film composite solvent resistant nanofiltration membrane via interfacial polymerization on an engineered polyethylene membrane support coated with polydopamine. *Journal of Membrane Science* 634: 119406.

Hermans, S., Mariën, H., Van Goethem, C., and Vankelecom, I.F. (2015). Recent developments in thin film (nano) composite membranes for solvent resistant nanofiltration. *Current Opinion in Chemical Engineering* 8: 45–54.

Homayoonfal, M., Mehrnia, M.R., Mojtahedi, Y.M., and Ismail, A.F. (2013). Effect of metal and metal oxide nanoparticle impregnation route on structure and liquid filtration performance of polymeric nanocomposite membranes: a comprehensive review. *Desalination and Water Treatment* 51 (16–18): 3295–3316.

Huang, Z.H., Zhang, X., Wang, Y.X. et al. (2020). Fe_3O_4/PVDF catalytic membrane treatment organic wastewater with simultaneously improved permeability, catalytic property and anti-fouling. *Environmental Research* 187: 109617.

Jiang, J.X., Su, F., Trewin, A. et al. (2007). Conjugated microporous poly (aryleneethynylene) networks. *Angewandte Chemie* 119 (45): 8728–8732.

Joseph, N., Thomas, J., Ahmadiannamini, P. et al. (2017). Ultrathin single bilayer separation membranes based on hyperbranched sulfonated poly (aryleneoxindole). *Advanced Functional Materials* 27 (9): 1605068.

Joshi, D.R. and Adhikari, N. (2019). An overview on common organic solvents and their toxicity. *Journal of Pharmaceutical Research International* 28 (3): 1–18.

Julian, H., Yaohanny, F., Devina, A. et al. (2020). Apple juice concentration using submerged direct contact membrane distillation (SDCMD). *Journal of Food Engineering* 272: 109807.

Junaidi, N.F.D., Othman, N.H., Fuzil, N.S. et al. (2021). Recent development of graphene oxide-based membranes for oil–water separation: a review. *Separation and Purification Technology* 258: 118000.

Jung, D.H., Lee, J., Lee, Y.G. et al. (2011). Simulation of forward osmosis membrane process: effect of membrane orientation and flow direction of feed and draw solutions. *Desalination* 277 (1–3): 83–91.

Kosaraju, P.B. and Sirkar, K.K. (2008). Interfacially polymerized thin film composite membranes on microporous polypropylene supports for solvent-resistant nanofiltration. *Journal of Membrane Science* 321 (2): 155–161.

Lalia, B.S., Kochkodan, V., Hashaikeh, R., and Hilal, N. (2013). A review on membrane fabrication: structure, properties and performance relationship. *Desalination* 326: 77–95.

Lee, J.S.M. and Cooper, A.I. (2020). Advances in conjugated microporous polymers. *Chemical Reviews* 120 (4): 2171–2214.

Lee, K.P., Arnot, T.C., and Mattia, D. (2011). A review of reverse osmosis membrane materials for desalination–development to date and future potential. *Journal of Membrane Science* 370 (1–2): 1–22.

Li, C.Y. (2019). Multi-index evaluation of applicability of nanofiltration separation of salvianolic acid B in organic phase. *Chinese Traditional and Herbal Drugs* 24: 1328–1333.

Li, X., De Feyter, S., Chen, D. et al. (2008). Solvent-resistant nanofiltration membranes based on multilayered polyelectrolyte complexes. *Chemistry of Materials* 20 (12): 3876–3883.

Li, S., Qin, F., Qin, P. et al. (2013). Preparation of PDMS membrane using water as solvent for pervaporation separation of butanol–water mixture. *Green Chemistry* 15 (8): 2180–2190.

Li, Y., Verbiest, T., and Vankelecom, I. (2013). Improving the flux of PDMS membranes via localized heating through incorporation of gold nanoparticles. *Journal of Membrane Science* 428: 63–69.

Li, D., Yan, Y., and Wang, H. (2016). Recent advances in polymer and polymer composite membranes for reverse and forward osmosis processes. *Progress in Polymer Science* 61: 104–155.

Li, C., Li, S., Lv, L. et al. (2018). High solvent-resistant and integrally crosslinked polyimide-based composite membranes for organic solvent nanofiltration. *Journal of Membrane Science* 564: 10–21.

Li, B., Cui, Y., Japip, S. et al. (2018). Graphene oxide (GO) laminar membranes for concentrating pharmaceuticals and food additives in organic solvents. *Carbon* 130: 503–514.

Li, Y., Wong, E., Volodine, A. et al. (2019). Nanofibrous hydrogel composite membranes with ultrafast transport performance for molecular separation in organic solvents. *Journal of Materials Chemistry A* 7 (33): 19269–19279.

Li, S., Gao, B., Wang, Y. et al. (2019). Antibacterial thin film nanocomposite reverse osmosis membrane by doping silver phosphate loaded graphene oxide quantum dots in polyamide layer. *Desalination* 464: 94–104.

Li, B., Japip, S., and Chung, T.S. (2020). Molecularly tunable thin-film nanocomposite membranes with enhanced molecular sieving for organic solvent forward osmosis. *Nature Communications* 11 (1): 1198.

Lim, S.K., Goh, K., Bae, T.H., and Wang, R. (2017). Polymer-based membranes for solvent-resistant nanofiltration: a review. *Chinese Journal of Chemical Engineering* 25 (11): 1653–1675.

Lindman, B., Karlström, G., and Stigsson, L. (2010). On the mechanism of dissolution of cellulose. *Journal of Molecular Liquids* 156 (1): 76–81.

Liu, Y.N., Qu, R., Zhang, W. et al. (2019). Lotus-and mussel-inspired PDA–PET/PTFE Janus membrane: toward integrated separation of light and heavy oils from water. *ACS Applied Materials & Interfaces* 11 (22): 20545–20556.

Liu, Q., Xu, N., Fan, L. et al. (2020). Polyacrylonitrile (PAN)/TiO_2 mixed matrix membrane synthesis by thermally induced self-crosslinking for thermal and organic-solvent resistant filtration. *Chemical Engineering Science* 228: 115993.

Liu, C., Takagi, R., Shintani, T. et al. (2020). Organic liquid mixture separation using an aliphatic polyketone-supported polyamide organic solvent reverse osmosis (OSRO) membrane. *ACS Applied Materials & Interfaces* 12 (6): 7586–7594.

Liu, B., Zhu, T., Liu, W. et al. (2020). Ultrafiltration pre-oxidation by boron-doped diamond anode for algae-laden water treatment: membrane fouling mitigation, interface characteristics and cake layer organic release. *Water Research* 187: 116435.

Livingston, A.G., Bhole, Y.S. and Jimenez Solomon, M.F.U.S. (2013). Solvent resistant polyamide nanofiltration membranes. US Patent 0, 112,619 A1, Issued 9 May 2013.

Loeb, S. and Manjikian, S. (1965). Six-month field test of reverse osmosis desalination membrane. *Industrial & Engineering Chemistry Process Design and Development* 4 (2): 207–212.

Marchetti, P. and Livingston, A.G. (2015). Predictive membrane transport models for organic solvent nanofiltration: how complex do we need to be? *Journal of Membrane Science* 476: 530–553.

Marchetti, P., Jimenez Solomon, M.F., Szekely, G., and Livingston, A.G. (2014). Molecular separation with organic solvent nanofiltration: a critical review. *Chemical Reviews* 114 (21): 10735–10806.

McCutcheon, J.R. and Elimelech, M. (2006). Influence of concentrative and dilutive internal concentration polarization on flux behavior in forward osmosis. *Journal of Membrane Science* 284 (1–2): 237–247.

Morshed, M., Simonaire, H., Alem, H., and Roizard, D. (2020). Investigation of OSN properties of PDMS membrane for the retention of dilute solutes with potential industrial applications. *Journal of Applied Polymer Science* 137 (6): 48359.

Muñoz-Almagro, N., Prodanov, M., Wilde, P.J. et al. (2020). Obtainment and characterisation of pectin from sunflower heads purified by membrane separation techniques. *Food Chemistry* 318: 126476.

Musale, D.A. and Kumar, A. (2000). Solvent and pH resistance of surface crosslinked chitosan/poly (acrylonitrile) composite nanofiltration membranes. *Journal of Applied Polymer Science* 77 (8): 1782–1793.

Namvar-Mahboub, M. and Pakizeh, M. (2013). Development of a novel thin film composite membrane by interfacial polymerization on polyetherimide/modified SiO_2 support for organic solvent nanofiltration. *Separation and Purification Technology* 119: 35–45.

Nunes, S.P., Culfaz-Emecen, P.Z., Ramon, G.Z. et al. (2020). Thinking the future of membranes: perspectives for advanced and new membrane materials and manufacturing processes. *Journal of Membrane Science* 598: 117761.

Omar, N.M.A., Othman, M.H.D., Tai, Z.S. et al. (2022). Bottlenecks and recent improvement strategies of ceramic membranes in membrane distillation applications: a review. *Journal of the European Ceramic Society* 42: 5179–5194.

Paredes, D.A.F., Laoretani, D.S., Zelin, J. et al. (2017). Screening of pervaporation membranes for the separation of methanol-methyl acetate mixtures: An approach based on the conceptual design of the pervaporation-distillation hybrid process. *Separation and Purification Technology* 189: 296–309.

Park, E. and Barnett, S.M. (2001). Oil/water separation using nanofiltration membrane technology. *Separation Science and Technology* 36 (7): 1527–1542.

Park, J.S., Kim, S.K., and Lee, K.H. (2000). Effect of $ZnCl_2$ on formation of asymmetric PEI membrane by phase inversion process. *Journal of Industrial and Engineering Chemistry* 6 (2): 93–99.

Park, S.H., Kwon, S.J., Shin, M.G. et al. (2018). Polyethylene-supported high performance reverse osmosis membranes with enhanced mechanical and chemical durability. *Desalination* 436: 28–38.

Park, S.H., Yang, C., Ayaril, N., and Szekely, G. (2021). Solvent-resistant thin-film composite membranes from biomass-derived building blocks: chitosan and 2,5-furandicarboxaldehyde. *ACS Sustainable Chemistry & Engineering* 10 (2): 998–1007.

Park, S.H., Alammar, A., Fulop, Z. et al. (2021). Hydrophobic thin film composite nanofiltration membranes derived solely from sustainable sources. *Green Chemistry* 23 (3): 1175–1184.

Paz, A.I., Blanco, C.A., Andrés-Iglesias, C. et al. (2017). Aroma recovery of beer flavors by pervaporation through polydimethylsiloxane membranes. *Journal of Food Process Engineering* 40 (6): e12556.

Peyravi, M., Rahimpour, A., and Jahanshahi, M. (2012). Thin film composite membranes with modified polysulfone supports for organic solvent nanofiltration. *Journal of Membrane Science* 423: 225–237.

Polotskaya, G.A., Meleshko, T.K., Gofman, I.V. et al. (2009). Polyimide ultrafiltration membranes with high thermal stability and chemical durability. *Separation Science and Technology* 44 (16): 3814–3831.

Rao, Y., Ni, F., Sun, Y. et al. (2020). Efficient recovery of the volatile aroma components from blackberry juice using a ZIF-8/PDMS hybrid membrane. *Separation and Purification Technology* 230: 115844.

Rdzanek, P., Marszałek, J., and Kamiński, W. (2018). Biobutanol concentration by pervaporation using supported ionic liquid membranes. *Separation and Purification Technology* 196: 124–131.

Rehman, W.U., Muhammad, A., Khan, Q.A. et al. (2019). Pomegranate juice concentration using osmotic distillation with membrane contactor. *Separation and Purification Technology* 224: 481–489.

Ren, D., Ren, S., Lin, Y. et al. (2021). Recent developments of organic solvent resistant materials for membrane separations. *Chemosphere* 271: 129425.

Sairam, M., Loh, X.X., Bhole, Y. et al. (2010). Spiral-wound polyaniline membrane modules for organic solvent nanofiltration (OSN). *Journal of Membrane Science* 349 (1–2): 123–129.

Santoro, S., Galiano, F., Jansen, J.C., and Figoli, A. (2017). Strategy for scale-up of SBS pervaporation membranes for ethanol recovery from diluted aqueous solutions. *Separation and Purification Technology* 176: 252–261.

Saththasivam, J., Yiming, W., Wang, K. et al. (2018). A novel architecture for carbon nanotube membranes towards fast and efficient oil/water separation. *Scientific Reports* 8 (1): 1–6.

Shen, J., Beale, K., Amura, I., and Emanuelsson, E.A. (2020). Ligand and solvent selection for enhanced separation of palladium catalysts by organic solvent nanofiltration. *Frontiers in Chemistry* 8: 375.

Sheth, J.P., Qin, Y., Sirkar, K.K., and Baltzis, B.C. (2003). Nanofiltration-based diafiltration process for solvent exchange in pharmaceutical manufacturing. *Journal of Membrane Science* 211 (2): 251–261.

Shi, G.M. and Chung, T.S. (2020). Teflon AF2400/polyethylene membranes for organic solvent nanofiltration (OSN). *Journal of Membrane Science* 602: 117972.

Shi, G.M., Farahani, M.H.D.A., Liu, J.Y., and Chung, T.S. (2019). Separation of vegetable oil compounds and solvent recovery using commercial organic solvent nanofiltration membranes. *Journal of Membrane Science* 588: 117202.

Si, Z., Wu, H., Qin, P., and Van der Bruggen, B. (2022). Polydimethylsiloxane based membranes for biofuels pervaporation. *Separation and Purification Technology* 298: 121612.

Siddique, H., Peeva, L.G., Stoikos, K. et al. (2013). Membranes for organic solvent nanofiltration based on preassembled nanoparticles. *Industrial & Engineering Chemistry Research* 52 (3): 1109–1121.

Solomon, M.F.J., Bhole, Y., and Livingston, A.G. (2012). High flux membranes for organic solvent nanofiltration (OSN)–interfacial polymerization with solvent activation. *Journal of Membrane Science* 423: 371–382.

Song, H.M., Zhu, L.J., Zeng, Z.X., and Xue, Q.J. (2018). High performance forward osmosis cellulose acetate (CA) membrane modified by polyvinyl alcohol and polydopamine. *Journal of Polymer Research* 25: 1–8.

Soroko, I. and Livingston, A. (2009). Impact of TiO_2 nanoparticles on morphology and performance of crosslinked polyimide organic solvent nanofiltration (OSN) membranes. *Journal of Membrane Science* 343 (1–2): 189–198.

Sourirajan, S. (1964). Separation of hydrocarbon liquids by flow under pressure through porous membranes. *Nature* 203: 1348–1349.

Szekely, G., Jimenez-Solomon, M.F., Marchetti, P. et al. (2014). Sustainability assessment of organic solvent nanofiltration: from fabrication to application. *Green Chemistry* 16 (10): 4440–4473.

Tandel, A.M., Guo, W., Bye, K. et al. (2021). Designing organic solvent separation membranes: polymers, porous structures, 2D materials, and their combinations. *Materials Advances* 2 (14): 4574–4603.

Tashvigh, A.A. and Chung, T.S. (2019). Robust polybenzimidazole (PBI) hollow fiber membranes for organic solvent nanofiltration. *Journal of Membrane Science* 572: 580–587.

Teodorescu, M., Bercea, M., and Morariu, S. (2018). Biomaterials of poly (vinyl alcohol) and natural polymers. *Polymer Reviews* 58 (2): 247–287.

Thakur, V.K. and Voicu, S.I. (2016). Recent advances in cellulose and chitosan-based membranes for water purification: a concise review. *Carbohydrate Polymers* 146: 148–165.

Tham, H.M. and Chung, T.S. (2020). One-step cross-linking and tannic acid modification of polyacrylonitrile hollow fibers for organic solvent nanofiltration. *Journal of Membrane Science* 610: 118294.

Tsarkov, S., Khotimskiy, V., Budd, P.M. et al. (2012). Solvent nanofiltration through high permeability glassy polymers: effect of polymer and solute nature. *Journal of Membrane Science* 423: 65–72.

Ulbricht, M. (2020). Design and synthesis of organic polymers for molecular separation membranes. *Current Opinion in Chemical Engineering* 28: 60–65.

Van Hecke, W. and De Wever, H. (2017). High-flux POMS organophilic pervaporation for ABE recovery applied in fed-batch and continuous set-ups. *Journal of Membrane Science* 540: 321–332.

Vandezande, P., Gevers, L.E., and Vankelecom, I.F. (2008). Solvent resistant nanofiltration: separating on a molecular level. *Chemical Society Reviews* 37 (2): 365–405.

Vanherck, K., Hermans, S., Verbiest, T., and Vankelecom, I. (2011). Using the photothermal effect to improve membrane separations via localized heating. *Journal of Materials Chemistry* 21 (16): 6079–6087.

Vanherck, K., Vankelecom, I., and Verbiest, T. (2011). Improving fluxes of polyimide membranes containing gold nanoparticles by photothermal heating. *Journal of Membrane Science* 373 (1–2): 5–13.

Vanherck, K., Verbiest, T., and Vankelecom, I. (2012). Comparison of two synthesis routes to obtain gold nanoparticles in polyimide. *The Journal of Physical Chemistry C* 116 (1): 115–125.

Vankelecom, I.F.J., De Smet, K., Gevers, L.E.M., and Jacobs, P.A. (2005). Nanofiltration membrane materials and preparation. In: *Nanofiltration, Principles and Applications* (ed. A.I. Schaefer, A.G. Fane, and T.D. Waite), 33–65. Oxford: Elsevier.

Volkov, A.V., Parashchuk, V.V., Stamatialis, D.F. et al. (2009). High permeable PTMSP/PAN composite membranes for solvent nanofiltration. *Journal of Membrane Science* 333 (1–2): 88–93.

Wei, X.Z., Yang, J., and Zhang, G.L. (2012). Preparation and characterization of nanofiltration membranes synthesized by hyperbranched polyester and terephthaloyl chloride (TPC). *Polymers and Polymer Composites* 20 (3): 261–270.

Werhan, H., Farshori, A., and von Rohr, P.R. (2012). Separation of lignin oxidation products by organic solvent nanofiltration. *Journal of membrane science* 423: 404–412.

Wu, Y., Hao, J., Wu, C. et al. (2012). Cation exchange PVA/SPPO/SiO_2 membranes with double organic phases for alkali recovery. *Journal of Membrane Science* 423: 383–391.

Xu, Y., Jin, S., Xu, H. et al. (2013). Conjugated microporous polymers: design, synthesis and application. *Chemical Society Reviews* 42 (20): 8012–8031.

Xu, Y.M., Le, N.L., Zuo, J., and Chung, T.S. (2016). Aromatic polyimide and crosslinked thermally rearranged poly (benzoxazole-co-imide) membranes for isopropanol dehydration via pervaporation. *Journal of Membrane Science* 499: 317–325.

Xue, Y., Liu, Y., Lu, F. et al. (2012). Functionalization of graphene oxide with polyhedral oligomeric silsesquioxane (POSS) for multifunctional applications. *The Journal of Physical Chemistry Letters* 3 (12): 1607–1612.

Yang, C., Li, S., Lv, X. et al. (2021). Effectively regulating interfacial polymerization process via in-situ constructed 2D COFs interlayer for fabricating organic solvent nanofiltration membranes. *Journal of Membrane Science* 637: 119618.

Yeo, K.E., Oh, W., Jeon, E. et al. (2020). One-pot single-step route toward bicontinuous nanoporous membranes of an organic–inorganic core–shell network. *Chemistry of Materials* 32 (19): 8318–8324.

Yun, M.A., Chen, J.M., and LI, Y.H. and Peng, P.G. (2017). Nanofiltration separation mechanism of phenolic acids from *Salvia miltiorrhiza* based on regulating dissociated state. *Chinese Traditional Patent Medicine* 39 (12): 2517–2520.

Zeng, G., Ye, Z., He, Y. et al. (2017). Application of dopamine-modified halloysite nanotubes/PVDF blend membranes for direct dyes removal from wastewater. *Chemical Engineering Journal* 323: 572–583.

Zhang, Y., Zhong, M., Luo, B. et al. (2017). The performance of integrally skinned polyetherimide asymmetric nanofiltration membranes with organic solvents. *Journal of Membrane Science* 544: 119–125.

Zhang, L., He, Y., Ma, L. et al. (2019). Hierarchically stabilized PAN/β-FeOOH nanofibrous membrane for efficient water purification with excellent antifouling performance and robust solvent resistance. *ACS Applied Materials & Interfaces* 11 (37): 34487–34496.

Zhang, L., Guan, H., Zhang, N. et al. (2019). A loose NF membrane by grafting TiO_2-HMDI nanoparticles on PES/β-CD substrate for dye/salt separation. *Separation and Purification Technology* 218: 8–19.

Zhang, Y., Lin, L., Wang, Q. et al. (2020). Polyrotaxane crosslinked modified EC/PVDF composite membrane displaying simultaneously enhanced pervaporation performance and solvent resistance for benzene/cyclohexane separation. *Journal of Materials Science* 55 (19): 8403–8419.

Zhu, Y., Xu, S., and Yi, D. (2010). Photocatalytic degradation of methyl orange using polythiophene/titanium dioxide composites. *Reactive and Functional Polymers* 70 (5): 282–287.

17

Biodegradable Organic Polymers for Environmental Protection and Remediation

Adhithya S. Santhosh, Sreehari Suresh, Mridul Umesh, Liya M. Stanly, Sneha Grigary, and Nilina James

CHRIST (Deemed to be University), Department of Life Sciences, Bangalore 560029, Karnataka, India

17.1 Introduction

The main source of environmental pollution is the waste of long-lasting plastics. Most of these materials are single-use packages. The extent of pollution is unimaginable; the debris of plastic is being reported even in the deep oceans. The damage done to the environment cannot be undone; however, further destruction can be controlled by replacing the synthetic plastic with bio-based, biodegradable polymers.

17.1.1 Environmental Concerns Raised by Synthetic Materials

The large rise in the production of plastic and the increased usage in daily life have influenced the environment to a great extent (Pan et al. 2020). The accumulation of these synthetic materials has endangered life on the earth (Geyer et al. 2017). Approximately, 390.7 million metric tons of plastic was produced in 2021, which is 4% increase from the previous year (Statista n.d.). Close to 44% of all plastics are used for packaging consumer goods, with the remainder plastics being utilized for manufacturing. Only 173 metric ton of the total plastic waste produced after the consumption is collected for recycling and landfill use, remaining are either burnt or disposed of in the environment. The disposed materials, due to lack of management and segregation, leaks into the ecosystems. Among the leached waste from the disposed plastic roughly 3% is carried to the oceans (Rai et al. 2021). In the ocean water the mechanical waves and the prolonged UV radiation break down the macroplastic to microplastic (<5 mm), which are potential hazards to aquatic life (Lebreton and Andrady 2019). The currents disperse these microplastics throughout the ocean, where they accumulate in aquatic life and travel up the food chain. The consumed plastic continues to exist in the digestive tract of the animals, which affects their feeding behavior, hormone production and causes the overexudation of gastric enzymes which leads to gastrointestinal impediment. The accumulation of plastic also affects the reproduction of these organisms (Ciriminna and Pagliaro 2020).

Organic Polymers in Energy-Environmental Applications, First Edition.
Edited by Ramesh Oraon, Pardeep Singh, Sanchayita Rajkhowa, Sangita Agarwal, and Ravindra Pratap Singh.

The carbon dioxide emissions caused by the burning of the non-biodegradable polymers like polyethylene, polyvinyl chloride, and polypropylene possess an increasing risk of global warming. Although the "reuse, reduce, and recycle" strategy provides a relief from the burdensome problem, it is insufficient to limit the excessive use of plastics in society and the environment as a whole (Borrelle et al. 2020).

17.1.2 Biodegradable Polymers: Significance and Prospects

More focus is being given to the growing need for bio-based materials for immediate applications as a substituent to non-biodegradable synthetic polymers. Their application spans huge areas including tissue engineering, wound healing, and drug delivery. These bioplastics are mostly formed of polysaccharides, proteins, and fibers which are either naturally occurring, synthetic or semi synthetic. These biopolymers lack some of the material properties which enables their wide application as synthetic plastics. In order to have a widespread application in different fields, extensive research and development has been done to improve the properties of biopolymers. The goal is to bring their mechanical and physical capabilities up to par with those of non-biodegradable polymers (Abioye et al. 2019). These polymers can quickly disintegrate into the simple compounds, when discarded into the environment because of the enzymatic activity of natural microbial flora (Ruggero et al. 2019). As a result of the breakdown of natural polymers, humic matter, methane, biomass, water, carbon dioxide and a number of other natural substances are released (RameshKumar et al. 2020).

Depending on the source, biopolymers are divided into three categories: microbial (polyhydroxyalkanoates, levan), animal (chitin, chitosan, keratin), and plant-based (starch, cellulose) (Siracusa 2019). Unique properties of these polymers allow them to replace synthetic plastic in a variety of applications.

17.1.3 Waste Valorization, a Crucial Strategy for Sustainable Production

Biodegradable polymers are an alternative to the synthetic polymers, but their production is extremely challenging. A waste valorization strategy should be used to make it more cost-effective. Primary generators of waste which can be used as substrate for biopolymer production are the agro-industry. The agro-industry alone generates one billion metric tons of waste (Tiseo 2023). The waste includes solids, liquids, and slurries generated from the processing of cereals, vegetables, seafood, dairy, fruits, poultry, and meat. These can serve as a low-cost source for biopolymer production because of their non-toxic content and abundance (Bolaji et al. 2021; Tomaszewska et al. 2018).

Applying waste valorization toward the production of bioplastics can significantly lower the carbon footprint of agricultural biomass like corn stalks, vegetable trash, jute, wood chips, stubble, grain husks, and banana peels. The raw waste can be treated chemically to obtain polymers like cellulose, chitin, and gelatin. By employing microbes for the fermentation of these raw waste materials, they can be converted into natural polyesters like polyhydroxyalkanoates (PHA), polyhydroxybutyrate (PHB), and poly lactic acid (PLA) (Rai et al. 2021). The biopolymers after synthesis/extraction are subjected to life cycle assessments in order to determine their potential application for replacing non-biodegradable plastics and to minimize their environmental impact. These polymers have a lot of potential since their properties can easily be manipulated by adding desirable characteristics by blending with other polymers or chemical modifications (Rijpkema et al. 2020).

17.2 Role of Organic Biodegradable Polymers in Environmental Bioremediation

The popularity of biopolymers is attributed to the application of biopolymers in various industrial sectors such as the medical sector, food and packaging sector, cosmetics, agriculture, automotives, electronics, and construction fields (Balaji et al. 2018). Apart from these applications, biopolymers are now being extensively studied and experimented for its usefulness in bioremediation processes. Conventional methods for the removal of environmental pollutants like metals from the environment are often expensive and also create secondary pollutants (Ayangbenro and Babalola 2018). Utilization of biodegradable biopolymers for bioremediation is a pragmatic way, as it not only cleans up the environment but will reduce the pile up.

The competition of biopolymers in regard to the efficiency and scalability is with synthetic polymers. In regard to this, the work of Prakash et al. (2016) is important as they prepared a chitosan/starch-based blend which was found to be very much effective in removing chromium (Cr) from waste. Blending of chitosan and starch improved the quality of the film and provided a more adsorbable site for heavy metal (Prakash et al. 2016). Synthetic dyes that are commonly present in effluents released from textile and leather industry pose major environmental risks. To remove dyes from industrial effluents, cellulose, chitin, and chitosan-based biopolymers have been found to be effective and are now in use (Dassanayake et al. 2021). For the treatment of water polluted with heavy metals, dyes and oils, one of the naturally occurring polymers called keratin has been found to be helpful (Liya and Mridul 2022).

The biopolymers are quite flexible to chemical and physical modifications and versatile products like films, gels, fibers, sponges, nanoparticles (NPs), and biocomposites could be designed from their raw structure (Vinod et al. 2020). Being abundantly found in nature, divergent sources of biopolymers include plants, animals, and microorganisms. Different types of biopolymers and its role in bioremediation are being discussed further.

17.2.1 Plant-Based Polymers for Environmental Remediation

17.2.1.1 Cellulose and Cellulose Derivatives

Cellulose is an organic biopolymer which is primarily produced by plants as cell wall material. It is also produced by some forms of bacteria. The basic building block of cellulose is D-glucose, which is polymerized by β (1 → 4) linkages. In plants cellulose fibers act as structural support for the cells, they maintain the rigidity of tissues and make up most of the hard components of the plants. The purest form of naturally available cellulose is cotton (90%) (Khan et al. 2022). The cellulose microfibrils in the plant cell wall provide rigidity to the cell; this physical property is derived from the crystalline nature of cellulose in aqueous state. Such crystallinity is very rare in polysaccharides, which make cellulose a unique compound (He et al. 2021).

Different chemical modifications can be done on cellulose to obtain derivatives of cellulose with a wide variety of properties suitable for multiple applications (He et al. 2021). Oxycellulose derivatives are the derivatives which have a carboxyl group in the position of the terminal alcohol group. They can be water soluble. It's used to prepare gauzes which can be applied on wounds at the time of surgery. Ethers of cellulose are used in drug delivery due to their hydrogel-forming ability when comes in contact with water. The hydrogel will help to protect the drug from water by forming a barrier (Khan et al. 2022). Cellulose ether can be water soluble based upon the derived chemical structure, as well as degree and arrangement of the substituted groups. Common cellulose ethers are methyl cellulose, hydroxyethyl cellulose, and carboxymethyl cellulose (CMC). Methyl cellulose is synthesized when cellulose is treated with methylating agents (methyl chloride, dimethyl sulfate) under alkaline conditions (Içten et al. 2017). Ethyl cellulose is formed by converting the hydroxyl groups in cellulose molecules into ethyl ether groups (Jiang et al. 2022). This derivative is formed by the reaction between alkali cellulose and ethyl chloride. CMC is a highly stable derivative of cellulose, which is formed by a substitution reaction. The hydroxyl group gets substituted with the carboxymethyl group when treated with monochloroacetic acid (Seddiqi et al. 2021).

Among the esters of cellulose acetate is the common one. Cellulose acetate is synthesized by ring-opening esterification and transesterification of cellulose. The reaction can take place under heterogeneous or homogeneous conditions (Khoshnevisan et al. 2018). Another derivative, nitrocellulose is a nitric ester of cellulose which is highly reactive; hence, it's called gun cotton. It is made by treating cotton with a mixture of sulfuric acid and nitric acid. The compound is highly explosive and has a flash point of 4.4 °C (El Nemr et al. 2021).Cellulose nitrate is the first semisynthetic cellulose polymer synthesized during the emergence of the plastic industry. It is produced by the substitution of hydroxyl groups with nitrate groups (Neves et al. 2019). Cellulose sulfate is another unique derivative of cellulose prepared by sulfation of cellulose. Sulfuric acid, sulfur trioxide, and chlorosulfonic acid are used for the sulfation (Seddiqi et al. 2021).

Cellulose sheets can be prepared from different waste sources like *Pistia* (water weed) (Umesh et al. 2022), corn husk, sugarcane bagasse (Bhardwaj et al. 2022). These sheets can be degraded easily when exposed in environmental soil for about four weeks, which is much faster than the currently used packaging materials. These findings are a beacon for cellulose sheets to be used as an alternative for plastic in the packaging industry. In addition to the biodegradable packaging Table 17.1 lists additional uses for various cellulose types.

Table 17.1 Environmental applications of cellulose.

Cellulose form	Application	Key findings	References
Cellulose powder	Phosphate removal and anti-microbial activity	TiO_2/cellulose nanocomposites were able to remove 96% of phosphate from the solution	Zong et al. (2021)
Cellulose powder	Dye removal	Removal of 139.6 mg of Congo red per g of cellulose beads loaded with graphene nanoplatelets was obtained	González-López et al. (2021)
Cellulose powder	Removal of heavy metal from water	Removal of 23.5 ± 2.5 mg of Co(II) per g of chitosan/cellulose composite	Zhuang et al. (2021)
CMC	Dye removal	671 mg of dye was removed per g of carboxymethyl cellulose hybrid	Salama (2016)
CMC	Removal of heavy metal from water	Removal of 141.1 mg of Zn^{2+} and 150.2 mg of Cd^{2+} per g of CMC bridged chlorapatite	Li et al. (2021)
CMC	Extracting toxic material	Absorption of bisphenol A	Azzaoui et al. (2017)
Cellulose nanocrystals	Filtration of particulate matter (PM) from air	99.1% of $PM_{2.5}$ removal using poly(vinyl alcohol) (PVA)/cellulose nano crystals (CNC)	Zhang et al. (2019)
Cellulose nanocrystal	Removal of methyl ethyl ketone and cyclohexane	Polyacrylonitrile (PAN)/cellulose nanocrystal with activated carbon nanofibers showed three times better adsorption than the neat PAN membrane	Awad et al. (2021)
Cellulose nanocrystal	Dye removal	131.58 mg of methylene blue was removed per g of jute cellulose nanocrystal/poly(*N*,*N*-dimethylacrylamide-*co*-3-methacryloxypropyltrimethoxysilane) hybrid hydrogels	Hossain et al. (2021)
Microcrystalline cellulose	Dye removal	95% removal of methylene blue	Salama (2017)
Cellulose nanofibers	Filtration of particulate matter and removal of humidity from air	Super-hydrophobicity (Contact Angle [CA] = 152.4°), 99.31% and 99.75% removal of $PM_{1.0}$ and $PM_{2.5}$, respectively, when cellulose nanofibers modified with methyltrimethoxysilane were used	Liu et al. (2021)
Cellulose nanofibers	Filtration of PM and removal of CO_2 from air	Cellulose nanofibers – chitosan nanofilter removed 99.56% of $PM_{2.5}$ and showed efficient adsorption of CO_2	Sepahvand et al. (2021)
Cellulose acetate	Pesticide removal	99% removal of dimethoate pesticide when metal organic framework–cellulose acetate was used	Abdelhameed et al. (2021)

17.2.1.2 Starch and Starch Derivatives

Starch is one of the substantially applied organic biopolymers for bioremediation due to its uncomplicated structure, its availability, renewability, non-toxicity, and low cost. Starch is made up of glucose molecules that are efficiently broken down by amylases and amylase-producing microorganisms. Hence, starch and its derivatives are employed in bioremediation (Amaraweera et al. 2021).

One of the sole starch-based derivatives used in bioremediation is biosorbents like corn starch and potato starch. They effectively remove heavy metals from polluted water, air, and soil. In a work by Akinterinwa et al. (2020) corn starch cross-linked with $PoCl_3$ was used in the removal of Pb, Cd, Hg, and Cu from water. It was observed that 1% starch could remove the heavy metals (Pb, Cd, Hg) effectively and reduce the amount of Cu from the contaminated water. The starch complex was also reusable after it had undergone weak acid treatment. Another work by (Gao et al. 2022) highlights a liquid mulch based on corn starch-sodium alginate which was made as an alternative to non-biodegradable liquid mulch. It was degraded within 25 days and thereafter improved soil fertility by increasing organic matter.

Dextrins and maltodextrins are products obtained from partial and complete hydrolysis of starch, respectively. These products are indirectly employed in bioremediation by providing them to bacteria as a substrate medium for growth, and the bacteria will degrade certain unfavorable chemicals like (polychlorinated biphenyls) PCBs (Jing et al. 2018).

Another starch derivative is starch-based nanomaterials which include nanoparticles, nanogels, and nanomicelles. Starch nanomaterials are utilized as adsorbents for removal of pollutants from water. Starch hybrid nanomaterials like starch/SnO_2, starch Fe-Pd, starch xanthane, carboxymethyl starch, etc. have application in water treatment to remove targeted heavy metals like Hg^{2+}, trichloroethane, (Ni^{2+}, Cu^{2+}, Cr^{3+}) and (Pb^{2+}, Cd^{2+}, Cu^{2+}), respectively. Modified cassava starch coated with nanoscale zero-valent iron (nZVI) was used in aqueous nitrate remediation. Starch/PVA was employed in textile wastewater treatment for both cationic and anionic dyes. Starch treatment improved the hydrophilicity of CNTs (carbon nanotubes) efficiently, and it was observed that multi walled carbon nano tube (MWCNT)–starch–iron oxide was better absorbent that MWCNT–iron oxide for removing methylene blue and methyl orange dyes (Gamage et al. 2022). A work by Salam et al. (2018) suggested that corn starch modified with diethylenetriamine penta-acetic acid (DTPA) along with chitosan improved interfiber bonding of recycled fiber paper sheets and showed antibacterial properties up to 97%.

Although the applications of starch and starch derivatives are widely being studied and utilized, these techniques have their limitations. Ingestion of nanomaterials by humans can harmfully affect our well-being, as nano-sized materials can interact with cells and intracellular structures inside the human body. Comprehensively, starch and starch derivatives play a critical part in environmental bioremediation. Being affordable and readily available have uplifted their usage by enzyme producing microorganisms as a source of energy, therefore reducing the pollutants. Therefore, starch and its derivatives are valuable tools for environmental bioremediation which in turn benefits human well-being.

17.2.2 Animal-Based Polymers for Environmental Remediation

17.2.2.1 Chitin and Chitosan

Chitin is the second most common biopolymer on earth after cellulose, and chitosan is created by deacetylating chitin, where both these biopolymers exhibit specific chemical, mechanical, optical, and physical properties as a result of their structural features, which give them adjustable traits like high porosity, low density, renewability, and biodegradability (Shaumbwa et al. 2021). Chitin and chitosan have similar characteristics, because of their microstructural similarity. These biopolymers can be a potential replacement for their natural and synthetic equivalents in environmental remediation due to their simplicity in processing, non-toxicity, and intrinsic antibacterial nature (Sirajudheen et al. 2021).

Chitin is a fibrous material composed of repeating glucose units (acetylglucosamine) connected by β-(1 → 4) glycosidic bonds, which is the primary component of fungal cell wall and crustacean exoskeletons. The exoskeleton of crustaceans is abundant in components like protein (20–40%), calcium carbonate (20–50%), and chitin (15–40%) which are advantageous for industrial purposes (Yan and Chen 2015). The exceptional properties of chitosan offer extensive applications in many fields including biotechnology, food industry, chemical industries, textiles, membranes, wastewater treatment, pharmaceutical, and biomedicine (El Knidri et al. 2018). Chitin and chitosan can be extracted by chemical and biological processes, where both methods involve three major steps: demineralization, deproteinization, and deacetylation (Mohan et al. 2020). Compared to chitin, chitosan has higher industrial significance and applications due to the presence of various functional groups with high reactivity and the polymer chain has higher flexibility. The free amino groups found in each D-glucosamine unit give chitosan its key characteristics and the hydroxyl groups present on the polysaccharide chain opens the possibility of structural modification of chitosan (Joseph et al. 2021). Chitosan in its pure form has limited applications compared to modified chitosan, where chitosan is modified through physical and chemical processes in order to enhance its original qualities and expand the range of applications, with a focus on their use in environmental remediation. Chitosan can be modified to prepare numerous adsorbent materials like films, hydrogels, membranes, aerogels, and composites which have various environmental applications such as removal of environmental pollutants like herbicides, industrial dyes, heavy metals, insecticide, and fungicides (Pal et al. 2021). Different methods of modification of chitin and chitosan-based adsorbents are given in Table 17.2.

Chitin and chitosan are used as environmental remediation agents in both water and soil treatment. Water pollution is brought on by toxic compounds that are easily dissolved in water when they come into touch with it from farms, cities, and factories. The majority of the industry uses dyes for commercial purposes, and the unprocessed industrial effluents contain large quantities of various organic dyes. People who live close to industry are in severe danger since they are frequently exposed to the contaminants, which is why the elimination of such pollutants from wastewaters is absolutely essential. Dyes are one of the common pollutants found in

Table 17.2 Pros and cons of chitin and chitosan modification.

Modification method	Advantage	Disadvantage	References
Co-precipitation	Controllability of particle size and contents Easy preparation The surface particle state and overall homogeneity can be altered in a wide range of ways Minimal temperature and usage of less energy	Time consuming process Batch-to-batch reproducibility problems Chances of precipitation of trace contaminants along with the product	Rane et al. (2018)
Hydrothermal synthesis	Eco-Friendly nature One-step synthetic route Better dispersion in solution	Difficulty in monitoring crystal formation during the reaction High cost of equipment	da Silva et al. (2020)
Lyophilization	High-quality physical structure of the dry products Potential to regulate the final product's moisture content Quick reconstitution of the dry product with a high specific surface area	Loss of volatile compounds from the model matrix during the process Need complex and expensive equipment	Wu et al. (2021)
Melt extrusion method	Less manufacture times Uniform diffusion of nanofillers	Need high-energy input	Ghaleb et al. (2019)
Polymerization	Products with uniform size and shape can be obtained	Long reaction time Require huge solvent volume Difficulty in removing the solvent from the product	Deng et al. (2019)

Simple blending	Improves physical and thermal properties of the polymer	—	Ambaye et al. (2022)
Solvent/solution casting method	Provides enhanced thermal and mechanical properties Provides better water barrier properties	Chemical bonds containing the binder gets affected during the process of solvent casting	Gopi et al. (2019)
Spray drying	The consistent dispersion of particle size Dry powder form of product	Vast distribution of particle dimension Long processing time	Amalraj et al. (2020)
Sol-gel method	Low-temperature of reaction High level of purity Better composition control	High price of the raw materials Chances of shrinkage during drying	Yilmaz and Soylak (2020)
Phase inversion	Morphology of the scaffold can be controlled by changing parameters	No large-scale studies	Chaudhary et al. (2020)
Electrospinning	Simplicity and effectiveness of the procedure Potential to control properties like fiber orientation and diameter	Need organic solvents	Wu et al. (2021)

water bodies as they are widely used in various sectors including textile, pharmaceutical, food, agriculture, leather tanning, and paper production. In addition to being exceptionally mutagenic and carcinogenic, these dyes have been discovered to be deadly wastes (Sirajudheen et al. 2020). Methyl orange (MO), methylene blue (MB), reactive yellow 145, malachite green (MG), methyl violet (MV), organic dye sunset yellow (SY), alizarin yellow, and congo red are some common dyes that can have negative impacts on both the environment and people. Chitin and chitosan's limited adsorption capacity in powder or flake form is due to their crystalline structure and modified chitin and chitosan have shown higher adsorption potential. The interactions between some divalent heavy metals and chitosan/graphene oxide (Cs/GO) improve the stability of chitosan and its adsorption potential toward heavy metals like Ni, Cu, As, Cd, and Pb (Menazea et al. 2020). Bioaugmentation, biostimulation, and degradation of xenobiotic compounds are some methods involved in the remediation of soil. The immobilization property of chitosan makes it an efficient applicant for bioaugmentation, and this procedure of remediation is quick and effective, and it can be applied in both in vitro and ex vitro.

17.2.2.2 Keratin and Keratin Derivatives

Keratin is a protein which is found in nature as components of feathers, hair, wool, claws, nails, horns, hooves, and beaks (Sharma et al. 2019). Keratin is also one of the most exploited biopolymers for different bioremediation applications. The building blocks of keratin include various amino acids which have functional and charged groups in its chemical structure. These functional groups and charged groups are the major sites where adsorption and sequestration of environmental contaminants take place (Saha et al. 2019). The tough structure of keratin protein and its resistance toward many organic solvents and peptidase enzymes contributed by the recurring presence of cysteine amino acid and disulfide bonds makes them a good option as a stable biopolymer. Apart from the structure, the resistance of keratinous materials against microbial attack and its slow response to natural degradation renders them to be used for a long time and over several recycling cycles also (Lange et al. 2016).

Many well-studied works on the application of keratin for bioremediation have been reported in the past like the work of Abba et al. (2020) for removal of heavy metals from aqueous solution. Two commonly occurring types of chicken feathers, melanized and non-melanized chicken feathers, were experimented for the removal of cadmium (Cd) and lead (Pb) from aqueous solutions. The feathers were able to remove 40% of cadmium (Cd) and 30% of lead (Pb) concentration from the solution within a time period of seven days (Abba et al. 2020). Similarly, the ability of guinea fowl feathers to remove copper (Cu) and zinc (Zn) heavy metal from aqueous solution was studied by Yusuf and Sharu (2022). They used raw feathers as well as autoclave-pretreated and acetone-pretreated feathers for the removal of heavy metals. All three forms of feathers were able to remove heavy metal from aqueous solution with subsequent degradation of feather-by-feather degrading *Pseudochrobactrum* sp. IY-BUKI (Yusuf and Sharu 2022). Hydrocarbons and oils discharged from industries, mining practices, and spilled from oil tankers have been polluting both land and water for a long time. Although activated carbon has been in use for the removal of oils for the past years, Okoya et al. (2020) were successful in proving the efficiency of chicken feather waste as an adsorbent for the removal of oil from aqueous solution with 99.95% of adsorption ability which was 4% higher than commercially used activated carbon (Okoya et al. 2020). Not only chicken feathers but also dog fur and human hair were experimented over peat moss and polypropylene for adsorption of crude oil and was found to be as efficient as peat moss and polypropylene for the removal of oil even after recycling also (Murray et al. 2020).

The applicability of keratin can be improved by modifying its functional groups and side chains and such modified forms of keratin are called keratin derivatives. Major modification methods for the preparation of keratin derivatives include oxidation, reduction, coupling, using ionic solutions, succinylation, quaternization, sulfitolysis, and enzymatic modifications (Maier et al. 2019). The process of keratin structural modification by oxidation and reduction is depicted in Figure 17.1. Such modified keratin can be used for the preparation of hydrogels, scaffolds, films, and fibers (Jaiswar et al. 2022).

Adsorption of oil and naphthalic acid that come up the surface during bitumen extraction was removed by a modified keratin biopolymer as mentioned in one of the works of Arshad et al. (2016). The modification of keratin was based on opening the structure of keratin using polyhedral oligomeric silsesquioxanes (POSS) nanocages and goethite dopant (Arshad et al. 2016). Similarly, to clean up spilled oil and organic solvents, keratin/cellulose based cryogel was prepared by adding ionic solvents. This cryogel efficiently adsorbed oil with the cryogel attaining a weight of 6.9–17.7 times more than its initial weight. Through adsorption distillation or adsorption desorption method, the gel could be recycled and used again (Guiza et al. 2021). Separation of oil and water from oil contaminated water can be brought about by preferentially absorbing water from oil. In view of this, a study was conducted by Gao et al. (2020) to increase the wettability of chicken feathers. Through modification of keratin in chicken feathers by reduction and coupling, they were able to make a keratin derivative which was hydrophilic and oleophobic in nature. These modified feathers could act as a surface which can effectively repel oil and thus can be used for oil water separation (Gao et al. 2020). A well-known method of modification of keratins includes hydrolysis wherein the bonds of keratin will be broken forming a hydrolyzed keratin as product. Utilizing this method, a study was conducted by Tissera et al. (2020)

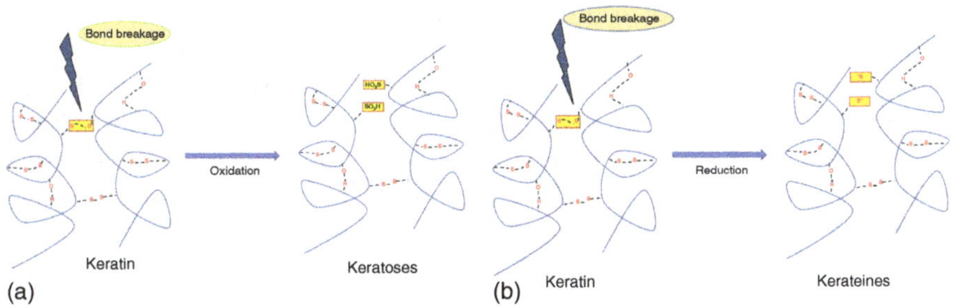

Figure 17.1 Formation of keratin derivatives through chemical modifications: (a) oxidation and (b) reduction.

where controlled surface hydrolysis of Merino wool fibers resulted in the removal of cuticle from wool fiber exposing cortical fibers which have exceptionally high amounts of amino acids. Such modified fibers were able to adsorb carcinogenic Rhodamine B dye up to 294 mg g^{-1} of the fiber in presence of 3.5% acetic acid (Tissera et al. 2020). Likewise for the absorption of anionic and cationic dyes, keratin sponges were prepared by freeze drying and chemical cross-linking. The sponges provided a good absorbing surface for Azure A and methyl orange dye, two of those dyes which are toxic and are most frequently found in waste waters (Posati et al. 2020). Wastewater from tannery is one of the most polluted discharges from industries as it contains a lot of toxic dyes, heavy metals, and harmful microbes (Liya et al. 2023). It is very important to treat this water before discharging it for disposal. In regard to this problem, Roy et al. (2022) worked out to prepare a nanocomposite based on graphene oxide, chitosan, and keratin biopolymer (GO-CS-K) which could reduce the turbidity and thereby toxin loads of tannery wastewater. Tannery effluent treated with this nanocomposite was also found to reduce the biochemical oxygen demand (BOD), chemical oxygen demand (COD) total suspended solids (TSS), and salinity in wastewater and made the effluent safe to be released to the environment (Roy et al. 2022). Environment pollutant load is increasing day by day. Organic biodegradable biopolymers and their derivatives can be a promising solution for bioremediation.

17.2.3 Microbe-Based Polymers for Environmental Remediation

17.2.3.1 Polyhydroxyalkanoates

Microorganisms produce microbial polyhydroxyalkanoates as intracellular granules under stress conditions. These biopolyesters are harmless and biodegradable. Microbial polyhydroxyalkanoates are mostly produced by prokaryotic organisms as granules inside the cytoplasm. The size of these water-insoluble granules ranges from 0.2 to 0.5 μm (Sedlacek et al. 2019). Most of the microorganisms produce these granular inclusions as a source of energy when the nutrients are scarce and carbon sources are abundant (Sagong et al. 2018), and studies have shown that these granules actually protect the bacterial cell from osmotic stress and other worse conditions (Obruca et al. 2018).

PHAs are important mainly because of their polymeric properties which make them environmentally friendly and can be considered as a replacement for the synthetic plastics (Amelia et al. 2019). These eco-friendly polymers can be used as biomaterials with zero impact on the environment, in agriculture fields, packaging materials, medical implants, drug-delivery carriers, 3D printing materials, etc. When it comes to the industrial sector a major application of PHAs is in using it as a packaging material. Pollution occurring in the oceans as a result of usage of plastics can be compensated by the use of bioplastics made using polyhydroxyalkanoates; these include the use of biodegradable plastic utensils like spoons, cups, and plates made by using PHAs (Mangaraj et al. 2019). A PHA copolymer named as "Metabolix" which is actually a blend of PHB and P(3HO) can be used in the production of food additives and packaging materials and has properties similar to petrochemical-derived plastics (Mangaraj et al. 2019).

Another field which uses PHA biopolyesters is in the production of biofuels. Here the used bioplastic products are methyl esterified to produce biofuels. During this esterification process the short-chain and medium-chain length PHAs can be esterified with methanol in order to produce hydroxyalkanoate methyl esters. This biofuel production using PHA is feasible as the PHA can be produced from activated sludge and nutrient-rich wastewater (Muneer et al. 2020). PHAs can be used for the another environmental remediation purposes such as in the removal of heavy metals. Hungund et al. (2018) used PHA composites comprising PHB-starch and PHB polyethylene glycol (PEG) for removing heavy metals from aqueous solutions, and it was observed that these composites showed better adsorption of Pb^{2+} ions (Hungund et al. 2018).

PHA-based nanoparticles and fibers are used in medical applications as well. Studies have shown that 3HB monomers from PHA can actually fight against Parkinson's and Alzheimer disease which can be done by blocking the neural cell death (Camberos-Luna et al. 2016). Another application in the medical field is its use as an anti-osteoporosis agent; here the grooved P(3HB-*co*-3HHx) film induced osteogenic initiation in bone-marrow-derived mesenchymal stem cells. This finding supports the use of PHAs in the treatment of osteoporosis (Ali and Jamil 2016). Certain studies have also shown the ability of PHA as an anti-cancerous agent by actually combining the 3HAs with D peptide (Sangsanoh et al. 2017).

In the agricultural sector PHAs can be used to make agricultural grow bags, neta and mulch films (Amelia et al. 2019). Mulching crops with PHA-made biofilms or bioplastics not only increase yield but also protect them. Agriculture mulching is made with copolymers like P(3HB-*co*-3HHx) (Kalia et al. 2019). NodaxTM and MirelTM are two examples of PHA-based agriculture mulch used for commercial purposes. When buried or burned, synthetic mulch pollutes the environment and has led to an increase in the number of landfills. As PHAs-based mulch films are being developed on a research scale and are anticipated to aid in the reduction of synthetic mulch films, bioplastics have addressed the issue (Amelia et al. 2019). In nanotechnology, PHA nanomaterials in the form of microspheres have been studied to increase herbicide activity by acting as nanocarriers of the herbicides with lower genotoxicity and higher biodegradability rates (Kalia et al. 2019).

PHAs, as bioplastic materials, can help conserve ecosystems by removing synthetic plastics from the environment. These amazing granular inclusions, which function as microbial polyesters, can assist us in combating environmental pollution by addressing current plastic debris and waste mismanagement scenarios.

17.2.3.2 Levan

Levan is a homopolysaccharide generated by a variety of microorganisms as exopolysaccharides (EPS) and a minor amount of plant species as non-structural storage carbohydrates. Levan is a β-(2,6)-linked fructose polymer, and this fructose backbone makes levan a fascinating carbohydrate polymer. Bacterial levan contributes to the formation of the exopolysaccharides matrix and helps in microbial biofilm formation which prevents the microorganisms from drying out as the water level fluctuates, aids in influencing the environment of their proximity, and protects them from other organisms. Levan has key biomedical qualities including antioxidant, anti-inflammatory, anti-carcinogenic, and hyperglycemic inhibitor in addition to general qualities like biocompatibility, biodegradability, renewability, flexibility, and ecofriendliness. In recent years, scientists have explored the potential of biopolymers as an applicant in environmental remediation due to its natural occurrence, non-toxicity, environmental friendliness, and biocompatibility. Biopolymer-based sorbents have several advantages over other materials in environmental remediation, and levan is an excellent candidate for molecular interactions with organic micropollutants due to the furanose rings in its structure (Hacıosmanoğlu et al. 2019). Due to its β-(2,6) linkage, the water-soluble carbohydrate polymer levan is soluble in both water and oil. However, it does not immediately settle out in a water suspension and has variable solubility in cold water but is entirely soluble in hot water. Levan is mostly produced by bacteria utilizing submerged fermentation processes and for higher yields, environmental factors including pH, temperature, aeration, and agitation must be developed and optimized for each organism (de Siqueira et al. 2020) (Figure 17.2).

Levan is different from other exopolysaccharides due to its high-water retention capacity and it is heat stable, with a glass transition temperature of 141 °C and a decomposition point of about 225 °C. Levan is extensively employed as an encapsulating and emulsifying agent, but the modification of levan enhances the naturally present properties of levan. These modifications involve blending of levan with other biopolymers such as chitin and chitosan to improve its mechanical properties (Wang et al. 2022). A study reported by (Lončarević et al. 2019) highlights the use of levan and pullulan as potential bio-detoxification agents for reducing adverse effects of copper on *Daphnia magna* and *Vibrio fischeri*. Phosphonated Halomonas Levan (PhHL) as an adsorbent for bisphenol, one of the highest volume chemicals

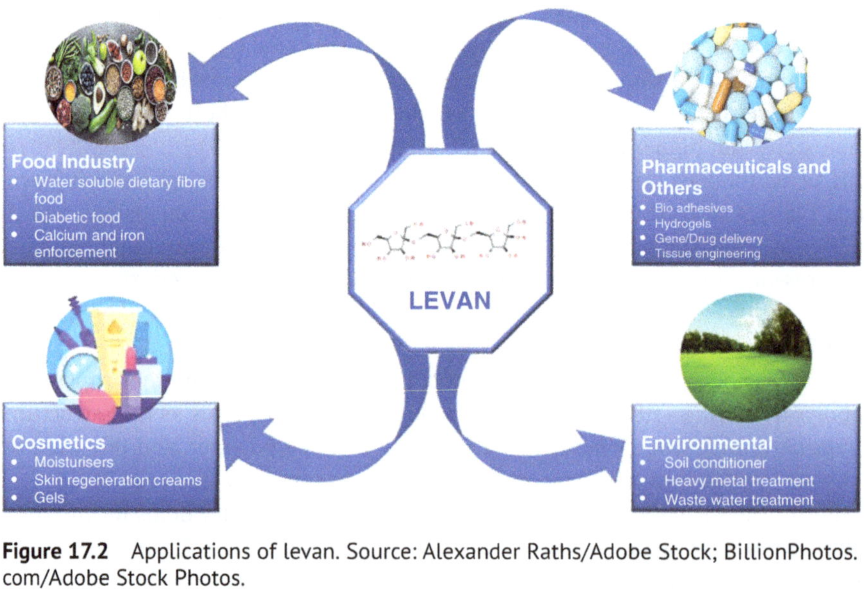

Figure 17.2 Applications of levan. Source: Alexander Raths/Adobe Stock; BillionPhotos.com/Adobe Stock Photos.

produced worldwide, showed maximum adsorption capacity of 126.6 $mg\,g^{-1}$ (Hacıosmanoğlu et al. 2019). Levan have been widely utilized in nanostructured systems as excipients numerous techniques have been used to create levan-based nanoparticles (NPs), including flame pyrolysis, co-precipitation, direct enzymatic synthesis, thermal and chemical reduction, self-assembly, and nanoprecipitation. Levan synthesized by *Azotobacter vinelandii* D-08 showed higher capacity for the binding of Na, Mg, S, Ca, and Fe in the cultivation medium underlines the usage of levan as sorbents for wastewater treatment to remove heavy metals and radionuclides (Shutova et al. 2021).

17.3 Conclusion

Despite the versatile material properties and high degree of biocompatibility of biopolymers, they still fail to compete with their synthetic counterparts majorly due to the cost associated with their production and recovery. Lack of understanding among the public and policy makers is the major reason that hinders the commercialization of biopolymers in the market. Although extensive research was conducted in lab and pilot scale, the industrial level scale up of these polymers is yet a long journey. Development of cost-effective methods for production and purification of these biopolymers and integration of waste valorization methods with biopolymer production can clearly revolutionize this field of eco-friendly product development.

References

Abba, Z.A., Yahaya, S., Ahmad, S.A. et al. (2020). Bioremediation of heavy metals by melanised and non-melanised feathers and heavy metal resistant feather-degrading bacteria. *Revista Mexicana de Ingeniería Química* 19 (1): 243–252.

Abdelhameed, R.M., Abdel-Gawad, H., and Emam, H.E. (2021). Macroporous Cu-MOF@cellulose acetate membrane serviceable in selective removal of dimethoate pesticide from wastewater. *Journal of Environmental Chemical Engineering* 9 (2): 105121.

Abioye, A.A., Fasanmi, O.O., Rotimi, D.O. et al. (2019). Review of the development of biodegradable plastic from synthetic polymers and selected synthesized nanoparticle starches. *Journal of Physics: Conference Series* 1378 (4): 042064.

Akinterinwa, A., Oladele, E., Adebayo, A. et al. (2020). Cross-linked-substituted (esterified/etherified) starch derivatives as aqueous heavy metal ion adsorbent: a review. *Water Science and Technology: A Journal of the International Association on Water Pollution Research* 82 (1): 1–26.

Ali, I. and Jamil, N. (2016). Polyhydroxyalkanoates: current applications in the medical field. *Frontiers in Biology* 11 (1): 19–27.

Amalraj, A., Jude, S., and Gopi, S. (2020). Polymer blends, composites and nanocomposites from chitin and chitosan; manufacturing, characterization and applications. In: *Handbook of Chitin and Chitosan* (ed. S. Gopi, S. Thomas, and A. Pius), 1–46. Elsevier.

Amaraweera, S.M., Gunathilake, C., Gunawardene, O.H.P. et al. (2021). Development of starch-based materials using current modification techniques and their applications: a review. *Molecules* 26 (22): 6880.

Ambaye, T.G., Vaccari, M., Prasad, S. et al. (2022). Preparation and applications of chitosan and cellulose composite materials. *Journal of Environmental Management* 301: 113850.

Amelia, T.S.M., Govindasamy, S., Tamothran, A.M. et al. (2019). Applications of PHA in agriculture. In: *Biotechnological Applications of Polyhydroxyalkanoates* (ed. V. Kalia), 347–361. Springer.

Arshad, M., Khosa, M.A., Siddique, T., and Ullah, A. (2016). Modified biopolymers as sorbents for the removal of naphthenic acids from oil sands process affected water (OSPW). *Chemosphere* 163: 334–341.

Awad, R., Mamaghani, A.H., Boluk, Y., and Hashisho, Z. (2021). Synthesis and characterization of electrospun PAN-based activated carbon nanofibers reinforced with cellulose nanocrystals for adsorption of VOCs. *Chemical Engineering Journal* 410: 128412.

Ayangbenro, A. and Babalola, O. (2018). Metal(loid) bioremediation: strategies employed by microbial polymers. *Sustainability* 10 (9): 3028.

Azzaoui, K., Mejdoubi, E., Lamhamdi, A. et al. (2017). Preparation and characterization of biodegradable nanocomposites derived from carboxymethyl cellulose and hydroxyapatite. *Carbohydrate Polymers* 167: 59–69.

Balaji, A.B., Pakalapati, H., Khalid, M. et al. (2018). Natural and synthetic biocompatible and biodegradable polymers. In: *Biodegradable and Biocompatible Polymer Composites* (ed. N.G. Shimpi), 3–32. Woodhead Publishing.

Bhardwaj, A., Sharma, N., Alam, T. et al. (2022). Development and characterization of chitosan and beeswax–chitosan coated biodegradable corn husk and sugarcane bagasse-based cellulose paper. *Waste and Biomass Valorization*. https://doi.org/10.1007/s12649-022-01952-1.

Bolaji, I., Nejad, B., Billham, M. et al. (2021). Multi-criteria decision analysis of agri-food waste as a feedstock for biopolymer production. *Resources, Conservation and Recycling* 172: 105671.

Borrelle, S.B., Ringma, J., Law, K.L. et al. (2020). Predicted growth in plastic waste exceeds efforts to mitigate plastic pollution. *Science* 369 (6510): 1515–1518.

Camberos-Luna, L., Gerónimo-Olvera, C., Montiel, T. et al. (2016). The ketone body, β-hydroxybutyrate stimulates the autophagic flux and prevents neuronal death induced by glucose deprivation in cortical cultured neurons. *Neurochemical Research* 41 (3): 600–609.

Chaudhary, S., Kumar, S., Kumar, V., and Sharma, R. (2020). Chitosan nanoemulsions as advanced edible coatings for fruits and vegetables: composition, fabrication and developments in last decade. *International Journal of Biological Macromolecules* 152: 154–170.

Ciriminna, R. and Pagliaro, M. (2020). Biodegradable and compostable plastics: a critical perspective on the dawn of their global adoption. *ChemistryOpen* 9 (1): 8–13.

Dassanayake, R.S., Acharya, S., and Abidi, N. (2021). Recent advances in biopolymer-based dye removal technologies. *Molecules* 26 (15). https://doi.org/10.3390/molecules26154697.

Deng, F., Luo, X.-B., Ding, L., and Luo, S.-L. (2019). Application of nanomaterials and nanotechnology in the reutilization of metal ion from wastewater. In: *Nanomaterials for the Removal of Pollutants and Resource Reutilization*, 149–178. Elsevier.

El Knidri, H., Belaabed, R., Addaou, A. et al. (2018). Extraction, chemical modification and characterization of chitin and chitosan. *International Journal of Biological Macromolecules* 120 (Pt A): 1181–1189.

El Nemr, A., Eleryan, A., Mashaly, M., and Khaled, A. (2021). Rapid synthesis of cellulose propionate and its conversion to cellulose nitrate propionate. *Polymer Bulletin* 78 (8): 4149–4182.

Gamage, A., Punniamoorthy, T., and Madhujith, T. (2022). Starch-based hybrid nanomaterials for environmental remediation. *Starch - Evolution and Recent Advances*. https://doi.org/10.5772/intechopen.101697.

Gao, D., Chen, J., Qian, W. et al. (2020). Improving wettability of feather fiber by surface modification. *Waste and Biomass Valorization* 11 (12): 6993–7003.

Gao, X., Fu, C., Li, M. et al. (2022). Effects of biodegradation of corn-starch–sodium-alginate-based liquid mulch film on soil microbial functions. *International Journal of Environmental Research and Public Health* 19 (14): 8631.

Geyer, R., Jambeck, J.R., and Law, K.L. (2017). Production, use, and fate of all plastics ever made. *Science Advances* 3 (7): e1700782.

Ghaleb, Z.A.A., Jaafar, M., and Rashid, A.A. (2019). Fabrication methods of carbon-based rubber nanocomposites and their applications. In: *Carbon-Based Nanofillers and Their Rubber Nanocomposites* (ed. S. Yaragalla, R.K. Mishra, S. Thomas, et al.), 49–63. Elsevier.

González-López, M.E., Laureano-Anzaldo, C.M., Pérez-Fonseca, A.A. et al. (2021). Congo red adsorption with cellulose-graphene nanoplatelets beads by differential column batch reactor. *Journal of Environmental Chemical Engineering* 9 (2): 105029.

Gopi, S., Pius, A., Kargl, R. et al. (2019). Fabrication of cellulose acetate/chitosan blend films as efficient adsorbent for anionic water pollutants. *Polymer Bulletin* 76 (3): 1557–1571.

Guiza, K., Ben Arfi, R., Mougin, K. et al. (2021). Development of novel and ecological keratin/cellulose-based composites for absorption of oils and organic solvents. *Environmental Science and Pollution Research International* 28 (34): 46655–46668.

Hacıosmanoğlu, G.G., Doğruel, T., Genç, S. et al. (2019). Adsorptive removal of bisphenol A from aqueous solutions using phosphonated levan. *Journal of Hazardous Materials* 374: 43–49.

He, X., Lu, W., Sun, C. et al. (2021). Cellulose and cellulose derivatives: different colloidal states and food-related applications. *Carbohydrate Polymers* 255: 117334.

Hossain, S., Shahruzzaman, Kabir, S.F. et al. (2021). Jute cellulose nanocrystal/poly(*N*,*N*-dimethylacrylamide-*co*-3-methacryloxypropyltrimethoxysilane) hybrid hydrogels for removing methylene blue dye from aqueous solution. *Journal of Science: Advanced Materials and Devices* 6 (2): 254–263.

Hungund, B.S., Umloti, S.G., Upadhyaya, K.P. et al. (2018). Development and characterization of polyhydroxybutyrate biocomposites and their application in the removal of heavy metals. *Materials Today: Proceedings* 5 (10): 21023–21029.

Içten, E., Purohit, H.S., Wallace, C. et al. (2017). Dropwise additive manufacturing of pharmaceutical products for amorphous and self emulsifying drug delivery systems. *International Journal of Pharmaceutics* 524 (1–2): 424–432.

Jaiswar, G., Modak, S., Singh, R., and Dabas, N. (2022). Functionalization of biopolymer keratin-based biomaterials and their absorption properties for healthcare application. In: *Polymeric Biomaterials for Healthcare Applications* (ed. K. Varaprasad), 257–270. Woodhead Publishing.

Jiang, Y.-H., Zhang, Y.-Q., Wang, Z.-H. et al. (2022). Cotton-derived green sustainable membrane with tailored wettability interface: synergy of lignin and ethyl cellulose. *Industrial Crops and Products* 183: 114993.

Jing, R., Fusi, S., and Kjellerup, B.V. (2018). Remediation of polychlorinated biphenyls (PCBs) in contaminated soils and sediment: state of knowledge and perspectives. *Frontiers in Environmental Science* 6. https://doi.org/10.3389/fenvs.2018.00079.

Joseph, S.M., Krishnamoorthy, S., Paranthaman, R. et al. (2021). A review on source-specific chemistry, functionality, and applications of chitin and chitosan. *Carbohydrate Polymer Technologies and Applications* 2: 100036.

Kalia, V.C., Ray, S., Patel, S.K.S. et al. (2019). The dawn of novel biotechnological applications of polyhydroxyalkanoates. In: *Biotechnological Applications of Polyhydroxyalkanoates* (ed. V.C. Kalia), 1–11. Singapore: Springer.

Khan, R., Jolly, R., Fatima, T., and Shakir, M. (2022). Extraction processes for deriving cellulose: A comprehensive review on green approaches. *Polymers for Advanced Technologies* 33 (7): 2069–2090.

Khoshnevisan, K., Maleki, H., Samadian, H. et al. (2018). Cellulose acetate electrospun nanofibers for drug delivery systems: applications and recent advances. *Carbohydrate Polymers* 198: 131–141.

Lange, L., Huang, Y., and Busk, P.K. (2016). Microbial decomposition of keratin in nature—a new hypothesis of industrial relevance. *Applied Microbiology and Biotechnology* 100 (5): 2083–2096.

Lebreton, L. and Andrady, A. (2019). Future scenarios of global plastic waste generation and disposal. *Palgrave Communications* 5 (1): 1–11.

Li, Z., Gong, Y., Zhao, D. et al. (2021). Enhanced removal of zinc and cadmium from water using carboxymethyl cellulose-bridged chlorapatite nanoparticles. *Chemosphere* 263: 128038.

Liu, T., Cai, C., Ma, R. et al. (2021). Super-hydrophobic cellulose nanofiber air filter with highly efficient filtration and humidity resistance. *ACS Applied Materials & Interfaces* 13 (20): 24032–24041.

Liya, M.S. and Mridul, U. (2022). Keratin as a sustainable biopolymer for waste water treatment. *Research Journal of Chemistry and Environment* 26 (9): 157–166.

Liya, S.M., Umesh, M., Nag, A. et al. (2023). Optimized production of keratinolytic proteases from *Bacillus tropicus* LS27 and its application as a sustainable alternative for dehairing, destaining and metal recovery. *Environmental Research* 221: 115283.

Lončarević, B., Lješević, M., Marković, M. et al. (2019). Microbial levan and pullulan as potential protective agents for reducing adverse effects of copper on *Daphnia magna* and *Vibrio fischeri*. *Ecotoxicology and Environmental Safety* 181: 187–193.

Maier, S.S., Rajabinejad, H., and Buciscanu, I.-I. (2019). Current approaches for raw wool waste management and unconventional valorization: a review. *Environmental Engineering and Management Journal* 18 (7): 1439–1456.

Mangaraj, S., Yadav, A., Bal, L.M. et al. (2019). Application of biodegradable polymers in food packaging industry: a comprehensive review. *Journal of Packaging Technology and Research* 3 (1): 77–96.

Menazea, A.A., Ezzat, H.A., Omara, W. et al. (2020). Chitosan/graphene oxide composite as an effective removal of Ni, Cu, As, Cd and Pb from wastewater. *Computational and Theoretical Chemistry* 1189: 112980.

Mohan, K., Ganesan, A.R., Muralisankar, T. et al. (2020). Recent insights into the extraction, characterization, and bioactivities of chitin and chitosan from insects. *Trends in Food Science & Technology* 105: 17–42.

Muneer, F., Rasul, I., Azeem, F. et al. (2020). Microbial polyhydroxyalkanoates (PHAs): efficient replacement of synthetic polymers. *Journal of Polymers and the Environment* 28 (9): 2301–2323.

Murray, M.L., Poulsen, S.M., and Murray, B.R. (2020). Decontaminating terrestrial oil spills: a comparative assessment of dog fur, human hair, peat moss and polypropylene sorbents. *Environments* 7 (7): 52.

Neves, A., Angelin, E.M., Roldão, É., and Melo, M.J. (2019). New insights into the degradation mechanism of cellulose nitrate in cinematographic films by Raman microscopy. *Journal of Raman Spectroscopy: JRS* 50 (2): 202–212.

Obruca, S., Sedlacek, P., Koller, M. et al. (2018). Involvement of polyhydroxyalkanoates in stress resistance of microbial cells: biotechnological consequences and applications. *Biotechnology Advances* 36 (3): 856–870.

Okoya, A.A., Ochor, N.O., Akinyele, A.B., and Olaiya, O.O. (2020). The use of chicken feather waste as an adsorbent for crude oil clean up from polluted water. *Journal of Agriculture and Ecology Research International* 21: 43–53.

Pal, P., Pal, A., Nakashima, K., and Yadav, B.K. (2021). Applications of chitosan in environmental remediation: a review. *Chemosphere* 266: 128934.

Pan, D., Su, F., Liu, C., and Guo, Z. (2020). Research progress for plastic waste management and manufacture of value-added products. *Advanced Composites and Hybrid Materials* 3 (4): 443–461.

Posati, T., Listwan, A., Sotgiu, G. et al. (2020). Keratin/hydrotalcites hybrid sponges as promising adsorbents for cationic and anionic dyes. *Frontiers in Bioengineering and Biotechnology* 8: 68.

Prakash, N., Arungalai Vendan, S., Sudha, P.N., and Renganathan, N.G. (2016). Biodegradable polymer-based ternary blends for adsorption of heavy metal from simulated industrial wastewater. *Synthesis and Reactivity in Inorganic, Metal-Organic, and Nano-Metal Chemistry* 46 (11): 1664–1674.

Rai, P., Mehrotra, S., Priya, S. et al. (2021). Recent advances in the sustainable design and applications of biodegradable polymers. *Bioresource Technology* 325: 124739.

RameshKumar, S., Shaiju, P., O'Connor, K.E., and Ramesh, B.P. (2020). Bio-based and biodegradable polymers – state-of-the-art, challenges and emerging trends. *Current Opinion in Green and Sustainable Chemistry* 21: 75–81.

Rane, A.V., Kanny, K., Abitha, V.K., and Thomas, S. (2018). Methods for synthesis of nanoparticles and fabrication of nanocomposites. In: *Synthesis of Inorganic Nanomaterials* (ed. S. Mohan Bhagyaraj, O.S. Oluwafemi, N. Kalarikkal, and S. Thomas), 121–139. Woodhead Publishing.

Rijpkema, S.J., Langens, S.G.H., van der Kolk, M.R. et al. (2020). Modular approach to the functionalization of polymersomes. *Biomacromolecules* 21 (5): 1853–1864.

Roy, K., Dey, T.K., Jamal, M. et al. (2022). Fabrication of graphene oxide–keratin–chitosan nanocomposite as an adsorbent to remove turbidity from tannery wastewater. *Water Science and Engineering*. https://doi.org/10.1016/j.wse.2022.12.003.

Ruggero, F., Gori, R., and Lubello, C. (2019). Methodologies to assess biodegradation of bioplastics during aerobic composting and anaerobic digestion: a review. *Waste Management & Research: The Journal of the International Solid Wastes and Public Cleansing Association, ISWA* 37 (10): 959–975.

Sagong, H.-Y., Son, H.F., Choi, S.Y. et al. (2018). Structural insights into polyhydroxyalkanoates biosynthesis. *Trends in Biochemical Sciences* 43 (10): 790–805.

Saha, S., Zubair, M., Khosa, M.A. et al. (2019). Keratin and chitosan biosorbents for wastewater treatment: a review. *Journal of Polymers and the Environment* 27 (7): 1389–1403.

Salam, A., Lucia, L., and Jameel, H. (2018). Starch derivatives that contribute significantly to the bonding and antibacterial character of recycled fibers. *ACS Omega* 3 (5): 5260–5265.

Salama, A. (2016). Functionalized hybrid materials assisted organic dyes removal from aqueous solutions. *Environmental Nanotechnology, Monitoring & Management* 6: 159–163.

Salama, A. (2017). New sustainable hybrid material as adsorbent for dye removal from aqueous solutions. *Journal of Colloid and Interface Science* 487: 348–353.

Sangsanoh, P., Israsena, N., Suwantong, O., and Supaphol, P. (2017). Effect of the surface topography and chemistry of poly(3-hydroxybutyrate) substrates on cellular behavior of the murine neuroblastoma Neuro2a cell line. *Polymer Bulletin* 74 (10): 4101–4118.

Seddiqi, H., Oliaei, E., Honarkar, H. et al. (2021). Cellulose and its derivatives: towards biomedical applications. *Cellulose* 28 (4): 1893–1931.

Sedlacek, P., Slaninova, E., Koller, M. et al. (2019). PHA granules help bacterial cells to preserve cell integrity when exposed to sudden osmotic imbalances. *New Biotechnology* 49: 129–136.

Sepahvand, S., Bahmani, M., Ashori, A. et al. (2021). Preparation and characterization of air nanofilters based on cellulose nanofibers. *International Journal of Biological Macromolecules* 182: 1392–1398.

Sharma, S., Gupta, A., and Kumar, A. (2019). Keratin: an introduction. In: *Keratin as a Protein Biopolymer: Extraction from Waste Biomass and Applications* (ed. S. Sharma and A. Kumar), 1–18. Springer.

Shaumbwa, V.R., Liu, D., Archer, B. et al. (2021). Preparation and application of magnetic chitosan in environmental remediation and other fields: a review. *Journal of Applied Polymer Science* 138 (42): 51241.

Shutova, V.V., Revin, V.V., Kalinkina, E.A. et al. (2021). Levan from *Azotobacter vinelandii* as a component of biosorbents for heavy metals and radionuclides. *Applied Biochemistry and Microbiology* 57 (1): 102–109.

da Silva, A.K., Ricci, T.G., de Toffoli, A.L. et al. (2020). The role of magnetic nanomaterials in miniaturized sample preparation techniques. In: *Handbook on Miniaturization in Analytical Chemistry* (ed. C.M. Hussain), 77–98. Elsevier.

de Siqueira, E.C., de Rebouças, J., S., Pinheiro, I.O., and Formiga, F.R. (2020). Levan-based nanostructured systems: an overview. *International Journal of Pharmaceutics* 580: 119242.

Siracusa, V. (2019). Microbial degradation of synthetic biopolymers waste. *Polymers* 11 (6). https://doi.org/10.3390/polym11061066.

Sirajudheen, P., Karthikeyan, P., Basheer, M.C., and Meenakshi, S. (2020). Adsorptive removal of anionic azo dyes from effluent water using Zr(IV) encapsulated carboxymethyl cellulose-montmorillonite composite. *Environmental Chemistry and Ecotoxicology* 2: 73–82.

Sirajudheen, P., Poovathumkuzhi, N.C., Vigneshwaran, S. et al. (2021). Applications of chitin and chitosan based biomaterials for the adsorptive removal of textile dyes from water – a comprehensive review. *Carbohydrate Polymers* 273: 118604.

Statista (n.d.). Plastic production worldwide 2021. https://www.statista.com/statistics/282732/global-production-of-plastics-since-1950/ (accessed 15 December 2022).

Tiseo, I. (2023). Total food waste produced worldwide in 2019, by sector. Statista. https://www.statista.com/statistics/1219836/global-food-waste-by-sector/ (accessed 15 December 2022).

Tissera, N.D., Wijesena, R.N., Yasasri, H. et al. (2020). Fibrous keratin protein bio micro structure for efficient removal of hazardous dye waste from water: surface charge mediated interfaces for multiple adsorption desorption cycles. *Materials Chemistry and Physics* 246: 122790.

Tomaszewska, J., Bieliński, D., Binczarski, M. et al. (2018). Products of sugar beet processing as raw materials for chemicals and biodegradable polymers. *RSC Advances* 8 (6): 3161–3177.

Umesh, M., Santhosh, A.S., Shanmugam, S. et al. (2022). Extraction, characterization, and fabrication of cellulose biopolymer sheets from *Pistia stratiotes* as a biodegradative coating material: an unique strategy for the conversion of invasive weeds into value-added products. *Journal of Polymers and the Environment*. https://doi.org/10.1007/s10924-022-02511-4.

Vinod, A., Sanjay, M.R., Suchart, S., and Jyotishkumar, P. (2020). Renewable and sustainable biobased materials: an assessment on biofibers, biofilms, biopolymers and biocomposites. *Journal of Cleaner Production* 258: 120978.

Wang, Y., Zhou, Z., and Han, Y. (2022). Levan-chitosan blend films: preparation, structural, physical properties and application in pork packaging. *International Journal of Biological Macromolecules* 217: 624–632.

Wu, Y.-X., Ma, H., Wang, J.-L., and Qu, W. (2021). Production of chitosan scaffolds by lyophilization or electrospinning: which is better for peripheral nerve regeneration? *Neural Regeneration Research* 16 (6): 1093–1098.

Yan, N. and Chen, X. (2015). Sustainability: don't waste seafood waste. *Nature* 524 (7564): 155–157.

Yilmaz, E. and Soylak, M. (2020). Functionalized nanomaterials for sample preparation methods. In: *Handbook of Nanomaterials in Analytical Chemistry* (ed. C. Mustansar Hussain), 375–413. Elsevier.

Yusuf, I. and Sharu, A. (2022). Bioremediation of Cu and Zn by guinea fowl feather wastes and biodegradation of Cu- and Zn-polluted feathers by free and immobilized heavy-metal-tolerant bacterium *Pseudochrobactrum* sp. IY-BUK1. *Biotechnologia* 103 (3): 221–234.

Zhang, Q., Li, Q., Young, T.M. et al. (2019). A novel method for fabricating an electrospun poly(vinyl alcohol)/cellulose nanocrystals composite nanofibrous filter with low air resistance for high-efficiency filtration of particulate matter. *ACS Sustainable Chemistry & Engineering* 7 (9): 8706–8714.

Zhuang, S., Zhu, K., and Wang, J. (2021). Fibrous chitosan/cellulose composite as an efficient adsorbent for Co(II) removal. *Journal of Cleaner Production* 285: 124911.

Zong, E., Wang, C., Yang, J. et al. (2021). Preparation of TiO_2/cellulose nanocomposites as antibacterial bio-adsorbents for effective phosphate removal from aqueous medium. *International Journal of Biological Macromolecules* 182: 434–444.

18

Application of Organic Polymers in Agriculture

Deepak Kumar[1], Nandni Sharma[1], Raman Tikoria[1,2], Sandeep Kour[1], Mohd. Ali[1], Parkirti[1], Roohi Sharma[1], and Puja Ohri[1]

[1] *Guru Nanak Dev University, Department of Zoology, Amritsar, Punjab 143005, India*
[2] *Central University of Punjab, School of Basic Sciences, Department of Zoology, Bathinda, Punjab 151401, India*

18.1 Introduction

Agriculture is one of the significant factors in good health, competent nutrition, environmental pollution, and economic development (Fasusi et al. 2021). Enhancing productivity per unit of the land utilized for crop production was once the main driving force for agricultural development around the world. Over time, cultivating, gathering, and processing food in agriculture have been notably accomplished by overusing agrochemicals and exploiting natural resources like water and soil (Tilman et al. 2002; Chopin et al. 2021). The unsustainability of these methods has been brought to light by factors such as the strong demand for freshwater, rising agrochemical manufacturing costs, and increased awareness of their negative environmental effects (Lykogianni et al. 2021). Recently, the globe has started working toward more sustainable farming methods for a brighter tomorrow in response to the mounting environmental repercussions (Altabbaa et al. 2022). Some of the contemporary challenges faced by global agriculture are controlled supply of agrochemicals, dealing with soil erosion, water pollution, and climate change. To prevent crop loss, synthetic agrochemicals are frequently utilized (Tikoria et al. 2022). Both humans and the environment have suffered greatly as a result of their poisonous leftovers attributed to their uncontrolled release (Himani et al. 2022). Further, the widespread distribution of heavy metals in the environment as a result of numerous residential, agricultural, medicinal, industrial, and technical uses has also raised worries about their possible consequences on human health, the environment, and agriculture (Ahmed et al. 2022). Furthermore, global climate change has hastened the flux of various abiotic challenges such as flooding, drought, extreme temperatures, and salt and biotic stresses including ever-evolving phytopathogens and diseases that reduce agricultural production (Zandalinas et al. 2021). Generally, plants have a built-in defense mechanism that can only withstand certain levels of stress. The increasing environmental

Organic Polymers in Energy-Environmental Applications, First Edition.
Edited by Ramesh Oraon, Pardeep Singh, Sanchayita Rajkhowa, Sangita Agarwal, and Ravindra Pratap Singh.

stressors have detrimental effects on plant growth, development, and crop output. These unfavorable circumstances alter the metabolism of reactive species (Chaki et al. 2020). High levels of these reactive species, which are more numerous than antioxidant defense enzymes can handle, disrupt redox equilibrium, which may cause macromolecules like lipids and proteins of membrane and nucleic acids to be damaged (Janků et al. 2019). The disruption of the natural defense system under extreme stress results in damage and abnormal physiological conditions in the plant cell, ultimately leading to plant cell death and yield reductions (Raza et al. 2022).

Researchers are attempting to address all of these challenges by utilizing a variety of environmentally sound solutions, such as the utilization of biological units, drainage, and water and soil management. Application of polymers can be considered as one such approach. Polymers play a significant role in daily life; they are used to create a variety of high-tech items, including biomaterials (prosthetic hip and knee joints) and aerospace materials (Guo et al. 2020). Polymers' enormous popularity can be attributed to their simplicity in processing. By adjusting the "atomic make-up" of repetitive component and changing the molecular weight, the properties of polymers can be adjusted to match particular needs (Puoci et al. 2008). Recent studies have begun investigating customized and stimulus-responsive smart polymeric systems for diverse agricultural applications (Stuart et al. 2010). Among different types of polymers organic polymers are extensively studied as they can be utilized as soil conditioners, carriers for controlled agrochemical release, and for mitigating

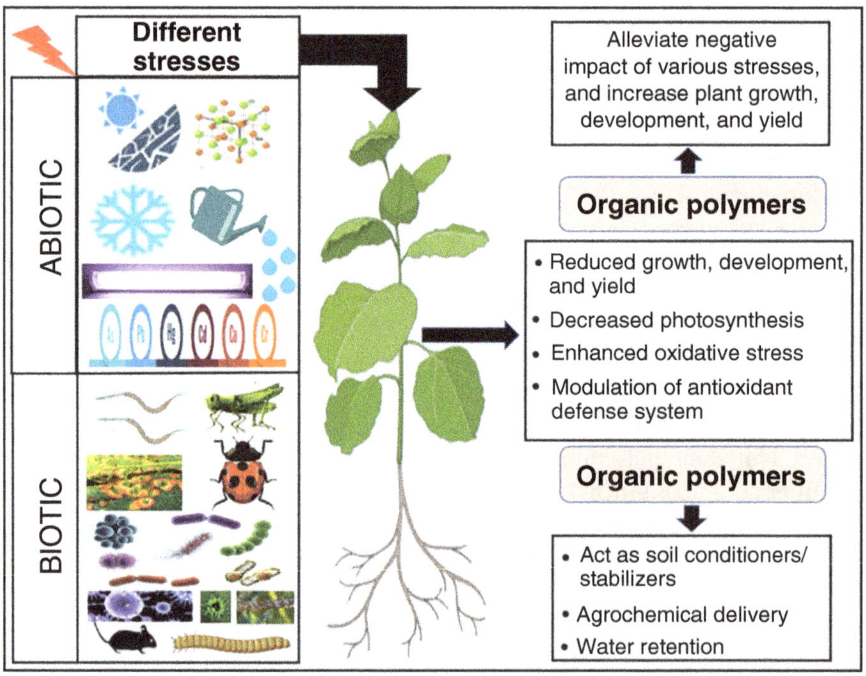

Figure 18.1 General illustration of different plant stressors and application of organic polymers in agriculture.

various biotic and abiotic stresses in plants (Figure 18.1). Further, a family of functional polymer materials known as organic super-absorbent polymers (SAPs) has a superior-water absorption function. It has great water retention capabilities and can absorb water that is hundreds to thousands of times heftier than it. Even under pressure, it is impossible to separate the water if the water absorption has expanded into a hydrogel (Ai et al. 2021). Several studies have exploited this property of SAPs for enhancing the soil water-holding capacity and mitigating drought stress in plants, thereby increasing agricultural production. In the present chapter, we have concentrated on various applications of organic polymers in agriculture.

18.2 Organic Polymers as Soil Conditioners/Stabilizers

The heavy use of fertilizers from scarce resources in modern agriculture has resulted in significant soil degradation due to the loss of soil organic matter and fertility that raises emission of carbon dioxide (Kopittke et al. 2019). In addition, better crop development and productivity require soils treatment very frequently with soil conditioners (such as coal fly ash pellets containing extra Ca, Mg, K, S, P, Si, Al, Fe, and Cu) and soil improvers (such as waste from agriculture). The utilization of fertilizers and chemicals, however, has received a lot of attention recently, while soil conditioners and soil amendments have received less (Nkebiwe et al. 2016). When these soil conditioners are used to improve the soil, more precise dosing, a longer-lasting release of nutrients were observed that are advantageous for soil health (Sarlaki et al. 2021).

As soil conditioners, natural and chemically synthesized polymers have been used since the 1950s. To date a variety of organic polymers are available in the market which is used for soil stabilization. Organic polymers are essentially substances having necessary nutrients to enhance physical, chemical, and biological characteristics of soil (Karamina and Fikrinda 2020). Among organic polymers, alginate, cellulose, chitin, chitosan, collagen, and dextran are natural organic polymers, whereas polymers like acrylic and methacrylic acid, polyethylene glycol, polyvinyl alcohol, and vinyl acetate are classified as synthetic organic polymers (Behera and Mahanwar 2019). Besides these compounds, organic polymers such as pectin and carboxymethyl cellulose are also used for soil improvement (Roy et al. 2014). Regarding their effectiveness as soil conditioners, various studies have been done on the synthetic organic polymers polyacrylamide and polysaccharide. Identical acrylamide and similar monomers are polymerized to generate the homopolymer complex known as polyacrylamide (Yu et al. 2003). The anionic form of polyacrylamide is most frequently employed in soil conditioning. Polysaccharide is derived from guar, which is categorized as a galactomannan made up of galactose and mannose and galactose subunits. Cationic density polysaccharide having low–medium charge is most frequently applied to soil preparation (Graber et al. 2006). Recently, organic polymer substances have been investigated as potential novel chemical soil stabilizers. Examples include polyacrylamide, polyurethane, resins, and xanthan gum. Such polymeric substances encompass huge number

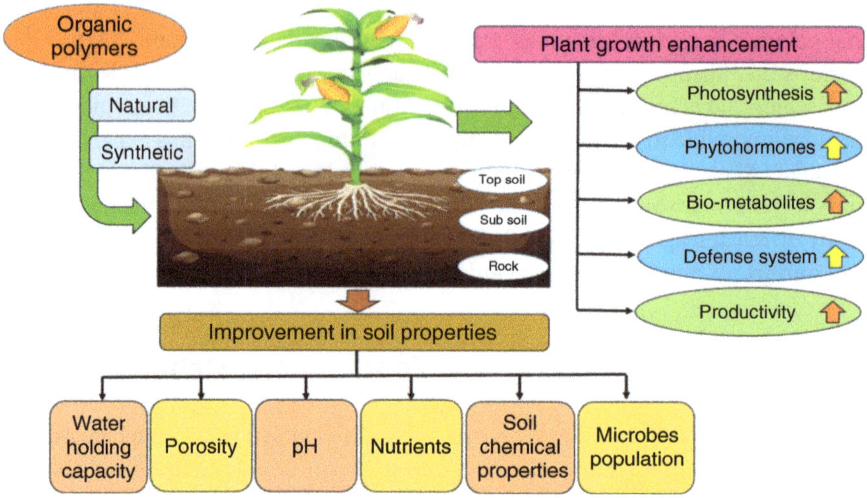

Figure 18.2 General applications of organic polymers as soil conditioners.

of macromolecular/chains and hydrophilic groups (Mohsin and Attia 2015). An elastic film structure is created when soil and an organic polymer stabilizer are combined, improving the soil's strength and stability properties. The main benefits of polymer reinforcements are strong soil–particle bonds and superior structural stability (Rezaeimalek et al. 2017).

The physical, chemical, and biological properties of soil can both be significantly influenced by these soil conditioners (Figure 18.2). In addition, it raises the availability of nitrate while reducing bulk density and extractile salt content of the soil (Babla et al. 2022). By preventing soil deterioration, enhancement of soil–air–water interactions and modification of soil aggregation and soil drainage to lessen soil compaction and combat water repellence, these organic polymers aided in the improvement of soil and production of food (Shinde et al. 2019). This ultimately results in better growth and development of plants and increased nutrient availability in the soil (Karamina and Fikrinda 2020). Along with increasing the availability of nutrients for crops, soil conditioners also promote weed competition and insect and disease defense. The main elements affecting polymers functions include charge density, composition, concentration, conformation, molecule size, molecular weight, and pH (Huang et al. 2021).

Compared to other organic polymers or fiber reinforcement alone, integrating augmentation of fiber and organic polymer may be thought of as a novel efficacious approach to enhance soil, providing it with higher solidity and refined structural stability (Liu et al. 2018). Furthermore, when a polymer is applied along with an electrolyte source, such as phosphogypsum, its benefits of preserving increased infiltration rate and decreased runoff and erosion are significantly enhanced. Furthermore, the presence of electrolytes enhances the flocculation of the soil clay, which is necessary for the successful cementing and stability of aggregates at the soil surface by anionic polymers (Graber et al. 2006). One important method for polymer stabilization is electrostatic interaction, particularly between clays and cationic polymers since the clay surfaces often have negative charge. The clay minerals' surfaces and edges readily accept the polycations, which alter water sorption close to the surface of clay and produce big and firm flocculated particles (Huang et al. 2021). Most of the soils stabilized with polymers show an increase in strength right away, even though the primary organic polymers under investigation have different fundamental features.

18.3 Organic Polymers and Agrochemicals Delivery

Providing plants with the nutrients including micronutrients like cobalt, nickel, molybdenum, boron, chlorine, copper, zinc, iron, and manganese as well as macronutrients like calcium, nitrogen, phosphorus, potassium, magnesium, and sulfur they require, is a crucial aspect of agriculture (Reetz 2016). However, these nutrients are habitually not present in environment in quantities that are enough for healthy plant development. As a result, the use of agrochemicals including fertilizers, herbicides, and insecticides is crucial in agricultural sector. On the contrary, as our understanding of the agrochemical lifecycle has escalated, so too has our concern over some unfavorable effects on environment like bioaccumulation in the food chain and potential ecosystem pollution (Puoci et al. 2008).

To relinquish agrochemicals in soil to immediately meet the nutritional demands of plants without incurring pollution, production and the usage of agrochemicals mixed with polymeric materials has been a solution to this issue (Puoci et al. 2008). The amalgamation of polymer and agrochemical can be carried out by the combination of two processes, viz., physical and chemical processes. For instance, physical processes result in control over the rate of release of agrochemicals in contrast with a chemical mechanism that is required for chemical stabilization of agrochemicals within the polymer matrix. The optimum system to release the active ingredient is chosen based on the drug's biological and chemical characteristics as well as on its physiochemical interactions (Puoci et al. 2008). Additionally, there are two categories for the release profile of agrochemicals: controlled- and slow-release systems. In contrast to controlled release, which involves the pervasion-controlled transfer of active components to a targeted surface from a reservoir to sustain predefined concentration for given amount of time, slow-release fertilizer pertains to the mechanism itself which remits the release (Trenkel 2010; Pereira et al. 2015). Since the parameters that affect the release rate and pattern of controlled-release fertilizer are known, they can be managed (Trenkel 1997, 2010; Shaviv 2001). Moreover, controlled nutrient delivery is not an entirely new idea. Oertli and Lunt (1962) published a study on the membrane-encapsulated controlled release of fertilizer minerals, while Allan and coworkers documented the controlled release of pesticides in 1971. Their pioneering research has imposed a significant impact on the industry, and the technology's advancement has still been increasing steadily.

A commonly used material for controlled release is acrylamide (acrylic-based organic polymer) (Allan et al. 1971). Furthermore, nanotechnology concepts have recently been used in studies to enhance the release mechanism by using nanoparticles. In terms of chemistry, this technology is being developed using organic polymers and inorganic polymers, and composites (Pérez de Luque and Hermosín 2013).

Natural rubber, polyethylene, and cyclopentadiene having glyceryl esters of unsaturated fatty acid and materials derived from cellulose are examples of other polymers (Puoci et al. 2008). The most widely utilized agrochemicals, i.e. herbicides, are typically used as pennant groups in polymers. Various categories of agrochemicals are pesticides like atrazine, bifenthrin, chlorpyrifos and tebuconazole, various herbicides including pentachlorophenol, 4-chloro-2-methylphenoxyacetic acid and 2,4-dichlorophenoxyacetic acid, and fertilizers having nitrogen, phosphorus, potassium, urea, potassium nitrate, calcium carbonate and various fertilizer with micro- and macro-nutrients (Roy et al. 2014; Kashyap et al. 2015). Moreover, numerous biodegradable polymers that are used for the fabrication of agrochemicals for release into the agriculture sector have been mentioned in Table 18.1.

Table 18.1 Studies depicting the use of organic polymers for the delivery of agrochemicals.

Polymer used for fabrication	Agrochemical	References
Xanthan	Potassium nitrate	Messa et al. (2016)
Alginate, chitosan	Paraquet	Grillo et al. (2015)
Chitosan-clay	Potassium nitrate	dos Santos et al. (2015)
Chitosan	Monocalcium phosphate	Ahmad et al. (2015)
Paraffins, urethane composites, ester copolymers, polyolefins	Nitrogen-phosphorus-potassium	Senna et al. (2015)
Chitosan	Nitrogen-phosphorus-potassium	Noppakundilograt et al. (2015)
Gellan gum, chitosan	Potassium dihydrogen phosphate	Sabadini et al. (2015)
Chitosan	Potassium nitrate	Melaj and Daraio (2014)
Polyacetic acid, polyethylene, polyvinyl acetate	Urea	Azeem et al. (2014)
Ethyl cellulose, polyhydroxybutyrate	Urea	Costa et al. (2013)
Chitosan-clay	Hexazinone	Celis et al. (2012)
Chitosan	Urea	Hussain et al. (2012)
Waxes, natural gum, resins	Nitrogen-phosphorus-potassium	Lubkowski and Grzmil (2007)

18.4 Organic Polymer and Heavy Metal Toxicity

Hazardous metals like As, Cd, Cr, Co, Hg, Ni, Pb, and Zn, valuable metals like Ag, Au, Pd, Pt, and Ru, and radioactivity metals such as Am, Th, Ra, and U can all be categorized as metal ions (Wang and Chen 2009). They help in supplying micronutrients and macronutrients to the soil, but intensive agriculture has resulted in massive contaminants in the form of metal ions (Singh et al. 2020). Such deposition has led to environmental degradation and ecological damage to agriculture (Alengebawy et al. 2021). Since time immemorial, this heavy metal toxification is a major concern. Likewise other bioremediations, organic polymers are utilized (Table 18.2) to improve soil quality and remove harmful heavy metal ions from the soil and water, enhancing overall plant growth (Malik et al. 2019).

Among organic polymers, polymeric biosorption is one way of cleaning up damaged regions; among which biosorption is a natural ability of biomass that has been incorporated into biomimicry kind of procedures for binding heavy metals even from the much-diluted sample (Shen et al. 2015; Milani et al. 2017). As a strategy for the elimination of heavy metals, biosorption method of environmental remediation has been explored for use in soil and water systems for agricultural practices. Using ionic exchange, this technology may eliminate both organic and inorganic contaminants (Rangabhashiyam et al. 2014). Findings demonstrate this technique's minimal prices and great sorption capability. Extracellular polymeric compounds of sewage sludge are recovered and converted into bio-based products. A study investigated the use of extracellular polymeric substances of granular waste sludge from anammox were used as a biosorbent to remove heavy metals, notably Pb, Cu, Ni, and Zn (Pagliaccia et al. 2022). Similarly, interpenetrating polymer network hydrogels based on biopolymeric substances including starch, chitosan, alginate, and other polysaccharides such as exopolysaccharides were also studied (Dragan 2014). For sorption of metal ion from an aqueous medium, Chauhan and Mahajan (2002) and Dragan (2014) demonstrated usage of an interpenetrating polymer network utilizing a mixture of biopolymers and poly(methacrylamide). They discovered that adding cellulose compounds to the polymer network significantly enhanced the material's sorption capacity (Chauhan and Mahajan 2002).

To clean up water and soils that have been contaminated by hazardous substances from agricultural inputs in plants, polymers have also frequently been utilized as sorbent materials where the functional groups and the chemical affinity in the sorbent materials are considered for understanding the mechanisms involved in this process (Hua et al. 2021; Demirbas 2008). Due to the presence of electroactive amine, amide, and carboxylic acid as functional groups for ionic bonding, polysaccharides including cellulose, chitin, chitosan, and exobiopolymers derived from organic sources are the type of polymers that are well-high frequently used for the aim of metal sorption in the plants (Milani et al. 2017; Udayakumar et al. 2021; Díez-Pascual 2022). These form complexes with heavy metals and thus help in alleviating phytotoxicity (Hidangmayum et al. 2019). Green coconut fiber, bamboo fiber, and natural fibers obtained from sugarcane bagasse have each been used as adsorbent materials. The fundamental benefit from these materials is that they are

Table 18.2 Organic polymer in mitigation of heavy metal toxicity.

Organic polymer	Heavy metal	Test crop/soil	Reported effect	References
Humic acid	Cd	Soil and soil aggregates	Modified soil properties like pH, moisture, organic carbon, and overall plant conditions	Zhao et al. (2022)
Ethyl cellulose, keratin-based superabsorbent	Cr	Wheat	Used as multifunctional ecological fertilizer for overall increased nutrient uptake	Chen et al. (2021)
Polysuccinimide nanoparticles	Cu	*Zea mays*	Promoted seed germination and growth of plant by increasing antioxidative enzymes activity	Xin et al. (2020)
Humic acid	Pb	*Camellia sinensis* (Tea plant)	Increased pectin and pectic acid in cell wall	Duan et al. (2020)
Chitosan	Ni	*Calendula tripterocarpa*	Reduced MDA content	Heidari et al. (2020)
Starch	Zn, Cu, Mn, Fe, Co, Pb, Cd, and As	*Triticum aestivum*	Increased growth and nutrient status of plants	Singh and Rathore (2020)
Xanthan gum, β-glucan	Zn	*Camelina sativa*	Enhanced plant growth and reduced zinc absorption	Shin et al. (2020)
Exopolysaccharide	Cd, Cu, Cr, Ni, Pb, Zn, and Mn	*Linum usitatissimum*	Enhanced nutrients bioavailability and plant growth parameters	Zainab et al. (2020)
Exopolysaccharide	As	Rice	Enhanced plant growth and plant growth parameters	Mukherjee et al. (2019)
Chitosan	Cd	*Zea mays*	Modulated gene expression and enhanced growth	Qu et al. (2019)
Humic acid	Cu, Cd	Triticale	Stress markers gene activation, enhanced plant growth and yield	Sergiev et al. (2013)
Chitosan and humic acid	Cd	Radish	Resulted increase in plant growth and yield and modulated physiological process	Farouk et al. (2011)
Alginate derived oligosaccharide	Cd	*Vicia faba*	Activated antioxidant enzyme system and enhanced photosynthetic rate	Ma et al. (2010b)
Oligo-alginates	Cd	Wheat	Decreased MDA content and increased overall yield	Ma et al. (2010a)

frequently leftovers from commercial or agricultural operations, are unrefined and inexpensive, and are simple to extract. They are mostly made of cellulose and hemicellulose, both of which include acidic groups that allow metal ions to bind to them (Khoo et al. 2018; Sarker et al. 2017). Research on such type of organic polymers (such as polysaccharides) as adsorbents is still underway with an emphasis of soil decontamination from metals by bioremediation and rhizoremediation (Bhatnagar and Kumari 2013). Polymers suitable for decontaminating impacted areas with high efficiencies are being developed (Kavamura and Esposito 2010). According to the findings of Pal and Paul, exopolymer films were used for water treatments in soil. The metals that were absorbed most effectively by such biofilm-like microbes through electrostatic interactions were Cd, Co, Cr, Cu, Ni, and CrO (Pal and Paul 2008). Furthermore, organic polymers can lessen the burden of heavy metals in plants growing under stressed conditions. For instance, administration of exopolysaccharide increased nutrient availability and plant growth parameters of *Linum usitatissimum* exposed to heavy metals (Zainab et al. 2020). Similarly, application of chitosan resulted decrease in the content of malondialdehyde (MDA) and modulated the enzymatic activity of catalase and superoxide dismutase in *Calendula tripterocarpa* under Ni stress (Heidari et al. 2020). Another study reported that Cd-stressed radish plants exhibited enhanced growth and yield after treatment with humic acid. Another study depicted that humic-acid-treated radish plants also showed modulation of various physiological activities under Cd stress (Farouk et al. 2011). Similar findings were observed by Mukherjee et al. (2019), who reported that application of exopolysaccharide enhanced the growth of rice exposed to As stress.

18.5 Organic Polymers and Other Plant Stress

In addition to heavy metal toxicity, organic polymers have the potential to help the plants to overcome other abiotic stressors including salinity, temperature, and flooding and biotic stressors such as fungi, bacteria, and nematodes stress (Table 18.3).

18.5.1 Salinity Stress

Salinity is one of the significant abiotic stresses turning most of the agricultural lands into unfruitful wastelands, causing a dearth of food. It influences plant's ability to absorb water, which causes an overall decline in growth and development (Bartwal and Arora 2020). Roots are the first part of the plants to encounter salt stress, which disrupts the cells' ionic balance and produces an excessive amount of reactive oxygen species which damages genetic material and significantly oxidizes important biomolecules like carbohydrates, lipids, and proteins of membrane. This results in the alteration of redox balance in plant cells and prevents growth of plants at various stages of stress-prompted deterioration (Hussain et al. 2022). Several reports have demonstrated that exogenously applied organic polymers like chitosan, xanthan, and alginate can alleviate the salinity-prompted detrimental effects on plant growth and metabolism (Hidangmayum et al. 2019; Campobenedetto et al. 2021;

Table 18.3 Role of organic polymers in mitigation of abiotic and biotic stresses.

Organic polymer	Type of stress	Test crop	Reported effect	References
Alginate	*Fusarium oxysporum*	Date palm	Stimulated activity of the enzyme phenylalanine ammonia-lyase and expression of genes involved in phenolic metabolism	Bouissil et al. (2022)
Chitosan	*Cercospora beticola*	*Beta vulgaris*	Upregulated synthesis of phytoalexin and stress-related genes and reduced incidence of cercospora leaf spot	Kappel et al. (2022)
Chitosan and EDTA	*Meloidogyne incognita*	Eggplant	Increased proline content and antioxidative enzymes	Attia et al. (2021a)
Chitosan	Salinity	Tomato	Minimized oxidative damage by improving	Attia et al. (2021b)
Humic acid	Salt	*Arabidopsis*	Promoted plant growth and induced stress tolerance	Cha et al. (2021)
Alginate	*Penicillium expansum*	Pear	Increased activity of antioxidants, phenol and flavonoid content and reduced fungal spore germination	Han et al. (2021)
Chitosan	Pinewood nematode	*Pinus pinaster*	Provided defense against pine wood nematode by modulating plant antioxidative enzymes	Nunes da Silva et al. (2021)
Chitosan	Water stress	Tomato	Minimized fruit juice pH by enhancing ascorbic acid content and percent juice content	Basit et al. (2020)
Humic acid	Temperature	Summer squash	Increased protein and proline content	Jia-yue et al. (2020)
Exopolysaccharide	Heat	Tomato	Increased length, fresh weight and dry weight of plants	Mukhtar et al. (2020)
Chitosan	Salinity	Tomato	Increased growth of tomato plants and content of photosynthetic pigments	Ullah et al. (2020)
Chitosan	Salinity	Tomato	Enhanced antioxidative enzymes and proline content	Chanratana et al. (2019)
Sulfated polysaccharide	*Phytophthora palmivora*	*Hevea brasiliensis*	Induced salicylic acid (SA) accumulation and expression of SA-responsive genes	Pettongkhao et al. (2019)
Alginate	*Verticillium dahlia*	Olive	Enhanced activity of PAL and reduced mycelial growth of fungus	Ben Salah et al. (2018)
Chitosan	Salt	Milk thistle	Protected plants from salt stress by elevating antioxidative enzymes and photosynthetic system	Safikhan et al. (2018)
Chitosan	Temperature	*Phaseolus vulgaris*	Mitigated temperature stress and increased shoot fresh weight	Ibrahim and Ramadan (2015)
Humic acid	Salt	Bean	Increased photosynthetic pigments and proline content	Adil et al. (2012)

Salachna et al. 2018). Chitosan, a second-generation natural biopolymer, is a linear polysaccharide that is derived from chitin, which is found in crustaceans shells, insect exoskeleton, and cell wall of fungus (Younes and Rinaudo 2015) and is significantly applied for providing resistance against salt stress. For instance, the exogenously applied chitosan enhanced the salinity resistance in tomato, and increased growth and yield by refining the quality of photosynthetic pigments and different morphological characters under salinity stress (Ullah et al. 2020). Another study conducted by Gu (2012) also investigated the role of chitosan to furnish stress resistance under salinity stress in tomato seedlings. Gu reported that exogenous application of chitosan reduced the content of H_2O_2 (hydrogen peroxide) and MDA and delayed decline of chlorophyll contents. Similarly, exogenously applied chitosan enhanced the germination rate, root length and weight, shoot length and weight and protein content but decreased proline, MDA, and H_2O_2 in sunflowers under salt stress (Jabeen and Ahmad 2013). Another study demonstrated that exogenously applied chitosan altered the antioxidative as well as non-antioxidative enzyme activity and decreased oxidative stress in tomato receiving salinity stress (Attia et al. 2021b). In another experiment, it was observed that treatment of oligo-alginate mitigates the negative impacts induced by salt stress in *Eucomis autumnalis* by increasing plant height, chlorophyll content, activities of antioxidative enzymes, proline content and stimulating the production of phenolic compound (Salachna et al. 2018). Application of bio-stimulant-based tannin enhances salt tolerance and plant growth by upregulating the stress response genes expression and essential transcription factors (Campobenedetto et al. 2021). Similarly, the application of humic acid and phosphorus mitigated the salt stress and increased plant growth parameters including root length, shoot length, shoot weight, and root weight of pepper plant (Çimrin et al. 2010). Furthermore, organic polymer compounds regulated biochemical and physiological indicators and maintained the K^+ and Na^+ homeostasis in the cotton root and leaf by modifying the transport of Na^+ and K^+ in cotton plants under alkaline and salinity stress (Wang et al. 2021).

18.5.2 Temperature Stress

In the current situation, change in climatic conditions and rise in greenhouse effect led to variations in temperature, which is emerging as one of the detrimental stresses in the context of continually changing environmental factors (Karwa et al. 2020). Increase in the temperature by 10–15 °C over ambient is considered as heat. From germination to harvest, heat stress has adverse effects on growth and development of crops. Sedentary by nature, plants are vulnerable to a wide range of temperatures 60–75 °F is the ideal range for plant growth. Extreme temperatures (high as well as low) can be hazardous to the ecosystem and cause abiotic stress to plants, which lowers crop productivity. Heat stress has worst impact on plant growth and metabolism out of all abiotic stimuli including drought, salinity, exposure to heavy metals, and low temperature (Shaffique et al. 2022). These negative effects of temperature stress can be reduced by the administration of organic polymers. For instance, seed priming with chitosan increased length and weight and decreased oxidative stress by improving antioxidative system in maize plants under temperature stress (Guan et al. 2009). Additionally, use of humic acid decreased MDA and H_2O_2 content by improving antioxidative system and increasing the endogenous phytohormones content in tomato plants exposed to heat stress (Khan et al. 2020). Similarly, treatment with chitosan oligosaccharide can mitigate cold stress by enhancing the activity of enzymatic antioxidants including polyphenol peroxidase and superoxide dismutase along with levels of chlorophyll in tea plants under cold stress (Li et al. 2020).

18.5.3 Flooding Stress

Flooding is a most prevalent and pervasive abiotic stressor having negative impact on growth, development, and physiology of various plant species (Sharma et al. 2022). Flooding causes chlorosis, necrosis, abscission, and foliar senescence in addition to a loss in shoot and root biomass and hydraulic conductivity (Lukić et al. 2021). Though flooding is regarded as a complex stress, advanced shoot elongation, alteration in leaf morphology, adventitious root growth, aerenchyma production, pressurized gas flow, and deposition of gas coatings on surfaces of leaf are some of the key responses in escape plants (Bispo and Vieira 2022). Flooding substantially restricts exchange of gases between the plants and surroundings, which leads to various internal modifications in plants. Submerged plant parts lack oxygen and/or carbon dioxide and may build up a lot of the gaseous hormone ethylene. While some crop species can only survive a brief inundation of the soil, some flood-resistant plants can endure partial or total flooding for some days or even months (Park et al. 2022). Studies have proved that the utilization of organic polymers can alleviate the harmful effect of flooding stress. For example, administration of humic acid increased water stress resistance and decreased oxidative stress like MDA and H_2O_2 by improving antioxidative enzymes phenolic compounds and proline content in rice seedlings (García et al. 2012). Similarly, foliar administration of chitosan increased water resistance by improving growth, physiological constituents, and yield of cowpea plants exposed to water stress (Farouk and Amany 2012).

18.5.4 Biotic Stress

Plants have to deal with various types of biotic stressors like bacteria, fungi, nematodes, insects, parasites, weeds, and viruses. To effectively cope up with these biotic stressors plants utilize an inducible defense mechanism which is close to the defense which is affected during abiotic stress (Kaur et al. 2022; Kumar and Ohri 2022). Plant growth regulators including salicylic acid, brassinosteroids, cytokinin, auxin, ethylene, and jasmonic acid are primarily responsible for regulating the activation of defense responses and associated signals in plants; in addition to phytohormones organic polymers such as algal polysaccharides, chitosan, alginate, humic acid, and xanthan are some of natural inducers that can activate plant defense mechanisms and promote growth and development. Eco-friendly substances that can promote plant development while also causing pathogen resistance and activating the expression profiling of defense pathway include polysaccharides and oligosaccharides (Saberi Riseh et al. 2022). Algal polysaccharides are one of the common and varied types of organic compounds, which can enhance plant resistance (Popletaeva et al. 2018). Alginate, a linear biopolysaccharide co-polymer composed of L-guluronate and β-D-mannuronate, can be organized in blocks that are both heteropolymeric and homopolymeric. Alginate can promote growth and development of plant along with induced resistance against biotic and abiotic stress resistance in plants (González et al. 2013). Lemic et al. (2021) performed an experiment for the management of pests by using apitoxin-loaded calcium-alginate microspheres. It has been observed that microspheres laden with apitoxin have a negative impact on insects and can be used as an eco-friendly method for pest management. Similarly, application of chitosan and oxalic acid to fungal-stressed strawberry plants reduced spore germination, linear growth and elongation of germ tube of fungi, *Rhizopus stolonifer* and *Botrytis cinerea*, and enhanced the enzymatic activity and levels of phenolic compounds of strawberry plants (Thabet 2019). In another experiment administration of humic acid increased the percent mortality of infective juveniles, decreased egg masses and population of *Meloidogyne incognita*; treatment with humic acid also increased root and shoot length and root and shoot weight of banana plants exposed to the nematode stress (Seenivasan and Senthilnathan 2018). Another study reported similar results where it was observed that chitosan has a crucial role in the activation of defense system along with protecting tomato plants from the diseases caused by *Alternaria solani* (Sathiyabama et al. 2014). Similarly, chitosan oligosaccharide increased the mortality of the infective stage of *M. incognita,* enhanced photosynthetic pigments, and altered the immune system in tomato plants (Fan et al. 2020).

18.6 Superabsorbent Organic Polymer and Agriculture

Superabsorbent organic polymers are polyelectrolytes having 3D structure where the polymer chains are attached by cross-linking. These hydrophilic polymers have deep range of implementations and have proved to be an asset in the field of agriculture. Their popularity in the agricultural sector is attributed to their biodegradability, high water-absorbing ability, and low-cost production (Sinha 2018; Behera and Mahanwar 2019). Since superabsorbent organic polymers have very high water-absorption as well as retention properties, they can be utilized, especially in arid areas, to keep moisture in soil and deal with the issues of water scarcity for irrigation purposes (Souza et al. 2016; Milani et al. 2017; Thombare et al. 2018). They serve as excellent water-saving and soil-conditioning materials and help to ameliorate drought stress in water-deficit regions (Dabhi et al. 2014). Along with revamping the water use proficiency, superabsorbent polymers can also be used in coalition with fertilizers to attain their controlled release in fields for sustainable production (Tubert et al. 2018; Li et al. 2018). They enhance growth of the plants by improving mineral absorption by plants and also reduce environmental pollution (Chang et al. 2021).

18.6.1 Superabsorbent Hydrogels and Water Retention

In agriculture, irrigation plays a central role in the plant growth by supplying essential nutrients to soil. The high cost of irrigation processes in arid regions has become a matter of concern. In such water-deficit regions, superabsorbent hydrogels (SAH) provide a feasible and cost-effective alternative. SAH is a 3D polymer network of crosslinked chains comprising ionic monomers that tend to swell up when aqueous solutions are provided and retain moisture in soil (Ullah et al. 2015; Milani et al. 2017). SAH has a vast array of applications in agriculture like reduced consumption of water and cost of irrigation, improved health and growth of plants, stabilization of fertilizers in the soil, and prevention of seeping out of essential nutrients into the groundwater (Behera and Mahanwar 2019).

SAH, owing to the existence of several hydrophilic groups along with a vast quantity of free volume between polymer chains, can soak up and hold large quantities of water, approximately 10–1000 g/g (Zohuriaan-Mehr et al. 2010). Various reports have depicted that the water-holding properties of the soil were amplified to a significant level by the inclusion of SAH (El-Tohamy et al. 2014; Liao et al. 2016; Mohawesh and Durner 2019). At the time of rainfall and/or irrigation after being incorporated in the moistened soil, SAH granules swell by soaking up water. In the act of soil dryness, these granules steadily release water through diffusion, thus preventing evaporation and loss of irrigation water (Behera and Mahanwar 2019). Moreover, when watered, SAH can swell up and inflate, thereby increasing the porosity of the clay and aeration capacity of soil and hence supplying better oxygen to the roots of the plant for their proper growth and development (Figure 18.3).

18.6.2 Superabsorbent Polymers and Drought Stress

Drought has adverse effects on sustainable agriculture. It negatively affects the plant growth and development ultimately having an impact on the yield. It has noteworthy effects on plant metabolism and has potential to alter physiological

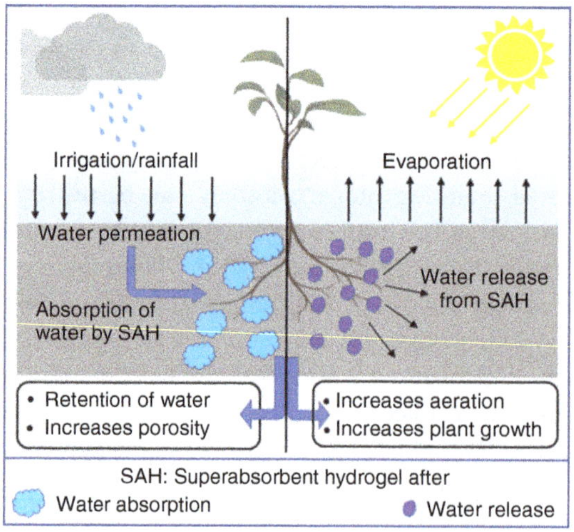

Figure 18.3 Superabsorbent hydrogels and water retention.

and biochemical properties of the plant (Khan et al. 2019). Drought decreases the soil-nutrient-utilizing ability of plants and can induce free radicals. Moreover, it reduces chlorophyll content and accentuates ethylene biosynthesis. It negatively affects the subcellular compartments of plants to the tissue and organ levels (Vurukonda et al. 2016). Plants have various natural inbuilt strategies to survive under drought stress using various physiological processes like osmotic adjustment, stomatal regulation, and development of root system. But to enhance their drought tolerance capability artificially, various strategies are also available like application of nitric oxide, proline, plant growth-promoting microorganisms, or hormones (Ilyas et al. 2021). Trending drought-coping strategies to secure agricultural production involve several methods like the addition of nutrients, fertilizers, composting, mulching, giving water to soil, seedbed procedure, shedding, tillage, and maturing (Azadi et al. 2018). Other strategies include use of integrated rhizosphere management that includes association of drought-tolerant microorganisms, nanoparticles, and plant-growth-promoting microorganisms which alter the effect of drought by improving morphological characters, physiological characters, adjusting biochemical processes (Zia et al. 2021). Use of organic polymers is one such method that prevents the adverse effects of drought in agriculture field. These polymers have the ability to enhance the soil-water-holding capacity. Among various organic polymers like polysaccharides, proteins, and other molecules, some organic molecules have a great tendency to hold water in the soil. These are called SAPs. SAP is a 3D network of polymers which can help in retention of abundant water. These SAPs can be polyelectrolyte gel. Their molecules can absorb water and swell because of electrostatic repulsion that takes place between charged chains of polymers and osmotic imbalance (Ono et al. 2007; Klein and Poverenov 2020). SAP can provide nutrients and pesticides containing water to the plant via the absorb–release–absorb process near the roots. The resistance to expansion caused by hydrogen bonds and crossed-linked structure as well as the expansion due to a repulsive ionic charge of macromolecular electrolytes determines the major water-absorbing capacity of SAPs (Bo et al. 2012). Natural as well as synthetic polymers can be used to make SAPs. Natural polymers include collagen, cellulose, chitosan, dextran, alginate, and chitin, while synthetic polymers can be vinyl acetate, acrylic and methacrylic acid, polyvinyl alcohol, polyethylene glycol, and various acrylates. These molecules are highly flexible enabling their polymers to hold more water (Behera and Mahanwar 2019). One vastly studied example of SAP is superabsorbent resin; the mechanism behind its water-absorbing capability involves its network-like microstructure of polymeric chains made of starch or cellulose or any synthetic compound like polydactyly acid and the osmotic pressure created by hydrophilic groups including hydroxyl and carboxyl groups that forms the dynamic bases of water absorption (Ai et al. 2021). In SAP, crosslinking is of two types; bulk crosslinking occurs in that polymerization in which some co-monomers act as crosslinking agents having higher functionality than other polymers and surface crosslinking in which surface treatment is necessary during polymerization that includes post-treatment of electrophilic polysubstituted compounds or polyvalent metal salts resulting in super crosslinked shell with higher absorption and water retention capacity of the polymers (Behera and Mahanwar 2019). This helps SAP absorb a thousand times more water than the weight of its molecule without dissolving itself; they absorb water and swell to retain water in them (Chang et al. 2021; Esteves 2011). The addition of these SAP to soil can combat the water scarcity problem in drought-prone areas enhancing better crop cultivation strategies. SAP present in soil can help in storing a sufficient amount of water; thus, less amount of water is required in the field, and it also reduces the loss of soil water via evaporation in deserted hot areas as it can release water stored in it (Cannazza et al. 2014). SAP can even release the pre-absorbed water rapidly to the plant roots when required, the respective water is released up to an equilibrium limit to that of respective molecule absorbing water from the SAP molecule (Yang et al. 2020a). Moreover, it is a good soil conditioner and the addition of superabsorbent polymer in the soil increases biomass, panicle number, grain yield along with relative water amount, and fertility rate of soil. It also monitors the temperature variations in soil, transpiration rate, chlorophyll content, photosynthetic rate, growth attributes of plants and yield (Yang et al. 2020b; El Idrissi et al. 2023). Use of these polymers is also eco-friendly, the degradation of superabsorbent polymers in soil in the presence of microorganisms was investigated, and it was found that these soil microorganisms can solubilize the polymers followed by their mineralization (Stahl et al. 2000). The outcomes of using SAP on various plants under drought stress have been reported during various studies. For instance, use of SAP for better growth also enhanced water retention capacity of soil and exhibited no adverse effects on soil microbiota (Li et al. 2013). Another study demonstrated the positive effects of SAP application in field of green pepper by increasing the soil porosity and the water-holding capacity from 59.5% to 81.3% with increased nutrient levels in plants (Fan et al. 2019b). Additionally, SAP as a soil amendment resulted in improved tolerance of tomato plants toward drought; it also enhanced the yield of tomato fruit by 16% (Cerasola et al. 2022). Fruit of mango orchards also showed a gain in fruit quality, and in yield of mango and soil water content in the presence of superabsorbent polymers which also saved 40% water application to the plants thus reducing irrigation volume to obtain better fruits (Liu et al. 2019). The use of superabsorbent along with salicylic acid was also reported to neutralize the consequences of drought in mung bean (Balli et al. 2019). In one study a lab prepared SAP resin and used it under drought conditions which ensured improved seed germination and seedlings growth with better root and stem height in *Artemisia desertorum* (Zhang et al. 2020). Further, SAP as a mixture is found to be more effective on plants, for example, a mixture of SAP along with fertilizer was found effective in increasing growth and chlorophyll content strengthening antioxidant activities and enhancing sugar and protein content in bamboo species for better development (Fan et al. 2019a). Additionally, SAP comprising attapulgite, straw biochar, polyacrylamide, and humic acid exhibited better results in holding soil water and germination of maize seeds (Yu et al. 2019). Even SAP-embedded bacteria are also used for successful establishment and survival of plants. A study in this regard has been done on eucalyptus seedlings whose drought tolerance was enhanced due to application of two *Pseudomonas* strains embedded in acrylic-hydrocellulosic SAP (Chaín et al. 2020). A combination of SAP with exopolysaccharides-secreting bacteria in a study was found to improve the tolerance of maize plant with maize plant; it also improved growth of plant along with maize yield (Yaseen et al. 2020). Moreover, application SAP (A200), another kind of water-absorbing SAP along with manure, was found effective against drought and induced resistance in wheat plants (Gholinezhad and Eivazi 2021).

18.7 Conclusion

It is a deep-rooted fact that excessive use of agrochemicals, soil contamination, and various environmental stresses, abiotic as well as biotic stress, pose detrimental impact on the agriculture-based economy by directly lowering the crop production. Additionally, this drop in crop production is getting worse day by day. The use of polymers is one of the many ways that have been explored to combat this productivity drop and may be a workable, ecologically secure substitute. In this continuation, organic polymers have been reported to act as soil conditioners and carriers for the controlled release of agrochemicals leading to improved properties of soil and reduced soil pollution leading to the better growth of plants. Further, organic polymers also efficiently alleviate different environmental stresses, resulting in increased crop yield. Furthermore, superabsorbent polymers also aid in increased productivity by increasing drought resistance in plants, though further research is required to explore and decode the additional facets of application and functioning of organic polymers in agriculture.

References

Adil, A., Canan, K., and Metin, T. (2012). Humic acid application alleviate salinity stress of bean (*Phaseolus vulgaris* L.) plants decreasing membrane leakage. *African Journal of Agricultural Research* 7 (7): 1073–1086. https://doi.org/10.5897/AJAR10.274.

Ahmad, N.N.R., Fernando, W.J.N., and Uzir, M.H. (2015). Parametric evaluation using mechanistic model for release rate of phosphate ions from chitosan-coated phosphorus fertiliser pellets. *Biosystems Engineering* 129: 78–86. https://doi.org/10.1016/j.biosystemseng.2014.09.015.

Ahmed, S.F., Kumar, P.S., Rozbu, M.R. et al. (2022). Heavy metal toxicity, sources, and remediation techniques for contaminated water and soil. *Environmental Technology & Innovation* 25: 102114. https://doi.org/10.1016/j.eti.2021.102114.

Ai, F., Yin, X., Hu, R. et al. (2021). Research into the super-absorbent polymers on agricultural water. *Agricultural Water Management* 245: 106513. https://doi.org/10.1016/j.agwat.2020.106513.

Alengebawy, A., Abdelkhalek, S.T., Qureshi, S.R., and Wang, M.Q. (2021). Heavy metals and pesticides toxicity in agricultural soil and plants: ecological risks and human health implications. *Toxics* 9 (3): 42. https://doi.org/10.3390/toxics9030042.

Allan, G.G., Chopra, C.S., Neogi, A.N., and Wilkins, R.M. (1971). Design and synthesis of controlled release pesticide-polymer combinations. *Nature* 234 (5328): 349–351. https://doi.org/10.1038/234349a0.

Altabbaa, S., Mann, N.A., Chauhan, N. et al. (2022). Era connecting nanotechnology with agricultural sustainability: issues and challenges. *Nanotechnology for Environmental Engineering* 8: 481–498. https://doi.org/10.1007/s41204-022-00289-3.

Attia, M.S., El-Sayyad, G.S., Abd Elkodous, M. et al. (2021a). Chitosan and EDTA conjugated graphene oxide antinematodes in eggplant: toward improving plant immune response. *International Journal of Biological Macromolecules* 179: 333–344. https://doi.org/10.1016/j.ijbiomac.2021.03.005.

Attia, M.S., Osman, M.S., Mohamed, A.S. et al. (2021b). Impact of foliar application of chitosan dissolved in different organic acids on isozymes, protein patterns and physio-biochemical characteristics of tomato grown under salinity stress. *Plants* 10 (2): 388. https://doi.org/10.3390/plants10020388.

Azadi, H., Keramati, P., Taheri, F. et al. (2018). Agricultural land conversion: reviewing drought impacts and coping strategies. *International Journal of Disaster Risk Reduction* 31: 184–195. https://doi.org/10.1016/j.ijdrr.2018.05.003.

Azeem, B., KuShaari, K., Man, Z.B. et al. (2014). Review on materials & methods to produce controlled release coated urea fertilizer. *Journal of Controlled Release* 181: 11–21. https://doi.org/10.1016/j.jconrel.2014.02.020.

Babla, M., Katwal, U., Yong, M.T. et al. (2022). Value-added products as soil conditioners for sustainable agriculture. *Resources, Conservation and Recycling* 178: 106079. https://doi.org/10.1016/j.resconrec.2021.106079.

Balli, Z., Amirinejad, A.A., and Ghobadi, M. (2019). Salicylic acid and superabsorbent polymer interaction effects on yield and yield components in Mung bean (*Vigna radiata* Wilczek) under different water regimes. *Iranian Journal of Soil and Water Research* 50 (5): 1193–1205.

Bartwal, A. and Arora, S. (2020). Brassinosteroids: molecules with myriad roles. In: *Co-Evolution of Secondary Metabolites* (ed. J.M. Mérillon and K. Ramawat), 869–895. Cham: Springer. https://doi.org/10.1007/978-3-319-96397-6_18.

Basit, A., Hassnain, M.A., Ullah, I. et al. (2020). Quality indices of tomato plant as affected by water stress conditions and chitosan application. *Pure and Applied Biology (PAB)* 9 (2): 1364–1375. https://doi.org/10.19045/bspab.2020.90143.

Behera, S. and Mahanwar, P.A. (2019). Superabsorbent polymers in agriculture and other applications: a review. *Polymer-Plastics Technology and Materials* 59 (4): 341–356. https://doi.org/10.1080/25740881.2019.1647239.

Ben Salah, I., Aghrouss, S., Douira, A. et al. (2018). Seaweed polysaccharides as bio-elicitors of natural defenses in olive trees against *Verticillium* wilt of olive. *Journal of Plant Interactions* 13 (1): 248–255. https://doi.org/10.1080/17429145.2018.1471528.

Bhatnagar, S. and Kumari, R. (2013). Bioremediation: a sustainable tool for environmental management–a review. *Annual Research & Review in Biology* 3 (4): 974–993.

Bispo, T.M. and Vieira, E.A. (2022). Assimilatory deficit and energy regulation in young *Handroanthus chrysotrichus* plants under flooding stress. *Journal of Plant Research* 135 (2): 323–336. https://doi.org/10.1007/s10265-022-01370-3.

Bo, Z., Renkuan, L., Yunkai, L. et al. (2012). Water-absorption characteristics of organic–inorganic composite superabsorbent polymers and its effect on summer maize root growth. *Journal of Applied Polymer Science* 126 (2): 423–435. https://doi.org/10.1002/app.36652.

Bouissil, S., Guerin, C., Roche, J. et al. (2022). Induction of defense gene expression and the resistance of date palm to *Fusarium oxysporum* f. sp. albedinis in response to alginate extracted from *Bifurcaria bifurcata*. *Marine Drugs* 20 (2): 88. https://doi.org/10.3390/md20020088.

Campobenedetto, C., Mannino, G., Beekwilder, J. et al. (2021). The application of a biostimulant based on tannins affects root architecture and improves tolerance to salinity in tomato plants. *Scientific Reports* 11 (1): 1–15. https://doi.org/10.1038/s41598-020-79770-5.

Cannazza, G., Cataldo, A., De Benedetto, E. et al. (2014). Experimental assessment of the use of a novel superabsorbent polymer (SAP) for the optimization of water consumption in agricultural irrigation process. *Water* 6 (7): 2056–2069. https://doi.org/10.3390/w6072056.

Celis, R., Adelino, M.A., Hermosín, M.C., and Cornejo, J.J.J.O.H.M. (2012). Montmorillonite–chitosan bionanocomposites as adsorbents of the herbicide clopyralid in aqueous solution and soil/water suspensions. *Journal of Hazardous Materials* 209: 67–76. https://doi.org/10.1016/j.jhazmat.2011.12.074.

Cerasola, V.A., Perlotti, L., Pennisi, G. et al. (2022). Potential use of superabsorbent polymer on drought-stressed processing tomato (*Solanum lycopersicum* L.) in a mediterranean climate. *Horticulturae* 8 (8): 718. https://doi.org/10.3390/horticulturae8080718.

Cha, J.Y., Kang, S.H., Ji, M.G. et al. (2021). Transcriptome changes reveal the molecular mechanisms of humic acid-induced salt stress tolerance in *Arabidopsis*. *Molecules* 26 (4): 782. https://doi.org/10.3390/molecules26040782.

Chaín, J.M., Tubert, E., Graciano, C. et al. (2020). Growth promotion and protection from drought in *Eucalyptus grandis* seedlings inoculated with beneficial bacteria embedded in a superabsorbent polymer. *Scientific Reports* 10 (1): 1–17.

Chaki, M., Begara-Morales, J.C., and Barroso, J.B. (2020). Oxidative stress in plants. *Antioxidants* 9 (6): 481. https://doi.org/10.3390/antiox9060481.

Chang, L., Xu, L., Liu, Y., and Qiu, D. (2021). Superabsorbent polymers used for agricultural water retention. *Polymer Testing* 94: 107021. https://doi.org/10.1016/j.polymertesting.2020.107021.

Chanratana, M., Joe, M.M., Roy Choudhury, A. et al. (2019). Physiological response of tomato plant to chitosan-immobilized aggregated *Methylobacterium oryzae* CBMB20 inoculation under salinity stress. *3 Biotech* 9 (11): 1–13. https://doi.org/10.1007/s13205-019-1923-1.

Chauhan, G.S. and Mahajan, S. (2002). Use of novel hydrogels based on modified cellulosics and methacrylamide for separation of metal ions from water systems. *Journal of Applied Polymer Science* 86 (3): 667–671. https://doi.org/10.1002/app.10943.

Chen, Y., Li, W., and Zhang, S. (2021). A multifunctional eco-friendly fertilizer used keratin-based superabsorbent as coatings for slow-release urea and remediation of contaminated soil. *Progress in Organic Coatings* 154: 106158. https://doi.org/10.1016/j.porgcoat.2021.106158.

Chopin, P., Mubaya, C.P., Descheemaeker, K. et al. (2021). Avenues for improving farming sustainability assessment with upgraded tools, sustainability framing and indicators. A review. *Agronomy for Sustainable Development* 41: 19. https://doi.org/10.1007/s13593-021-00674-3.

Çimrin, K.M., Türkmen, Ö., Turan, M., and Tuncer, B. (2010). Phosphorus and humic acid application alleviate salinity stress of pepper seedling. *African Journal of Biotechnology* 9 (36): 5845–5851.

Costa, M.M., Cabral-Albuquerque, E.C., Alves, T.L. et al. (2013). Use of polyhydroxybutyrate and ethyl cellulose for coating of urea granules. *Journal of Agricultural and Food Chemistry* 61 (42): 9984–9991. https://doi.org/10.1021/jf401185y.

Dabhi, R., Bhatt, N., and Pandit, B. (2014). Effect on the absorption rate of agricultural super absorbent polymers under the mixer of soil and different quality of irrigation water. *International Journal of Emerging Technology Advanced Engineering–IJETAE* 3 (1): 1402–1406.

Demirbas, A. (2008). Heavy metal adsorption onto agro-based waste materials: a review. *Journal of Hazardous Materials* 157 (2–3): 220–229. https://doi.org/10.1016/j.jhazmat.2008.01.024.

Díez-Pascual, A.M. (2022). Biopolymer composites: synthesis, properties, and applications. *International Journal of Molecular Sciences* 23 (4): 2257. https://doi.org/10.3390/ijms23042257.

Dragan, E.S. (2014). Design and applications of interpenetrating polymer network hydrogels. A review. *Chemical Engineering Journal* 243: 572–590. https://doi.org/10.1016/j.cej.2014.01.065.

Duan, D., Tong, J., Xu, Q. et al. (2020). Regulation mechanisms of humic acid on Pb stress in tea plant (*Camellia sinensis* L.). *Environmental Pollution* 267: 115546. https://doi.org/10.1016/j.envpol.2020.115546.

El Idrissi, A., Dardari, O., Metomo, F.N.N.N. et al. (2023). Effect of sodium alginate-based superabsorbent hydrogel on tomato growth under different water deficit conditions. *International Journal of Biological Macromolecules* 253: 127229. https://doi.org/10.1016/j.ijbiomac.2023.127229.

El-Tohamy, W.A., El-Abagy, H.M., Ahmed, E.M. et al. (2014). Application of super absorbent hydrogel poly(acrylate/acrylic acid) for water conservation in sandy soil. *Transaction of the Egyptian Society of Chemical Engineering* 40 (2): 1–8.

Esteves, L.P. (2011). Superabsorbent polymers: on their interaction with water and pore fluid. *Cement and Concrete Composites* 33 (7): 717–724. https://doi.org/10.1016/j.cemconcomp.2011.04.006.

Fan, L., Xue, L., Rong, J. et al. (2019a). Effects of super absorbent polymers and nitrogen fertilizer mixture on resistance physiology and chlorophyll fluorescence parameters of 3 bamboo species in coastal sand. *Journal of Fujian Agriculture and Forestry University (Natural Science Edition)* 48 (3): 378–385.

Fan, R., Zhang, Z., and Luo, J. (2019b). Study on effect of super absorbent polymer (SAP) on fertilizer retention and nutrient absorption of green pepper. *Jiangsu Agricultural Sciences* 47 (3): 107–109.

Fan, Z., Qin, Y., Liu, S. et al. (2020). Chitosan oligosaccharide fluorinated derivative control root-knot nematode (*Meloidogyne incognita*) disease based on the multi-efficacy strategy. *Marine Drugs* 18 (5): 273.

Farouk, S. and Amany, A.R. (2012). Improving growth and yield of cowpea by foliar application of chitosan under water stress. *Egyptian Journal of Biology* 14: 14–16. https://doi.org/10.4314/ejb.v14i1.2.

Farouk, S., Mosa, A.A., Taha, A.A., and El-Gahmery, A.M. (2011). Protective effect of humic acid and chitosan on radish (*Raphanus sativus*, L. var. sativus) plants subjected to cadmium stress. *Journal of Stress Physiology & Biochemistry* 7 (2): 99–116.

Fasusi, O.A., Cruz, C., and Babalola, O.O. (2021). Agricultural sustainability: microbial biofertilizers in rhizosphere management. *Agriculture* 11 (2): 163. https://doi.org/10.3390/agriculture11020163.

García, A.C., Berbara, R.L.L., Farías, L.P. et al. (2012). Humic acids of vermicompost as an ecological pathway to increase resistance of rice seedlings to water stress. *African Journal of Biotechnology* 11 (13): 3125–3134. https://doi.org/10.5897/AJB11.1960.

Gholinezhad, E. and Eivazi, A. (2021). The effect of super absorbent polymer A200 and manure on reducing drought stress in wheat (*Triticum aestivum* L.) different cultivars. *Water and Soil Science* 31: 69–86.

González, A., Castro, J., Vera, J., and Moenne, A. (2013). Seaweed oligosaccharides stimulate plant growth by enhancing carbon and nitrogen assimilation, basal metabolism, and cell division. *Journal of Plant Growth Regulation* 32 (2): 443–448. https://doi.org/10.1007/s00344-012-9309-1.

Graber, E.R., Fine, P., and Levy, G.J. (2006). Soil stabilization in semiarid and arid land agriculture. *Journal of Materials in Civil Engineering* 18 (2): 190–205. https://doi.org/10.1061/(ASCE)0899-1561(2006)18:2(190).

Grillo, R., Clemente, Z., de Oliveira, J.L. et al. (2015). Chitosan nanoparticles loaded the herbicide paraquat: the influence of the aquatic humic substances on the colloidal stability and toxicity. *Journal of Hazardous Materials* 286: 562–572. https://doi.org/10.1016/j.jhazmat.2014.12.021.

Gu, L. (2012). Effects of chitosan on physiological characteristics of tomato seedlings under salt stress. *Agricultural Science & Technology-Hunan* 13 (3): 551–553.

Guan, Y.J., Hu, J., Wang, X.J., and Shao, C.X. (2009). Seed priming with chitosan improves maize germination and seedling growth in relation to physiological changes under low temperature stress. *Journal of Zhejiang University Science B* 10 (6): 427–433. https://doi.org/10.1631/jzus.B0820373.

Guo, Y., Ruan, K., Shi, X. et al. (2020). Factors affecting thermal conductivities of the polymers and polymer composites: a review. *Composites Science and Technology* 193: 108134. https://doi.org/10.1016/j.compscitech.2020.108134.

Han, J., Zhao, L., Zhu, H. et al. (2021). Study on the effect of alginate oligosaccharide combined with *Meyerozyma guilliermondii* against *Penicillium expansum* in pears and the possible mechanisms involved. *Physiological and Molecular Plant Pathology* 115: 101654. https://doi.org/10.1016/j.pmpp.2021.101654.

Heidari, J., Amooaghaie, R., and Kiani, S. (2020). Impact of chitosan on nickel bioavailability in soil, the accumulation and tolerance of nickel in *Calendula tripterocarpa*. *International Journal of Phytoremediation* 22 (11): 1175–1184. https://doi.org/10.1080/15226514.2020.1748564.

Hidangmayum, A., Dwivedi, P., Katiyar, D., and Hemantaranjan, A. (2019). Application of chitosan on plant responses with special reference to abiotic stress. *Physiology and Molecular Biology of Plants* 25 (2): 313–326. https://doi.org/10.1007/s12298-018-0633-1.

Himani, P.U., Mahawer, S.K., Kumar, R., and Prakash, O. (2022). Plant protection through agrochemicals and its consequences. In: *Plant Protection: From Chemicals to Biologicals* (ed. R. Soni, R. Goel, and R. Syal), 25. Berlin: Walter de Gruyter GmbH.

Hua, Y., Kumar, V., and Kim, K.H. (2021). Recent progress on hollow porous molecular imprinted polymers as sorbents of environmental samples. *Microchemical Journal* 171: 106848. https://doi.org/10.1016/j.microc.2021.106848.

Huang, J., Kogbara, R.B., Hariharan, N. et al. (2021). A state-of-the-art review of polymers used in soil stabilization. *Construction and Building Materials* 305: 124685. https://doi.org/10.1016/j.conbuildmat.2021.124685.

Hussain, M., Devi, R.R., and Maji, T.K. (2012). Controlled release of urea from chitosan microspheres prepared by emulsification and cross-linking method. *Iranian Polymer Journal* 21 (8): 473–479. https://doi.org/10.1007/s13726-012-0051-0.

Hussain, S., Zhang, R., Liu, S. et al. (2022). Methyl jasmonate alleviates the deleterious effects of salinity stress by augmenting antioxidant enzyme activity and ion homeostasis in rice (*Oryza sativa* L.). *Agronomy* 12 (10): 2343. https://doi.org/10.3390/agronomy12102343.

Ibrahim, E.A. and Ramadan, W.A. (2015). Effect of zinc foliar spray alone and combined with humic acid or/and chitosan on growth, nutrient elements content and yield of dry bean (*Phaseolus vulgaris* L.) plants sown at different dates. *Scientia Horticulturae* 184: 101–105. https://doi.org/10.1016/j.scienta.2014.11.010.

Ilyas, M., Nisar, M., Khan, N. et al. (2021). Drought tolerance strategies in plants: a mechanistic approach. *Journal of Plant Growth Regulation* 40 (3): 926–944.

Jabeen, N. and Ahmad, R. (2013). The activity of antioxidant enzymes in response to salt stress in safflower (*Carthamus tinctorius* L.) and sunflower (*Helianthus annuus* L.) seedlings raised from seed treated with chitosan. *Journal of the Science of Food and Agriculture* 93 (7): 1699–1705. https://doi.org/10.1002/jsfa.5953.

Janků, M., Luhová, L., and Petřivalský, M. (2019). On the origin and fate of reactive oxygen species in plant cell compartments. *Antioxidants* 8 (4): 105. https://doi.org/10.3390/antiox8040105.

Jia-yue, C., Ling-zhi, L., Hai-ping, L. et al. (2020). Effect of humic acid as initiating solution on seed germination of summer squash under low temperature stress. *Chinese Journal of Agrometeorology* 41 (8): 483. https://doi.org/10.3969/j.issn.1000-6362.2020.08.002.

Kappel, L., Kosa, N., and Gruber, S. (2022). The multilateral efficacy of chitosan and *Trichoderma* on sugar beet. *Journal of Fungi* 8 (2): 137. https://doi.org/10.3390/jof8020137.

Karamina, H. and Fikrinda, W. (2020). Soil amendment impact to soil organic matter and physical properties on the three soil types after second corn cultivation. *AIMS Agriculture and Food* 5 (1): 150–169.

Karwa, S., Bahuguna, R.N., Chaturvedi, A.K. et al. (2020). Phenotyping and characterization of heat stress tolerance at reproductive stage in rice (*Oryza sativa* L.). *Acta Physiologiae Plantarum* 42 (2): 1–16.

Kashyap, P.L., Xiang, X., and Heiden, P. (2015). Chitosan nanoparticle based delivery systems for sustainable agriculture. *International Journal of Biological Macromolecules* 77: 36–51. https://doi.org/10.1016/j.ijbiomac.2015.02.039.

Kaur, R., Sharma, N., Tikoria, R. et al. (2022). Insights into biosynthesis and signaling of cytokinins during plant growth, development and stress tolerance. In: *Auxins, Cytokinin and Gibberellins Signaling in Plants* (ed. T. Aftab), 153–187. London: Springer Nature. https://doi.org/10.1007/978-3-031-05427-3_7.

Kavamura, V.N. and Esposito, E. (2010). Biotechnological strategies applied to the decontamination of soils polluted with heavy metals. *Biotechnology Advances* 28 (1): 61–69. https://doi.org/10.1016/j.biotechadv.2009.09.002.

Khan, N., Bano, A., Rahman, M.A. et al. (2019). Comparative physiological and metabolic analysis reveals a complex mechanism involved in drought tolerance in chickpea (*Cicer arietinum* L.) induced by PGPR and PGRs. *Scientific Reports* 9 (1): 1–19.

Khan, M.A., Asaf, S., Khan, A.L. et al. (2020). Extending thermotolerance to tomato seedlings by inoculation with SA1 isolate of *Bacillus cereus* and comparison with exogenous humic acid application. *PLoS One* 15 (4): e0232228. https://doi.org/10.1371/journal.pone.0232228.

Khoo, R.Z., Chow, W.S., and Ismail, H. (2018). Sugarcane bagasse fiber and its cellulose nanocrystals for polymer reinforcement and heavy metal adsorbent: a review. *Cellulose* 25 (8): 4303–4330. https://doi.org/10.1007/s10570-018-1879-z.

Klein, M. and Poverenov, E. (2020). Natural biopolymer-based hydrogels for use in food and agriculture. *Journal of the Science of Food and Agriculture* 100 (6): 2337–2347. https://doi.org/10.1002/jsfa.10274.

Kopittke, P.M., Menzies, N.W., Wang, P. et al. (2019). Soil and the intensification of agriculture for global food security. *Environment International* 132: 105078. https://doi.org/10.1016/j.envint.2019.105078.

Kumar, D. and Ohri, P. (2022). Say "NO" to plant stresses: unravelling the role of nitric oxide under abiotic and biotic stress. *Nitric Oxide* 130: 36–57. https://doi.org/10.1016/j.niox.2022.11.004.

Lemic, D., Orešković, M., Mikac, K.M. et al. (2021). Sustainable pest management using biodegradable apitoxin-loaded calcium-alginate microspheres. *Sustainability* 13 (11): 6167. https://doi.org/10.3390/su13116167.

Li, X., He, J.Z., Liu, Y.R., and Zheng, Y.M. (2013). Effects of super absorbent polymers on soil microbial properties and Chinese cabbage (*Brassica chinensis*) growth. *Journal of Soils and Sediments* 13 (4): 711–719.

Li, H., Sui, L., and Niu, Y. (2018). Preparation and properties of a double-coated slow-release urea fertilizer with poly(propylene carbonate), a sodium polyacrylate hydroscopicity resin and sodium alginate. *Chemistry Select* 3 (26): 7643–7647. https://doi.org/10.1002/slct.201800913.

Li, Y., Zhang, Q., Ou, L. et al. (2020). Response to the cold stress signaling of the tea plant (*Camellia sinensis*) elicited by chitosan oligosaccharide. *Agronomy* 10 (6): 915. https://doi.org/10.3390/agronomy10060915.

Liao, R., Wu, W., Ren, S., and Yang, P. (2016). Effects of superabsorbent polymers on the hydraulic parameters and water retention properties of soil. *Journal of Nanomaterials* 2016: 5403976. https://doi.org/10.1155/2016/5403976.

Liu, J., Bai, Y., Feng, Q. et al. (2018). Strength properties of sand reinforced with a mixture of organic polymer stabilizer and polypropylene fiber. *Journal of Materials in Civil Engineering* 30 (12): 04018330. https://doi.org/10.1061/(ASCE)MT.1943-5533.0002541.

Liu, W., Zhong, Q., and Luo, L. (2019). Effects of different irrigation volume co-applied with super absorbent polymers on soil moisture and fruit of mango orchards. *Southwest China Journal of Agricultural Sciences* 32 (10): 2378–2382. https://doi.org/10.16213/j.cnki.scjas.2019.10.020.

Lubkowski, K. and Grzmil, B. (2007). Controlled release fertilizers. *Polish Journal of Chemical Technology* 9 (4): 81–84. https://doi.org/10.2478/v10026-007-0096-6.

Lukić, N., Trifković, T., Kojić, D., and Kukavica, B. (2021). Modulations of the antioxidants defence system in two maize hybrids during flooding stress. *Journal of Plant Research* 134 (2): 237–248. https://doi.org/10.1007/s10265-021-01264-w.

Lykogianni, M., Bempelou, E., Karamaouna, F., and Aliferis, K.A. (2021). Do pesticides promote or hinder sustainability in agriculture? The challenge of sustainable use of pesticides in modern agriculture. *Science of the Total Environment* 795: 148625. https://doi.org/10.1016/j.scitotenv.2021.148625.

Ma, L.J., Li, X.M., Bu, N., and Li, N. (2010a). An alginate-derived oligosaccharide enhanced wheat tolerance to cadmium stress. *Plant Growth Regulation* 62 (1): 71–76. https://doi.org/10.1007/s10725-010-9489-2.

Ma, L.J., Zhang, Y., Bu, N., and Wang, S.H. (2010b). Alleviation effect of alginate-derived oligosaccharides on *Vicia faba* root tip cells damaged by cadmium. *Bulletin of Environmental Contamination and Toxicology* 84 (2): 161–164. https://doi.org/10.1007/s00128-009-9914-2.

Malik, L.A., Bashir, A., Qureashi, A., and Pandith, A.H. (2019). Detection and removal of heavy metal ions: a review. *Environmental Chemistry Letters* 17 (4): 1495–1521. https://doi.org/10.1007/s10311-019-00891-z.

Melaj, M.A. and Daraio, M.E. (2014). HPMC layered tablets modified with chitosan and xanthan as matrices for controlled-release fertilizers. *Journal of Applied Polymer Science* 131: 40839. https://doi.org/10.1002/app.40839.

Messa, L.L., Froes, J.D., Souza, C.F., and Faez, R. (2016). Chitosan-clay hybrid for encapsulation of fertilizers and release sustained of potassium nitrate fertilizer. *Quimica Nova* 39 (10): 1215–1220. http://dx.doi.org/10.21577/0100-4042.20160133.

Milani, P., França, D., Balieiro, A.G., and Faez, R. (2017). Polymers and its applications in agriculture. *Polímeros* 27: 256–266. https://doi.org/10.1590/0104-1428.09316.

Mohawesh, O. and Durner, W. (2019). Effects of bentonite, hydrogel and biochar amendments on soil hydraulic properties from saturation to oven dryness. *Pedosphere* 29 (5): 598–607. https://doi.org/10.1016/S1002-0160(17)60426-0.

Mohsin, M.A. and Attia, N.F. (2015). Inverse emulsion polymerization for the synthesis of high molecular weight polyacrylamide and its application as sand stabilizer. *International Journal of Polymer Science* 2015: 436583. https://doi.org/10.1155/2015/436583.

Mukherjee, P., Mitra, A., and Roy, M. (2019). *Halomonas* rhizobacteria of *Avicennia marina* of Indian Sundarbans promote rice growth under saline and heavy metal stresses through exopolysaccharide production. *Frontiers in Microbiology* 10: 1207. https://doi.org/10.3389/fmicb.2019.01207.

Mukhtar, T., Rehman, S.U., Smith, D. et al. (2020). Mitigation of heat stress in *Solanum lycopersicum* L. by ACC-deaminase and exopolysaccharide producing *Bacillus cereus*: effects on biochemical profiling. *Sustainability* 12 (6): 2159. https://doi.org/10.3390/su12062159.

Nkebiwe, P.M., Weinmann, M., Bar-Tal, A., and Müller, T. (2016). Fertilizer placement to improve crop nutrient acquisition and yield: a review and meta-analysis. *Field Crops Research* 196: 389–401. https://doi.org/10.1016/j.fcr.2016.07.018.

Noppakundilograt, S., Pheatcharat, N., and Kiatkamjornwong, S. (2015). Multilayer-coated NPK compound fertilizer hydrogel with controlled nutrient release and water absorbency. *Journal of Applied Polymer Science* 132 (2). https://doi.org/10.1002/app.41249.

Nunes da Silva, M., Santos, C.S., Cruz, A. et al. (2021). Chitosan increases Pinus pinaster tolerance to the pinewood nematode (*Bursaphelenchus xylophilus*) by promoting plant antioxidative metabolism. *Scientific Reports* 11 (1): 1–10. https://doi.org/10.1038/s41598-021-83445-0.

Oertli, J.J. and Lunt, O.R. (1962). Controlled release of fertilizer minerals by incapsulating membranes: I. Factors influencing the rate of release. *Soil Science Society of America Journal* 26 (6): 579–583. https://doi.org/10.2136/sssaj1962.03615995002600060019x.

Ono, T., Sugimoto, T., Shinkai, S., and Sada, K. (2007). Lipophilic polyelectrolyte gels as super-absorbent polymers for nonpolar organic solvents. *Nature Materials* 6 (6): 429–433.

Pagliaccia, B., Carretti, E., Severi, M. et al. (2022). Heavy metal biosorption by extracellular polymeric substances (EPS) recovered from anammox granular sludge. *Journal of Hazardous Materials* 424: 126661. https://doi.org/10.1016/j.jhazmat.2021.126661.

Pal, A. and Paul, A. (2008). Microbial extracellular polymeric substances: central elements in heavy metal bioremediation. *Indian Journal of Microbiology* 48 (1): 49–64. https://doi.org/10.1007/s12088-008-0006-5.

Park, S.U., Lee, C.J., Park, S.C. et al. (2022). Flooding tolerance in sweet potato (*Ipomoea batatas* (L.) Lam) is mediated by reactive oxygen species and nitric oxide. *Antioxidants* 11 (5): 878. https://doi.org/10.3390/antiox11050878.

Pereira, E.I., Giroto, A.S., Bortolin, A. et al. (2015). Perspectives in nanocomposites for the slow and controlled-release of agrochemicals: fertilizers and pesticides. In: *Nanotechnologies in Food and Agriculture* (ed. M. Rai, C. Ribeiro, L. Mattoso, and N. Duran), 241–265. Minneapolis: Springer. https://doi.org/10.1007/978-3-319-14024-7_11.

Pérez de Luque, A. and Hermosín, M.C. (2013). Nanotechnology and its use in agriculture. In: *Bio-Nanotechnology: A Revolution in Food, Biomedical and Health Sciences* (ed. D. Bagchi), 383–398. Oxford: Blackwell Publishing Ltd. https://doi.org/10.1002/9781118451915.ch20.

Pettongkhao, S., Bilanglod, A., Khompatara, K., and Churngchow, N. (2019). Sulphated polysaccharide from *Acanthophora spicifera* induced *Hevea brasiliensis* defense responses against *Phytophthora palmivora* infection. *Plants* 8 (3): 73. https://doi.org/10.3390/plants8030073.

Popletaeva, S.B., Statsyuk, N.V., Voinova, T.M. et al. (2018). Evaluation of eliciting activity of peptidil prolyl cys/trans isomerase from *Pseudonomas fluorescens* encapsulated in sodium alginate regarding plant resistance to viral and fungal pahogens. *AIMS Microbiology* 4 (1): 192. https://doi.org/10.3934/2Fmicrobiol.2018.1.192.

Puoci, F., Iemma, F., Spizzirri, U.G. et al. (2008). Polymer in agriculture: a review. *American Journal of Agricultural and Biological Sciences* 3 (1): 299–314. https://doi.org/10.3844/ajabssp.2008.299.314.

Qu, D.Y., Gu, W.R., Zhang, L.G. et al. (2019). Role of chitosan in the regulation of the growth, antioxidant system and photosynthetic characteristics of maize seedlings under cadmium stress. *Russian Journal of Plant Physiology* 66 (1): 140–151. https://doi.org/10.1134/S102144371901014X.

Rangabhashiyam, S., Anu, N., Giri Nandagopal, M.S., and Selvaraju, N. (2014). Relevance of isotherm models in biosorption of pollutants by agricultural byproducts. *Journal of Environmental Chemical Engineering* 2 (1): 398–414. https://doi.org/10.1016/j.jece.2014.01.014.

Raza, A., Salehi, H., Rahman, M.A. et al. (2022). Plant hormones and neurotransmitter interactions mediate antioxidant defenses under induced oxidative stress in plants. *Frontiers in Plant Science* 13: 961872. https://doi.org/10.3389/fpls.2022.961872.

Reetz, H.F. (2016). *Fertilizers and their Efficient Use*. International Fertilizer Industry Association.

Rezaeimalek, S., Huang, J., and Bin-Shafique, S. (2017). Evaluation of curing method and mix design of a moisture activated polymer for sand stabilization. *Construction and Building Materials* 146: 210–220. https://doi.org/10.1016/j.conbuildmat.2017.04.093.

Roy, A., Singh, S.K., Bajpai, J., and Bajpai, A.K. (2014). Controlled pesticide release from biodegradable polymers. *Central European Journal of Chemistry* 12 (4): 453–469. https://doi.org/10.2478/s11532-013-0405-2.

Sabadini, R.C., Martins, V.C., and Pawlicka, A. (2015). Synthesis and characterization of gellan gum: chitosan biohydrogels for soil humidity control and fertilizer release. *Cellulose* 22 (3): 2045–2054. https://doi.org/10.1007/s10570-015-0590-6.

Saberi Riseh, R., Gholizadeh Vazvani, M., Ebrahimi-Zarandi, M., and Skorik, Y.A. (2022). Alginate-induced disease resistance in plants. *Polymers* 14: 661. https://doi.org/10.3390/polym14040661.

Safikhan, S., Khoshbakht, K., Chaichi, M.R. et al. (2018). Role of chitosan on the growth, physiological parameters and enzymatic activity of milk thistle (*Silybum marianum* (L.) Gaertn.) in a pot experiment. *Journal of Applied Research on Medicinal and Aromatic Plants* 10: 49–58. https://doi.org/10.1016/j.jarmap.2018.06.002.

Salachna, P., Grzeszczuk, M., Meller, E., and Soból, M. (2018). Oligo-alginate with low molecular mass improves growth and physiological activity of *Eucomis autumnalis* under salinity stress. *Molecules* 23 (4): 812. https://doi.org/10.3390/molecules23040812.

dos Santos, B.R., Bacalhau, F.B., dos Santos Pereira, T. et al. (2015). Chitosan-montmorillonite microspheres: a sustainable fertilizer delivery system. *Carbohydrate Polymers* 127: 340–346. https://doi.org/10.1016/j.carbpol.2015.03.064.

Sarker, T.C., Azam, S.M.G.G., El-Gawad, A.M.A. et al. (2017). Sugarcane bagasse: a potential low-cost biosorbent for the removal of hazardous materials. *Clean Technologies and Environmental Policy* 19 (10): 2343–2362. https://doi.org/10.1007/s10098-017-1429-7.

Sarlaki, E., Kermani, A.M., Kianmehr, M.H. et al. (2021). Improving sustainability and mitigating environmental impacts of agro-biowaste compost fertilizer by pelletizing-drying. *Environmental Pollution* 285: 117412. https://doi.org/10.1016/j.envpol.2021.117412.

Sathiyabama, M., Akila, G., and Charles, R.E. (2014). Chitosan-induced defence responses in tomato plants against early blight disease caused by *Alternaria solani* (Ellis and Martin) Sorauer. *Archives of Phytopathology and Plant Protection* 47 (16): 1963–1973. https://doi.org/10.1080/03235408.2013.863497.

Seenivasan, N. and Senthilnathan, S. (2018). Effect of humic acid on *Meloidogyne incognita* (Kofoid & White) Chitwood infecting banana (*Musa* spp.). *International Journal of Pest Management* 64 (2): 110–118. https://doi.org/10.1080/09670874.2017.1344743.

Senna, A.M., do Carmo, J.B., da Silva, J.M.S., and Botaro, V.R. (2015). Synthesis, characterization and application of hydrogel derived from cellulose acetate as a substrate for slow-release NPK fertilizer and water retention in soil. *Journal of Environmental Chemical Engineering* 3 (2): 996–1002. https://doi.org/10.1016/j.jece.2015.03.008.

Sergiev, I., Todorova, D., Moskova, I. et al. (2013). Protective effect of humic acids against heavy metal stress in triticale. *Comptes Rendus de L'Academie Bulgare des Sciences* 66: 53–60.

Shaffique, S., Khan, M.A., Wani, S.H. et al. (2022). A review on the role of endophytes and plant growth promoting rhizobacteria in mitigating heat stress in plants. *Microorganisms* 10 (7): 1286. https://doi.org/10.3390/microorganisms10071286.

Sharma, N., Kour, S., Kumar, D. et al. (2022). Role of brassinosteroids (BRs) in modulating antioxidative defense mechanism in plants growing under abiotic and biotic stress conditions. In: *Antioxidant Defense in Plants* (ed. T. Aftab and K.R. Hakeem), 325–367. Singapore: Springer. https://doi.org/10.1007/978-981-16-7981-0_15.

Shaviv, A. (2001). Advances in controlled-release fertilizers. *Advances in Agronomy* 71: 1–49. https://doi.org/10.1016/S0065-2113(01)71011-5.

Shen, L.C., Nguyen, X.T., and Hankins, N.P. (2015). Removal of heavy metal ions from dilute aqueous solutions by polymer–surfactant aggregates: a novel effluent treatment process. *Separation and Purification Technology* 152: 101–107. https://doi.org/10.1016/j.seppur.2015.07.065.

Shin, J.H., Kim, H.S., Kim, E., and Ahn, S.J. (2020). Biopolymer amended soil reduces the damages of Zn excess in *Camlina sativa* L. *Ecology and Resilient Infrastructure* 7 (4): 262–273. https://doi.org/10.17820/eri.2020.7.4.262.

Shinde, R., Sarkar, P.K., and Thombare, N. (2019). Soil conditioners. *Agriculture & Food: e-Newsletter* 1: 1–5.

Singh, R. and Rathore, D. (2020). Role of transitory starch on growth, development and metal accumulation of *Triticum aestivum* cultivars grown under textile effluent fertilization. *Environmental Science and Pollution Research* 27 (19): 24201–24217. https://doi.org/10.1007/s11356-020-08735-y.

Singh, S.B., Singh, K., Butola, S.S. et al. (2020). Determination of macronutrients, micronutrients and heavy metals present in *Spilanthes acmella* Hutch and Dalz: possible health effects. *Natural Product Sciences* 26 (1): 50–58. https://doi.org/10.20307/nps.2020.26.1.50.

Sinha, S. (2018). Biodegradable superabsorbents: Methods of preparation and application—a review. *Fundamental Biomaterials: Polymers* 307–322. https://doi.org/10.1016/B978-0-08-102194-1.00014-1.

Souza, A.J.J., Guimarães, R.J., Dominghetti, A.W. et al. (2016). Water-retaining polymer and seedling type when planting irrigated coffee. *Revista Ciência Agronômica* 47: 334–343. https://doi.org/10.5935/1806-6690.20160039.

Stahl, J.D., Cameron, M.D., Haselbach, J., and Aust, S.D. (2000). Biodegradation of superabsorbent polymers in soil. *Environmental Science and Pollution Research* 7 (2): 83–88. https://doi.org/10.1065/espr199912.014.

Stuart, M.A.C., Huck, W.T., Genzer, J. et al. (2010). Emerging applications of stimuli-responsive polymer materials. *Nature Materials* 9 (2): 101–113.

Thabet, M. (2019). Application of chitosan and oxalic acid combined with hot water to control postharvest decay of strawberry fruits caused by *Botrytis inerea* and *Rhizopus stolonifer*. *Sciences* 9 (1): 63–77.

Thombare, N., Mishra, S., Siddiqui, M.Z. et al. (2018). Design and development of guar gum based novel, superabsorbent and moisture retaining hydrogels for agricultural applications. *Carbohydrate Polymers* 185: 169–178. https://doi.org/10.1016/j.carbpol.2018.01.018.

Tikoria, R., Sharma, N., Kour, S. et al. (2022). Vermicomposting: an effective alternative in integrated pest management. In: *Earthworm Engineering and Applications* (ed. A.P. Vig, S.S. Suthar, and J. Singh), 103–118. New York: Nova Science Publishers.

Tilman, D., Cassman, K.G., Matson, P.A. et al. (2002). Agricultural sustainability and intensive production practices. *Nature* 418 (6898): 671–677. https://doi.org/10.1038/nature01014.

Trenkel, M.E. (1997). *Controlled-Release and Stabilized Fertilizers in Agriculture*, vol. 11. Paris: International Fertilizer Industry Association.

Trenkel, M.E. (2010). *Slow- and Controlled-Release and Stabilized Fertilizers: An Option for Enhancing Nutrient Use Efficiency in Agriculture*. Paris: International Fertilizer Industry Association.

Tubert, E., Vitali, V.A., Alvarez, M.S. et al. (2018). Synthesis and evaluation of a superabsorbent-fertilizer composite for maximizing the nutrient and water use efficiency in forestry plantations. *Journal of Environmental Management* 210: 239–254. https://doi.org/10.1016/j.jenvman.2017.12.062.

Udayakumar, G.P., Muthusamy, S., Selvaganesh, B. et al. (2021). Ecofriendly biopolymers and composites: preparation and their applications in water-treatment. *Biotechnology Advances* 52: 107815. https://doi.org/10.1016/j.biotechadv.2021.107815.

Ullah, F., Othman, M.B.H., Javed, F. et al. (2015). Classification, processing and application of hydrogels: a review. *Materials Science and Engineering: C* 57: 414–433. https://doi.org/10.1016/j.msec.2015.07.053.

Ullah, N., Basit, A., Ahmad, I. et al. (2020). Mitigation the adverse effect of salinity stress on the performance of the tomato crop by exogenous application of chitosan. *Bulletin of the National Research Centre* 44 (1): 1–11. https://doi.org/10.1186/s42269-020-00435-4.

Vurukonda, S.S.K.P., Vardharajula, S., Shrivastava, M., and SkZ, A. (2016). Enhancement of drought stress tolerance in crops by plant growth promoting rhizobacteria. *Microbiological Research* 184: 13–24. https://doi.org/10.1016/j.micres.2015.12.003.

Wang, J. and Chen, C. (2009). Biosorbents for heavy metals removal and their future. *Biotechnology Advances* 27 (2): 195–226. https://doi.org/10.1016/j.biotechadv.2008.11.002.

Wang, X., An, M., Wang, K. et al. (2021). Effects of organic polymer compound material on K^+ and Na^+ distribution and physiological characteristics of cotton under saline and alkaline stresses. *Frontiers in Plant Science* 12: 636536. https://doi.org/10.3389/fpls.2021.636536.

Xin, X., Zhao, F., Rho, J.Y. et al. (2020). Use of polymeric nanoparticles to improve seed germination and plant growth under copper stress. *Science of the Total Environment* 745: 141055. https://doi.org/10.1016/j.scitotenv.2020.141055.

Yang, J., Sun, Z., Zhao, Y. et al. (2020a). The water absorption-release of superabsorbent polymers in fresh cement paste: an NMR study. *Journal of Advanced Concrete Technology* 18 (3): 139–145. https://doi.org/10.3151/jact.18.139.

Yang, F., Cen, R., Feng, W. et al. (2020b). Effects of super-absorbent polymer on soil remediation and crop growth in arid and semi-arid areas. *Sustainability* 12 (18): 7825. https://doi.org/10.3390/su12187825.

Yaseen, R., Hegab, R., Kenawey, M., and Eissa, D. (2020). Effect of super absorbent polymer and bio fertilization on Maize productivity and soil fertility under drought stress conditions. *Egyptian Journal of Soil Science* 60 (4): 377–395. https://doi.org/10.21608/ejss.2020.35386.1372.

Younes, I. and Rinaudo, M. (2015). Chitin and chitosan preparation from marine sources. Structure, properties and applications. *Marine Drugs* 13 (3): 1133–1174. https://doi.org/10.3390/md13031133.

Yu, J., Lei, T.W., Shainberg, I. et al. (2003). Infiltration and erosion in soils treated with dry PAM and gypsum. *Soil Science Society of America Journal* 67 (2): 630–636. https://doi.org/10.2136/sssaj2003.6300.

Yu, G., Chen, F., Xie, Y. et al. (2019). Effects of organic-inorganic superabsorbent polymer composites on soil moisture and growth of maize at seedling stage. *Journal of Henan Agricultural Sciences* 48 (10): 23–29.

Zainab, N., Din, B.U., Javed, M.T. et al. (2020). Deciphering metal toxicity responses of flax (*Linum usitatissimum* L.) with exopolysaccharide and ACC-deaminase producing bacteria in industrially contaminated soils. *Plant Physiology and Biochemistry* 152: 90–99. https://doi.org/10.1016/j.plaphy.2020.04.039.

Zandalinas, S.I., Fritschi, F.B., and Mittler, R. (2021). Global warming, climate change, and environmental pollution: recipe for a multifactorial stress combination disaster. *Trends in Plant Science* 26 (6): 588–599. https://doi.org/10.1016/j.tplants.2021.02.011.

Zhang, W., Liu, Q., Liu, S. et al. (2020). Study on seed emergence and seedling growth of *Artemisia desertorum* with superabsorbent polymers. *Polymers* 12 (12): 2873. https://doi.org/10.3390/polym12122873.

Zhao, K., Yang, Y., Peng, H. et al. (2022). Silicon fertilizers, humic acid and their impact on physicochemical properties, availability and distribution of heavy metals in soil and soil aggregates. *Science of the Total Environment* 822: 153483. https://doi.org/10.1016/j.scitotenv.2022.153483.

Zia, R., Nawaz, M.S., Siddique, M.J. et al. (2021). Plant survival under drought stress: implications, adaptive responses, and integrated rhizosphere management strategy for stress mitigation. *Microbiological Research* 242: 126626. https://doi.org/10.1016/j.micres.2020.126626.

Zohuriaan-Mehr, M.J., Omidian, H., Doroudiani, S., and Kabiri, K. (2010). Advances in non-hygienic applications of superabsorbent hydrogel materials. *Journal of Materials Science* 45 (21): 5711–5735. https://doi.org/10.1007/s10853-010-4780-1.

19

Porous Organic Polymers as Potential Catalysts

Manisha Sharma[1], Chanchal Bhardwaj[1], Pinki R. Agrawal[1,2], Chankit Kaushik[1], Rahul Sharma[1,2], and Ashok. K. Sharma[1]

[1] *Thin Film Laboratory, Deenbandhu Chhotu Ram University of Science and Technology, Department of Chemistry, Murthal, Sonipat. 131039, India*
[2] *AcSIR, CSIR-National Physical Laboratory, New Delhi, India*

19.1 Introduction

Humans are facing serious issues due to the scarcity of resources and the quick consumption of fossil fuels, such as an energy shortage, greenhouse effects, and environmental damage. Thus, to overcome these obstacles, numerous efforts have been made to pursue alternate forms of clean, sustainable energy and technology (Chakraborty et al. 2019). To achieve this goal, the use of recyclable catalysts for organic synthesis to reduce waste production and maximize catalyst efficiency has been thoroughly researched during the past few decades. Photocatalyst development has exploded during the past few decades, with an emphasis on inorganic semiconductors (such as TiO_2, CdSe, WO_3, ZnS, and ZnO), as well as metal complexes (such as ruthenium and iridium complexes), organic dyes, and others (e.g. fluorescein and rose Bengal). In any case, the down-to-earth utilizations of homogeneous photocatalysts are compelled by the noxious nature, significant expense, and innate disservice of uncommon metal parts. Also, a significant number of them are just initiated by UV light, making it hard to use the sun's energy completely. POPs have in this way turned into a promising stage for viable photocatalysis. Ongoing advancements in the blend of nanoporous polymers, like covalent-natural system (COF), and other permeable natural polymers (POPs) present new open doors for heterogeneous catalysis.

Large pores, high specific surface areas, superior stability, and durability in a variety of solvents are characteristics of porous organic polymers (POPs), which are advantageous for mass transfer, supplying more active sites, and the photocatalytic recycling process (Liu and Liu 2020).

POPs are a family of multi-dimensional porous network materials made of various organic building blocks with varying geometries and topologies linked together via strong covalent bonds (Zhang and Riduan 2012). POPs can be classified into the category of amorphous or crystalline POPs. Amorphous POPs mostly incorporate polymers of intrinsic microporosity (PIMs), hyper-cross connected polymers (HCPs),

Organic Polymers in Energy-Environmental Applications, First Edition.
Edited by Ramesh Oraon, Pardeep Singh, Sanchayita Rajkhowa, Sangita Agarwal, and Ravindra Pratap Singh.

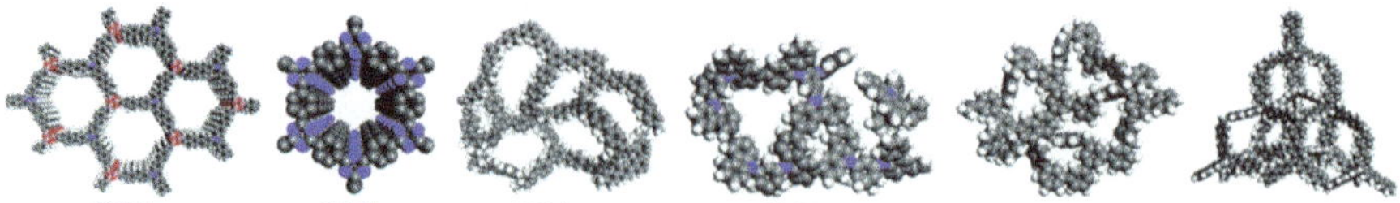

Figure 19.1 Example of POPs. Source: Yang et al. (2022)/Royal Society of Chemistry.

and formed microporous polymers (CMPs) which elements expanded pi formation framework in contrast with other undefined POPs, and permeable sweet-smelling systems (PAFs), while glasslike POPs are addressed by covalent organic framework systems (COFs) and covalent triazine structure (CTFs), have requested structure. The polymers can be synthesized in a variety of particle morphologies, including nanosphere, nanosheet, and nanotube, by carefully balancing reaction conditions, monomer structure, and synthesis technique. POPs are analogous to other porous materials like porous carbon and molecular organic framework (MOFs) in that they have surface areas per gram that range from hundreds to thousands of square meters (Figure 19.1).

POPs' advantages of a large specific surface area and structural (morphological, compositional) tailor ability make them favorable for chemical transformation reactions in terms of locally segregated exclusiveness for particular reactions and catalytic sites. POPs with active mononuclear metal centers work more efficiently than molecular metal catalysts as single-site catalysts for synthetic organic chemistry. POPs' large surface areas, regularly spaced structural components (particularly for COFs), and generally uniform pore sizes enable the materials to move through and interact with one another with ease within the channels and networks of POPs during the catalytic process. POPs can achieve high catalytic efficiency per unit mass or specific surface area, serving as more than just heterogeneous substrates for molecular catalysts.

19.2 Synthesis of POP Catalyst

19.2.1 Radical Polymerization

The radical mechanism-assisted polymerization of alkene groups is a popular technique for creating single-site catalysts. Therefore, for alkene functionalization, ligands that exhibit the appropriate selectivity and activity in homogeneous catalysis are frequently the initial choice, followed by creating novel heterogeneous catalysts by polymerization. Free radical chain development polymerization, which is begun by an extreme initiator, is the system for polymerization. While utilizing solvothermal union, the polymerization step is as often as possible did by whirling the monomers and extremist initiator together at room temperature first. *N,N*-dimethylformamide (DMF), tetrahydrofuran (THF), are chlorobenzene is the most often utilized solvents, while different choices incorporate ethyl acetic acid derivation, 1-methyl-2-pyrrolidinone, and blends with benzo trifluoride and water. This kind of polymerization regularly utilizes monomers that are roused by phosphine or nitrogen-based ligands like PPh_3, BINAP, or bipyridine. Rather than most other routinely utilized polymerization strategies, extremist polymerization has the advantage of occurring without the utilization of metal impetuses. As a result, this method eliminates the possibility of metal contamination in the final metal–POP catalyst; hence, no safety measures are necessary. Radical polymerization produces polymers that are frequently quite flexible and produce environments that are practically quasi-homogeneous. The high frequency of benzylic C–H groups in the polymer backbone, despite the robustness of these polymers, may make them vulnerable to oxidation or functionalization in the presence of radical-generating reaction conditions, such as certain oxidations and cross-coupling reactions.

19.2.2 Free Radical Reactions

Another crucial technique for creating single-site catalysts on polymer substrates is Friedel–Crafts alkylation/acylation polymerization. $FeCl_3$ or $AlCl_3$ is used in stoichiometric or excess amounts, together with a linker precursor such as dichloromethane or formaldehyde dimethyl acetal, to perform the polymerizations. Since the Friedel–Crafts reaction can occur directly on aryl moieties without any pre-functionalization, it has the advantage of not requiring the same kind of synthetic alteration on the building blocks as the other approaches described for the creation of POPs. This implies, however, that if the ligand sites contain nucleophilic aryl groups, cross-linking may be unselective and may even change the ligand sites. Despite the low cost of the metals employed for polymerization, the requirement for stoichiometric quantities during polymerization once again highlights the significance of efficient strategies for metal removal from POP.

19.2.3 Sonogashira Coupling

Further polymerization reaction that produces POPs for single-site catalysis involves the coupling of a terminal alkyne and an aryl halide, or Sonogashira coupling. Usually, in this process, THF or DMF and PPh_3 and NEt_3 are present and the process was catalyzed by a catalyst combination of palladium/copper. A helpful characteristic in the synthesis of POPs is the coupling of two distinct functions, which gives complete control over the polymerization process and allows for the formation of exclusively ABAB-composed polymers. This is different from the polymerization processes mentioned above, when the building pieces are incorporated into the polymer backbone at random.

Unfortunately, a precious metal catalyst is needed for the polymerization. A trustworthy procedure is also required because the Sonogashira coupling involves the two transition metals.

19.2.4 Miscellaneous

It has additionally been demonstrated the way that polymerization can happen by build-up, with water discharge going as the driving variable, a method for the polymerization of 1,3,5-triformyl-phloroglucinol and amino-functionalized porphyrin by a keto enamine condensation was used. In contrast to the other methods described above, this sort of polymerization approach does not call for additions in the form of radical initiators, metal catalysts, or stoichiometric concentrations of metal. Although theoretically possible, the condensations can be reversed, and the resulting polymer could not be as strong as those joined by carbon-carbon bonds.

There have also been reports of polymerization using oxidative coupling. In the study by Han et al., two distinct tris(2-phenylpyridine) iridium(III)-complexes were polymerized to create iridium-based photocatalysts. The related polymer, which had a high iridium loading, turned out to be ideal for photocatalysis. In contrast to many of other examples given here, this type of polymerization approach does not call for a linker. BET surface areas for two iridium-based polymers, however, are still between 460 and 480 $m^2\ g^{-1}$ and the polymers have a distinct microporous structure even in the absence of a stiff linker.

Other processes for producing metal-POPs include Suzuki cross-coupling between bromide-functionalized porphyrins and aryl boronic esters, nucleophilic aromatic substitution between fluoride-functionalized porphyrins and linkers containing hydroxy-groups, nickel-catalyzed Ullmann coupling for polymerizing porphyrin moieties, and oxidative Eglinton coupling for producing Ru-bpy-POPs.

19.3 Advantageous Features of POPs

The area of porous materials has greatly advanced, moving from inorganic zeolites to coordination polymers or metal–organic frameworks (MOFs), and in the present, POPs and covalent organic frameworks (COFs). Heterogeneous catalysts must have had both chemical and thermal stability. Although MOFs and COFs are known to have a large surface area and porosity, the majority of these materials frequently exhibit significant moisture and pH sensitivity, which can cause structural collapse. POPs, in contrast, have exceptional stability in practically all common solvents, including water, as well as throughout a wide range of pH values (Kaur et al. 2011). Additionally, POPs, particularly CMPs, are very stable at high temperature and hydrothermal conditions and continue to be a good option for catalysts in challenging circumstances. Zeolites have good thermal and chemical stability as well; however, they are a poor candidate for catalysis due to their difficulty incorporating specified functionality into the skeleton for particular applications. However, the user-friendliness of these materials is further enhanced by the straightforward synthesis of POPs and structural alterations.

19.4 Principle

The electrons at the valence bands (VB) are typically stimulated to the conduction bands (CB) and simultaneously depart the VB, resulting in excited electrons and holes in the CB and VB, respectively, when POPs absorb photons with energy equal

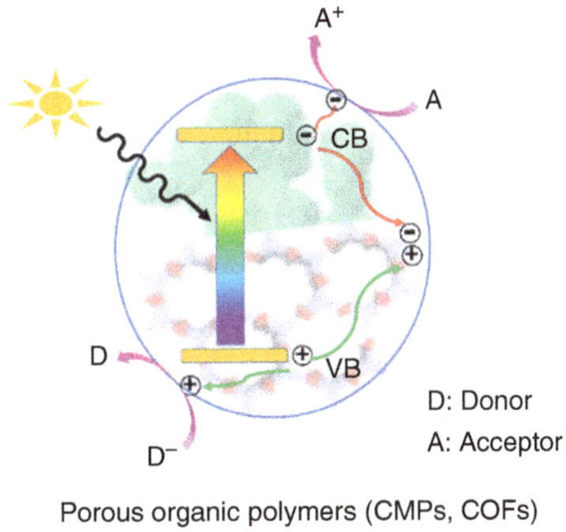

Figure 19.2 Mechanism for porous organic polymers for photocatalysis. Source: Zhang et al. (2020)/Royal Society of Chemistry.

to or higher than their band gaps (Zhang et al. 2020). They then proceed to the surface of the photocatalyst for a redox reaction. Many reactive chemicals, including the superoxide radical (O_2) and singlet oxygen ($^1O_2$), hydroxyl radical (OH), hydrogen peroxide (H_2O_2), photogenerated hole, and electrons, may be involved in these photocatalytic processes. Control experiments are typically performed to pinpoint the likely reactive species and the proper scavenger (Figure 19.2).

19.5 Properties and Functions

The structural, physical, and functional properties of porous organic materials have a significant impact on how they are used. The initial stage in managing the structure–property–function of porous organic materials is accurate synthesis of organic molecules since molecular architectures determine processing strategies, structures, features, and end functions, including self-assembly. Pore size is the primary structural component. The relationship between pore size and surface area and volume ultimately affects end functions including size-based permeability and selectivity for visitor species. Hard templating and other approaches have been used to manipulate macropore diameters, which is a very simple process. Similar to that, soft templates can be used to control mesopore diameters. Through molecular self-assembly and guided assembly, the successful development of block copolymers over the past few decades has made it possible to have exceptional control over the size and arrangement of mesopores at the molecular level. However, the manipulation of the organic polymers' micropore diameters has largely remained some art. Organic frameworks with naturally small pore size distributions are one exception. There have been some accomplishments in manipulating the micropore diameters of porous polymers thanks to mechanisms that have been discovered to control the structures, configuration, and conformation of organic molecules. There is still much potential for improvement when it comes to controlling the spatial arrangement of the micropores, which affects the gas separation and filtration performances. It is frequently insufficient to have perfect control over a single pore size for mass movement with high flux and rapidity. Multiple types of pores must be prepared at various length scales, but more significantly, their interconnectedness must be guaranteed. Reduced structural tortuosity is still a significant obstacle for porous membranes that transport guest molecules, in addition to pore interconnectivity.

Interfacial characteristics, one of the many features, control the wettability to liquids, the binding affinity to molecules, and eventually the molecular permeability and adsorption selectivity. Surface functional groups are primarily responsible for controlling these features. Surface functions must be adjusted in order to maximize the capabilities of porous organic materials in separation, gas storage, and biological applications. For instance, functional group-decorated pore surfaces in gas separation membranes can interact with specific gases in a selective manner. The molecules that make up the porous matrix's building blocks can be designed to have surface functional groups. At some time, the organic matrix starts to incorporate these functional groups naturally. As an alternative, surface functional groups can be added through post-processing, which involves adding moieties to functionalize porous organic materials and change the surface chemistries. Porous organic materials can gain functional groups through chemical changes carried out under precise control of time, temperature, and pressure; however, care must be taken when post-treating porous organic materials using techniques like thermal annealing. Polymers have defined thermal stability and working temperature ranges, and handling them outside of these limits will permanently change or even damage their molecular structures. For precise tunability in this regard, techniques that directly alter the surface functional groups are crucial. Notably, porous organic materials can be used in innovative ways due to functional groups of variable compositions, densities, and spatial distributions. The use of tactics from other communities may help achieve this objective. For instance, spatial confinement guides the production of planar N-dopants (pyridinic- and pyrrolic-nitrogen) over nonplanar quaternary N-dopants to regulate the makeup of the surface functions of graphene electrocatalysts.

19.6 Porous Organic Polymer as the Catalyst

19.6.1 Degradation of Organic Pollutants

In addition to endangering human health, water pollution has an impact on the growth of aquatic creatures. The two main categories of water pollution include organic dyes and heavy metal ions that are chemically stable and non-biodegradable. To date, several techniques, including sorption, ion exchange, photocatalysis, and sorption, have been used to eliminate or degrade organic colors in an aqueous solution (Zhang and Riduan 2012). One of the most efficient, economical, and environmentally friendly processes is adsorption. Numerous porous substances, such as zeolites and activated carbon, have been utilized extensively as sorbents, while CMPs and COFs, which are emerging novel POPs, have just recently been investigated for the removal of pollutants from water. However, these sorbents usually have poor adsorption capacities. It is advantageous that many CMPs and COFs can act as sorbents and photocatalysts for the removal of dyes or the photodetoxification of heavy metal ions for water clean-up. The huge electron-delocalized network structure, effective light absorption, and charge carrier movement are advantageous to them.

Methyl orange (MO), rhodamine B (RhB), methylene blue (MB), and Congo Red (CR) are common organic dyes that are released into the ecosystem by the paper, leather, and food industries. They are exceedingly dangerous to aquatic life and present significant environmental dangers. One or more active species, such as the superoxide radical (O_2), singlet oxygen ($^1O_2$), photogenerated hole (h^+), and OH, may be implicated in the photodegradation of dyes in water when CMPs or COFs are utilized as photocatalysts. The following processes occur when light is irradiated: (i) the electron in the VB is excited to the CB, leaving holes in the VB; (ii) Two active oxygen species, O_2 and $^1O_2$, are produced through singlet electron transfer (SET) and energy transfer (ET) from the excited photocatalyst to molecular oxygen; (iii) O_2 reacts with water to form OH, another active species; (iv) the dyes are degraded by one or more of the reactive oxidative species produced; and (v) the holes spread throughout the entire polymer and further trap pollutants, facilitating the separation of exactions. CMPs showed broad UV/vis absorption, which is necessary for photocatalysis. Because of the fascinating photochemical properties of polyoxometalates (POMs), the hybrid materials' superior photocatalytic activity outperformed that of either of their component parts. This is because CMPs have a high degree of reusability and high porosity. Due to the powerful electrostatic interactions between the positively charged Anderson–POMs and the negatively charged dyes (RhB and MB), these POM–CMPs' photocatalytic activities were also comparable to or even greater than those of benzothiadiazole-based CMP nanoparticles and porous carbon nitride under the same reaction conditions (C_3N_4). Because Bn–Anderson–CMP has a larger BET surface area than Th–Anderson–CMP, it also exhibits more photocatalytic activity and takes less time to achieve the same RhB degradation efficiency. This, when exposed to light, gives additional sites that are receptive for the photocatalytic activity. $^1O_2$ and H_2O_2 were discovered to be the principal active species in the photodegradation of RhB, and POM–CMPs maintained a significant catalytic activity throughout this reaction.

19.6.2 Hydrogen Evolution

As a potential replacement for fossil fuels, hydrogen fuel is anticipated to only create safe, pollution-free water and a vast quantity of heat when it is burned. This makes hydrogen fuel a sort of clean energy. The earth's plentiful water supplies offer a variety of possibilities for the generation of hydrogen.

Catalysts are required to lower the reaction's activation barrier because the hydrogen produced from water undergoes an energy-rising (uphill) reaction that needs 1.23 eV of energy and a high activation barrier to overcome. The majority of the reported photocatalysts for hydrogen evolution are inorganic semiconductors like TiO_2, and noble metal catalysts; but they have drawbacks such as pricey metal

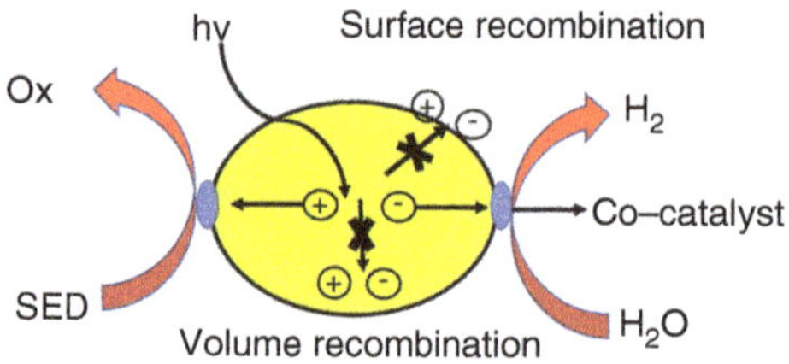

Figure 19.3 Photocatalytic hydrogen evolution process. Source: Zhang et al. (2020)/Royal Society of Chemistry.

components and limited visible-light catalytic activity. Organic semiconductors have received much scientific interest due to their various synthesis methods, customizable electrical structures, extensive visible-light absorption, and high carrier mobility. As a photocatalyst for the creation of hydrogen, C_3N_4 has typically been utilized extensively. However, its structure and energy levels are difficult to modify, and the synthetic temperature is often high. Contrarily, CMPs and COFs are readily made and have tunable structural properties, and as a result, an increasing number of CMPs and COFs have lately been described as effective photocatalysts for hydrogen evolution (Figure 19.3).

Due to their outstanding stability, adaptable chemical structures, and unique electron properties, CMPs have been recognized as promising candidates for hydrogen evolution. Cooper et al. revealed how they used a library of pyrene-based CMPs with different building block ratios to photocatalyze the production of hydrogen. The compositional alterations led to significant changes in their absorbance (420–640 nm), emission spectra (445–588 nm, from blue to red emission), BET surface areas (600–1700 $m^2\ g^{-1}$), and band gaps (1.94–2.95 eV). Under visible-light (4420 nm) irradiation using diethylamine as a sacrificial electron donor, the photocatalytic hydrogen generation rate of CP-CMP1-15 initially increased and then decreased as the optical gap expanded. This can be explained by the fact that electron–hole recombination grew easier and gradually took the lead in restricting the photocatalytic activity of the CMPs as pyrene concentration rose while the energy level for charge transfer declined. COFs with an organized framework, excellent photoelectric properties, and varying energy levels have also been used as photocatalysts for the evolution of hydrogen.

19.6.3 Reduction of Carbon Dioxide

Recently, research has focused more on CO_2 Photocatalytic conversion of CO_2 into fine C_1 compounds such as CO, CH_4, CH_3OH, and HCOOH as well as C_2 or higher-order hydrocarbons as fuels. Recently, CO_2 has been successfully reduced into CO and H_2 using visible-light active POPs. PEosinY-1, -2, and 3 were created through Sonogashira coupling of eosin Y and the corresponding terminal alkynes TEB, DEBP, and TEB, respectively, according to Yu et al findings. The resulting polymers showed significant micro porosity and had a large BET surface area of up to 610 $m^2\ g^{-1}$ and evaluated the VB and CB locations and band gap of the networks, using electrochemical investigation. The VB potentials for PEosinY-1, -2, and -3 were found at 1.70, 1.77, and 1.68 V (vs. NHE), respectively, whereas CB ranged from −0.74 to −0.76 V (vs. NHE). Given that the NHE CO_2/CO reduction potential was 0.48 V and the NHE H_2O/O_2 oxidation potential was 0.82 V, it is said that all three POPs would be able to convert CO_2 to CO when exposed to water.

19.6.4 Water Splitting

Typically, aromatic nitriles or triazine precursors are used to create covalent triazine frameworks (CTFs), which have a high level of thermal and chemical stability (Xu et al. 2018). The electron-withdrawing feature of the nitrogen-rich conjugated unit triazine makes it a potential redox-active site for photocatalytic water splitting (Schwinghammer et al. 2015). These characteristics make CTFs useful as heterogeneous catalysts and have drawn attention for their photocatalytic properties when exposed to visible light (Puthiaraj et al. 2016). CTFs with significant photocatalytic activity are amorphous. The crystallinity makes it simpler to reveal the photocatalytic activity of polymeric semiconductors due to their open-pore arrays (Lan et al. 2018). The potent Bronsted acid catalyst was used to limit carbonization, although at the expense of CTFs' crystallinity. Some novel synthesis methods, such as low-temperature polycondensation methods, have recently been developed (Jiang et al. 2018).

19.6.5 Oxidation

The oxidation of alkenes is one of the most studied reactions involving POPs with an active mononuclear metal core. In 2011, Hupp, Nguyen, and colleagues successfully epoxidized styrene using a catalyst consisting of a manganese porphyrin complex encased in a POP. A similar iron catalyst was likewise found to have less activity after investigation. The related manganese-porphyrin-role POPs involved in the epoxidation of styrene and divinylbenzene were examined in greater detail by the same research team in 2015. Tetrabromo-spirobifluorene and bromide-functionalized porphyrins underwent an Ullmann coupling to produce the catalyst. A 2017 paper on the epoxidation of styrene employing a manganese-porphyrin-POP was written by Gu and colleagues. Although PhIO was used as an oxidant, the polymer's structural makeup was very different. Five repeated runs revealed a very slight decrease in yield, proving that these changes enabled the catalyst to be recycled.

According to Zou, Liu, and colleagues, styrene can be selectively oxidized to phenylacetaldehyde. The catalyst was constructed of manganese porphyrins coupled by an organic backbone, similar to the one Hupp, Nguyen, et al. used for styrene epoxidation. The POP was created using a Suzuki–Miyaura cross-coupling as opposed to a Sonogashira cross-coupling. While the manganese-free porphyrin-POP (>99% selectivity for phenylacetaldehyde) displayed significant background activity, the manganese-metalated POP displayed a higher reaction rate.

19.6.6 Miscellaneous

Numerous additional electrocatalytic processes, such as the oxidation of benzyl alcohol and the nitrate/nitrite reduction of nitrobenzene and N_2, also use POPs

Table 19.1 List of other electrocatalytic processes using POP-based electrocatalysts.

Electrocatalyst	Electrocatalytic reaction	Design approach	Active center	Faradaic efficiency	Potential application	References
Pt-CTF with Ketjen black	Hydrogen oxidation reaction	Pt(II) modification and hybrid system	Pt site	—	Hydrogen fuel cells	Kamai et al. (2016)
B-rich COF on N-doped carbon nanosheets	N_2 reduction	Heteroatom incorporation	B site	45.43% (NH_3)	Agrochemical industry	Liu et al. (2019)
$PMo_{10}V_2$@CTF	Benzyl alcohol oxidation	$[PMo_{10}V_2O_{40}]^{5-}$ incorporation	$\{PMo_{10}V_2\}$ site	96% (benzyl aldehyde)	Pharmaceutical industry	Li et al. (2021)
Pt-CTF with Ketjen black	Nitrite reduction	Pt(II) modification and hybrid system	Pt site	—	Water purification	Kamai et al. (2017)
Cu-CTF with Ketjen black	Nitrate reduction	Cu(II) modification and hybrid system	Cu site	18% (N_2O)	Water purification	Yoshioka et al. (2016)
Ru-CTF with Ketjen black	Benzyl alcohol oxidation	Ru(III) modification and hybrid system	Ru site	75% (PhCOOH)	Pharmaceutical industry	Yamaguchi et al. (2017)
Cu-CTF	Nitrobenzene reduction	Cu(II) modification	CTF and Cu site	65% (aniline)	Hazardous waste treatment	Iwase et al. (2018)

Source: Yang et al. (2022)/Royal Society of Chemistry.

as catalysts. The efficient use of clean energy and the reduction of environmental pollution are both significantly aided by all of these electrocatalytic reactions, which primarily target the agrochemical industry, hydrogen fuel cells, water purification, hazardous waste treatment, and the pharmaceutical industry (Yang et al. 2022). The most important chemical, for example, is ammonia (NH_3), which is used to create chemical fertilizers for use in agriculture. Additionally, it works well as a hydrogen storage and energy transporter. The electrochemical reduction of N_2 to NH_3 is viewed as a potential substitute to the traditional Haber–Bosch process of manufacturing NH_3 due to its mild reaction conditions, obvious energy conservation benefits, and environmental protection advantages. Yan and colleagues demonstrated that a boron-rich COF exhibits very effective activity for N_2 reduction with a faradaic efficiency of 45.43%. The high activity is attributed to the large number of B sites, whose strong Lewis acidity facilitates N_2's (a weak Lewis base) easier adsorption in the COF. According to DFT calculations, the enhanced COF electronic structure greatly decreases the energy barrier for N_2 dissociation on B sites at a sufficient voltage. A collision between B sites and N_2 molecules is more likely due to the concentrated high N_2 concentration, which boosts the reaction activity. Su and colleagues have proposed a potential N_2 reduction catalyst, the conductive Mo-COF, which is composed of 2,3,9,10,16,17,23,24-octaamino-metallophthalocyanine and pyrene-4,5,9,10-tetraone. It has a high faradaic efficiency of about 100% and a low over potential of 0.16 V. The extraordinary activity is principally due to the large positive charge and spin moment of the Mo core, as well as to the suitable adsorption strength for reaction species (Table 19.1).

19.7 Conclusion

A fascinating area of research is the development and use of suitable POPs as possible photocatalysts. Due to their sustainability, accessibility, and processing ability, these POPs can frequently be produced at lower costs than other homogeneous photosensitizers and typical inorganic transition metal-based photocatalysts. This is particularly accurate when taking into account the ease of POP manufacturing and potential for long-term impacts. Furthermore, it has been shown that the structure and morphology of POPs can be dramatically altered by a modest alteration in the synthetic conditions or the application of templating agents. There is a significant likelihood that interesting migration and carrier movement over the skeleton will be feasible for a number of photo-responsive applications because conjugated POPs have sustained conjugation throughout their three-dimensional structure. This distinctive electronic arrangement rationally proves its efficient application in solar energy conversion when combined with simple structural modifications.

References

Chakraborty, J., Nath, I., Song, S. et al. (2019). Porous organic polymer composites as surging catalysts for visible-light-driven chemical transformations and pollutant degradation. *Journal of Photochemistry and Photobiology C: Photochemistry Reviews* 41: 100319. https://doi.org/10.1016/j.jphotochemrev.2019.100319.

Iwase, K., Fujinami, N., Hashimoto, K. et al. (2018). Cooperative electrocatalytic reduction of nitrobenzene to aniline in aqueous solution by copper-modified covalent triazine framework. *Chemistry Letters* 47: 304–307. https://doi.org/10.1246/cl.171117.

Jiang, Q., Sun, L., Bi, J. et al. (2018). MoS_2 quantum dots-modified covalent triazine-based frameworks for enhanced photocatalytic hydrogen evolution. *ChemSusChem* 11: 1108–1113. https://doi.org/10.1002/cssc.201702220.

Kamai, R., Kamiya, K., Hashimoto, K., and Nakanishi, S. (2016). Oxygen-tolerant electrodes with platinum-loaded covalent triazine frameworks for the hydrogen oxidation reaction. *Angewandte Chemie - International Edition* 55: 13184–13188. https://doi.org/10.1002/anie.201607741.

Kamai, R., Nakanishi, S., Hashimoto, K., and Kamiya, K. (2017). Selective electrochemical reduction of nitrogen oxides by covalent triazine frameworks modified with single Pt atoms. *Journal of Electroanalytical Chemistry* 800: 54–59. https://doi.org/10.1016/j.jelechem.2016.09.027.

Kaur, P., Hupp, J.T., and Nguyen, S.T. (2011). Porous organic polymers in catalysis: opportunities and challenges. *ACS Catalysis* 1: 819–835. https://doi.org/10.1021/cs200131g.

Lan, Z.-A., Fang, Y., Zhang, Y., and Wang, X. (2018). Photocatalytic oxygen evolution from functional triazine-based polymers with tunable band structures. *Angewandte Chemie* 130: 479–483. https://doi.org/10.1002/ange.201711155.

Li, Z., Zhang, J., Jing, X. et al. (2021). A polyoxometalate@covalent triazine framework as a robust electrocatalyst for selective benzyl alcohol oxidation coupled with hydrogen production. *Journal of Materials Chemistry A* 9: 6152–6159. https://doi.org/10.1039/d0ta09421h.

Liu, T. and Liu, G. (2020). Porous organic materials offer vast future opportunities. *Nature Communications* 11: 10–12. https://doi.org/10.1038/s41467-020-15911-8.

Liu, S., Wang, M., Qian, T. et al. (2019). Facilitating nitrogen accessibility to boron-rich covalent organic frameworks via electrochemical excitation for efficient nitrogen fixation. *Nature Communications* 10: 1–9. https://doi.org/10.1038/s41467-019-11846-x.

Puthiaraj, P., Lee, Y.R., Zhang, S., and Ahn, W.S. (2016). Triazine-based covalent organic polymers: design, synthesis and applications in heterogeneous catalysis. *Journal of Materials Chemistry A* 4: 16288–16311. https://doi.org/10.1039/c6ta06089g.

Schwinghammer, K., Hug, S., Mesch, M.B. et al. (2015). Phenyl-triazine oligomers for light-driven hydrogen evolution. *Energy & Environmental Science* 8: 3345–3353. https://doi.org/10.1039/c5ee02574e.

Xu, C., Zhang, W., Tang, J. et al. (2018). Porous organic polymers: an emerged platform for photocatalytic water splitting. *Frontiers in Chemistry* 6: 1–12. https://doi.org/10.3389/fchem.2018.00592.

Yamaguchi, S., Kamiya, K., Hashimoto, K., and Nakanishi, S. (2017). Ru atom-modified covalent triazine framework as a robust electrocatalyst for selective alcohol oxidation in aqueous electrolytes. *Chemical Communications* 53: 10437–10440. https://doi.org/10.1039/c7cc05841a.

Yang, D.H., Tao, Y., Ding, X., and Han, B.H. (2022). Porous organic polymers for electrocatalysis. *Chemical Society Reviews* 51: 761–791. https://doi.org/10.1039/d1cs00887k.

Yoshioka, T., Iwase, K., Nakanishi, S. et al. (2016). Electrocatalytic reduction of nitrate to nitrous oxide by a copper-modified covalent triazine framework. *Journal of Physical Chemistry C* 120: 15729–15734. https://doi.org/10.1021/acs.jpcc.5b10962.

Zhang, Y. and Riduan, S.N. (2012). Functional porous organic polymers for heterogeneous catalysis. *Chemical Society Reviews* 41: 2083–2094. https://doi.org/10.1039/c1cs15227k.

Zhang, T., Xing, G., Chen, W., and Chen, L. (2020). Porous organic polymers: a promising platform for efficient photocatalysis. *Materials Chemistry Frontiers* 4: 332–353. https://doi.org/10.1039/c9qm00633h.

20

Developing Trend in Organic Polymer Science

Jyoti Sarwan, Arfat P. Dar, Deeksha, Komal Mittal, and Jagadeesh C. Bose K

University Institute of Biotechnology, Chandigarh University, Mohali, Punjab 140413, India

20.1 Introduction

Today, polymer chemistry and physics serve as a foundation for modern technology, enabling the manufacturing of plastics, rubber, fibers, varnishes, glues, as well as potential composite materials. They also represent a large autonomous field of natural science that goes by the name "polymer science." Energy and resource availability, food security, human health, mobility and infrastructure, and communication will be the top issues for people in the future. Without any doubt, the development of polymers will be essential to overcoming these obstacles. The producer of polymeric parts that are green, energy efficient, high quality, reasonably priced, etc. will be a major trend in the coming millennium. Modern polymer research is very interdisciplinary and integrated, and as a result, it does not easily fit into the conventional natural science structural models. These days, organic polymers are used in many different technical applications, including solar cells, sensors, and LEDs. Typically, natural polymer carriers are weak, safe for microbes to consume, and quickly degraded by them in anaerobic environments. Agar, sponges, chitin, carrageenan, alginate, gelatin, and chitosan are examples of common natural polymer transporters. Synthetic organic polymer gel carriers provide excellent mechanical strength, resistance to microbiological deterioration, and chemical stability. Polyvinyl alcohol, polyacrylamides, light-curing resins, and polyacrylic acid gel are some of the most used synthetic organic polymer gel carriers. There are many ways to measure the importance of a specific discipline. One is to consider its pervasiveness. Proteins, nucleic acids, and polysaccharides are examples of polymers that are the building blocks of life.

By the end of 2050, the global population is projected to exceed seven billion people due to the alarming rate of growth. Huge-scale environmental pressures have been put on our environment as a result of population increase, the growth of massive, densely populated cities and corresponding global industrialization, which could jeopardize food security and environmental sustainability. This has

Organic Polymers in Energy-Environmental Applications, First Edition.
Edited by Ramesh Oraon, Pardeep Singh, Sanchayita Rajkhowa, Sangita Agarwal, and Ravindra Pratap Singh.

caused the biosphere to warm and the accumulation of chemical and biological pollutants, particularly in soils and sediments (Novais et al. 2019).

In order to feed the world's growing population, the primary goal of agricultural growth during the twentieth century was primarily to increase productivity per area of land used for food production. Pesticides and fertilizers were used excessively, which was essentially how it was accomplished, as well as overuse of water and plant life, as well as other natural resources. Although this method significantly increased crop production in the near area, the ability of arable land to produce decreased and natural water resources was exhausted, contaminated with dangerous pesticides and artificial fertilizers, endangering the existence and wellbeing of all life forms on Earth. As a result, the focus of food production in the twenty-first generation is primarily on the ecological use of plant, water, and land resources in agriculture. Modern agriculture's main objective is to safeguard the environment and natural resources while increasing land and water productivity. Food supply and security are agriculture's primary goals, but it also has other crucial roles to play. Emerging trends, however, point to a growing importance in terms of management and conservation of soil and water. Concerning anomalies in the world's climate, throughout human history, droughts of different intensity and duration have occurred in various regions of the world as a result of prolonged absence, insufficiency, or poor distribution of precipitation. Mass starvation (and even famines) and a suspension of economic activity have occurred in many nations as a result of drought events, especially in the developing world, where agriculture is fundamentally and closely related to economies. By concentrating on macromolecular substances that are both naturally occurring and artificially produced, such as proteins, polyacrylates, polyacrylamides, and polysaccharides, discussion of these polymers has aided in the development of polymers for agricultural use. Many commercially available natural/synthetic polymers have been luckily used as a diverse of agricultural applications because of the collaborative efforts of polymer chemists.

One of the main issues affecting human health is environmental pollution, which is caused by the rapid expansion of the chemical, battery, steel, fertilizer, and pesticide industries as well as mining operations. Metal(loid) pollution puts the health of living things at risk. The ecosystem has undergone major changes due to the build-up of metal(loids) due to anthropogenic and natural activity (Luna et al. 2016). Environmentally harmful substances such as cadmium, chromium, lead, and mercury have also been connected to a number of diseases, including birth deformities, damage to internal organs like the kidney and also liver, specific learning difficulties, a range of different ailments. These metals can be found in wastewater and soil along with other metal ions. Because of their toxic, mutagenic, genotoxic, or carcinogenic effects, their presence in the environment puts human health at risk. Metal(loid) are both naturally occurring parts of the earth's crust and man-made products. While some metal(loid) serve as vital micronutrients for living things, they can also be extremely poisonous in high amounts. Because they are not biodegradable, metal(loid) ions stay in the ecosystem. As the surface tension of the water at polluted areas decreases, they become bioavailable for removal (Ayangbenro and Babalola 2017; Das 2019; Fashola et al. 2016; Sriram et al. 2011).

Table 20.1 Name and application of polymers.

S. No.	Name of polymers	Applications	References
1	Teflon	Nonstick cook ware plastic	Plunkett (1938, 1986)
2	Polyvinyl chloride (PVC)	Pipes, tubes	
3	Rubber	Making tires, elastic material	Charles Goodyear (1839)
4	Nylon-6	Fabric	
5	BUNA-S	Synthetic rubber	
6	Tarylene	Fabric	Jewkes et al. (1969)
7	Bakelite	Plastic switches, mugs, buckets	
8	Melamine	Crockery	
9	Lexan	Bulletproof glasses	
10	Antigen	Substances that can stimulate the formation of antibodies	Karl Landsteiner (1920)

20.1.1 What Are Polymers?

Polymers are organic macromolecules which are essentially composed of a number of smaller molecules are called as monomer. Polymer is the Greek word for "many parts." We see polymers everywhere we turn. From polypropylene, a substance that is used all over the world, to the human DNA strand, a biopolymer that naturally occurs.

Natural polymers are those that can be found in animals and plants, while synthetic polymers are those that are created by people. Due to their distinct physical and chemical characteristics, many polymers are employed in daily life (Table 20.1).

20.1.2 Types of Polymers

Polymers can be classified as

1) Organic Polymers
2) Inorganic polymers

20.1.3 What Are Organic Polymers?

Macromolecules known as organic polymers are made up of many monomer units that repeat. Both man-made and natural polymers are essential to daily living. The three primary categories of biopolymers found in living cells are polysaccharides, polypeptides, and polynucleotides. These polymers are created in cells by enzyme-mediated mechanisms. Synthetic polymers are often made from monomers that include two or more functional groups, a multiple bond, or a ring with three to seven members. While the physical characteristics of polymers vary, their chemical

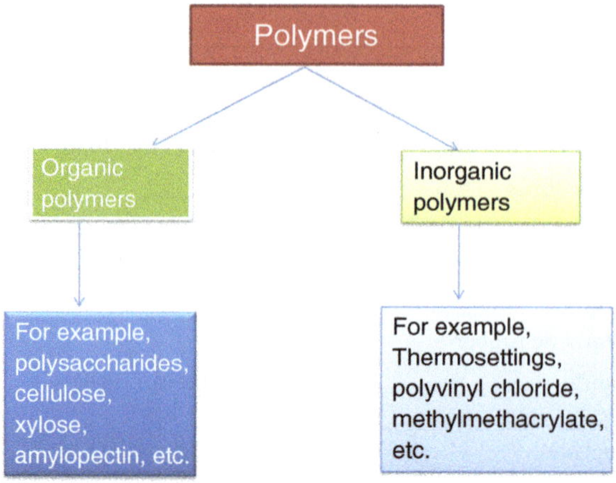

Figure 20.1 An illustration of different types of polymers.

characteristics are obtained from their monomer units. The physical characteristics of polymers determine whether they are thermoplastics, thermosets, elastomers, or fibers.

There are numerous uses for polymers. In the biomedical field, biodegradable polymers are frequently employed for tissue engineering and drug delivery. Wet suits and shoe soles both contain neoprene. Teflon is utilized in non-stick pans, polyvinyl chloride is used in pipes, and other polymers are widely employed in textiles and fabrics. Polyethylene terephthalate, a synthetic polymer, is used to make the well-known polyethylene terephthalate (PET) bottles. Appliances and household gadgets are made with polystyrene resins. The textile and plastics industries both make substantial use of nylon-6. Many consumer goods contain the glass-like polymer poly(methyl methacrylate) (Figure 20.1).

20.1.4 Advantages and Disadvantages of Organic Polymers

Over the past 200 years, the swift rise in the world's population and consumption patterns has had a number of negative repercussions on the environment. When developing a society with more sustainable production/consumption systems, considerations for scenarios like deforestation, water pollution, soil silting, and solid waste accumulation should be taken into account. In relation to plastic wastes (Kaza et al. 2018), they account for 12% of the solid waste produced worldwide, and since 1950, their annual output has increased. They generated more than 6 billion tons of trash between 1950 and 2015 (Geyer et al. 2017).

It is unlikely that synthetic plastics with petrochemical origins can ever totally replace them in the short or even long term, despite the fact that technology and bioproducts, such as bioplastics, can help ameliorate these environmental difficulties. However, some applications for bioplastics may eventually be largely replaced (Tharanathan 2003). For instance, because they account for a sizeable amount of the world's plastics output, biodegradable polymers for packaging and other short-lived use sectors are practicable (Hopewell et al. 2009; Li et al. 2016).

Bioplastics include, for instance, materials made from cellulose and starch, polymers created through microbial fermentation, such as polyhydroxyalkanoates (PHAs), and biomaterials created chemically from renewable raw materials, such as polylactic acid (PLA), bio-polyethylene (BPE), bio-nylons, and bio-polyurethanes. The production of BPE involves polymerizing ethylene from bio-ethanol, the production of bio-nylon involves synthesizing diacids from biomass, and the production of bio-polyurethanes involves adding polyols from plants (Queiroz and Collares-Queiroz 2009). Even oils, though, can serve as a feedstock for the production of bioplastics (Abe et al. 2021).

Bioplastics are made from agro-polymers, which include proteins (casein, zein, gluten, and gelatin) and agro-polymers including starch, cellulose, pectins, and hemicellulose. In order to create the semi-rigid three-dimensional polymeric network that contains the solvent, these natural polymers often entail intra- and intermolecular interactions as well as cross-links (crosslinking) between polymeric elements (Tharanathan 2003; Guilbert et al. 1995).

However, exorbitant costs compared to synthetic polymers made from fossil fuels and worries about functionality are preventing the large-scale manufacture of bioplastics for many uses (Peelman et al. 2013). Numerous biopolymers are used, but they have disadvantages such as high water vapor permeability, oxygen permeability, fragility, low temperature resistance, poor tensile characteristics, sensitivity to degradation, and poor processability (Cabedo et al. 2006; Joshi 2008; Jamshidian et al. 2010).

By producing and using bioplastics rather than synthetic plastics, harmful gas radiation is reduced, raw materials are readily available, and a viable alternative for disposal of solid biomass leftovers is provided. According to the Intergovernmental Panel on Climate Change (IPCC), radiation of greenhouse gases, which contribute to environmental issues like the greenhouse effect, is a growing global concern. By 2050, emissions of greenhouse gas (GHG) must be reduced by 50% in order to prevent a 2 °C rise in the global temperature. One of the ways to combat global warming is through the use of biomaterials like biofuels and bioplastics (de Paula et al. 2018; Naik et al. 2010).

20.2 Applications of Organic Polymers

20.2.1 In Biomedical Field

The therapeutic potential of polymers and polymeric self-assemblies in the treatment of a wide range of illnesses and medical requirements has long been recognized. By encouraging cellular regeneration and medication delivery, safety, efficacy, and absorption into disease regions, polymer-coated biomaterials, tissue engineering scaffolds, wound dressings, and drug delivery devices, for instance, can all improve therapy outcomes. However, as research on this topic has progressed over the past few decades, it has become clear that the physical characteristics of the employed polymers, such as chemistry, charges, composition, and biodegradation, as well as the characteristics of self-assembles, such as size, shape, and surface chemistry, all play a significant role in determining how they behave in biological environments.

Realizing nanostructures that can explore the body, infect and modify cells, or find and mend damaged cells requires an understanding of how the physicochemical qualities of polymeric materials affect these interactions. Our general understanding of how material chemistry and hydrophobicity influence interactions with tissues is still in its infancy, despite intensive research efforts devoted to studying correlations between polymer molecular weights, mechanical properties, self-assembly, solubility, and degradation behaviors. Intriguing publications have shown that the hydrophobicity of drug-delivery systems, implants, or scaffolds for tissue regeneration mediates the interactions of materials with tissues (together with other parameters like chemistry and molecular weights). With an emphasis on the attribute of polymer hydrophobicity, we synthesize studies from the fields of antimicrobial polymers, gene transfer, tissue engineering, and lipid membrane interactions in this perspective. In addition to our focus on experimental investigations, we specifically cover the many computational models of structure–property connections that have been utilized to verify experimental results. In order to combine existing knowledge and suggest prospective future study fields, we aim to identify concepts that are shared by these distinct professions.

20.2.1.1 Applications of Polymers in the COVID-19 Pandemic Crisis for Medical Care

For a very long time, people have recognized the value of chemistry in daily life. There are numerous instances some of which are even emblematic where this discipline of science has significantly improved our quality of life. However, occasionally, even if subtly, concerns regarding its advantages for modern society surface. Environmental or social activism has questioned the value of this science and of all those who practice it without giving it adequate thought. We need to promote a thoughtful and constructive dialogue regarding chemistry's legacy.

There are countless instances where chemists have contributed to the development of science and technology. Since the Curie couple's (Curie 1904) experimentation with radioactive elements until Goodyear's discovery (Korman 2002) of rubber formulas that would later be utilized in the production of tires – via the professors Ziegler and Natta's (Boor 1979) production of catalysts which was used in process called polymerization there has been chemistry. And in each of those instances, it makes a mark, demonstrating that science is a force for good in the world. From this angle, modern society has benefited from polymer chemistry. Many polymers have been created and/or modified using complex methods of organic synthesis, becoming adaptable to different uses and circumstances. The pandemic catastrophe that began in 2020 has demonstrated the importance of polymers in daily living, even beyond consumption (Mitra and Wieswg 2020). By investing in mask producers, the Japanese government was able to guarantee a monthly supply of more than 600 million masks (Park et al. 2020). In addition, businesses whose core competencies do not include producing polyphenylene ether (PPE) have instructed their manufacturers to produce masks and other medical supplies. In this instance, 14 mask production lines with a daily output of 1.7 million parts were built up by the SAIC-GM-Wuling JV of the General Motors Company. A section of their industrial park in Italy has been put up for this special use by many businesses that have committed themselves to the creation of masks (Amendola et al. 2020). The current scenario necessitates concerted action by medical experts, physical therapists, materials engineers, and the industrial sector to reevaluate techniques of preparation of accessible supplies, as also mentioned by Dargaville et al. (2020).

There is no denying that the way of life that society currently leads to situations where pollution and environmental deterioration are visible. It makes sense that the environment is a topic of discussion at conferences and forums where the major international actors (politicians and business leaders) gather. However, the current humanitarian crisis has come to a quality of polymers, up until now, had gone unnoticed: their potential for diversifying their uses and adapting as necessary. As fairly extensively discussed by Czigány and Ronkay (2020), there are numerous examples that show how polymers have aided in the protection of lives on numerous fronts in the struggle against the invisible foe, the coronavirus. The purpose of the current work is to demonstrate how polymers can be employed with already-accessible techniques to create lab test swabs and personal protective utensils.

The COVID-19 outbreak has had an effect on each productive sector, albeit to differing degrees. Due to a lack of PPE, especially masks (Corrêa & Corrêa 2020), which were previously only necessary for health professionals but are now critical for the general people as a major strategy of preventing the virus' transmission, efforts are being made to develop PPE from other materials. Companies and volunteers use unconventional resources to supply the demand for these products, resulting in a movement that provides strong evidence of economic circularity. One thing they had in common was the usage of polymer materials, whether they were employed as virgin materials (raw materials that had not yet been treated) or as recycled or used materials.

20.2.2 In Agriculture Field

Population growth, the emergence of huge, densely populated metropolises, and concurrent global industrialization have all put enormous environmental stresses on our environment, possibly endangering food security and environmental sustainability. Due to this, the biosphere has warmed and contaminants, particularly in soils, have accumulated (Yazdani et al. 2007).

In order to feed the world's burgeoning population, agricultural growth during the twentieth century placed a major emphasis on enhancing productivity per area of land used for the food preparation. The majority of this was done by overusing fertilizers and insecticides, as well as overusing other natural which includes water and plant life (Bhat et al. 2009).

20.2.2.1 Organic Polymers and Their Applications in Agriculture

In agriculture, synthetic polymers are frequently used as building materials for mulches, greenhouses, and other structures that work to distribute water and control its movement while fostering a favorable environment for plant growth. However, the primary requirement for the polymers utilized in these applications is their physical qualities as inert materials, such as transmission, reliability, permeability, or weather ability, rather than being active molecules. Reactive functionalized polymer science and technology is one of the most fascinating fields of polymer chemistry for the creation of superior materials. Polymeric molecules have been widely exploited as reactive materials because of the potential advantages of the numerous active functional groups and the unique characteristics of the molecules. They are used in a wide range of technological fields, such as solid-phase synthesis, physiologically active systems, and others (Akelah 1990).

Polymeric Biocides and Herbicides Recently, technology has advanced in the usage of controlled-release formulations, which are meant to prevent or reduce any potential negative effects associated with the use of biologically active medications. The objectives of this method are to maintain the agent supply, enable automatic delivery of the agent to the target at controlled rates, and maintain the agent concentration within the required ranges for a predetermined period of time. There are two techniques to combine biological agents and polymeric materials function as a rate-controlling device and serve as an agent carrier by physical combination (encapsulation or heterogeneous dispersion) or chemical combination (Akelah 1990).

The polymeric biocide has numerous advantages, some of which include the following: it can be applied at lower concentrations than conventional biocides because it releases the required amount of active agent over a prolonged period of time; the number of applications is decreased due to the prolonged period of activity caused by a single application; and it does not require repeated applications. Its macromolecular design reduces the requirement for extensive dispersion of high biocide levels in the environment, as well as evaporation and degradation losses caused by environmental causes or leaching into the soil or waterways. Additionally, because fewer active substances are required, it reduces the toxicity, duration, and cost of repeated treatments. It lessens phytotoxicity by reducing the biocides' high soil mobility and, as a result, their presence in the food chain. Additionally, regular herbicide use reduces phytotoxicity by lowering the high mobility of the biocides in the soil and, as a result, minimizes their residue in the food chain. This increases the herbicide's selectivity to other crops. How quickly the active group is released from the polymer matrix, the chemical makeup of the polymer backbone, and the rate of breakdown of the connection between the active ingredient and the polymer by chemical, biological, and physical processes are all influenced by the chemical properties of the active agent's structure, the type of active–agent–polymer bond (such as esters, amides, ureas, urethanes, or acetals), the distance between the active agent and the polymer. The commoner content, solubility, degree of cross-linking, stereochemistry, degree of polymerization, and chemical make-up with the active site are all factors that affect the polymer molecule's size, shape, and composition.

Because they are unstable in water life environment, conventional herbicides with less persistence and greater specificity must be used in larger quantities than are actually necessary to control the plant. Due to their lesser persistence, they also require multiple sprays, which are costly and very harmful to farm workers. However, because persistent herbicides are frequently ingested by animals, large-scale applications of them are not desired.

Attempting to draw a connection between the issues and the use of traditional herbicides. In order to make certain working polymers with pesticide moieties as pendant groups, vinyl monomer-type free radical polymerization was used, and the rates of their hydrolysis under various circumstances were investigated. Its hydroxyl group also provides a workable way to covalently bind to a variety of units that can be polymerized.

A number of vinyl monomers with ester linkages that include polychloropene (PCP) have been created (Sun et al. 1994). Styrene and 4-vinylpridine were used to homo- and copolymerize these monomers to give the polymers a hydrophobic and hydrophilic character, respectively (Water, pH = 4; pH = 10, dioxane-water, and dioxane-water) are the four distinct mediums that were used to study the polymer rates of PCP release at 30 °C. When different polymers' release rates are compared, it can be seen that the rates rise as the hydrophilicity rises. This means that copolymers hydrolyze more quickly than homo-polymers, which in turn hydrolyze more quickly than hydrophobic copolymers.

In order to illustrate the polymers' responsiveness by altering the chemical composition of linkage bonds and the chemical properties of the active agent structure, researchers have studied polymer herbicides with active molecules ionically attached to ammonium salt groups (Akelah 1990). PCP has been used in conjunction with conventional herbicides including 2,4-dichlorophenoxyacetic acid, 2-methyl-4-chlorophenoxyacetic acid, and 2,4-dinitro-6-methyl phenol because they feature functional groups for binding to the polymer matrix.

The huge number of controls that were used as a means of transporting herbicides and for the removal of the leftover herbicide pose the main obstacles to the cost-effective use of these polymeric herbicides, which may be hazardous to plants and soil. The idea of transferring pesticides to biodegradable carriers comprising polysaccharide-based agricultural wastes like bark, sawdust, cellulose, and other cellulosic wastes has been put up in various attempts to minimize this issue. The main disadvantages, however, are the challenges involved in attempted chemical modification of such naturally occurring polymers. Large concentrations of these bioactive polymers, which are normally needed for herb control, will therefore unavoidably be utilized. The pendant pesticide hydrolyzes more quickly as a result of these polymers' hydrophilic and cross-linking characteristics. Another factor is their quick biodegradation in soil, which causes the polymeric matrix to be swiftly dissolved and reduces the herbicide's duration of action.

In order to reduce the detrimental effects of using too many inert polymers as carriers and the detrimental effects of using dissolved nitrogen fertilizers, the concept of a controlled emitted herbicide and fertilizer duel application has recently been used. The foundation of this concept is the use of appropriate polymers as carriers, where the residual byproducts of polymer degradation act as fertilizer for the soil and plant growth. For instance, condensation polymerization has been used to produce and polymerize herbicide derivatives of bifunctional chemicals conditions. The idea behind the second attempt is to modify the fundamental characteristics of sandy soil by connecting herbicides to polymeric hydrogels. These polymers' principal purpose is to regulate the pace at which herbicides are delivered, but they can also be very helpful in preventing quick leaching, which will improve the amount of water that sandy soil can hold onto. Because of this, using a combined strategy of water saving and controlled release herbicides can help transform traditional agricultural irrigation, particularly for sandy soil (Sun et al. 1994).

20.2.3 Bioremediation

Bio-surfactant is one of these microbial polymers that has proven useful for complexing metal(loids). They are metabolic chemicals with surface activity and amphiphilicity. This amphiphilic metabolite has both hydrophobic and hydrophilic moieties, and as a result, it induces the formation of micelles or the combination of these moieties at the interfaces of fluids with different polarities, thereby lowering the interfacial surface tension (Mazaheri Assadi and Tabatabaee 2010). Common hydrophilic groups found in bio-surfactants include anions, cations, mono-, di-, or polysaccharides, amino acids, and peptides. Fatty acids, which can be branched, unsaturated, saturated, or hydroxylated, make up the hydrophobic groups (Kitamoto et al. 2009; Singh and Cameotra 2004).

Bio-surfactants are components with varying structural characteristics that are taken out of yeast or bacterial cell membranes. Both high- and low-molecular-weight polymers, which are potent emulsifiers and can dramatically lower surface tension, are present in these secondary metabolites. By emulsifying and dissolving these hydrophobic chemicals or by changing the cell's surface properties, hydrophobic molecules' bioavailability can be enhanced. Their production is regarded as a crucial tactic to help microorganisms recover from the stress caused by pollutants (Bezza and Chirwa 2016). They operate as mediators, increasing the pace at which hydrophobic chemicals are transferred mass-wise by making them easily accessible to bacteria. Additionally, they change the characteristics of cell membranes, increasing microbial adhesion. This mechanism is essential when there are two immiscible liquids and substrate uptake is likely (Dadrasnia and Ismail 2015), therefore assist in overcoming the mass transfer limitations brought on by diffusion (Bezza and Chirwa 2016).

Bio-surfactants, which have different metal(loid) binding capabilities and selectivity, are produced by various microorganisms (Arjoon et al. 2013). Species-specific selectivity in metal(loid) binding (Miller 1995) is essential to remove the target metal(loid) since it competes with other metal(loid)s in solution (loid). Ionic radius and surfactant-metal(loid) complex stability have an impact on selectivity as well (Yuan et al. 2008). *Rhamnolipids* from *Pseudomonas aeruginosa* have been found to have a greater affinity for lead and cadmium than any other soil component to which they are bonded. When compared to macronutrient cations, rhamnolipids have a greater affinity for non-essential and necessary metallic cations, which could be dangerous if present in excess in exposure pathways (Torrens et al. 1998; Singh and Cameotra 2004).

Neutral, cationic, or anionic bio-surfactants are employed as washing agents. In a washing unit, metalloids from the excavated soils or sediments can be dispersed and desorb (Dahrazma and Mulligan 2007). It has been proposed that anionic biosurfactants use electrostatic interactions to bind metal(loid) ions. The apparent metal(loid) solubility is increased as a result of these complexations (Rufino et al. 2012). This biopolymer reduces metal(loid)s, resulting in the precipitation of less soluble metal salts such as phosphate and sulfide, which decreases the bioavailability of the metals (Mosa et al. 2016). By sticking to the surface of sediments, drawing metal(loid) ions from the sediment, and releasing them into the pore water, bio-surfactants produce complexes with metal(loid) ions. When coupled with surfactant micelles, they can therefore remove metal(loid)s that have been adsorbed to the soil (Franzetti et al. 2014; Mazaheri and Tabatabaei 2010).

Actinobacteria, bacteria, and fungi of various types all create biosurfactants. The classifications include particulate and polymeric surfactants, phospholipids, glycolipids, fatty acids, lipopeptides, lipoproteins, and lipopeptides (Sriram et al. 2011; Singh and Cameotra 2004). With their actions and structural traits, they have attained a substantial level of importance. It has been observed that several bacterial species, including *Arthrobacter*, *Bacillus*, and *Pseudomonas*, create surfactants with metal(loid) binding properties.

The reductive process, rather than the oxidative pathway, is used by the majority of microbes to convert metal(loid)s. This is because metal(loid)s in their reduced form are less hazardous. Tolerating exposure to metal(loids) is done by processes such as accumulation, complexation, the efflux system, and reduction mechanisms. Gene clustering on plasmids under stressful conditions can also provide resistance to metal(loid) ions (Gnanamani et al. 2010). Due to their amphiphilic quality, biosurfactants serve as a bridge between the fluid surfaces and lower surface tension. Increased metal(loid) mobilization from unsaturated soils is caused by a decrease in water surface tension, making it feasible to remove metal(loid)s as a result.

Two key ideas have been offered as a way for biosurfactants to remove metal(loid)s from solutions. Through complexation of the free metal ion, biosurfactant-mediated metal sorption is increased. The Le Chatelier principle is used in this procedure, and the activity of the metal(loid) ion in the solution phase decreases to encourage metal desorption (loid). Direct contact with a metal(loid) is another mechanism that operates at the solid–solution interface, where there is less interfacial tension. The biosurfactant can then build up at the solid–solution contact, which facilitates the binding of metal(loid) (Chakraborty and Das 2014; Singh and Cameotra 2004).

The biosurfactant's method for removing metal consists of ion exchange, counter binding, and precipitation (loids). As part of the ion-exchange process, negatively charged anionic biosurfactants interact with positively charged metal ions to form a stronger bond than the one between a metal ion and soil. The cell surface receives a negative charge from the bio surfactant. This mechanism supports the electrostatic interaction with the metal ion. The polar head moieties of surfactant micelles bind the metal ions, making them water soluble. The metal ions are subsequently transferred into solution by the surfactant micelles, making it simpler to recover the metal by flushing, washing, or pumping (Mulligan 2005; Aşçı et al. 2010). The metal(loid) – biosurfactant complex is removed when the soil is flushed with water because of its strong link to the soil matrix (Mulligan 2005). The same charged ions are replaced by cationic biosurfactants, which compete with negatively charged surfaces. The metalloid ion can then be taken out of the complex and put to use in a variety of ways. The biosurfactant mono-rhamnolipid, which is made by *P. aeruginosa*, favors metals like Cd^{2+}, Pb^{2+}, and Zn^{2+} for its carboxyl group. The combination of the metal and biosurfactant poses no threat to bacteria. Biosurfactants are utilized for the precipitation process when cleaning dirt. Due to the biosurfactants' ability to create foam, the biosurfactant–metal complex can be extracted from soil by foaming it with air. The biosurfactant can then be gathered and produced through precipitation by reducing pH.

Dirt and metal(loids) are closely connected. This reduces the bioavailability of bacteria. The activities of the biosurfactant–metal(loid) ion complex in the soil are influenced by the cation-exchange capacity of the soil, its composition and pH, the size of the particles, the kind and timing of contamination, and the geological pattern. Persistent metal(loid) contamination causes soil bonding and metal(loid) ion stabilization, which makes remediation challenging. A frequently used product for removing arsenic and other metals (copper, lead, and zinc) from mine tailings is the rhamnolipid biosurfactant JBR425 (Wang and Mulligan 2009). They found that 0.1% of the surfactant removed 148, 74, 2379, and 259 mg kg^{-1} of Pb, As, Zn, and Cu, respectively, in an alkaline environment. They came to the conclusion that the presence of another metal in an alkaline environment facilitated the formation of the biosurfactant–metal complex, which allows metal bridging via the mobilization of arsenic by the rhamnolipid surfactant (Figure 20.2).

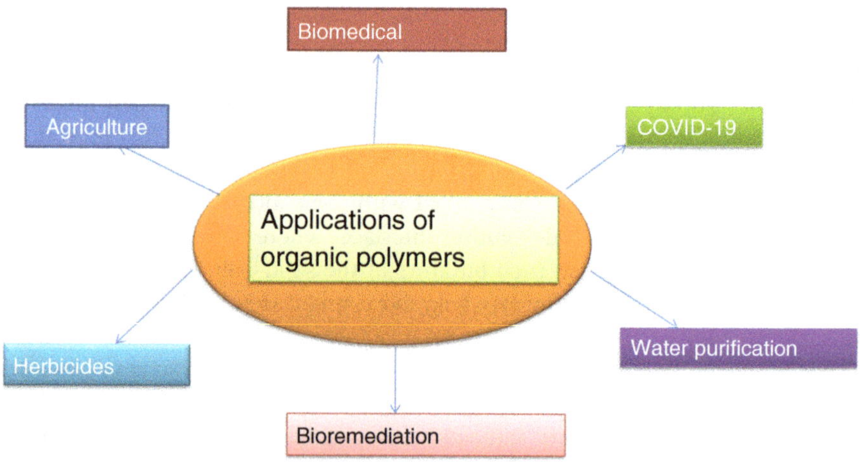

Figure 20.2 Different applications of organic polymers.

20.3 Purification of Drinking Water

Organic polymers, which are long-chained macromolecules made up of multiple repeating monomer units, are often utilized in modern water filtration. Their typical application is in the primary coagulants or coagulant aids phases of the purification of drinking water. In an aqueous suspension, these compounds have a tendency to strongly adsorb the particles on their surface. A faster rate of settling, cheaper costs, better-quality treated water, and sludge dewatering capabilities are just a few benefits of using organic polymers.

The main applications of organic polyelectrolytes in the creation of drinkable water are coagulation, flocculation, and the dewatering of treatment plant sludge. Normal water production procedures are followed by sedimentation and filtration, while the sedimentation stage may be bypassed in the event of water that is only mildly contaminated. In waterways with a lot of algae, flotation can take the place of sedimentation. Due to its extremely high water content and requirement to be further concentrated to cut down on transportation costs, polymers are used in the conditioning of the sludge after it has been acquired from the various separation procedures. The following are a few benefits of using polymers in water treatment:

- Less requirements for coagulant dosage,
- Lesser amount of deposit,
- Lesser addition in the treated water's ionic load
- Reduced aluminum content in water,
- Savings up to 25–30% on costs (Rout et al. 1999; Nozaic et al. 2001).

There are drawbacks, of course; the main ones are environmental factors and higher costs in specific circumstances. Incorrect dosage can cause more sensitivity, and in some cases, turbidity and the removal of natural organics can be less effective.

20.4 Conclusion

As organic polymers more efficiently works than synthetic polymers, the organic polymers can be obtained in the nature from its pure form and showing a natural approach with least of their side effects. The organic polymers are designed by the universe so that they can easily decompose in the environment and do not harm the ecosystem of that particular area. Despite the various applications of the synthetic or composites, organic polymers are still in priority. Not only in the terms of easy decomposition abilities, but also numerous fields rely on the organic polymers. There are number of applications of the organic polymers like bioremediation, COVID-19 crisis, agricultural field, herbicides, weedicides, and waste water treatment. In conclusion this chapter is quite close enough to explain various applications of organic polymers over synthetic polymers.

References

Abe, M.M., Martins, J.R., Sanvezzo, P.B. et al. (2021). Advantages and Disadvantages of Bioplastics Production from Starch and Lignocellulosic Components. *Polymers.* 13 (15): 2484. https://doi.org/10.3390/polym13152484.

Akelah, A. (1990). Applications of functionalized polymers in agriculture. *Journal of Islamic Academy of Sciences* 3 (1): 49–61.

Amendola, L., Saurini, M.T., Di Girolamo, F., and Arduini, F. (2020). A rapid screening method for testing the efficiency of masks in breaking down aerosols. *Microchemical Journal* 157: 104928.

Arjoon, A., Olaniran, A.O., and Pillay, B. (2013). Co-contamination of water with chlorinated hydrocarbons and heavy metals: challenges and current bioremediation strategies. *International Journal of Environmental Science and Technology* 10: 395–412.

Aşçı, Y., Nurbaş, M., and Açıkel, Y.S. (2010). Investigation of sorption/desorption equilibria of heavy metal ions on/from quartz using rhamnolipid biosurfactant. *Journal of Environmental Management* 91 (3): 724–731.

Ayangbenro, A.S. and Babalola, O.O. (2017). A new strategy for heavy metal polluted environments: a review of microbial biosorbents. *International Journal of Environmental Research and Public Health* 14 (1): 94.

Bezza, F.A. and Chirwa, E.M.N. (2016). Biosurfactant-enhanced bioremediation of aged polycyclic aromatic hydrocarbons (PAHs) in creosote contaminated soil. *Chemosphere* 144: 635–644.

Bhat, N.R., Suleiman, M.K., and Abdal, M. (2009). Selection of crops for sustainable utilization of land and water resources in Kuwait. *World Journal of Agricultural Sciences* 5 (2): 201–206.

Boor, J. (1979). *Ziegler-Natta Catalysts and Polymerizations*. Cambridge, MA: Academic Press.

Cabedo, L., Luis Feijoo, J., Pilar Villanueva, M. et al. (2006). Optimization of biodegradable nanocomposites based on aPLA/PCL blends for food packaging applications. In: *Macromolecular Symposia* (Vol. 233, No. 1), 191–197. Weinheim: Wiley-VCH Verlag.

Chakraborty, J. and Das, S. (2014). Biosurfactant-based bioremediation of toxic metals. In: *Microbial Biodegradation and Bioremediation*, 167–201. Elsevier.

Corrêa, H.L. and Corrêa, D.G. (2020). Polymer applications for medical care in the COVID-19 pandemic crisis: will we still speak ill of these materials? *Frontiers in Materials* 7: 283.

Curie, M. (1904). *Recherches sur les Substances Radioactives*. Gauthier-Villars.

Czigány, T. and Ronkay, F. (2020). The coronavirus and plastics. *Express Polymer Letters* 14 (6): 510–511.

Dadrasnia, A. and Ismail, S. (2015). Biosurfactant production by *Bacillus salmalaya* for lubricating oil solubilization and biodegradation. *International Journal of Environmental Research and Public Health* 12 (8): 9848–9863.

Dahrazma, B. and Mulligan, C.N. (2007). Investigation of the removal of heavy metals from sediments using rhamnolipid in a continuous flow configuration. *Chemosphere* 69 (5): 705–711.

Dargaville, T., Spann, K., and Celina, M. (2020). Opinion to address the personal protective equipment shortage in the global community during the COVID-19 outbreak. *Polymer Degradation and Stability* 176: 109162.

Fashola, M.O., Ngole-Jeme, V.M., and Babalola, O.O. (2016). Heavy metal pollution from gold mines: environmental effects and bacterial strategies for resistance. *International Journal of Environmental Research and Public Health* 13 (11): 1047.

Franzetti, A., Gandolfi, I., Fracchia, L. et al. (2014). Biosurfactant use in heavy metal removal from industrial effluents and contaminated sites. *Biosurfactants: Production and Utilization—Processes, Technologies, and Economics* 159: 361.

Geyer, R., Jambeck, J.R., and Law, K.L. (2017). Production, use, and fate of all plastics ever made. *Science Advances* 3 (7): e1700782.

Gnanamani, A., Kavitha, V., Radhakrishnan, N. et al. (2010). Microbial products (biosurfactant and extracellular chromate reductase) of marine microorganism are the potential agents reduce the oxidative stress induced by toxic heavy metals. *Colloids and Surfaces B: Biointerfaces* 79 (2): 334–339.

Goodyear, C. (1839). Specification of a Patent for an Improvement in the Manufacture of Gum Elastic Shoes. Granted to CHARLES GOODYEAR, Roxbury, Norfolk county, Massachusetts July 24th, 1838. *Journal of the Franklin Institute, of the State of Pennsylvania, for the Promotion of the Mechanic Arts; Devoted to Mechanical and Physical Science, Civil Engineering, the Arts and Manufactures, and the Recording of American and Other Patent Inventions (1828–1851)*, 24 (1): 24.

Guilbert, S., Gontard, N., and Cuq, B. (1995). Technology and applications of edible protective films. *Packaging Technology and Science* 8 (6): 339–346.

Hopewell, J., Dvorak, R., and Kosior, E. (2009). Plastics recycling: challenges and opportunities. *Philosophical Transactions of the Royal Society B: Biological Sciences* 364 (1526): 2115–2126.

Jamshidian, M., Tehrany, E.A., Imran, M. et al. (2010). Poly-lactic acid: production, applications, nanocomposites, and release studies. *Comprehensive Reviews in Food Science and Food Safety* 9 (5): 552–571.

Jewkes, J., Sawers, D., Stillerman, R. et al. (1969). 'Terylene' polyester fibre. In: *The Sources of Invention*, 310–312.

Joshi, S. (2008). Can nanotechnology improve the sustainability of biobased products? The case of layered silicate biopolymer nanocomposites. *Journal of Industrial Ecology* 12 (3): 474–489.

Landsteiner, K. and Wiener, A.S. (1920). An agglutinable factor in human blood recognized by immune sera for rhesus blood. *Proceedings of the Society for Experimental Biology and Medicine* 43 (1): 223–223.

Kaza, S., Yao, L., Bhada-Tata, P., and Van Woerden, F. (2018). *What a Waste 2.0: A Global Snapshot of Solid Waste Management to 2050*. World Bank Publications.

Kitamoto, D., Morita, T., Fukuoka, T. et al. (2009). Self-assembling properties of glycolipid biosurfactants and their potential applications. *Current Opinion in Colloid & Interface Science* 14 (5): 315–328.

Korman, R. (2002). *The Goodyear Story: An Inventor's Obsession and the Struggle for a Rubber Monopoly*. Encounter Books.

Li, W.C., Tse, H.F., and Fok, L. (2016). Plastic waste in the marine environment: a review of sources, occurrence and effects. *Science of the Total Environment* 566: 333–349.

Luna, J.M., Rufino, R.D., and Sarubbo, L.A. (2016). Biosurfactant from *Candida sphaerica* UCP0995 exhibiting heavy metal remediation properties. *Process Safety and Environmental Protection* 102: 558–566.

Mazaheri Assadi, M. and Tabatabaee, M.S. (2010). Biosurfactants and their use in upgrading petroleum vacuum distillation residue: a review. *International Journal of Environmental Research* 4 (4): 549–572.

Miller, R.M. (1995). Biosurfactant-facilitated remediation of metal-contaminated soils. *Environmental Health Perspectives* 103 (Suppl. 1): 59–62.

Mitra, K. and Wieswg, M. (2020). *How is Polymer Demand Impacted by the COVID–19 Pandemic*. London: IHS Markit.

Mosa, K.A., Saadoun, I., Kumar, K. et al. (2016). Potential biotechnological strategies for the cleanup of heavy metals and metalloids. *Frontiers in Plant Science* 7: 303.

Mulligan, C.N. (2005). Environmental applications for biosurfactants. *Environmental Pollution* 133 (2): 183–198.

Naik, S.N., Goud, V.V., Rout, P.K., and Dalai, A.K. (2010). Production of first and second generation biofuels: a comprehensive review. *Renewable and Sustainable Energy Reviews* 14 (2): 578–597.

Novais, R.M., Senff, L., Carvalheiras, J. et al. (2019). Sustainable and efficient cork-inorganic polymer composites: An innovative and eco-friendly approach to produce ultra-lightweight and low thermal conductivity materials. *Cement and Concrete Composites* 97: 107–117.

Nozaic, D.J., Freese, S.D., and Thompson, P. (2001). Longterm experience in the use of polymeric coagulants at Umgeni water. *Water Science and Technology: Water Supply* 1 (1): 43–50.

Park, C.Y., Kim, K., and Roth, S. (2020). *Global Shortage of Personal Protective Equipment Amid COVID-19: Supply Chains, Bottlenecks, and Policy Implications* (No. 130). Asian Development Bank.

de Paula, F.C., De Paula, C.B., and Contiero, J. (2018). Prospective biodegradable plastics from biomass conversion processes. In: *Biofuels-State of Development*, 245–272.

Peelman, N., Ragaert, P., De Meulenaer, B. et al. (2013). Application of bioplastics for food packaging. *Trends in Food Science & Technology* 32 (2): 128–141.

Plunkett, R.J. and Evans, W.L. (1938). The Mechanism of Carbohydrate Oxidation. XXIV. The Action of aldehydo-d-Glucose and of aldehydo-d-Galactose in Alkaline Solutions. *Journal of the American Chemical Society* 60 (12): 2847–2852.

Plunkett, R.J. (1986). The history of polytetrafluoroethylene: discovery and development. In *High Performance Polymers: Their Origin and Development: Proceedings of the Symposium on the History of High Performance Polymers at the American Chemical Society Meeting held in New York, April 15–18, 1986* (pp. 261–266). Dordrecht: Springer Netherlands.

Queiroz, A.U. and Collares-Queiroz, F.P. (2009). Innovation and industrial trends in bioplastics. *Journal of Macromolecular Science®, Part C: Polymer Reviews* 49 (2): 65–78.

Rout, D., Verma, R., and Agarwal, S.K. (1999). Polyelectrolyte treatment—an approach for water quality improvement. *Water Science and Technology* 40 (2): 137–141.

Rufino, R.D., Luna, J.M., Campos-Takaki, G.M. et al. (2012). Application of the biosurfactant produced by *Candida lipolytica* in the remediation of heavy metals. *Chemical Engineer* 27: 61–66.

Singh, P. and Cameotra, S.S. (2004). Enhancement of metal bioremediation by use of microbial surfactants. *Biochemical and Biophysical Research Communications* 319 (2): 291–297.

Sriram, M.I., Kalishwaralal, K., Deepak, V. et al. (2011). Biofilm inhibition and antimicrobial action of lipopeptide biosurfactant produced by heavy metal tolerant strain *Bacillus cereus* NK1. *Colloids and Surfaces B: Biointerfaces* 85 (2): 174–181.

Sun, H., Mumby, S.J., Maple, J.R., and Hagler, A.T. (1994). An ab initio CFF93 all-atom force field for polycarbonates. *Journal of the American Chemical Society* 116 (7): 2978–2987.

Tharanathan, R.N. (2003). Biodegradable films and composite coatings: past, present and future. *Trends in Food Science & Technology* 14 (3): 71–78.

Torrens, J.L., Herman, D.C., and Miller-Maier, R.M. (1998). Biosurfactant (rhamnolipid) sorption and the impact on rhamnolipid-facilitated removal of cadmium from various soils under saturated flow conditions. *Environmental Science & Technology* 32 (6): 776–781.

Wang, S. and Mulligan, C.N. (2009). Rhamnolipid biosurfactant-enhanced soil flushing for the removal of arsenic and heavy metals from mine tailings. *Process Biochemistry* 44 (3): 296–301.

Yazdani, F., Allahdadi, I., and Akbari, G.A. (2007). Impact of superabsorbent polymer on yield and growth analysis of soybean (*Glycine max* L.) under drought stress condition. *Pakistan Journal of Biological Sciences: PJBS* 10 (23): 4190–4196.

Yuan, X.Z., Meng, Y.T., Zeng, G.M. et al. (2008). Evaluation of tea-derived biosurfactant on removing heavy metal ions from dilute wastewater by ion flotation. *Colloids and Surfaces A: Physicochemical and Engineering Aspects* 317 (1–3): 256–261.

21

Functionalization and Characterization of Organic Polymers

Ritu Raj[1], *Subhashri Dutta*[2], *Rajan Singh*[1], *Vikash Kumar*[3], *and Gajendra P. Singh*[1]

[1] *Central University of Jharkhand, Department of Metallurgical & Materials Engineering, Manatu, Ranchi 835222, India*
[2] *Central University of Jharkhand, Department of Chemistry, Manatu, Ranchi 835222, India*
[3] *RV Institute of Technology and Management, Department of Electronics and Communication Engineering, Bangalore 560076, Karnataka, India*

21.1 Introduction

Some organic monomers are radically polymerized in an aqueous solution. Some organic monomers' functional polymers are converted via ring-opening polymerization, particularly ethylene oxide, ethylene imine, and oxazolines. Organic fuctionalized materials show better properties in comparison to non-functionalized organic materials like properties as stronger contact, separation, and reactivity. Well-designed organic polymer with an in-chain or chain-end structure group or classes may be changed by popular living anion methods (Storey et al. 2013). Such types of functional organic polymer groups may be controlled by reversible ionic activity, chain extension, branching, or crosslinking reactions. Functionalized organic polymers are polymers that have been modified with functional groups. These functional groups can impart new properties to the polymer, such as optical, electronic, catalytic, or biological properties. Organic polymer functionalization is defined by an organic functional polymer synthesis process consisting of three main categories: direct organic functional polymer synthesis (in situ), post-functionalization (post-polymer modification), and functional group transformation.

21.2 Synthesis of Functionalized Organic Polymer

21.2.1 Direct Polymerization

21.2.1.1 End-Functionalized Polymerization

End-functional polymers have been produced through vinyl moieties with functionalized terminal groups, such as carboxylic acid, hydroxyl, amine, thiol, aldehyde,

Organic Polymers in Energy-Environmental Applications, First Edition.
Edited by Ramesh Oraon, Pardeep Singh, Sanchayita Rajkhowa, Sangita Agarwal, and Ravindra Pratap Singh.

and acetyl, epoxy group (Coessens et al. 2001). The following factors can affect the yield and functionality of a polymer synthesis: Chain-end structure: The structure of the chain ends can affect the reactivity of the polymer and its ability to interact with other molecules, Solvent: The solvent can affect the solubility of the polymer and the rate of the reaction, Temperature: The temperature can affect the rate of the reaction and the morphology of the polymer, Concentration: The concentration of the reactants can affect the rate of the reaction and the uniformity of the polymer, Stoichiometry: The stoichiometry of the reactants can affect the yield of the polymer and the distribution of its molecular weights, Manner and order in which reagents are added: The manner and order in which the reagents are added can affect the rate of the reaction and the morphology of the polymer, Polar additives: Polar additives can affect the solubility of the polymer and the rate of the reaction. By optimizing the method of termination, side reaction can be reduced. In the termination process, many functional groups may be used without protection because the reaction of haloalkyl moieties with living chain polymers is more rapid than functional groups. Halide-functionalized polymers, such as polystyrenes and polyacrylates, can participate in nucleophilic substitution reactions. This is because the halide end-groups are nucleophilic and can be displaced by a nucleophile. Numerous end-functional, clearly defined polymeric materials have been created using this method. One of the benefits is the capability to integrate features that cannot be accommodated during the polymerization procedure. Moreover, when required, it enables the substance to be characterized prior to further functionalization. Through this process, polymers can be created, and functionality can be chosen that is ideal for additional specialized reactions, such as the attachment of materials to biomaterials or the immobilization of materials on surfaces (Matyjaszewski et al. 1997; Coessens et al. 2001; Hoskins and Grayson 2010) (Figure 21.1).

21.2.1.2 In-Chain Functionalized Polymerization

Considering that most functional groupings are harmful to life, so anionic polymerization, using a protective group in the production of polymeric materials, has come to be increasingly more crucial. The protected functionality that monomers can offer can be converted into reactive or associative pendant groups. The preparation of anionic poly(methacrylic acid) serves as a highly significant example (Donini et al. 2002). Since carboxylic acid cannot be created by direct anionic polymerization of the appropriate monomer (acrylic acid), the carboxylic acid groups in the backbone are often covered by tert-butyl esters. Instead, carboxylic acid is synthesized during polymerization hydrolysis (Morton and Fetters 1975). Similar to this, a variety of design polymers and copolymers can be made by sulfonating homo-polymers or styrene copolymers (Yang et al. 2002). A separate protective system takes a different path. A well-designed material is created by an anionically polymerizing 1-alkoxyethyl methacrylate, and polyacids are created by acetyl group hydrolysis (Zhang and Ruckenstein 1998). Functional groups with active hydrogen-containing chemical molecules, such as alcohol and primary and secondary amines, require protecting groups in particular. Dimethyl-tert-butylsilyl groups will also provide protection to alcohols (R-OHs), trimethylsilyl groups

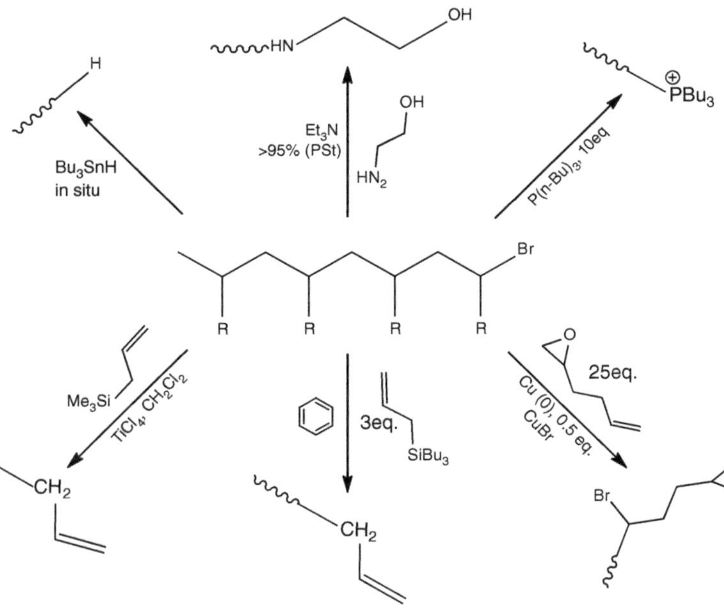

Figure 21.1 Some examples of end-functionalized polymerization.

commonly function as protective layers for primary and secondary amines, namely RNH_2 and R_2NH. Halogen is one example of a reactive polymer site that can be employed as an attachment point for premade side chains or as an initiation site. Smart functionalized polymers have been used to create a variety of functionalized monomers based on methacrylates, styrene, and conjugated dienes with diverse pendant groups for applications ranging from electronics to biomedicine.

21.2.2 Post-polymerization

When direct polymerization is challenging due to the interference of certain functional groups, post-polymerization or post-modification, a method that has been widely employed for many years is preferred (Günay et al. 2013; Gauthier et al. 2009). To create the appropriate functionalized polymer, the polymer may undergo thiol-ene click reaction, sulfonation, chloromethylation, carboxylation, formylation, esterification, alkylation, acylation, or amine functionality. When monomers containing the desired functionality are incompatible with multiple components, one approach is to remove protective groups through the utilization of the polymerization process. The functional groups are protected by polymerization of functional monomers; when the reaction is complete, the desired functionality gets deprotected. Some examples of acidic monomers that undergo protonation and subsequently degrade N-based ligands in transition metal complexes used as ATRP catalysts include (meth)acrylic acid, isomeric vinylbenzoic acids, and unsaturated sulfonic or phosphonic acids. Protected acids, such as t-bu(1,2) for the functionality of acrylic acid or the TMS group, are preferred for polymerizing methacrylic acid (MAA) using atom transfer radical polymerization (ATRP).

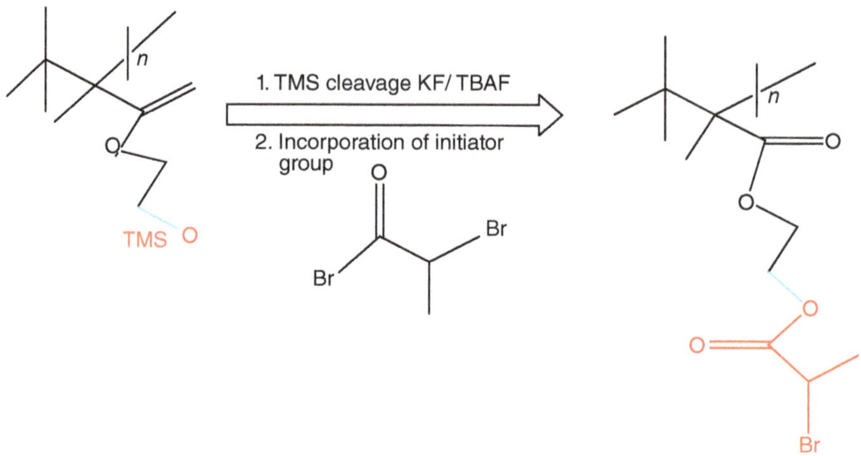

The formation of copolymers with distributed tetrazole units is by the post polymerization reaction of functionality-distributed acid. Tetrazoles are both acidic, resembling carboxylic acids, and coordinating compounds and hence the direct ATRP of vinyl tetrazoles has not been reported. The preparation of well-defined tetrazole-containing polymers can be accomplished using ATRP to (co)polymerize AN followed by a "click" chemistry (Lummerstorfer and Hoffmann 2004; Gissibl et al. 2007) chemical modification of the produced polyAN to yield the desired materials (Tsarevsky et al. 2004).

21.3 Transformation of Functional Group

If functionalization of the polymer via direct and post-polymerization techniques is challenging, functional group transformation may be used. In this process, the desired functional polymer is first modified utilizing a functional group transformation, such as oxidation or reduction, and then the undesirable functional polymer is synthesized using direct or post-functionalization.

It is necessary to establish a network of interconnected channels in order to create a pore structure that can accommodate guest molecules, such as gases. It is often necessary to use stiff building pieces that stop the networks from collapsing and, as a result, fill space more effectively in order for these pores to be stable. At least one of the monomers (the node) needs to have connectivity greater than two in order for a three-dimensional network to form. Network creation is conceivable when these nodes are connected by monomers with at least two functionalities (the struts). Only if the structure is unable to collapse and seal the pores will the network remain permanently porous when the solvent is removed. There are two subclasses of porous polymer networks: crystalline networks and amorphous networks. The diameters of the monomer struts are related to the pore sizes of crystalline networks because they feature well-ordered structures and homogeneous pore sizes. While certain amorphous polymer networks really have relatively tight pore size distributions, amorphous networks often have a statistically chaotic structure and may have a greater pore size range. Similar to crystalline networks, amorphous networks' pore size may likewise be statistically related to the length of the monomer (Jiang et al. 2008). For example, a series of substances known as covalent organic frameworks (COFs) were created either by the self-condensation of boronic acids or by the condensation reaction of boronic acids with diols to create boroxine rings. Crystalline porous polymers are typically created using reversible-bond-forming chemistry.

Characterization of porous polymer networks is more difficult in comparison to small molecules and linear, soluble polymers because of their inherent insolubility. The utilization of conventional polymer analysis techniques such as solution NMR, gel permeation chromatography, or MALDI-TOF mass spectroscopy is not possible. Infrared (IR) spectroscopy can provide qualitative information on the composition of the materials. IR has been used, for example, to demonstrate monomer condensation and the depletion of the early functionalities in the synthesis of COF, where a decrease in the O–H band's strength indicates condensation (Cote et al. 2005). Similar to this, internal alkynes and a decrease in the intensity of terminal alkynes are indicators of efficient Sonogashira–Hagihara coupling for conjugated microporous polymers (CMPs) (Jiang et al. 2007). Similar to carbonyl networks, triazine networks also experience an increase in the triazine ring at 1507 and 352 cm^{-1} and a decrease in the carbonyl bond at 2228 cm^{-1} as a result of the reaction (Kuhn et al. 2008). It is frequently more difficult to evaluate the results of microanalysis of polymer networks. Solid-state NMR (ssNMR) is a crucial analytical tool for insoluble polymer networks because it can be used to confirm the final material's structure and provide quantitative or semi-quantitative structural information (Jiang et al. 2007). Small-angle X-ray scattering (SAXS) is a less popular but very useful technique for determining the porosity in polymer networks (Weber et al. 2008). In comparison to gas sorption techniques, SAXS can determine surfaces with higher areas for a given material. SAXS can measure occluded volume, or isolated pores, which cannot be found by sorption. Positron annihilation lifetime spectroscopy (PALS) and ^{129}Xe-NMR are two other techniques that have been used to calculate the free volume in microporous materials (Yampolskii 2007). The free volume in PIMs has been calculated using PALS (Budd et al. 2008). PALS was used to show that the nitrogen adsorption results were caused by the increased externally accessible surface area of the powder, rather than its internal pore structure. ^{129}Xe-NMR has been used to calculate the pore size based on the chemical shift of confined xenon atoms.

21.3.1 Transforming in Conducting Polymers

Intrinsically conducting polymers (ICP), also referred to as "synthetic metals," are organic polymers that have the mechanical, processibility, electrical, electronic, magnetic, and optical properties of metal while retaining those of a typical polymer. Its characteristics are inherent to the polymer's "doped" state. Contrary to "conducting polymers," which are essentially a physical amalgamation of a nonconductive polymer with a conducting ingredient, such as a metal or carbon powder scattered throughout the material, this class of polymer is entirely distinct from those "conducting polymers." Electrical insulators or semiconductors are both types of conjugated organic polymers. Electronic polymers are polymers whose conductivity can be increased by several orders of magnitude from the semiconductor regime. Electronic polymers have been used in light-emitting diodes since 1990, electronic polymers have gained tremendous scientific and technological significance (MacDiarmid 2001). To demonstrate the several orders of magnitude increases in electrical conductivity that can be obtained by doping, trans-$(CH)_x$ and polyaniline in its emeraldine base form are used. The recent finding of superconductivity in regioregular poly(3-hexylthiophene) has exponentially boosted the conductivity that an electronic polymer is capable of superconductivity of an organic polymer constitutes a historical quantum leap (Schön et al. 2001). The discovery of superconductivity in regioregular poly(3-hexylthiophene) was only observed in a very thin layer of the polymer in a field-effect transistor (FET) arrangement at a very low temperature (~2 K). All conducting polymers, such as poly-(para-phenylene), can be chemically or electrochemically doped to modify the number of electrons attached to the polymer backbone. This process, known as p- and/or n-redox doping, can significantly change the electrical properties of the polymer (Kanatzidis et al. 1990). The instances of various forms of doping that are shown below are only a few. The discovery of p-doping in organic polymers was made by treating trans-$(CH)_x$ with an oxidizing agent, such as iodine. P-doping is a process that involves the partial oxidation of the polymer backbone, which results in an increase in the polymer's conductivity (Chiang et al. 1978).

$$\text{trans-}[\mathrm{CH}]_x + 1.5xy\mathrm{I}_2 \rightarrow \left[\mathrm{CH}^{+y}\,(\mathrm{I}_3)^{y-}\right]_x\ (y \leq 0.07)$$

where trans-$(CH)_x$ was used to discover n doping, or the partial reduction of the backbone pi system of an organic polymer, by treating it with a reducing agent such as liquid sodium amalgam or, preferable, sodium naphthalide (Chiang et al. 1978).

$$\text{trans-}[\mathrm{CH}]_x + (xy)\mathrm{Na}^+ + (\mathrm{Nphth})^- \rightarrow \left[\mathrm{Na}_y{}^+\,(\mathrm{CH})^{-y}\right] + \mathrm{Nphth}^-\ (y \leq 0.1)$$

Polyanilines are a particularly significant class of electronic/conducting polymers. They can be thought of as being derived from a polymer with the general composition of aniline, $C_6H_5NH_2$.

21.4 Characterization of Functional Organic Polymer

An analytical method called chromatography is used to separate molecular mixtures by distributing them among two or more phases. Chromatography comes in a variety of forms, including gel-permeation, gas, and high-performance liquid chromatography (HPLC), as well as ion-exchange and GC (GPC). Liquid chromatography (LC) is a powerful analytical tool that can be used to characterize the molecular weight distribution, chain conformation, chemical content, and microstructure of functionalized organic polymers (Storey et al. 2013). HPLC is an improved method of liquid chromatography that uses high pressure to force the mobile phase through a column packed with chromatographic packing material. Size-exclusion chromatography (SEC) and interaction chromatography (IC) are the two main categories of chromatographic techniques used to analyze macromolecules. These techniques offer high resolution, speed, and sensitivity (Chang et al. 1997). SEC is the most widely used method for analyzing the molecular weight distribution of polymers. The organic polymer molecules have been divided in the SEC method based on the length of the organic polymer chain. Chain size in chemically homogeneous linear organic polymers varies with molecular weight. As a result, SEC may not be regarded as an effective approach for separating organic polymer molecules where the molecular features, such as chemical composition, molecular conformation, and functionality, do not simply correlate with chain size (Chang 2005). GPC, a kind of SEC, is used for analyzing the composition of copolymers as well as for determining molecular-weight distributions. Branched copolymer analysis often calls for a variety of characterization techniques. By using the appropriate detectors and analysis techniques, it is possible to obtain qualitative information about long-chain branching or identify the distribution of a copolymer's constituents (Rowland and Striegel 2012). SEC was used to characterize the thermal degradation of poly(lactic acid) (PLA), poly(butylene adipate-*co*-terephthalate) (PBAT), and their blends before and after reactive extrusion with a chain extension/branching agent containing nine glycidyl methacrylate (GMA) groups (Hoskins and Grayson 2010). In fact, the decrease in molecular weight is a sign of heat breakdown in PLA and PBAT. The measurements of the average molecular weight were conducted to fully understand the chain extension/branching balance that occurs during reactive processing. The SEC of modified PLA and PBAT shows that the polymers have developed extended and branched chains, which has increased their molecular weight. The incorporation of this functionalized epoxy into PLA and PBAT also significantly increases and improves their Mw and thermal stability (Al-Itry et al. 2012). Donini et al. (2002) used SEC to characterize the derivatization procedure of poly(caprolactone) (PCL). They discussed the development and characterization of PCL modified with acrylate and azide functional groups. SEC analysis showed that the molecular weight of the copolyesters decreased slightly when azide replaced chloride. The SEC curves also show that the cycloaddition was successful. The SEC study also showed that the cycloaddition conditions were mild enough to protect the polymer chains from significant degradation. The polydispersity index (PDI) increased from 1.5 to 1.6 after the cycloaddition reaction, while the apparent molar mass (AM) decreased from 20,000 to 18,000 g mol^{-1} (Riva et al. 2007). GPC was used to characterize the star polymers and supramolecular structures that were formed when polystyrene (polySt) and poly(polyethylene glycol) acrylate (polyPEG-A) were used to make three-armed biodegradable star polymers using a trifunctional reversible addition-fragmentation chain-transfer (RAFT) agent. GPC was used to characterize the star polymers and supramolecular structures that were formed when polystyrene (polySt) and poly(polyethylene glycol) acrylate (polyPEG-A) were used to make three-armed biodegradable star polymers using a trifunctional RAFT agent (Setijadi et al. 2009).

In 2011, a study was published that established the critical conditions for liquid chromatography at critical conditions (LCCC) analysis of poly(2-ethyl-2-oxazoline) (PEtOx) (Tsarevsky et al. 2004). Liquid chromatography under critical conditions (LCCC) was successfully used to analyze a number of customized poly(2-ethyl-2-oxazoline) (PEtOx) homopolymers with end groups of different polarity, as well as a block copolymer of PEtOx with a fluorinated block. The results of the study were also supported by combining tandem mass spectrometry (MS/MS) and liquid chromatography at critical conditions (LCCC). This combination of techniques provides a powerful tool for the detailed analysis of complex polymers, such as PEtOx copolymers (Baumgaertel et al. 2011). LCCC was used in combination with other liquid chromatographic techniques to characterize block copolymers of 2-ethylhexyl acrylate (2EHA) and methyl acrylate (MA) produced using RAFT-mediated dispersion polymerization. SEC and HPLC are two different analytical methods used to characterize polymers. These techniques use different separation concepts and are used for various objectives. In the investigation of the reaction products, it was observed that both SEC and HPLC techniques revealed certain characteristics of the polymer samples. SEC analysis displayed a multimodal molar-mass distribution, indicating the presence of different polymer sizes or molecular weights within the sample. This could be attributed to the formation of the predicted block copolymer, as well as the existence of surviving P2EHA and PMA homopolymer fractions that were not fully reacted or incorporated into the block copolymer structure. The multimodal distribution suggests the coexistence of different polymer species with distinct molecular weights.

On the other hand, HPLC analysis provided insights into the chemical composition distribution of the polymer samples. It showed the presence of the predicted block copolymer, which confirms the successful formation of the desired copolymer structure. Additionally, the detection of the surviving P2EHA and PMA homopolymer fractions indicated that the reaction did not go to completion, and some unreacted monomers remained in the mixture. This could be due to various factors such as incomplete reaction kinetics, side reactions, or limitations in the reaction conditions. The combination of SEC and HPLC analyses allows for a comprehensive characterization of the polymer samples. SEC provides information about the molecular weight and size distribution, while HPLC reveals the chemical composition distribution. Together, these techniques enable a deeper understanding of the reaction products, including the formation of the predicted block copolymer and the presence of any unreacted homopolymer fractions. LCCC measurements were instrumental in monitoring the concentration of unreacted macro-RAFT agent during the initial hours of polymerization. Additionally, LCCC tests allowed for the determination of the relative molar mass of the PMA blocks in the copolymers, providing valuable information about the structural characteristics of the copolymer system (Raust et al. 2010).

21.4.1 Nuclear Magnetic Resonance Spectroscopy

21.4.1.1 NMR-^{1}H

Nuclear magnetic resonance spectroscopy can be used to determine the proton environment in the molecule. The molecule's proton exhibits a distinct chemical shift *d*. It is typically stated by frequency in parts per million (ppm). Based on the following equation, it is calculated:

$$\delta = \frac{\Delta\nu \times 10^6}{\text{oscillator frequency (cps)}} \tag{21.1}$$

where the oscillator frequency is a defining property of the instrument and is the difference in frequencies of absorption between the sample and the reference in cps. A combination of chemical shift information and spin–spin splitting, resulting from proton–proton contact, can be used to identify chemical structures. In order to analyze stereochemistry of polymer and monomer sequencing, the NMR is helpful. The ^{1}HNMR spectra of a wide variety of substances with known, relatively structures have been measured in order to determine the chemical shifts of ^{1}H-nuclei in varied chemical environments. The chemical shifts for several popular forms of ^{1}H nuclei are provided in Table 21.1.

Table 21.1 Organic functional groups and their 1H NMR Chemical Shift Values (ppm).

S. No.	Types	*d* (ppm)
1	C–CH_3	0.85–0.95
2	C–CH_2–C	1.20–1.35
3	C–CH(C)–C	1.40–1.65
4	CH_3–C=C	1.6–1.9
5	CH_3–Ar	2.2–2.5
6	CH_3–C=O	2.1–2.6
7	CH_3–N	2.1–3.0
8	CH_3–O–	3.5–3.8
9	–CH_2–CI	3.6–3.8
10	–CH–CI_2	5.8–5.9
11	–CH_2–F	4.3–4.4
12	–CH_2–Br	3.4–3.6
13	–CH_2–I	3.1–3.3
14	CH_2=C	4.6–5.0
15	Ar–H	6.6–8.0
16	H–C≡C–H	2.4–2.7
17	–COOH	10–13
18	–COH	9.5–9.7
19	R–OH	0.5–5.5
20	Ar–OH	4.8

Currently, polymer scientists have access to very-high-resolution NMR tools to learn more about the stereochemistry and monomer sequencing of polymers. For instance, the potential configurations for each repeat unit in pure isotactic poly(methyl methacrylate) are depicted in Figure 21.2. As illustrated in Figure 21.3, 500-MHz NMR analysis of a poly(methyl methacrylate) sample revealed that it is primarily (95%) an isotactic isomer.

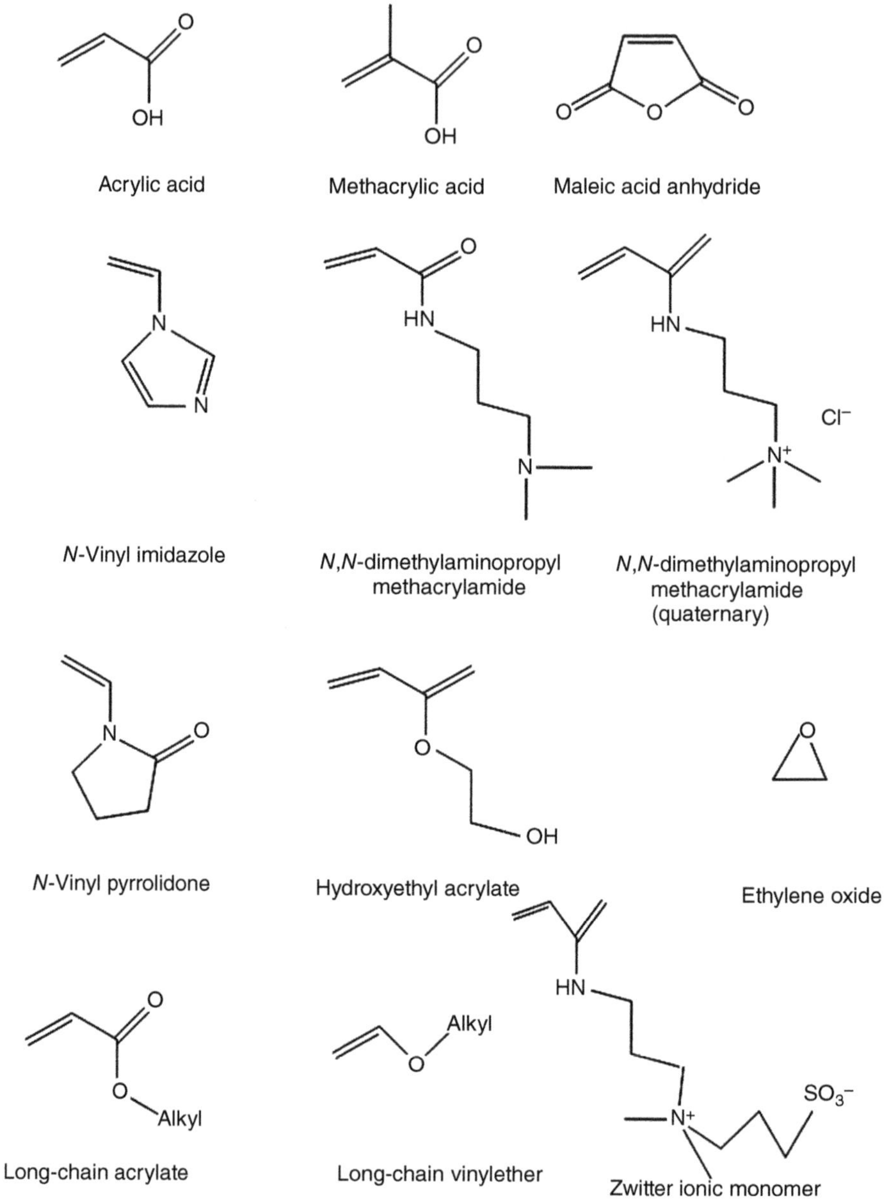

Figure 21.2 Some examples of monomers of functional polymers.

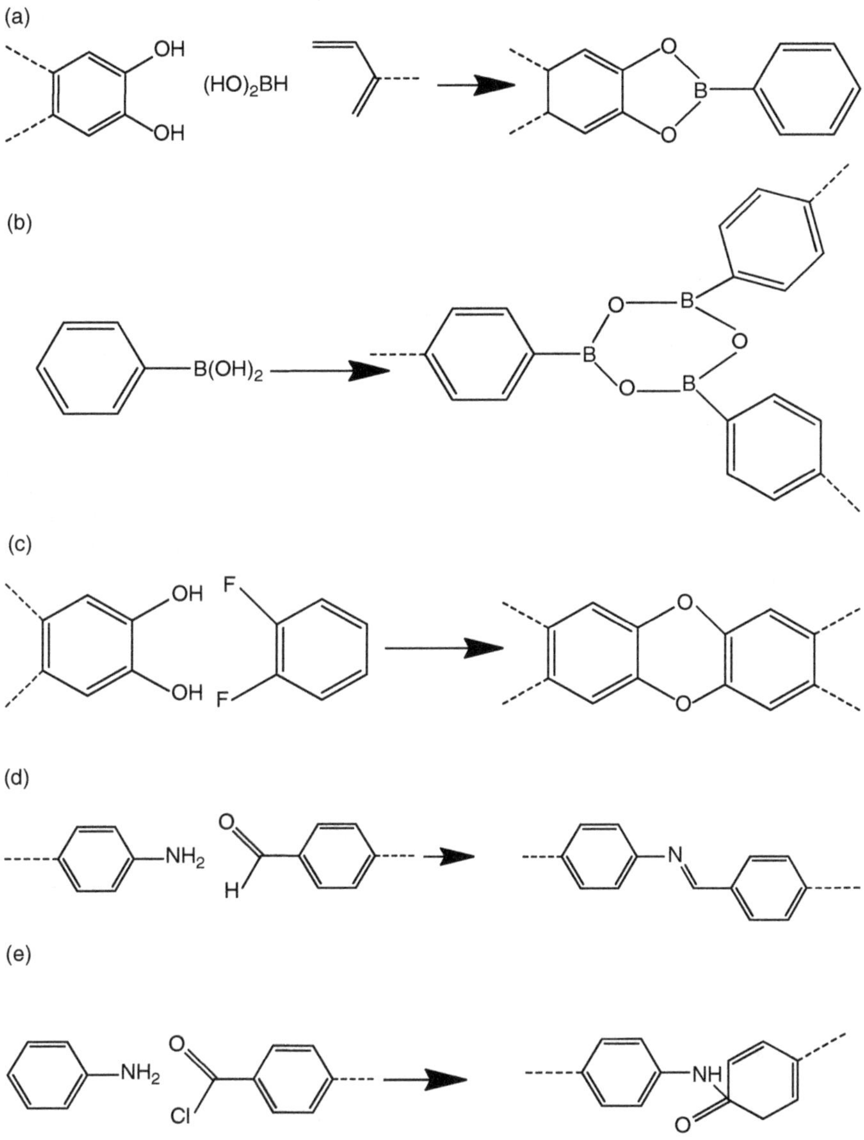

Figure 21.3 Example of condensation reactions used to form nanoporous polymer networks. (a) BO_2C_2 ring formation, (b) boroxine ring formation, (c) dioxane ring formation, (d) imine reaction, (e) amide reaction.

The sample's methyl protons can be resolved to the pentad level and the methylene protons to the hexad level. Annotations like erythro and threo are used to denote the respective placements. Because of the ester group's deshielding effect, the erythro resonance of the methylene protons is farther downfield than that of the ester groups. With the exception of the minuscule pentad at roughly about 1.16 ppm, the spectra at the minimal sensitivity (1×) are almost entirely that for an isotactic polymer. Higher gain (5 and 100×) makes it easier to see departures from purely isotactic behavior.

21.4.1.2 NMR-^{13}C

High resolution of proton-decoupled ^{13}C spectra of polypropylene with changing tactility is shown in Figure 21.4. Pentad sequences may once more be associated with fine structure in the spectra. Crosspolarization, magic angle spinning, and proton decoupling together produce solids spectra that are close to those obtained with solutions in terms of resolution. As an illustration, the solid-state ^{13}C-NMR spectrum of polycarbonate including the carbonyl and ring carbon resonances is shown in Figure 21.5 presence and in absence of crosspolarization magic angle spinning.

21.4.2 Characterization and Physical Structure

The characterization of functionalized polymers requires the use of spectroscopic methods that provide accurate details about the molecular structure of various polymers. IR spectroscopy and NMR have been used, and the reported results show their potential for analyzing the functionalization of polymers. For studying the chemical

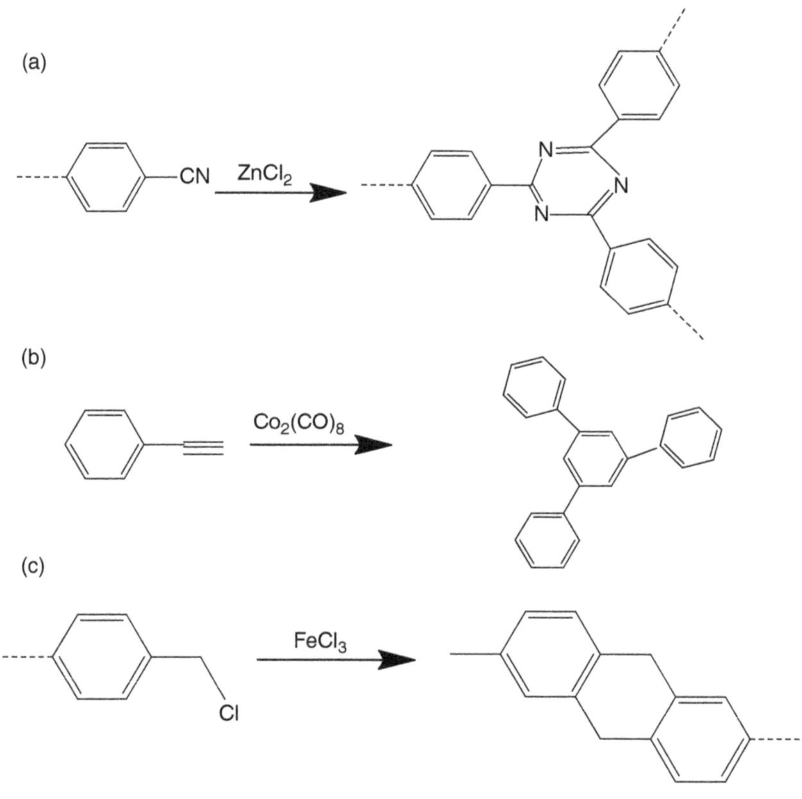

Figure 21.4 Metal-catalyzed reactions used to form nanoporous polymer networks. (a,b) Cyclotrimerization. (c) Friedel–Craft acylation.

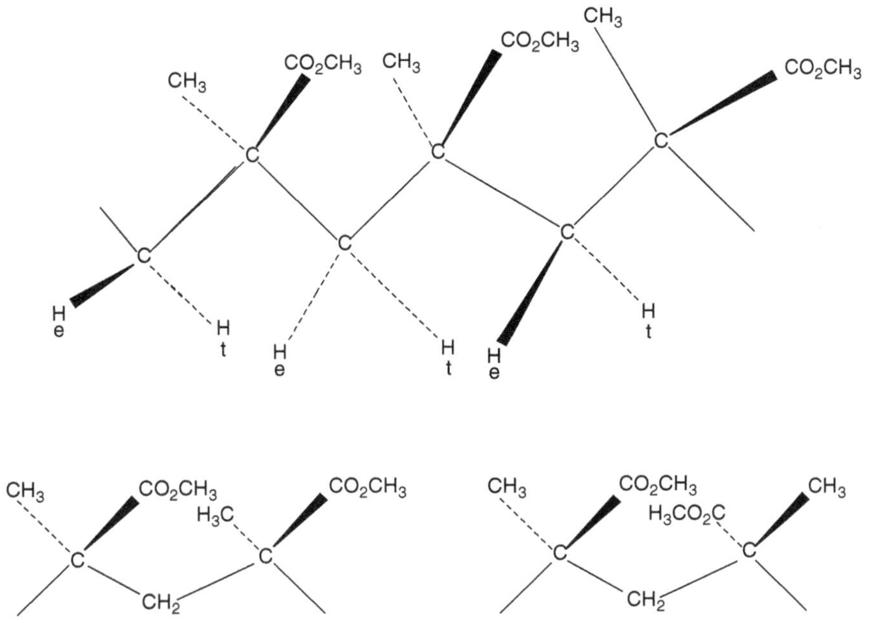

Figure 21.5 Poly(methyl methacrylate) (PMMA) can exhibit different isotactic repeating unit arrangements.

structures and dynamics of various macromolecular species, NMR is a potent spectrometric tool. It gives information on the polymers' stability and degradability, including their levels of acetylation, amination, and sulfonation (Mulijani et al. 2014; Nuñez and Hickner 2013; Parrondo et al. 2014; Sankir et al. 2004). Because 1H and 13C NMR are sensitive nuclei of the elements in organic materials, they are the most often investigated nuclei by NMR. NMR has the ability to characterize a broad range of functional groups that cannot be acquired by any other method, in addition to providing absolute identification of certain isomers in simple mixes. Quantitative data is obtained through NMR. The various functions present in the sample, as well as their contents, can be identified by analyzing a 13C or 1H spectrum. The structures of several functionalized poly(-caprolactone) copolymers, including poly(-caprolactone-*co*-dimethyl vinylphosphonate) (P(MDOVPE)) and poly(-caprolactone-*co*-vinylphosphonic acid) (P(MDOVPA)), were determined by 1H and 13C NMR in different study. In this study, two vinyl monomers, dimethyl vinylphosphonate (VPE) and vinylphosphonic acid (VPA), are copolymerized with MDO (2-methylene-1,3-dioxepane), through introduction of phosphonic acid and dimethylphosphonate groups in order to create versatile polymeric inorganic hybrid materials. Copolymers with a differing ester percentage or of functional groups were synthesized when the feed ratio of the MDO co-monomers and the vinyl monomer was changed accordingly. According to this study's NMR findings, hydrogen was transferred during copolymerization. The copolymers also developed a branching ester structure. The NMR spectra of P(MDOVPE), which have a high ester content, which have a high 1H and 13C NMR spectra, support this.

The distinctive methyl peaks coming from hydrogen transfer in the ester unit (0.94 ppm, Figure 17.5a, 13.9 ppm, Figure 17.5b) were seen when these spectra were contrasted with the 1H and 13C NMR spectra of P(MDO) (Jin and Gonsalves 1998). The chemical makeup of the surfaces of the substrate in presence of and absence of vapor-phase grafting of the most commonly used as biodegradable biomaterials, poly(L-lactide) (PLLA), poly(lactideco-glycolide) (PLGA), poly(-caprolactone) (PCL), and poly(trimethylene carbonate) (PTMC) was studied using FTIR analysis with an ATR (attenuated total reflectance) probe. The findings are for the original substrates, PLLA, PCL, and PLGA, respectively, and the characteristic ester CQO bands are located at 1747, 1718, and 1746 cm^{-1}, while PTMC exhibits a carbonate CQO at 1734 cm^{-1}. New peaks develop when VP is grafted covalently onto the surfaces.

21.4.3 FTIR

The varied IR absorption of chemical bonds by polymers (vibrational transition) in response to IR radiation produces the infrared spectra of polymers (Scheirs et al. 1991). The vibrational transition's magnitude can be expressed by

$$v = \frac{1}{2\pi C}\sqrt{\frac{k}{M_xM_y/(M_x + M_y)}} \tag{21.2}$$

where k (force constant) in dynes cm^{-1}; single bond (=5×10^5 dynes cm^{-1}), double bond (=10×10^5 dynes cm^{-1}), and triple bond (=15×10^5 dynes cm^{-1}); m is the stretching frequency in cm^{-1}; M_x and M_y are the masses of two atoms involved in stretching; C is the velocity of light, 2.998×10^{10} cm s^{-1}; and the number of waves cm^{-1} is used to measure infrared frequency, which is often represented in wavenumber units.

Approximately 700–5000 cm^{-1} is the scanning range of common instruments. This frequency range is associated with energies between 2 and 12 kcal mol^{-1}. The provided energy is more than enough to cause vibrations of bond (motions like bond bending or stretching), but it is remarkably less than that would be necessary to dissolve the bonds. Figure 21.6 provides an illustration of these actions.

Certain types of organic molecule bonds typically stretch within specific, very small frequency ranges, which is highly helpful to ascertain the chemical structure of the molecule. The stretching frequency ranges for several bonds that are frequently observed in organic compounds are provided in Table 21.2. Most often, the distinctive IR absorption frequency of unknown polymers can be used to identify their chemical structures. However, except they can be compared to experimental data or from the data of IR spectra of their respective monomers, it is impossible to determine their precise chemical structures only from IR spectra. The p-substituting linkage between the aromatic structure of monomers and the repeating unit model is where the IR spectra of polyimide and its repeating unit compound differ.

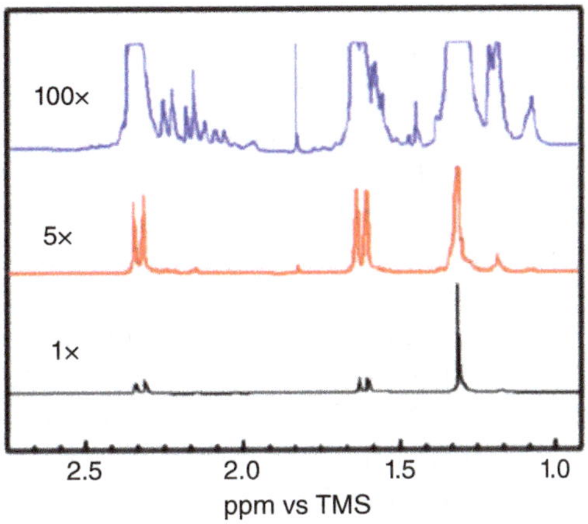

Figure 21.6 500-MHz 1H nuclear magnetic resonance spectrum of isotactic poly(methyl methacrylate). Source: Stevens (1990)/Oxford University Press.

Table 21.2 Organic functional groups and their corresponding FTIR Bands (cm^{-1}).

Bond type	Class compound	Frequency range (cm^{-1})
Single bond	Alkanes	2850–3000
	Alkenes and aromatic	3030–3140
	Alkynes	3300
	Alcohols and phenols	3500–3700 (free)
		3200–3500 (hydrogen bonded)
	Carboxylic acids	2500–3000
	Amines	3200–3600
	Thiols	2550–2600
Double bonds	Alkenes	1600–1680
	Amines, oximes	1500–1680
	Aldehydes, ketones, esters, acids	1650–1780
Triple bond	Alkynes	2100–2260
	Nitriles	2200–2400
Substitute aromatics	Ortho	750
	Para	830
Halide	C-F	2962–2853
	C-Cl	800–600
	C-Br	600–500

21.4.4 Raman Spectroscopy

Raman is derived from molecular vibrational transitions, just like IR spectroscopy. Light is dispersed when visible light strikes molecules. The various modes of vibration in the scattering molecules affect the frequency of the light that is scattered.

The Raman effect is used to describe this. Raman effect reacts to the symmetric modes of vibration, whereas IR spectra reveal un-symmetric bond stretching and bending modes of vibration. A molecule's polar groups produce the strongest IR signals, whereas its nonpolar groups produce the strongest Raman signals.

As a result, IR and Raman spectroscopy complement one another. The interpretation of structural information is enhanced when the same compound's paired IR and Raman spectra are used (Kuptsov and Zhizhin 1998).

In case of C—C bonds, symmetrical stretching is what makes Raman most sensitive. By contrasting the spectra of polymer chains with those of bulky-chained "model" alkanes, it is possible to investigate the structure of conformation in polymer chains. Raman may be used to analyze polymer deformations, sulfur crosslinks in rubber, and cis–trans isomerism in elastomers since stereoisomers of polymers are created by rotation around single bonds (Figure 21.7).

Raman spectra is particularly useful in conformational investigations of several biopolymers in an aqueous medium since Raman scattering by water is low in comparison to water's high IR absorption. The comparison of trans-poly pentenamer's IR and Raman is shown in Figure 21.8. Caused by the symmetrical structure of alkane (–C–C– at 2900 cm^{-1}) and alkene (–C=C– at 3000 cm^{-1}), the Raman exhibits more features than the IR. Figure 21.9 compares the poly(3-hexyl thiophene) (P3HTIR)'s and Raman spectra. While C–H stretching prevalent in the IR band spectra, symmetrical C–C stretching dominates the Raman spectrum. According to Figure 21.10, the Grignard metathesis reaction produces P3HT (as shown in Figure 21.13). The yield might exceed 70%.

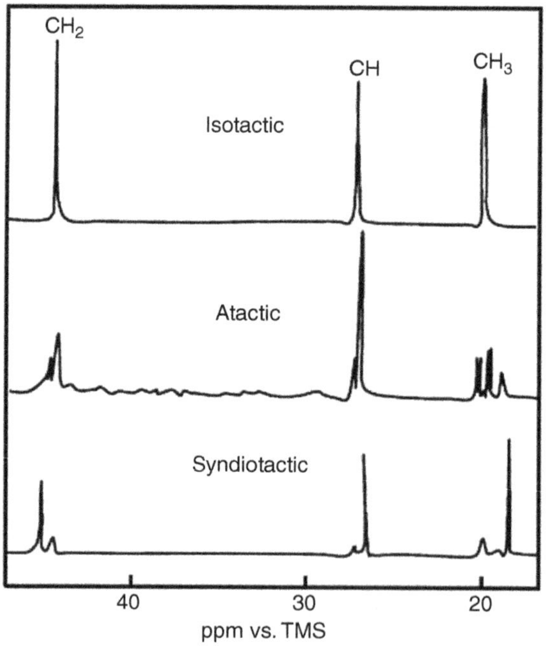

Figure 21.7 Here is the spectrum of polypropylene at 24-MHz 13C nuclear magnetic resonance (3.5% weight/volume in 1,2,4-trichlorobenzene at 135 °C).

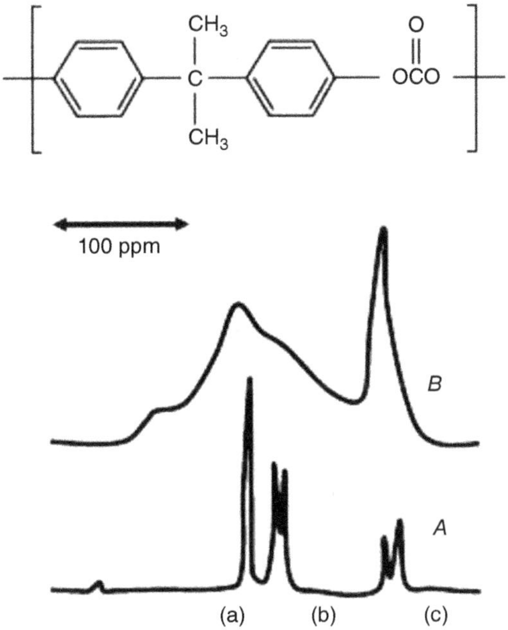

Figure 21.8 Here are the solid-state crosspolarization magic angle spinning 13C nuclear magnetic resonance spectra of polycarbonate, (a) with and (b) without. Carbonyl carbon, substituted ring carbons, and unsubstituted ring carbons were assigned as the peaks, respectively. Source: Stevens (1990)/Oxford University Press.

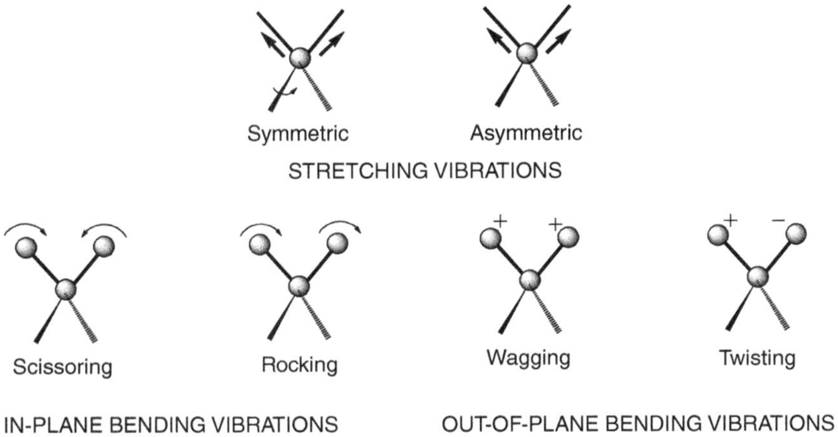

Figure 21.9 Atomic group vibrations are the vibrations of atoms within a molecule. They can be classified into three types: stretching vibrations, bending vibrations, and torsional vibrations.

21.4.5 UV-Visible Spectroscopy

When matter undergoes the $n \rightarrow \pi^*$ and $\pi \rightarrow \pi^*$ transition upon light irradiation, UV-Vis (ultraviolet-visible) spectroscopy (Dyer 1978) is used to discover the chromophores qualitatively as well as quantitatively. It has proved helpful in finding contaminants in polymers, similar to residual monomer, antioxidants, inhibitors, and so on, because to its sensitivity (10-5 mol). For instance, the styrene at λ_{max} = 292 nm can be used to quantitatively measure the amount of styrene

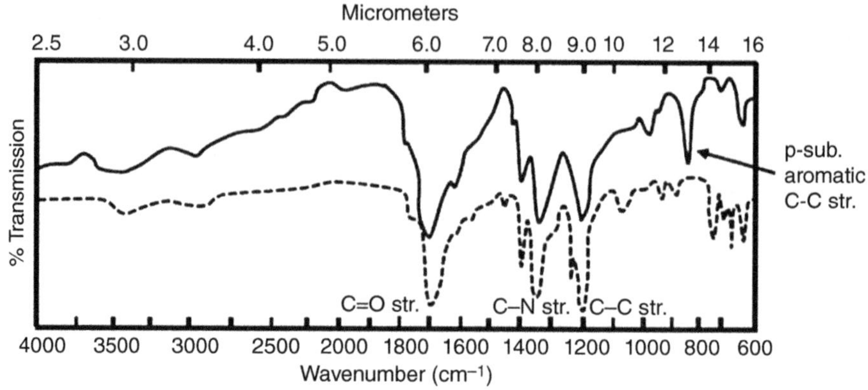

Figure 21.10 Shows the infrared spectra of polyimide (–) and a model molecule (- - - -). Source: Stevens (1990)/Oxford University Press.

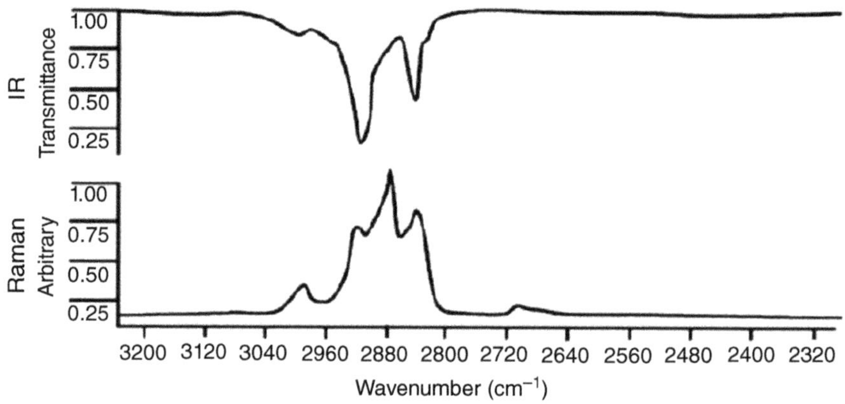

Figure 21.11 Comparative research of trans-poly pentenamer IR and Raman spectroscopy.

monomer in polystyrene. After styrene is polymerized, the peak at $\lambda_{max} = 292$ nm vanishes and is replaced by the λ_{max} of benzene for polymer at 203 and 254 nm. A typical organic compounds UV absorption spectrum is shown in Figure 21.14. Contrary to infrared spectra (IR), UV-Vis spectra are considerably broader and typically only exhibit a few peaks. The wavelengths of the maxima are reported as the peaks. The conjugated, and unsaturated ketone spectrum is depicted in Figure 21.14 absorbs strongly at $\lambda_{max} = 232$ nm and just little at $\lambda_{max} = 330$ nm. While the longer wavelength, weaker intensity band corresponds to a transition of the nonbonding electrons on the carbonyl oxygen atom, the shorter wavelength band correlates to a transition of the π^* electrons.

A quantitative expression for an absorption band's intensity is possible. The specific chemical structure of the band and the quantity of absorbing molecules in the path of light both affect band intensity. Beer's law provides the following as an example of absorbance, which is described as the log of the ratio of light intensities entering as well as leaving the sample.

$$A = \varepsilon c l \tag{21.3}$$

Molar absorptivity location (sometimes called the extinction coefficient), l is the length in centimeters, and c is the concentration of the solution in moles per liter. Any peak's value in a compound's spectrum is a property (constant) of that specific chemical structure. As an instance, the peak values for the given unsaturated ketone in Figure 21.11 spectra are $\lambda_{max} = 233$ nm ($\varepsilon = 12{,}600$) and $\lambda_{max} = 331$ nm ($\varepsilon = 78$). The effect of the conjugate length on the wavelength and extinction coefficient is presented in Figure 21.15.

The most frequent method for detecting conjugation uses UV-Vis spectra. In extensive, molecules with one double bond or none does not absorb in the UV to visible spectrum (~200–800 nm). However, the conjugated systems absorb here, and as shown in Figure 21.12, the longer the wavelength of maximum absorption, higher the conjugation.

21.4.6 Thermal Analysis

Research on polymers and the creation of new polymeric systems require the use of thermal analysis techniques. Accurate thermoanalytical measurements are

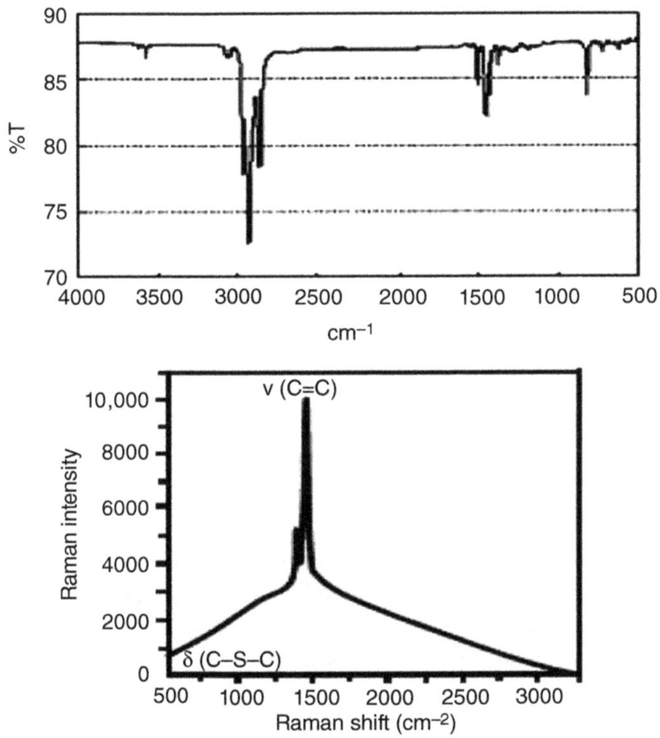

Figure 21.12 Shows the IR spectrum (top) and Raman spectrum (bottom) of P3HT.

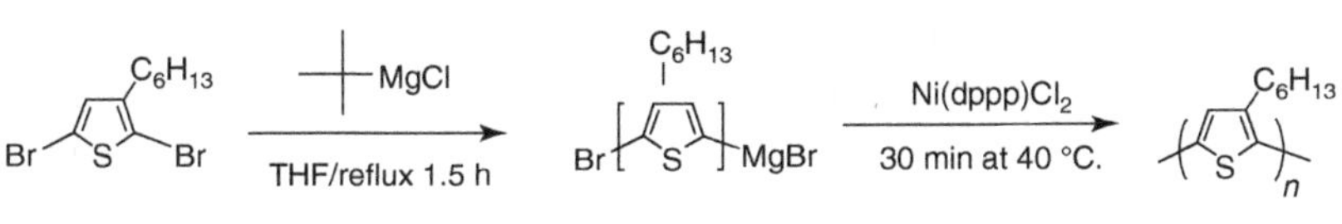

Figure 21.13 Synthesis of P3HT using Grignard metathesis reaction.

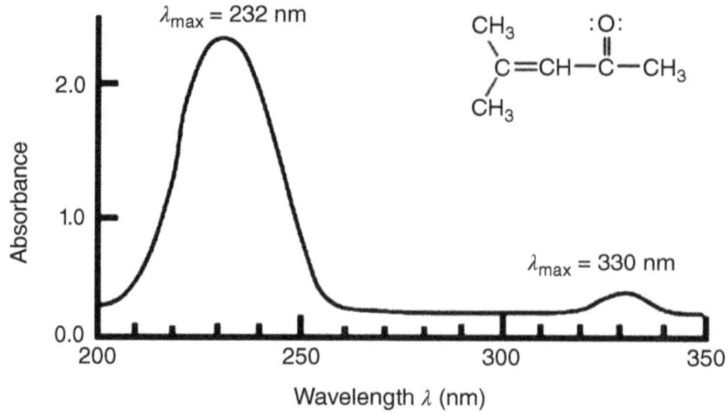

Figure 21.14 The absorption spectrum of 4-methyl-3-penten-2-one. Source: Su and Su (2013)/Springer Nature.

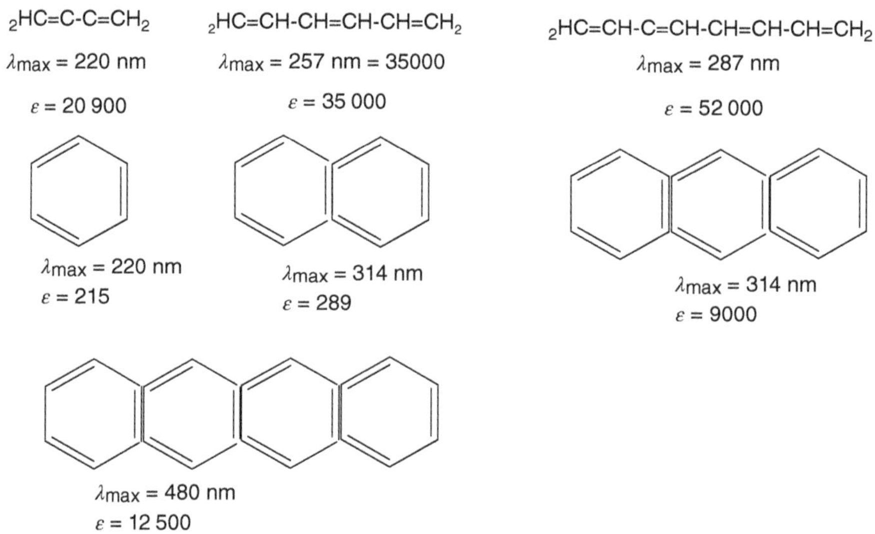

Figure 21.15 Effect of conjugation length on the λ_{max} and ε of organic compounds.

crucial because chemical structure has a major impact on the physical properties of polymeric materials. To assess the functionalized biopolymer's potential for biomedical applications, it is also required to assess its temperature resistance in detail. For instance, the thermal characteristics of polymeric materials used in restorative applications are crucial because they may expand or contract in response to temperature changes, which may affect their suitability for a particular application (such as a prosthetic) (Flores-Ramírez et al. 2005). It could be thought of as a collection of methods that analyze a sample's physical characteristics in relation to its temperature profile. Thermal characterization is mostly used to ascertain the temperature at which polymer melting, crystallization, and glass transition (T_g) occur in biomedical applications (Parthasarathy and Sethuraman

2014). Widely used thermal analysis instruments for the characterization of functionalized polymers are differential scanning calorimetry (DSC) and thermal gravimetric analysis (TGA). In DSC, the sample is susceptible to a managed temperature programme in order to determine the quantity of energy absorbed and released, but in TGA, the loss of mass is measured instead of raising the sample's temperature. The application of DSC and TGA techniques, as well as how to interpret the findings, are shown in the examples below. In order to study the thermal characteristics of copolymers with varying functional pendant groups produced by the ring-opening polymerization of L-lactide (L-LA) and RS-*b*-benzyl malate, Zhang and Ruckenstein (1998) utilized DSC analysis (MA). They also investigated how the thermal characteristics of the copolymers were affected by hydrogenolysis, which removes the benzyl groups. The findings of the DSC study (Figure 21.16a) demonstrated that the copolymers' crystallinity reduced as the MA concentration rose. As the MA concentration rose, so did the copolymers'

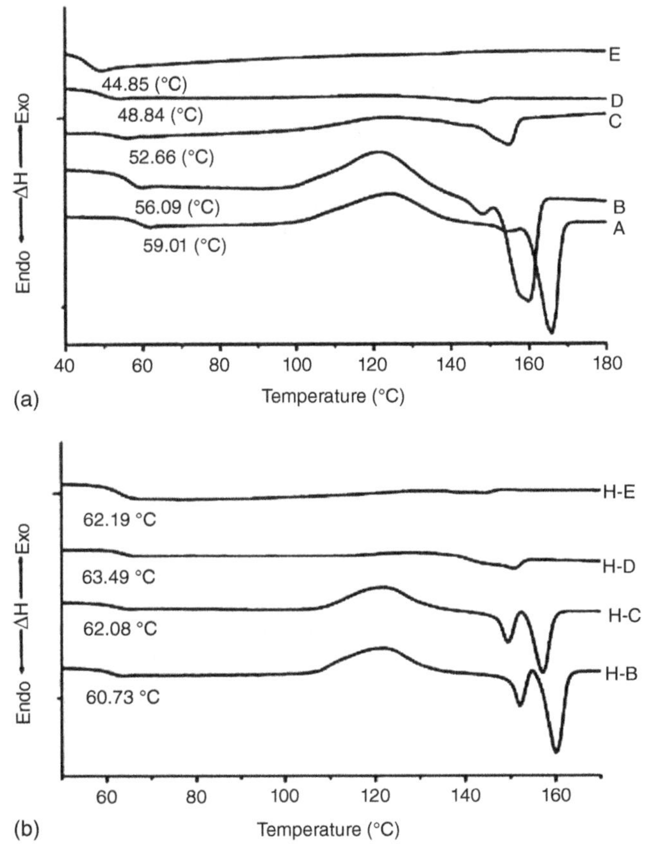

Figure 21.16 (a) DSC spectra of PLLA and poly(L-lactide-*co*-RS-*b*-benzyl malate). Composition of L-LA/MA was (A) 100/0, (B) 92/8, (C) 82/18, (D) 70/30, (E) 59/41; (b) DSC spectra of poly(L-lactide-*co*-RS-*b*-malic acid). Composition of L-LA/MA was (H-B) 92/8, (H-C) 82/18, (H-D) 70/30, (H-E) 59/41. Source: He et al. (2003)/with permission of Elsevier.

crystal-melting temperature (T_m) and glass transition temperature (T_g). However, all of the deprotected (hydrogenolyzed) copolymers had T_g values that were larger than the comparable protected copolymers (Figure 21.16b). This was explained by the intermolecular hydrogen connections that developed between the pendant carboxyl groups (Zhang and Ruckenstein 1998). Thermal analysis techniques were used in the study by Flores-Ramírez et al. (2005) to analyze the functionalization of (CTS)/GMA material at two sharp stoichiometric molar ratios, 1 : 1 and 1 : 4. These hybrid composite functionalized materials are softer than the precursor material at room temperature, according to DSC analysis of CTS/GMA, which reveals that T_g in both of the functionalized polymers is lower than the value observed for pure CTS. Results from TGA show that as functionalization increases, degradation temperatures likewise fall (Flores-Ramírez et al. 2005).

21.4.7 Mechanical Testing and Rheometric Analysis

Mechanical characteristics of polymeric biomaterials may have an impact on their biocompatibility (Mamaghani et al. 2018). Tensile testing can be used to assess the most crucial mechanical properties for biomedical applications, such as tensile strength, compressive strength, and elastic or Young's modulus. The impact of these characteristics will vary depending on how biomaterials are used. Rheometry, which measures viscosity and storage modulus, can be used to assess the viscoelastic behavior and rheological characteristics of polymeric biomaterial.

The degree of polymerization, polymer structure, crosslinking, and functionality are all aspects of polymeric materials that affect their mechanical properties. Examples are given here to show how functionalizing polymers affect their rheological and mechanical properties. It was investigated (Cai et al. 2016) how functionalizing nanofillers with their hyperbranched polyglycerol (PG) improved the mechanical properties of composite nanofibers. In the present study, poly(vinyl alcohol) (PVA) was chosen as the model polymer matrix and nanodiamonds (ND) were used as the nanofillers. By using the electrospinning technique, various unmodified ND/PVA and PG-modified ND/PVA composite nanofibers are created, and tensile tests were used to assess their mechanical characteristics. The mechanical properties of membranes of ND/PVA composite nanofiber were improved by the grafting of PG on the ND surface, as evidenced by a 60% increased tensile strength, a 51% increase in Young's modulus, and a 59% increase in toughness at a 3 weight percent filler content. Interactions with matrix (Nie et al. 2007). In another study (Baumgaertel et al. 2011), crosslinking maleimide-functionalized of high-molecular-mass heparin (HMWH) with thiolderivatized PEGs of diverse molecular weights and polymer structures were used to analyze the gelation time and elasticity of the synthesized hydrogels. The findings of this work demonstrate that the generation of hydrogels and their rheological properties was significantly influenced by the alteration of carboxylates in HMWH with various maleimide groups and with thiol derivatization of PEG crosslinker. Elastic moduli (G′) in hydrogels were measurable impacts of variations in HMWH functioning, as expected.

References

Al-Itry, R., Lamnawar, K., and Maazouz, A. (2012). Improvement of thermal stability, rheological and mechanical properties of PLA, PBAT and their blends by reactive extrusion with functionalized epoxy. *Polymer Degradation and Stability* 97 (10): 1898–1914.

Baumgaertel, A., Weber, C., Fritz, N. et al. (2011). Characterization of poly(2-oxazoline) homo-and copolymers by liquid chromatography at critical conditions. *Journal of Chromatography A* 1218 (46): 8370–8378.

Budd, P.M., McKeown, N.B., Ghanem, B.S. et al. (2008). Gas permeation parameters and other physicochemical properties of a polymer of intrinsic microporosity: polybenzodioxane PIM-1. *Journal of Membrane Science* 325 (2): 851–860.

Cai, N., Li, C., Luo, X. et al. (2016). A strategy for improving mechanical properties of composite nanofibers through surface functionalization of fillers with hyperbranched polyglycerol. *Journal of Materials Science* 51: 797–808.

Chang, T. (2005). Polymer characterization by interaction chromatography. *Journal of Polymer Science Part B: Polymer Physics* 43 (13): 1591–1607.

Chang, T., Lee, H.C., and Lee, W. (1997). High performance liquid chromatography characterization of macromolecules. In: *Macromolecular Symposia* (Vol. 118, No. 1), 261–265. Basel: Hüthig & Wepf Verlag.

Chiang, C.K., Druy, M.A., Gau, S.C. et al. (1978). Synthesis of highly conducting films of derivatives of polyacetylene, (CH)x. *Journal of the American Chemical Society* 100 (3): 1013–1015.

Coessens, V., Pintauer, T., and Matyjaszewski, K. (2001). Functional polymers by atom transfer radical polymerization. *Progress in Polymer Science* 26 (3): 337–377.

Cote, A.P., Benin, A.I., Ockwig, N.W. et al. (2005). Porous, crystalline, covalent organic frameworks. *Science* 310 (5751): 1166–1170.

Donini, C., Robinson, D.N., Colombo, P. et al. (2002). Preparation of poly(methacrylic acid-g-poly(ethylene glycol)) nanospheres from methacrylic monomers for pharmaceutical applications. *International Journal of Pharmaceutics* 245 (1–2): 83–91.

Dyer, J.R. (1978). *Applications of Absorption Spectroscopy of Organic Compounds.* Prentice Hall India Learning Private Limited.

Flores-Ramírez, N., Elizalde-Peña, E.A., Vásquez-García, S.R. et al. (2005). Characterization and degradation of functionalized chitosan with glycidyl methacrylate. *Journal of Biomaterials Science, Polymer Edition* 16 (4): 473–488.

Gauthier, M.A., Gibson, M.I., and Klok, H.A. (2009). Synthesis of functional polymers by post-polymerization modification. *Angewandte Chemie International Edition* 48 (1): 48–58.

Gissibl, A., Padié, C., Hager, M. et al. (2007). Synthesis and application of phosphorus dendrimer immobilized azabis (oxazolines). *Organic Letters* 9 (15): 2895–2898.

Günay, K.A., Theato, P., and Klok, H.A. (2013). Standing on the shoulders of Hermann Staudinger: post-polymerization modification from past to present. *Journal of Polymer Science Part A: Polymer Chemistry* 51 (1): 1–28.

He, B., Bei, J., and Wang, S. (2003). Synthesis and characterization of a functionalized biodegradable copolymer: poly(l-lactide-co-RS-β-malic acid). *Polymer* 44 (4): 989–994.

Hoskins, J.N. and Grayson, S.M. (2010). MALDI-TOF mass spectral characterization of polymers containing an azide group: evidence of metastable ions. *Macromolecules* 43 (14): 6225–6228.

Jiang, J.X., Su, F., Trewin, A. et al. (2007). Conjugated microporous poly(aryleneethynylene) networks. *Angewandte Chemie International Edition* 46 (45): 8574–8578.

Jiang, J.X., Su, F., Trewin, A. et al. (2008). Synthetic control of the pore dimension and surface area in conjugated microporous polymer and copolymer networks. *Journal of the American Chemical Society* 130 (24): 7710–7720.

Jin, S. and Gonsalves, K.E. (1998). Synthesis and characterization of functionalized poly(ε-caprolactone) copolymers by free-radical polymerization. *Macromolecules* 31 (4): 1010–1015.

Kanatzidis, M.G., Wu, C.G., Marcy, H.O. et al. (1990). Conductive polymer/oxide bronze nanocomposites. Intercalated polythiophene in vanadium pentoxide (V_2O_5) xerogels. *Chemistry of Materials* 2 (3): 222–224.

Kuhn, P., Antonietti, M., and Thomas, A. (2008). Porous, covalent triazine-based frameworks prepared by ionothermal synthesis. *Angewandte Chemie International Edition* 47 (18): 3450–3453.

Kuptsov, A.H. and Zhizhin, G.N. (1998). *Handbook of Fourier Transform Raman and Infrared Spectra of Polymers*. Elsevier.

Lummerstorfer, T. and Hoffmann, H. (2004). Click chemistry on surfaces: 1,3-dipolar cycloaddition reactions of azide-terminated monolayers on silica. *The Journal of Physical Chemistry B* 108 (13): 3963–3966.

MacDiarmid, A.G. (2001). Nobel lecture: "synthetic metals": a novel role for organic polymers. *Reviews of Modern Physics* 73 (3): 701.

Mamaghani, K.R., Naghib, S.M., Zahedi, A. et al. (2018). GelMa/PEGDA containing graphene oxide as an IPN hydrogel with superior mechanical performance. *Materials Today: Proceedings* 5 (7): 15790–15799.

Matyjaszewski, K., Nakagawa, Y., and Gaynor, S.G. (1997). Synthesis of well-defined azido and amino end-functionalized polystyrene by atom transfer radical polymerization. *Macromolecular Rapid Communications* 18 (12): 1057–1066.

Morton, M. and Fetters, L.J. (1975). Anionic polymerization of vinyl monomers. *Rubber Chemistry and Technology* 48 (3): 359–409.

Mulijani, S., Dahlan, K., and Wulanawati, A. (2014). Sulfonated polystyrene copolymer: synthesis, characterization and its application of membrane for direct methanol fuel cell (DMFC). *International Journal of Materials, Mechanics and Manufacturing* 2 (1): 36–40.

Nie, T., Baldwin, A., Yamaguchi, N., and Kiick, K.L. (2007). Production of heparin-functionalized hydrogels for the development of responsive and controlled growth factor delivery systems. *Journal of Controlled Release* 122 (3): 287–296.

Nuñez, S.A. and Hickner, M.A. (2013). Quantitative 1H NMR analysis of chemical stabilities in anion-exchange membranes. *ACS Macro Letters* 2 (1): 49–52.

Parrondo, J., Arges, C.G., Niedzwiecki, M. et al. (2014). Degradation of anion exchange membranes used for hydrogen production by ultrapure water electrolysis. *RSC Advances* 4 (19): 9875–9879.

Parthasarathy, M. and Sethuraman, S. (2014). Hierarchical characterization of biomedical polymers. In: *Natural and Synthetic Biomedical Polymers* (ed. S.G. Kumbar, C.T. Laurencin, and M. Deng), 33–42. Elsevier.

Raust, J.A., Houillot, L., Save, M. et al. (2010). Two dimensional chromatographic characterization of block copolymers of 2-ethylhexyl acrylate and methyl acrylate, P2EHA-b-PMA, produced via RAFT-mediated polymerization in organic dispersion. *Macromolecules* 43 (21): 8755–8765.

Riva, R., Schmeits, S., Jérôme, C. et al. (2007). Combination of ring-opening polymerization and "click chemistry": toward functionalization and grafting of poly(ε-caprolactone). *Macromolecules* 40 (4): 796–803.

Rowland, S.M. and Striegel, A.M. (2012). Characterization of copolymers and blends by quintuple-detector size-exclusion chromatography. *Analytical Chemistry* 84 (11): 4812–4820.

Sankir, M., Harrison, W.L., Wiles, K.B. et al. (2004). Proton exchange membrane fuel cells: synthesis and characterization of disulfonated poly(arylene ether benzonitrile) copolymers. In: *Abstracts of Papers of the American Chemical Society*, vol. 228, U656–U656. Washington, DC: American Chemical Society.

Scheirs, J., Bigger, S.W., and Delatycki, O. (1991). Characterizing the solid-state thermal oxidation of poly(ethylene oxide) powder. *Polymer* 32 (11): 2014–2019.

Schön, J.H., Meng, H., and Bao, Z. (2001). Self-assembled monolayer organic field-effect transistors. *Nature* 413 (6857): 713–716.

Setijadi, E., Tao, L., Liu, J. et al. (2009). Biodegradable star polymers functionalized with β-cyclodextrin inclusion complexes. *Biomacromolecules* 10 (9): 2699–2707.

Stevens, M.P. (1990). *Polymer Chemistry*. New York: Oxford University Press.

Storey, R., Vana, P., and Yagci, Y. (2013). *Fundamentals of Controlled/Living Radical Polymerization*. Royal Society of Chemistry.

Su, W.F. (2013). Characterization of polymer. In: *Principles of Polymer Design and Synthesis*, 89–110. Springer.

Tsarevsky, N.V., Bernaerts, K.V., Dufour, B. et al. (2004). Well-defined (co)polymers with 5-vinyltetrazole units via combination of atom transfer radical (co)polymerization of acrylonitrile and "click chemistry"-type postpolymerization modification. *Macromolecules* 37 (25): 9308–9313.

Weber, J., Antonietti, M., and Thomas, A. (2008). Microporous networks of high-performance polymers: elastic deformations and gas sorption properties. *Macromolecules* 41 (8): 2880–2885.

Yampolskii, Y.P. (2007). Methods for investigation of the free volume in polymers. *Russian Chemical Reviews* 76 (1): 59.

Yang, J.C., Jablonsky, M.J., and Mays, J.W. (2002). NMR and FT-IR studies of sulfonated styrene-based homopolymers and copolymers. *Polymer* 43 (19): 5125–5132.

Zhang, H. and Ruckenstein, E. (1998). Graft copolymers by combined anionic and cationic polymerizations based on the homopolymerization of a bifunctional monomer. *Macromolecules* 31 (3): 746–752.

22

Organic Polymers for Adhesive Applications: History, Progress, and the Future

Samiran Upadhyaya[1], Madhabi Devi[2], Saponjeet Borah[3], and Neelotpal S. Sarma[4]

[1]Moridhal College (affiliated to Dibrugarh University), Department of Chemistry, Moridhal, Dhemaji, Assam, 787057, India
[2]Moridhal College (affiliated to Dibrugarh University), Department of Physics, Moridhal, Dhemaji, Assam, 787057, India
[3]Gauhati University, Department of Physics, Jalukbari, Guwahati, Assam, 781014, India
[4]Institute of Advanced Study in Science and Technology, Advanced Materials Laboratory, Physical Sciences Division, Paschim Boragaon, Guwahati, Assam, 781035, India

22.1 Introduction

Polymer materials are of great use in our daily life for various applications. Among all others, polymeric materials as adhesives have been in demand due to their broad range of applicability. Adhesives are a class of essential materials that bind similar or different kinds of items in strong and mostly permanent manners. Most of the adhesives used in daily life need some specific pressure to be applied for binding two surfaces and hence are also known as pressure-sensitive adhesives (PSAs). Tapes, labels, and protective films can also be classified into the class of bonds. Undoubtedly, adhesives are among the primary areas of research in the current era. They are non-metallic, polymeric substances applied to one of the sides of substances to be bound. They are more advantageous than the other contemporary binding techniques, such as sewing, welding, and fastenings. They are engineered such that they can perform instant operations, like veneering, jointing, or edge-banding oils, and plastic films. The adhesives themselves undergo a metamorphosis in the course of their matchmaking for binding substrates. When applied to the binding materials, they must be fluidic in nature, so that they can spread rapidly, to achieve molecular contact. However, on bringing the two substrates together with adhesive, the adhesives must become hard and develop the required strength so that the bonded composite resists disruption. The adhesive is among the fastest-growing sectors in the world, leading to a highly competitive market. Adhesives render services in designing items, which range from decoratives to automobiles, textiles, electronics, and space industries (Miller et al. 2013; Barnes and Pashby 2000; Zhang et al. 2008). In fact, adhesives, both natural and synthetic, are among the many industrial sources that help in the economic growth of a nation. Natural adhesives are eco-friendly and

Organic Polymers in Energy-Environmental Applications, First Edition.
Edited by Ramesh Oraon, Pardeep Singh, Sanchayita Rajkhowa, Sangita Agarwal, and Ravindra Pratap Singh.

less toxic, which might not always be the perfect material of choice, while synthetic adhesives are designed in the laboratory to fulfill our requirements. Among the natural organic polymers, adhesives are formed by using starch, animal glue, casein, and rubber, in water or organic solvents. Even the crosslinking of adhesives has ancient beginnings – the egg albumin coagulation and the drying of blood by heat or light, the treatment of lime and casein. The earliest human use of adhesives could be estimated at approximately about 200,000 years ago for binding wooden handles and stone tools, using tar produced by the dry distillation of birch bark (Wadley 2010). During the early ages, adhesives formulations were derived from eggshells, plant extracts, lime, etc. for construction, design of hunting tools, and crafts. The first plant used to obtain animal glue for commercial purposes was produced at the end of the 1600s, and the since then adhesives have been in use for decorative purposes and joint production (Tout 2000).

Nails and pegs were employed initially to join two substrates, initially. However, in the nineteenth century, with the enhancement in the quality of animal glue furniture makers started to depend solely on the glue bonds' strength to ensure joint security. The commercially known glue was developed in 1690 initially in Holland; the casein glue, during the early nineteenth century, was manufactured in Switzerland (Mittal and Pizzi 2003; Delmonte 1947). The first patent on glue was issued in Britain in 1750, which reported an adhesive made from fish. The first US patent (No. 183,024) for glue was published in 1876. Synthetic adhesives have been in focus since the mid-1900s. In recent years, the adhesive synthesis is performed using different polymers, like acrylates, epoxy, polyurethanes, and cyanoacrylates (Tenorio-Alfonso et al. 2020; Jin et al. 2015; Maurya et al. 2018; Raja 2016). Polymerization processes, such as hot melt synthesis, emulsion technique, radiation curing, and solution-phase synthesis, are employed to obtain adhesives. Research has been in progress around the globe to formulate efficient adhesives to meet the growing demands. In addition, continuous efforts have been made around the globe to formulate adhesives that could contribute toward the future energy demand, along with safeguarding the environment. This chapter focuses on the various aspects of adhesives, including their applications in energy and the environment and their prospects. The past and ongoing research and the market scenario have also been presented in this chapter. The detailed study of adhesives will undoubtedly lead to advancements in the research and progress in adhesives/glues and sealants.

22.2 History and Developments

Adhesives could be estimated to be discovered in the Pleistocene era around 200,000 years back, obtained from the tar obtained from birch bark. The Africans used glue obtained from tree sap to protect their caves around 75,000 BC. Adhesives obtained from natural sources were utilized by people to preserve the native culture and craft tools, weapons, and much more since prehistoric times. The earliest known adhesive applications date back to 6000 years ago, applied for designing decoration materials. Archaeological investigations have also revealed the use of adhesives in the Babylonian temples dating to approximately 4000 BC (Chattopadhyay 2014). Glue became commonplace across central and southern Europe with the rise of the Greek and Roman empires, when glues were being used for small daily jobs to massive building constructions that have survived to this day. Adhesives were used to create laminated woodwork some 3500 years ago, while they were also used to design the tombs of Egyptian Pharaohs 2500 years ago; the adhesives used were obtained from animal products, which used extracts from bones, skins, blood, and brain. The first reference to adhesives in literature has been traced to approximately 2000 BC; to design artifacts, which include paintings by wood gluing and caskets using animal glue. The design of bows, arrows, and furniture using casein-based glues was much favored in ancient times. Wood marquetry and veneering were developed, followed by the production of animal and fish glues by the Romans from AD 1 to 500. Egg-based pastes were used via the incorporation of natural ingredients, which include milk, vegetables, bones, cheese, blood, and grains to bond materials. Over time, slaked lime was utilized by the Greeks as mortar, while the Romans mixed sand and volcanic ash with lime, thus further extending the mortar development. Romans are known to be the first ones who used beeswax and tar as sealants for wood to be used in boats and ships (Muhammad et al. 2019). On the other hand, the Mongols in 1100 AD constructed bows using a new adhesive with laminated lemonwood and bullhorn. Interestingly, the great Ahoms in 1200–1400 AD used duck-egg shells and yolk, rice husk, Bora chaul (a special rice variety), and black gram to construct monuments, temples, and war tunnels, which are still prominent tourist spots in the various locations of Assam, India (Verma et al. 2014).

In Europe, adhesive use started and subsequently in growth during the period from AD 1500 to 1700. Thomas Chippendale and Duncan Phyfe, the two renowned furniture makers, began to use adhesives to bind their products during this period. With human civilization development, the first-ever adhesive industry was started in Holland in 1690, termed Horse glue, which became extensively popular (Anderson 1990). Old horses were disposed to the glue factories to produce the materials. The horse glue continued to be used for a more extended period, with different variations. In 1750, the first British patent was granted on an adhesive that was obtained from fish. In the following year, casein glue production was started in Germany and Switzerland. In 1808, the first animal glue factory was set started in the United States. In 1845, the natural-rubber-based adhesive was developed by Henry Day, which was later used widely for surgical tapes. In 1847, the first postal stamp was designed, which required the use of glue. In 1876, the Ross brothers received the first US patent for developing casein glue.

Rubber cement, cellulose nitrate dopes, laminating veneers, and waterproof adhesives were other prominent developments in the nineteenth century. By 1825, the PSA tap industry started, which is found in various forms in today's world, such as sticky notes, scotch tapes, and fluorescent tapes. The invention of thermosetting plastics called Bakelite in 1910 was a crucial step in the development of adhesives (Jewkes et al. 1969). Bakelite-resins-based adhesives were used in plywood as coatings. Today, plywood bonding by phenol-formaldehyde resins is the most extensive for adhesive material end-product use. The phenolic resin is cost-effective, and when applied on the plywood material does not delaminate in the hot dampness of June–August. The importance of these resins was severely realized in the late 1930s. Since 1930, urea-formaldehyde resins started to be significantly used as adhesives, which are still being widely used as binders in particleboard, and utilized in flooring and furnishing (Rozelsky 2021). The adhesives sector's growth reached a great height with the First and the Second World Wars from the 1920s to 1940s. With the advancements of civilizations, a substitute for animal-based glue was felt necessary. America's leading food and beverages company, Borden, made Elmer's all-glue in 1932, which used animal products in the formulations, with many restrictions and precautions. Many countries framed some laws restricting the use of animals for adhesives. In 1947, the first polyvinyl alcohol (PVA)-based glue reported was made of milk protein. The synthetic adhesive composed of PVA was found to be more effective for multi-purpose applications. In 2003, Borden was acquired by Berwind Corporation, which started producing Elmer's all-glue by using PVA, which did not involve the use of animal products. With the growing interest in nanotechnology, many adhesive formulations were made, with much improved adhesive efficiency. For example, Gecko-tape, a nanomaterial-based adhesive tape, was 1000 times more effective than regular adhesive tapes, with no harmful materials up to cellular levels. With further advancements in research, many adhesives have been synthesized, which are presently available in the markets. Different adhesives are known today as per the usage and requirements, such as acrylic, polyurethanes, epoxy, and PVA-based adhesives.

22.3 Classifications of Adhesives

Several methods are known according to which adhesives can be classified. Several aspects are considered for the classifications of adhesives. Among all, the major category is structural and non-structural adhesive classification. Structural adhesives provide crucial services for daily life utility; they provide shear strength of more than 6900 kPa and have stable properties that do not alter with aging. Non-structural adhesives are mostly used as packaging adhesives, pressure-sensitive tapes, or hot melts. Their shear strength is less than 6900 kPa, and they are durable under relatively mild environment (Petrie 2007). These adhesives are chemically composed of polyurethanes, resorcinol, phenols, melamine, epoxy, urea-formaldehyde, acrylates, and polyimides. Non-structural adhesives are chemically composed of butyl/styrene-butadiene rubbers, polyesters, polyvinyl acetate, silicones, acrylics, ethylene-vinyl acetate copolymers, cellulosic resins, polyolefins, and naturally polymers. Considering the chemical/physical behavior, adhesives are categorized as thermosetting, thermoplastic, elastomeric, or hybrids. Thermosetting adhesives are crosslinked structures that do not soften in the presence of solvents or heat. Thermoplastic glues are not crosslinked; they turn to a gel-like state on heating. Elastomeric adhesives shows low tensile strength and high elongation usually and are formulated via natural or synthetic rubbers. Hybrid adhesives are compounds formed via polymeric mixtures, with specific property adjustments, as required.

Another route to classify adhesives involves application and subsequent curing methods. Several steps are involved during the process of adhesion: application, wetting, and setting. The setting up of adhesive proceeds in numerous ways, such as by cooling, the release of solvent/carrier, or via in situ polymerization reaction (Jovanović and Dubé 2004).

Below are the classifications of adhesives, discussed briefly.

22.3.1 Based on the Origins

22.3.1.1 Natural Adhesives

As the name suggests, adhesives are obtained from natural sources, viz. vegetable starch, natural resins, or animal sources. These adhesives are also commonly referred to as bio-adhesives. Common examples found in daily life include cooking flour paste in water, boiled rice paste, and plant resins.

Starch-based adhesives, albumen, casein glue, lignin, etc. can be categorized under this class. Starch-derived glues are used in production of paper sack, wallpaper adhesives, and paper tube winding. Albumen finds application commonly in the ply industry and is composed of the blood–protein component. Casein glue is primarily applied to bind labels of glass bottles. Animal glues were used traditionally in book-binding, wood joining, and several other fields; however, synthetic glues have largely replaced them. In certain specific applications, such as stringed instruments (production and repair), animal glues are still prominently used.

22.3.1.2 Synthetic Adhesives

The adhesives synthesized in the laboratory by using different organic precursors are known as synthetic adhesives. Naturally produced adhesives were extensively used until, in the 1920s, the first synthetic adhesive was produced by Karlsons Klister. There are different types of synthetic adhesives, based on the precursors used. Common types of synthetic adhesives are epoxy, polyurethanes, acrylic, cyanoacrylate, and polyvinyl-acetate-based adhesives.

Epoxy-based adhesives are highly used for binding different substances, like polymeric composites, metals, and concrete structures. They form highly crosslinked adhesive materials with numerous advantages such as various curing techniques, low shrinkage during the curing process, high resistance to corrosion, good mechanical strength, and excellent chemical/thermal resistance. The good wettability of these adhesives provides an excellent joint for most substrates. In addition, epoxy can be easily conjugated with nanomaterials, thus enhancing the characteristics of the adhesives. Thus, epoxy-based adhesives are among the most used materials for multiple purposes (Ahmadi 2019).

Due to the polar urethane group in polyurethane, the polyurethane-based adhesives form strong adhesion to different substrates. They are obtained by reacting isocyanates with different polyols, leading to chemically hardened adhesive formulations, due to the availability of reactive terminal groups.

Acrylic-based adhesives have been a topic of discussion in recent times. Copolymers of acrylic acids have been an essential aspect of many researchers around the globe. Due to their ability to cross-link extensively, acrylic-based adhesives are widely applicable in applications like drug delivery, wound dressing, and environment safety. One of the trending fields of interest in acrylic-based polymers is the development of adhesives. In coatings and furnishing, they are considered the most suitable adhesives due to their durability, cure speed, strength, and low toxicity (Smeets 2013). Other major sectors in which acrylic-based adhesives are widely are construction, dentistry, engineering, and textiles.

Cyanoacrylate or Superglue, also called Krazy Glue, was discovered by Dr. Harry Coover in 1942, although it was later rejected by Coover as it was too sticky. In 1951, Coover, while supervising research at the Eastman Company with Dr. Fred Joyner, rediscovered cyanoacrylate. Coover discovered that cyanoacrylate was a useful material to glue together the prisms and finally in 1958, the Eastman compound was marketed, which later was packaged as Superglue (Baig 2009).

Polyvinyl acetate (PVAc) is formed by the polymerization of vinyl monomers and is most widely used as an emulsion. The emulsion is the familiar "white glue" used for household purposes. The PVAc adhesive has replaced natural casein-made adhesives due to its low cost and ability to resist the attack by microorganisms. Thus PVAc-derived adhesives are used in multiple fields, for example, in the manufacture of furniture. Due to the easy diffusion of water into the wood, the adhesive gets set easily on the substrate and the setting times are much more rapid at room temperature.

22.3.2 Based on the Dispersion Medium

Based on the dispersion medium, adhesives can be mainly categorized into two classes: solvent-borne and water-borne adhesives. A medium of dispersion in the form of a solvent is always necessary for the gelation of the polymeric resins to form adhesives. Sometimes the water is used as the dispersion medium to form the adhesive gel; in this case, the glue may be termed water-borne adhesive. On the other hand, in some adhesives, organic solvents are used as a medium of dispersion for the formation of adhesives. Such a category of adhesives may be termed solvent-borne adhesives.

Solvent-based adhesives might be toxic to an extent due to the release of volatile solvents utilized in adhesives. But at the same time, they provide a wide range of utilities for different binding types of materials. For PSAs, solvent-borne adhesives are often ideal, as they take much shorter curing time compared to water-borne adhesives. On the other hand, water-borne adhesives are most desirable for binding cellulose-based materials, like wood, paper, ply, and thermocols, which can easily absorb excess water.

22.4 Adhesive Characterization Techniques

Adhesives can be characterized using various spectroscopic and analytical techniques. Adhesives are polymeric resins, so the characterization techniques for the adhesives are those involved in the characterization of polymers, with few additions (Hon 2003).

Fourier transformed infrared (FT-IR) spectroscopy: Structural elucidation of adhesives is started with FT-IR analysis. FT-IR spectra determine the presence of different functional groups in the polymeric framework of an adhesive.

Nuclear magnetic resonance (NMR) analysis: NMR is a critical tool to analyze the structure of a synthetic polymeric adhesive. The linkage of C-C bonds and the environment of the hydrogens present can be studied by the ^{13}C and ^{1}H NMR, respectively. The type and functionality of an adhesive largely depend on its structural framework, which can be studied with the help of NMR spectroscopy.

Gel permeation chromatography (GPC): GPC is an important technique to determine the total molecular mass of a polymeric material. The polydispersity index (PDI) is also calculated by GPC. The total molecular mass and PDI determine the nature of the adhesive and the overall chain linkage.

X-ray diffraction (XRD): XRD helps us to determine the crystalline and/or amorphous characteristics of an adhesive. The overall type of bonding can be analyzed by determining the miller planes of the adhesives. Interestingly, the analysis of XRD also gives an idea of the extent of crosslinking in a resin.

Differential scanning calorimetry (DSC): DSC is an important characterization technique for the analysis of adhesives. The properties of adhesives are largely dependent on the glass transition temperature (T_g) and the melting and crystalline temperature. With the help of DSC, we can find the T_g, which helps us analyze the type and utility of a glue. The range of stability and utility can also be analyzed via DSC.

Thermogravimetry analysis (TGA): TGA is an analytical technique that helps us to determine thermal stability with respect to the weight of a polymer or resin. The analysis of the thermal stability of an adhesive helps us to predict the range up to which an adhesive can be conveniently used. The more the thermal stability of an adhesive, the more it can withstand the bonding.

Viscosity: The viscosity is a fluid's resistance to flow. An adhesive is a gel-like polymeric resin and hence has lesser fluidity. Viscosity is an important characteristic for the analysis of an adhesive. The more viscous an adhesive is, the less time and temperature are required for curing. Viscometer and/or rheometer are used to determine the viscosity of the glue.

22.5 Adhesive Efficiency and Strength Test

The utmost essential characteristic of an adhesive is the test for the adhesive strength or the tensile strength, which is defined as the total force required per unit cm^2

to detach the adhesive joint applied for the binding material. The force applied is always perpendicular to the substrate. The lap shear strength, also known as the tensile strength or adhesive strength, determines the scope of utility of an adhesive, starting from decorative to buildings. The adhesive strength measurements are based on some set of rules, as laid down by the American Society for Testing and Materials (ASTM) (ASTM 1958). For every specific material, a different set of rules and measurements have been designated as per the ASTM to determine the adhesive strength. All the adhesive strength test methods, as per the ASTM standard, are explained briefly in Table 22.1.

22.5.1 Specimen Failures During Adhesive Strength Tests

Sometimes the samples undergo breakdown during the process of the adhesive strength test. There are two types of failures that may take place in PSAs, viz. adhesive failure and cohesive failure. The failure occurs at the interface of the substrate and the adhesive. On the other hand, cohesive failure takes place within the center of the adhesive. Both failures depend on the viscoelastic characteristics of PSAs. Certain chemical compounds, known as tackifiers, are used to formulate adhesives that can specifically increase the stickiness and the tack of the surface of the adhesive. Tackifiers are generally aliphatic cyclo-aliphatic or aromatic resins, hydrogenated hydrocarbon resins, terpenes and modified terpenes, and terpene-phenol resins.

22.6 Applications of Adhesives

Adhesives are majorly used materials in different applications, starting from minor household daily use to industrial and medicines. Apart from their role in adhesion, adhesives are also used for multiple other applications, such as sealing agents, to nullify the effect of self-loosening due to dynamic loads, to seal certain parts to prevent oxidation and corrosion, waterproofing, etc. Adhesives as sealants can also be applied as electrical or thermal insulators, barriers to fire, and also as products for smoothing, filleting, etc. The adhesives and sealants are mainly used to bond substrates such as plastics (thermosets and thermoplastics), wood and wood products, composites, metals, glass and ceramics, elastomers, and sandwich and honeycomb structures. Two major sectors where adhesives are widely used are the industrial sector and pharmaceuticals and biomedicines, although the application of adhesives is not limited. The various applications of adhesives are depicted in Figure 22.1.

22.6.1 Use of Adhesives in Industrial Sectors

Many industries, such as construction, packaging, and transportation, require adhesives for their functioning. Adhesives are among the basic requirements considered for the industrial sectors. A few industrial applications of adhesives are mentioned below.

Table 22.1 ASTM methods for the analysis of adhesive strength, for different materials

Sl. No.	Standard	Description
1	ASTM C1135	It describes a method for the quantitative measurement of the tensile adhesion properties of sealants.
2	ASTM C1265	It includes a method to quantitatively measure the tensile strength and adhesion properties of insulating glass edge seals applied in structural sealant glazing applications.
3	ASTM C1382	It describes a procedure to measure the tensile adhesion properties of sealants to EIFS (exterior insulation and finishes systems) under various environmental conditions.
4	ASTM C1523	This process covers a procedure for analyzing the modulus, joint movement ability, tear, and adhesion properties of elastomeric joint sealants.
5	ASTM C906	This procedure involves the test to determine the peel strength of a hot-applied sealant when applied between flexible metal samples of a T-type shape.
6	ASTM C907	This method describes a methodology to determine the tensile adhesive strength of a tape sealant. It also enables us to determine the degree of cohesive/adhesive failure.
7	ASTM C908	This test includes a method to calculate the yield strength of preformed tape sealants.
8	ASTM C961	It is a procedure used to determine the lap shear strength of hot-applied sealants. In addition, this method also describes the adhesive bond of the sealants to the tested materials.
9	ASTM D1002	This procedure helps to determine the apparent strengths of adhesives for bonding metals when tested on a standard single-lap-joint specimen and under specified conditions of preparation and test.
10	ASTM D1062	It determines the cleavage properties of adhesive bonds when tested on metal specimens of specific dimensions under specified conditions of preparation and testing.
11	ASTM D1184	This test determines the comparative properties of the adhesive-bonded assemblies of either wood or metal when subjected to rupture stresses with specimens of standard dimensions, under certain conditions of pretreatment, relative humidity, temperature, and testing technique.
12	ASTM D1623	It helps to determine the tensile adhesion properties of rigid cellular materials of standard shape as test specimens, under the optimum humidity, temperature, and speed of the testing machine.
13	ASTM D1780	This process helps to determine the creep formed as a result of the metal-to-metal adhesive bonds due to the summed up effects of tensile shear stress, temperature, and time.
14	ASTM D1781	It describes the calculation of the peel resistance of adhesive bonds between a rigid and a relatively flexible adherend and its core when the test is carried out under specified conditions.

(Continued)

Table 22.1 (Continued)

Sl. No.	Standard	Description
15	ASTM D1876	It is a primarily designed method to analyze the relative peel resistance of adhesive bonds between flexible adherends taking a T-type test sample.
16	ASTM D1995	These tests include the techniques to test adhesives in four popularly practiced stressing modes, viz. compression shear, climbing drum peel, cleavage, and T-peel, and in different combinations of rigid and flexible adherends.
17	ASTM D2095	This procedure deals with the calculation of the relative tensile strength of adhesives, using bar and rod-shaped butt-joined specimens the specific preparation, and testing conditions.
18	ASTM D2293	This method describes the analysis of the creep properties of adhesives in metals bonding while testing on a standard sample and subjected to certain conditions of temperature and compressive stress in a spring-loaded testing instrument.
19	ASTM D2294	The procedure deals with the analysis of the creep properties of adhesives for metals adhesion when testing is done using a standard sample in certain conditions of temperature and tensile stress in a spring-loaded testing device.
20	ASTM D2295	It determines the comparative shear strengths of adhesives for adhesion of metals when a standard specimen is used for testing under specified preparation and testing conditions at elevated temperatures.
21	ASTM D2339	It includes the calculation of the comparative shear strengths of adhesives when tested on a standard sample specimen, under specified preparation, conditioning, and testing conditions. This method is designed to be applied only to adhesives applied in wood-wood bonding.
22	ASTM D2918	The test gives information to assess the shelf-life of adhesive joints using T-peel-type sample specimens stressed in contact with air, water, aqueous solutions, or other environmental factors at different temperatures.
23	ASTM D2919	It provides data to assess the durability of adhesive lap-shear joints during stressed contact with air, the air in equilibrium with certain solutions, water, or other environmental aspects at different temperatures.
24	ASTM D3163	This method provides a complement to D1002 with extended application in determining single-lap shear adhesive joints of rigid plastic substrates.
25	ASTM D3164	This procedure is a complement to Test Method D 1002 and Test Method D 3163 for extended application to single-lap-shear adhesive joints using plastic adherends.

Table 22.1 (Continued)

Sl. No.	Standard	Description
26	ASTM D3165	The method intends to determine the comparative shear strengths of adhesives in joints with large area while the testing is done using a standard single-lap-joint sample, under specific preparation and testing conditions. Adhesives show different responses to small versus large area joints.
27	ASTM D3166	This procedure deals with the fatigue strength measurement in shear by tension loading of adhesives on a specific specimen and under certain preparation, loading, and testing conditions.
28	ASTM D3167	This test involves the determination of the relative peel resistance of adhesive bonds between a flexible and a rigid adherend when tested under certain specific conditions of preparation and testing.
29	ASTM D3330	These procedures assist in the determination of the peel adhesion of pressure-sensitive tapes.
30	ASTM D3433	This method determines the fracture strength of adhesives' cleavage when tested on standard specimens and under specified conditions of preparation and testing.
31	ASTM D3759	In this method the tensile strength and stretches properties (elongation and "F" value) of the pressure-sensitive tapes are measured.
32	ASTM D3762	This test provides the qualitative simulation of the forces and effects on an adhesive bond joint at the interface of metal–adhesive/primer.
33	ASTM D3807	The test covers the analysis of the comparative cleavage/peel strengths of adhesives for engineering plastics bonding when tested on a standard sample specimen under specified preparation and testing conditions.
34	ASTM D3931	This process determines the comparative shear properties of gap-filling adhesives in wood-to-wood joints taking certain thicknesses of bonding in the dry condition when standard specimens are used for testing under specified conditions of preparation, conditioning, and loading in compression.
35	ASTM D3983	This process describes a procedure for calculating the shear modulus and rupture stress in the shear of adhesives in bonded joints.
36	ASTM D4501	This test describes a method and fixture for determining the shear strengths of adhesives used to bond samples having higher moduli than that of the adhesive.
37	ASTM D4562	This test determines the shear strength of curing liquid adhesives utilized to retain cylindrical assemblies or to lock and seal threaded fasteners.
38	ASTM D4680	This process determines the time-dependent characteristics of structural adhesives in wood-to-wood bonds, the samples of which are subjected to shearing stresses at varying static load levels, relative humidity, and constant temperature.

(Continued)

Table 22.1 (Continued)

Sl. No.	Standard	Description
39	ASTM D5656	This procedure deals with the preparation and testing of thick-adherend lap-shear specimens to determine the stress–strain behavior of adhesives.
40	ASTM D6004	This method describes a procedure to measure shear strength development for adhesives used for bonding carpet adhesives under layments.
41	ASTM D6195	These methods determine the properties of a pressure-sensitive adhesive, and they apply to those adhesives that rapidly form a bond of measurable strength when kept in contact with another surface.
42	ASTM D6862	This test provides a procedure to determine the resistance-to-peel strength of an adhesive bond between one flexible and one rigid adherend when the test is performed at an angle of approximately 90 degrees, considering specified preparation and testing conditions.

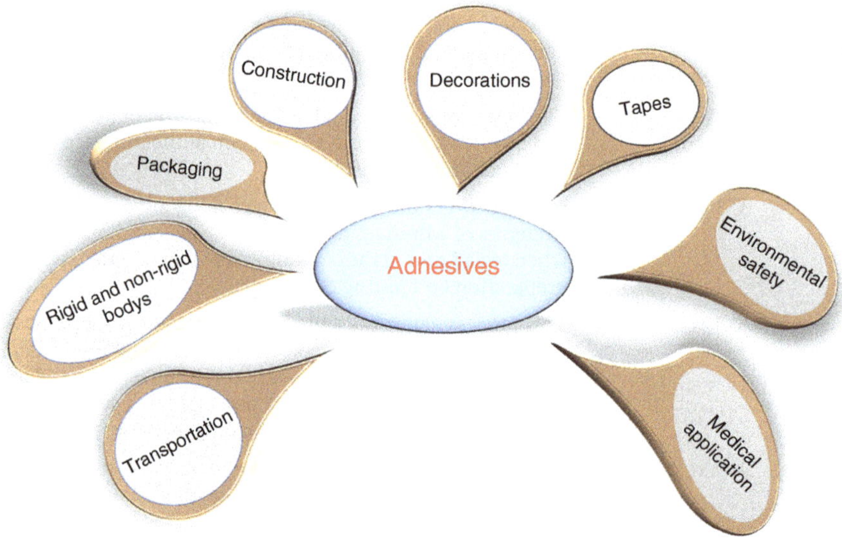

Figure 22.1 Applications of adhesives in different fields.

1) *Construction*: Adhesives are used for continuous flooring installation, ceramic tile installation, countertop lamination, manufacture of prefabricated beams and trusses, carpet adhesives, flooring adhesives, installation of prefinished panels, joint cement, drywall lamination adhesives, and covering installations (Dinte and Sylvester 2018).
2) *Consumer adhesives*: The adhesives consumed for, decoratives model supplies, stationery products, etc. fall under this category.

3) *Packaging*: Adhesives are widely used as carton closures, bonding disposable materials, cigarettes, bags, cups, and for manufacturing filters, containers for composites, and tubes.
4) *Tapes*: This segment of adhesives is used in packaging, masking, surgical consumer tapes, and industrials.
5) *Transportation*: The adhesives in this category include assemblies of vehicles (auto, truck, bus), structural assemblies of aircraft/aerospace, weather-strip, and gasket bonding.
6) *Other rigid bondings*: These include furniture manufacture, millwork, doors, kitchen cabinets, and vanities; appliance assembly and trim attachment; T.V., radio, and electronics assembly and machinery manufacture and assembly.
7) *Other non-rigid bondings*: This category includes shoe assembly, apparel laminates, book-binding, sports equipment, flock cement, and for manufacturing air and liquid filter.

22.6.2 Application of Adhesives in Medical and Pharmaceuticals

The pressure-sensitive adhesives initially used to develop bandage materials were composed of natural rubber. With further development, synthetic rubber, and, polyacrylic acid ester-derived adhesives have also been significantly used nowadays. A variety of adhesives are used as medical dressings/tapes/devices for wound healing, or for supporting an injured portion of the body. Advanced adhesives are widely used in various biomedical and pharmaceutical applications, which have been reviewed in detail by Webster (1997). Adhesives are an integral part of multiple medical procedures, such as medical apparatus/devices: to treat and restore dental filings, anesthetic administration, intravenous delivery of drugs, bypass surgery of the heart, and complex surgeries, like urological plastic surgery. Below are the common uses of adhesives for medicinal purposes.

1) Bioadhesives for drug-delivery applications are emerging systems for delivering various important biomedical substances, ranging from popular ones to probiotics, proteins, and herbal medicines. The drug-delivery systems may exist as solids, semisolids, and/or liquids in the common formulations or as nanoparticle-based systems designated for different routes such as oral, nasal, ocular, vaginal, skin, and esophageal routes (Brahmbhatt 2017).
2) Apart from their use as adhesives, bioadhesive polymers are also known to perform other pharmaceutical activities, such as disintegrant, diluent, and enhance viscosity. Their nature, quantity, and association with other excipients highly influence the final drug product quality.
3) Bioadhesive polymers are suitable for drug-delivery systems, and to design of medical devices. Transdermal patch technology is an emerging field of biomedical adhesives, which ensure drug administration to the bloodstream via the skin; it is a highly efficient process of drug delivery, in which the incorporation of the drug takes place into an adhesive membrane that binds the patch to the skin and controls the rate of absorption of the drug. The systems are designed to ensure

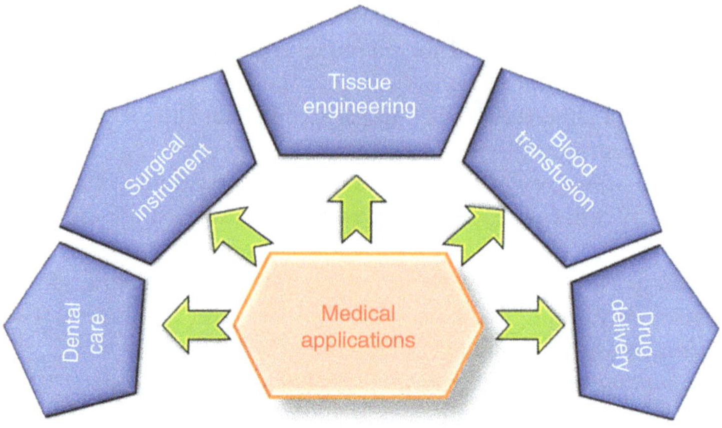

Figure 22.2 Applications of organic adhesives in medicals and pharmaceuticals.

continuous administration of drugs throughout the day, and thus avoid any alterations in plasmatic concentration, generally linked with oral administration of drugs (Padula et al. 2007).

4) Surgical adhesives are conveniently used methods for wound dressing that have multiple advantages, such as minimum pain, no removal of suture material, and localized release of drugs. Tissue adhesives are those classes of materials with properties that allow adherence of tissues with other tissue/non-tissue surfaces via in situ polymerization. They may be applied for various functions, such as prostheses, and bleeding control (hemostats), and also act as a barrier to liquids and gas (sealants). Over the past 30–40 years, various tissue adhesives and hemostats have been developed, and the work is still in progress (Duarte et al. 2012).
5) Dental adhesives are designed to help in the retention of composite fillings and are defined as monomers resin (having hydrophilic as well as hydrophobic entities) that enable the interaction between the resin and dental substrate. A few of them are also applied to provide protective against enamel erosion. Most often, inadequate sealing may lead to the failure of restorations. The recently developed dental adhesives have dramatically simplified the application procedure, as compared to classical multistep bonding methods, thus minimizing the technique sensitivity and application/healing time. Recently, fluorescent adhesives have been found to do wonders for dental care, owing to the easy visualization of tooth decay and tooth protection under UV light (Milia et al. 2012).

The most common applications of adhesives in medical fields are depicted in Figure 22.2.

22.7 Commercial Aspects of Adhesives

The glue industry is one of the fastest-growing sectors, and the continuous development of stronger and more flexible glues is being explored. Although considered

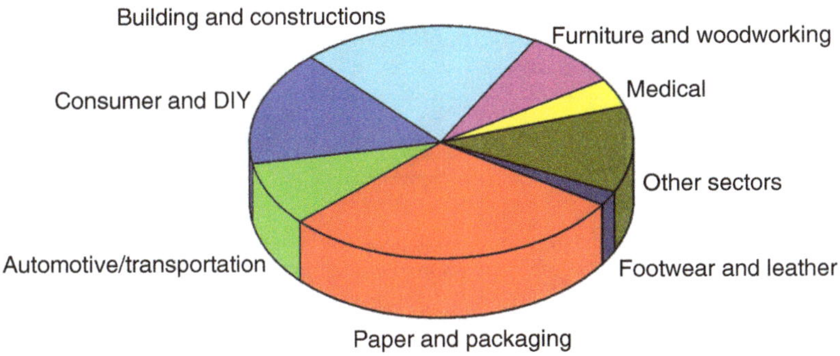

Figure 22.3 The adhesive market scenario as per their applications (as of 2021).

a simple household material, glue will continuously be considered essential for the sustained development of the world. The current market size of global adhesives and sealants is around USD 67.5 billion in 2022 and with the expected compound annual growth rate (CAGR) of 5.9%, this value is estimated to reach 107.27 by 2030, due to the increasing demand from various sectors, such as packaging, furniture construction, and footwear.

The adhesives are of different types, which find their applications in wide areas, as depicted in Figure 22.3.

The paper and packaging industry has been found to be the largest segment of the adhesive market, having a revenue share of more than 29%. Another major sector occupying the adhesive market is building/construction, which has been increasing in the last few years, due to architectural and technological innovations (20%). Consumer and DIY products also include a large share of the adhesive market (14%). Other important sectors contributing to the growth of the adhesive market are footwear industries, transportation, medical, etc.

22.8 Advanced Adhesive Formulations for Environment Sustainability and Applications in the Energy Sector

Adhesives are one of the most important classes of materials, which are an essential part of many industries. Every sector, including the energy and environment sectors, directly or indirectly are affected by the adhesives and their market share. With further enhancements in research and development, the next decade will certainly see higher growth in the application of adhesives in energy and environmental applications.

22.8.1 Use of Organic Adhesives for Environment Sustainability

Environment safety and sustainability have been major concerns for all researchers around the globe. Biodegradable polymers have been considered important replacements to traditional plastics and polythenes for every sector (Upadhyaya and Sen

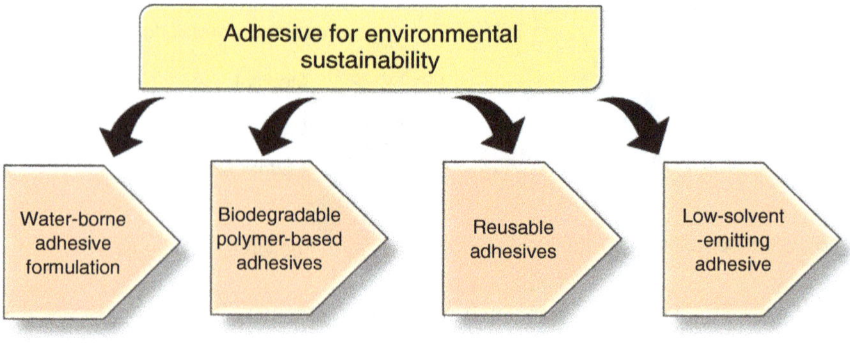

Figure 22.4 Various means of obtaining adhesives for environmental sustainability.

Sarma 2022). Continuous use of synthetic adhesives can have negative impacts on both humans and the environment. Various regulations have been set up by the competent authorities for the development of sustainable environment-friendly adhesives and packaging materials. Efforts are being made to develop eco-friendly adhesives, without compromising the strength. Considering environmental safety, a considerable decrease in the use of solvent-based adhesives has been observed in recent times, to minimize the effect of volatile solvents. The most suitable way to preserve the environment from the toxic effects of organic adhesives is to use suitable biopolymer-derived organic adhesives, such as polyhydroxyalkanoates (PHAs), bio-epoxy resins, and bio-acrylates (Packham 2009). The various aspects through which environmentally sustainable adhesives can be obtained are depicted in Figure 22.4.

To obtain adhesives of desired characteristics, without compromising on environmental safety, various research groups have made numerous efforts. Few successful adhesives have been derived from bio-compatible sources, thus safeguarding the environment. Among all the bio-derived adhesives, starch-based biocompatible polymers have been known for their utility as eco-friendly adhesives for a safe environment. The starch-based eco-friendly polymers are found to be renewable, non-toxic, and low-solvent emitting, and thus could help safeguard the environment (Maulana et al. 2022). Further, the use of starch-modified polymer composites is also known for their application in environment-friendly, water-resistant, bio-compatible adhesives (Watcharakitti et al. 2022). Thus, starch can be used as a substitute for harsh chemicals for adhesive development. In addition, milk-protein-derived polymers are also used to derive environment-friendly adhesives. Caseins and whey are known to be important milk proteins, which according to the reports can be potentially used as precursors to eco-friendly adhesives. The polymerization of the milk protein is an important aspect, which leads to the formation of aggregates to be used as adhesives (Guo and Wang 2016). Another work by Bhakare et al. reports the development of Arjuna-bark derived reusable composite for dual applications: as an adhesive and also as a flame retardant, as a substitute to traditional halogen-based flame retardants (Bhakare et al. 2023).

Thus, with progress in research, special focus has been given to green methods of adhesive development for the protection of the environment.

Interestingly, certain recently invented adhesives have been found to be used for the separation of waste from the environment. A recent interesting report by Rieland et al. presents the use of organic pressure-sensitive adhesives for the quantification of microplastics and their isolation from water bodies, thus helping in the environmental remediation process (Rieland et al. 2023). Other notable works related to the applications of environment-friendly adhesives include the following:

Applications of silk-fibroin/tannic-acid-based adhesive for underwater use (Kim et al. 2023), water-borne powder adhesive with additional applications in crafts and anti-counterfeiting (Upadhyaya et al. 2020), bio-inspired adhesives for environment safety and next-gen bone repair (Shokri et al. 2022), real-time deployable, high strength adhesive for application in harsh environment (Cui et al. 2022), etc.

22.8.2 Applications of Organic Adhesives in the Energy Sector

With the development of a wide variety of adhesives, due to the growing demand, adhesives have also been recently known to be used in energy sectors, specifically for energy conservation. Although at the very initial stage now, researchers have been going on to utilize the adhesives, with the viewpoint of energy conservation.

Epoxy adhesives have been in use for large wind turbine blades to obtain maximum output from a windmill, with minimum energy requirement. The major role of epoxy adhesives is to fabricate rotor blades, to meet the required mechanical strength, and to increase the manufacturing speed (Thomas and Ramachandra 2018). Certain bio-compatible adhesives are also used against biofouling, for environmental remediation (Vinagre et al. 2020). Considering the aspect of energy minimization, apart from cost and environmental balance, the use of adhesives with high strength is being considered as a technical alternative to welding, riveting, soldering, etc. (Wegner 2010). In addition, emerging techniques are being used, for curing the adhesives applied in electronics packaging. Controlled microwave frequency can be used in this method, to reduce the time frame of curing (Fathi et al. 1996). Recently, biomimetic adhesives have been in use for multiple applications, apart from the bonding substrates, thus bringing the natural adhesion phenomena to the market. By studying the structural properties, the polymers were designed in such a way that they provide the characteristics features of being an adhesive of choice, such as underwater adhesion, minimum toxicity, low cost, and high strength (Putnam 2020). Fiber-reinforced polymer-based adhesives have been recently developed for reversible applications. The reversibility helps in recycling and also contributes towards a circular economy. The bonding/debonding technology uses mechanical methods, thermal methods, use of foaming agents, etc. (Goodenough et al. 2023).

Keeping in mind the concern of energy conservation, recently many industries have started to develop cementitious adhesives, as per the standard of the ETICS (external thermal insulation composite system). ETICS has been in use for the last 60 years, and it has been found to be suitable for households. The use of ETICS has

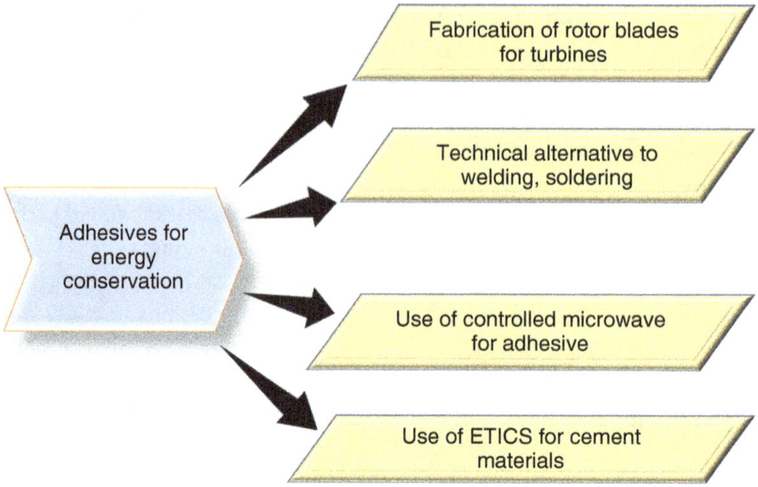

Figure 22.5 The various applications of adhesives for energy conservation.

shown a great response toward maximizing efficiency, in addition to energy conservation and environment safety (Czernik et al. 2020). The various applications of adhesives for energy conservation are depicted in Figure 22.5.

Although adhesives are prominently used for bonding various materials, however with progress in research, many adhesives have been synthesized which can be used for energy conservation. A special focus has been laid on the development of adhesives, with minimum energy requirement, thus contributing toward sustainable development.

22.9 Disadvantages of Organic Adhesives

1) Most of the advanced adhesive formulations require the use of toxic reagents, which cause ill effects on the health and environment. The starting materials are often toxic, and the dispersion medium (organic) often emits vapors, which are harmful to the environment, and causes various respiratory diseases in humans and animals.
2) The World Health Organization (WHO) and government agencies have been emphasizing minimizing the use of organic-solvent-based adhesives. Thus, the need for water-borne adhesives is increasing, owing to safety concerns. Although safe to use, water-borne adhesives require a much longer time to cure and are limited to be used only on some specific materials, mainly cellulose-based, such as paper and wood.
3) The disposal of unused adhesives is also a matter of concern for both environment and health. Most of the adhesives come with limited shelf life and are prone to dry up too soon, thus left unused and dumped.

4) Most of the adhesives available are difficult to carry, as the glue fluid may get spilled. Although few convenient adhesives have been developed for paper use, most adhesives with higher strength are often difficult to store or transport.
5) Convenient adhesives would be those, which can be stored in powder form for a longer time. These adhesives can be mixed with the required solvent medium and used when needed. But, at the present stage, there has been very little work done to develop powder adhesives.
6) The necessity of a fluorescent water-borne has been felt for a long time. Fluorescent glues find great applications in crafts and designs to easily detect fractured joints and dentistry. But unfortunately, most of the reported adhesives today are not fluorescent, and further development in fluorescent adhesive production is needed.

22.10 Conclusion

Adhesive or glue is essential for our daily use and has been used by humans since ancient times in different forms. Many advanced synthetic adhesives have been under various services, characterized by different types and groups. The adhesive industry is one of the fastest-growing sectors in the current era. A country's economy depends on adhesives' market prospects and has become a leading market. The present market value of the adhesive industry can be estimated to be around 67.5 billion USD. Adhesives are essential for various applications, which include packaging, transportation, construction, and medicines. Different types of adhesives are used for different purposes, which are used based on their properties and the nature of the adherends. Standard testing protocols have been set up by the ASTM, according to which the adhesive strengths of the materials are determined. Although a material of utmost necessity, certain ill effects of adhesives cannot be ignored. Research around the globe has been in progress toward the formulation of advanced adhesives, which will eventually lead to a boost in the adhesive/glue industry.

References

Ahmadi, Z. (2019). Nanostructured epoxy adhesives: a review. *Progress in Organic Coatings* 135: 449–453.

Anderson, K.J. (1990). Adhesive materials. *MRS Bulletin* 15 (8): 78–79.

ASTM, C. (1958). *ASTM Standards*. Philadelphia: American Society for Testing Materials.

Baig, S. (2009). History of cyanoacrylate Bristol University, UK Molecule of the Month, July 2009.

Barnes, T. and Pashby, I. (2000). Joining techniques for aluminium spaceframes used in automobiles: part II–adhesive bonding and mechanical fasteners. *Journal of Materials Processing Technology* 99 (1–3): 72–79.

Bhakare, M.A., Lokhande, K.D., Bondarde, M.P. et al. (2023). Dual functions of bioinspired, water-based, reusable composite as a highly efficient flame retardant and strong adhesive. *Chemical Engineering Journal* 454: 140421.

Brahmbhatt, D. (2017). Bioadhesive drug delivery systems: overview and recent advances. *International Journal of Chemical and Life Sciences* 6 (3): 2016–2024.

Chattopadhyay, A.K. (2014). 'Boring Boron' and adhesives. *Adhesives & Sealants Industry* 21 (9): 32.

Cui, C., Gu, R., Wu, T. et al. (2022). Zwitterion-initiated spontaneously polymerized super adhesive showing real-time deployable and long-term high-strength adhesion against various harsh environments. *Advanced Functional Materials* 32 (4): 2109144.

Czernik, S., Marcinek, M., Michałowski, B. et al. (2020). Environmental footprint of cementitious adhesives–components of ETICS. *Sustainability* 12 (21): 8998.

Delmonte, J. (1947). *Technology of Adhesives*. Reinhold Publishing.

Dinte, E. and Sylvester, B. (2018). Adhesives: applications and recent advances. In: *Applied Adhesive Bonding in Science and Technology* (ed. H. Özer), 119–134. IntechOpen.

Duarte, A., Coelho, J., Bordado, J. et al. (2012). Surgical adhesives: systematic review of the main types and development forecast. *Progress in Polymer Science* 37 (8): 1031–1050.

Fathi, Z., Tucker, D.A., Wei, B.J. et al. (1996). Rapid cure of liquid encapsulants and structural adhesives for electronics packaging using variable frequency microwave (VFM) energy. *MRS Online Proceedings Library (OPL)* 445: 125.

Goodenough, J., Fitzgerald, A., Bean, K. et al. (2023). Reversible adhesives and debondable joints for fibre-reinforced plastics: characteristics, capabilities, and opportunities. *Materials Chemistry and Physics* 299: 127464.

Guo, M. and Wang, G. (2016). Milk protein polymer and its application in environmentally safe adhesives. *Polymers* 8 (9): 324.

Hon, D.N.-S. (2003). Analysis of adhesives. In: *Handbook of Adhesive Technology* (ed. A. Pizzi and K.L. Mittal), 293–324. New York: Marcel Dekker, Inc.

Jewkes, J., Sawers, D., Stillerman, R. et al. (1969). Bakelite. *The Sources of Invention* 233–234.

Jin, F.-L., Li, X., and Park, S.-J. (2015). Synthesis and application of epoxy resins: a review. *Journal of Industrial and Engineering Chemistry* 29: 1–11.

Jovanović, R. and Dubé, M. (2004). Emulsion-based pressure-sensitive adhesives: a review. *Journal of Macromolecular Science, Part C: Polymer Reviews* 44 (1): 1–51.

Kim, E., Jung, J.-S., Yoon, S.-G., and Park, W.H. (2023). Eco-friendly silk fibroin/tannic acid coacervates for humid and underwater wood adhesives. *Journal of Colloid and Interface Science* 632: 151–160.

Maulana, M.I., Lubis, M.A.R., Febrianto, F. et al. (2022). Environmentally friendly starch-based adhesives for bonding high-performance wood composites: a review. *Forests* 13 (10): 1614.

Maurya, S.D., Kurmvanshi, S., Mohanty, S., and Nayak, S.K. (2018). A review on acrylate-terminated urethane oligomers and polymers: synthesis and applications. *Polymer-Plastics Technology and Engineering* 57 (7): 625–656.

Milia, E., Cumbo, E., Cardoso, J.A., and Gallina, G. (2012). Current dental adhesives systems. A narrative review. *Current Pharmaceutical Design* 18 (34): 5542–5552.

Miller, M.S., O'Kane, J.C., Niec, A. et al. (2013). Silver nanowire/optical adhesive coatings as transparent electrodes for flexible electronics. *ACS Applied Materials & Interfaces* 5 (20): 10165–10172.

Mittal, K.L. and Pizzi, A. (2003). Historical development of adhesives and adhesive bonding. In: *Handbook of Adhesive Technology, Revised and Expanded* (ed. A. Pizzi and K.L. Mittal), 9–20. CRC Press.

Muhammad, A.E., Abd-Elgadir, H.G., Abdalla, M.E., and Barakat, T.A.M. (2019). Adhesion properties of commercial wood glue. *South Asian Research Journal of Engineering and Technology* 1 (1): 1–6.

Packham, D. (2009). Adhesive technology and sustainability. *International Journal of Adhesion and Adhesives* 29 (3): 248–252.

Padula, C., Nicoli, S., Aversa, V. et al. (2007). Bioadhesive film for dermal and transdermal drug delivery. *European Journal of Dermatology* 17 (4): 309–312.

Petrie, E.M. (2007). *Handbook of Adhesives and Sealants*. McGraw-Hill Education.

Putnam, A.A. (2020). *Designing Functional Biomimetic Adhesives: Bringing Nature's Methods to Market*. Purdue University Graduate School.

Raja, P.R. (2016). Cyanoacrylate adhesives: a critical review. *Reviews of Adhesion and Adhesives* 4 (4): 398–416.

Rieland, J.M., Hu, Z., Deese, J.S., and Love, B.J. (2023). Pressure sensitive adhesives for quantifying microplastic isolation. *Separation and Purification Technology* 307: 122819.

Rozelsky, E. (2021). *Thermoplastic-Thermoset Bonding via Reactive Interfaces*. Clemson University.

Shokri, M., Kharaziha, M., Dalili, F. et al. (2022). Strong and bioactive bioinspired biomaterials, next generation of bone adhesives. *Advances in Colloid and Interface Science* 305: 102706.

Smeets, N.M. (2013). Amphiphilic hyperbranched polymers from the copolymerization of a vinyl and divinyl monomer: the potential of catalytic chain transfer polymerization. *European Polymer Journal* 49 (9): 2528–2544.

Tenorio-Alfonso, A., Sánchez, M.C., and Franco, J.M. (2020). A review of the sustainable approaches in the production of bio-based polyurethanes and their applications in the adhesive field. *Journal of Polymers and the Environment* 28: 749–774.

Thomas, L. and Ramachandra, M. (2018). Advanced materials for wind turbine blade – a review. *Materials Today: Proceedings* 5 (1): 2635–2640.

Tout, R. (2000). A review of adhesives for furniture. *International Journal of Adhesion and Adhesives* 20 (4): 269–272.

Upadhyaya, S. and Sen Sarma, N. (2022). Biodegradable plastics as a substitute to traditional polythenes: a step toward a safer environment. In: *Environmental Sustainability and Industries* (ed. P. Singh, J.P. Bassin, S. Rajkhowa, et al.), 193–215. Elsevier.

Upadhyaya, S., Konwar, A., Chowdhury, D., and Sen Sarma, N. (2020). High-performance water-borne fluorescent acrylic-based adhesive: synthesis and application. *RSC Advances* 10 (43): 25408–25417.

Verma, P.K., Kumar, V., Kaushik, P., and Yadav, A. (2014). Bryophyte invasion on famous archaeological site of Ahom Dynasty 'Talatal Ghar' of Sibsagar, Assam (India). *Proceedings of the National Academy of Sciences, India Section B: Biological Sciences* 84: 71–74.

Vinagre, P.A., Simas, T., Cruz, E. et al. (2020). Marine biofouling: a European database for the marine renewable energy sector. *Journal of Marine Science and Engineering* 8 (7): 495.

Wadley, L. (2010). Compound-adhesive manufacture as a behavioral proxy for complex cognition in the Middle Stone Age. *Current Anthropology* 51 (S1): S111–S119.

Watcharakitti, J., Win, E.E., Nimnuan, J., and Smith, S.M. (2022). Modified starch-based adhesives: a review. *Polymers* 14 (10): 2023.

Webster, I. (1997). Recent developments in pressure-sensitive adhesives for medical applications. *International Journal of Adhesion and Adhesives* 17 (1): 69–73.

Wegner, J. (2010). "The ongoing substitution process will speed up" sustainability in the adhesives and sealants industry. *Adhesion Adhesives & Sealants* 7: 7–9.

Zhang, J.-t., Lu, H.-n., and Cao, F.-q. (2008). Ultrasonic image diagnoses for multi-layered adhesive structure in solid rocket motor. *Journal of Propulsion Technology-Beijing* 29 (1): 13.

23

Remediation of Environmental Toxins with Porous Organic Molecules

Shambhawi Singh, Jyoti Sarwan, Shilpa Shalini, Nazim Uddin, Komal Mittal, and Jagadeesh C. Bose K

University Institute of Biotechnology, Chandigarh University, Gharuan, Mohali, Punjab, 140413, India

23.1 Introduction

There are so many environmental impacts and effects issues that two of the United Nations' sustainable development objectives have been set aside for them: clean water and excellent health and well-being. Researchers in several disciplines are working on solutions to clean up pollution in the environment. By creating materials that may act as catchers of hazardous chemicals existing in polluted soil, water, or air, material scientists have made a significant contribution to this objective. An efficient adsorbent material must react non-covalently with contaminants to avoid dissociation throughout adsorption, but the contact must be weak enough to be reversible. Some of the best prospects in this approach are organic macrocycles (Skorjanc et al. 2021). They offer strength, insolubility, and stability to covalent polymeric structures, providing additional benefits (Beuerle and Gole 2018). While some of these systems have been made available for purchase, others are yet to be fully investigated. Air damages soil, and water destroys people's health and various ecosystems. Adsorption has been a widely used method for removing pollutants, but modern substances scientific research is contributed to developing sorbent materials that are more effective, affordable, and scalable (Zia et al. 2020). Porous organic polymers (POPs) contain functional groups and voids that can build supramolecular interactions to sequester contaminants (Isaeva et al. 2021). The insolubility of such compounds greatly enhances their potential for regeneration and recycling.

POPs are penetrable or permeable materials constructed by the process of polymerization in which organic building blocks show strong covalent interaction. Due to their unique stability, POPs hold tremendous interest (Sarkar et al. 2021). They have attributes including large surface area, and configurable porosity, are simple to synthesize, and are easy to process for technological applications giving them an edge compared to all other porous materials (Wang et al. 2020). Advances in organic synthesis enable the fabrication of a wide range of porous polymers with much chemical functionality incorporated inside the microporous or on the surfaces. The

Organic Polymers in Energy-Environmental Applications, First Edition.
Edited by Ramesh Oraon, Pardeep Singh, Sanchayita Rajkhowa, Sangita Agarwal, and Ravindra Pratap Singh.

application of POPs in several areas, such as selective permeation, catalysis, clean energy storage, gas storage and adsorption, sensing, molecular motors, and photoconductors, is now being extensively investigated (Wang et al. 2021). According to their pore diameters, POPs are often divided into three classes: macroporous (pores larger than 450 nm), mesoporous (2–50 nm), and microporous (2 nm) (Ranganatha 2020). The study of porous materials with customized functions generated from basic molecular building blocks is now a key area of research in material science. POPs can be classified as either crystalline or amorphous analogs, depending on the level of crystallinity. The crystalline form of POPs includes covalent organic frameworks (COFs) that are standard, while amorphous POPs primarily comprise conjugated microporous polymers (CMPs), polymers of intrinsic microporosity (PIMs), hyper-crosslinked polymers (HCPs), porous aromatic frameworks (PAFs) (Luo et al. 2020). Usually, in the process of their production, reversible processes often result in organized framework structures, whereas irreversible reactions invariably result in amorphous porous polymers due to their advantages of inherent porosity, low weight, resilience, tunable architectures, and custom functionality as well as their prospective applications in environmental remediation (Korde and Kandasubramanian 2019).

POPs are distinctive due to a combination of attributes not present in other materials:

1) Complaisant production permits uptake performance for compositional engineering by varying the cluster of composition (Valencia et al. 2020).
2) Stability across a variety of chemical circumstances (such as very reductive or oxidative environments or acidic or basic environments), leading to long-term steady performance and possible recycling to streamline workup procedures (Ushani et al. 2020).
3) The large surface area and adjustable pore size provide high absorption capacity and controlled mass transfer (Xu et al. 2015).
4) POPs have featured the composition of a lightweight element with the potential for strong gravimetric performance in terms of energy and molecular absorption (Liu et al. 2020).
5) They have minimal density. As light components (C, H, N, O, B, etc.) comprise the majority of POPs, the comparatively low density can culminate in a remarkable gravimetric performance, which raises the adsorption capacity (Ding et al. 2022).
6) POPs show structural adaptability due to the variety of building blocks and the rise of reticular chemistry, and a wide range of POPs with rational functional groups may be easily developed and synthesized.
7) POPs are made without the involvement of metal ions, unlike metal–organic frameworks (MOFs), preventing the danger of metal ion emission and secondary pollution (Karmakar et al. 2022).

With these characteristics, POPs represent the most recent development in high-performance adsorbents, showing significant value as potentially "game-changing" advancements in the area of selective sequestration.

Polymers have a wide range of applications. Biodegradable polymers are commonly used in the biomedical industry for tissue engineering and drug delivery.

As support, organic polymers are frequently used because of their stable, cheap, nontoxic, and processable qualities. High efficiency and activity are constrained by limited dispersion in solvents solution, traditional polymers' active sites are exposed, and there is an interaction between substrates and catalysts. Single-atom catalysis may be a feature of hierarchically porous materials made of monomers with coordination sites or catalytic functional groups.

A straightforward POPs synthesis often yields a network with a high microspore concentration leading to a broad surface area value. Figure 23.1 is an illustration of the different applications of POPs. These micropores are restricted to the catalysis of gaseous molecules and are helpful for gas adsorption applications. However, by generating steric limitations leading to constraining catalytic efficiency, these pores frequently lead to ineffective reactant and substrate transport. Meso- and/or macropores are hence more advantageous for superior POP catalytic applications that achieve broad substrate scopes. POPs are superior porous molecules for wastewater treatment because of their distinctive crystalline properties, diverse topologies, and compositions. POPs with configurable functionalization, programmable structures, and adjustable pore size may be produced as a result of organic chemistry's adaptability. These characteristics, along with POPs' outstanding thermal and chemical stability, have given them the power to survive harsh conditions for future effective adsorption applications. POPs are also projected to be of significant importance for wastewater treatment shortly due to their simple separation processes and low energy needs. It will be appealing to combine with other approaches like

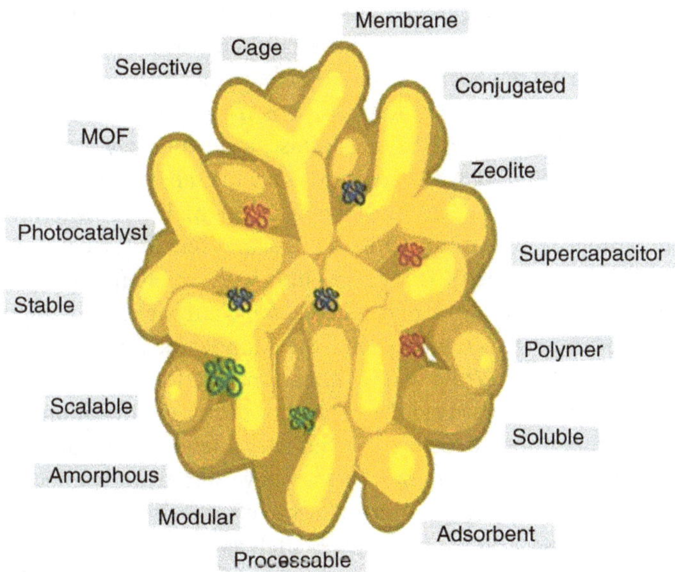

Figure 23.1 POPs and their functions. Source: Adapted from https://images.app.goo.gl/gWkVbT7FHWm7yLmG8. Kewley et al. (2015).

photocatalysis, which converts extremely hazardous pollutants into non-toxic or less toxic chemicals while also having a high removal efficiency, to achieve effective wastewater management. Due to the continued efforts of researchers from around the world, it is projected that POPs will be able to function as adaptive adsorbent materials for the efficient removal of various metallic ions in actual water samples in the next years.

Organic polymers are mostly microporous varying with pore sizes. Their design ability has to be performed in a controlled manner. Amorphous and ionic organic networks are amorphous, and cannot be crystallized, while metal–organic frameworks are highly crystalline. Instability trend out of the four, amorphous porous organic polymer shows the highest, followed by covalent and nanoporous ionic organic polymer leaving metal–organic framework with the least stability in comparison. Metal and covalent organic polymers can be self-assembled based on their intermolecular interactions and bonding. However, amorphous porous polymer requires condensation and nanoporous ionic networks require template synthesis A decrease in micropore concentration and a rise in macropore porosity are typical causes of a decrease in BET surface area (Kreisberg and Antropova 2014). It should be mentioned that a considerable surface area is required for catalysis activity at the interfaces of catalysts, as in the degradation of dyes since it enhances the interfacial interaction between substrate and catalyst. Therefore, POP substrates with a considerable surface area and hierarchical porosity are advantageous for catalytic applications. POPs have advantages over other porous materials. The exploration of porous materials has significantly advanced, progressing from inorganic zeolites to coordination polymers or metal–organic frameworks (MOFs), and more recently covalent organic frameworks (COFs) (Ma et al. 2022). It is currently needed to analyze the advantages and drawbacks of POPs against all other porous materials. Heterogeneous catalysts must have both chemical and thermal stability. Although COFs and MOFs have a high surface area as well as porosity, they frequently exhibit significant moisture and pH sensitivity, which can cause structural collapse (Yoon et al. 2021). POPs, in contrast, are extremely stable in practically all major solvents, including water, and across a wide range of pH values. Additionally, POPs, particularly CMPs, are very stable under high hydrothermal conditions, and temperatures and continue to be a great alternative for catalysts in tough circumstances (Tao et al. 2020). Though zeolites have excellent chemical and thermal stability as well, they are a poor choice for catalysis due to their difficulty incorporating specified functionality into the framework for particular applications (Ding and Wang 2013). While on the other hand, POPs are even more user-friendly due to their straightforward synthesis and structural modification. As a result of the growth of contemporary society and industry, metallic ions, organic micro-pollutants, radioactive elements, and other toxins have significantly affected our natural environment. Therefore, the question of how to effectively remove these contaminants is crucial. Because of POP's advantages such as low cost, easy operation, and high efficiency, porous materials have been extensively explored for the separation and absorption of pollutants at this time. The performance of traditional porous materials, such as activated carbon and molecular sieves, can't be improved by further enhancement of their segregation and adsorption performance. Even though MOFs have been studied as potential pollution removers, they exhibit poor chemical stability, and the hazardous metals produced after the structural disintegration of MOFs cause secondary contamination (Dias and Petit 2016). Therefore, POPs are exciting alternatives to their traditional equivalents. Due to their porosity hierarchies, host–guest interactions, ease of structure-and-function tuning, and high chemical stabilities, macrocycle-based POPs have shown exceptional and great performance in pollutant removal. Figure 23.2 describes the various applications in bioremediation with POPs. Here, we focus on the removal of micropollutants, charged species, metal ions, oils and organic solvents, perfluorinated substances, iodine, and volatile organic compounds.

23.2 Parameters of Oxygen-Rich POPs in Dye Adsorption from Water Sources

One of the biggest causes of water pollution is organic dyes in effluent from the paper, textile, plastic, and food sectors. They are extremely dangerous to the aquatic system and may be harmful to human health. The most straightforward and effective method of purifying water of pollutants is adsorption (Rashid et al. 2021). Activated carbons, zeolites, and porous organic polymers (POPs), among other porous materials with high specific surface area and rich porous architectures, have all been employed as adsorbent materials to effectively remove dyes from wastewater. POPs are one of the most attractive and interesting fields, because of their distinct benefits, including ease of functionalization, adjustable porosity, and structural diversity.

When POPs are added to oxygen-containing functional groups it could result in certain unique features. Firstly, the occurrence of oxygen-containing compounds can make POPs more hydrophilic and increase the region in which they come into contact with dye molecules in water (Yang et al. 2021). Because of its notable characteristics, like huge pore size, accurate balance, and broad surface area robustness in various solvents, it recently turned out to be a growing study subject in porous substances. These features of POPs are favorable for mass transfer, providing more lively sites, and emerging as a promising platform for the recycling process in photocatalysis (Zhang et al. 2019). POPs can be fabricated reversibly or irreversibly. For reversible synthesis, Koil and organic frameworks are utilized. However, for a reversible we can perform Friedel–Crafts alkylation, coupling of carbon bonds, imine formation, or cyclotrimerization additionally, the oxygen-rich subunits can make the functionalized POPs more electronegative, affording additional electrostatic attraction with cationic dye molecules. Adding hydrophilic hydroxyl groups can increase the ability of CMPs to adsorb dye (Fan et al. 2022) and developed the novel triterpene and crown ether-based highly permeable copolymer POP-TCE-15, and they recognized that the methylene blue which is a positively charged cationic dye, and the crown-ether-15 which contains the lone pair of electrons on O-atom have electrostatically interacted (Xu et al. 2018). POP-O exhibits a hysteresis loop and both Type I and Type IV absorption isotherms. The BET model determines

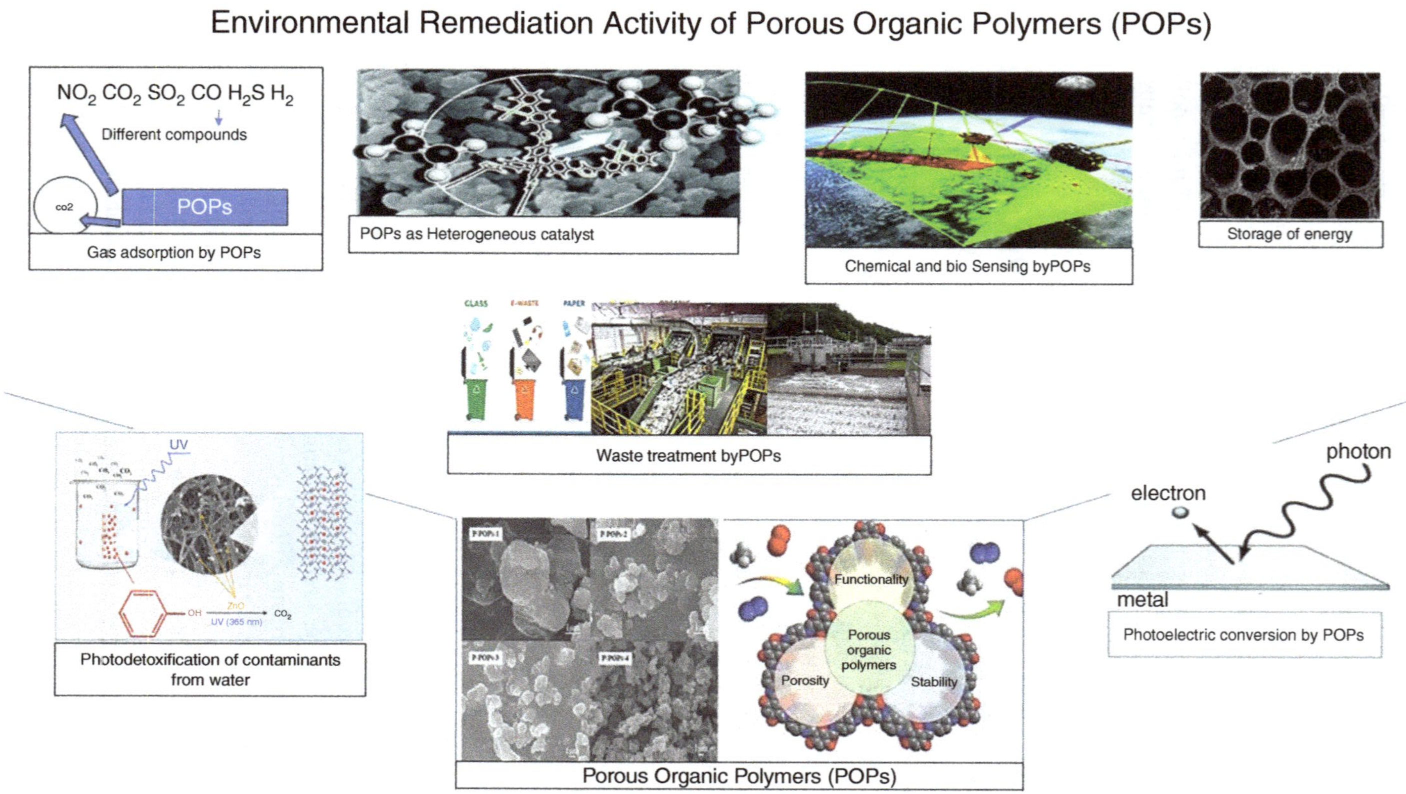

Figure 23.2 POPs including various methods of environmental remediation. Source: Gudellaphoto/Adobe Stock; Elena Istomina/Getty Images; nordroden/Adobe Stock.

that the POP-O's surface area is 619 $m^2 g^{-1}$. POP-O mostly contains micropores that are 0.9 nm in size, mesopores that are 2.2 nm in size, and a few mesopores that are 5–40 nm in size (Yang et al. 2021). The outcomes show that the polymer has a hierarchical pore structure. The hierarchy of pore structure incorporates the strengths of micropores and mesopores, which could accelerate the mass transfer rate of the substance and make it simpler for dye molecules to enter POP-O, expanding its potential for application in the adsorption industry. The porosity of POP affects its ability to adsorb substances, but it also significantly affects both the size of the particles and the porosity in the solution. The size distributions of the POP-O particles are 100–180 nm and 1000–1700 nm. The large-sized molecules originating from the clustering of POP-B nanoparticles show only one point focused at 1414 nm. The small particulate matter of POP-O is advantageous for dye molecule adsorption. POPs soak up more or equal amounts of energy from photons than their gaps between bands. In gasoline repositories and their detachment, catalysis, photoelectric alteration, energy conversion and storage, and many others, POPs have acquired an interest in research and attention due to their advantages of pre-designable structures, lightweight, higher essential permeability, and first-rate stability.

POP-O may have effective adsorption properties for organic dye contaminants from water due to its abundance of oxygen-rich units and hierarchical pore structure. To investigate the adsorption capabilities of oxygen-enriched POP for water-soluble contaminants, rhodamine B (Rh B) is utilized as the adsorbate. One of the most important characteristics of an adsorbent that directly affects its actual adsorption effectiveness is fast kinetic adsorption capacity. The strong adsorption capability of oxygen-enriched POP for RH B is primarily attributed to various aspects. First, the POP-O has strong hydrophilicity because it contains the hydroxyl and carbonyl groups, which makes it simpler to produce a homogeneous dispersion in Rh B aqueous solution (Pech-Canul et al. 2020). Second, POP-O becomes electronegative due to the oxygen-rich units, allowing for even stronger electrostatic linkage with cationic dye molecules (Rh B). Last, but not least, the huge unique surface area of POP-O offers a substantial volume for the storage of dye molecules during adsorption. Thus, the structural characteristics of the POP-O, large surface area, numerous oxygen-containing groups, and hierarchical pore architectures, combine to provide good adsorption capability of Rh B dyes extracted from aqueous solutions. Figure 23.3 describes the artificial pathway of synthesis of polymers.

23.3 TALPOPs Include Reversible Iodine Detection and Extraction

Numerous adsorbent substances have been investigated and evaluated for the efficient sensing and sequestration of iodine. According to some reports, the adsorbent's surface area, particularly high-affinity binding sites, pore size, conjugated units, and polar groups all have a role in determining how effectively iodine is sequestered (Yildirim et al. 2022). Thus, in addition to the expanded surface area, boosting the host's affinity for iodine may also have a significant impact on iodine collection. It

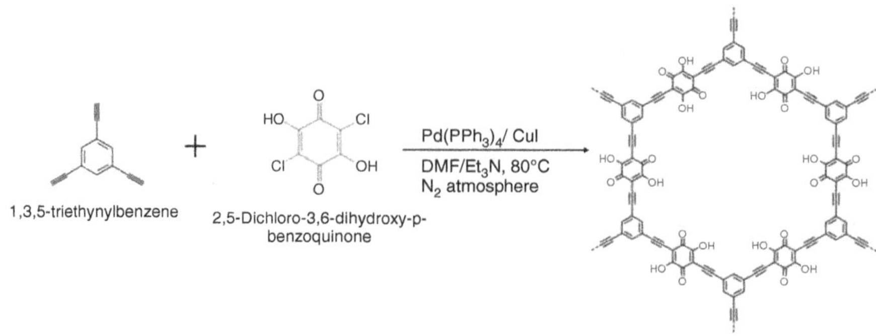

Figure 23.3 Artificial pathway for synthesis of the polymer POP-O. Adapted from https://images.app.goo.gl/NocLovSuyCAPmXYB8. Ong et al. (2020).

has been observed that substances such as silica, activated carbon, silver-doped zeolites, microporous polymers, and chalcogenide aerogels have effective radioiodine adsorption capabilities (Chee et al. 2020).

Using the metal-independent-Schiff base polycondensation technique, a new TALPOP (triazene-anthracene-based fluorescent animal-linked porous organic polymer) is created by combining 2-furaldehyde, 9,10-bis-(4,6-diamino-S-triazin2-yl) anthracene (DTA), and the core of the envisioned POP. 2-Furaldehyde serves as an organic linker in this structure. The polymer has excellent porosity and exceptional thermal and chemical stabilities, with a Brunauer–Emmett–Teller surface area of 401 $m^2 g^{-1}$ (Sabri et al. 2020). Such porosity combined with the high concentration of conjugate heteroatom-rich structure allows the polymer molecule to produce a remarkable vapor of iodine, which is uptake to 314 wt% and also found reversible adsorption of iodine in solution (Lustig et al. 2017). Due to the presence of anthracene moieties, the TALPOP shows iodine detection with excellent sensitivity through fluorescence quenching with $2.9*10^3$ l mol^{-1} of K_vs value. Because of its low cost and excellent iodine absorption and sensing, TALPOP is a perfect material for environmental restoration. Nitrogen adsorption–desorption isotherm at 77 K is used to determine the precise surface area and pore characteristics of TALPOP. The polymer is generally microporous with some mesoporous pore size distribution, as evidenced by the steep increase of isotherm in nitrogen gas absorption at a low relative pressure (P/P_0 0.01) and lower uptake at the range of 0.01–0.8 P/P_0 (Sabri et al. 2020). Due to their specific characteristics of electron-rich nature and large surface area, these POPs can acquire a significant quantity of iodine. Figure 23.4 is an illustration of the synthetic route for the preparation of the synthetic polymers including TALPOP.

23.4 Sulfur- and Nitrogen-Rich Hierarchically POPs for Adsorptive Expulsion of Mercury

Mercury poisoning is a critical environmental problem because the high concentration of mercury in the air, water, or soil causes harm to humans. The best method to

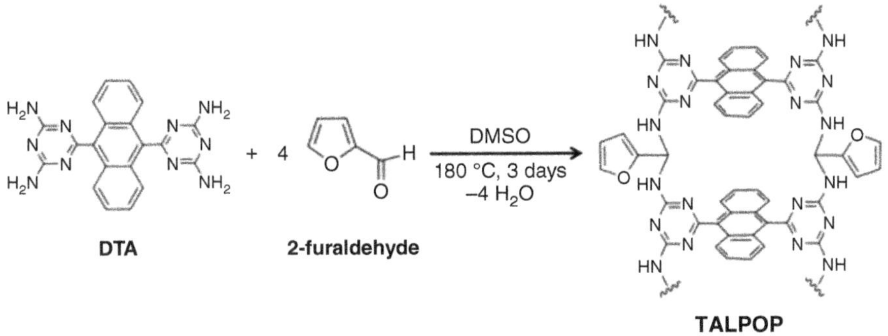

Figure 23.4 Synthetic route for the preparation of the TALPOP. Source: Sabri et al. (2020)/Springer Nature.

resolve this problem is mercury adsorption removal (Chowdhury et al. 2021). Firstly, the adsorptive removal of mercury is quite demanding for nitrogen- and sulfur-rich large surface area POPs, because high surface area improves the adsorption capacity of Hg(II) and these elements present at the pore structural surface of the material provide strong linkages with Hg(II) ions (Zhang et al. 2019). Second, TPP-1 (secondary amine-linked, S-containing porous organic polymer), which has a very large surface area and hierarchical porosity, also removes mercury (Kostova et al. 2013). It was created by combining a monoaldehyde and thiophene-2-carbaldehyde or tetramine and the tetrapodal amine 6,6-(1,3,5-triazine-2,4-diamine) with a Schiff base, and it showed extremely high uptake for organic methylmercury (691 mg g^{-1}) and Hg^{2+} (3106 mg g^{-1}) (Chowdhury et al. 2021). Theoretical calculation or DFT (density functional theory) calculation and AIM analysis support the strong Hg absorption and mercury connection between the S and N-sites of TTP-1. There is a very strong bond between the S and N lone pair through noncovalent interactions. Therefore, the occurrence of strong-to-moderately stronger non-covalent linkages between the ligand sites of the polymer surface and Hg^{2+} can cause significant Hg^{2+} adsorption by this porous organic polymer material (Modak et al. 2020). The great potential of TTP-1 for the adsorptive elimination of both inorganic and organic mercury can have considerable positive effects on the environment and promote the development of a long-lasting adsorbent for the treatment of wastewater.

23.5 Novel N-Enriched Covalent Crystalline POPs for Efficient Removal of Cadmium

The development of effective adsorbents with highly abundant active sites, to enable fast absorption and high capacity for pollutants like cadmium, is one of the rising concerns in environmental remediation (Algarni and Al-Mohaimeed 2022). Cadmium and its constituents are more soluble than other heavy metals, and they can easily enter biological systems. In contrast, the high toxicity of Cd(II) promotes serious illnesses such as hypertension, nausea, skeletal deformation patterns, and

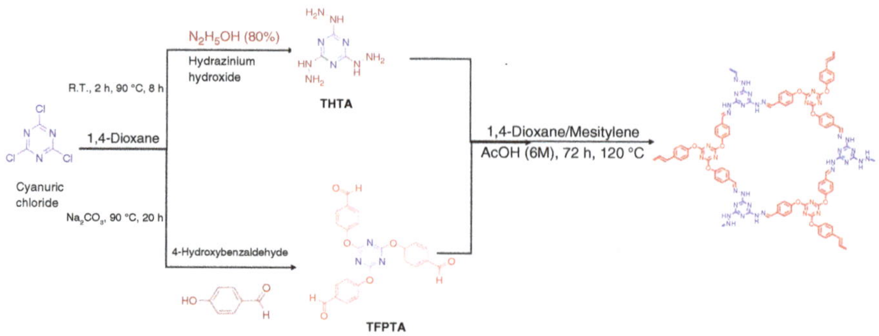

Figure 23.5 Synthesis of N-enriched covalent crystalline POPs. Source: Dinari and Hatami (2019)/with permission of Elsevier.

kidney stone formation. The research of cadmium removal has thus received a lot of attention over the past decade, and a variety of techniques including nanofiltration, ultrafiltration, chemical precipitation, reverse osmosis, ion exchange, and biological treatment have all been developed (Awual 2019). However, these systems have drawbacks such as complicated operations, high-energy consumption, and expensive running costs. The suitable adsorbent for eliminating cadmium from aqueous solutions is N-enriched crystalline COF, which is generated via condensation of 2,4,6-tris(p-formyl phenoxy)-1,3,5-triazine and 2,4,6-tris(hydrazino)-1,3,5-triazine. An imine-linked porous COF with sufficient nitrogen that provided greater performance in cadmium removal was rationally developed and synthesized, and it is shortly referred to as N-enriched COF (Dinari and Hatami 2019). After that, a flexible coupling block and a monomer are derived from a triazine, which causes the network resonance following polymerization. Lone pair electrons are so freely accessible and able to interact with the metal ions. By increasing the pH of the solution and making the functional centers of the adsorbent accessible, an anionic form of COF is created. At pH = 7, the optimum option is the maximum elimination of cadmium (98%) (Pérez-Marín et al. 2007). Figure 23.5 describes the synthesis of N-enriched covalent crystalline porous organic polymers. The elimination of cadmium is examined at various intervals between 5 and 60 minutes under 80 ppm concentration and optimal pH = 7.

23.6 Novel Phenyl-Phosphate-Based POPs for the Elimination of Pharmaceutical Water Contaminants

Living species depend on clean water systems to survive; thus, removing new toxins from aquatic systems is a fundamental research concern to restore the equilibrium of the ecosystem regarding a sustainable society. Due to rising manufacturing and utilization of synthetic chemicals, water system poses a serious issue to people and aquatic life. Many pollutants, commonly known as micropollutants, increase toxicological effects even at a trace level. Besides these, pharmaceuticals and personal care products (PPCPS) and pharmaceuticals such as antiepileptic drugs, anti-inflammatory medications, and antibiotics are claimed to be frequently observed in surface water (Liu and Wong 2013), in fresh vegetables, in wild animals, and even in sediments. Different types of therapeutic strategies have been investigated to remediate PPCPs in water. These include membrane filtration, adsorption, photocatalytic degradation, and enhanced oxidation. To chemically destroy PPCPs and mineralize them, photocatalytic degradation and advanced oxidation utilize highly reactive agents such as hydroxyl radicals; however, these pathways frequently have large energy demands or exhibit deficient effectiveness, primarily when background organics are abundant (Rosman et al. 2018). The reaction between 2,3,4-hydroxybenzaldehyde and phenyl trichlorosilane in the presence of a base and further reacting with benzidine with glacial acetic acid is to obtain an organic polymer. Various phases of clinical trials have been done to check the efficacy of therapeutics and determine whether the formulation of organic polymer is safe or not. Phase 1 is generally associated with the dosage and safety of polymer for its utilization in the treatment of disease. In this phase, tolerance and side effects are observed to decide whether it is safe to use or not. Under phase 2, the effectiveness of the molecule is observed and evaluated in correspondence to that particular disease. If no side effect is observed, then it is considered as good to go for the third phase. In the third phase, we determine the efficacy and reactions of the molecule as a treatment aid, and its long-term effectiveness is measured and observed. During this undesirable by-product and chemical breakdown could also develop. Because most PPCPs may move through larger high-porosity membranes, the elimination of PPCPs via membrane needs reverse osmosis or nanofiltration membranes (Shin et al. 2022). Because of its convenience in the method design and ease of usage, adsorption has generated a lot of attention for the removal of PPCP. POPs have the potential to be an effective adsorbent for the specific elimination of water pollutants like PPCPs. To maintain structural stability over prolonged usage, such polymers must be developed with caution so that the functional group does not disrupt the heterogeneous character of the adsorbent. According to a prior study, adsorbents containing phosphate functional groups may be useful for removing pharmaceutical pollutants and Sewage Treatment plants.

Particularly for weakly basic pharmacological or biological pollutants, phosphate sites inside the polymer structure should diffuse a high sorption property. It happens due to the weakly acidic nature of phosphates. The objective of this study is to construct phosphate-based POP and examine their function for the extraction of pharmaceutical components from water. Diphenyl phosphate serves as the primary functional unit precursor in the synthesis of phosphate-based porous organic polymers (P-POP), whereas 1,3,5-triphenyl benzene and 1,1,2,2-tetraphenylethylene serve as linkers. By utilizing a Friedel–Crafts procedure, two significant phosphate-based POPs, P-POP-1, and P-POP-2 are produced. Diphenyl phosphate as precursors are for P-POP-1 and 1,1,2,2-tetraphenylethylene for P-POP-2 (Ravi et al. 2020). The phenyl ring structures in the porous polymer are linked at various substitution sites. In contrast, one linkage may join two phenyl rings at their ortho locations, while another would connect them at their para positions. P-POPs are most likely to be polymeric materials with repeating molecules composed of phenyl phosphate and aromatic hydrocarbons (Zhou et al. 2022), but they have an amorphous polymer matrix, according to a combination of NMR, XPS, FTIR, and

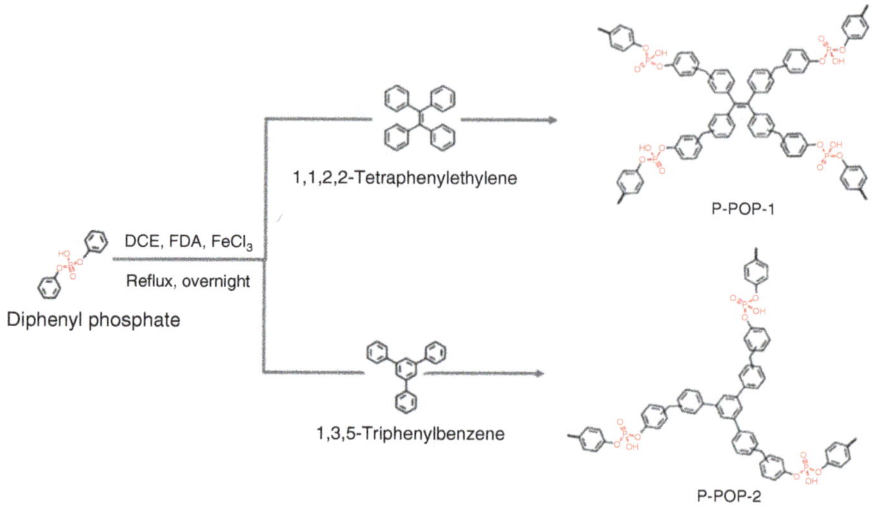

Figure 23.6 Synthesis route of phosphate-based POPs. Source: Adapted from https://images.app.goo.gl/mNXqnKPdjkxrGwyT6. Yang et al. (2018).

XRD data. N_2-adsorption desorption isotherms are used to assess pore volume, surface area, and permanent pore structure (Vinu et al. 2008). N_2-adsorption desorption isotherms are used to evaluate the pore volume, surface area, and permanent pore structure of the polymer. At relative pressures of 0.1–0.2, P-POP-1 exhibits type IV N_2 adsorption isotherms with steeper gas uptakes and illustrates the microporous nature of the polymer materials (Ravi et al. 2020). Phosphate-based POP-2 are further distinguished by type IV isotherms with strong nitrogen absorption at low pressures and weak nitrogen uptake at high pressures, i.e. 0.7–1.0 exhibits a major hysteresis loop generated by mesoporous channels of these networks. Moreover, the P-POP-2 has a mesoporous percentage that is much larger (32.5%) than the P-POP-1 (19.6%). Such mesoporosity greatly facilitates interaction between target components and the adsorbate. In contrast, the P-POP-1 adsorbs > 70% of medicinal compounds within 20 min of contact time, while the P-POP-2 absorbs > 75% over the same time frame. In the context of common cations (such as Mg^{2+}, Na^+, and Ca^{2+}) in water, P-POP-1 and P-POP-2, both demonstrate great selectivity (more than 98%) for the medicinal compounds. Figure 23.6 describes the route for the synthesis of phosphate-based POPs. Additionally, they exhibit excellent chemical stability upon repeated usage with no reduction of adsorption capability.

23.7 Nano-architecture POPs in the Elimination of Toxic Metal Ions

POPs are based on strong covalent bonds and have been involved as a potential class of porous materials. Due to their unique properties, they have been frequently utilized as adsorbents for the elimination of radioactive/toxic metal ions. Figure 23.7

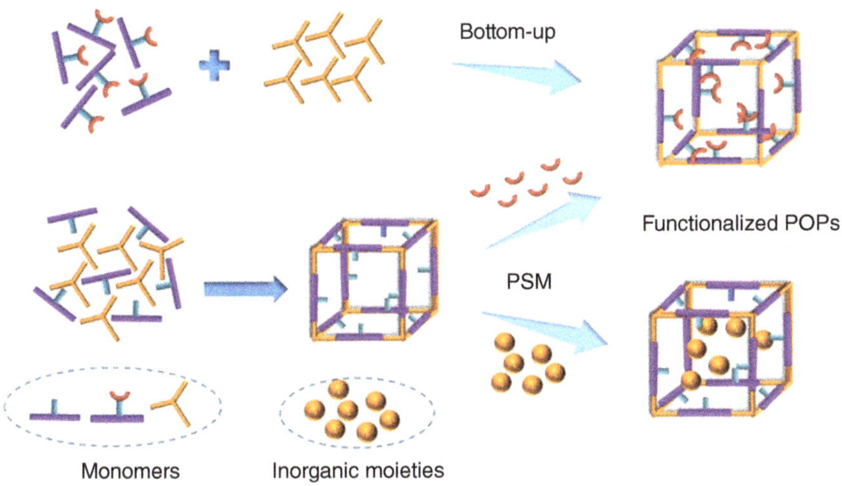

Figure 23.7 Artificial route for the formation of functionalized POPs. Source: Adapted from https://images.app.goo.gl/tLp1PSZGaP6xgzeB9. Jin et al. (2022).

shows the pathway for the synthesis of artificial pathways of functionalized POPs. Diverse heterotopic atom-doped functional groups have been introduced into the polymer backbone of POPs using post-synthetic modification (PSM) synthetic and bottom-up techniques to improve the accumulation of metal ions like chromium (Cr), mercury (Hg), lead (Pb), and arsenic (As) (Subramaniam et al. 2019), which are non-biodegradable, poisonous, and likely to accumulate in aquatic organisms, living organisms, and human beings.

23.8 PSM (Post-synthetic Modification)

PSM offers a versatile approach for manufacturing functionalized POPs with highly customizable moieties than straightforward synthesis techniques. This strategy is also adaptable and very useful for introducing specialized functional groupings, particularly for those tough to directly prepare. Through the controlled PSM approach, many functional groups, including $-NH_2$, $-COOH$, and $-SH$, have so far been effectively introduced into POPs (Raza et al. 2019). Generally, the introduction of either inorganic or organic moieties is frequently used to perform PSM. POPs have undergone a reduction reaction, esterification, click reaction, hydrolysis of esters, and other procedures to include different organic functional groups into their structure (Liu et al. 2021). Doping inorganic moieties into the host backbone, such as metal nanoparticles, metals, or metal cations, has also been effective in demonstrating the PSM approach.

23.8.1 Bottom-Up Method

The “bottom-up” process is utilized for developing heterotopic atom-doped porous organic polymer, in which functional groups are involved via methodically constructing and applying unique building blocks. To fulfill the specific requirements of synthetic materials, there are two appealing techniques. One is to customize pure monomers with numerous functional groups before the high and effective yielding distillation. This is an excellent method for bottom-up synthesis. Multiple functional groups, including –OH, $-SO_3H$, and –COOH, have been effectively involved in the skeleton of POPs in this respect (Sun et al. 2015). The most apparent benefit of this method is the pore characteristics of the well-maintained source material, avoiding pore blocking. Another wonderful bottom-up method is to utilize monomer-containing hybrid components like carbazole, thiophene, and triphenylamine. Due to the enhanced affinities or sensitivity, the inclusion of several elements (such as S, N, O, and P) not just increases the architectural complexity of POPs but is also extremely advantageous for the capture of metal ions. Tetra(thiophene-2-yl) silane is used as a functional monomer to produce a new sulfur-rich microporous polymer (SMP) with a sulfur content of 31.4 weight percent. Due to the powerful coordination reaction between soft acid mercury and soft basic sulfur, the resultant SMP with the evenly distributed sulfur atom and the extremely porous structure demonstrated a significant adsorption capability toward Hg^{2+} ($595.2\,mg\,g^{-1}$).

23.9 POPs for Gas Extraction, Segregation, and Transformation

POPs are distinctive because they are conjugated and nonporous, providing for molecular design and synthetic control over their structures. POPs have evolved into a novel category of porous material for gas storage and adsorption activities because of their huge surface areas and microporous properties. For the storage and adsorption of gases including methane, carbon dioxide, and hydrogen, POPs have been extensively investigated (Bae and Snurr 2011). A functional group can be introduced to the wall surface to change the pore habitat; in contrast, synthetic processes provide control regarding porosity. The selectivity and capability of particular gas may be improved by using this surface modification, which is a significant and efficient method (Low et al. 2017).

23.10 Hydrogen

It has been proven that porous materials' capacity for hydrogen absorption may be improved by having a wide surface area, which is good for the material's capacity for absorption. Therefore, POPs with wide surface regions have a variety of significant benefits as sorbents during H_2 storage. The first benefit is the ability to build polymers entirely of light components (Broom et al. 2019). Second, there are several synthetic pathways available by which a variety of functionality may be added. Third, polymers are a modular technology, and there are existing instances of systems that are manufactured on a global scale for use in industry (such as macroporous polymer composites for separations) (Savina et al. 2016). The ability to manufacture POPs in a sculpted "monolithic" configuration has a lesser evident benefit (Salmean and Dimartino 2019). By performing so, volumetric storage problems with the packaging of porous particulate components may be avoided.

More endeavors have been made to enhance the heat of adsorption, which is particularly necessary for potential implementation, to further maximize the hydrogen absorption of POP materials. One effective strategy to improve hydrogen efficiency could be to dope metal into the materials. By metalizing POP materials, coordinately unsaturated metal complexes (Li, Mn, Ni) may be provided. Several POPs, including F-CTF$_3$, HCP, Ni-Por-1, TNP4, N-PEINK, PCTF-4, and SHCP-3, permit the separation and storage of H_2 (Zhang et al. 2017).

23.11 Methane

Natural gas, which contains mostly methane gas, is a plentiful source of energy 180 cm^3 (STP) at 298 K and 35 bar is the DOE's target range for methane storage (Knerelman et al. 2022). The investigation of POPs for methane adsorption has received less attention than that of other gases. Organic porous materials are indeed very promising because the isosteric sorption temperatures for methane in carbonaceous materials range within the optimal condition for storage around room temperature (Thomas et al. 2009). Out of all the materials presented, Friedel–Crafts self-condensation of 50 bis chloromethyl monomers, including 4,4′-bis(chloromethyl)-1,10-biphenyl (BCMBP) and DCX (dichlorosilane), produced several hyper crosslinked polymers in 2008 that Cooper and colleagues evaluated for their CH_4 storage capacity. The BCMBP-based compound exhibits the estimated Brunauer–Emmett–Teller (BET) surface area (1904 m^2 g^{-1}) (Makal et al. 2012), and greatest CH_4 uptake (116 cm^3 g^{-1}, 5.2 mmol g^{-1}). In MOPs (microporous organic polymer), PPN-4 had the maximum methane absorption. At 55 and 35 bar, the uptake would approach 28.0 and 21.5 wt%, respectively. PPN-13 recently showed a very high 3420 m^2 g^{-1} BET surface area. At 77 K and 40 bar, the whole hydrogen absorption is 52 mg g^{-1}, whereas at 65 bar and 298 K, the overall methane absorption is 179 mg g^{-1} (Lu et al. 2014).

Some more POPs contributed to methane uptakes such as CTF-TPC, Ni-Por-1, FCDTPA-K-700, PIN1, SHCP-3, PCTF-4, BILP-12, and P5-CMP-1 (Zhang et al. 2017).

23.12 Carbon Dioxide

The greater utilization of fossil fuels has increased carbon dioxide (CO_2) emissions into the environment, which has a detrimental effect on the attempt to combat global warming. While ethanol-based hydrophilic amines have not yet been assessed as a viable choice for CO_2 exfoliation, several POPs have lower energy penalties and provide superior physiochemical properties. Adsorption using porous molecules is one of several methods for CO_2 collection that is both technically and energetically effective (Zhang 2013). In parameters relevant to either pre-combustion or post-combustion capture, a variety of POPs have just been identified for the uptake of CO_2. Because of their significant porosity and considerable tenability of pore interface functioning, POPs are incredibly exciting and can improve interactions with CO_2 significantly.

Several POPs, including PAF-60, CTF-TPC, BILP-2, HCP-BDM, ThPOP-1, HOMP, BILP-6-NH_2, etc., have been observed to absorb CO_2. These POPs exhibit great selectivity and capacity for CO_2 uptake. They can contribute to the solution and encourage scientists in this sector to use comparable material design techniques to combat climate change. The materials are ecologically beneficial due to their metal-free composition and may be carefully tailored to trap toxic gases using inexpensive catalysts, like copper (Modak et al. 2020). Yavuz and Coskun have described it as a sequence of abridged nanoporous covalent organic polymers with exceptional growth in CO_2/N_2 selectivity with temperature rise (Patel et al. 2014). It also gives the researcher a way to regulate POPs' surface characteristics, which may create novel opportunities for the logical design and synthesis of new nanocomposite materials for CO_2 uptake. This creative alternative not only allows the researcher to attain an astonishingly high CO_2 adsorption capacity but also regulates the surface characteristics of POPs.

23.13 Conclusion

In this chapter, we have outlined the main techniques used to generate POPs and divided the functions of these POPs into categories for environmental remediation such as dye adsorption, iodine detection and extraction, adsorption of mercury, elimination of cadmium, removal of pharmaceutical water contaminants, toxic metal ion eradication, gas extraction, separation, and transformation. For the storage and adsorption of gases including methane, carbon dioxide, and hydrogen, POPs have been extensively investigated. POPs are penetrable or permeable materials constructed by the process of polymerization in which organic building blocks show strong covalent interaction. A thorough analysis of the data suggests that POP's extraordinary performance in both chemical adsorptions as well as heterogeneous catalytic processes is due to the unique host–guest interactions offered by supramolecular macrocycles. POPs showed much greater efficacy in comparison to pristine materials. POPs have strong mechanical strength and can be subsequently manufactured into membranes for non-polar and polar organic chemical separation. Even though surface areas of POPs can periodically be reduced by the addition of macrocycles, their gas adsorption affinity and sensitivity are surprisingly increased. Therefore, more studies should have been focused on developing unique, high-performance POPs using eco-friendly, effective techniques. In conclusion, we provide a thorough description of the synthetic approaches and utilization of hierarchical POPs formed from macrocycles. POPs in several areas, such as selective permeation, catalysis, clean energy storage, gas storage and adsorption, sensing, molecular motors, and photoconductors, are now being extensively investigated. Simple POPs synthesis frequently results in a network with a high microspore concentration, which results in a large surface area value. An instance of the various uses for POPs is shown in Figure 23.1. These micropores are useful for gas adsorption applications and are only used to catalyze gaseous molecules. However, these holes typically result in inefficient reactant and substrate transport because they create steric restrictions that reduce catalytic efficiency. For better POP catalytic applications that reach large substrate scopes, meso- and/or macropores are hence more beneficial. A reduction in BET surface area is frequently caused by an increase in macropore porosity and a decrease in micropore concentration.

Using the Schiff base reaction, extremely porous polymers rich in nitrogen were synthesized. One of the highest values for POPs recorded, the extraordinary CO_2 absorption of 7.06 mmol g^{-1}, is the result of the large surface area, nitrogen-adorned pores, and electron-rich aromatic structures. POPs are a fascinating platform for the conversion, separation, and collection of CO_2. Although many POPs have already been found to absorb CO_2, only a small number have now been investigated for the conversion of CO_2 into value-added compounds. POPs for methane adsorption have received less attention than those of other gases. With our contribution, we want to inspire the development of novel POPs for the conversion and collection of CO_2. In addition, COFs have proven to be very effective at relocating heavy metals because of their distinctive characteristics like stability, high porosity, large surface area, and even more accessible binding sites. With a great substantial adsorption capacity of 396 mg^{-1} and outstanding efficacy in the removal of cadmium, the synthetic N-enriched COF performed well. The N-enriched COFs have demonstrated additional benefits including easy fabrication and no by-products, as well as high efficiency, cost effectiveness, and environmental friendliness. Hence, it might be a potential adsorbent in the remediation of environmental contamination. Due to their distinct crystalline characteristics, varied topologies, and compositions, POPs are excellent porous molecules for wastewater treatment. POPs with wide surface regions have a variety of significant benefits as sorbents during H_2 storage. The first benefit is the ability to build polymers entirely of light components. Due to the adaptability of organic chemistry, POPs with customizable functionalization, designable architectures, and changeable pore sizes may be generated. These properties have given POPs the strong capacity to withstand hostile conditions for future effective adsorption applications, combined with their exceptional thermal/chemical stability. POPs are also anticipated to be of great relevance for wastewater treatment shortly due to their low-energy requirements and straightforward separation procedures. To accomplish efficient wastewater management, it will be appealing to combine with other techniques like photocatalysis, which transforms highly hazardous pollutants into non-toxic or less toxic compounds while also having a high removal efficiency. In the upcoming years, it is anticipated that POPs will be able to function as adaptable adsorbent materials for the effective removal of diverse metallic ions in actual water samples due to the ongoing efforts of researchers from all over the world.

References

Algarni, T.S. and Al-Mohaimeed, A.M. (2022). Water purification by adsorption of pigments or pollutants via metal oxides. *Journal of King Saud University - Science* 34 (8): 102339. https://doi.org/10.1016/j.jksus.2022.102339.

Awual, M.R. (2019). A facile composite material for enhanced cadmium(II) ion capturing from wastewater. *Journal of Environmental Chemical Engineering* 7 (5): 103378. https://doi.org/10.1016/j.jece.2019.103378.

Bae, Y.S. and Snurr, R.Q. (2011). Development and evaluation of porous materials for carbon dioxide separation and capture. *Angewandte Chemie - International Edition* 50 (49): 11586–11596. https://doi.org/10.1002/anie.201101891.

Beuerle, F. and Gole, B. (2018). Covalent organic frameworks and cage compounds: design and applications of polymeric and discrete organic scaffolds. *Angewandte Chemie - International Edition* 57 (18): 4850–4878. https://doi.org/10.1002/anie.201710190.

Broom, D.P., Webb, C.J., Fanourgakis, G.S. et al. (2019). Concepts for improving hydrogen storage in nanoporous materials. *International Journal of Hydrogen Energy* 44 (15): 7768–7779. https://doi.org/10.1016/j.ijhydene.2019.01.224.

Chee, T.S., Tian, Z., Zhang, X. et al. (2020). Efficient capture of radioactive iodine by a new bismuth-decorated electrospinning carbon nanofiber. *Journal of Nuclear Materials* 542: 152526. https://doi.org/10.1016/j.jnucmat.2020.152526.

Chowdhury, A., Das, S.K., Mondal, S. et al. (2021). Sulfur-containing nitrogen-rich robust hierarchically porous organic polymer for adsorptive removal of mercury: experimental and theoretical insights. *Environmental Science: Nano* 8 (9): 2641–2649. https://doi.org/10.1039/d1en00448d.

Dias, E.M. and Petit, C. (2016). Erratum: Towards the use of metal-organic frameworks for water reuse: a review of the recent advances in the field of organic pollutants removal and degradation and the next steps in the field (Journal of Materials Chemistry A (2015) 3 (22484–22506)). *Journal of Materials Chemistry A* 4 (9): 3565. https://doi.org/10.1039/c6ta90036d.

Dinari, M. and Hatami, M. (2019). Novel N-riched crystalline covalent organic framework as a highly porous adsorbent for effective cadmium removal. *Journal of Environmental Chemical Engineering* 7 (1): 102907. https://doi.org/10.1016/j.jece.2019.102907.

Ding, M., Liu, X., Ma, P., and Yao, J. (2022). Porous materials for capture and catalytic conversion of CO_2 at low concentration. *Coordination Chemistry Reviews* 465: 214576. https://doi.org/10.1016/j.ccr.2022.214576.

Ding, S.Y. and Wang, W. (2013). Covalent organic frameworks (COFs): from design to applications. *Chemical Society Reviews* 42 (2): 548–568. https://doi.org/10.1039/c2cs35072f.

Fan, K., Zhang, T., Xiao, S. et al. (2022). Preparation and adsorption performance of functionalization cellulose-based composite aerogel. *International Journal of Biological Macromolecules* 211: 1–14. https://doi.org/10.1016/j.ijbiomac.2022.05.042.

Isaeva, V.I., Vedenyapina, M.D., Kurmysheva, A.Y. et al. (2021). Modern carbon–based materials for adsorptive removal of organic and inorganic pollutants from water and wastewater. *Molecules* 26 (21): 1–95. https://doi.org/10.3390/molecules26216628.

Jin, J., Xue, J., Wu, D. et al. (2022). Improved performance of the pyrimidine-modified porous In-MOF and an in situ prepared composite Ag@ In-MOF material. *Chemical Communications* 58 (56): 7749–7752.

Karmakar, A., Velasco, E., and Li, J. (2022). Metal-organic frameworks as effective sensors and scavengers for toxic environmental pollutants. *National Science Review* 9 (7): https://doi.org/10.1093/nsr/nwac091.

Kewley, A., Stephenson, A., Chen, L. et al. (2015). Porous organic cages for gas chromatography separations. *Chemistry of Materials* 27 (9): 3207–3210.

Knerelman, E.I., Karozina, Y.A., Shunina, I.G., and Sedov, I.V. (2022). Highly porous materials as potential components of natural gas storage systems: part 1 (a review). *Petroleum Chemistry* 62 (6): 561–582. https://doi.org/10.1134/S0965544122040077.

Korde, J.M. and Kandasubramanian, B. (2019). Fundamentals and effects of biomimicking stimuli-responsive polymers for engineering functions. *Industrial and Engineering Chemistry Research* 58 (23): 9709–9757. https://doi.org/10.1021/acs.iecr.9b00683.

Kostova, I., Vassileva, C., Dai, S. et al. (2013). Influence of surface area properties on mercury capture behavior of coal fly ashes from some Bulgarian power plants. *International Journal of Coal Geology* 116–117: 227–235. https://doi.org/10.1016/j.coal.2013.03.008.

Kreisberg, V.A. and Antropova, T.V. (2014). Changing the relation between micro- and mesoporosity in porous glasses: the effect of different factors. *Microporous and Mesoporous Materials* 190: 128–138. https://doi.org/10.1016/j.micromeso.2014.02.002.

Liu, J.L. and Wong, M.H. (2013). Pharmaceuticals and personal care products (PPCPs): a review on environmental contamination in China. *Environment International* 59: 208–224. https://doi.org/10.1016/j.envint.2013.06.012.

Liu, X., Liu, C.F., Lai, W.Y., and Huang, W. (2020). Porous organic polymers as promising electrode materials for energy storage devices. *Advanced Materials Technologies* 5 (9): 1–20. https://doi.org/10.1002/admt.202000154.

Liu, X., Pang, H., Liu, X. et al. (2021). Orderly porous covalent organic frameworks-based materials: superior adsorbents for pollutants removal from aqueous solutions. *The Innovation* 2 (1): 100076. https://doi.org/10.1016/j.xinn.2021.100076.

Low, J., Cheng, B., and Yu, J. (2017). Surface modification and enhanced photocatalytic CO_2 reduction performance of TiO_2: a review. *Applied Surface Science* 392: 658–686. https://doi.org/10.1016/j.apsusc.2016.09.093.

Lu, W., Wei, Z., Yuan, D. et al. (2014). Rational design and synthesis of porous polymer networks: toward high surface area. *Chemistry of Materials* 26 (15): 4589–4597. https://doi.org/10.1021/cm501922h.

Luo, R., Xu, W., Chen, M. et al. (2020). Covalent triazine frameworks obtained from nitrile monomers for sustainable CO_2 catalysis. *ChemSusChem* 13 (24): 6509–6522. https://doi.org/10.1002/cssc.202002422.

Lustig, W.P., Mukherjee, S., Rudd, N.D. et al. (2017). Metal-organic frameworks: functional luminescent and photonic materials for sensing applications. *Chemical Society Reviews* 46 (11): 3242–3285. https://doi.org/10.1039/c6cs00930a.

Ma, M., Lu, X., Guo, Y. et al. (2022). Combination of metal-organic frameworks (MOFs) and covalent organic frameworks (COFs): recent advances in synthesis and analytical applications of MOF/COF composites. *TrAC - Trends in Analytical Chemistry* 157: 116741. https://doi.org/10.1016/j.trac.2022.116741.

Makal, T.A., Li, J.R., Lu, W., and Zhou, H.C. (2012). Methane storage in advanced porous materials. *Chemical Society Reviews* 41 (23): 7761–7779. https://doi.org/10.1039/c2cs35251f.

Modak, A., Bhanja, P., Selvaraj, M., and Bhaumik, A. (2020). Functionalized porous organic materials as efficient media for the adsorptive removal of Hg(II) ions. *Environmental Science: Nano* 7 (10): 2887–2923. https://doi.org/10.1039/d0en00714e.

Ong, W.S., Smaldone, R.A., and Dodani, S.C. (2020). A neutral porous organic polymer host for the recognition of anionic dyes in water. *Chemical Science* 11 (29): 7716–7721.

Patel, H.A., Je, S.H., Park, J. et al. (2014). Directing the structural features of N_2-phobic nanoporous covalent organic polymers for CO_2 capture and separation. *Chemistry - A European Journal* 20 (3): 772–780. https://doi.org/10.1002/chem.201303493.

Pech-Canul, A.D.C., Ortega, D., García-Triana, A. et al. (2020). A brief review of edible coating materials for the microencapsulation of probiotics. *Coatings* 10 (3): 197.

Pérez-Marín, A.B., Zapata, V.M., Ortuno, J.F. et al. (2007). Removal of cadmium from aqueous solutions by adsorption onto orange waste. *Journal of Hazardous Materials* 139 (1): 122–131. https://doi.org/10.1016/j.jhazmat.2006.06.008.

Ranganatha, S. (2020). 2D nanostructured materials for high-performance electrochemical supercapacitors. *ACS Symposium Series* 1353: 79–92.

Rashid, R., Shafiq, I., Akhter, P. et al. (2021). A state-of-the-art review on wastewater treatment techniques: the effectiveness of adsorption method. *Environmental Science and Pollution Research* 28 (8): 9050–9066. https://doi.org/10.1007/s11356-021-12395-x.

Ravi, S., Choi, Y., and Choe, J.K. (2020). Achieving effective fructose-to-5-hydroxymethylfurfural conversion via facile synthesis of large surface phosphate-functionalized porous organic polymers. *Applied Catalysis B: Environmental* 271: 118942. https://doi.org/10.1016/j.apcatb.2020.118942.

Raza, W., Kukkar, D., Saulat, H. et al. (2019). Metal-organic frameworks as an emerging tool for sensing various targets in aqueous and biological media. *TrAC - Trends in Analytical Chemistry* 120: 115654. https://doi.org/10.1016/j.trac.2019.115654.

Rosman, N., Salleh, W.N.W., Mohamed, M.A. et al. (2018). Hybrid membrane filtration-advanced oxidation processes for removal of pharmaceutical residue. *Journal of Colloid and Interface Science* 532: 236–260. https://doi.org/10.1016/j.jcis.2018.07.118.

Sabri, M.A., Al-Sayah, M.H., Sen, S. et al. (2020). Fluorescent aminal linked porous organic polymer for reversible iodine capture and sensing. *Scientific Reports* 10 (1): 1–11. https://doi.org/10.1038/s41598-020-72697-x.

Salmean, C. and Dimartino, S. (2019). 3D-printed stationary phases with ordered morphology: state of the art and future development in liquid chromatography. *Chromatographia* 82 (1): 443–463. https://doi.org/10.1007/s10337-018-3671-5.

Sarkar, C., Shit, S.C., Das, N., and Mondal, J. (2021). Presenting porous-organic-polymers as next-generation invigorating materials for nanoreactors. *Chemical Communications* 57 (69): 8550–8567. https://doi.org/10.1039/d1cc02616j.

Savina, I.N., Ingavle, G.C., Cundy, A.B., and Mikhalovsky, S.V. (2016). A simple method for the production of large volume 3D macroporous hydrogels for advanced biotechnological, medical and environmental applications. *Scientific Reports* 6: 1–9. https://doi.org/10.1038/srep21154.

Shin, M.G., Choi, W., Park, S.J. et al. (2022). Critical review and comprehensive analysis of trace organic compound (TOrC) removal with polyamide RO/NF membranes: mechanisms and materials. *Chemical Engineering Journal* 427: 130957. https://doi.org/10.1016/j.cej.2021.130957.

Skorjanc, T., Shetty, D., and Trabolsi, A. (2021). Pollutant removal with organic macrocycle-based covalent organic polymers and frameworks. *Chem* 7 (4): 882–918. https://doi.org/10.1016/j.chempr.2021.01.002.

Subramaniam, M.N., Goh, P.S., Lau, W.J., and Ismail, A.F. (2019). The roles of nanomaterials in conventional and emerging technologies for heavy metal removal: a state-of-the-art review. *Nanomaterials* 9 (4): 625. https://doi.org/10.3390/nano9040625.

Sun, Q., Dai, Z., Meng, X., and Xiao, F.S. (2015). Porous polymer catalysts with hierarchical structures. *Chemical Society Reviews* 44 (17): 6018–6034. https://doi.org/10.1039/c5cs00198f.

Tao, R., Ma, X., Wei, X. et al. (2020). Porous organic polymer material supported palladium nanoparticles. *Journal of Materials Chemistry A* 8 (34): 17360–17391. https://doi.org/10.1039/d0ta05175f.

Thomas, A., Kuhn, P., Weber, J. et al. (2009). Porous polymers: enabling solutions for energy applications. *Macromolecular Rapid Communications* 30 (4–5): 221–236. https://doi.org/10.1002/marc.200800642.

Ushani, U., Lu, X., Wang, J. et al. (2020). Sulfate radicals-based advanced oxidation technology in various environmental remediation: a state-of-the-art review. *Chemical Engineering Journal* 402: 126232. https://doi.org/10.1016/j.cej.2020.126232.

Valencia, P.M., Farokhzad, O.C., Karnik, R., and Langer, R. (2020). Microfluidic technologies for accelerating the clinical translation of nanoparticles. *Nano-Enabled Medical Applications* 7 (10): 93–112. https://doi.org/10.1201/9780429399039-3.

Vinu, A. (2008). Two-dimensional hexagonally-ordered mesoporous carbon nitrides with tunable pore diameter, surface area, and nitrogen content. *Advanced Functional Materials* 18 (5): 816–827. https://doi.org/10.1002/adfm.200700783.

Wang, H., Shao, Y., Mei, S. et al. (2020). Polymer-derived heteroatom-doped porous carbon materials. *Chemical Reviews* 120 (17): 9363–9419. https://doi.org/10.1021/acs.chemrev.0c00080.

Wang, J., Li, Z., and Gu, Z. (2021). A comprehensive review of template-synthesized multi-component nanowires: from interfacial design to sensing and actuation applications. *Sensors and Actuators Reports* 3: 100029. https://doi.org/10.1016/j.snr.2021.100029.

Xu, F., Tang, Z., Huang, S. et al. (2015). Facile synthesis of ultrahigh-surface-area hollow carbon nanospheres for enhanced adsorption and energy storage. *Nature Communications* 6: 7221. https://doi.org/10.1038/ncomms8221.

Xu, T., He, Y., Qin, Y. et al. (2018). Facile preparation of porous organic copolymer based on triptycene and crown ether for efficient organic dye adsorption. *RSC Advances* 8 (9): 4963–4968. https://doi.org/10.1039/c7ra12495c.

Yang, J., Xu, M., Wang, J. et al. (2018). A facile approach to prepare multiple heteroatom-doped carbon materials from imine-linked porous organic polymers. *Scientific Reports* 8 (1): 4200.

Yang, G., Gao, H., Li, Q., and Ren, S. (2021). Preparation and dye adsorption properties of an oxygen-rich porous organic polymer. *RSC Advances* 11 (26): 15921–15926. https://doi.org/10.1039/d1ra01382c.

Yildirim, O., Tsaturyan, A., Damin, A. et al. (2022). Quinoid-thiophene-based covalent organic polymers for high iodine uptake: when rational chemical design counterbalances the low surface area and pore volume. *ACS Applied Materials and Interfaces* https://doi.org/10.1021/acsami.2c20853.

Yoon, J.W., Kim, J.H., Kim, C. et al. (2021). MOF-based hybrids for solar fuel production. *Advanced Energy Materials* 11 (27): 2003052. https://doi.org/10.1002/aenm.202003052.

Zhang, D., Wang, L., Zeng, H. et al. (2019). A three-dimensional macroporous network structured chitosan/cellulose biocomposite sponge for rapid and selective removal of mercury(II) ions from aqueous solution. *Chemical Engineering Journal* 363: 192–202. https://doi.org/10.1016/j.cej.2019.01.127.

Zhang, J. (2013). Study on CO_2 capture using theriomorphic biphasic solvents with energy-efficient regeneration, 1–142.

Zhang, W., Aguila, B., and Ma, S. (2017). Potential applications of functional porous organic polymer materials. *Journal of Materials Chemistry A* 5 (19): 8795–8824. https://doi.org/10.1039/c6ta11168h.

Zhou, Y., Klinger, G.E., Hegg, E.L. et al. (2022). Skeletal Ni electrode-catalyzed C-O cleavage of diaryl ethers entails direct elimination via benzyne intermediates. *Nature Communications* 13 (1): https://doi.org/10.1038/s41467-022-29555-3.

Zia, Z., Hartland, A., and Mucalo, M.R. (2020). Use of low-cost biopolymers and biopolymeric composite systems for heavy metal removal from water. *International Journal of Environmental Science and Technology* 17 (10): 4389–4406. https://doi.org/10.1007/s13762-020-02764-3.

24

Versatile Applications of Organic Polymer and Their Prospects

Shreya Sahani, Jyoti Sarwan, Komal Mittal, Richa Chaurasia, Kanchan Kumari, Nazim Uddin, and Jagadeesh C. Bose K

University Institute of Biotechnology, Chandigarh University, Gharuan, Mohali, Punjab, 140413, India

24.1 Introduction

Traditional industrial manufacturing processes entail significant energy consumption and environmental degradation, and synthetic chemicals have greatly benefited human civilization. Modernizing manufacturing processes and their energy sources are necessary to address these problems; in this context, photocatalysis, mainly visible-light-catalysts, is a potent instrument for the transformation of almost limitless solar energy to chemical energy (Schneider et al. 2014). A polymer, which is a huge macromolecule, is essentially made up of a lot of smaller molecules. The Greek term "polymer" means "many parts." Everywhere we look, polymers are present – from the widely used biopolymer polypropylene to the naturally occurring biopolymer found in the human DNA strand. Polymers have a wide range of applications. Biodegradable polymers are commonly used in the biomedical industry for tissue engineering and drug delivery. Neoprene is used in both shoe soles and wet suits. Non-stick pans use Teflon; pipelines use polyvinyl chloride, while textiles and fabrics frequently use different polymers. The well-known PET bottles as called as polyethylene terephthalate are produced using synthetic polymer polyethylene terephthalate. Polystyrene resins are used to manufacture household goods and appliances. Both the plastics and textiles sectors utilize nylon-6 extensively. The glass-like polymer poly(methyl methacrylate) is present in many consumer goods. Due to their porosity and high specific area organic polymers are used widely in the world. Not only the surface area but the presence of a designable structure, low density, and easily available for functionalization, organic polymers are usually the source of applications in numerous of fields. There is another advantage of the organic polymers to use them in different application is to have metal-free structure present in them. The metal-free structure of the organic polymer does not let them interfere during the making derivatives of the synthesis process. Nowadays, porous organic polymers are taking the attention of several researchers. There is an advantage of using porous organic polymers due to the ability of biodegradation in

Organic Polymers in Energy-Environmental Applications, First Edition.
Edited by Ramesh Oraon, Pardeep Singh, Sanchayita Rajkhowa, Sangita Agarwal, and Ravindra Pratap Singh.

their polymeric chains. They are not persisting in the environment and cause harm to nature. This article provides an in-depth analysis of natural biopolymer research and a greater comprehension of the trends that have emerged in the industry over the past 22 years and can also act as a resource for upcoming studies. Organic polymers are macromolecules consisting of many monomer units that repeat. Natural and synthetic polymers are both necessary for daily life. Polysaccharides, polypeptides, and polynucleotides are the three basic types of biopolymers that are present in living cells. Organic polymers are frequently utilized in the electronics sector as packaging materials, insulators, dielectrics, and thermal transfer media, and to shield delicate joints and components from shock and vibration, mechanical and thermal damage, and airborne contaminants. Organic polymers are also necessary for basic structural elements and participation in vital life processes in living beings. For instance, polymers make up all of the solid parts of plants. Among them are cellulose, lignin, and different resins. For a very long time, people have recognized the value of chemistry in daily life. There are numerous instances some of which are even emblematic where this discipline of science has significantly improved our quality of life.

Natural polymers can be found in animals and plants, as opposed to synthetic polymers, which are created by humans. Due to their unique physical and chemical characteristics, several polymers are used in daily life. Macromolecules known as organic polymers are made up of many monomer units that repeat. Both natural and artificial polymers are essential for daily life. The three primary categories of biopolymers that are present in living cells are polysaccharides, polypeptides, and polynucleotides. Enzyme-mediated processes in cells produce these polymers. The producer of polymeric parts that are green, energy-efficient, high-quality, reasonably priced, etc. will be a major trend in the coming millennium. Modern polymer research is very interdisciplinary and integrated, and as a result, it does not easily fit into the conventional natural science structural models. These days, organic polymers are used in many different technical applications, including solar cells, sensors, and LEDs. Typically, natural polymer-carriers are weak, safe for microbes to consume, and quickly degraded by them in anaerobic environments Monomers containing two or more functional groups, numerous bonds, or rings with three to seven members are frequently used to create synthetic polymers. While polymers' physical properties can vary, their chemical properties can be determined by the monomer units that make up each polymer. Polymers are classified as thermoplastics, thermosets, elastomers, or fibers based on their physical properties. One of the main issues affecting human health is environmental pollution, which is caused by the rapid expansion of the chemical, battery, steel, fertilizer, and pesticide industries as well as mining operations. Metal(loid) pollution puts the health of living things at risk. The ecosystem has undergone major changes due to the build-up of metal(loids) due to anthropogenic and natural activity. Environmentally harmful substances such as cadmium, chromium, lead, and mercury have also been connected to several diseases, including birth deformities, damage to internal organs like the kidney and also liver, specific learning difficulties, and a range of different ailments. These metals can be found in wastewater and soil along with other metal ions.

Polymers have a wide range of applications. Biodegradable polymers are commonly used in the biomedical industry for tissue engineering and drug delivery.

As support, organic polymers are frequently used because of their stable, cheap, nontoxic, and processable qualities. High efficiency and activity are constrained by limited dispersion in solvents solution, traditional polymers' active sites are exposed, and there is the interaction between substrates and catalysts. Single-atom catalysis may be a feature of hierarchically porous materials made of monomers with coordination sites or catalytic functional groups (Singh et al. 2013).

24.2 Characteristics of Organic Polymer

Organic polymer with more than 40 carbons does not exist because large macromolecules cannot be crystallized to that extent. The development of organic polymers has been aided by a variety of techniques over the years, including coupling reactions between multi-halogen aromatic and multi-boric acid or multi-alkynyl compounds, the Schiff base reaction between aromatic multi-amino monomers and a dialdehyde or tri aldehyde, the Knoevenagel reaction of multi-active methylene groups and active aldehyde groups, and the cyclo trimerization of multi-ethyls (Xu and Hedin 2013; Uribe-Romo et al. 2009). New methods that specify the production of organic polymers in a trustworthy and adaptable approach that are thought to be the opposite for expanding sustainable cycle for organic polymers to get over these difficulties. A "two-in-one" method for creating covalent organic frameworks has recently garnered a lot of interest. The basic idea behind the process is that two distinct functional groups are combined into a single practical building block to pre-prepare two-in-one monomers. As a result, covalent organic framework (COF) films can be produced by self-condensing without considering solvent adaptability and stoichiometry (Li et al. 2019; Hao et al. 2019). Today, *the* purification of water is frequently accomplished using organic polymers, which are lengthy macromolecules composed of several repetitive monomer units. Their typical use in the coagulant aids or primary coagulants stages of drinking water purification. In an aqueous suspension, these chemicals have a propensity to significantly adsorb the particles on their surface. Increased rate of settling, lower costs, enhanced quality of treated water, and sludge dewatering properties are just a few of the benefits of using organic polymers. A family of very porous, crosslinked materials with a covalent bond between the monomers is referred to as organic polymers. Metals that are heavy including cadmium, zinc, and copper can be removed from wastewater using COP-63, a disulfide-linked covalent polymer that was created. As per report, COPs are a potential sorbent for water treatment since they exhibit a capacity for cadmium sorption that is seven times higher and also sixteen times faster than that of activated carbon (Maiti et al. 2021). Porous organic polymers (POPs) are materials that can be penetrated or are permeable and are created through the polymerization process, in which the organic building blocks exhibit strong covalent interactions. POPs are of great interest because of their distinct stability. They have advantages over all other porous materials, including a large surface area, customizable porosity, ease of synthesis, and ease of processing for technological applications. With numerous chemical functions included inside the microporous or on the surfaces, a variety of porous polymers may now be created thanks to advancements in organic synthesis. It is unlikely that synthetic plastics with petrochemical origins can ever totally replace them in the short or even long term, even though technology and bio-products, such as bioplastics, can help ameliorate these environmental difficulties. However, some applications for bioplastics may eventually be largely replaced. For instance, because they account for a sizeable amount of the world's plastics output, biodegradable polymers for packaging and other short-lived use sectors are practicable.

POPs are currently the subject of intensive research in several fields, including selective permeation, catalysis, sustainable energy storage, gas storage and adsorption, sensing, molecular motors, and photoconductors, among others (Wang et al. 2021). POPs are frequently classified into three kinds based on the size of their pores: macroporous (pores greater than 450 nm), mesoporous (2–50 nm), and microporous (2 nm). A major field of research in material science today is the study of porous materials with customized functionality created from fundamental molecular building blocks. Depending on their degree of crystallinity, POPs can be categorized as either crystalline or amorphous analogs. COFs are a common component of POPs in their crystalline form, while conjugated microporous polymers (CMPs), polymers with intrinsic microporosity (PIMs), hyper-crosslinked polymers (HCPs), and porous aromatic frameworks (PAFs) make up the majority of POPs in their amorphous form. Reversible processes typically lead to organized framework structures during their formation, but irreversible reactions usually lead to amorphous porous polymers. Due to their potential uses in environmental clean-up as well as their benefits of inherent porosity, low weight, resilience, tunable structures, and customized functionality.

Natural and artificial organic polymer carriers are both types of organic polymer carriers. Natural polymer carriers typically have low strength, good mass transfer capability, and are not harmful to microorganisms. They can also be quickly degraded by bacteria in anaerobic environments. Agar, gelatine, sponges, chitin, alginate, and chitosan are typical natural polymer carriers. Figure 24.1 shows various qualities of polymers. Agar-embedding cell procedure is easy to use and also non-toxic to cells. It also has a wider clearance, permitting the passage of high molecular compounds, but it has poor mechanical and chemical stability. These polymers have high catalytic reaction rates per unit volume achieved by high internal surface areas and active site densities. Possibilities for post-synthesis modification of MOFs with complementary catalytic groups, enabling catalysis by a variety of functional groups as well as tandem or bifunctional catalysis; shape-selective catalysis, confinement effects for stabilizing encapsulated catalytic species, placement of different catalytic groups next to each other or in separate,

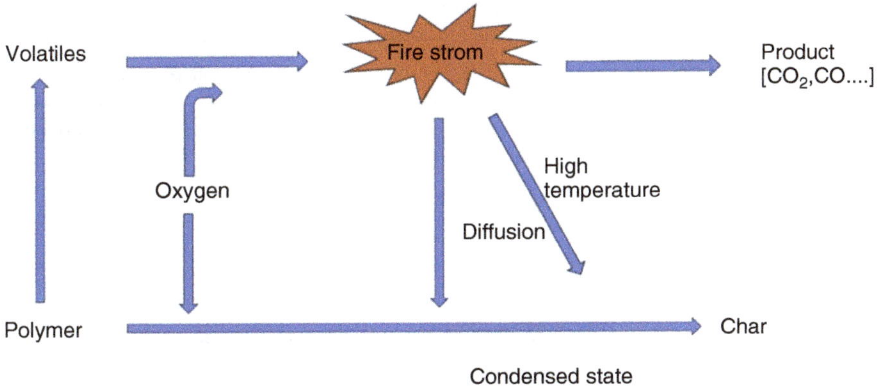

Figure 24.1 Organic polymer utilization in the formation of fuel.

nearby environments are all possibilities for tailoring MOF pore structures to allow for these things.

These appealing characteristics and the potential for large-scale catalytic applications, albeit none have yet been discovered to our knowledge, help to explain in part the rapid expansion of research on MOF catalysis.

Synthetic organic polymer gel carriers exhibit excellent mechanical strength, strong chemical stability, and good resistance to antimicrobial degradation. However, transfer practice is subpar, and the embedding procedure will result in decreased cell activity. Polyacrylamides, light-curing resins, polyvinyl alcohol, and polyacrylic acid gel are typical synthetic organic polymer gel carriers (Chen and Wang 2017). When heated over a particular critical temperature, which in turn depends on their chemical structures, ORGANIC polymers break to give volatile flammable compounds. If the gaseous mixture created when deteriorating volatiles are mixed with air is within the range of flammability and the temperature is greater than the ignition temperature, combustion will begin. An extremely complicated process, the combustion of the polymeric substance involves a few interconnected and/or independent steps that take place in the condensed phase, the gaseous phase, and at the interfaces between the two phases (Cullis and Hirschler 1981). Fires utilizing organic polymers present substantially identical risks to those brought on by fires fueled by other fuels. Other issues exist in addition to flammability and property damage. Fire fatalities brought on by toxic gases and evolved smokes are more significant, and they are sometimes made worse by toxic fumes released by synthetic organic polymers (Levin 1992). To create stationary phases for HPLC, organic polymers are utilized in a variety of ways.

The polymers can be created as monoliths or used as tiny particles, comparable to those formed from hydrated silica. Additionally, some polymers can be utilized as solid support that is further derivative to produce an active stationary phase, or they can be employed as stationary phases as they are produced. Table 24.1 shows the application of organic polymers in fuel utilization. The use of organic polymers as stationary phase is quite prevalent for several particular forms of chromatography,

Table 24.1 Classification of different types of POPs.

Amorphous organic polymer	Crystalline organic polymer
Hyper-crosslinked polymer (HCPs)	Covalent organic frameworks (COFs)
Conjugated micro-porous polymer (CMPs)	
Polymers of intrinsic microporosity (PIMS)	
Porous aromatic framework (PAF)	

such as size exclusion or ion exchange. The polymers used as the stationary phase in other HPLC methods, including reversed-phase, are less significant.

24.3 Methods and Preparation

Organic polymers can be prepared with the aid of diverse polymerization tactics. The synthesis with the aid of mass polymerization includes a uniform system wherein the testing agents, blended inside the medium of reaction, polymerize with the help of radical reactions. The process ought to be done in an oxygen-free environment since the occurrence of oxygen can put off the process of polymerization. The polymerization is caused with the aid of UV or heat, within the process it relies on the employed monomer (da Silva et al. 2019). Numerous units can be used, according to the desired software and interface with the analyte. Mass polymerization harvests a stiff inflexible polymer grid, commonly, the product received is softened dehydrated, and, depending on its use, isolated with the aid of particle dimensions. Usually, organic polymers are providing constancy over an extensive pH range and low density. To avoid the grinding step and firm matrix formation, it is conceivable to embrace other synthesis procedures consisting of polymerization through suspension or via precipitation (da Silva et al. 2019). In the "bottom-up" method, functional groups are included in heterotopic atom-doped porous organic polymers by carefully creating and using special building blocks. There are two enticing ways to satisfy the particular needs of synthetic materials. One is to add a variety of functional groups to pure monomers before high-yielding distillation. This is a very effective bottom-up synthesis technique. According to Sun et al. (2015) several functional groups, such as –OH, $-SO_3H$, and –COOH, have been productively implicated in the skeleton of POPs. The well-maintained source material's pore properties, which prevent pore clogging, are the method's most obvious advantage. Utilizing monomer-containing hybrid components like carbazole, thiophene, and triphenylamine is another fantastic bottom-up approach. The addition of various elements (such as S, N, O, and P) improves the architectural complexity of POPs and is also very helpful for the capture of metal ions because of the increased affinities or sensitivity. To create a new sulfur-rich microporous polymer (SMP) with a sulfur content of 31.4 weight percent, tetra(thiophene-2-yl) silane is utilized as a functional monomer. The resultant SMP with the evenly dispersed sulfur atom and the very porous structure revealed a significant adsorption capability toward Hg^{2+} (595.2 $mg\,g^{-1}$) as a result of the strong coordination interaction between soft acid mercury and soft basic sulfur.

24.4 Properties and Characteristics of Organic Polymers

Organic polymers have been applied to every aspect of human life and play a crucial role. They are now recognized as critical materials in the modern day, supplying basic structural substances and participating in critical survival techniques. For instance, the stable components of all plant life are made of polymers. These consist of lignin, numerous resins, and cellulose. Lignin, a complex organic polymer, plays a very important role in many biological functions, such as mechanical support, formation of the cell wall, and water transport. Cellulose is a type of homopolysaccharide found in plant cell walls that are composed of monosaccharide units containing oxygen, carbon, and hydrogen atoms. Wood resins are plant materials that are highly viscous and solid. It is a polymer of an easy hydrocarbon, isoprene. Rubber is other known isoprene polymer.

Different prime natural polymers, including proteins (a polymer composed of amino acids), serve a wide range of functions in living organisms, such as transportation, defense, metabolism, and organization. In the nucleic acids (composed of nucleotides) it is the complicated molecules made up of sugars, nitrogen-containing bases, and phosphoric acid. Nucleic acid (DNA and RNA) contains genetic material and plays a very significant role in protein synthesis. Starch, a carbohydrate, is one of the primary sources of food for humans. For the conversion, separation, and collection of CO_2, POPs provide an intriguing platform. Only a small number of POPs have so far been studied for the conversion of CO_2 into value-added chemicals, even though numerous POPs have already been identified to absorb CO_2. With our contribution, we hope to stimulate the creation of fresh POPs for CO_2 conversion and capture. Furthermore, due to their specific qualities including stability, high porosity, vast surface area, and even more accessible binding sites, COFs have been demonstrated to be particularly successful at moving heavy metals. The synthesized N-enriched COF functioned admirably, with a great significant adsorption capacity of 396 mg^{-1} and exceptional efficacy in the elimination of cadmium. Additional advantages of the N-enriched COFs have included ease of manufacturing, the absence of by-products, high efficiency, cost effectiveness, and environmental friendliness. As a result, it could serve as a possible adsorbent in the clean-up of environmental contaminants. POPs are superior porous molecules for wastewater treatment because of their distinctive crystalline properties, diverse topologies, and compositions. POPs with configurable functionalization, programmable structures, and adjustable pore size may be produced as a result of organic chemistry's adaptability. These characteristics, along with POPs' outstanding thermal and chemical stability, have given them the power to survive harsh conditions for future effective adsorption applications. POPs are also projected to be of significant importance for wastewater treatment shortly due to their simple separation processes and low energy needs. It will be appealing to combine with other approaches like photocatalysis, which converts extremely hazardous pollutants into non-toxic or less toxic chemicals while also having a high removal efficiency, to achieve effective wastewater management. Due to the continued efforts of researchers from around the world, it is projected that POPs will be able to function as adaptive adsorbent materials for the efficient removal of various metallic ions in actual water samples in the next years.

Organic polymers are mostly microporous varying with pore sizes. Their design ability has to be performed in a controlled manner. Amorphous and Ionic organic networks are amorphous, cannot be crystallized, while metal–organic frameworks are highly crystalline. In the stability trend out of the four, amorphous porous organic polymer is highest, followed by covalent and nanoporous ionic organic polymer leaving metal–organic framework with the least stability in comparison. Metal and covalent organic polymers can be self-assembled based on their intermolecular interactions and bonding, However, amorphous porous polymer requires condensation and nanoporous ionic networks require template synthesis.

Whereas porous organic polymers have another advantage too, such as being used in water purification or upgrading the water conditioning, and escalating the tempo of settlement (Yadav and Sinha 2022).

24.5 Types of Organic Polymers

24.5.1 Porous Organic Polymers (POPs)

Porous organic polymers or porous polymers are multifaceted in nature and composed of sturdy covalent linkages between various organic building blocks. Because of their notable characteristics, like huge pore size, accurate balance, and broad surface area robustness in various solvents, they recently turned out to be a growing study subject in porous substances. These features of POPs are favorable for mass transfer, providing more lively sites, and emerging as a promising platform for the recycling process in photocatalysis. Porous organic polymers can be fabricated reversibly or irreversibly (Figure 24.2). For reversible synthesis, Koil and organic frameworks are utilized. However, for a reversible, we can perform Friedel–Crafts alkylation, coupling of carbon bonds, imine formation, or cyclotrimerization (Figure 24.3).

POPs are generally classified into two categories: amorphous polymers and crystalline polymers. Further amorphous and crystalline organic polymers include, under amorphous we have HCPs, CMPs, PIMS, and PAF (Table 24.2).

POPs soak up more or equal amounts of energy from photons than their gaps between bands (Zhang et al. 2020a). In gasoline repositories and their detachment, catalysis, photoelectric alteration, energy conversion and storage, and many others, POPs have acquired an interest in research and attention due to their advantages of

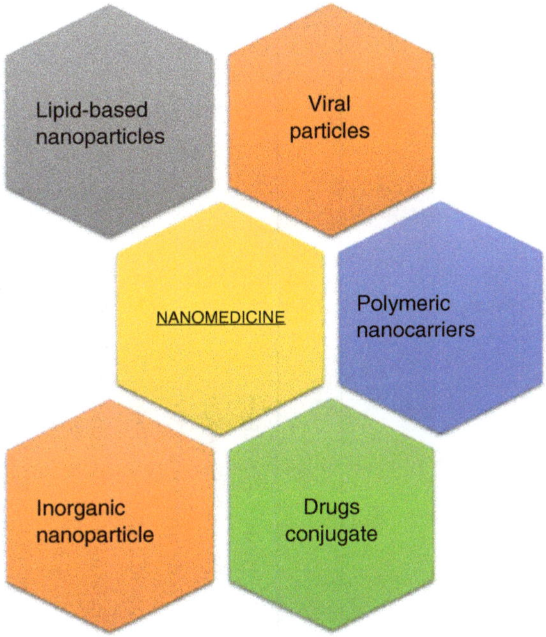

Figure 24.2 Various types of nanomedicines in which organic polymers are utilized as an important aspect.

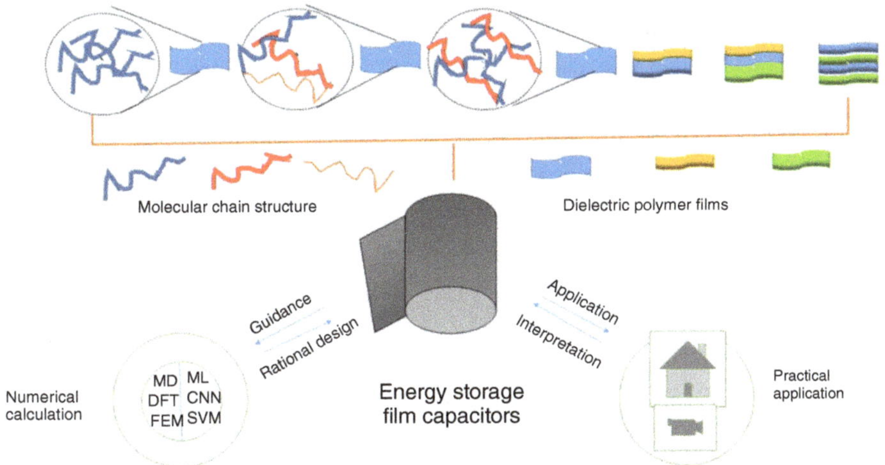

Figure 24.3 The fields of biomedical, environmental, and chemical engineering of organic polymers.

pre-designable structures, light-weight, higher essential permeability, and first-rate stability (Zhang et al. 2020b).

24.5.2 Covalent Organic Polymer (COPs)

Covalent organic polymers are porous community polymers with permanent pores. These permanent voids (in general, in a nano-porous place, with pore sizes under

Table 24.2 Types of polymer dosage form.

Drug 1	Drug 2
PDEA-PDMA-PEGMsiBcl-2DOXHep	PEG-PLL-PAspNPsiBcl-2DOXHep
Trimethyl ChitosaNPIL-2DOXSMMC7721	PEI-PLA/PEG-PAspNPsiSurvivinPCTA549
PEO-b-PCLMsiMDR-1DOXMDA-MB-435	PDP-PDHANPshSurvivinDOXMCF7/ADR
HT1080 mPEG-PCL-SS-DOX/mPEG-PCL	Poly(acrylic acid)HDOXCDDPMCF7/ADR
Poly(amido amine)sMDTXMMP9shRNAMCF7	PEG-PLG-PDMAPMANPsiMDR1DOXMCF7
PLGA-PEG-PLGAHDOX	MTXCDDPSaos-2
miRNA-34a-S-S-PEMMmiRNA-34aDOXMCF7	HT1080PEG-pp-PEI-DOPEMsiSurvivinPCTA549T24PEG-PLA
G0-C14NPVEGF siRNAPCTHT-1080	A375
PC-3 Magnetic responsive PLGA/TPGS/OANPTPGSDOXMCF7	MCF7/ADR-ASA-MNPs-CDDP/mPEG-PLL-FANPCDDPTFPI2 DNAHNE-1
NP69 PCL(NIPAAm-co-HEMA-co-MAA-co-TMSPMA)	NPDOXMTXMCF7

100 nm) convey porous material properties including gasoline seizing and separation, water treatment, and catalysis (Zhang et al. 2021).

In general, COP grew accustomed to inspecting its multipurpose luminescent programs. Taking benefit of the excessive permeability and fluorescence, the COP became utilized as an especially sensitive and selective probe for the uncovering of MnO_4, TNP, and Fe_3O_4 (Zhang et al. 2021).

24.5.3 Nanomedicines

The convergence of nanotechnology and medicine has led to the interdisciplinary field of nanomedicine. Improvement in molecular and cellular biology, genetics, material science, proteomics, and bioengineering, all these sectors have contributed to this emerging field, which deals on the nanoscale level with physiological processes (Sara Soares et al. 2018). Nanoparticles (NPs) are key factors of nanomedicine, and currently, a giant range of nanoparticle kinds exist.

An extensive style of nanoparticles and substances are utilized in nanomedicine, the various maximum widely used are polymers, quantum dots (QDs), liposomes, iron oxide (IO) debris, and carbon nanotubes.

24.5.4 Applications of Organic Polymer

Wide ranges of applications often take place in organic polymers in several fields (Taheri et al. 2022).

1) Originally, natural polymers were used. For example wood, rubber, cotton, wool, leather, silk.
2) Oldest known uses: Building materials, amber used for jewelry 13,000 years earlier, clothing, rubber balls used by Mesoamericans 36,000 years earlier.
3) Energy storage film capacitors can also be fabricated from molecular chains or dielectric polymeric chains of organic polymers. Along with this, these organic polymers can be used for computerized calculations in concepts like machine learning and genome materials shows fabrication of energy storage film capacitors through organic polymeric chains. These polymers are heat resistant and hence are suitable for this.
4) Applications for synthetic organic polymers are increasing: Automobiles, jet engines, when low weight is required, the addition of fibers, composites of glass or carbon.
5) Nucleic acid organic polymer can be utilized for diagnostics, bio-fabrication of biomedicines, etc. (Kwak and Herrmann 2010).
6) MOPs: Microporous organic polymer can be used for the application of hydrogen storage, methane storage, capturing, and separation of carbon dioxide (Chang et al. 2013).
7) These have also been reported to be used in the field of agriculture, Massive environmental strains brought on by population increase, the development of massive, densely populated metropolises, and concurrent global industrialization may jeopardize environmental sustainability and food security.

8) Bioremediation: One of these microbial polymers that has shown promise for complexing metal(loids) is bio-surfactant. They are metabolic compounds that have amphiphilicity and surface activity. The formation of micelles or the combination of these moieties at the interfaces of fluids with different polarities is induced by this amphiphilic metabolite's dual hydrophobic and hydrophilic moieties, which lowers the interfacial surface tension. Bio-surfactants frequently contain hydrophilic groups such as anions, cations, mono-, di-, or peptides, amino acids, and polysaccharides. The hydrophobic groups are made up of fatty acids, which can be branched, hydroxylated, saturated, or unsaturated.
9) These have also been reported to be utilized in water treatment. Organic polymers, which are long-chained macromolecules made up of multiple repeating monomer units, are often utilized in modern water filtration. Their typical application is in the primary coagulants or coagulant aids phases of the purification of drinking water. In an aqueous suspension, these compounds tend to strongly adsorb the particles on their surface. A faster rate of settling, cheaper costs, better-quality treated water, and sludge dewatering capabilities are just a few benefits of using organic polymers.
10) Nonetheless, the physical properties of the polymers used in these applications are the main requirement as inert materials, such as transmission, reliability, permeability, or weather ability, rather than being active molecules. Reactive functionalized polymer science and technology is one of the most fascinating fields of polymer chemistry for the creation of superior materials. Polymeric molecules have been widely exploited as reactive materials because of the various active functional groups' potential benefits and the compounds' distinctive properties. They are used in a wide range of technological fields, such as solid-phase synthesis and physiologically active systems.
11) The polymeric biocide has numerous advantages, some of which include the following: it can be applied at lower concentrations than conventional biocides. The number of applications is reduced as a result of the prolonged duration of activity brought on by a single application, which releases the required amount of active agent over a lengthy period; and it does not require repeated applications. Its macromolecular design reduces the requirement for extensive dispersion of high levels of biocides in the environment, as well as degradation and evaporation losses caused by environmental causes or waterways or leaching into the soil. Additionally, because fewer active substances are required, it reduces the toxicity, duration, and cost of repeated treatments. The foundation of this concept is the use of appropriate polymers as carriers, where the residual by-products of polymer degradation act as fertilizer for the soil and plant growth. For instance, condensation polymerization has been used to produce and polymerize herbicide derivatives of bifunctional chemical conditions. The idea behind the second attempt is to modify the fundamental characteristics of sandy soil by connecting herbicides to polymeric hydrogels. These polymers' principal purpose is to regulate the pace at which herbicides are delivered, but

they can also be very helpful in preventing quick leaching, which will improve the amount of water that sandy soil can hold onto.

Bio-surfactants are components with varying structural characteristics that are taken out of yeast or bacterial cell membranes. Both high- and low-molecular-weight polymers, which are potent emulsifiers and can dramatically lower surface tension, are present in these secondary metabolites. By emulsifying and dissolving these hydrophobic chemicals or by changing the cell's surface properties, hydrophobic molecules' bioavailability can be enhanced. Their production is regarded as a crucial tactic to help microorganisms recover from the stress caused by pollutants. They operate as mediators, increasing the pace at which hydrophobic chemicals are transferred mass-wise by making them easily accessible to bacteria. Additionally, they change the characteristics of cell membranes, increasing microbial adhesion. This mechanism is essential when there are two immiscible liquids and substrate uptake is likely.

The huge number of control that is used as a means of transporting herbicides and for the removal of the leftover herbicide poses the main obstacles to the cost-effective use of these polymeric herbicides, which may be hazardous to plants and soil. The idea of transferring pesticides to biodegradable carriers comprising polysaccharide-based agricultural wastes like bark, sawdust, cellulose, and other cellulosic wastes has been put up in various attempts to minimize this issue. The main disadvantages, however, are the challenges involved in the attempted chemical modification of such naturally occurring polymers. Large concentrations of these bioactive polymers, which are normally needed for herb control, will therefore unavoidably be utilized. The pendant pesticide hydrolyzes more quickly as a result of these polymers' hydrophilic and cross-linking characteristics. Another factor is their quick biodegradation in soil, which causes the polymeric matrix to be swiftly dissolved and reduces the herbicide's duration of action.

Recently the development of new strategies is made visible for more accuracy in the quantitative establishment of polyelectrolytes in the range of 0.00002 N–0.005 N concentrations (Wang et al. 1978). Synthesis of the polymer has key factors toward CO_2 to cancer treatment and vaccines, 3D printing, and non-flammable polymer electrolytes for lithium-ion batteries (DeSimone et al. 2016). Biomedical applications involve the identification of tumor cells, and drug delivery using organic polymer nano- and microparticles. The amalgamation of several modifications scheme allows the fabrication multifunctional formation which contributes to performing various chores at a specific time (Krug et al. 2019). Nanotechnology including nanomedicines formed with organic polymer to help in therapeutic modalities.

Standard methods of the polymers in drug-delivery system (Harris et al. 2019).

1) *Diffusion-controlled system*: The system includes the encapsulation, spray, and dip coating in the pharmaceutical industry from years ago. The polymers used in the system are rich in cellulose imitative such as polyethylene glycol (PEG) and poly N-vinyl pyrrolidone. To initiate drug delivery different devices swelling or osmotically controlled devices, monolithic devices come under function.

2) *Solvent-activated system*: The system included two concomitantly moving fronts, swelling and diffusion in the erosion front using cylindrical hydroxypropyl methylcellulose (HPMC). The devices operate drug delivery as controlled swelling under glassy to a rubbery state.
3) *Biodegradable system*: The important class of drug delivery is represented by biodegradable and bio-erodible polymers. Surface or bulk processes take place in both degradation and erosion. Ester derivatives like poly lactic/glycolic acid or hydrolytic cleavage of ester bonds relied on the maximum biodegradable synthetic polymers.

Dirt and metal(loids) are closely connected. This reduces the bioavailability of bacteria; the cation exchange capacity, composition, pH, and particle size of the soil all have an impact on the bio-surfactant–metal(loid) ion complex's activities in the soil, the kind and timing of contamination, and the geological pattern. Persistent metal(loid) contamination causes soil bonding and metal(loid) ion stabilization, which makes remediation challenging. A frequently used product for removing arsenic and other metals (Cu, Pb, and Zn) from mine tailings is the rhamnolipid bio-surfactant JBR425. In an alkaline environment, they discovered that 0.1% of the surfactant eliminated 148, 74, 2379, and 259 mg kg^{-1} of Pb, As, Zn, and Cu, respectively.

24.6 Polymeric Co-delivery Systems in Cancer Treatment

Cancer is the second most death-causing disease next to heart disease spread worldwide with an increase in mortality rates. The pharmacological agents used to treat disease are chemotherapy and radiotherapy. The issues regarding these therapies are poor stability which causes rapid blood/renal clearance, low bioavailability, and damage to the healthy tissues. The non-uniform biodistribution, difference in injection schedules, pharmacokinetic properties are involved to take benefit of drugs to form molar ratios (Pan et al. 2019). The activated moieties include stimuli-responsive drugs to increase the efficacy of the therapeutic system (Mahira et al. 2019).

The interconnection between neurons and external devices performs the function of the nervous system which is diffused during disease. The rigidity of Parkinson's disease is used to implant basal ganglia for the treatment of brain blockage (Fattahi et al. 2014). The nervous system contains neurons, extracellular matrix (ECM), and glial cells including nanoscale dimensions; the new intrinsic properties of nanomaterials show efficiency with the features and function of neural tissue. The central and peripheral nervous system involved electrically active nanomaterials which include silicon nanowires, carbon nanotubes, conducting polymer nanotubes, and gallium phosphide nanowires. To identify the disorder in humans electroencephalogram (EEG) signals range from 5 to 300 μV, <100 Hz, electrocorticogram (ECoG) signals range from 0.01 to 5 mV, <200 Hz are largely used. The examples include biological uses for hydrogels, bioartificial organs, and tissue engineering.

Additionally, granular activated carbon's surface was functionalized with COP-19 and examined for its ability to remove ions of heavy metals and organic dyes. Within 30 minutes of contact, the activated carbon-covalent organic polymer (AC-COP) and COP-19 demonstrated a strong affinity for azo dye adsorption. The way that metal ions and sulfur groups interact in COP suspended in water.

There is no denying that the way of life that society currently leads to situations where pollution and environmental deterioration are visible. It makes sense that the environment is a topic of discussion at conferences and forums where the major international actors (politicians and business leaders) gather. However, the current humanitarian crisis has come to a quality of polymers, up until now, had gone unnoticed: their potential for diversifying their uses and adapting as necessary. As fairly extensively discussed by, numerous examples show how polymers have aided in the protection of lives on numerous fronts in the struggle against the invisible foe, the coronavirus. The purpose of the current work is to demonstrate how polymers can be employed with already-accessible techniques to create lab test swabs and personal protective utensils.

The COVID-19 outbreak has affected each productive sector, albeit to differing degrees. Due to a lack of PPE, especially masks, which were previously only necessary for health professionals but are now critical for the general people as a major strategy of preventing the virus' transmission, efforts are being made to develop PPE from other materials. Companies and volunteers use unconventional resources to supply the demand for these products, resulting in a movement that provides strong evidence of economic circularity. One thing they had in common was the usage of polymer materials, whether they were employed as virgin materials (raw materials that had not yet been treated) or as recycled or used materials. Many polymers have been created and/or modified using complex methods of organic synthesis, becoming adaptable to different uses and circumstances. The pandemic catastrophe that began in 2020 has demonstrated the importance of polymers in daily living, even beyond consumption. By investing in mask producers, the Japanese government was able to guarantee a monthly supply of more than 600 million masks. In addition, businesses whose core competencies do not include producing PPE have instructed their manufacturers to produce masks and other medical supplies. In this case, 1.7 million components are produced each day on 14 mask production lines built up by the SAIC-GM-Wuling JV of the General Motors Company.

24.7 Recent Clinical Research of Organic Polymers

The repeating monomer units comprise macromolecules, i.e. organic polymers. The main supremacy of organic polymer is its biosensing attribute in electrochemical applications to identify biological fragments from human body (Kim et al. 2013). The recent use of polymers in divergent fields of present-day medicine has created significant ways in tissue engineering, diagnostic and therapeutic methods. Polymers play important role to check the rate of drug delivery, avert toxicity, saving drugs from demeaning before release, targeting drug to the place of action, and enhance absorption and therapeutic efficacy. Over the years the structure and synthesis of multifunctional organic polymer makes them appropriate for the storage of carbon dioxide. The reaction between 2,3,4-hydroxybenzaldehyde and phenyl trichlorosilane in the presence of a base and further reacting with benzidine with glacial acetic acid to obtain an organic polymer. Various phases of clinical trials have been done to check the efficacy of therapeutics and determine whether the formulation of organic polymer is safe or not. Phase 1 is generally associated with the dosage and safety of polymer for its utilization in treatment of

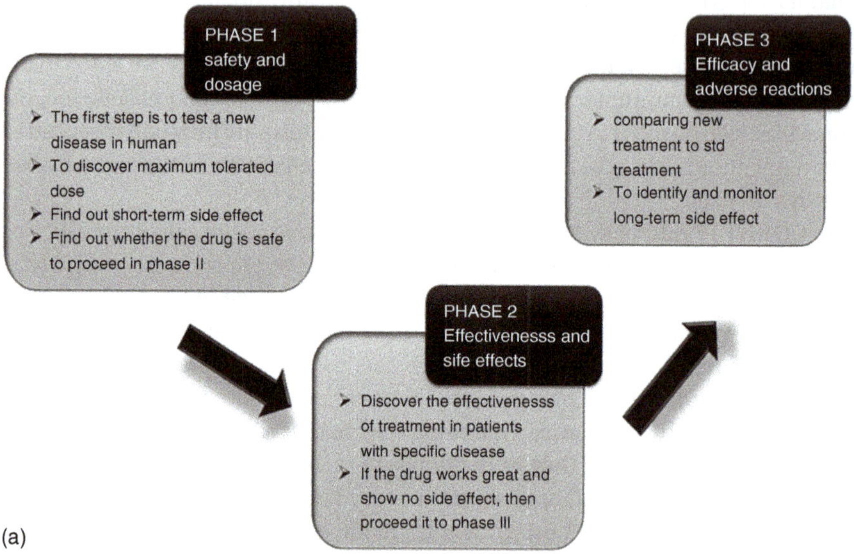

(a)

Phases of clinical research

Phase and purpose	Participants	Length of the study	Reseachers questions
Determine a medication device's safety	20 TO 100 healthy volunteers participated with disease	Take few months	• How it works on body? • Side effects of taking high dosage? • Might the IMP work in patients?
Test a product's or device's effectiveness	Up to few hundred people were participated with disease	Few months to 2 years	• Is IMP safe In patients?
Monitoring of adverse reaction	300–3,000 people participated with disease	1–4 years	• Is IMP really safe in patients?
Safety and effectiveness	Thousand volunteers participated with disease	During the post - market safety after FDA approval	• How safe is the new medicine? • Does the medicine work in real world?

(b)

Figure 24.4 Discussion about the different Phases of clinical trials.

disease. In this phase, tolerance and side effects are observed to make a decision whether it is safe to use or not. Under Phase 2, the effectiveness of molecule is observed and evaluated in correspondence to that particular disease. If no side effect is observed, then it is considered as good to go for the third phase. In the third phase we determine the efficacy and reactions of the molecule as a treatment aid, and its long-term effectiveness is measured and observed (Figure 11). These phases require time and consistent work under trials. It could take several years to complete. These phases and purpose of clinical trials can be encapsulated in Figure 24.4.

24.8 Current Scenario of Organic Polymers

In the current scenario polymer science is inescapable and relevant to emphasize its role. The two Nobel Prizes awarded in the physical chemistry of polymeric systems. Evolution still exists as classical thermodynamics approaches are examined and fully authorized, while major rectification is being initiated during H-bonding systems. Self and cross-interconnection are being ardently investigated by lengthening available mean-field models to describe the system. Several efforts have been committed in the way of research development of nanocomposites (classical polymer-composite) (Kango et al. 2013). The highly cross-linked POPs, having micropores, currently appear as a versatile program for disposition of catalysts (Kaur et al. 2011).

24.9 Future Perspectives of Organic Polymer

The importance of polymers increases with the development of advanced electronic gadgets and electric power system that hinge on dielectric film capacitors maximum energy storage. Several perspectives of organic polymers in different ways are as follows-

1) Porous organic polymers in solar cells: These are the unique materials of solar cells. It describes the challenges for POPs from the photoactive layer to transporting layer in the appearance of perovskite and dye-sensitized solar cells.
2) Porous organic polymers for light-driven organic transformations: The high specific surface area, less density, better stability, uniform pore size, and structural versatility led to the strongest platform of heterogeneous photocatalysis.
3) Emerging porous organic polymers for biomedical applications. This shows the vast applications in the biomedical department such as drug insertion, phototherapy, bioimaging, and antibacterial applications.

24.10 Conclusion

Organic polymers have already been applied and are utilized in various applications of our day-to-day lives. These versatile applications of organic polymer are because of their exceptional characteristics like porosity, design-ability, crystallinity, stability, and bio-compatibility. These properties have enabled their applications and enhancement in today's world. Now these organic polymers are widely used in chemical engineering, therapeutics, medical devices, and more. In addition to this, their properties and some of applications are further explored in clinical research. From the widely used biopolymer polypropylene to the naturally occurring biopolymer found in the human DNA strand. Moreover, many scientists have even received Nobel Prizes and various other recognitions while exploring and working on the potential of these organic polymers. It has been estimated that these polymers may even revolutionize the solar energy systems and can cure many other dreadful diseases of the world by their exceptional properties.

References

Chang, Z., Zhang, D.S., Chen, Q., and Bu, X.H. (2013). Microporous organic polymers for gas storage and separation applications. *Physical Chemistry Chemical Physics* 15 (15): 5430–5442.

Chen, H. and Wang, L. (2017). Chapter 5—microbial cell refining for biomass conversion. In: *Technologies for Biochemical Conversion of Biomass*, 101–135. Apple Academic Press.

Cullis, C.F. and Hirschler, M.M. (1981). *The Combustion of Organic Polymers*, vol. 5. Oxford University Press.

DeSimone, J.M., Mecham, S.J., and Farrell, C.L. (2016). Organic polymer chemistry in the context of novel processes. *ACS Central Science* 2 (9): 588–597.

Fattahi, P., Yang, G., Kim, G., and Abidian, M.R. (2014). A review of organic and inorganic biomaterials for neural interfaces. *Advanced Materials* 26 (12): 1846–1885.

Hao, W., Chen, D., Li, Y. et al. (2019). Facile synthesis of porphyrin based covalent organic frameworks via an A2B2 monomer for highly efficient heterogeneous catalysis. *Chemistry of Materials* 31 (19): 8100–8105.

Harris, A.R., Molino, P.J., Wang, C. et al. (2019). Biomedical applications of organic conducting polymers. In: *Conjugated Polymers*, 783–812. CRC Press.

Kango, S., Kalia, S., Celli, A. et al. (2013). Surface modification of inorganic nanoparticles for development of organic–inorganic nanocomposites—a review. *Progress in Polymer Science* 38 (8): 1232–1261.

Kaur, P., Hupp, J.T., and Nguyen, S.T. (2011). Porous organic polymers in catalysis: opportunities and challenges. *ACS Catalysis* 1 (7): 819–835.

Kim, G.B., Fattahi, P., and Abidian, M.R. (2013). Electrode-neural tissue interactions: immune responses, current technologies, and future directions. In: *Biomaterials Surface Science*, 539–565. Wiley-VCH Verlag.

Krug, P., Kwiatkowska, M., Mojzych, I. et al. (2019). Polypyrrole microcapsules loaded with gold nanoparticles: perspectives for biomedical imaging. *Synthetic Metals* 248: 27–34.

Kwak, M. and Herrmann, A. (2010). Nucleic acid/organic polymer hybrid materials: synthesis, superstructures, and applications. *Angewandte Chemie International Edition* 49 (46): 8574–8587.

Levin, B.C. (1992). The development of a new small-scale smoke toxicity test method and its comparison with real-scale fire tests. *Toxicology Letters* 64: 257–264.

Li, Y., Chen, Q., Xu, T. et al. (2019). De novo design and facile syntheses of 2D covalent organic frameworks: a two-in-one strategy. *Journal of the American Chemical Society* 141: 13822–13828.

Mahira, S., Kommineni, N., Husain, G.M., and Khan, W. (2019). Cabazitaxel and silibinin co-encapsulated cationic liposomes for CD44 targeted delivery: a new insight into nanomedicine based combinational chemotherapy for prostate cancer. *Biomedicine & Pharmacotherapy* 110: 803–817.

Maiti, A., Mule, A. A., Kumar, A., Bhatnagar, A., & Mondal, P. (2021). *00148 Polymers in Wastewater Treatment*.

Pan, J.H., Zhou, H., Cooper, L. et al. (2019). LAYN is a prognostic biomarker and is correlated with immune infiltrates in gastric and colon cancers. *Frontiers in Immunology* 10: 6.

Schneider, J., Matsuoka, M., Takeuchi, M. et al. (2014). Understanding TiO_2 photocatalysis: mechanisms and materials. *Chemical Reviews* 114 (19): 9919–9986.

da Silva, F.L., Scheel, G.L., Segatelli, M.G., and Tarley, C.R.T. (2019). Redox Preconcentration/speciation of chromium by using Nanocomposites based on carbon nanotubes and functional polymers. In: *Composite Nanoadsorbents*, 139–180. Elsevier.

Singh, S., Mahalingam, H., and Singh, P.K. (2013). Polymer-supported titanium dioxide photocatalysts for environmental remediation: a review. *Applied Catalysis A: General* 462–463: 178–195.

Soares, S., Sousa, J., Pais, A., and Vitorino, C. (2018). Nanomedicine: principles, properties, and regulatory issues. *Frontiers in Chemistry* 20 (6): 360. https://doi.org/10.3389/fchem.2018.00360.

Sun, Q., Dai, Z., Meng, X. et al. (2015). Porous polymer catalysts with hierarchical structures. *Chemical Society Reviews* 44 (17): 6018–6034. https://doi.org/10.1039/c5cs00198f.

Taheri, N., Dinari, M., and Asgari, M. (2022). Recent applications of porous organic polymers prepared via Friedel–Crafts reaction under the catalysis of $AlCl_3$: a review. *ACS Applied Polymer Materials* 4 (9): 6288–6302.

Uribe-Romo, F.J., Hunt, J.R., Furukawa, H. et al. (2009). A crystalline imine-linked 3-D porous covalent organic framework. *Journal of the American Chemical Society* 131: 4570–4571.

Wang, L.K., Wang, M.H., and Kao, J.F. (1978). Application and determination of organic polymers. *Water, Air, and Soil Pollution* 9 (3): 337–348.

Wang, J., Li, Z., and Gu, Z. (2021). A comprehensive review of template-synthesized multi-component nanowires: from interfacial design to sensing and actuation applications. *Sensors and Actuators Reports* 3: 100029. https://doi.org/10.1016/j.snr.2021.100029.

Xu, C. and Hedin, N. (2013). Synthesis of microporous organic polymers with high CO_2-over-N_2 selectivity and CO_2 adsorption. *Journal of Materials Chemistry A* 1: 3406–3414.

Yadav, Anshul, and Niraj Sinha. "Organic polymers for drinking water purification." (2022): 997-1003.

Zhang, W., Mohamed, A.R., and Ong, W.J. (2020a). Z-scheme photocatalytic systems for carbon dioxide reduction: where are we now? *Angewandte Chemie International Edition* 59 (51): 22894–22915.

Zhang, T., Xing, G., Chen, W. et al. (2020b). Porous organic polymers: a promising platform for efficient photocatalysis. *Materials Chemistry Frontiers* 4 (2): 332–353. https://doi.org/10.1039/C9QM00633H.

Zhang, H., Ding, G.Y., Yousaf, A. et al. (2021). A typical 2D covalent organic polymer as multifunctional sensor and assemble a WLED. *Journal of Solid State Chemistry* 297: 122101.

25

Transdermal Drug Delivery and Organic Polymers: Current Scenario and Future Prospects

Sukanta Roy[1], Sourav Das[1], Dibya Das[3], Himangshu S. Maji[3], Anirbandeep Bose[2], Senjuti Bhattacharjee[4], Sudeepto Biswas[4], and Subas Chandra Dinda[1]

[1] *The Neotia University, School of Pharmacy, Diamond Harbour Road, Jhinga, Sarisha, West Bengal 743368, India*
[2] *Adamas University, Department of Pharmaceutical Technology, Barrackkpore Road, 24 Paraganas North, West Bengal 700126, India*
[3] *JIS University, Department of Pharmaceutical Technology, 81, Nilgunj Road, Agarpara, Kolkata, West Bengal 700109, India*
[4] *Hola Nutriment, Garalgacha Bazar, Baluhati Main Road, Ganguli Para, Chanditala II, Hoogly, West Bengal 712311, India*

25.1 Transdermal Drug Delivery System (TDDS) and Current Scenario

A comprehensive assessment of the global TDDS market reveals a projected growth trajectory of 6.39% CAGR from 2021 to 2027, leading up to a total market size of US$ 8.7 billion by 2027. This promising growth rate stands in stark contrast to other more traditional routes of drug administration such as oral, intravascular, subcutaneous, and transmucosal. TDDS offers advantages that could gain greater attention as researchers continue to explore novel ways of distributing drugs (Grand View Research 2021). Transdermal drug delivery (TDD) is an advanced drug-delivery technique, which enables the safe and consistent administration of biologically active agents, such as medications or cosmeceuticals, via the skin. TDD has been demonstrated to offer several beneficial outcomes in comparison to traditional oral or intravenous formulations; primarily, the controlled release of drugs over time, decreased first-pass metabolism, and fewer side effects. Additionally, the direct administration of drugs via TDDS avoids the use of needle-based injections, which can be painful for patients. TDDS has been proven with significant impact on the delivery of Analgesics, Hormones, Cardiovascular, and central nervous systems agents with therapeutic value (Jeong et al. 2021). TDDS provides a non-invasive, safe, easy-to-use, and effective method for drug delivery, making it particularly suitable for child and elderly patients (Sabbagh and Kim 2022; Prausnitz et al. 2004).

The inception of transdermal drug delivery systems (TDDS) in the United States occurred in 1979. In 1996, nicotine patches were the first system to be implemented

Organic Polymers in Energy-Environmental Applications, First Edition.
Edited by Ramesh Oraon, Pardeep Singh, Sanchayita Rajkhowa, Sangita Agarwal, and Ravindra Pratap Singh.

extensively. Currently, there are nineteen TDDS available for such substances as estradiol, fentanyl, lidocaine, and testosterone. Various drug combination patches, iontophoretic, and ultrasonic delivery systems are available for contraception and hormone replacement, as well as for the application of analgesia. TDDS exist for a variety of substances including estradiol, fentanyl, lidocaine, and testosterone. The average rate of new transdermal delivery systems being approved has more than tripled in the past 5 years, with an average of one new system being approved every 7.5 months (Jeong et al. 2021).

25.2 Navigating the Complexities of Transdermal Delivery

The skin is a complex biological barrier and presents many limitations in regard to percutaneous drug delivery (Pastore et al. 2015). The outermost layer of the skin, the stratum corneum (SC), is composed of a lattice of lipids and corneocytes, with an approximate thickness of 20 μm of adult skin, which is roughly 3 mm thick in total (see Figure 25.1). This renders the skin largely impermeable to most biopharmaceuticals and small molecules (Erdő et al. 2016) with only moderately lipophilic compounds being able to penetrate through the barrier.

In order for a substance to traverse the stratum corneum and enter into the viable epidermis, it must possess suitable hydrophilicity to enable adequate transdermal delivery into the bloodstream (Prausnitz and Langer 2008; Sloan et al. 2011; Bos and Meinardi 2000). However, employing TDDS may increase the possibility of local side effects by changing the skin barrier. In some circumstances, these effects can vary from skin irritation to allergic reactions. Managing the danger of unfavorable local reactions is thus a significant concern in TDDS (Barry 1991; Murphy and Carmichael 2000; Johnson 2007).

The range of scientific techniques employed to develop a device capable of electroporation, sonophoresis, or iontophoresis may vary extensively on the basis of application requirements. Depending on the specifics of the delivery problem at hand, the complexity of the techniques chosen to develop a suitable device can range from

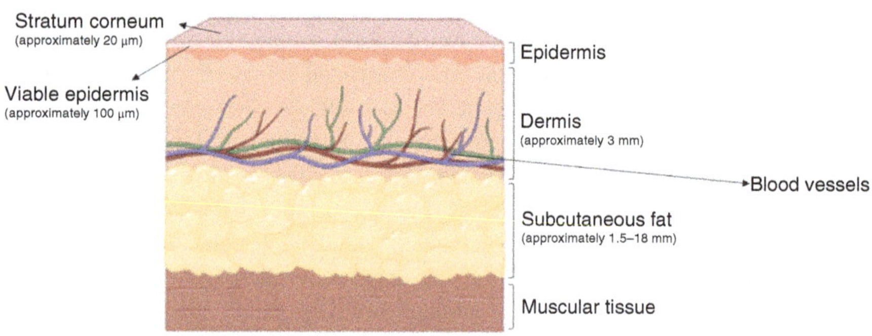

Figure 25.1 Schematic view of skin.

simple conductive material construction to the implementation of intricate neural networks and advanced sensing technologies. In addition, developing a stable and non-irritating formulation to deliver a specific molecule, leads to further obstacles in TDD (Prausnitz and Langer 2008).

Furthermore, there are additional obstacles to achieving satisfactory patient usability with TDDS. For instance, when creating matrix or reservoir patches, it is crucial to design them to have minimal detachment and patch loss rates, which can be particularly difficult to achieve in the context of once-weekly TDDS. On the other hand, transdermal gels must be created in a way that limits the possibility of the active ingredient being transferred to unintended recipients. In the design of complicated devices, patient comfort and reducing skin irritation are of highest importance. These are just a few instances of the challenges that a research team could encounter while developing a novel transdermal medication (Goffin et al. 2000; Prausnitz et al. 2004; Karande et al. 2005).

Despite the difficulties, the development of TDDS throughout the previous five decades has made progress. It becomes evident that there have been two key levers used to create prospects for new transdermal drugs:

1) The process of identifying appropriate molecules from the collection of drugs that have been approved.
2) Lowering the skin's barrier capacity, particularly the SC layer, through several methods such as chemical agents (penetration enhancers), biological techniques (hydration/occlusion), or physical means (heat, ultrasound, etc.)

A closer examination of the current utilization of transdermal medicines reveals that drug development research is currently focused on developing products with low- or non-invasive administration of therapeutic molecules. Such development involves multiple levers such as active, minimal and non-invasive approaches, though details of their effectiveness remain uncertain. The recent emergence of microneedle technology provides an avenue for novel transdermal products to be realized through minimal disruption. Therefore, it is worthwhile to invest in such technology and identify appropriate macromolecules to capitalize on, in order to generate valuable new products (Paudel et al. 2010; Tiwary et al. 2007; Prausnitz et al. 2004).

25.3 Structural Overview of Transdermal Delivery Systems

The objective of TDDS relies on polymeric layers to form the system. TDDS are composed of an outer impermeable backing to ward off drug leakage, an inner adhesive or rate-controlling membrane, along with additional layers of polymer material. Depending upon the components involved, the systems are divided into three categories: drug reservoirs, drug–polymer matrices, and polymeric laminates (see Figure 25.2).

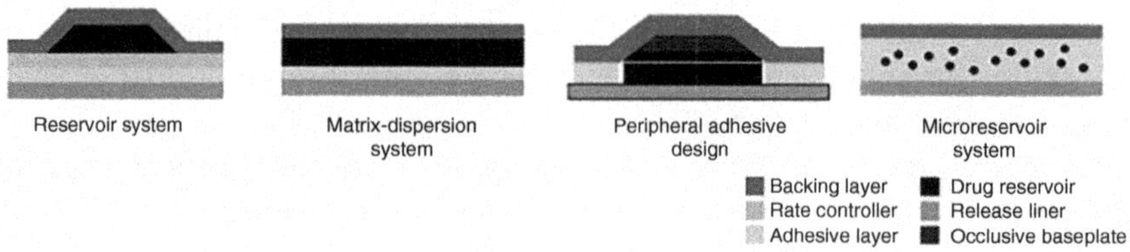

Figure 25.2 Schematic representation of different TDDS. Source: Khandavilli et al. (2002)/Pharmaceutical Technology.

25.3.1 Key Optimizing Principles for Designing a Transdermal Delivery System

25.3.1.1 Reservoir Systems

This system utilizes a pharmacological agent contained within an impermeable backing layer and a rate-controlling membrane. The rate-controlling membrane, which may be either microporous or nonporous in nature, is the only passage for the release of the pharmacological agent. This reservoir of drug is filled with a solution, suspension, or gel, or derived from a solid polymer matrix. An additional hypoallergenic adhesive polymer helps to augment the outermost layer of the reservoir (Stevenson et al. 2012).

25.3.1.2 Microreservoir Systems

A TDDS is developed by combining a drug reservoir and matrix-dispersion system. The drug reservoir utilizes a water-soluble polymer to solubilize the medication, forming an aqueous solution that is spread throughout a lipophilic polymer to produce microscopic spheres that contain the drug. The thermodynamic instability of the suspension is stabilized via polymer cross-linking in situ to form a stable TDDS (Omray et al. 2008).

25.3.1.3 Matrix Systems

The drug-in-adhesive or matrix delivery system takes advantage of a polymer-based adhesive to create a drug-containing reservoir. This is achieved through the uniform distribution of drugs into the adhesive matrix, which is then applied onto an impervious base layer through either solvent casting or melt processing. A composite adhesive structure is then formed by layering additional non-medicated adhesive matrix on top of the reservoir, thus forming a suitable matrix based drug-delivery system (Aqil et al. 2003).

25.3.1.4 Matrix-Dispersion System

A polymeric matrix-dispersion system in either a lipophilic or a hydrophilic variation is utilized to evenly distribute the drug. This matrix is then affixed to an occlusive base plate via a separately produced drug-impermeable backing layer. This ensures the secure placement of the drug depot. To complete the process, a strip of adhesive encaging the edge of the reservoir holds the entire ensemble in place (Akram et al. 2018).

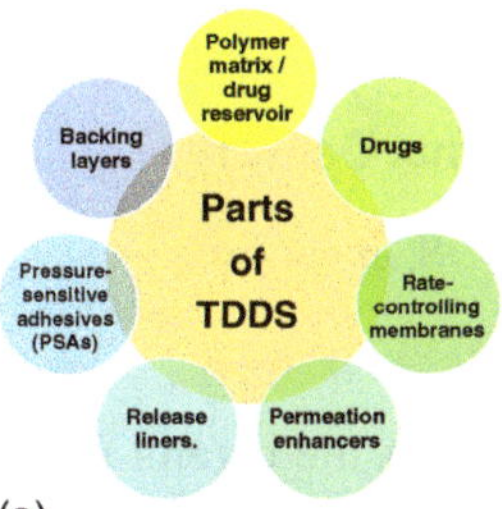

(a)

Components	Ideal properties for TDDS
Drugs	Dose <25 mg/day; half-life- ≤10 h; molecular weight < 500 Dalton; log P (octanol-water)- between 1 and 3; melting point <200 °C; non irritating and non-sensitizing to skin; oral bioavailability- low.
Permeation enhancer	Inert; harmless; non-allergenic; and non-irritating.
PSAs	hypoallergenic, non-irritating, and easily removable without leaving a residue or causing any pain
Backing membrane	lowest modulus; high flexibility; good oxygen transmission; and a high moisture vapor transmission rate.
Release liner/	Non-occlusive; inert; harmless; non-allergenic; and non-irritating; easily removable.
Rate-controlling membranes	Inert; harmless; non-allergenic; and non-irritating;
Polymers	Inert; harmless; non-allergenic; and non-irritating; microporous or nonporous in nature; highly soluble; non-reactive; biodegradable

(b)

Figure 25.3 (a) Describes different parts of TDDS. (b) Ideal properties of components suitable for TDDS. Source: Adapted from Kathe and Kathpalia (2017); Yu et al. (2021).

The area of transdermal drug delivery is currently experiencing an exponential growth, owing largely to the advances in polymer science. This branch of drug delivery is rapidly advancing, and this chapter is dedicated to the selection of polymeric materials used in such systems.

Polymers have been identified as essential components in the development of effective transdermal drug-delivery systems due to their multifaceted physical and chemical characteristics. Their flexibility, permeability, viscosity, and mechanical strength provide desirable traits for formulators looking to create efficient drug-delivery systems. A thorough analysis of the respective polymers enables the formulator to identify the most suitable structural choices for their desired application (see Figure 25.3).

25.3.1.5 Polymer Matrix/Drug Reservoir

Polymers are essential for the successful fabrication of a TDDS. These materials must possess high stability, biocompatibility, and a low rate of reaction with the drug itself and other components of the device. Additionally, easy fabrication of the polymer into the desired shape and structure can be ensured by its non-toxic degradation products that are not antigenic to the host organism (Kamaly et al. 2016). Based on the sources, the polymers used for TDDS can be classified into two categories (Gutierrez Cisneros et al. 2021; Liechty et al. 2010; Bangar et al. 2014) as follows.

Potential Natural Polymers in Transdermal Delivery Hydroxypropyl methylcellulose (HPMC), sodium carboxy methyl cellulose (sodium CMC), cellulose acetate, methylcellulose, ethyl cellulose, gelatin, chitosan, sodium carboxymethyl guar, sodium alginate, polymerized rosin, etc.

Potential Synthetic Polymers in Transdermal Delivery Polyvinyl alcohol, polyethylene, polyethylene glycol, polyvinylpyrrolidone, eudragits, ethylene vinyl acetate copolymer, ethyl vinyl acetate, silicon rubber, etc.

25.3.2 Characterization of Drug Molecules for Transdermal Delivery

Drugs that possess specific characteristics are chosen for TDDS, as follows.

25.3.2.1 Characterization of Physicochemical Properties

The successful implementation of TDD treatments depends on the careful selection of pharmacologically active molecules with diverse criteria (see Figure 25.3). For instance, the solubility of the molecules must be at least $1\ \mathrm{mg\ ml^{-1}}$ in both oil and water, and the melting point should be below 200 °F to improve the membrane concentration gradient. Additionally, the molecular weight should remain under 500 kD, while the pH of the aqueous solution should be between 5 and 9. Lastly, for optimal results, the molecules should feature less than two hydrogen-bonding groups. Systematically selecting molecules with these specific properties can ensure the success of TDD treatments (Choudhary and Singh 2021; Prausnitz and Langer 2008).

25.3.2.2 Characterization of Biological Properties

In order to maximize efficacy, an ideal drug should have a potency of less than 25 mg per day, a short biological half-life, and should neither irritate nor cause allergic reactions on contact with human skin. Additionally, it is desired for the drug to have stability upon exposure, not elicit any immune responses, display a near zero-order release profile in transdermal delivery and should not bind permanently to the subcutaneous tissue or undergo metabolic changes in the skin. Moreover, to prevent tolerance, the drug should be optimally formulated, and should not bind permanently to the subcutaneous tissue or undergo any metabolic changes in the skin (Choudhary and Singh 2021; Prausnitz and Langer 2008).

25.3.3 Investigating the Properties of Various Permeation Enhancers for Transdermal Delivery

25.3.3.1 Chemical Permeation Enhancers

The stratum corneum consists of highly ordered intercellular lipid bilayers, which can be disrupted by the introduction of amphiphilic molecules or the removal of lipids. This interruption reversibly reduces the hindrance of resistance, permitting the entry of drugs that are administered in unison (Prausnitz and Langer 2008). In order for an enhancer to be effective, it should meet certain criteria. The desired material must be inert, harmless, non-allergenic, and non-irritating. Moreover, it should be able to operate in a unidirectional fashion and interact harmoniously with the drugs and excipients it is associated with. The potency of the enhancer appears to be affected by the type of drug, skin, and concentration (Williams and Barry 2004). Permeants, also known as permeation enhancers, are molecules that facilitate the passage of drugs or other substances through biological membranes, such as skin or the gastrointestinal tract. Examples of permeants which are used to enhance permeation include ethanol, which is the most common, as well as essential oils and terpenes such as cineole, carveol, menthone, citral, menthol, and d-limonene. In addition, dimethyl sulfoxide, propylene glycol, N-methyl-2-pyrrolidine, ethyl pyrrolidine, polyethylene glycol 400, isopropyl myristate, myristic acid, succinic acid, laurocapram(azone), methyl laureate, lauric acid, sodium lauryl sulfate, non-ionic surfactants (spans, tweens), pluronic, oleic acid, diethylene glycol monoethyl ether, and urea are also known to have permeating properties (Yu et al. 2021; Arora et al. 2017; Harrison et al. 1996; Walters and Hadgraft 1993; Wang et al. 1998).

25.3.3.2 Physical Permeation Enhancers

The physical approach of TDDS involves removing the stratum corneum and hydrating it. Iontophoresis, sonophoresis, thermal energy, and electroporation are the physical approaches used. Iontophoresis in particular holds a great potential for drug delivery by using a low-voltage electric current to increase and regulate drug permeation. This is achieved by two mechanisms: electrorepulsion, where the electric field applied to the permeant causes it to disperse, and electroosmosis, due to the negative charge of the skin at physiological pH (Dhote et al. 2012).

Stratum corneum (SC) is the foremost layer of the skin, playing the essential role of a protective barrier between the body's interior and the exterior environment, limiting the transdermal exchange of molecules between the two. This layer ranges in thickness from 10 to 100 μm and is created of numerous layers of corneocytes (cells crammed with keratin) and a lipid matrix, which presents a continuous medium. The intercellular lipids consist of nearly identical proportions of ceramides, cholesterol, and free fatty acids.

The skin is capable of absorbing substances through three distinct pathways: intercellular transport via corneocytes and lamellar lipids, intracellular uptake through individual cells, and appendageal uptake through hair follicles, sweat ducts, and exocrine glands (Park et al. 2019).

Sonophoresis is an innovative technique of permeating semi-solid topicals into the epidermis. This is achieved by harnessing the power of sound waves and converting ultrasonic vibrations into micro-vibrations. These micro-vibrations are effective in raising the kinetic energy of topical agent molecules, resulting in increased permeation rates of compounds through the epidermis. It currently has many potential applications across medical fields such as drug delivery through the skin, as well as in physical therapy and iontophoresis. The ultrasonic energy is transferred to the skin through coupling agents such as gels, creams, and ointments. It is believed that this process is enhanced by the effects of cavitation, microstreaming, and heating (Park et al. 2019; Tawfik et al. 2021).

In the process of electroporation, high-voltage electrical pulses are applied to the SC of the skin, creating temporary pores that enable increased transdermal delivery of therapeutic drugs. This novel delivery method efficiently creates aqueous trails in SC for drugs to permeate and significantly improving drug transfer into underlying tissue (Gehl 2003). Erbium: Yttrium-Aluminum-Garnet (Er: YAG) laser is implied efficiently remove the SC layers of the skin through the application of a single pulse of low energy (Arora et al. 2012). Ultrasound and microneedle treatments effectively create minuscule pathways or channels within the external most layer of the epidermis, the stratum corneum. This allows drug products to penetrate the skin's barrier for therapeutic purposes, increasing the absorption of the medication (Singh et al. 2013).

25.3.3.3 Other Permeation Enhancers

Various strategies can be employed to enhance the permeability of the skin, including ethanolic liposomes, niosomes, protransferosome gels, and prodrug approaches, which facilitate the solubilization and diffusion of drugs into the skin (Dubey et al. 2007; Rai et al. 2017; Fang and Leu 2006).

25.3.4 Pressure-Sensitive Adhesives (PSAs)

The attachment of TDDS to the skin must be secure to ensure a regulated release of the drug. To achieve this, pressure-sensitive adhesives (PSAs) are used. These PSAs are hypoallergenic and non-irritating, allowing for easy and painless removal without any residue. Polyacrylate, polyisobutylene, and silicones are the most commonly used PSAs in TDDS, providing a necessary skin–drug interface for long-term drug delivery (Tan and Pfister 1999; Lobo et al. 2016).

25.3.5 Optimization of Suitable Backing Membrane

To ensure optimal performance, it is essential to select the right backing materials for use in skin adhesives. Polyolefins, polyesters and elastomers are the most common material used for this purpose, and are available in clear, pigmented, or metalized form. Low-density polyethylene (LDPE) has been identified as the optimal material for backing materials used in many skin-related applications. This is owing to its elastomeric nature, which far surpasses that of polyester in regard to its contouring and adhesion properties, while enabling enhanced skin intricacies. To further enhance the performance of the backing material, it must exert a low water-vapor-permeable rate, so as to ensure improved skin hydration and permeation.

The selection of backing materials for topical drug-delivery systems is a critical step in enabling suitable performance for the desired application. Low-density polyolefins, polyesters, and elastomers are commonly employed, often in clear, pigmented, or metalized forms that possess the flexibility and tensile strength required for proper functioning. Elastomeric low-density polyethylene is often favored due to its superior adhesion and greater flexibility in comparison to other materials such as polyester. The backing material used for any transdermal skin delivery system should provide several beneficial properties, such as low modulus, enhanced flexibility, high oxygen transmission, and a high moisture vapor transmission rate. These properties will enable the device to transfer the corresponding drug efficiently and provide excellent comfort. Vinyl, polyester films, polyester-polypropylene films, polypropylene and polyethylene resins, polyurethane, Co Tran 9722 film, ethylene-vinyl acetate, and aluminized plastic laminate are some of the materials which have ideal characteristics designed to optimize the performance of the transdermal patch. Furthermore, these materials should also have low water vapor transmission to retain skin hydration and ensure proper permeability. In addition, the backing material should be heat-sealable to ensure form–fill–seal packaging of the drug reservoir (Zhan et al. 2015; Sharma 2018; Liu et al. 2014).

25.3.6 Release Liner

The TDDS patch incorporates a release liner for secure fixation on the skin. The liner material commonly incorporates non-occlusive substrates such as paper fabric, or occlusive substances including polyethylene and polyvinylchloride, with a release layer crafted from silicon. To facilitate the patch attachment, this liner must be carefully removed before applicability (Wokovich et al. 2010).

25.3.7 Other Excipients

The combination of water, ethanol, isopropyl myristate, isopropyl alcohol, and dichloromethane can be utilized for the production of a drug-containing reservoir for transdermal drug delivery. This combination of solvents can either be used in individual form or in combination to achieve optimized results (Prodduturi et al. 2009; Rastogi and Yadav 2012). The efficacy of transdermal delivery systems can be improved by adding permeation enhancers in combination with two co-solvents, such as propylene glycol and ethanol. Moreover, the incorporation of plasticizers, such as diethyl phthalate, dibutyl phthalate, glycerol, triethyl citrate, polyethylene glycol 400, eudraflex, and propylene glycol, can improve the flexibility of the patch. The Androderm product by TheraTech Inc./SmithKline Beecham is formulated with a drug reservoir of alcohol and glyceryl monooleate, which is gelled with an acrylic acid copolymer. This is backed by a metalized polyester/ethylene methacrylic acid copolymer/EVA polyethylene microporous membrane, with a peripheral acrylic adhesive and a release liner of silicone-coated polyester. Estraderm by Alza/Ciba Geigy utilizes a drug reservoir compounded with a combination of hydroxypropyl cellulose, alcohol, and light mineral oil, which is held in place by a polyester/polyethylene composite EVA copolymer with 5% vinyl acetate and PIB adhesive. The reservoir is then lined with a silicone-coated polyethylene terephthalate release liner for optimal release. The Testoderm TTS product by Alza contains a drug reservoir of drug and alcohol gelled with HPC, and is backed by a polyester/EVA copolymer, with a PIB adhesive and a silicone-coated polyester release liner. The Transderm Nitro product by Alza/Ciba Geigy is a patch containing a drug reservoir made up of a combination of lactose, colloidal silica, and silicone oil, which is backed by a polyfoil EVA copolymer with a flesh-like color. The patch is further equipped with a silicone adhesive and a fluorocarbon polyester film release liner (Hanumanaik et al. 2012).

Table 25.1 Different approaches for transdermal delivery and potential drug candidates.

Types	Approaches	Potential molecules for delivery	References
Active	Iontophoresis	Local anesthetics, opioids, steroids, non-steroidal anti-inflammatory drugs, antibacterial drugs, antifungal drugs, antiviral drugs, anti-cancer drugs, fluorides, vitamins	Ansari et al. (2022)
	Sonophoresis	Hydrocortisone, betamethasone dipropionate, oligodeoxynucleotides, chymotrypsin alpha, dexamethasone, fluocinonide, heparin, hyaluronidase, ibuprofen, indomethacin, insulin, iodine, ketoprofen, lidocaine, mecholyl, naproxen, piroxicam, sodium salicylate, trypsin, and zinc	Vranić (2004)
	Electroporation	Vaccine delivery (HIV, cancer); gene delivery (IL-2, IL-12, IL-15, IL-18, GM-CSF, IFN-gamma); drug delivery (Bleomycin); analgesic and anti-inflammatory drugs; anti-diuretic drugs; anti-viral drugs; beta-blocker agents; insulin; photosensitizers; folic acid antagonists	Bolhassani et al. (2014), Escobar-Chávez et al. (2009), and Gehl (2008)
	Photomechanical waves	Protein delivery (insulin)	Cross and Roberts (2004)
	Microneedle	Hormones (para thyroid hormone), insulin, β-blockers, antidiabetic drugs (TZDs), antifungal drugs	Ita (2015), Jana and Wadhwani (2019), and Kelchen and Brogden (2018)
	Thermal ablation	Lipophilic and hydrophilic drugs including biomacromolecules such as peptides, proteins, vaccines, and DNAs	Amjadi et al. (2017), Li et al. (2013), and Szunerits and Boukherroub (2018)
Passive	Vesicles • Emulsomes • Enzymosomes • Virosomes • Ethosomes • Liposomes • Pharmacosomes • Transferosomes • Niosomes • Sphingosomes	Topical treatment of skin diseases, peptides, or proteins	Palac et al. (2014), Rajan et al. (2011), Gill et al. (2014), and Singh et al. (2015)
	Polymeric nanoparticles	Pilocarpine HCL; Retinol; Antisense oligonucleotides; Aciclovir; Plasmid DNA; Clobetasol-17-propionate; Quercetin	Zhang et al. (2013)
	Nanoemulsion	Camphor, menthol; methyl salicylate; Clobetasol Propionate; Amphotericin B; Meloxicam; Curcumin; Finasteride; 5-ALA; Ropinirole	Rai et al. (2018)

The TDDS is broadly categorized mainly in two types depending upon their permeability pathway through the skin, (i) Active Pathway and (ii) Passive Pathway. The active transdermal drug-delivery system necessitates external stimulation, which can include electrical, mechanical, or physical means, to enables the drug to permeate through the different layers of the skin. In contrast, the passive TDDS does not necessitate the utilization of external stimuli. However, it entirely depends upon the physicochemical property of the drug molecule. These two classes are also further classified depending upon the techniques implied to improve skin permeation are shown in Table 25.1 (Jeong et al. 2021).

25.4 Exploring Different Organic and Synthetic Polymer-Based Strategies for Transdermal Drug Delivery

25.4.1 Cellulose

Cellulose is a natural polymer found in many organisms, yet its insolubility in water and other solvents has hindered its use in hydrogel-based applications. To tackle this issue, numerous cellulose derivatives have been synthesized and incorporated with cross-linking agents to enhance solubility. Examples of such derivatives include hydroxypropyl methylcellulose (HPMC), methylcellulose, carboxymethylcellulose, carboxyethyl cellulose, hydroxyethyl cellulose, and ethyl cellulose. HPMC is a widely used cellulose derivative that has been found to be advantageous in the pharmaceutical industry due to its biodegradability, availability, and ease of use. HPMC is often employed as a matrix for drug delivery in both films and gels, with its remarkable swell ability influencing the drug release kinetics. To confirm this, Prajapati et al. (2011) conducted a study to investigate the potential of using a transdermal patch with solvent-casting technique and HPMC as the polymer matrix to deliver Repaglinide. The results of the study revealed the effectiveness of the patch in transporting the drug. This patch was demonstrated to be efficient in minimizing recrystallization probabilities and promoting the drug loading ratio (Esuendale and Gabriel 2016).

25.4.2 Dextran

Dextran is a branched polysaccharide composed of glucose molecules that is synthesized either chemically or through microbial fermentation. It is widely employed in medical applications, such as in the treatment of hypoglycemia and hemorrhage, due to its high water solubility and ability to form stable colloidal solutions. Its structure is composed of a large number of hydroxyl groups, which, when combined with a specific cross-linker, form hydrogels. In a study Nematpour et al. (2019) incorporated electrospinning to craft a transdermal patch. This was achieved through the utilization of dextran, graphene oxide (GO), and polycaprolactone (PCL). The composition of the patch includes an inner layer loaded with tetracycline-incorporated GO atop dextran nanofibers, shielded by an external hydrophobic PCL layer to restrict moisture penetration. Despite its advantageous components, dextran is not exempt from certain drawbacks such as its elevated cost and susceptibility to induce more serious anaphylactic reactions compared to starch and gelatin. These responses of dextran-reactive antibodies to elicit vasoactive mediator release can be attenuated through pre-treatment with a hapten. This is due to the ability of the hapten to interfere with the antibody–antigen binding, thereby preventing the instigation of the immunological response (Ahn et al. 2020).

25.4.3 Chitosan

Chitosan, a naturally occurring polysaccharide derived from chitin, is a highly versatile biopolymer that exhibits a plethora of advantageous characteristics. Its bifunctional and biodegradable nature make it an attractive material for a gamut of biomedical and pharmaceutical applications, such as tissue engineering, drug delivery, and membranes. Furthermore, due to its abundance in nature, chitosan is the second most abundant biopolymer after cellulose. Nevertheless, its solubility in acidic solvents due to the protonation of its free amino groups, weak mechanical properties, water resistance and low thermal stability, limit its potential applications. To improve these features, chitosan is often blended with other synthetic or natural polymers, and this combination has been shown to be beneficial (Nair 2019). Castilla-Casadiego et al. (2022) used chitosan and meloxicam for the fabrication of microneedle patches with uniform measurements and precise arrangement.

25.4.4 Silicone

Silicone-based coatings can be applied to film substrates to create release liners, which impart a protective layer to adhesive surfaces. These liners facilitate handling of the substrate and optimize the delivery of materials for coating. The application of the silicone coating can be performed on a single side or both sides of the film substrate (Aliyar and Schalau 2015). Panchaxari et al. (2013) developed a transdermal patch containing diclofenac diethylamine (DDEA) and studied in detail. This patch is composed of both silicone and acrylic adhesives, and is known to have a successful effect in transporting the active ingredient through the skin and has been shown to be effective in delivering the active ingredient transdermally. The patch has been tested and characterized in order to determine its efficacy and safety in use. This new patch could provide an effective and safe alternative to other forms of administration of DDEA, such as oral or injectable forms.

25.4.5 Acrylic Polymers

The composition of acrylic polymers consists of various elements, such as acrylic acid, cyanoacrylic acid, methacrylic acid, and their respective esters, including acrylamide and acrylonitrile. Among the esters of methacrylic acid, methyl methacrylate is the most widely used in the fabrication of medicinal devices. Although acrylics are highly resistant to lipids, they provide no protection against organic solvents. To address this drawback, the use of acrylic blends and copolymers can be employed to enhance resilience to soaps and salt water. By adjusting the proportions of the two polymers or comonomers in the mix, it is possible to mitigate this issue (Khandavilli et al. 2002). Verma have explored the potential of drugs like nicotine, scopolamine, trimegestone, isosorbide dinitrate, tacrine, and hydrocortisone to be delivered via the skin by using acrylic polymers such as Eudragit R, Eudragit NE, Eudragit E100 and Eudragit L100 (Verma 2016).

25.4.6 Styrene-Block-(Ethylene-*co*-Butylene)-Block-Styrene (SEBS) Copolymers

SEBS copolymers are renowned for their biocompatibility, as well as their range of advantageous properties. Without the need for vulcanization post-molding, these materials are able to achieve high levels of tensile strength, with values up to 3000 psi attainable without the use of fillers. An additional benefit of SEBS copolymers is their small domain size, which allows for both opaqueness and transparency. However, their physical cross-links enabled by hard blocks are subject to mobility in higher temperatures (Paglicawan and Celorico 2021). Gennari et al. conducted a study to evaluate the possibility of employing elastomers for the fabrication of transdermal patches. The formulations employed were evaluated for biopharmaceutical and adhesive properties, and those that were deemed suitable for this purpose were further explored. The study revealed the potential of utilizing SEBS block copolymers for engineering of transdermal patches (Gennari et al. 2020).

25.4.7 Polycaprolactone (PCL)

Polycaprolactone (PCL) is a low-cost, non-toxic synthetic polyester that has been widely utilized in biomedical applications. It offers a glass transition temperature of −60 °C. The melting point of PCL was found to be 60 °C. Typically PCL synthesized via a ring-opening polymerization reaction of ε-caprolactone and can be degraded by enzymes such as esterases and lipases found in microorganisms. However, the limited degradation rate and lack of bioactivity restrict its applications. In order to improve the degradation rate and bioactivity of PCL, researchers have implemented strategies such as incorporating inorganic materials like calcium silicate, hydroxyapatite, bioglass, and calcium phosphate into PCL-based composites and blending natural polymers like gelatin, starch, and zein into the same. These approaches have been shown to result in enhanced degradability and bioactivity of PCL-based composites (McKeen 2017). The development of nanocomposites made from tetracycline/graphene oxide (GO) loaded dextran/polycaprolactone (PCL) has opened up new biomedical applications, such as wound dressings. The composite materials are designed with a hydrophobic PCL layer which prevents moisture from entering the inner layers. Graphene has attracted attention for its remarkable electrical, thermal, and mechanical properties and has been used to create nanocomposites with impressive flexural strength and mechanical properties. Additionally, nanocomposites consisting of GO and PCL prepared with an in situ grafting approach have been found to be highly soluble in polymer solvent (Nematpour et al. 2019).

25.4.8 Polyvinylpyrrolidone (PVP)

Polyvinylpyrrolidone (PVP) are chemically composed of 1-vinyl-2-pyrrolidinone linear molecules and characterized as synthetic polymer. Its non-ionic and water-soluble character allows it to be highly suitable for variety of industries, like cosmetics, medicine, and pharmaceutical industries. PVP featured with several beneficial attributes including excellent water solubility, exceptional binding capacity with resins and polymers, and non-toxic biocompatibility. Additionally, PVP is hygroscopic, meaning it has the ability to absorb water from the atmosphere (Hiremath et al. 2019). Despite its many benefits, there are some potential reported events of allergy, pseudorheumatoid nodules a distinct type of subcutaneous granulomas common in children, altering the pulmonary vascular arrangement, and accumulation in the reticuloendothelial system have all been linked to this polymer and its derivatives. This can limit the amount of this ingredient that can be used in products. It is important to not use a high molecular weight of PVP and to avoid multiple applications in the same area. One of the major drawbacks of PVP is its hygroscopicity, making it challenging to store. To improve the tackiness of PVP, acetylated monoglycerides can be added to the coating solution. PVPs are not biodegradable, which makes them unsuitable for parenteral administration. To mitigate this issue, only pyrogen-free PVPs of low molecular weight can be used as they can be eliminated by the kidneys (Kurakula and Rao 2020). Swetha et al. formulated a transdermal patch composed of rivastigmine tartrate by using polyvinylpyrrolidone (PVP) and ethylcellulose (EC) in different molar ratios of one to one, one to three and one to five, respectively, with 30% dibutyl phthalate as a plasticizer (Vennapoosa and Sandeep 2013).

25.4.9 Sodium Alginate

Sodium alginate is a biodegradable, non-toxic, biocompatible, and water-soluble anionic macromolecule with a range of advantageous properties. It can be engineered through a combination of physical process and chemical processes. To modify the sodium alginate hydrophilic, vinyl monomers can be used for graft copolymerization or through blending with functional components. Additionally, when sodium alginate reacted with divalent ions like Ca^{2+}, it shows liquid gel transformation making it a popular choice in tissue engineering and drug delivery. Despite its advantages, alginate suffers from some drawbacks, such as limited mechanical properties and ion leaching, leading to instability. Moreover, due to the exchange of freely available monovalent cations in the surrounding medium, ionically cross-linked alginate gels are not suitable for long-term stability in physiological conditions (Lee and Mooney 2012). Recently developed transdermal lamivudine patch (LAM) by Ramadan et al. reported using a tri-polymeric combination of sodium alginate, methylcellulose, and chitosan to form matrix-like construction to facilitate effective transdermal drug delivery, as it enabled successful penetration of LAM through rat skin and into the bloodstream. It resulted with a sustained drug release in the bloodstream for up to six days, suggesting that this transdermal delivery system has the potential to extend the biological half-life of LAM (Ramadan et al. 2018).

25.4.10 Pullulan

Pullulan, a polysaccharide-based organic polymer, namely structured with maltotriose units covalently attached with other three maltotriose units by α-1,6 glycosidic bonds. It is extensively utilized due to its impressive water solubility. The terminally active functional bonds allow it to exert a range of chemical properties. These functional groups can be manipulated to produce a partially soluble or completely insoluble product in water. Such versatility makes it suitable for numerous industrial applications as adhesive materials or as adhesive film due its low-viscosity and highly water-soluble properties. Additionally, its solubility can be engineered to regulate drug-delivery system and the release kinetics of drugs molecules in plasma to obtain desired drug release profile in pre-determined rate. However, the use of pullulan films possesses some drawbacks, such as a difficult processing process, brittleness, and a lack of water resistance. To enhance its performance for various purposes, a cross-linking reaction is commonly adopted (Prajapati et al. 2013). Lalitkumar et al. have recently engineered dissolvable microneedle patches constructed from pullulan. In vitro examinations have indicated that these microneedles can be proficiently used to delivering low- to high-molecular-weight drugs through the epidermis. Additionally, the biomolecules were established to be secure in the gels used to construct the pullulan microneedles, and the physical properties of these microneedles were deemed to be sufficiently promising. These discoveries suggest that pullulan microneedles could be a feasible option for transdermal delivery of biomolecules (Vora et al. 2020).

25.5 Conclusion and Future Prospects

The chapter "Transdermal Drug Delivery and Organic Polymers" has elucidated the growing prevalence of transdermal drug delivery in the medical sector and its capacity to revolutionize the manner in which drugs are administered. Additionally, the examination has also highlighted the significance of organic polymers in the formulation of effective transdermal systems. Organic polymers provide a number of advantageous features such as biocompatibility, convenience of biodegradability, and cost-effectiveness, all of which make them highly desirable for transdermal drug-delivery systems. Ultimately, the review has identified a diversity of obstacles that must be addressed to ensure the successful development and utilization of TDDS.

Going forward, the development of TDD and organic polymer technologies will remain paramount. To achieve this, it is essential to further refine the characteristics of numerous organic polymers and apply them to the creation of reliable transdermal drug-delivery systems. Moreover, research must be conducted to create novel materials that are capable of delivering a broad spectrum of medications safely and efficiently. With sustained effort, it is expected that transdermal drug delivery and organic polymers will keep contributing to the progress of a gamut of medical treatments.

References

Ahn, G.Y., Eo, H.S., Kim, D., and Choi, S.W. (2020). Transdermal delivery of FITC-dextrans with different molecular weights using radiofrequency microporation. *Biomaterials Research* 24 (1): 22.

Akram, M.R., Ahmad, M., Abrar, A. et al. (2018). Formulation design and development of matrix diffusion controlled transdermal drug delivery of glimepiride. *Drug Design, Development and Therapy* 12 (1): 349–364.

Aliyar, H. and Schalau, G. (2015). Recent developments in silicones for topical and transdermal drug delivery. *Therapeutic Delivery* 6 (7): 827–839.

Amjadi, M., Mostaghaci, B., and Sitti, M. (2017). Recent advances in skin penetration enhancers for transdermal gene and drug delivery. *Current Gene Therapy* 17 (2): 139–146.

Ansari, T.M., Srivastava, D., Kaliyappan, D. et al. (2022). Iontophoretic drug delivery systems. In: *Systems of Nanovesicular Drug Delivery* (ed. A.K. Nayak, H. Ms, T.M. Aminabhavi, and V.P. Torchilin), 387–401. Academic Press.

Aqil, M., Sultana, Y., and Ali, A. (2003). Matrix type transdermal drug delivery systems of metoprolol tartrate: in vitro characterization. *Acta Pharmaceutica (Zagreb, Croatia)* 53 (2): 119–125.

Arora, P., Sarkar, R., Garg, V.K., and Arya, L. (2012). Lasers for treatment of melasma and post-inflammatory hyperpigmentation. *Journal of Cutaneous and Aesthetic Surgery* 5 (2): 93–103.

Arora, R., Katiyar, S.S., Kushwah, V., and Jain, S. (2017). Solid lipid nanoparticles and nanostructured lipid carrier-based nanotherapeutics in treatment of psoriasis: a comparative study. *Expert Opinion on Drug Delivery* 14 (2): 165–177.

Bangar, B., Shinde, N., Deshmukh, S., and Kale, B. (2014). Natural polymers in drug delivery development. *Research Journal of Pharmaceutical Dosage Forms and Technology* 6 (1): 54–57.

Barry, B.W. (1991). Lipid-protein-partitioning theory of skin penetration enhancement. *Journal of Controlled Release* 15 (3): 237–248.

Bolhassani, A., Khavari, A., and Zahra, O. (2014). Electroporation – advantages and drawbacks for delivery of drug. In: *Gene and Vaccine* (ed. A.D. Sezer). IntechOpen.

Bos, J.D. and Meinardi, M.M. (2000). The 500 Dalton rule for the skin penetration of chemical compounds and drugs. *Experimental Dermatology* 9 (3): 165–169.

Castilla-Casadiego, D.A., Miranda-Muñoz, K.A., Roberts, J.L. et al. (2022). Biodegradable microneedle patch for delivery of meloxicam for managing pain in cattle. *PLoS One* 17 (8): e0272169.

Choudhary, N. and Singh, A. (2021). Transdermal drug delivery system: a review. *Indian Journal of Pharmacy and Pharmacology* 8 (1): 5–9.

Cross, S.E. and Roberts, M.S. (2004). Physical enhancement of transdermal drug application: is delivery technology keeping up with pharmaceutical development? *Current Drug Delivery* 1 (1): 81–92.

Dhote, V., Bhatnagar, P., Mishra, P.K. et al. (2012). Iontophoresis: a potential emergence of a transdermal drug delivery system. *Scientia Pharmaceutica* 80 (1): 1–28.

Dubey, V., Mishra, D., and Jain, N.K. (2007). Melatonin loaded ethanolic liposomes: physicochemical characterization and enhanced transdermal delivery. *European Journal of Pharmaceutics and Biopharmaceutics: Official Journal of Arbeitsgemeinschaft fur Pharmazeutische Verfahrenstechnik e.V* 67 (2): 398–405.

Erdő, F., Hashimoto, N., Karvaly, G. et al. (2016). Critical evaluation and methodological positioning of the transdermal microdialysis technique. A review. *Journal of Controlled Release: Official Journal of the Controlled Release Society* 233: 147–161.

Escobar-Chávez, J.J., Bonilla-Martínez, D., Villegas-González, M.A., and Revilla-Vázquez, A.L. (2009). Electroporation as an efficient physical enhancer for skin drug delivery. *Journal of Clinical Pharmacology* 49 (11): 1262–1283.

Esuendale, D. and Gabriel, T. (2016). Cellulosic on transdermal drug delivery system: a review. *Journal of Drug Delivery and Therapeutics* 6 (5): 57–64.

Fang, J.Y. and Leu, Y.L. (2006). Prodrug strategy for enhancing drug delivery via skin. *Current Drug Discovery Technologies* 3 (3): 211–224.

Gehl, J. (2003). Electroporation: theory and methods, perspectives for drug delivery, gene therapy and research. *Acta Physiologica Scandinavica* 177 (4): 437–447.

Gehl, J. (2008). Electroporation for drug and gene delivery in the clinic: doctors go electric. In: *Methods in Molecular Biology*, vol. 423 (ed. S. Li), 351–359. (Clifton, N.J.): Humana Press. https://doi.org/10.1007/978-1-59745-194-9_27.

Gennari, C.G.M., Quaroni, G.M.G., Creton, C. et al. (2020). SEBS block copolymers as novel materials to design transdermal patches. *International Journal of Pharmaceutics* 575: 118975.

Gill, B., Singh, J., Sharma, V., and Kumar, S.L.H. (2014). Emulsomes: an emerging vesicular drug delivery system. *Asian Journal of Pharmaceutics (AJP)* 6 (2): 87–94.

Goffin, V., Henry, F., Piérard-Franchimont, C., and Piérard, G.E. (2000). Penetration enhancers assessed by corneoxenometry. *Skin Pharmacology and Applied Skin Physiology* 13 (5): 280–284.

Gutierrez Cisneros, C., Bloemen, V., and Mignon, A. (2021). Synthetic, natural, and semisynthetic polymer carriers for controlled nitric oxide release in dermal applications: a review. *Polymers* 13 (5): 760.

Grand View Research. (2021). Transdermal drug delivery system market size report, 2030. https://www.grandviewresearch.com/industry-analysis/transdermal-drug-delivery-systems-industry (accessed 1 July 2023).

Hanumanaik, M., Patil, U., Kumar, G. et al. (2012). Design, evaluation and recent trends in transdermal drug delivery system: an overview. *International Journal of Pharmaceutical Sciences and Research* 3 (8): 2393–2406.

Harrison, J.E., Watkinson, A.C., Green, D.M. et al. (1996). The relative effect of Azone and Transcutol on permeant diffusivity and solubility in human stratum corneum. *Pharmaceutical Research* 13 (4): 542–546.

Hiremath, P., Nuguru, K., and Agrahari, V. (2019). *Material Attributes and Their Impact on Wet Granulation Process Performance*. Elsevier eBooks.

Ita, K. (2015). Transdermal delivery of drugs with microneedles-potential and challenges. *Pharmaceutics* 7 (3): 90–105.

Jana, B.A. and Wadhwani, A.D. (2019). Microneedle - future prospect for efficient drug delivery in diabetes management. *Indian Journal of Pharmacology* 51 (1): 4–10. https://doi.org/10.4103/ijp.IJP_16_18.

Jeong, W.Y., Kwon, M., Choi, H.E., and Kim, K.S. (2021). Recent advances in transdermal drug delivery systems: a review. *Biomaterials Research* 25 (1): 24.

Johnson, W. (2007). Final report on the safety assessment of capsicum annuum extract, capsicum annuum fruit extract, capsicum annuum resin, capsicum annuum fruit powder, capsicum frutescens fruit, capsicum frutescens fruit extract, capsicum frutescens resin, and capsaicin. *International Journal of Toxicology* 26 (3): 106.

Kamaly, N., Yameen, B., Wu, J., and Farokhzad, O.C. (2016). Degradable controlled-release polymers and polymeric nanoparticles: mechanisms of controlling drug release. *Chemical Reviews* 116 (4): 2602–2663.

Karande, P., Jain, A., Ergun, K. et al. (2005). Design principles of chemical penetration enhancers for transdermal drug delivery. *Proceedings of the National Academy of Sciences of the United States of America* 102 (13): 4688–4693.

Kathe, K. and Kathpalia, H. (2017). Film forming systems for topical and transdermal drug delivery. *Asian Journal of Pharmaceutical Sciences* 12 (6): 487–497.

Kelchen, M.N. and Brogden, N.K. (2018). Effect of dosing regimen and microneedle pretreatment on in vitro skin retention of topically applied beta-blockers. *Biomedical Microdevices* 20 (4): 100.

Khandavilli, S., Nair, V., and Panchagnula, R. (2002). Polymers in transdermal drug delivery systems. *Pharmaceutical Technology* 26: 62–81.

Kurakula, M. and Rao, G.S.N.K. (2020). Pharmaceutical assessment of polyvinylpyrrolidone (PVP): as excipient from conventional to controlled delivery systems with a spotlight on COVID-19 inhibition. *Journal of Drug Delivery Science and Technology* 60: 102046.

Lee, K.Y. and Mooney, D.J. (2012). Alginate: properties and biomedical applications. *Progress in Polymer Science* 37 (1): 106–126.

Li, Y., Guo, L., and Lu, W. (2013). Laser ablation-enhanced transdermal drug delivery. *Photonics and Lasers in Medicine* 2 (4): 315–322.

Liechty, W.B., Kryscio, D.R., Slaughter, B.V., and Peppas, N.A. (2010). Polymers for drug delivery systems. *Annual Review of Chemical and Biomolecular Engineering* 1: 149–173.

Liu, N., Zhang, Y., Cun, D. et al. (2014). Effect of backing films on the transdermal delivery of donepezil from patches. *AAPS PharmSciTech* 15 (6): 1569–1573.

Lobo, S., Sachdeva, S., and Goswami, T. (2016). Role of pressure-sensitive adhesives in transdermal drug delivery systems. *Therapeutic Delivery* 7 (1): 33–48.

McKeen, L.W. (2017). Environmentally friendly polymers. In: *Plastics Design Library, Permeability Properties of Plastics and Elastomers (Fourth Edition)* (ed. L.W. McKeen), 305–323. Norwich, NY: William Andrew Publishing.

Murphy, M. and Carmichael, A.J. (2000). Transdermal drug delivery systems and skin sensitivity reactions. Incidence and management. *American Journal of Clinical Dermatology* 1 (6): 361–368.

Nair, S.S. (2019). Chitosan-based transdermal drug delivery systems to overcome skin barrier functions. *Journal of Drug Delivery and Therapeutics* 9 (1): 266–270.

Nematpour, N., Farhadian, N., Ebrahimi, K.S. et al. (2019). Sustained release nanofibrous composite patch for transdermal antibiotic delivery. *Colloids and Surfaces A: Physicochemical and Engineering Aspects* 586. https://doi.org/10.1016/j.colsurfa.2019.124267.

Omray, L.K., Kohli, S., Khopade, A.J. et al. (2008). Development of mesophasic microreservoir-based transdermal drug delivery system of propranolol. *Indian Journal of Pharmaceutical Sciences* 70 (5): 578–584. https://doi.org/10.4103/0250-474X.45394.

Paglicawan, M.A. and Celorico, J.R. (2021). Properties of styrene-ethylene-butylene styrene block copolymer/exfoliated graphite nanoplatelets nanocomposites. *Polymers and Polymer Composites* 29 (9): 154–165. https://doi.org/10.1177/0967391121991290.

Palac, Z., Engesland, A., Flaten, G.E. et al. (2014). Liposomes for (trans)dermal drug delivery: the skin-PVPA as a novel in vitro stratum corneum model in formulation development. *Journal of Liposome Research* 24 (4): 313–322. https://doi.org/10.3109/08982104.2014.899368.

Panchaxari, D.M., Pampana, S., Pal, T. et al. (2013). Design and characterization of diclofenac diethylamine transdermal patch using silicone and acrylic adhesives combination. *DARU: Journal of Faculty of Pharmacy, Tehran University of Medical Sciences* 21 (1): 6. https://doi.org/10.1186/2008-2231-21-6.

Park, J., Lee, H., Lim, G.S. et al. (2019). Enhanced transdermal drug delivery by sonophoresis and simultaneous application of sonophoresis and iontophoresis. *AAPS PharmSciTech* 20 (3): 96. https://doi.org/10.1208/s12249-019-1309-z.

Pastore, M.N., Kalia, Y.N., Horstmann, M., and Roberts, M.S. (2015). Transdermal patches: history, development and pharmacology. *British Journal of Pharmacology* 172 (9): 2179–2209. https://doi.org/10.1111/bph.13059.

Paudel, K.S., Milewski, M., Swadley, C.L. et al. (2010). Challenges and opportunities in dermal/transdermal delivery. *Therapeutic Delivery* 1 (1): 109–131. https://doi.org/10.4155/tde.10.16.

Prajapati, S.T., Patel, C.G., and Patel, C.N. (2011). Formulation and evaluation of transdermal patch of repaglinide. *ISRN Pharmaceutics* 2011: 651909. https://doi.org/10.5402/2011/651909.

Prajapati, V.D., Jani, G.K., and Khanda, S.M. (2013). Pullulan: an exopolysaccharide and its various applications. *Carbohydrate Polymers* 95 (1): 540–549. https://doi.org/10.1016/j.carbpol.2013.02.082.

Prausnitz, M.R. and Langer, R. (2008). Transdermal drug delivery. *Nature Biotechnology* 26 (11): 1261–1268. https://doi.org/10.1038/nbt.1504.

Prausnitz, M.R., Mitragotri, S., and Langer, R. (2004). Current status and future potential of transdermal drug delivery. *Nature Reviews. Drug Discovery* 3 (2): 115–124. https://doi.org/10.1038/nrd1304.

Prodduturi, S., Smith, G.J., Wokovich, A.M. et al. (2009). Reservoir based fentanyl transdermal drug delivery systems: effect of patch age on drug release and skin permeation. *Pharmaceutical Research* 26 (6): 1344–1352. https://doi.org/10.1007/s11095-009-9843-0.

Rai, S., Pandey, V., and Rai, G. (2017). Transfersomes as versatile and flexible nano-vesicular carriers in skin cancer therapy: the state of the art. *Nano Reviews & Experiments* 8 (1): 1325708. https://doi.org/10.1080/20022727.2017.1325708.

Rai, V.K., Mishra, N., Yadav, K.S., and Yadav, N.P. (2018). Nanoemulsion as pharmaceutical carrier for dermal and transdermal drug delivery: formulation development, stability issues, basic considerations and applications. *Journal of Controlled Release: Official Journal of the Controlled Release Society* 270: 203–225. https://doi.org/10.1016/j.jconrel.2017.11.049.

Rajan, R., Jose, S., Mukund, V.P., and Vasudevan, D.T. (2011). Transferosomes - a vesicular transdermal delivery system for enhanced drug permeation. *Journal of Advanced Pharmaceutical Technology & Research* 2 (3): 138–143. https://doi.org/10.4103/2231-4040.85524.

Ramadan, E., Borg, T., Abdelghani, G.M., and Saleh, N. (2018). Design and in vivo pharmacokinetic study of a newly developed lamivudine transdermal patch. *Future Journal of Pharmaceutical Sciences* 4. https://doi.org/10.1016/j.fjps.2018.03.002.

Rastogi, V. and Yadav, P. (2012). Transdermal drug delivery system: an overview. *Asian Journal of Pharmaceutics* 6: 161. https://doi.org/10.4103/0973-8398.104828.

Sabbagh, F. and Kim, B.S. (2022). Recent advances in polymeric transdermal drug delivery systems. *Journal of Controlled Release: Official Journal of the Controlled Release Society* 341: 132–146. https://doi.org/10.1016/j.jconrel.2021.11.025.

Sharma, N. (2018). A brief review on transdermal patches. *Organic & Medicinal Chemistry* 7 (2). https://doi.org/10.19080/OMCIJ.2018.07.555707.

Singh, D., Pradhan, M., Nag, M., and Singh, M.R. (2015). Vesicular system: versatile carrier for transdermal delivery of bioactives. *Artificial Cells, Nanomedicine, and Biotechnology* 43 (4): 282–290. https://doi.org/10.3109/21691401.2014.883401.

Singh, T.R.R., McMillan, H.L., Mooney, K. et al. (2013). Microneedles for drug delivery and monitoring. *Microfluidic Devices for Biomedical Applications*. https://doi.org/10.1533/9780857097040.2.185.

Sloan, K.B., Devarajan-Ketha, H., and Wasdo, S.C. (2011). Dermal and transdermal delivery: prodrugs. *Therapeutic Delivery* 2 (1): 83–105. https://doi.org/10.4155/tde.10.83.

Stevenson, C.L., Santini, J.T. Jr., and Langer, R. (2012). Reservoir-based drug delivery systems utilizing microtechnology. *Advanced Drug Delivery Reviews* 64 (14): 1590–1602. https://doi.org/10.1016/j.addr.2012.02.005.

Szunerits, S. and Boukherroub, R. (2018). Heat: a highly efficient skin enhancer for transdermal drug delivery. *Frontiers in Bioengineering and Biotechnology* 6: 15. https://doi.org/10.3389/fbioe.2018.00015.

Tan, H.S. and Pfister, W.R. (1999). Pressure-sensitive adhesives for transdermal drug delivery systems. *Pharmaceutical Science & Technology Today* 2 (2): 60–69. https://doi.org/10.1016/s1461-5347(99)00119-4.

Tawfik, M.A., Mohamed, M.I., Tadros, M.I., and El-Helaly, S.N. (2021). Low-frequency sonophoresis as an active approach to potentiate the transdermal delivery of agomelatine-loaded novasomes: design, optimization, and pharmacokinetic profiling in rabbits. *AAPS PharmSciTech* 22 (8): 261. https://doi.org/10.1208/s12249-021-02147-y.

Tiwary, A.K., Sapra, B., and Jain, S. (2007). Innovations in transdermal drug delivery: formulations and techniques. *Recent Patents on Drug Delivery & Formulation* 1 (1): 23–36. https://doi.org/10.2174/187221107779814087.

Vennapoosa, S. and Sandeep, D. (2013). Formulation and evaluation of a matrix-type transdermal patch containing rivastigmine tartrate. *International Journal of Pharma Sciences and Research* 4 (2): 10–16.

Verma, N. (2016). Adhesive polymers in fabrication of transdermal drug delivery. *Research Journal of Pharmacy and Technology* 9 (7): 945–956. https://doi.org/10.5958/0974-360X.2016.00182.7.

Vora, L.K., Courtenay, A.J., Tekko, I.A. et al. (2020). Pullulan-based dissolving microneedle arrays for enhanced transdermal delivery of small and large biomolecules. *International Journal of Biological Macromolecules* 146: 290–298. https://doi.org/10.1016/j.ijbiomac.2019.12.184.

Vranić, E. (2004). Sonophoresis-mechanisms and application. *Bosnian Journal of Basic Medical Sciences* 4 (2): 25–32. https://doi.org/10.17305/bjbms.2004.3410.

Walters, K.A. and Hadgraft, J. (1993). *Pharmaceutical Skin Penetration Enhancement*. London: Informa Health Care.

Wang, S., Kara, M., and Krishnan, T.R. (1998). Transdermal delivery of cyclosporin-a using electroporation. *Journal of Controlled Release: Official Journal of the Controlled Release Society* 50 (1-3): 61–70. https://doi.org/10.1016/s0168-3659(97)00117-x.

Williams, A.C. and Barry, B.W. (2004). Penetration enhancers. *Advanced Drug Delivery Reviews* 56 (5): 603–618. https://doi.org/10.1016/j.addr.2003.10.025.

Wokovich, A.M., Shen, M., Doub, W.H. et al. (2010). Release liner removal method for transdermal drug delivery systems (TDDS). *Journal of Pharmaceutical Sciences* 99 (7): 3177–3187. https://doi.org/10.1002/jps.22067.

Yu, Y.Q., Yang, X., Wu, X.F., and Fan, Y.B. (2021). Enhancing permeation of drug molecules across the skin via delivery in nanocarriers: novel strategies for effective transdermal applications. *Frontiers in Bioengineering and Biotechnology* 9: 1–17. https://doi.org/10.3389/fbioe.2021.646554.

Zhan, X., Mao, Z., Chen, S. et al. (2015). Formulation and evaluation of transdermal drug-delivery system of isosorbide dinitrate. *Brazilian Journal of Pharmaceutical Sciences* 51 (2): 373–382.

Zhang, Z., Tsai, P.C., Ramezanli, T., and Michniak-Kohn, B.B. (2013). Polymeric nanoparticles-based topical delivery systems for the treatment of dermatological diseases. *Wiley Interdisciplinary Reviews. Nanomedicine and Nanobiotechnology* 5 (3): 205–218. https://doi.org/10.1002/wnan.1211.

Index

Organic Polymers in Energy-Environmental Applications, First Edition.
Edited by Ramesh Oraon, Pardeep Singh, Sanchayita Rajkhowa, Sangita Agarwal, and Ravindra Pratap Singh.

d

e

f

p

www.ingramcontent.com/pod-product-compliance
Lightning Source LLC
Chambersburg PA
CBHW080930240926
51849CB00003BA/425
9783527352371